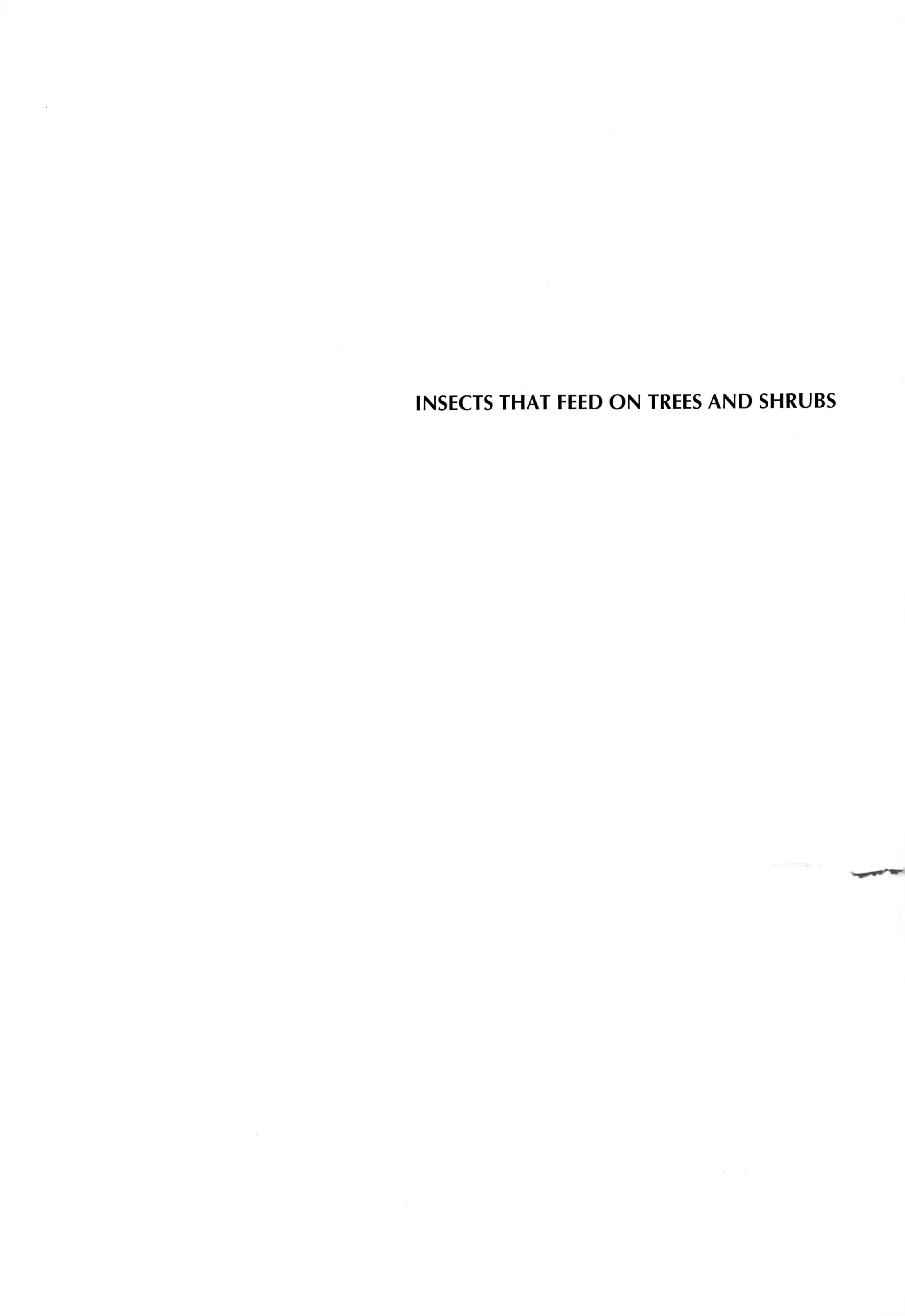

INSECTS THAT FEED ON TREES AND SHRUBS

INSECTS THAT FEED ON

COMSTOCK PUBLISHING ASSOCIATES, a division of
CORNELL UNIVERSITY PRESS | ITHACA AND LONDON

TREES AND SHRUBS

SECOND EDITION, REVISED

By WARREN T. JOHNSON
Professor of Entomology
Cornell University

and HOWARD H. LYON
Photographer, Department of Plant Pathology
Cornell University

Contributing Authors

C. S. KOEHLER
University of California, Berkeley

J. A. WEIDHAAS
Virginia Polytechnic Institute and State University

First edition published 1976 by Cornell University Press. Second edition 1988.
Second edition, revised, 1991. Second printing with corrections 1994.

Library of Congress Cataloging-in-Publication Data

Johnson, Warren T.
Insects that feed on trees and shrubs / by Warren T. Johnson and Howard H. Lyon : contributing authors, C.S. Koehler, J.A. Weidhaas. — 2nd ed., rev.
p. cm.
Includes bibliographical references and indexes.
ISBN 0-8014-2602-2
1. Ornamental trees—Diseases and pests—United States. 2. Ornamental shrubs—Diseases and pests—United States. 3. Insect pests—United States—Identification. 4. Insect pests—United States—Pictorial works. 5. Ornamental trees—Diseases and pests—Canada. 6. Ornamental shrubs—Diseases and pests—Canada. 7. Insect pests—Canada—Identification. 8. Insect pests—Canada—Pictorial works. I. Lyon, Howard H. II. Title.
SB762.J63 1991 635.9'77—dc20 90-46218

Printed in China

Cornell University Press strives to use environmentally responsible suppliers and materials to the fullest extent possible in the publishing of its books. Such materials include vegetable-based, low-VOC inks and acid-free papers that are recycled, totally chlorine-free, or partly composed of nonwood fibers. Books that bear the logo of the FSC (Forest Stewardship Council) use paper taken from forests that have been inspected and certified as meeting the highest standards for environmental and social responsibility. For further information, visit our website at www.cornellpress.cornell.edu.

Cloth printing 10 9 8 7 6 5

Contents

Preface to the First Edition 7
Preface to the Second Edition 8
Acknowledgments 9
Reader's Guide to Identification of Insects and Related Pests 10
Introduction 11
- The Book and Its Use 11
- Pest Control 11
- Pest Identification and Problem Diagnosis 11
- Natural History 13

Insects That Feed on Conifers **15**
(Numbers following titles refer to plates)

- Sawfly Defoliators, Bud Miners, and Shoot Borers, 1–3 16
- Webspinning Sawflies, Pine Webworm, Pine False Webworm, 4 22
- Moth and Butterfly Defoliators of Conifers, 5–6 24
- Spruce Bud Moth and Spruce Budworms, 7 28
- Juniper Webworm, 8 30
- Spruce Needleminers, 9 32
- Douglas-fir Tussock Moth and Silverspotted Tiger Moth, 10 34
- Larch Casebearer, 11 36
- Needle Miners and Webbers, 12–13 38
- Cypress Tip Miners and Arborvitae Leaf Miners, 14 42
- Needle and Twig Midges, 15 44
- Pine Tube Moth, 16 46
- Pine Tip or Shoot Moths, 17–18 48
- Pine Shoot Borers, 19 52
- Conifer Twig Weevils, 20 54
- Trunk and Root Collar Weevils, 21 56
- Pine Needle Weevils and Juniper Twig Girdler, 22 58
- Cypress Bark Moth and Douglas-fir Twig Weevil, 23 60
- Conifer Bark Beetles, 24 62
- Eastern Pine Bark and Shoot Beetles, 25 64
- Western Conifer Bark Beetles, 26 66
- Cedar Bark Beetles, 27 68
- Cypress Bark Beetles, 28 70
- Pitch Moths, 29 72
- Balsam Woolly Adelgid, 30 74
- Adelgids of Pine and Spruce, 31 76
- Adelgids of Larch, Spruce, and Hemlock, 32 78
- Balsam Twig Aphid, 33 80
- *Cinara* Aphids and Other Aphids of Conifers, 34–35 82
- Conifer Spittlebugs, 36 86
- Mealybugs of Yew and Pine, and the Woolly Pine Scale, 37 88
- Golden Mealybug, 38 90
- *Matsucoccus* Scales, 39 92
- Irregular Pine Scale and Monterey Pine Scale, 40 94
- Pine Tortoise Scale, Spruce Bud Scale, and Virginia Pine Scale, 41 96
- Fletcher Scale, 42 98
- Black Pineleaf Scale, 43 100
- Hemlock Scale, 44 102
- Elongate Hemlock Scale, 45 104
- Juniper Scale, 46 106
- Pine Needle Scale, 47 108
- Armored Scales of Conifers, 48 110
- Cooley Spruce Gall Adelgid, 49 112
- Eastern Spruce Gall Adelgid, 50 114
- Conifer Gall Midges, 51 116
- Spider Mites of Conifers, 52–53 118
- Eriophyid Mites of Conifers, 54 122

Insects That Feed on Broad-leaved Trees and Shrubs **125**
(Numbers following titles refer to plates)

- Dogwood Sawfly, 55 126
- Mountain-ash Sawfly and Dusky Birch Sawfly, 56 128
- "Slug" Sawflies, 57 130
- Rose Sawflies, 58 132
- Sawflies, 59–60 134
- Gypsy Moth, 61–62 138
- Cankerworms, 63 142
- Loopers, 64 144
- California Oakworm, 65 146
- Fruitworms and Other Caterpillars, 66 148
- Walnut Caterpillar, 67 150
- Nymphalid Caterpillars, 68 152
- Notodontid Caterpillars, 69–70 154
- Tussock Moth Caterpillars, 71–72 158
- Subtropical Caterpillars, 73 162
- Armed Caterpillars, 74 164
- Webworms, 75 166
- Eastern Tent Caterpillar and Forest Tent Caterpillar, 76 168
- Western Tent Caterpillars, 77 170
- Uglynest Caterpillar, Oak Webworm, Rollers, and Tiers, 78 172
- Euonymus Caterpillar, 79 174
- Bagworms, 80–81 176
- Mimosa Webworm, 82 180
- A Bougainvillea Caterpillar, 83 182
- Birch Leaf-mining Sawflies, 84 184
- Sawfly Leaf Miners and Case Bearers, 85–86 186
- Leaf-mining Beetles, 87 190
- Oak Leaf Miners, 88 192
- Blotch and Serpentine Leaf Miners, 89–90 194
- Cherry Leaf Miner and Cambium Miners, 91 198
- Maple Petiole Borer, and Shoot and Twig Borers, 92 200
- Azalea Leaf Miners, 93 202
- Boxwood Leafminer, 94 204
- Leaf-mining Maggots and Holly Budmoth, 95 206
- Madrone Shield Bearer and Other Shield Bearers, 96 208
- Yellow Poplar Weevil, 97 210
- Maple and Other Trumpet Skeletonizers, and Maple Leafcutter, 98 212
- Leaf Rollers, Tiers, and Webbers, 99–101 214
- Skeletonizers of Oak, Birch, and Apple, 102 220
- Elm Leaf Beetles, 103 222
- Leaf Beetles, 104 224
- Cottonwood Leaf Beetles and Related Species, 105 226
- Imported Willow Leaf Beetle and Flea Beetles, 106 228
- Fuller Rose Beetle, 107 230
- Seagrape Borer and a Seagrape Gall Midge, 108 232
- Walnut and Hickory Insects, 109 234
- Japanese Beetle, European Chafer, and Rose Chafer, 110 236
- Baccharis Leaf Beetle and Rose Curculio, 111 238
- Weevils and Other Root-feeding Beetles, 112–114 240
- Vectors of Elm Diseases, 115 246
- Elm Bark Beetles, 116 248
- Shothole Borer and Other Bark Beetles, 117 250
- American Plum Borer, 118 252
- Wood-boring Caterpillars, 119–121 254
- Lilac Borer and Banded Ash Clearwing, 122 260
- Dogwood Borers, 123 262
- Twig Pruners and Twig Girdlers, 124 264
- Oak Twig Girdlers, 125 266

Poplar-and-Willow Borer, 126 268
Flatheaded Borers, 127 270
Bronze Birch Borer, 128 272
Locust Borers, 129 274
Roundheaded Borers, 130–131 276
Root-feeding Beetles, 132 280
Trunk Borers, 133 282
Wood-boring Beetles, 134 284
Shoot and Twig Borers, 135 286
Acacia Psyllid, 136 288
Psyllids, 137 290
Tuliptree Aphid and Other Foliage-feeding Aphids, 138 292
California Laurel Aphid and Podocarpus Aphid, 139 294
Aphids of Beech, Birch, and Apple, 140 296
Spirea Aphid and Aphids of Manzanita, 141 298
Snowball Aphid and Other Foliage-feeding Aphids, 142 300
Black Citrus Aphid and Other Foliage-feeding Aphids, 143 302
Woolly Aphids, 144–145 304
Leaf and Stem Aphids, 146–147 308
Aphids of Oleander and Bamboo, 148 312
Leaf, Shoot, and Twig Aphids, 149 314
Woolly Root and Twig Aphids, 150 316
Whiteflies of Rhododendron and Azalea, 151 318
Greenhouse Whitefly and Relatives, 152–153 320
Mealybugs and Whiteflies, 154 324
Comstock Mealybug, 155 326
Subtropical Mealybugs, 156 328
Noxious Bamboo Mealybug, 157 330
Beech Scale and Cypress Bark Mealybug, 158 332
Sycamore Scale, 159 334
Azalea Bark Scale, 160 336
Cottony Cushion Scale, 161 338
Cottony Maple Scale, 162 340
Cottony Scales and Mealybugs, 163 342
Cottony Camellia Scale, 164 344
Cottony Scales, 165 346
Pit-making Pittosporum Scale, 166 348
Bamboo Scale, 167 350
Pit-making Scales of Oak and Holly, 168 352
Magnolia Scale, Calico Scale, and Frosted Scale, 169 354
Wax Scales, 170 356
Black Scale, 171 358
Nigra Scale and Hemispherical Scale, 172 360
Tuliptree Scale, 173 362
Lecanium Scales and Kermes Scales, 174 364
Rose Scale and Globose Scale, 175 366
European Elm Scale and Other Bark-feeding Scales, 176 368
Oystershell Scale and Camellia Scale, 177 370
Greedy Scale and Peony Scale, 178 372
Oleander Scales, 179 374
Subtropical Armored Scales, 180–182 376
Dictyospermum Scale and Gall-forming Armored Scales, 183 382
Obscure Scale and Gloomy Scale, 184 384
San Jose Scale and Walnut Scale, 185 386
Euonymus Scales, 186 388
Armored Snow Scales, 187 390
White Peach Scale and White Prunicola Scale, 188 392
Fiorinia Scales, 189 394
Fourlined Plant Bug and *Lygocoris* Plant Bugs, 190 396
Tarnished Plant Bug and Boxelder Plant Bugs, 191 398
Plant Bugs and a Beetle Affecting Sycamore, 192 400
Ash Plant Bugs, 193 402
Plant Bugs, Leafhoppers, and a Treehopper of Honeylocust, 194 404
Treehoppers, 195 406
Thornbug and Relatives, 196 408
Twomarked Treehopper, 197 410
Rose Leafhopper, 198 412
Potato Leafhopper and Whitebanded Elm Leafhopper, 199 414
Foliage Injury by Leafhoppers, 200 416
Redbanded Leafhopper and Other Sharpshooter Leafhoppers, 201 418
Spittlebugs on Angiosperms, 202 420
Planthoppers and Plant Bugs, 203 422
Lace Bugs of Broad-leaved Evergreens and Sages, 204 424
Lace Bugs of Deciduous Plants, 205–206 426
Greenhouse Thrips, 207 430
Thrips, 208–209 432
Dogwood Club Gall, 210 436
Bud, Shoot, and Stem Galls, 211 438
Oak Leaf and Twig Galls, 212–216 440
Aphid and Psyllid Galls, 217 450
Hackberry Galls, 218 452
Yaupon Psyllid Gall, 219 454
Psyllid Gall of *Persea,* 220 456
Laurel, Eugenia, Ceanothus, and Pepper Tree Psyllids, 221 458
Petiole and Leaf Stipule Galls, 222 460
Eyespot Galls, 223 462
Leaf Galls of Ash and Elm, 224 464
Honeylocust Pod Gall, 225 466
Midge and Sawfly Galls of Poplar and Willow, 226 468
Rhododendron Gall Midge, Rose Midge, and Other Gall Midges, 227 470
Spider Mites, False Spider Mites, and Tarsonemid Mites, 228–229 472
Bud and Rust Mites, 230–231 478
Eriophyid Gall Makers, 232–235 482
Katydids, Grasshoppers, and Periodical Cicadas, 236 490
Crickets and Walkingstick, 237 492
Wasps and Bees as Plant Pests, 238 494
Ants, 239 496
Slugs and Snails, 240 498
Birds and Small Mammals That Injure Trees, 241 500

Sources of Information on Pests and Pest Control 503
Land-Grant Institutions and Agricultural Experiment Stations in the United States 503
U.S. Department of Agriculture, Forest Pest Control Offices 503
Regional Pest Control Information in Canada 504

Glossary 505
References 507
Index of Insects, Mites, and Other Animals 522
Index to Insects by Host Plants 533

Preface to the First Edition

Entomological investigations in the United States over the past century have been directed mainly toward the many pests that feed on food and fiber crops. The efficient production of foodstuffs and the resulting increased productivity of American agriculture today is attributable in no small way to the contributions of early entomologists and their successors. Efficiency of agricultural production has reached the point where only a small fraction of our population is now directly engaged in farming. But as the rural population diminishes, the urban and suburban populations acquire greater voting power. It seems inevitable that the increased strength of the urban and suburban vote will in the years ahead bring about changes in the research and extension orientation of land-grant institutions and other publicly funded agencies within which most entomologists operate. Clear signs of change have already been indicated in several states. Rather than continue to commit nearly all entomological talent to areas of traditional agriculture, state and federal governments can be expected to place greater emphasis on the quasiagricultural problems of the urban and suburban resident and his recreational lands.

This book is an attempt to add one brick to the foundation upon which urban entomology must be built. Entomology's success in the future—as has been the case in the past—will depend to a considerable degree on service to a clientele that expresses a need. The need is no less than it was a century ago; only the clientele has changed.

As cities and suburbs grow, increased value is being placed on greenbelts and parks and on the plants that occupy these spaces. People are becoming more aware that trees and other green plants are a part of their life-support system. Vegetation supplies oxygen, filters out dust from the air, and reduces noise pollution. Shade trees can lower the temperature of an urban street by 10 or more degrees (the cooling effect results from transpiration as well as the sun-shielding action of the leaves). A properly placed row of trees can serve as an effective windbreak, reducing either hot or cold air movement. Trees and shrubs increase the quality of life in other ways. The flowering cherry trees of Washington, D.C., attract visitors from all over the world, as do the fall-colored hardwoods of the Northeast and the redwoods of California. The benefits provided by trees and other woody ornamentals are available to everyone.

One of the greatest challenges confronting the urban entomologist is appraising the actual importance of the innumerable insects and mites that attack ornamental plants. In agriculture, a species is usually termed an economic pest when the expected monetary loss it may cause, if the crop attacked is left unprotected, is greater than the cost of the usual treatment needed to prevent or mitigate that loss. In the case of ornamentals, however, it is more difficult to identify "pests," since no economic loss (in the agricultural sense) is involved. Pest-caused damage to ornamentals, then, must be measured in terms of the reduced aesthetic value of the plants, the loss of shade, or the cost of replacement of the afflicted plant. Loss may entail inconvenience, as in the case of pest-induced premature leaf drop, requiring the frequent raking of leaves, or even embarrassment (falling caterpillars—or their droppings—at a social affair held under shade trees). The point is that, regardless of what constitutes loss in the case of ornamentals, it is extremely difficult to measure. Yet it is important that appraisal be made.

Current trends in agricultural pest-control philosophy favor reduction of pest numbers to a level just below the so-called economic injury level in cases in which the pest's natural enemies are thereafter capable of containing the infestation. The urban entomologist must not only establish the importance of a pest but also determine the level at which the pest is of no concern to those who make use of the tree or shrub in the landscape. This level must be expected to vary according to the location of the plant and the uses for which it is intended.

This reference work has been in preparation since 1965. Though its beginnings predate the present environmental crisis by several years, it is appropriate for the book to appear at this time because of the widespread attention that has recently been focused on pest control. It is hoped that this book will be regarded as contributing to a rational approach to pest control problems as they relate to woody ornamental vegetation.

C. S. Koehler

Berkeley, California

Preface to the Second Edition

There have been many important changes in the "Green Industry" since the preparation of the first edition of this book, published in 1976. People in urban and suburban areas are more aware and interested than ever before in gardening and landscaping. A recent Gallup poll ranked these activities first nationwide among outdoor pursuits undertaken for recreation, health, and economy. The nursery and arboriculture industry has grown markedly, as have professional landscape and other horticultural businesses, which prospered even through the recession of the late 1970s. The International Society of Arboriculture now has more than 4000 members, compared with fewer than 2400 in 1973.

The increased interest in ornamental horticulture and home gardening is reflected in changes in the U.S. land-grant universities. Research and extension activities on pests of trees, shrubs, and other urban vegetation have more than doubled. Formerly staffed with professionals trained and working in programs related to food and fiber production, research and extension programs now include specialists in ornamental horticulture and related sciences. Their work is an important part of agricultural and natural resource endeavors. New positions have been created, and established positions redirected, to accommodate growing advances in technology.

The resulting surge in new information, along with huge increases in urban and suburban audiences for extension programs, has created an urgent need for improved dissemination of knowledge and technology. County offices of the Cooperative Extension Service have redirected their resources to serve the needs and interests of urban and suburban ornamental horticulture. For example, Fresno County, California, the leading agricultural county in the United States, recently hired a full-time urban horticulture adviser. In Virginia the number of extension agents involved in urban-suburban horticultural programs has increased 10-fold since the late 1960s. Many counties in numerous states have developed Master Gardener or similar educational programs, which train thousands of volunteer horticulturists to advise homeowners about the proper growing, maintenance, and protection of garden and landscape plants. Such programs have upgraded the knowledge, competence, and skills of many people who continually seek better and more sophisticated technology and educational resources.

Broad, sweeping changes in federal and state laws that regulate pesticide use have markedly changed pest control practices. The public's growing resistance to the use of pesticides has been fed by constant and wide publicity from the news media on toxic spills, groundwater contamination by chemicals, insect eradication or suppression programs, and chemical hazards to health. The rapid spread of the gypsy moth across the United States and large-scale programs aimed at other pests have greatly expanded the number of people embroiled in confrontations about pest control practices. Litigation over pesticide legislation and use has increased dramatically. Recently there have been astronomical increases in the costs of liability insurance for firms that apply pesticides to urban and forest vegetation.

Since the mid-1970s the concept of Integrated Pest Management (IPM) has evolved in response to public concern about the use of pesticides. IPM utilizes all available methods of regulating populations of destructive pests and diseases. Biological control with parasites, predators, traps, attractants, cultural practices, resistant plants, and other nonchemical strategies has received major attention. With the IPM approach, the accurate identification of pest problems is more critical than ever, and more organisms are consequential in the diagnostic and decision-making processes. Arboricultural firms are beginning to offer such IPM services as pest detection, population monitoring, and plant protection rather than merely offering spray service. Consequently, professionals involved in shade tree and landscape maintenance and protection have a growing need for useful guides to the identification of hundreds of organisms that inhabit or attack a vast array of plant species.

This second edition provides expanded coverage of wood and bark feeders, sawflies, leaf rollers, leaf tiers, foliage webbers, and gall midges. There are 29 new plates, and many old plates have been reorganized, with new pictures inserted. In keeping with our desire to illustrate as many plant-feeding insects as possible we have increased, by threefold, the number of black-and-white illustrations on text pages. Again our objective is to assist arborists, landscape and grounds maintenance professionals, extension advisers, agents and specialists, garden store personnel, Master Gardener volunteers, nurserymen, park managers, county and urban foresters, and many others in identifying and diagnosing insect and mite problems on woody ornamental plants. The most important first step in dealing with any pest problem is to identify the causal agent, and it is to that goal this book is dedicated.

As in the first edition, chemical control measures are not included, for they are subject to constant change and vary from state to state. Discussion of complex IPM methods is beyond the scope of this book. In a few instances cultural controls are suggested. Attention is also given to volatile chemicals (semiochemicals) produced by arthropods and used for communication among or between species. Sex pheromones, one of many semiochemicals, have been used often to effectively monitor or control a number of pests. Although classical biological control is usually beyond the capacity of most IPM practitioners, common or important parasitoids and predators are mentioned to encourage appreciation for the role that natural enemies of insects and mites play in pest population management.

We have worked on this revision with the hope and conviction that it will be more useful than its predecessor, and we are especially pleased that the companion volume, *Diseases of Trees and Shrubs,* is now available. In combination, these two reference books should make it possible for the tree and shrub specialist trained in horticulture or forestry to identify most of the major insect, mite, and disease problems found in the United States and Canada.

The authors

Ithaca, New York

Acknowledgments

In the preparation of the second edition we incurred many debts. Acknowledgment alone seems inadequate, but it is, at least, a beginning. The breadth and quality of this project was made possible through many and diverse contributors; some gave money, others their professional expertise. Many persons reviewed and edited sections of the manuscript. Each has made the resulting book more authoritative and complete. Sectional reviewers included Douglas C. Allen, State University College of Environmental Science and Forestry, Syracuse, NY; Leland R. Brown, University of California, Riverside; John A. Davidson and Michael J. Raupp, University of Maryland, College Park; Raymond G. Gagné and Manya B. Stoetzel, USDA, Washington, DC, and Beltsville, MD, respectively; Fred P. Hain, North Carolina State University, Raleigh; W. W. Middlekauff, University of California, Berkeley; Laverne L. Pechuman, Cornell University, Ithaca, NY; James D. Solomon, U.S. Forest Service, Stoneville, MS; and A. G. Wheeler, Jr., Pennsylvania Department of Agriculture, Harrisburg. We are especially grateful to Douglas C. Allen for editing the entire manuscript before it was submitted for publication.

Others who supplied advice, technical information, or field assistance were Arthur L. Antonelli, Washington State University, Puyallup; Leslie W. Barclay, Barbara A. Barr, and Jerry A. Powell, University of California, Berkeley; Douglas L. Caldwell, Chemlawn Research Center, Columbus, OH; Joseph Capizzi, Jr., Oregon State University, Corvallis, who made the Kenneth Gray collection of insect photographs available to us; Sharon J. Collman, Washington State University, Seattle; John A. Davidson, University of Maryland, College Park; Robert F. DeBoo, Ministry of Forests, Victoria, B.C.; Harold A. Denmark, Avas B. Hamon, and Frank W. Mead, Florida Department of Agriculture and Consumer Services, Gainesville; John G. Franclemont, Cornell University, Ithaca, NY; Raymond J. Gill, California Department of Food and Agriculture, Sacramento; James E. Gillaspy, Texas A & I University, Kingsville; William H. Hoffard, U.S. Forest Service, Asheville, NC; Norman E. Johnson, Weyerhaeuser Company; Michael Kosztarab, Virginia Polytechnic Institute and State University, Blacksburg; Gerald N. Lanier, State University College of Environmental Science and Forestry, Syracuse, NY; David A. Leatherman, Colorado State Forestry Service, Fort Collins; John C. Moser, U.S. Forest Service, Pineville, LA; John W. Neel, USDA, Beltsville, MD; Roy D. Parker, Texas A & M University, Corpus Christi; Milo E. Richmond, Cornell University, Ithaca, NY; Peter B. Schultz, Ornamentals Research Station, Virginia Beach, VA; Joseph D. Shorthouse, Laurentian University, Sudbury, Ont.; and Robert D. Wolfe, U.S. Forest Service, Broomall, PA.

We thank Susan Pohl and Alice B. Johnson for word processing and editorial assistance, and we are especially grateful to Hélène Maddux, Robb Reavill, and Richard Rosenbaum, members of the editorial and production staff of Cornell University Press, for their dedication to excellence.

Preparation of this second edition and the companion volume, *Diseases of Trees and Shrubs*, were sponsored by the New York State College of Agriculture and Life Sciences at Cornell University. The project was started in the academic environment fostered and promoted by Dean Charles E. Palm and was completed some 20 years later.

Special thanks are due the administration, faculty, and staff of North Carolina State University for providing technical assistance while the senior author was on sabbatical leave there. In particular, we thank James R. Baker, Lewis L. Deitz, Maurice H. Farrier, Fred P. Hain, Ronald J. Kuhr, Herbert H. Neunzig, Carrol S. Parron, and Jane F. Stephens.

Money also made this book possible. Contributions toward the cost of printing have resulted in a book price within reach of students. These contributions came from the following organizations:

American Association of Nurserymen
Bartlett Tree Experts
Chemlawn Corporation
Ted Collins Associates
Davey Tree Expert Company
Greymont Tree Specialists
Grace Griswold Fund
Horticultural Research Institute
International Society of Arboriculture
Long Island Arborist Association
J. M. McDonald Foundation
Maryland Arborist Association
National Arborist Association
New York State Arborist Association
Kenneth Post Foundation
Readers Digest Association, Inc.
Rosedale Nurseries, C. Powers Taylor
Weyerhaeuser Company Foundation
Wisconsin Arborist Association

Most of the photographs were taken by the authors and are used by permission of the Departments of Entomology and Plant Pathology, Cornell University. Others who supplied photographs and drawings are acknowledged in the captions.

Reader's Guide to Identification of Insects and Related Pests

(Numbers refer to plates)

Pests That Feed on Conifers

- Leaf-consuming pests
  - Bagworms, 80–81
  - Gross defoliators, 1–7, 10, 22
  - Leaf tiers and webbers, 8, 12–13, 16
  - Needle miners and case bearers, 9, 11–14
- Borers of bark, twigs, and shoots
  - Bark feeders, 21, 23–29
  - Tip and shoot borers, 3, 17–20
  - Twig girdlers, 22
- Piercing/sucking pests
  - Aphids and adelgids, 30–35, 49, 50
  - Mealybugs, 37–38
  - Mites, 52–54
  - Needle and twig midges, 15
  - Scale insects, 37, 39–48
  - Spittlebugs, 36
- Gall makers, 30, 49–51

Pests That Feed on Broad-leaved Trees and Shrubs

- Leaf- or flower-consuming pests, 55–74, 83, 105–107, 110–114, 236–237, 240
  - Bagworms and case bearers, 80–81, 85
  - Leaf miners, 84–98
  - Skeletonizers, 98, 102–104, 110
  - Webworms and leaf rollers, 75–79, 82, 99–101
- Root-feeding and root-crown-feeding pests, 110, 112–114, 132, 150
- Borers of shoots, stem, bark, and wood
  - Bark beetles and borers, 115–118, 127–128
  - Stem and shoot borers, 91–92, 108–109, 122–125, 135
  - Wood borers, 119–123, 126–134
- Piercing/sucking/rasping pests
  - Aphids, 138–150
  - Lace bugs, 204–206
  - Leaf hoppers, 115, 198–201
  - Mealybugs, 154–158, 163
  - Mites, 228–235
  - Plant bugs, 190–194, 203
  - Plant hoppers and tree hoppers, 194–197, 203
  - Psyllids, 136–137, 217–221
  - Scale insects, 158–189
  - Spittlebugs, 202
  - Thrips, 207–209
  - Whiteflies, 151–154
- Gall makers, 108, 131, 210–227, 232–235

Miscellaneous Pests

- Ants, cicadas, crickets, walkingstick, and wasps, 236–239
- Birds, 241
- Mammals, 241
- Slugs and snails, 240

Introduction

The Book and Its Use. This book and its companion volume on diseases are reference manuals that cover the major biotic pests and environmental problems associated with woody ornamental plants. This volume covers arthropod pests, providing essential information about most of the more common and important insects, mites, and other animals that can damage woody ornamental plants. Its readership is intended to be the horticultural adviser, teacher, student, nurseryman, arborist, forester, gardener, scientist, as well as any other person having direct or peripheral responsibility for the maintenance of trees and shrubs. It deals in a pragmatic way with the science of entomology and provides visual assistance in the identification of insects and related animals often considered pests. Scientific information is given in nontechnical language to make it more useful to those interested in plant protection and natural history. The metric system (SI) of measurement has been used throughout. Transition to this international system has already started, and in a few years *millimeter, centimeter, liter,* and *Celsius* may well be a part of our everyday language.

The plants considered in this volume and in Sinclair et al., *Diseases of Trees and Shrubs,* are either aesthetically pleasing or serve some environmental function important to mankind. The number of species of insects and mites destructive to ornamental plants in the United States alone is conservatively estimated to be about 2500. Several of the species illustrated and discussed in this second edition are relatively new to North America. As might be expected, about 65% of newly introduced arthropods are first found in urban areas. We deal with about 900 species, both native and introduced. They represent our appraisal of which ones are the "most important" species in the United States and Canada or, in our opinion, which ones have the potential to spread and become important to the health of woody ornamental plants over a large geographic area. We have included some common species that are not pests (their damage is inconsequential) because of their unusual features such as galls or because they are frequently encountered and are often perceived to be threats to their hosts.

Although the color photographs show specific examples of insects and their damage, many are sufficiently generalized to enable one to recognize closely related species not illustrated. By comparing the actual arthropod specimen and injury symptoms with the photograph and by reading the text and caption, the reader should be able to identify the cause of plant injury.

In preparing this volume, we assumed that the user would have some horticultural or urban forestry background and at least general knowledge about the identification of ornamental plants. Where meaningful, common names of both plants and pests are used, but to avoid confusion both common (if officially recognized) and Latin names have been given for pests and most plants. Our authorities for plant names are McClintock and Leiser (504) and Little (475). It is important also to know what constitutes the normal and the abnormal—in many instances not as simple as it may seem. The more knowledge one has about trees and shrubs and their normal growing habits, the more useful this book will be. The process of elimination is an important means of identifying the cause or causes of plant disorders under field conditions; such an empirical method is the tool of the scientist as well as the lay person. The index is also an important tool. Make good use of it! The Reader's Guide to Identification of Insects and Related Pests is a simplified key, based in part on plant groups and in part on where or how an insect feeds.

A list of references is included for those desiring more detailed information about specific arthropods covered in this book. Entries are listed alphabetically by the author's last name and are numbered in an appendix. The references applicable to the discussion are listed by number at the end of each text/plate unit. Also, when the work of a specific scientist is referred to in the text, it is cited parenthetically by its number in the list of references. To avoid burdening the text with such references, we have often cited reviews rather than original papers. We regret that historical precedence and proper investigator credit is occasionally obscured by this procedure. There are many horticulturally important arthropod species and groups for which there are meager or no published accounts about their behavior and life histories. This is particularly true of many species of gall makers. The references cited can serve as a source for other illustrations and provide details that space would not allow in this volume. A glossary provides definitions of the more technical terms.

Pest Control. Pest control is a dynamic field, and changes in its technology come about frequently. New chemical compounds that are continually being made available usually displace older materials. Recommendations or other guidelines for the control of pests of ornamentals are prepared and distributed by agricultural agencies in most states and Canadian provinces (see "Sources of Information on Pests and Pest Control" at the back of the book). Such recommendations, based in most cases on local conditions and experience, are always to be preferred over those that attempt to generalize over a broad area. These statements are by no means intended to imply that chemical control is the sole means available to combat all pests that attack woody ornamental plants. Although insecticides and acaricides (miticides) have been heavily relied on in the past and their continued use in the foreseeable future is a certainty, many interesting new approaches to insect control are being investigated. Many of these were developed to control pests of traditional food and fiber agriculture yet were found to have application in the control of pests of ornamental plants. Since the mid-1970s increased research emphasis has been given to the control of ornamental pests. The effective use of insect pest-resistant plants, of biological control, of pest control through the application of plant nutrients on soil and in foliar sprays, and of monitoring ornamental pests by pheromone traps have become realities in a number of situations. Combining these and all other appropriate pest control tactics is known as Integrated Pest Management (IPM), and is becoming the standard procedure for control of pests of woody ornamentals.

Pest Identification and Problem Diagnosis. With each change of season a plant must adjust to a new set of living conditions. In fact, there are daily changes in the plant's environment—some subtle, some dramatic. The growing plant must accommodate to each new set of conditions if it is to continue to perform well. Failure to adjust often brings about distress symptoms. Sometimes the problem is obvious, but more often one must search for specific, subtle symptoms and signs in order to narrow the range of possible causes.

A *symptom* is an injury by, or a plant response to, a pest agent. A *sign* is the pest organism itself, its skeleton, or a product produced by the pest; a sign is helpful in identifying the cause of a symptom (Table 1). The process of problem diagnosis is based largely upon symptoms and signs. Symptoms are produced from two general sources: the living (biotic) agents and the physical (abiotic) environment. The living agents are parasites, that is, animal pests and plant pathogens (see Sinclair et al., *Diseases of Trees and Shrubs*). Physical environmental factors are soil, climate, local weather conditions, and human-modified atmospheric and soil conditions. All of these must be taken into account by those who attempt to diagnose plant disorders (Table 1). Symptomatic evidence of insect or mite attack is treated in Table 2.

These symptoms and signs are based on the fact that all insects and their relatives must feed in order to survive and reproduce. In doing so,

***Table 1.* Possible causes of injury to plants**

Biotic environment		Physical environment	
Animal	Plants (including microorganisms)	Human-controlled	Natural
Nematodes	Viruses and viroids	Industrial wastes	Mineral deficiencies
Insects	Mycoplasmalike organisms, *Spiroplasma*	Air pollution	Frost/freeze
Mites	Bacteria	Phytotoxicity (pesticides)	Sun scorch
Millipedes	Rickettsialike organisms	Salt (chemical) accumulations	Drought
Slugs and snails	Fungi	Concrete and macadam	Lightning
Birds (woodpeckers)	Dodder	Soil compaction	Wind
Rabbits, mice, moles, and other rodents	Mistletoe	Changes in soil level	Hail
Dogs	Algae	Changes in groundwater level	Blowing sand
Humans	Moss	Mechanical injury by auto, machinery, and vandalism	Flood
	Weeds		
	Strangling vines		
	Strangling roots		

they produce or leave behind symptomatic evidence of their having been on the plant. The symptomatic categories alone do not serve to identify or distinguish different groups of insects because a single pest may cause symptoms or signs in more than one category. Aphids, for example, cause symptoms of yellowing (category II), plant distortion (category III), and are responsible for products such as honeydew, sooty mold, and cottony wax (category V), and some species of aphids commonly cause several kinds of symptoms or signs concurrently. Scale insects may cause dieback of plant parts (category IV) and leave products on plants (category V). Snails and slugs cause tattered foliage (category I) and leave slime trails (category V).

To diagnose a problem, let the plant serve as an early indicator of what may be wrong. Then search for pests capable of causing the kinds of symptoms or signs observed. Keep in mind that the pests found must occur in numbers sufficient to cause the damage noted. Remember also that the simple association of insects found on a damaged host is not always a cause-and-effect situation.

Symptoms of injury resulting from nonliving environmental factors may resemble injury symptoms caused by living entities. A severe frost during the time a maple leaf bud is in a critical state of development will cause a series of holes to form in the developing leaf. Certain caterpillars can chew holes of similar size and shape.

The same symptom may be produced by several organisms that are unrelated, or by the physical environment. *Verticillium,* a disease-causing fungus, may induce wilting in a single maple branch or the entire tree may be involved. Cryptic wood borers, drought, or a lightning strike may also cause the same symptom. Sometimes, too, a tree may be damaged by several agents working together or in sequence. For example, a tree may be weakened by some human activity, or by climate, resulting in greater susceptibility to an insect, which, in feeding, may inoculate the tree with a pathogen that actually kills the plant. If sequential events are indicated, the diagnostician should follow the symptoms of injury back in time, to determine the event or pest that initiated the problem.

The plant, of course, must be identified first. Sometimes broad identification categories such as oak or pine are acceptable, and it is not necessary to identify the plant more specifically, for example, as a white oak or a red pine. The diagnostician must know, however, whether or not the plant is normal and healthy. Some years ago certain galls on plants were considered normal growth rather than the symptoms of gall maker attack. General health and vigor can often be determined by examining the distances between branch nodes and by comparing them with plants of the same species growing nearby. Short internodes indicate a history of stress or poor growing conditions (Figure 1).

The diagnostician must be aware that plants have a characteristic life expectancy, which may vary from place to place. Thus, the decline and death of trees and shrubs may come about as a result of old age, as with people and other animals. In certain locations gray birch or mimosa may be old at 40 years. A 50-year-old birch could be senile and therefore difficult, if not impossible, to save from the bronze birch borer. The coast redwood, at the other extreme, may live for 30 centuries.

The death of a tree or shrub is itself a symptom. In the case of certain bark beetles, symptoms from a dead plant are useful. But for the most part postmortem diagnosis is not very useful because many secondary organisms enter dead plants and erase, subdue, or confound the symptoms that are essential to understanding why the plant died.

Climate, of course, is a major limiting factor in the distribution of plant species. The orange tree and the palm, for example, will not survive the winter outdoors in Michigan. Likewise, many northern conifers do not survive or perform well in southern latitudes. People

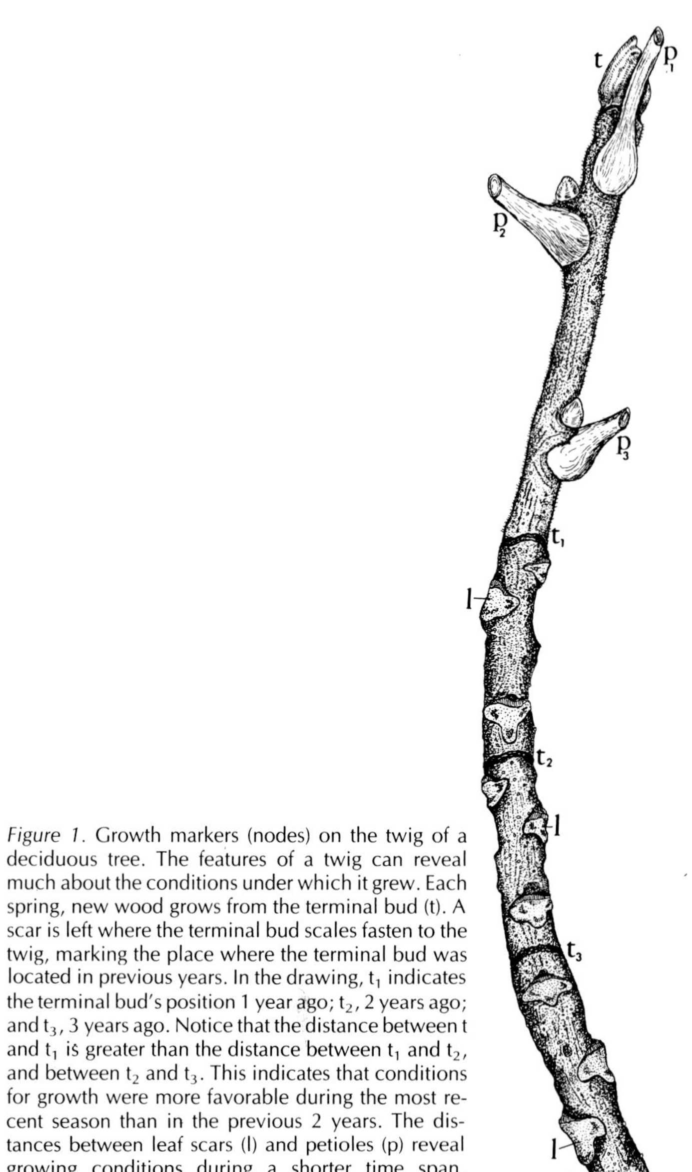

Figure 1. Growth markers (nodes) on the twig of a deciduous tree. The features of a twig can reveal much about the conditions under which it grew. Each spring, new wood grows from the terminal bud (t). A scar is left where the terminal bud scales fasten to the twig, marking the place where the terminal bud was located in previous years. In the drawing, t_1 indicates the terminal bud's position 1 year ago; t_2, 2 years ago; and t_3, 3 years ago. Notice that the distance between t and t_1 is greater than the distance between t_1 and t_2, and between t_2 and t_3. This indicates that conditions for growth were more favorable during the most recent season than in the previous 2 years. The distances between leaf scars (l) and petioles (p) reveal growing conditions during a shorter time span. (Drawing by R. Dirig.)

Table 2. **Symptoms or signs of insect attack**

Category	Pests often responsible
I. Chewed foliage or blossoms	Larvae of moths and butterflies Sawfly larvae Beetle larvae or adults Tree crickets, grasshoppers, and walkingsticks Snails and slugs
II. Bleached, bronzed, silvered, stippled (flecked), streaked, or mined leaves	Leafhoppers Lace bugs Plant bugs Thrips Aphids Psyllids Spider mites Leaf miners
III. Distortion (swelling, twisting, cupping) of plant parts	Thrips Aphids Eriophyid mites Gall makers Psyllids
IV. Dieback of twigs, shoots, or entire plant; stems, branches, and exposed roots sometimes with holes in bark; wood dust, frass, gum, or pitch may issue from holes	Wood borers Bark beetles Scale insects Gall makers Root-feeding beetle larvae
V. Presence of insect, or insect-related, products on plants:	
Honeydew and subsequent sooty mold	Aphids Soft scales Leafhoppers Mealybugs Psyllids Whiteflies
Fecal specks on leaves	Lace bugs Greenhouse thrips Some leaf beetles Some plant bugs Some sawfly adults
Tents, webs, silken mats	Tent caterpillars Leaf tiers Webworms
Bags and cases	Bagworms Case bearers
Spittle	Spittlebugs
Cottony fibrous material	Adelgids Mealybugs Some aphids Some scales Some whiteflies Flatids
Slime	Snails Slugs
Pitch tubes	Some bark beetles
Pitch or gum masses and sap flow	Larvae of certain moths Larvae of certain beetles Larvae of certain midges

are experimenters and continually attempt to introduce plants to areas where they normally will not grow. The plants may survive, and in some cases even thrive, until the 1-in-10-year freeze or drought occurs. Knowing the climatic zones that favor various plant species is an important aid in diagnosing disorders.

Trees or shrubs growing in situations where they are exposed to salt, fumes, natural gas, or other hydrocarbon vapors can produce symptoms difficult to diagnose. Drought symptoms may be delayed or not fully apparent until several years after the dry period has passed. Similarly, it may take more than a year for the symptoms of waterlogged soils to become evident.

The diagnostician should have access to the expertise of insect taxonomists. Even after the sign or signs have been characterized, or a sample collected, exact identification of the pest—if such is required—may take weeks or months. Some insects such as aphids, scales, and certain caterpillars cannot be precisely identified in their immature stages. If they are to be identified, they must be reared through to the adult stage. Certain species of leafhoppers can be positively determined only if the male insect is available and then only by a taxonomic specialist. The male must be killed and dissected to expose internal structures that will permit identification.

The information in this volume capitalizes on the diagnostic sleuthing, symptom interpretation, and taxonomic skills of many scientists. The plates are illustrations of signs and symptoms. The text and legends help to interpret the symptoms and identify the signs. This system, admittedly, is a shortcut to problem diagnosis in which there are many pitfalls. It will, however, permit identification of the more important pests.

In some cases the actions of disease organisms and arthropods are inexorably linked, as when arthropods act as disease vectors. Mechanical and practical considerations have prevented an amalgamation of the two subjects into one book; therefore, we present to our readers two volumes. Details about the diseases are found in Sinclair et al., *Diseases of Trees and Shrubs*; details about the arthropods are available herein.

Natural History. A pictorial study of plant or animal life is usually a nontechnical approach to natural history. Volumes on birds, mammals, fish, reptiles, and flowers are much used by the layperson, student, and professional. Pictorial representations reduce the use of, and help the nonprofessional understand, technical jargon. Insect natural history can be interesting and exciting to those who have economic reasons for learning it as well as to those who simply have a curiosity about animal life.

Insects affect human life and comfort daily. They are the most abundant group of animals on the earth. We need to go no farther than our own premises to explore their natural history. Their activities and behavior are nothing less than phenomenal, but one must be a careful observer to appreciate them.

Certain insects have evolved to the point where separate sexes are unnecessary for the multiplication of the species. Males of certain species are unknown. Thus the female produces more females without fertilization (see discussion of Asiatic weevils, Plate 114). A phenomenon common to gall-producing wasps and certain aphids is alternation of generations. One generation may be produced sexually, have a very characteristic appearance, and feed in a very precise way. Its offspring may be only female, may look entirely different, and may feed on a plant unrelated to the one used as a host by the parent (Plate 140). Insects have unique ways of communicating with one another. Chemical communication by odor is a means by which certain insects convey messages. These chemicals, called semiochemicals, are currently of much interest to those studying insect biochemistry and behavior. Pheromones, one kind of semiochemical, aid a species in finding food (Plate 25), marking a trail, and finding a mate (Plate 61).

The range of longevity of insects and related arthropods is great. The adult males of many species (scale insects) live only a few hours, just long enough for mating to take place. Likewise, some adult females live only a few hours or days. The bagworm female (Plate 80) has no functional mouthparts. Her only role is to produce eggs. When her eggs are laid, she dies. For the most part adult insects live for a short time—3 days to 8 weeks. Certain beetles such as the pales weevil (Plate 21) live 9–18 months. In warm regions a cycle from birth to death may occur in 4 days (Plates 228–229). In the temperate zone the pattern of many insects is a 1-year cycle with a long hibernation period. Life is longer for the wood-boring insects; some live as long as 4 years under normal conditions (Plate 133). The insect with the longest life span in North America is the periodical (17-year) cicada (Plate 236).

Protective coloration is illustrated in Plate 21F. Defense adaptation with stinging hairs is illustrated in Plate 74. Sexual dimorphism is illustrated in Plate 196. And so it goes—endless examples of strange and interesting insect natural history.

The reader will note a frequent statement in the text describing the plates: "Little is known about this insect's biology and habits." There is much that needs to be learned about insects for our own well-being. There is no shortage of work left to be done.

Insects and Injury

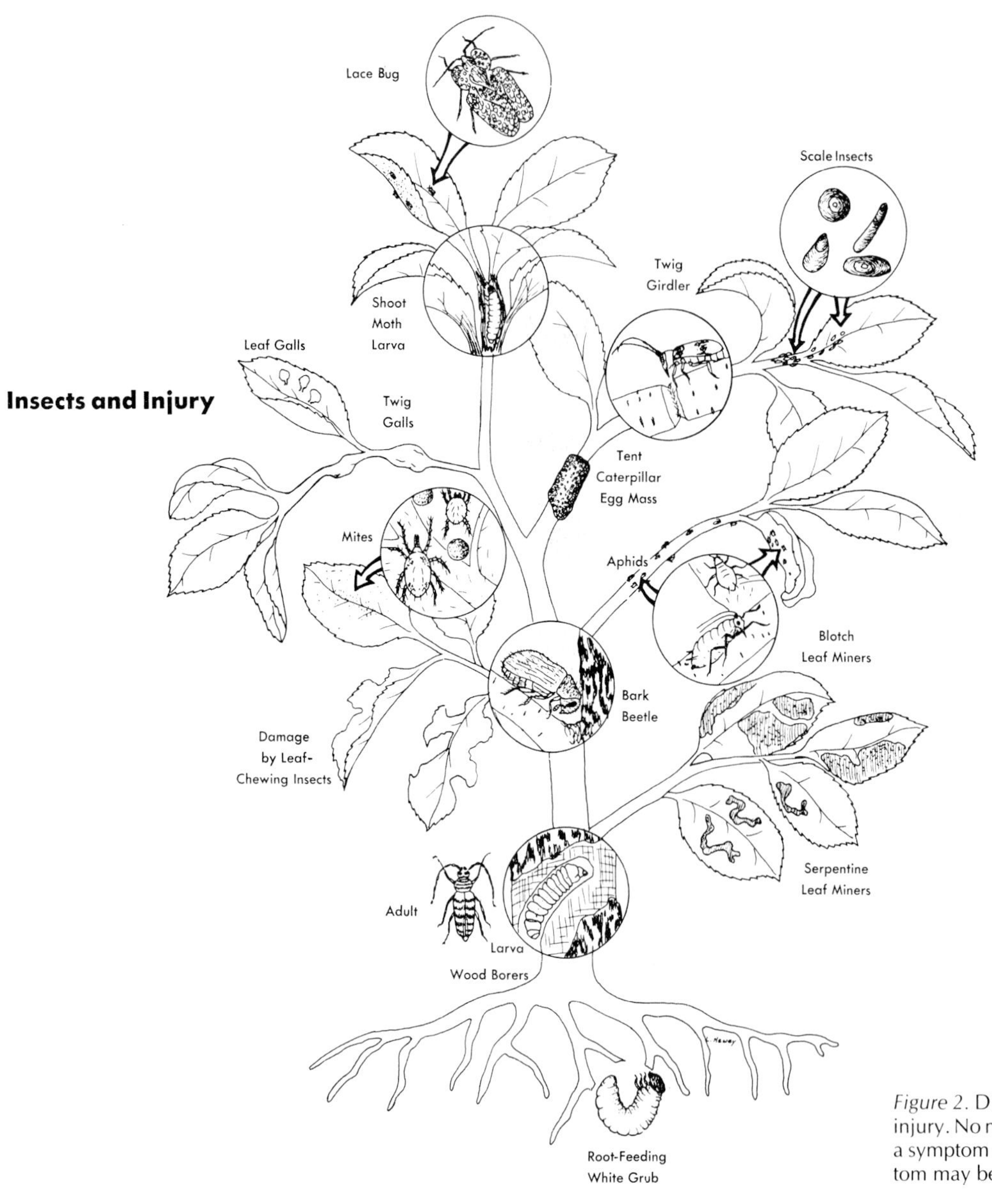

Figure 2. Diagram of the relationship of insects and plant injury. No matter how or where an insect feeds on a plant, a symptom is produced. Sometimes, however, the symptom may be difficult to see or interpret.

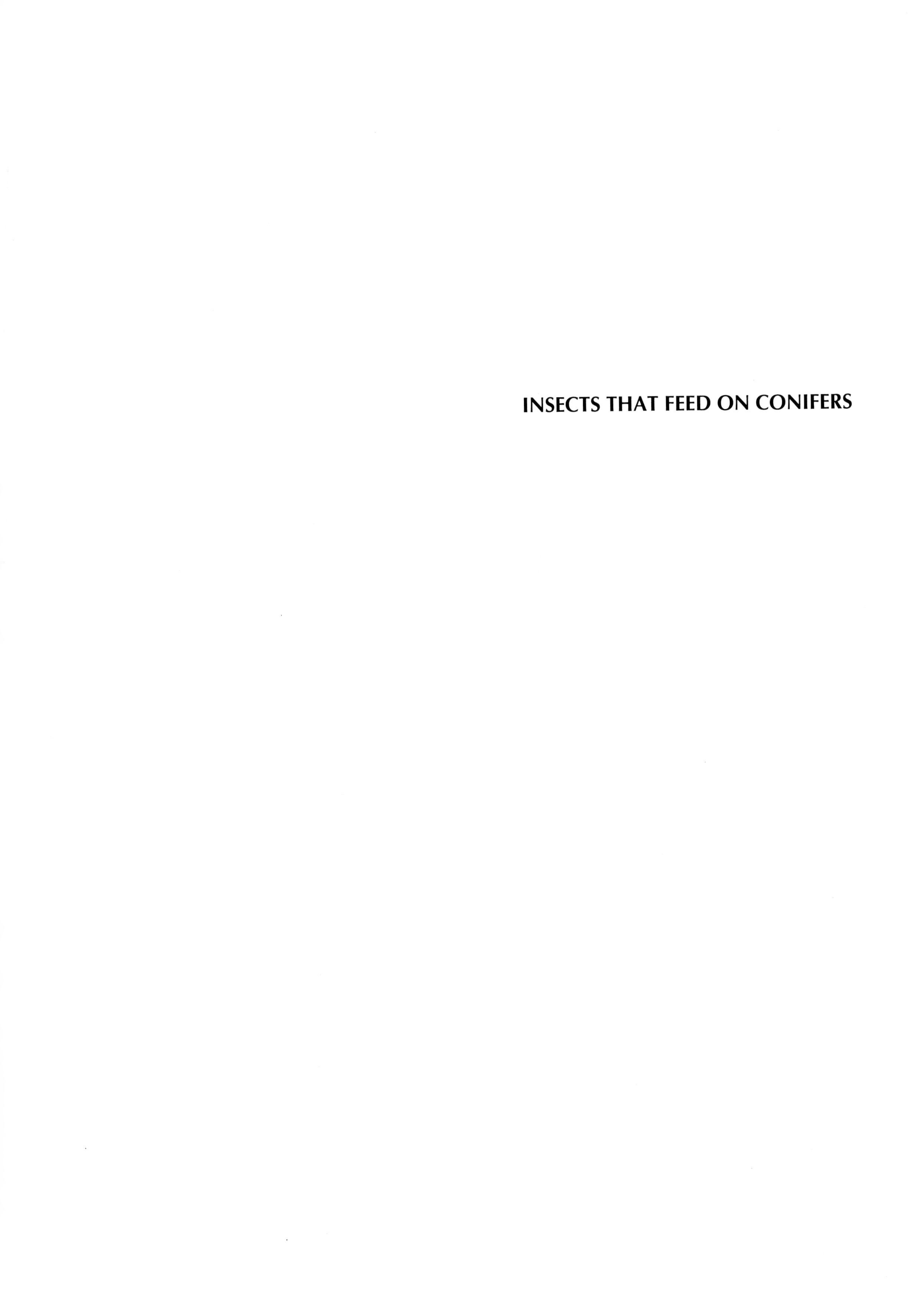

INSECTS THAT FEED ON CONIFERS

Sawfly Defoliators, Bud Miners, and Shoot Borers (Plates 1–3)

The conifer-feeding sawflies may feed on foliage, may mine buds, or may bore into the pith of shoots. Sawflies are not flies at all, but nonstinging wasps. The females of most economically important species have a sawlike apparatus (Figure 3) at the tip of their abdomens, which accounts for the name *sawfly*. The "saw" is used to slit or cut plant tissue and aids in the insertion of eggs into these slits (Plate 2C). Numerous species of sawflies occur in the United States. Probably 100 species, in the larval stage, feed on conifer foliage. Some of these species feed only on the old needles, others only on the new. Sawflies in the first larval stages eat only the outer portion of the needles; later the entire needle may be consumed. Partially eaten needles look like fine straw hanging on twigs (Plate 3A, B). Some species occur only in the spring, others in the summer, and others throughout the growing season. The larvae of typical defoliating sawflies look much like the caterpillars of moths and butterflies; they may be differentiated from them, however, by the number of prolegs on the abdomen. Sawfly larvae have more than five pairs, each of which lacks the hooked spines (crochets) typical of caterpillar larvae. Caterpillars have two to five pairs. Sawflies that attack broad-leaved trees may be defoliators, leaf rollers, web formers, skeletonizers, leaf miners, bud miners, stem borers, or gall makers (Plates 55–56, 84, 92, 226).

The life cycle of the redheaded pine sawfly, *Neodiprion lecontei* (Fitch) is typical of most sawflies. An important pest of ornamental, forest, and especially plantation trees, this species attacks jack, shortleaf, loblolly, slash, red, Scots, and other two- and three-needle pines. It may also feed on five-needle pines, Norway spruce, deodar cedar, and larch if these trees are growing among other, preferred hosts.

The winter is spent as a prepupa (stage between a larva and pupa) in a cocoon spun on the ground or in duff under host trees (Plate 3E). Pupation is completed in the spring and the adults emerge a few weeks later. Some prepupae may wait until the second or third season before transforming into adults. This delayed emergence is probably a mechanism to allow the species to survive should all those that emerge in a particular season die.

The female deposits more than 100 eggs in rows of slits in the edges of several needles. Egg laying may take place without mating, but unfertilized eggs produce only males. The fertilized eggs produce males and females.

Larvae hatch from the eggs after about a month and feed gregariously for another month or so, before dropping from the host to the ground to spin their cocoons. In Canada and the northern United States, a single generation occurs each year. In the latitude of southern Michigan and New York, a second generation may develop, and in parts of South Carolina, Georgia, and the Gulf states, three generations occur annually.

The rapid decline of a population of redheaded pine sawflies is often due to rodents that feed on the pupae and to diseases that kill large numbers of larvae. The redheaded pine sawfly is shown in panels D and E.

The introduced pine sawfly, *Diprion similis* (Hartig), is an unintentional import from Europe. It feeds on a wide range of pines, including white, red, Scots, and jack. Austrian pine is somewhat resistant. The sawfly is found from Maine to North Carolina, in the Central and Great Lakes states, in parts of Ontario and Quebec, and as far west as Minnesota. Two generations occur each year in its northern range, but three occur in the mountains of North Carolina. A mature larva is shown in panel F. Eggs are laid in series in old needles and are covered with a green frothy substance.

N. excitans Rohwer, the blackheaded pine sawfly, is an important pest in the South, where it attacks all the southern pines and has been known to cause widespread damage in young pine plantations. Four or five generations may occur in a year. The mature larvae have glossy black heads and olive-green bodies. Two longitudinal black stripes run along the back of the body, and a row of conspicuous black spots can be found on each side.

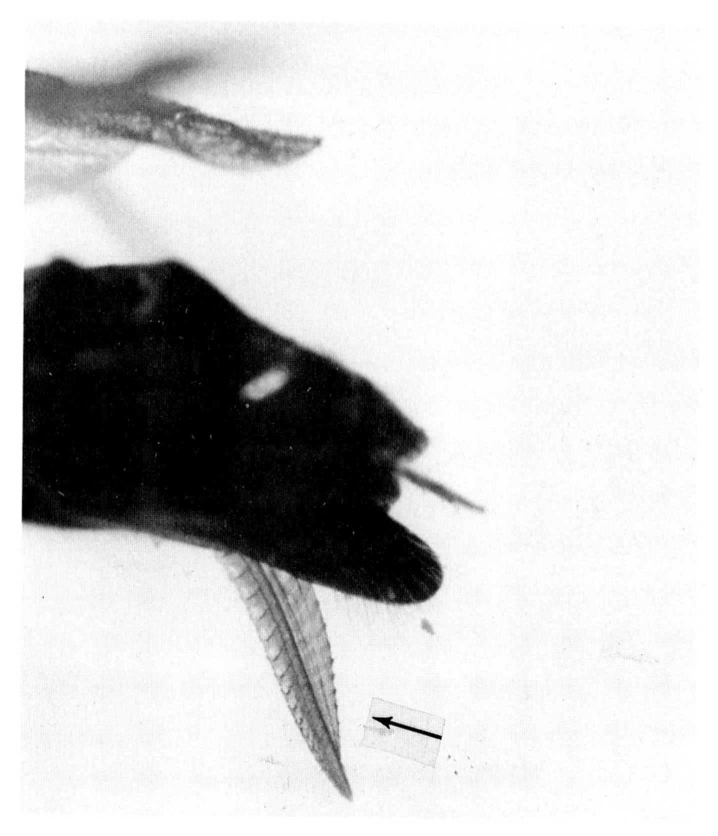

Figure 3. The abdomen of a female sawfly. Arrow points to the "saw," or ovipositor, used to cut plant tissue and to insert eggs.

The European pine sawfly, *N. sertifer* (Geoffroy), is an important pest of red, Scots, Japanese red, jack, Swiss mountain, and mugo pines. The larvae are gregarious and feed on mature foliage (Plate 2A). The mature larva is grayish green with markings that allow it to be distinguished from other species (Plate 2D, E).

The European pine sawfly is found in the northeastern United States and Canada. One generation occurs annually.

From Anderson (8) we have this list of common spring- and summer-feeding sawflies that attack pine.

1. Loblolly pine sawfly, *N. taedae linearis* Ross; south central United States; loblolly and shortleaf pines.
2. Spotted loblolly pine sawfly, *N. t. taedae* Ross; southeastern United States (Virginia); loblolly, shortleaf, and Virginia pines.
3. Jack pine sawfly, *N. pratti banksianae* Rohwer; eastern Canada and the Great Lakes region; jack pine and occasionally other pines.
4. Virginia pine sawfly, *N. p. pratti* (Dyar); southeastern United States; loblolly, shortleaf, and scrub pines.
5. *N. p. paradoxicus* Ross; northeastern United States, from Maine to North Carolina; jack, Virginia, pitch, and shortleaf pines.
6. Red pine sawfly, *N. nanulus nanulus* Schedl; southeastern Canada, northeastern United States and the Great Lakes states; red and jack pines.

A. A branch of *Larix decidua* partially defoliated by larch sawfly larvae.
B. Close-up of mature larvae of the larch sawfly.
C. An adult larch sawfly.
D. A branch of *Pinus nigra* partially defoliated by larvae of the redheaded pine sawfly. Larva at arrow.
E. Mature larvae of the redheaded pine sawfly.
F. Mature larva of the introduced pine sawfly.
G. Larch sawfly eggs embedded in larch twig. (Courtesy H. Tashiro, New York State Agricultural Experiment Station, Geneva.)

A
B
C
D
E
F
G

7. European pine sawfly, *N. sertifer* (Geoffroy); northeastern North America west to Illinois; red, pitch, shortleaf, and occasionally other pines.

8. Hetrick's sawfly, *N. hetricki* Ross; southeastern United States; loblolly and possibly other pines.

9. Redheaded pine sawfly, *N. lecontei* Fitch; eastern half of North America; all hard (two- or three-needle) pines (sometimes the foliage of other conifers is eaten by older larvae). Usually infests smaller trees up to about 5 meters tall.

10. White-pine sawfly, *N. pinetum* Norton; northeastern and north central North America; white pine. (Figures 4–6)

11. Pitch pine sawfly, *N. pinusrigidae* Norton; northeastern United States; pitch pine.

12. Lodgepole sawfly, *N. burkei* Middleton; western North America; lodgepole pine.

13. Abbott's sawfly, *N. abbottii* Leach; eastern North America; all hard pines.

14. Swaine jack pine sawfly, *N. swainei* Middleton; southeastern Canada and the Great Lakes states; jack pine.

15. *N. excitans* Rohwer; southeastern United States; most hard pines.

16. Introduced pine sawfly, *Diprion similis* (Hartig); northeastern North America west to the Great Lakes states and south to North Carolina; chiefly the soft five-needle pines.

17. *Gilpinia frutetorum* Fabricius; northeastern North America; red and Scots pines.

In addition, there are several Pamphiliidae in the genus *Acantholyda* that are called pine-webbing sawflies because they spin fine webs that become filled with excrement and needle fragments. These are discussed in the text accompanying Plate 4.

It might seem that only pines have sawfly enemies, but other conifers have sawfly pests, too, though not as many as do pines. The hemlock sawfly, *Neodiprion tsugae* Middleton, is an example. It is an important defoliator of western hemlock and Pacific silver fir in the Northwest (Montana, Idaho, Oregon, Washington, and British Columbia), damaging specimen trees as well as those in forests. In British Columbia the larvae may be present from June to August. Newly hatched larvae are almost black with a shiny black head. Mature larvae, 1.5–2 cm long, are green or yellow-green with two dark bands along each side. Larvae are gregarious and feed on old foliage. In autumn females lay eggs in needles, and the eggs overwinter there. Adults and larvae resemble other sawflies in the genus *Neodiprion* and exhibit many of the same behaviors.

Following is a list of some of the other more important sawfly pests of conifers other than pine, their ranges, and hosts:

1. Larch sawfly, *Pristiphora erichsonii* (Hartig); across North America and southern Canada; all species of larch. (Plate 1)
2. Twolined larch sawfly, *Anoplonyx occidens* Ross; northwestern United States; western larch.
3. Western larch sawfly, *A. laricivorus* Rohwer & Middleton; western United States; western larch.
4. European spruce sawfly, *G. hercyniae* (Hartig); northeastern United States, eastern Canada; spruce.
5. Yellowheaded spruce sawfly, *Pikonema alaskensis* (Rohwer); Canada, northern United States south to latitude of Colorado; spruce.
6. Cypress sawfly, *Susana cupressi* Rohwer & Middleton; California; Monterey cypress; *Cupressus, Juniperus,* and *Thuja.*

Sawflies are hosts to many kinds of parasites, and are subject to various viral and fungal diseases. Rodents and birds are important predators. But even with these natural controls, outbreaks of sawflies occur in their range every year and cause severe injury to forests and landscape ornamentals.

Figure 4 (left). An adult male white pine sawfly, *Neodiprion pinetum.* × 4. (Courtesy USDA.)

Figure 5 (right). An adult female white pine sawfly. × 4. Note difference in body size and antennae compared with the male shown in Figure 4. (Courtesy USDA.)

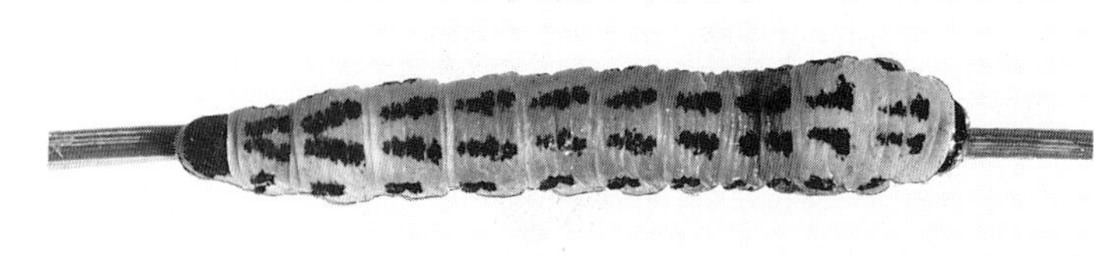

Figure 6. The larva of the white pine sawfly feeds on needles of eastern white pine. × 4. (Courtesy USDA.)

A. Scots pine needles being devoured by European pine sawfly larvae.
B. Eggs of European pine sawfly in Scots pine needles.
C. Close-up of eggs on a pine needle. Note that the eggs are partially embedded in the needle.
D,E. Fully grown European pine sawfly larvae.

A
B
C
D
E

Figure 7. An adult female *Pleroneura lutea* laying an egg into the bud of a white fir; about 5.5 mm long. (Courtesy C. P. Ohmart, CSIRO, Div. of Forest Research, Canberra, Australia.)

Figure 8. Damage to white fir buds and needles caused by *Pleroneura* larvae. [Xyelidae] (Courtesy C. P. Ohmart, CSIRO, Div. of Forest Research, Canberra, Australia.)

Within the genus *Pleroneura* are both bud miners and shoot borers that attack several species of fir. The species have similar life cycles and may occupy the same niche on the same host. The damage affects the host little, but it is undesirable and sometimes unacceptable on shade and ornamental trees.

In general, their life cycle develops as follows. Adults emerge in the spring soon after snow melts. Mating takes place and the female lays her eggs shortly thereafter. Eggs are laid *in* buds (Figure 7). By the time eggs hatch buds have opened, and needles and shoots begin to elongate. Larvae feed at the apex of the buds and fuse the tips of the needles together, but because the larvae bore down the central portion of the needle cluster, the needles in the outer layers are free along most of their length. As the needles elongate, they bow, forming a characteristic pattern (Figure 8). Larvae, as they develop, enter new shoots and form a hollow tube, but the shoot continues to grow. When mature, each larva chews a hole on the side of the shoot and drops to the ground, where it burrows into the soil and pupates. Winter is passed in either the larval or pupal stage. Full-grown larvae vary from 4.5 mm to 8.5 mm long, depending on the species. Most of these insects are found in the Rocky Mountains and west to the Pacific Coast. The sawflies illustrated in the figures below are western species. An eastern species found in Canada and the northern United States is called the balsam shootboring sawfly, *Pleroneura brunneicornis* Rohwer. Its larva tunnels in the new shoots, causing them to die and turn reddish brown, an effect similar to that caused by late frost. Such symptoms occur by early June. They are usually of concern only to Christmas tree growers or to people who own specimen trees. Hosts include balsam and white fir.

References: 8, 59a, 67, 191, 193, 194, 222, 275, 333, 498, 582, 659, 714, 778, 821, 855, 857, 860, 868a

A. Injury to Colorado blue spruce by the larvae of an undetermined sawfly. Note the strawlike appearance of the damaged needles.
B. A single needle from Colorado blue spruce injured by early instar sawfly larvae. Only the major vascular bundles (midrib) remain in the injured area.
C. A sawfly larva on white fir, *Abies concolor*.
D. Close-up of damage on white fir.
E. Typical sawfly cocoons.

A
B
C
D
E

Webspinning Sawflies, Pine Webworm, and Pine False Webworm (Plate 4)

Stripped foliage and large, ugly, and loosely webbed masses of needle fragments and frass are unmistakable symptoms and signs of webspinning sawflies and pine webworms (A, B, E). Although the symptoms and web masses produced by these two groups are similar, the animals producing them are distinguished by order characteristics (see text accompanying Plate 1) and by the lack of abdominal prolegs on webspinning sawfly larvae. These insects are of potential economic importance because they can severely defoliate forests, plantations, or ornamental plants.

The 21 or more species of webspinning sawflies that feed on conifer foliage are separated into two genera, *Acantholyda* and *Cephalcia,* which feed gregariously or singly. The solitary feeders build silk tube shelters along twigs or small branches.

The pine false webworm, *Acantholyda erythrocephala* (Linnaeus), originated in Europe and was first found in North America in Pennsylvania in 1925. An increasingly important defoliator, it now occurs from Pennsylvania northeast to New Brunswick and west to the Great Lakes states and Manitoba. It is considered to be the most common of the nest-building (web-spinning) sawflies. Females appear to favor *Pinus resinosa* (red pine) and *P. strobus* (eastern white pine) for laying their eggs. Larvae also feed on *P. mugo, P. montana, P. sylvestris, P. densiflora, P. nigra,* and *P. pungens.*

Figure 9. An egg and hatching larva of the pine false webworm. Eggs, attached to a pine needle, are about 2.8 mm long. (Photo by James F. Stimmel. Courtesy Karl Valley, Pennsylvania Dept. of Agriculture, Harrisburg.)

***Table 3.* Biological attributes of *Cephalcia* species**

Species	Distribution	Hosts	Adult emergence	Clutch size and form	Arboreal larvae (days)	Instars ♂	Instars ♀	Habit and shelter
californica	W	Pines: lodgepole, ponderosa	June, wk 1,2	7–11 eggs, 2 or 3 rows	33	5	6	Gregarious; frass nest
distincta	E	Balsam fir, eastern hemlock	May, wk 4; June, wk 1	—	—	—	—	Solitary; probably silk tube
fascipennis	T	Spruce: blue, Engelmann, red, white	May, wk 4; June, wk 1,2	1–3 eggs, single or zigzag rows	30	5	6	Primitively gregarious; frass nest
frontalis	E	Red pine	June, wk 2–4	To 10 eggs, 2 or 3 rows	48–52	5	6	Gregarious; frass nest
fulviceps	E	Pines: red, jack	June, wk 1,2	To 16 eggs, 2 or 3 rows	29	5	6	Gregarious; frass nest
hopkinsi	SW	Pines?	—	—	—	—	—	
marginata	E	Pines: red, jack, white	June, wk 1,2	4–15 eggs, 1 row	57	4	5	Gregarious; frass nest
nigra	E	White spruce	June, wk 4	—	—	—	—	Solitary; silk tube
provancheri	T	Spruce: black, Engelmann, red, Sitka, white	May, wk 3,4	1 egg	30	4	5	Solitary; silk tube
semidea	E	White spruce	June, wk 1	1 egg	37	No male	5	Solitary; silk tube

Note: E, eastern; SW, southwestern; T, transcontinental; W, western.
Source: Modified from Eidt (206).

Adults emerge from their earthen cells from mid-April to mid-May in the insect's southern range. Eggs are laid end to end on the flat surface of year-old needles in clusters of 3–10. Females produce an average of 16 eggs. The female makes a small slit in the needle, where she embeds a small portion of the chorion. After hatching (Figure 9), larvae spin loose webs at the base of needles. Within each web several larvae feed by cutting needles and pulling them into the web. Having no abdominal prolegs and only relatively short thoracic legs (F), the larvae are not equipped to move freely on the foliage. Without the web, they move unsteadily and they soon fall from the tree.

Larvae feed for about 18–20 days. Middlekauff (535) described the defoliation as unmistakable. "Defoliated twigs, stems, and branches are coated with large unsightly webbing to which much frass is attached [E]. Older needles are stripped and the current year's growth, with its needles, extends prominently beyond the naked frame of the tree. The new growth may be killed with no apparent feeding injury." Fully grown larvae drop to the ground and form earthen cells 5–8 cm beneath the surface. One generation occurs each year.

The bionomics of the several species of *Acantholyda* differ, but their web-spinning habit is consistent. Common species of limited significance include *A. verticalis* (Cresson), a reddish brown larva that may feed on all of the native northwestern pines, and *A. burkei* Middlekauff, which feeds primarily on *P. radiata* (Monterey pine). *A. zappei* (Rohwer) overlaps the geographic region of the pine false webworm.

The *Cephalcia* are native to North America and are easily confused with the *Acantholyda*. Eidt (206) stated that "eggs of all species are laid on needles of the host tree with a small knoblike part inserted into the host tissue. The larvae eat previous years' foliage, develop rapidly, and drop to the ground when through feeding. Larvae do not construct cocoons but overwinter in a cell formed in the soil." Pupation and emergence may occur the next spring or after two or more winters in the soil. Table 3 lists some of the bionomics of the more common species.

Because *Tetralopha robustella* Zeller, the pine webworm, is a caterpillar that attacks pines, it is important to homeowners and to growers of Christmas trees. It builds ugly web nests on terminal twigs; the loose webbing contains dead needles and brown frass (C). Usually the injury has been done and the nest vacated before the pest is noticed.

The biology of this insect was studied in Wisconsin in 1959. One generation occurs each year in its northern range; two generations are

---

A. A mugo pine with the abandoned nest of the pine webworm, *Tetralopha robustella.* [Pyralidae]
B. Close-up of frass, web, and dead needles—the leftovers from a colony of pine webworms.
C. A white pine twig with a nest of the pine webworm. The needles were removed and eaten by larvae.
D. The six instars of the pine webworm. (Courtesy D. M. Benjamin, Univ. of Wisconsin.)
E. A web mass of the pine false webworm on red pine.
F. A late-instar larva; about 20 mm long. [Pamphiliidae]

A
B
C
D
E
F

believed to occur each year from coastal Virginia to Florida. The larva overwinters inside a cocoon in the soil. The moths first appear in June, and emergence continues through July and part of August. Eggs are commonly laid in single rows along the pine needles. A needle may contain more than 20 eggs.

Upon hatching, the young larvae wander about, spinning silk threads among the needles. Then, selecting a needle, they begin feeding as miners. When they become too large to mine needles, the larvae, in colonies of a few to as many as 78 individuals, construct nests of frass and silk, covered by loosely webbed needles. The nests are usually 5 cm long but occasionally reach 15 cm, depending on the number of larvae in the colony. As the larvae devour the needles, the web becomes loosely filled with brown, nearly oblong, fecal pellets wrapped around the twig. Fully grown larvae, about 15 mm long, are yellowish brown with dark brown longitudinal stripes. They crawl or drop from the nest to pupate in the soil in late September. On Long Island, webs are found in late June and July; in Florida, webs of the year's first generation are found in April.

This insect occurs from New England to Florida and west to Wisconsin and Minnesota. The plant hosts by region are the following: Great Lakes states—jack pine, red pine, eastern white pine, Scots pine; Northeast—mugo pine, pitch pine, red pine, Scots pine, Virginia pine; South—slash pine, shortleaf pine, longleaf pine, loblolly pine. Parasites include *Apanteles* species, *Cryptus albitarsis* (Cresson), and *Mesostenus thoracicus* Cresson, all wasps, and *Phorocera* sp., a tachinid fly. A microsporidian, *Pleistophora schubergi,* is a known parasite of several species of sawflies. Birds such as chickadees and nuthatches tear at the nests to obtain the larvae.

References: 14, 46, 264, 535, 812

Moth and Butterfly Defoliators of Conifers (Plates 5–6)

On this and the following plates we illustrate three important groups of lepidopterous defoliators—loopers (or measuring worms), noctuid moth larvae, and pierid butterfly larvae—that may become abundant enough to defoliate pines and other conifers. Since evergreen conifers normally retain functional foliage for 2–5 years, defoliation is serious and often kills the tree.

The loopers (Geometridae and a few Noctuidae) are so called because of their method of locomotion. When crawling, they loop the abdomen as they bring the prolegs, at the end of the abdomen, forward toward the thoracic legs. Only caterpillars having a reduced number of abdominal legs walk with a looping motion.

Several species of loopers are important defoliators of conifers. *Lambdina pellucidaria* (Grote & Robinson), the eastern pine looper, can be a serious defoliator of pitch, shortleaf, red, and other pines. It overwinters in the pupal stage in the duff beneath its host tree. In Virginia the moths emerge in March; in New England they begin to lay eggs on the needles in May and June. The young larvae chew notches out of the needles, which turn straw brown (A) after a few days. This color change is often the first symptom of the presence of larvae. Later they devour the entire needle. Fully grown larvae are about 30 mm long. One generation occurs each year in Atlantic Coast states. On Cape Cod outbreaks have occurred in 10-year cycles for about 40 years. In 1970 an outbreak occurred on Long Island.

Related species, *L. fiscellaria* (Guenée), commonly called the hemlock looper, and *L. athasaria* (Walker), are defoliators of hemlock. More widely distributed than *L. pellucidaria,* they occur from Canada to Georgia and west to Wisconsin and Alberta. The hemlock looper is a pest of primary importance where hemlock and balsam fir are native species. It has a black lateral band with white teardrop-shaped spots along the length of its body. When it is viewed from above, small paired black spots appear to cover most of its body segments, including the head. Eggs overwinter and hatch in late May or early June. Young larvae feed on new foliage, but as they mature they accept old needles and foliage from other conifers, that is, larch, spruce, and arborvitae. They are wasteful feeders, rarely consuming the whole needle. Damaged needles turn brown, and by midsummer the outer shoots and twigs may be barren of needles. Larvae are slow growers, completing their development by late August or early September, by which time they have attained a length of about 30 mm. Adults are on the wing in September and October, laying their eggs singly on the bark of trunks or limbs.

The western hemlock looper, *L. fiscellaria lugubrosa* (Hulst), is periodically destructive to western hemlock and is limited to the Pacific northwest, Idaho, and Montana. In high populations the larvae readily accept Sitka spruce, Pacific silver fir, and Douglas-fir. The larva looks much like the hemlock looper but is generally yellowish to brown, lighter than its eastern counterpart. First and second instars feed in buds; older larvae feed on foliage of all age classes. Silken webs produced by the mature larvae as they descend from the tree become a nuisance, particularly on urban properties. Outbreaks of the western hemlock looper usually last about 3 years, after which they are generally brought under control by the natural action of parasites, predators, and disease. Of course, in an urban area or public park such a wait can kill a tree if human intervention is absent.

Phaeoura mexicanaria (Grote), sometimes called the pine looper, is found in the natural range of ponderosa pine, which appears to be its only host. Eggs are laid, unprotected, on needles or branches in linear clusters. Young larvae are smooth, hairless, and light brown. As the larva matures, the body becomes rough with tubercles and changes color to resemble a stubby pine twig. The fully grown larva is about 44 mm long. Tree injury first becomes evident in late July. The pupa overwinters on the soil, covered by duff.

Other loopers that attack conifers include *Semiothisa sexmaculata* (Packard), sometimes called the larch looper, and *Cingilia catenaria* (Drury), or chainspotted geometer. *S. sexmaculata,* which feeds on larch, is found from Newfoundland to Washington. The larvae have two color phases: dark, in a range of black, brown, and gray; and light, green with pale green stripes on the sides and back. The chainspotted geometer feeds on a wide range of both coniferous and deciduous trees. The colorful larva has numerous white, black, and yellow stripes extending from head to tail. When fully grown, it is about 45 mm long. Larvae are present from June to August.

The pine butterfly, *Neophasia menapia* (Felder & Felder) (D), is primarily a pest of ponderosa pine in the northwestern states and British Columbia. The butterfly is closely related to and resembles the better known cabbage butterfly. Adults fly in August, September, and October and lay their eggs (G) in rows on pine needles. Eggs overwinter and, in June or when the new needles begin to appear, hatch into tiny green caterpillars with black heads. Young larvae feed in clusters and prefer old needles. Older larvae (F) feed singly and mature by late July. Like other butterflies, they pupate by forming a chrysalis that is suspended by silk attached to needles or twigs. After about 2 weeks the adult butterfly emerges and flutters about pine foliage. When ground observations of about 24 per tree are made, an epidemic population can be expected the following year.

The caterpillar shown in panel H is a species of *Zale*. The larva of one of the noctuid moths, it has four pairs of abdominal prolegs. Solitary feeders, larvae of this genus are rarely abundant enough to cause economic injury, though they have a wide geographic range. The one whose photograph appears here was found on red pine in New York State in August. Larvae may be found from mid-June to late August. One generation is believed to occur each year.

A. Twigs of *Pinus rigida*. The strawlike needles are a symptom of attack by young caterpillars of *Lambdina pellucidaria*.
B. The measuring-worm posture of *L. pellucidaria*; 25–35 mm long.
C. Close-up of green needles notched by larvae. Later they will turn brown.
D. The pine butterfly. The wingspread is about 30 mm. (Courtesy Gray Collection, Oregon State Univ.)
E. The larva of a species of *Semiothisa*; about 22 mm long when fully grown.
F. The pine butterfly larva; about 25 mm long. (Courtesy Washington State Univ.)
G. Eggs of the pine butterfly attached to a ponderosa pine needle.
H. The larva of a species of *Zale* on pine needles.

A
B
C
D
E
F
G
H

Baldcypress (*Taxodium distichum*) is often thought of as a swamp tree, but it grows well on any good soil and is used successfully as a shade and specimen tree within its native range. One of its native defoliators, *Anacamptodes pergracilis* (Hulst), which could appropriately be called the cypress looper, is likely to be found in small numbers every year, held in check by natural forces. It has caused serious defoliation in Arkansas, Georgia, Maryland, and Florida. Because of its reproductive potential, the looper may pose problems in the future.

In Florida the moth stage, present throughout the year, flies at night and is attracted to lights. Adequate studies have not been made, but it is likely that overwintering or dormancy occurs in either the pupal or egg stage. The moth lays eggs (B) in small clusters in bark crevices or under loose bark flaps of the host tree. Upon hatching, larvae crawl to the foliage and feed for about 3 weeks. Early instars are the same color as foliage and may be described as nibblers, chewing notches in each of several leaves. The notched foliage usually dies. Where the insect is abundant, injured trees turn reddish brown. Older larvae consume everything but the petiole. In Florida injury becomes evident by July (A). Later instars (C) mimic the features of a baldcypress twig in both color and physical ornamentation. When ready to pupate, larvae move to the trunk and often chew a cavity along bark crevices or under bark flaps, making cells that are lined with bark fibers and silk. Pupation takes place here, although an equal number of individuals may pupate in exposed sites. In Florida six to eight generations are presumed to occur each year.

The imperial moth, *Eacles imperialis* (Drury), is a large, predominantly yellow moth that has several recognized subspecies. Illustrated in panel F are two color forms of the larva of *E. imperialis pini* Michener. Additional study may prove this insect to be a distinct species.

The larvae are defoliators of red, eastern white, jack, and Scots pines; they rarely occur on juniper and white spruce. Red and eastern white pines are the primary hosts. *E. imperialis pini* is found from eastern New York and Quebec west to Minnesota. This insect rarely becomes abundant, but because of its size just 10–20 individuals can defoliate large sections of a big tree, devouring entire needles. The large caterpillars are characterized by thoracic horns, spiracular white spots, and long, fine hair.

The pandora moth, *Coloradia pandora* Blake, is another in the group described as giant silkworm moths. It, too, has several recognized subspecies. Pandora moths infest pine forests, park lands, and occasionally specimen trees. Their range covers the southern Rocky Mountain region westward to California and Oregon.

The larvae are defoliators of ponderosa, Jeffrey, and lodgepole pines and, less frequently, Coulter and sugar pines. In parts of the West outbreaks occur at intervals of 20–30 years. Infestations were welcomed by American Indians before white colonists came because the insects were an important source of food. Today the Paiute Indians consider them a delicacy. They are cooked and processed in a way that allows for storage for up to 2 years. The Indians eat them like popcorn or in a stew with vegetables.

The pandora moth requires 2 years to complete its life cycle. Pupation typically lasts 1 year, but some individuals remain in diapause for up to 5 years before transforming into moths. Adults, which appear in late June or July, lay pale blue eggs, each about 3 mm long in clusters on pine needles or bark. Young caterpillars have shiny black heads and black or brownish bodies covered with short dark hairs. They feed in groups on new foliage, reaching a length of about 25 mm by the end of the trees' growing season. These half-grown larvae overwinter in clusters, attaching themselves at the base of needles. They resume feeding the following spring, reaching full size by the end of June. Pupation takes place in earthen cells fairly deep in the soil.

The larvae do the most severe damage when they come out of hibernation. The most obvious symptoms of infestation are thin crowns on the trees; signs of depredation by young larvae are stubs of needles and cast larval molted "skins." When moving from one feeding site to another, larvae travel together in single file.

At least four *Dasychira* species (tussock moths) defoliate conifers in Canada and the United States. *Dasychira plagiata* (Walker) and *D. pinicola* (Dyar) are more northern in distribution; *D. grisefacta* (Dyar) is widely distributed in the West; and *D. manto* (Strecker) occurs in the Southeast, including Texas. The larvae of these species have many similarities. *D. plagiata* and *D. pinicola* appear to be the most economically important. Their populations overlap in the Great Lakes states.

D. plagiata occurs from the Canadian maritimes to New York and the central Appalachians west to the Great Lakes states. It feeds on the foliage of jack, red, and eastern white pines, spruces, fir, larch, and hemlock. Eggs, which are laid from mid-July to early August, hatch shortly thereafter, and the larvae feed on the flat surface of needles, usually killing them. *D. plagiata* overwinters as immature larvae covered by a few silken threads attached either to rough bark or between the bases of needles. In the spring the larvae resume feeding on staminate flowers and young needles. As they mature (G), they consume the entire needle. Larval development is complete by May or June. Pupation occurs in cocoons attached to twigs or needles.

D. pinicola, also called the pine tussock moth, resembles *D. plagiata* in both its larval and adult stages. It has a seemingly disjunct distribution, for it is found in the Great Lakes states as well as in coastal areas of New Jersey north to Cape Cod. In New Jersey it is double brooded. *D. manto* has two or three broods each year, whereas the other species are single brooded.

References: 33, 65a, 104, 105, 175, 179, 193, 220, 237, 238, 257, 275, 499, 657–659, 748, 809, 862

A. Baldcypress defoliated by the cypress looper, *Anacamptodes pergracilis*. This gross symptom is evident by July. [Geometridae] (Courtesy W. Carothers, U.S. Forest Service.)

B. Eggs of *A. pergracilis* attached to the underside of a bark flap. Their color is often a deeper green than shown here. (Courtesy W. Carothers, U.S. Forest Service.)

C. A nearly fully grown cypress looper, about 25 mm long. (Courtesy W. Carothers, U.S. Forest Service.)

D. A late-instar larva of the "Black Hills" pandora moth, *Coloradia doris* (Barnes), feeding on pine. (Courtesy D. L. Leatherman, Colorado State Forest Service.)

E. A pandora moth larva; 5.5–7.5 mm long. (Courtesy Sharon J. Collman, Cooperative Extension, Washington State Univ.)

F. Two color phases of an imperial moth larva, *Eacles imperialis pini*, feeding on eastern white pine; 60–80 mm long. [Saturniidae]

G. The pine tussock moth, *Dasychira plagiata*; about 37 mm long. [Lymantriidae] (Courtesy J. Franclemont, Cornell Univ.)

A
B
C
D
E
F
G

Spruce Bud Moth and Spruce Budworms (Plate 7)

The variation in common names for the same species has unfortunately caused some confusion in discussions of the several caterpillars that feed on buds and shoots of spruce. *Zeiraphera canadensis* Mutuura & Freeman, commonly called the spruce bud moth, is one of the many insects introduced from Europe before effective plant quarantine laws were passed. It is now distributed throughout the eastern half of Canada and south to southern New York and probably Pennsylvania.

The larva of this insect feeds on the newly developing buds, needles, and shoots of white spruce (*Picea glauca*), black spruce (*P. mariana*), and balsam fir (*Abies balsamea*). Eggs are usually laid in late July under and between loose scales at the base of the current season's shoots. Eggs that overwinter successfully in this protected space hatch in early May in New York and in late May in eastern Quebec and New Brunswick. Upon hatching, the larva (C) moves to a developing bud and enters under a broken bud cap to feed on the tips and edges of the very young needles at the bud tip. The larva attaches the bud cap (B) to the developing needles with silk; here it remains, protected by this shell-like covering for most of its feeding period. The bud cap, which normally blows off in light winds, makes feeding sites conspicuous. By the fourth instar the larva, which has destroyed all of the foliage under the cap, begins to move down the new "candle" as it feeds at the base of needles and on tender bark. Late-instar feeding creates scars on the shoot (A) and weakens it; if the tree is being grown as a Christmas tree, its value is reduced. The larvae pupate in duff under the host tree, where the insects remain until adults emerge in early July. There is one generation a year. Egg and larval parasites are often numerous enough to control the bud moth in specimen plantings. The ant *Camponotus herculeanus pennsylvanicus* (De Geer) preys on pupae and on larvae that have dropped to the ground. A sex pheromone emitted by female moths has been identified, synthesized, and tested, and found to be effective bait for traps.

The spruce budworm, *Choristoneura fumiferana* (Clemens), and western spruce budworm, *C. occidentalis* Freeman, are the most important conifer defoliators in North America. They destroy billions of board feet of balsam fir and spruce timber in the East, and Douglas-fir, lowland white fir, and subalpine fir in the West. A closely related species, *C. pinus* Freeman, found in the Great Lakes states, prefers Scots and jack pine, and *C. lambertina* (Busck) in the West feeds on sugar pine. The eastern blackheaded budworm, *Acleris variana* (Fernald), injures hemlock, fir, and spruce in the western United States, Canada, and Alaska. Ornamental conifers are occasionally attacked, especially in regions where an outbreak occurs on forest trees.

The life cycle of the spruce budworm is completed in 12 months (Figure 10). The female moth is a strong flier but normally relies on wind currents for distant migration, which can exceed 80 km (50 miles) when airborne moths encounter a strong frontal system. Eggs, deposited in clusters (D) on needles in July, hatch in 1–2 weeks. Some very minor foliage feeding and dispersal occurs during the first instar. Most of the larva's energy at this time is used to spin a hibernaculum or cocoonlike shelter, in which it overwinters. Inside the hibernaculum the larva molts to the second instar and remains dormant until the following spring. Early in May it awakens and initially feeds on staminate flowers or mines the previous year's needles. Later, larvae enter expanding vegetative buds (G) and may destroy several of them before shoot elongation begins. Still later, the larvae feed on new foliage, and during the fourth or fifth instar they tie the tips of shoots together to form a "nest." They are wanton feeders, sometimes only severing needles, which are attached to the twig with silk and slowly turn reddish brown. From a distance a heavily infested tree appears to have needleless tips, and the remaining foliage looks scorched. Larvae (E) are about 25 mm long when fully grown.

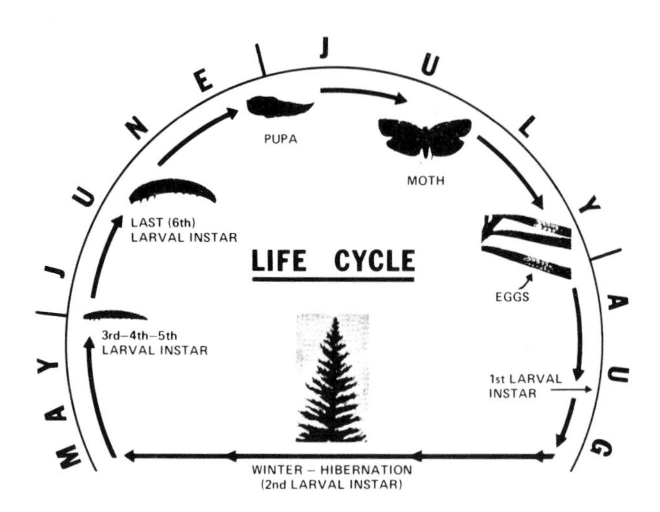

Figure 10. Life cycle of the spruce budworm. (Courtesy U.S. Forest and Canadian Forestry Service, CANUSA.)

Environmental pressures such as adverse climate and parasites play an important role in reducing budworm populations. The most common naturally occurring parasites are wasps, such as *Meteorus trachynotus* Vierech and *Phaeogenes hariolus* (Cresson), and a tiny fly, *Lypha setifacies* (West). High populations of budworms occur in cycles that are often linked to overmature stands of spruce and balsam fir. Forest management to remove stands before overmaturity or to discriminate against fir (the favored host) reduce the risk of outbreaks. The spruce budworm is rarely a problem on landscape trees.

References: 18, 448, 468, 498, 617, 658, 790

A, B. White spruce buds and shoots injured by the spruce bud moth. Arrow in A points to fifth-instar feeding damage. The presence of the bud cap, normally shed, is a symptom of early feeding injury (B).
C. A larva of the spruce bud moth. [Tortricidae]
D. A spruce budworm adult female. Note egg masses at arrows. (Courtesy Canadian Forest Service, Great Lakes Forest Research Center.)
E. A nearly fully grown (sixth-instar) spruce budworm larva. [Tortricidae] (Courtesy D. C. Allen.)
F. A balsam fir severely damaged by spruce budworm larvae.
G. A second-instar spruce budworm larva nearly covered by silk is attacking a newly developing spruce bud. (Courtesy Canadian Forest Service.)

A
B
C
D
E
F
G

Juniper Webworm (Plate 8)

The juniper webworm, *Dichomeris marginella* (Fabricius), is a moth belonging to the family Gelechiidae. European in origin, it was first reported in the United States in 1910. The species is known from Quebec and Maine to North Carolina, west to the Pacific Coast states and British Columbia. *D. marginella* has been found only on *Juniperus* species, especially the species or cultivars *hibernica, horizontalis, depressa, aurea, stricta, squamata,* and *suecica,* and also on Chinese juniper, although *Juniperus chinensis* var. *pfitzeriana* and *J. sabina* are apparently immune.

One generation is produced each year. Moths are active during May and June in the southern part of its range and during June and July in northern areas. In central Illinois peak moth emergence occurs about the third week of June. The moths live for about 15 days, during which time a female may produce 50–130 eggs. The moth can often be identified by its presence on the host plant and by the copper brown forewings, which are outlined with a band of white on the front and rear margins. Its wingspan is about 15 mm.

Eggs are laid singly on leaves at or near axils of the current year's shoots. When first laid, eggs are whitish but eventually turn yellow, then orange, and finally red. They are 5 mm long, broadly rounded, and sculptured. Newly hatched larvae feed as leaf miners, and they mine many leaves before becoming surface feeders. During this time they construct silken tubes around the feeding site. Early symptoms of infestation often go unnoticed because young larvae tend to inhabit the thick inner foliage of the host plant. By autumn, feeding sites may contain three to five larvae in webbed communities often described as nests. Winter may be passed in instars 5 through 7. Larvae commence feeding again in the spring, at this time producing most of the webbing. Larval feeding causes the foliage to turn brown (A, D), but the larvae are hidden by the naturally dense foliage, which is made more dense by the silk binding. Pupation occurs in whitish silken cases attached to the webbed foliage. The juniper webworm has several natural enemies including the mite predator *Pyemotes tritici* (Lagreze-Fossot & Montane) and several insect parasites, including *Tetrastichus* species, *Catoloccus aeneoviridis* (Girault), *Itoplectis conquisitor* (Say), *Coccygomimus aequalis* (Provancher), and *Bracon gelechiae* Ashmead.

Another species, *D. ligulella* Hübner, occurs on apple, cherry, pear, plum, hazel, and oak in the Northeast and west through Minnesota, as well as in parts of Canada.

References: 33, 453, 581, 659, 829, 862

A. Brown needles show the extent of damage done by juniper webworms on a large columnar juniper.
B,C. A mature larva; about 13 mm long. Strands of silk are evident in C.
D. Characteristic "nests," represented by dead foliage. Twigs and foliage are bound together with silk to make a cylindrical mass.
E. The mode of larval feeding and webbing tends to leave some foliage untouched.

A
B
C
D
E

Spruce Needleminers (Plate 9)

Several species of moth larvae mine spruce needles, causing similar damage. The spruce needleminer, *Endothenia albolineana* (Kearfott), which is widely distributed in the spruce-growing regions of the United States and Canada, is a problem primarily for Christmas tree growers and arborists. In the West it attacks Engelmann, blue, and Sitka spruce, and in the northeastern United States and eastern Canada it attacks primarily Norway, white, and blue spruce.

The moths emerge in May and June. Shortly thereafter, each female deposits a group of pale green eggs, shingle fashion, on the base of a needle. The greenish to greenish brown larva cuts a hole near the base of a needle and then mines the interior. Sometimes several larvae mine a single needle. When the interior of the needle has been consumed, the larva often cuts it off; the needle remains attached to the other dead needles by silk strands. Typical damage is shown in panels A–C and Figure 11. Feeding continues until the first heavy frost, when each larva enters a hollowed-out needle and spins a web over the opening to overwinter. It emerges again in the spring and resumes feeding until ready for pupation in mid-April. Larvae become social in the spring, with two or more living within each web.

Epinotia nanana (Treitschke) is a smoky brown moth that has several common names, for example, European spruce needleminer (E). Introduced from Europe, it occurs in the United States from Maine and Massachusetts to Ohio and Michigan, and in Canada in Quebec, Ontario, and British Columbia. Red, white, and blue spruce are common hosts, but this moth has a marked preference for Norway, Colorado, and Engelmann spruce. The moths fly during June, laying eggs singly on needles; eggs hatch in July. The newly hatched larva mines in the base of the previous year's needles. More than 10 needles may be mined by a single larva before cold weather stops its feeding. It spends winter in the last hollowed-out needle and resumes feeding in April, when it may kill an additional 15 or so needles before completing its growth. Pupation occurs within a mass of mined needles and excreta held loosely together by silk strands.

Even light infestations of *Epinotia nanana* and *Endothenia albolineana* can give ornamental spruce an unsightly appearance. At times they can cause severe defoliation of both ornamental and forest trees. These two species occupy essentially the same ecological niche, and their life cycles are similar, making diagnosis difficult. The best symptom for field diagnosis is the presence of densely webbed needles (Figures 11–12), always associated with *E. albolineana*.

Coleotechnites ducharmei Freeman, another leaf miner of red, white, and black spruce, occurs from Nova Scotia to Alberta and probably in the northern states within the same longitudinal boundaries. Trees situated in well-lighted habitats are most frequently attacked.

Figure 11 (left). Spruce twig illustrating nature of webbing and mining caused by the spruce needleminer, *Endothenia albolineana*. (From Freeman, ref. 265, courtesy *Canadian Entomologist*.)

Figure 12 (right). Spruce twig illustrating nature of webbing and mining caused by the green spruce leafminer, *Epinotia nanana*. (From Freeman, ref. 265, courtesy *Canadian Entomologist*.)

Eggs are laid singly in June and early July. Larvae develop through five instars, and all stages feed as leaf miners. The life cycle is completed in 1 year. Mature larvae, about 5 mm long, have distinctive transverse pink bands at the base of each segment on a creamy yellow background. They mature in late August and early September, dropping to the soil, where they pupate and overwinter.

C. piceaella (Kearfott), sometimes called the orange spruce needleminer because of its color, mines the needles of various spruce from eastern Canada and New England to Alberta and Colorado. Larvae overwinter in mined needles or in silken shelters at the base of mined needles. They are about 8 mm long when fully grown and complete their feeding by early July. Most damage by this species occurs on ornamental spruce.

References: 142, 193, 265, 519, 538a, 658, 862

A. Damage to blue spruce caused by the spruce needleminer, *Endothenia albolineana*.
B. Close-up of photograph in A.
C,D. Damage by *Epinotia nanana* to the foliage of Norway spruce.
E. A fully grown larva of *Epinotia nanana*; about 8 mm long.
F. An adult *Epinotia nanana*. [Tortricidae]
G. The moth of the spruce needleminer, *Endothenia albolineana*; about 9 mm long.

A
B
C
D
E
F
G

Douglas-fir Tussock Moth and Silverspotted Tiger Moth (Plate 10)

The Douglas-fir tussock moth, *Orgyia pseudotsugata* (McDunnough), is an enigma, even a source of fear to many people, largely because of its explosive outbreaks, which usually last for 3 years and then abruptly subside. This moth is a problem on the West Coast and the intermountain region from British Columbia to New Mexico. It defoliates a wide range of western conifers, but Douglas-fir, grand fir, and white fir are the preferred hosts. Larvae will feed on understory shrubs; subalpine and red fir; ponderosa, Jeffrey, sugar, and lodgepole pines; western hemlock; Engelmann spruce; and western larch after their preferred hosts have been stripped. Although forest trees take the brunt of their feeding orgies, ornamental and specimen trees are also subject to attack. When open-grown trees are attacked, the upper third of the crown is defoliated first (B). Defoliation is much more critical to a conifer than to a deciduous tree, for reasons that cannot be discussed here. Pest management becomes increasingly urgent as the outbreak continues. Moreover, allergic reactions of the respiratory system and skin may afflict people who touch the caterpillar or come in contact with floating "hair" fragments, which may lodge on skin or adhere to the mucous membranes of the nose or throat.

Tussock moths are so named because larvae possess unique tussocks, or compact tufts of hairs that resemble round brushes. Several species are economically important (Plate 71). The Douglas-fir tussock moth produces one generation each year. Adults appear from late July to November, depending on the climate and location. The male moth, a day flier, is brown to gray, with a wingspread of 25–30 mm. The furry female is wingless (F) and approaches 20 mm in length. Her eggs are often deposited in a single mass on the cocoon from which she emerged and are located most frequently on the trunk and lower side of larger limbs. The egg is the overwintering stage. Hatching coincides with bud break and the beginning of shoot elongation of host trees. The young larvae, sparsely clothed with long hairs and lacking the tufts, feed on new foliage. The four characteristic tufts of hair (A) on the abdomen are not obvious until the fourth larval stage. By this stage, larvae feed on all host foliage regardless of needle age. They go through six instars and pupate in cocoons (C) on the host tree.

The population is usually checked by a variety of natural controls. One of the most effective parasites is a fly, *Carcelia yalensis* Sellers, that lays its eggs on the back of a mature tussock moth larva, which eventually dies in its pupal stage. A naturally occurring virus that can be produced under laboratory conditions has been a highly effective control in both ground and aerially treated plots.

Two insect families (Arctiidae and Lymantriidae) have members described as tussock moths. The tiger moths (arctiids) are so named because many have distinct and attractive spots or stripes on their wings (H). The silverspotted tiger moth, *Lophocampa argentata* (Packard), has brushlike tufts of hairs on the abdomen in its advanced larval stage (G). Like the Douglas-fir tussock moth, the tiger moth feeds on conifers and is restricted to northern California and the Pacific northwest. Its principal host plant is Douglas-fir, but it also feeds on western hemlock, lodgepole pine, grand fir, Sitka spruce, western redcedar, and several other conifers. It is most important as a defoliator of urban and park trees. The larva makes ugly, dirty webs that temporarily destroy the appearance of specimen trees (E).

The moth, which may be seen during July and August, lays its eggs (Figure 13) either in clusters or rows on twigs and needles. The eggs hatch in late summer, and the larvae are gregarious (D), living in colonies under loose silky webs (E) that contain dead needles, frass, and other debris. With the onset of cold weather, the larvae become quiescent and remain inside the web. Up to and through the overwintering stage, they are rather sparsely clothed with short, fine hairs. The injury they do is much more evident in spring, when their tubular webs may be half a meter long. Buds are not harmed. Fully grown larvae, which may be up to 37 mm long, are reddish brown with many tufts of hair and many long, single body hairs (G). The tuft hairs contain urticating chemicals that may cause a skin rash. In June the larvae spin brownish cocoons that they attach to the defoliated trees or to debris on the ground.

Figure 13. A female tiger moth with her pearly eggs attached to Douglas-fir foliage. (Courtesy R. W. Duncan, Pacific Forest Research Centre, Canadian Forestry Service.)

One subspecies, *L. argentata sobrina* Stretch, feeds on Monterey pine in coastal California; another, *L. argentata subalpina* French, feeds on juniper and occasionally pinyon pine in the Rocky Mountain region.

References: 18a, 51, 198, 275, 697, 850

A. A fully grown Douglas-fir tussock moth larva; about 30 mm long. [Lymantriidae] (Courtesy Gray Collection, Oregon State Univ.)
B. Webbing and needle injury made by Douglas-fir tussock moth larvae on white fir. (Courtesy Sharon J. Collman, Cooperative Extension, Washington State Univ.)
C. Douglas-fir tussock moth cocoons and twig damage by larvae. (Courtesy Gray Collection, Oregon State Univ.)
D. A mass of young silverspotted tiger moth larvae on western white pine. (Courtesy Sharon J. Collman, Cooperative Extension, Washington State Univ.)
E. The tubular webbing and frass of silverspotted tiger moth larvae. Note cluster of young larvae at arrow point.
F. A female (wingless) Douglas-fir tussock moth and her egg mass attached to a twig. Females are 12–20 mm long.
G. A fully grown larva of the silverspotted tiger moth. Larvae may be as long as 37 mm. [Arctiidae] (Courtesy Gray Collection, Oregon State Univ.)
H. A female silverspotted tiger moth. Wingspan averages 47 mm.

A
B
C
D
E
F
G
H

Larch Casebearer (Plate 11)

Although larch is only occasionally used as an ornamental, in certain parts of the United States it grows into a beautiful yard tree. Larch has an uncommon trait for a conifer—it loses its foliage in the winter. Only baldcypress, pondcypress, and dawn redwood do the same. The casebearer sometimes causes larch to lose its foliage before winter comes.

Coleophora laricella (Hübner) derives its common name from the case constructed by the larva (C, D). The case provides camouflage that allows larvae to go undetected by nearly all but trained observers. The case, formed by part of a mined-out larch needle lined with silk spun by the larva, is the color of a dead needle and shaped somewhat like a miniature cigar. The casebearer overwinters in its case, which it firmly attaches to a branch, often near the base of a bud (C). The larva resumes feeding in the spring as soon as the foliage begins to appear; it grows to a length of about 6 mm. The needles it damages look as though they have been bleached or scorched (A, B).

When the larva finishes feeding in late May or June, it pupates within the case. Adults emerge during June and early July, and they disperse. The adult moths are silvery gray and have a wing expanse of only 8 mm. They deposit one or more distinctively shaped eggs to a needle. Under magnification, the reddish brown eggs resemble inverted jelly molds with 12–14 ridges. When the egg hatches, the larva burrows directly into the needle, where it feeds as a miner for about 2 months.

The most conspicuous damage inflicted by casebearers occurs in the spring, when the foliage of the host tree may shrivel and die. Heavy defoliation retards both height and diameter growth. If defoliation occurs during 2 or more consecutive years, the host may be killed.

The casebearer, an unfortunate introduction from Europe, was first reported in Massachusetts in 1886. It was known only in the eastern United States until the late 1950s, when it was found in the western United States. It is now ravaging larch forests in the northern Rocky Mountains, Oregon, Washington, and British Columbia. All species of North American *Larix* as well as exotic species may be severely injured, whether in the forest or urban landscape. A number of native and introduced parasites such as *Agathis pumila* (Ratzeburg), *Bassus pumilus* Nees, and *Chrysocharis laricinellae* (Ratzeburg) attack the casebearer but are unable to hold it completely in check. In some localities birds play an important role in the control of this insect.

References: 193, 275, 356, 359, 641, 789, 890

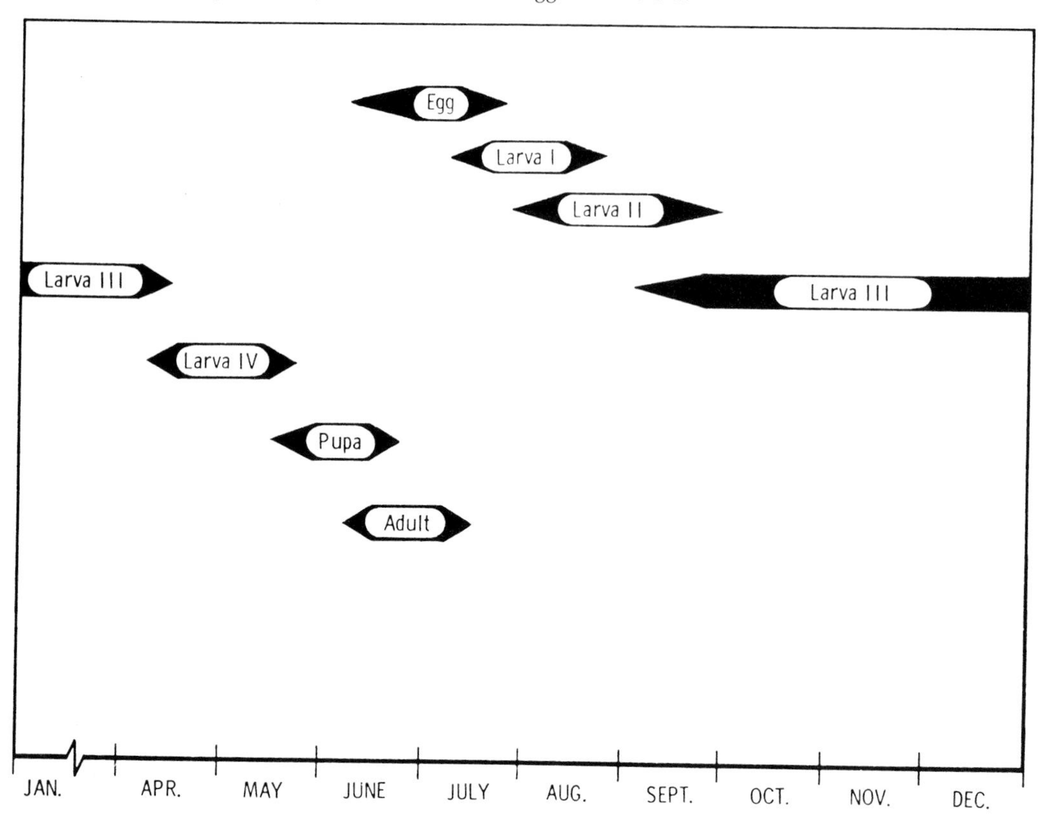

Figure 14. Occurrence of the life stages of the larch casebearer in western larch forests. (Courtesy U.S. Forest Service.)

A, B. Typical damage by the larch casebearer, *Coleophora laricella,* to foliage of *Larix decidua.*

C. Larch casebearer larvae overwintering in cases at the base of a bud.

D. A larch casebearer larva in its strange walking position. [Coleophoridae]

A
B
C
D

Needle Miners and Webbers (Plates 12–13)

The term *needle miner* describes the larval feeding habits of insects that bore into and feed on the soft internal tissue of conifer foliage. Some species have a very limited host range and feeding site. Others begin their larval stage by mining foliage or buds and later become gross feeders because of their size. A needle miner is a borer that has chosen the soft foliage tissue rather than twigs or stems. Pictures on this plate and those of Plates 9 and 14 serve as examples of some of the common leaf miners of conifers. Many seriously injure specimen plants as well as Christmas trees and forests. All of the economically important leaf-mining species affecting conifers belong to the Lepidoptera.

The genus *Coleotechnites* contains about 45 species of small, narrow-winged, and generally grayish moths. All are leaf miners in their caterpillar stage. *Coleotechnites apicitripunctella* (Clemens), known in Canada as the green hemlock needleminer, is both a leaf miner and a webber (A, C). The young larva enters a leaf near the base in mid to late July and continues mining through the growing season. The immature larva overwinters inside the mine and resumes feeding in the spring. When it becomes so large that it cannot be accommodated inside the leaf, it evacuates the mine and ties several leaves together with silk. Though it feeds openly on these leaves, it is yet partially protected, for it feeds on the underside by excavating a groove between the midvein and the leaf margin (C). Such feeding injury is sufficiently characteristic to identify the species as either *C. apicitripunctella* or *C. macleodi* (Freeman). By the time these symptoms are produced, the larva has reached a length of 6–8 mm. *C. apicitripunctella* larvae are pale green with light brown heads. The body is almost naked, and only with magnification can an occasional hair be seen on each body segment. Pupation occurs in early summer in a silk tube within the webbed needles (nest). *C. macleodi,* a closely related species with nearly the same life cycle and behavior, is known as the brown hemlock needleminer because of the larva's reddish brown color. Hemlock is the primary host for both species, and both occur in eastern Canada and the northeastern United States. Local outbreaks may be severe in both urban and forest areas.

The common name cypress leaftier has been given to *Epinotia subviridis* Heinrich by several West Coast entomologists. It is a significant pest of ornamental *Cupressus* and to a lesser degree *Thuja* and *Juniperus,* but is limited to the Pacific Coast states and British Columbia. The wings of the cypress leaftier moth have a patchwork of brownish orange lines and rectangular spots. The larva feeds from silk-lined tunnels, eating primarily foliage, but early instars scar and kill the bark of twigs and shoots. Larvae incorporate living twig tips, silk, frass, and foliage fragments into the tunnels. They inflict most of the injury in the spring, when many brown twigs flag the dense foliage (D). The fully grown larva (E) is about 10 mm long and blackish pink, owing in part to black, seta-bearing tubercles on each segment.

The white fir needleminer, *E. meritana* Heinrich, occasionally defoliates forest, park, and urban ornamental white firs (*Abies concolor*) and red fir (*A. magnifica*). On white fir, needleminer activity results in bleached yellow needles, noticeable from late spring to early autumn. Abandoned mined needles often become detached from the twig and dangle in bunches from webs created by larvae. A characteristic sign of needleminer infestation is the presence of single pupal cases protruding from mined needles. Large needleminer populations cause thin crowns and dead limbs, and they occasionally kill trees. The distribution of this insect is believed to be limited to the Rocky Mountain and Pacific Coast states and to western Canadian provinces.

The pine needle sheathminer, *Zelleria haimbachi* Busck, is a transcontinental species reported from New Jersey to British Columbia. It attacks many species and hybrids of two- and three-needle pines, for example, jack pine in the East and ponderosa and Jeffrey pines in the West. It is believed to occur wherever ponderosa and Jeffrey pines are native.

Biology and life history studies were conducted in northern California by Stevens (753). The larva is a typical needle miner in its first two instars. After that, the term *sheath miner* is appropriate because only part of the needle tissue in the sheath is devoured. Small larvae enter the sheath completely. When larvae reach fourth and fifth instars, they feed by inserting only the head and first few segments into the sheath. During its sheath-mining stage the larva protects itself with a tubelike

Figure 15. Needle damage by the pine needle sheathminer. *a,* needles devoured; *b,* white cocoon at the base of mined needle sheaths. Note holes (*arrows*) on fascicle sheaths. Webbing has been removed. Black needles represent foliage killed by the insect. (From Freeman, ref. 264, courtesy *Canadian Entomologist.*)

covering of silk across the bases of several needles. The larva pupates on the shoot tip, with its body held in place by silk.

Significant damage is caused by both needle and sheath mining (Figure 15). Mining begins on the current season's needles that have reached full size. The larva overwinters in the mine as a first instar, reaching the second instar in its mine the following spring. During this time frass is packed in the tunnel, and the needle becomes straw colored as it dries. After the shoots elongate and new needles begin to grow, the larva cuts a hole in the mined needle and crawls to the sheath of a new needle, where it begins to feed as a sheath miner (F). Some of the needles from mined sheaths droop sharply, providing another symptom of insect injury. One generation occurs each year.

In the West this insect seems to prefer trees less than 9 meters tall. The newly hatched larva is bright orange with a glossy black head. In its late instars, the larva fades to tan and has a dull orange stripe on each side. The fully developed larva is commonly 14 mm long.

References: 87, 91, 121, 264, 265, 275, 658, 753, 817

A, B. Injury in the form of dead needles and "nests" of hemlock needles tied with silk by larvae of the green hemlock needleminer. [Gelechiidae]

C. The underside of mined needles show damage caused by the hemlock needleminer.

D. A tunnel of frass and silk along with dead foliage and shoots caused by the cypress leaftier, *Epinotia subviridis*. The host plant is believed to be western redcedar. [Tortricidae] (Courtesy Oregon State Extension Service.)

E. A larva (about 10 mm long) of the cypress leaftier excised from its silk tunnel. (Courtesy Oregon State Extension Service.)

F. These dead needles were killed by the pine needle sheathminer. [Yponomeutidae] (Courtesy Oregon State Extension Service.)

A
B
C
D
E
F

The pine needleminer, *Exoteleia pinifoliella* (Chambers), is a native moth that in its caterpillar stage may mine the needles of jack, red, mugo, Virginia, loblolly, and perhaps other hard pines. Finnegan (244) believes that pitch pine is its primary host. This native insect occurs in the eastern half of both the United States and Canada; on occasion it has been a serious defoliator. Depredation by this insect causes the mined needle to die and turn brown (A). Usually only one needle in each fascicle or pair is attacked. The remaining healthy needle holds the dead one in place until late the following year.

Eggs are laid in late spring. Upon hatching, the larva crawls to a new fascicle and enters the flat side of one of the needles. The entrance hole becomes plugged with resin and is difficult to find. The larva feeds in this needle through the second instar, at which time it vacates the first needle, then enters the apical end of a second healthy needle. Here it feeds for the balance of the summer and overwinters in its mine. In the spring it mines toward the base of the needle, where it cuts a round exit hole (C). The larva then enters a third needle about midway and mines toward the apex. By this time, the insect has reached its full length of 6 mm. Pupation (E) occurs in the mine of the third needle (Figure 16).

Other eastern needleminers that cause similar symptoms include *Coleotechnites canusella* (Freeman) on jack pine and *C. resinosae* (Freeman) on red pine. The food of *Argyresthia pilatella* Braun, sometimes called the Monterey pine needleminer, appears to be limited to Monterey pine, *Pinus radiata*. Natural distribution of the pine is geographically restricted to a small region in coastal California near the city of Monterey. However, humans have transported the tree to other areas of California and to many southern states, and it has become a major timber tree in New Zealand, Australia, South Africa, and Spain.

C. canusella mines the current year's needles, and individuals limit their feeding to the needles of a single fascicle. The newly hatched larva migrates toward the distal end of the needle and bores in just above the sheath. After entering a needle, the larva mines toward the base. The larva has six developmental stages, and it alternately mines from apex to base, always packing the tunnel with frass. The fully grown larva is 3.6 mm long. During the final instar, it cuts a round exit hole and abandons the mine, pupating in duff under its host tree. Moth flight and egg laying occur from early to late spring, depending on the climate.

In the West the lodgepole needleminer, *C. milleri* (Busck), is a notorious native forest pest with a 2-year life cycle. Essentially limited to high-elevation forests west of the Sierra Nevada crest, it is not normally a pest of ornamental trees. The injury it does is similar to that caused by other lepidopterous needle miners on pine. Caterpillar feeding causes the foliage to turn yellow, then reddish brown, followed by needle drop. One or more holes in affected needles are plainly visible. Under good lighting a caterpillar or pupa may be seen within the needle. The fully grown larva is about 8 mm long and varies in color from lemon yellow to deep reddish orange. Each larva normally mines at least five needles before pupation.

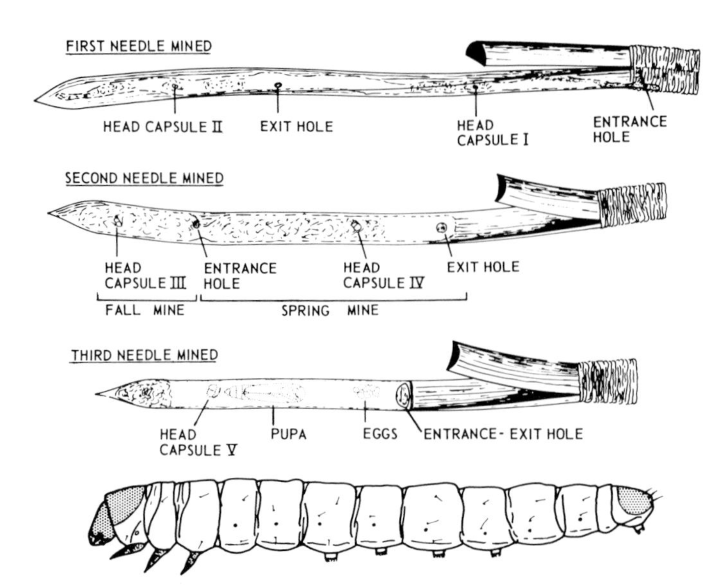

Figure 16. Sequence of injury in jack pine needles caused by the larva of the pine needleminer, *Exoteleia pinifoliella*. (From Finnegan, ref. 244, courtesy *Canadian Entomologist*.)

References: 244, 408, 439

A, B. Needles killed by the caterpillar mines of the pine needleminer, *Exoteleia pinifoliella*. [Gelechiidae]

C. Close-up of damaged needles, easily recognizable by the brown tips. Moths emerge through characteristic holes. (See also Figure 16.)

D. An exposed larval mine in a needle. Larva is near arrow.

E. A pupa of the pine needleminer inside a mined needle.

A
B
C
D
E

Cypress Tip Miners and Arborvitae Leaf Miners (Plate 14)

Several species of western *Argyresthia,* among them *Argyresthia cupressella* Walsingham and *A. franciscella* Busck, are commonly known as cypress tip miners or cypress tip moths. The first-named species attacks the tips of foliage of cypress (*Cupressus*), juniper (*Juniperus*), arborvitae (*Thuja*), and occasionally redwood (*Sequoia sempervirens*) in coastal regions of the western United States. *A. franciscella* also mines the tips of cypress and occasionally juniper in the San Francisco Bay area of California. Both insects tunnel within the growing points of the host, killing affected tips. Entire plants may appear brown as a consequence of heavy infestations, and repeated annual attacks result in dieback of twigs.

The moth of *A. cupressella* (I) is silvery tan with a wingspan of about 6 mm; it is active only in the spring and early summer. The exact time of its occurrence depends on how much the infested plant is exposed to the sun, and on latitude. In the vicinity of Corvallis, Oregon, moths occur from early May until late June, with the peak of activity in May and early June. In southern California adults are most numerous in April, but can be found from mid-March until late in July.

Eggs are laid on the green tips of 1- and 2-year twigs of susceptible host plants. Upon hatching, the tiny larvae tunnel into leaf scales and mine within the foliage until late winter or spring of the following year. Relatively little foliage discoloration is caused by young larvae. Beginning in late winter, however, a yellowing and then a browning of the infested tips become apparent. The dead tips can be readily broken because of their dried, shell-like condition.

After feeding has been completed, the larvae (6 mm long when mature) of *A. cupressella* leave their mines, and each spins a white paperlike cocoon among the living or dead foliage. The pupal stage is passed within this cocoon. *A. franciscella,* however, spends its pupal stage within the hollowed-out, slightly swollen dead tip. Several weeks later moths emerge, mate, and begin laying eggs. One generation occurs each year.

In the San Francisco region 13 members of the Cupressaceae were evaluated for their susceptibility to infestation by *A. cupressella*:

Highly resistant to infestation
Juniperus chinensis cv. Kaikuza
J. chinensis var. *sargentii* cv. Glauca
J. scopulorum cv. Erecta Glauca
Thuja plicata

Moderately resistant to infestation
J. chinensis cv. Pfitzerana Aurea
J. sabina cv. Arcadia
J. sabina cv. Tamariscifolia
J. virginiana cv. Prostrata

Highly susceptible to infestation
Chamaecyparis lawsoniana cv. Allumii
J. chinensis cv. Pfitzerana
J. chinensis cv. Robust Green
J. virginiana cv. Cupressifolia
T. occidentalis

Other western species that cause similar damage include *A. trifasciae* Braun, *A. libocedrella* Busck, and *A. arceuthobiella* Busck. The latter two mine the tips of incense cedar, *Libocedrus decurrens,* in Oregon; *A. trifasciae* attacks *Cupressus* species in California. *Stenolechia bathrodyas* Meyrick mines the tips of *Juniperus, Cupressus,* and *Cupressocyparis* species along the coast of southern California.

Several species of *Coleotechnites,* including *C. stanfordia* (Keifer) and *C. juniperella* (Kearfott), mine the tips of Monterey cypress, *Cupressus macrocarpa,* and western juniper, *Juniperus occidentalis,* respectively. The latter insect occurs also in the northeastern United States and adjacent Canada, where it mines the tips of *Juniperus* species.

In northeastern United States and adjacent eastern Canada four species of leaf miners attack arborvitae (*Thuja*): *A. thuiella* (Packard), *A. freyella* Walsingham, *A. aureoargentella* Brower, and *Coleotechnites thujaella* Kearfott. The arborvitae leafminer, *A. thuiella,* is the most abundant species and has a much greater known range than the other eastern species. *A. thuiella* is found not only in New England and eastern Canada but also in the mid-Atlantic states and as far west as Missouri.

Moths appear from mid-June to mid-July and lay their eggs in mixed populations over their respective ranges. The damage (A) caused by all eastern species is virtually identical to that described for *A. cupressella.* Trees may lose 80% of their foliage from leaf-miner attack and still survive. Twenty-seven parasites of larvae and pupae have been recovered in New Brunswick, Canada. *Pentacnemus bucculatricis* Howard, a wasp parasite, is the most widely abundant and important in natural control.

A. freyella infests eastern redcedar, *Juniperus virginiana,* as well as arborvitae (*Thuja*) in the northeastern United States and Canada.

References: 87, 88, 193, 204, 275, 437, 661, 696

A. An arborvitae heavily infested by the arborvitae leafminer, *Argyresthia thuiella.* The tree has lost about half its leaves.
B. Loss of foliage on arborvitae following attack by the arborvitae leafminer. The brown foliage eventually drops and leaves the twig bare.
C. An exposed arborvitae leafminer. Tunnels made by the larva are visible.
D. Healthy foliage adjacent to mined foliage.
E. Browning of foliage on Monterey cypress resulting from attack by *A. franciscella.*
F, G. Close-up of tip damage caused by *A. cupressella.* Dark spots in G are exit holes where the larvae emerged.
H. An adult arborvitae leafminer.
I. An *A. cupressella* adult. [Argyresthiidae]

A
B
C
D
E
F
G
H
I

Needle and Twig Midges (Plate 15)

The larvae of several gall midges (flies) feed in or near the vascular tissue of conifer shoots and twigs. Collectively, they may be described as pitch or resin midges because a resin mass is usually associated with the feeding site. Other gall midge larvae mine leaves, feed under the needle sheath, or cause galls (Plate 51). These larvae are typically orange or reddish orange and when fully grown usually measure 2–3.5 mm long.

Such a midge of transcontinental distribution is *Cecidomyia piniinopis* (Osten Sacken), most commonly found from the upper Ohio Valley to New York and east to New England and Virginia, and in all of the far western states. It feeds on pitch, Virginia, eastern white, and jack pines in the East, and ponderosa, Jeffrey, and lodgepole pines in the West. *C. piniinopis* selectively attacks established trees and causes needles to die in tufts that are often described as flags (A). Flags begin to appear in the summer but are most conspicuous in the dormant season. Brown needles drop to the ground, thereby revealing extensive twig distortion. Close examination of injured twigs reveals surface swelling. When the swollen area is shaved away, pits that may contain one to several larvae (B) immersed in fluid resin are exposed. Two or three pits have no noticeable effect on a shoot, but as the number increases plant growth processes are disrupted. Flagging results when pits are numerous enough to girdle the shoot. In 1972 *C. piniinopis* was considered one of the 10 most important forest insects in California.

The adult gall midge is small and delicate (Plate 226B shows a midge of similar form), and is active in the spring. Eggs are deposited singly or in groups attached to the surface of the new shoot. After hatching, the maggot crawls to a resin droplet at the base of a needle fascicle and soon becomes embedded in a resin pocket of the inner bark (B). Larvae feed and overwinter in this location. The following spring, they work their way to the bark surface and crawl to needles, where they spin light-colored cocoons. Each cocoon is securely attached to a needle. These cocoons, often with the dark pupal "skin" protruding through one end, are readily visible in the spring. One to several generations occur each year.

The severity of injury on ponderosa pine and perhaps other pines is related to the nature of the surface of spring shoots. Those that are sticky owing to a resinous exudation are likely to be severely injured. Those shoots that are smooth or have a powdery or waxy covering are much less susceptible. The shoot surface type is a variable character within pine species, the knowledge of which may be useful in the study of genetic resistance to the midge.

Other pitch midges that cause similar symptoms include *C. resinicola* (Osten Sacken), another transcontinental species; the Monterey pine resin midge, *C. resinicoloides* Williams, a western species; and *C. reeksi* Vockeroth, associated with jack pine. The latter two species are not considered economically important but could be of concern on ornamental plants.

Douglas-fir needles host three species of needle-mining gall midges. One species mines near the leaf petiole, another mines at the needle apex, and the third in the middle of the leaf. The injury illustrated in panel C represents a midleaf mine caused by *Contarinia pseudotsugae* Condrashoff, which is considered the most abundant and significant pest in Christmas tree plantations. Damaged needles usually drop, but some may remain on the tree for 2 years. These leaf miners occur in the Pacific states and British Columbia as well as Idaho and Montana.

Adults emerge about the time Douglas-fir buds are breaking. They lay eggs on new needles, and in a few days the hatched larvae penetrate the needle, where they feed throughout the summer (D). They vacate their mine in late autumn and drop to the ground. Maggots overwinter in the organic layer of soil. There is one generation each year.

Other midge larvae that live inside needles or the needle sheath include *Thecodiplosis piniresinosae* Kearby, *Resseliella pinifoliae* (Felt), and *Contarinia baeri* Prell (Figure 17). *T. piniresinosae* maggots live on red pine (*Pinus resinosa*) at the base of current-season needles. Their presence induces a bulbous swelling at the feeding site, where cavities or chambers develop. These structures are covered on one side by the needle sheath. Injured needles begin to turn brown in October (Wisconsin). One or more larvae may be found in the needle fascicle. *R. pinifoliae* causes the needles of white pine to drop before the foliage attains its normal length. As with most conifer-feeding midges, they overwinter as larvae in the soil. A proposed common name for *C. baeri* is pine needle midge. This insect can be a serious pest of Scots pine grown as Christmas trees and is also reported on *P. rigida* and *P. resinosa*. The pine needle midge is believed to be of European origin; its expanding distribution currently includes Quebec, eastern Ontario, the maritime provinces, New England, and New York. One of the readily noticeable symptoms of injury is needle droop. At first one needle of a pair will droop at a sharp angle (E). Unfortunately for growers, it is too late to apply chemical control measures for current-season protection after the droop symptom is observed. As foliage matures, injured needles turn brown, and by mid-July in Canada, they begin to drop. Although defoliation of new needles may occur, the tree's survival is not at stake. Christmas tree growers can successfully restore the appearance of damaged trees by shearing; however, a year's leader growth may be lost.

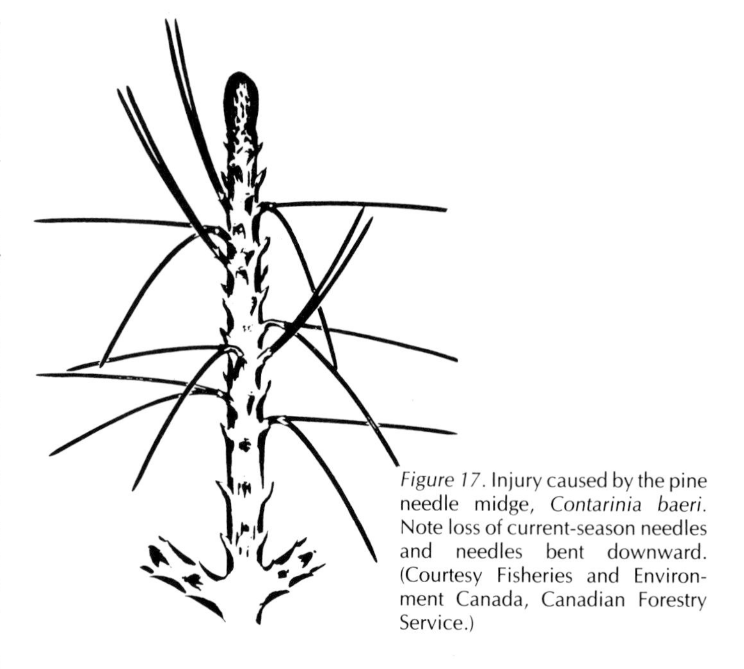

Figure 17. Injury caused by the pine needle midge, *Contarinia baeri*. Note loss of current-season needles and needles bent downward. (Courtesy Fisheries and Environment Canada, Canadian Forestry Service.)

In eastern Canada, pine needle midge eggs are laid from late June to early July on the foliage of new shoots. Larvae feed on the interior flat surface of needles within the sheath. The feeding action of the mouthparts scars and weakens the needle. The feeding site covers about 6 mm. If the vascular sieve tubes are severed, the needle will die. We have seen up to 23 larvae in a single bundle. Normally, only one or two dominant larvae survive in the sheath. When larvae mature, they spin cocoons in the lesion area, where additional protection is provided by the sheath. Injured needles along with the sheath usually drop to the ground bearing the cocoon in which the larva overwinters. Midge adults resembling the adults of the pine needle midge can be seen in Plate 226B. A small hymenopterous parasite, *Tetrastichus* sp.,

---

A. Dead needles on the new shoot of a lodgepole pine caused by the feeding of *Cecidomyia piniinopis* larvae (see Plate 18 for similar symptoms caused by a caterpillar). (Courtesy Gray Collection, Oregon State Univ.)
B. A feeding site opened to expose the maggots of *Cecidomyia piniinopis*. (Courtesy Gray Collection, Oregon State Univ.)
C. Leaf-mining injury to Douglas-fir needles caused by *Contarinia pseudotsugae*. (Courtesy Washington State Univ.)
D. Douglas-fir needle mines opened to expose the larva of *Contarinia pseudotsugae*. [Cecidomyiidae] (Courtesy Gray Collection, Oregon State Univ.)
E. Needle droop of Scots pine caused by *Contarinia baeri* larvae.
F. Needle droop of new Austrian pine needles caused by severe moisture deficit.

A
B
C
D
E
F

(Plate 15, continued)

is a major control factor. Where high midge populations occur, parasitism may exceed 90%.

Because droop (F) is a symptom that may be caused not only by midges, but also by the fungus *Aureobasidium pullulans* (deBary) and severe moisture deficit in late spring, each droop must be thoroughly examined to determine its cause.

Two vexing gall midges cause tip dieback in various species of *Juniperus*. The reddish orange larva of the juniper midge *Contarinia juniperina* Felt bores into twigs at the base of needles and produces a mine under the bark that may be as long as 25 mm. One to several larvae live and overwinter in each mine. By mid-May they have completed their development and make their way to the bark surface, drop to the ground, and burrow into the soil, where pupation occurs. The orange adults are gnatlike and emerge from the soil by early June, when eggs are laid on new foliage and the cycle is repeated.

A dead tip, which may be the result of maggot feeding, becomes evident near the end of the growing season. In early spring after the maggots awaken from diapause and resume feeding, additional twigs die. Dead tips may be up to 10 cm long. Residential plantings of Andorra, blue rug junipers (*Juniperus horizontalis*), and Hetzii (*J. chinensis*) are often severely injured if left unattended. Other cultivars of the named species may also be attacked, as are native redcedar and cultivars of *J. sabina*. Problems with this insect occur in the Midwest and as far east as New York and Pennsylvania.

The juniper tip midge, *Oligotrophus betheli* Felt, differs from the juniper midge by producing several successive generations throughout the growing season and by causing swollen tips best described as galls. After the larva vacates the gall, injured tips turn brown and die back for a distance of about 5 mm. This midge is presumed to occur over the same geographic area as the juniper midge and in British Columbia.

The dieback or blight symptoms produced by these midges so resemble the damage caused by the fungal pathogens in the genus *Phomopsis* (see Sinclair et al., *Diseases of Trees and Shrubs*) that the cause may be easily confused. Careful examination of the dead tips may reveal mine injury near or at the point where dead and live tissue meet or the swollen area terminates.

See Plate 51 for gall-forming midges.

References: 21, 124, 154a, 155, 195, 201, 276, 342, 420, 657, 701, 787

Pine Tube Moth (Plate 16)

Nature often provides its creatures with a home. If not, it provides them with the capability for building one. The pine tube moth, *Argyrotaenia pinatubana* (Kearfott), builds its own shelter by webbing 5–20 needles together into the form of a tube. The larva, nestled inside, surrounds itself with soft white silk. Eastern white pine, *Pinus strobus,* is its major, if not sole, host plant.

The pine tube moth's biology is not well known. It overwinters as a pupa in its tube. Moths emerge in early May and lay eggs on the needles. The young larvae feed on the tips of needles and become fully grown by mid-June (A). A new generation of moths appears in July. Eggs are laid soon thereafter. The moths, male (B) and female (C), are shown as pinned specimens. In fresh specimens the hind wings are pearl gray.

This insect is not considered a serious pest. When abundant, however, it can cause needles to turn brown, thus reducing the ornamental value of the tree. Injury may go unnoticed until the second generation appears.

The pine tube moth is most likely distributed throughout the native range of eastern white pine. It has been found from Florida north to Canada and west to Wisconsin.

References: 193, 359

A. A larva of the pine tube moth, *Argyrotaenia pinatubana,* and two bundles of needles showing the webbing inside the tube. The larva is about 12 mm long.
B. A male moth, pinned specimen.
C. A female moth, pinned specimen. Wingspan is about 14 mm.
D. Evidence of damage after the insect has vacated the tube.

A
B
C
D

Pine Tip or Shoot Moths (Plates 17–18)

Larvae of several species of small moths tunnel in the tips of pines wherever these conifers are grown in the United States. The most damaging species belong to the genus *Rhyacionia*. Some of the more important species and their hosts are listed in Table 4.

***Table 4.* Important pine tip and shoot moths in North American forests and urban landscapes**

Name	Principal hosts	Range
European pine shoot moth, *Rhyacionia buoliana*	Red, mugo, Scots, Austrian, Swiss mountain, Japanese black, ponderosa, lodgepole pines	Northern United States from coast to coast; and Ontario east to maritimes
Monterey pine tip moth, *R. pasadenana*	Monterey, Bishop pines	Coastal California
Nantucket pine tip moth, *R. frustrana*	All two- and three-needle pines except slash pine and longleaf pine	Eastern, southern, and south central United States; southern California
Pitch pine tip moth, *R. rigidana*	Pitch, Virginia, red, Scots, loblolly, slash pines	Eastern, southern, and south central United States
Ponderosa pine tip moth, *R. zozana*	Ponderosa, Jeffrey pines	California, Oregon, Washington, Idaho, Montana, Wyoming, South Dakota
R. montana	Lodgepole pine	Rocky Mountains
Southwestern pine tip moth, *R. neomexicana*	Ponderosa pine	Southwestern United States, Utah, Dakotas, Nebraska
Western pine tip moth, *R. bushnelli*	Ponderosa pine	Nebraska, Great Lakes states, Dakotas, Montana, New Mexico, Arizona

The damage inflicted on hosts by all species of *Rhyacionia* is similar: the tips of terminals and laterals are killed as a result of larval boring, initially into the base of the needles or buds, and then into the shoot itself. *R. frustrana* may kill back shoots of loblolly or shortleaf pine as much as 30 cm.

At times trees that are heavily infested by this species appear reddish as a consequence of the many dead tips. Small trees may be killed. Light infestations, however, may actually improve the form of some ornamental pines, giving them a bushier appearance. Both *R. frustrana* and *R. buoliana* are troublesome pests in nurseries and Christmas tree plantations. Repeated infestations leave the trees distorted, unsightly, and sometimes unmarketable.

A lure impregnated with sex pheromone is commercially available for monitoring the presence of male Nantucket pine tip moths. Traps baited with this lure are useful to determine the proper time for pesticide applications. Coupled with the degree-day model of both host and pest of Gargiullo et al. (280), a precise biological timing can assure tip moth control.

A single generation of European pine shoot moth occurs annually. The Nantucket pine tip moth may undergo four or more annual generations as far north as Oklahoma and Tennessee, with injury symptoms appearing numerous times during the growing season. The adult European pine shoot moth begins to lay small, flattened eggs on new shoots near the base of the needles or bud scales in late spring. The egg-laying period lasts for several weeks. When the eggs hatch, larvae bore into the needle sheaths and mine needles at or near their base. Mined needles die and turn brown, providing a summer season symptom. The wound area is always covered with resin-coated webs, providing another symptom. By midsummer, larvae move to buds and scar or burrow into them. In August they cease feeding for that growing season. They overwinter in the wound area, where the black larva, 2–3 mm long, is covered with resin-coated webs. Early indications of attack are yellowing of needles near the tips of twigs (A) and small, often clear deposits of pitch around and between new bud clusters (B). Later, dead buds and dead or deformed shoots result in striking deformity of the tree (Plate 18C). Some species of tip moths overwinter as larvae or pupae within the shoot; others drop to the ground and spend the winter there.

All species of tip and shoot moths are attacked by a number of parasites that, at times, seem to keep moth populations in check. Dry weather and poor soil conditions reportedly encourage damage by tip moths.

Moths of the genera *Eucosma, Petrova, Dioryctria,* and *Exoteleia* cause similar damage to pines in various parts of the United States. *Dioryctria abietivorella* (Grote) larvae frequently take up residence where other boring insects have been. They also feed on soft, smooth bark and bore into terminal or lateral shoots, cones, and galls on pine, spruce, Douglas-fir, and arborvitae. They do not feed on foliage. When twig bark is injured, the terminal portion beyond the site of injury usually dies. A dead flag remains as a symptom. Partly grown larvae overwinter inside the host and complete their growth in the spring. Fully grown larvae are about 18 mm long. There is little information about the biology and distribution of this species, but it is presumed to occur throughout the United States.

Larvae of related *Dioryctria* are among the most serious pests of cones in the South and West. In northeastern and Great Lakes states, larvae of the Zimmerman pine moth, *D. zimmermani* (Grote), and related species bore into the terminal shoots of Japanese black pine as well as Scots, Austrian, and native pines. Injured twigs often exude pitch mixed with sawdustlike frass at the entrance site. The affected shoot dies and turns brown. The larvae are dark with black heads and grow to 15 mm at maturity. The adult moth is similar to the one illustrated in Plate 18H.

Exoteleia burkei Keifer, sometimes called the Monterey pine shoot moth, is a common tip-killing insect of Monterey pine in California. Usually only 3–5 cm of growth are affected, however. The adults, which occur from late April until June, apparently lay their eggs at the bases of buds that will be the next year's points of growth. Newly hatched larvae tunnel into buds, yet the injured shoot develops to its normal length before its tip is killed the following year. The fully grown shoot moth larva is 5–6 mm long and is brownish yellow with a black head. Pupation takes place within the killed tip. One generation occurs each year.

References: 22, 59a, 60, 87, 97, 193, 210, 220, 275, 280, 359, 629, 754, 771, 878, 885

A. Brown, stunted needles at the tip of a pine twig, caused by the larva of the European pine shoot moth. [Tortricidae]
B. Pitch accumulation at the site of larval activity.
C. European pine shoot moth larvae exposed in a pitch mass.
D. A European pine shoot moth pupa in a tunneled twig.
E. The European pine shoot moth.

A
B
C
D
E

(Pine Tip or Shoot Moths, continued, Plate 18)

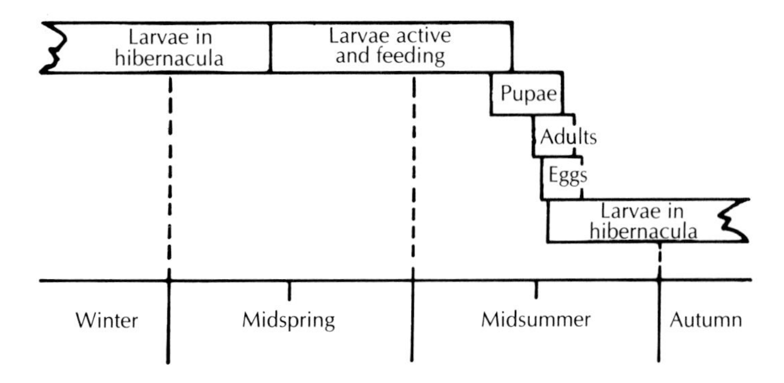

Figure 18. Seasonal development of the Zimmerman pine moth in Illinois. (Courtesy J. E. Appleby, Illinois Natural History Survey.)

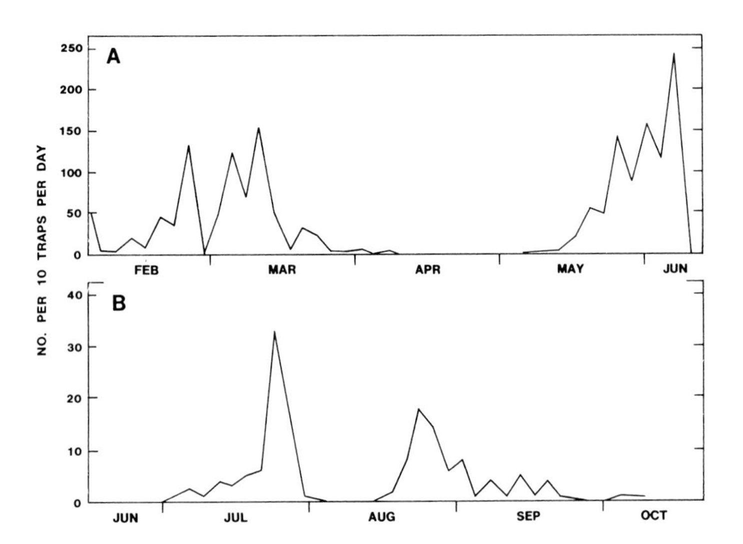

Figure 19. Incidence of Nantucket pine tip moths, *Rhyacionia frustrana*, caught in sex pheromone traps in 1983, Effingham County, GA. *A*, generations 1 and 2; *B*, generations 3 and 4. This information may be useful for planning a control strategy. (Redrawn from Gargiullo et al., ref. 280, courtesy C. W. Berisford and *Journal of Economic Entomology*.)

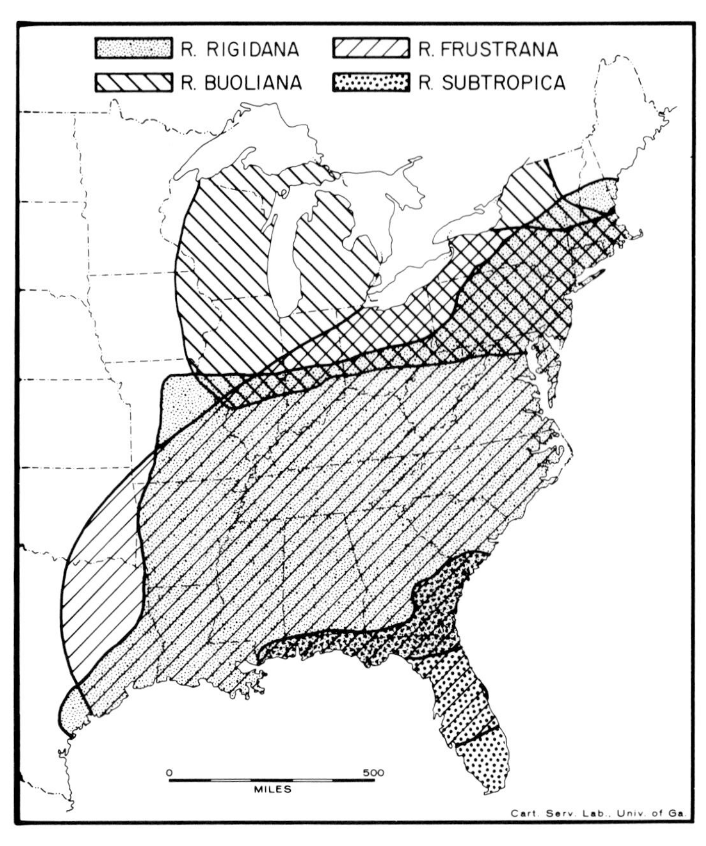

Figure 20. Approximate ranges in the eastern United States of the European pine shoot moth, *Rhyacionia buoliana*; the Nantucket pine tip moth, *R. frustrana*; the pitch pine tip moth, *R. rigidana*; and the subtropical pine tip moth, *R. subtropica*. (From Berisford et al., ref. 885, courtesy *Journal of Chemical Ecology*.)

A. A mugo pine "candle" (shoot) damaged by the European pine shoot moth, *Rhyacionia buoliana*.
B. Several new pine shoots showing symptoms of injury by a European pine shoot moth larva.
C. The "crook" symptom of European pine shoot moth injury. The leader recovered but the tree will remain disfigured.
D. A black pine showing characteristic bushy growth after repeated attack by the European pine shoot moth.
E. A twig of *Thuja occidentalis* showing bark injury by the larva of *Dioryctria abietivorella*.
F. A tip killed by the European pine shoot moth larva.
G. A mature *Dioryctria* larva.
H. A *D. abietivorella* adult; approximately 15 mm long.

A
B
C
D
E
F
G
H

Pine Shoot Borers (Plate 19)

Nine or more species of tip and shoot moths attack pine. The eastern pine shoot borer, *Eucosma gloriola* Heinrich, is a native of North America. It may be found throughout the natural range of white pine, including eastern Canada south to New Jersey and Virginia and west to Minnesota and Manitoba. The major host plants are 4- to 10-year-old white, Scots, and red pines. Other hosts include the Austrian, jack, mugo, and pitch pines and, on occasion, Douglas-fir and other conifers.

Damage to the plant is caused by larval feeding and tunneling in the pith of new shoots (C). Either the leader or lateral shoots are attacked, but more often the latter. Foliage on mined shoots fades and turns red in the summer, although smaller twigs may redden earlier (G). Such injury may be confused with that of the white pine weevil (Plate 20).

As the larva develops, it cuts into the wood at the base of the burrow but leaves the bark intact. The weakened shoot droops or breaks—this is often the first symptom of attack by the pest.

The loss of needles and the reduced size of the terminal leader are quite important to the Christmas tree grower. A leader size of 20–35 cm is desirable. After an eastern pine shoot borer attacks its host, only a 7.5- to 10-cm stub may remain.

The shoot borer overwinters as a pupa in the soil. Adults emerge any time from late April to early June, depending on the geographic location. Moth emergence in some areas seems to occur shortly after the Scots pine buds burst open. The adult moth has a wingspan of 14–15 mm (D). During the day, the insect rests on the shaded inner foliage of the pine. Eggs are laid on twigs or on needle sheaths, and can be found singly or in twos and threes. These eggs are round, flattened, pale yellow, 0.5 mm in diameter, and hatch in 10–15 days. The newly hatched larva enters the shoot behind a needle fascicle and develops through five instars within the shoot over a period of 42–55 days. The larva is dirty white and has a dark head. It can be as long as 13 mm. When fully developed, the larva chews an exit hole and drops to the ground. A light brown pupa is then formed in a loosely woven cocoon either in the top few inches of the soil or in debris.

The western pine shoot borer, *Eucosma sonomana* Kearfott, is a western species that occurs throughout the range of ponderosa pine. Its feeding habits and the damage it causes resemble that of *E. gloriola*. *E. sonomana* has a wingspan of 20–21 mm. In some areas of the West, larvae of this species reduce growth in the leaders of ponderosa and lodgepole pines without causing evident tissue injury. Some insects belonging to the genus *Eucosma* attack cones, thus interfering with seed production.

One parasite is known to overwinter as a mature larva in its own cocoon, which in turn is located inside the host's cocoon. This parasite is an ichneumon wasp in the genus *Glypta*. Synthetic sex pheromones have been developed for attracting the males of both *Eucosma* species.

References: 153, 154, 157, 192, 220, 275, 302

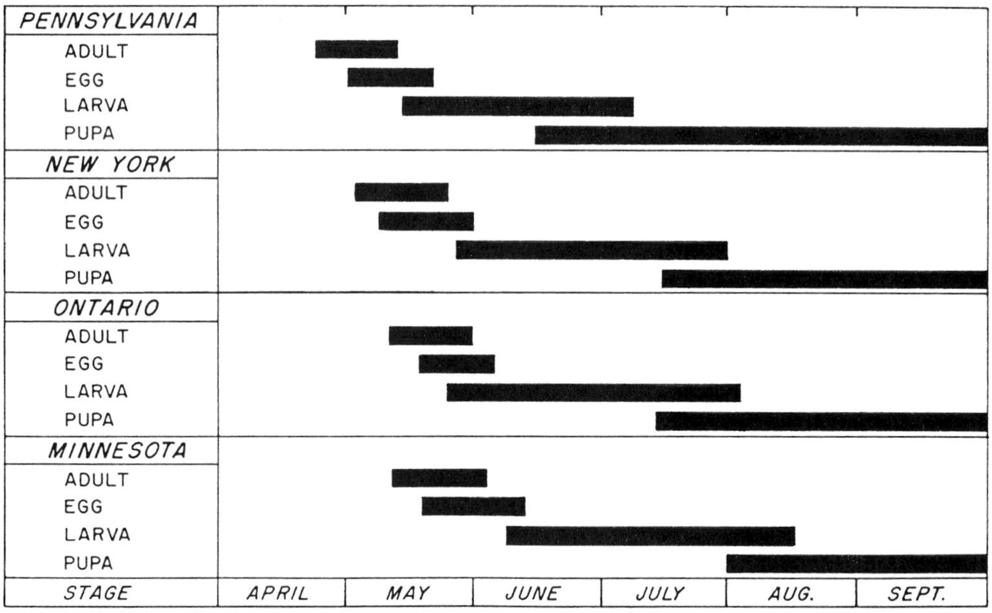

PENNSYLVANIA
ADULT
EGG
LARVA
PUPA
NEW YORK
ADULT
EGG
LARVA
PUPA
ONTARIO
ADULT
EGG
LARVA
PUPA
MINNESOTA
ADULT
EGG
LARVA
PUPA
STAGE | APRIL | MAY | JUNE | JULY | AUG. | SEPT.

Figure 21. Life cycle of the eastern pine shoot borer, *Eucosma gloriola,* in Pennsylvania, New York, Ontario, and Minnesota. (From DeBoo, ref. 154, courtesy R. F. DeBoo.)

Figure 22. Late symptoms of shoot injury to Scots pine by the eastern pine shoot borer. The shoot always droops, sometimes at a sharp angle. Needle length varies but is always shorter than that of normal needles. (From DeBoo, ref. 154, courtesy R. F. DeBoo.)

A. Damage inflicted by the larva of the eastern pine shoot borer, *Eucosma gloriola,* to the terminal leader of eastern white pine.
B. The leader of an eastern white pine with the "shepherd's crook" symptom.
C. An exposed tunnel in the pith of a new shoot with a shoot borer larva.
D. Pinned specimens of the eastern pine shoot borer (*top*) and the western pine shoot borer, *E. sonomana.* Note the similar color pattern and difference in size. [Tortricidae]
E. Pupae and a larva of the eastern pine shoot borer.
F, H. Infested Scots pine shoots that have dropped their needles because of shoot borer injury.
G. Reddish brown needles on a shoot indicate the presence of a shoot borer larva.

A
B
C
D
E
F
G
H

Conifer Twig Weevils (Plate 20)

Taxonomists now consider the white pine weevil, the Engelmann spruce weevil, and the Sitka spruce weevil as a single species, *Pissodes strobi* (Peck), a name that formerly referred to only the white pine weevil. This species is without doubt the most important factor limiting the production of timber from eastern white pine, *Pinus strobus*, and Sitka spruce, *Picea sitchensis*. It also causes serious damage to landscape plantings and Christmas tree plantations of Norway spruce, *Picea abies*. It is known to attack a wide variety of pines and spruce in addition to those listed above, occasionally even Douglas-fir.

This weevil, which kills the top 2- to 3-year's growth of the host tree, causes homeowners a great deal of concern. Weevil attack does not, however, kill the entire tree, although it may become crooked and limby as a result of repeated top killing. To some, the affected tree may be more aesthetically pleasing than a normal one.

A number of related weevils attack the tips and lateral branches of conifers. In the West, *Pissodes terminalis* Hopping attacks various species of pine, occasionally causing extensive damage to young lodgepole pine (1.5–3 meters tall) in the central Rocky Mountain region. This weevil prefers the most vigorous, open-grown trees and causes dieback of the current year's growth. The first symptoms are small punctures at the base of a new leader. Resin droplets ooze from these wounds and solidify. By midsummer the new foliage is chlorotic, its color gradually changing from yellow to red to brown. Some very small weevils (about 4 mm) of the genus *Cylindrocopturus* attack and kill small branches and tips of pines and Douglas-fir. They are particularly troublesome to Christmas tree plantings. Weevils of the genera *Magdalis* and *Scythropus* (Plate 22) feed on the needles of various conifers. The black vine weevil, *Otiorhynchus sulcatus* (Fabricius), feeds on the foliage and causes considerable damage to yew as well as to a number of other conifer and broad-leaved ornamentals (Plate 113). In the South and East, the eastern pine weevil (=deodar weevil), *Pissodes nemorensis* Germar, attacks deodar cedars, Atlas cedars, and cedar of Lebanon, all well-established exotic tree species. A native insect, this weevil is also considered a pest of the southern pines *Pinus taeda*, *P. echinata*, and *P. palustris*. Adult weevils look much like the white pine weevil (compare Plate 20D and Plate 21F) and cause injury during both adult and larval stages. Adult weevils feed on the inner bark, often girdling a stem or twig. Needles turn brownish red and may curl on lightly damaged trees. Adult weevils appear in April or May.

Although the biology of the twig weevils varies by species, a description of that of *Pissodes strobi* follows here as an example.

Adults usually overwinter in the litter on the ground. In the spring, adults fly or crawl to the leaders of suitable host trees. Mating takes place on the leader, where from one to a dozen or more pairs may gather. The female excavates a round hole in the bark and deposits from one to five eggs into the cavity (H). She then fills the hole with a plug of mascerated bark. Hundreds of eggs may be deposited in a single leader. The eggs hatch and the larvae feed in the bark, killing the leader of the previous year. While they are feeding, the current year's flush of growth starts but soon droops, producing the symptom known as "shepherd's crook" (Plate 19). The larvae bore into the wood and produce pupal chambers filled with shredded wood and bark. The new adults leave the host by late summer and do some feeding on twigs prior to overwintering. In British Columbia some adult weevils live up to 4 years.

References: 56, 193, 274, 586, 725

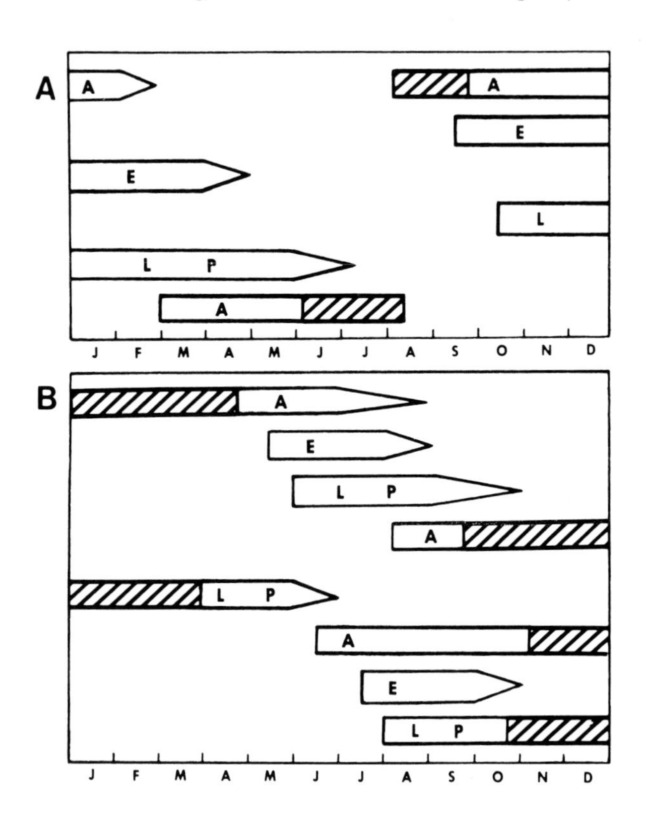

Figure 23. Generalized life cycles of southern (*A*) and northern (*B*) populations of the eastern pine weevil, *Pissodes nemorensis*. Cross-hatching indicates periods of inactivity. *E*, eggs; *L*, larvae; *P*, pupae; *A*, adults. (From Phillips et al., ref. 615a, courtesy T. W. Phillips.)

A. Drooping tip of Norway spruce, an early indication of attack by the white pine weevil, *Pissodes strobi*.
B. Dead top in pine resulting from weevil attack.
C. Hardened pitch exuded from weevil punctures in the leader of Norway spruce.
D. An adult white pine weevil.
E. Monterey pine twig has been split to show damage by the Monterey pine weevil, *Pissodes radiatae*.
F. Evidence of external feeding of the Monterey pine weevil on Monterey pine.
G. A weevil larva within spruce leader. In larger leaders the larvae do not bore into the pith.
H. Egg-laying punctures at arrow 1; punctures open to show eggs at arrow 2.

A
B
C
D
E
F
G
H
2
1

Trunk and Root Collar Weevils (Plate 21)

The pales weevil, *Hylobius pales* (Herbst), is the most serious pest of seedling pines in the eastern half of the United States and Canada. It is also a serious problem in forest and ornamental nurseries, Christmas tree plantations, and in freshly cutover pine forests. This insect is seldom a problem in landscape plantings because there is little breeding material available.

The adult weevil is the injurious stage because it feeds on the bark of twigs and small branches of conifers (I). The insect first chews a series of small holes in the bark. If the infestation is light, the holes fill with resin and heal. If the infestation is heavy, the holes soon become patches, and patches may extend to an entire branch or seedling. Most if not all of the bark may be removed. Whitish crystallized resin often covers the debarked areas. On large trees, feeding is generally restricted to the bark of twigs near the ends of branches. Twigs may be girdled, and if so needles die distal to the girdle, producing a flag of red needles. Feeding takes place mostly at night; adult weevils take refuge in leaf litter during daylight. This description of pales weevil injury applies to that caused by the pitch-eating weevil, *Pachylobius picivorus* (Germar), most commonly found in the Gulf states from Alabama to Texas.

Pales weevils overwinter as adults in the soil. In northern areas they become active in the spring between late April and early June, congregating on or near fresh cut stumps. In southern states adults may be active throughout the year. Eggs are usually laid in roots of recently cut stumps. The female burrows underground and chews small niches in root bark, where she lays one to three eggs. Upon hatching, the young larvae feed in bark phloem at the wood-bark interface. Here they make irregular tunnels that score the wood. Tunnels may be more than 1 meter long on stump buttress roots. Larvae pass through five or six stages and attain a length of about 12 mm. They are legless and look like the illustration in panel E. When it is time to pupate, the larva plugs its gallery with wood shavings and goes through metamorphosis in a chamber in the bark or within sapwood. The spring brood of adults usually emerges by the end of September, and they feed for a short period before diapause. Adults are long lived. In southern Ontario adults may live through two consecutive winters.

Hosts of the pales weevil include most native pines, spruce, fir, hemlock, Douglas-fir, juniper, larch, and cedar. They are a continuing pest of Christmas tree plantations and may threaten the life of young trees. Their geographic range includes all of the eastern half of the United States and neighboring Canada.

Hylobius radicis Buchanan, known as the pine root collar weevil, looks much like the pales weevil. It attacks both young and old pines, especially those growing on poor sites from Minnesota east to the Atlantic Ocean and south to Virginia and Kentucky. Scots, jack, red, Austrian, Corsican, and mugo pines are among those species that are attacked by the insect. Pitch pine is somewhat resistant and white pine is rarely attacked.

The pine root collar weevil feeds on healthy open-grown trees, a habitat found in nurseries and pine plantations. Both adults and larvae feed at the root crown on the inner bark. Adults may occasionally take nocturnal excursions to the crown of the tree and feed on the tender bark of twigs and small branches. Beginning in May young overwintered females emerge from diapause to lay eggs; they continue to oviposit into September in southern Ontario. Eggs are usually placed in a feeding wound made by an adult in the inner bark of the root collar. Some eggs may be laid in the soil a few centimeters from a root. Larvae feed in the bark phloem. A flow of resin from the injured root is the first external symptom. Vigorous trees respond to larval injury by producing copious amounts of resin under the bark at and near the injury site. The resin is drained off through tunnels made in both outer and inner bark and into the surrounding soil. A layer of pitch-infiltrated soil 40–50 cm thick may form near the feeding site. All stages of this weevil may be found throughout the growing season. Affected trees usually live 2–4 years after initial colonization.

The eastern pine weevil (=deodar weevil), *Pissodes nemorensis* Germar, is very closely related to the white pine weevil (Plate 20). It normally selects freshly cut stumps, blown-down trees, and standing trees of any size that are in a weakened or dying condition to feed on and lay its eggs. When large numbers of this insect are present, healthy trees and recently transplanted trees may be attacked and sometimes killed. The first symptom of attack is usually a discoloration and browning of foliage on twigs and small branches (A). The adult weevils (F) bore holes or pits in the bark of the trunk, branches, or shoots. Here they insert their curved snout to feed on the inner bark and hollow out a larger area around entrance holes. Such damage to the bole of a young tree may weaken it enough to make it attractive to ovipositing females. Adult maturation feeding in shoots may kill young terminals. Feeding pits and egg pits look the same except that those with eggs are plugged with macerated phloem. Eggs are most frequently laid low in the tree and in the deepest shade. After eggs hatch, larvae feed on the inner bark, all the while making tunnels between the bark and wood. The larvae may be found in the root collar (G) or in branches as small as 13 mm in diameter. Pupation occurs in their tunnel inside a "chip" cocoon (D) made of wood fibers that each larva cuts from the xylem by using its mandibles. The life cycle may be completed in 1–2 years. Larval stages may be found throughout most of the year (see Figure 23, accompanying Plate 20).

The eastern pine weevil is a secondary insect that rarely weakens and kills healthy trees. However, it is known as a vector of a pitch canker disease caused by the fungus *Fusarium moniliforme* var. *subglutinans* that causes twig and branch dieback. Sclerroderris canker caused by *Ascocalyx abietina* is not carried by eastern pine weevils, but weevil feeding on susceptible pines certainly makes it a facilitator of the disease (see Sinclair et al., *Diseases of Trees and Shrubs*). The weevil's distribution is believed to be limited to the eastern half of the United States and adjacent Canada. Red, eastern white, Scots, and Virginia pines are species commonly attacked in the north; exotic cedars such as cedar of Lebanon as well as southern pines are commonly attacked in the South. A synthetic sex pheromone is used to monitor populations, which aids in developing control strategies.

There are many species of *Pissodes* in the west. *P. radiatae* Hopkins, the Monterey pine weevil (Plate 20E, F) develops in the leader, trunk, and root collar of knobcone, lodgepole, Bishop, and Monterey pines. *P. terminalis* Hopping, the lodgepole terminal weevil, attacks open-grown young pines and kills developing terminals down to the uppermost whorl or branches. It feeds primarily on lodgepole and jack pines. *P. schwarzi* Hopkins, the Yosemite bark weevil, attacks the trunk of many species of *Picea, Larix,* and *Pinus.*

References: 193, 242, 243, 252, 275, 300, 351, 359, 456, 579, 586, 615a, 863

A. A Douglas-fir twig killed by the eastern pine weevil, *Pissodes nemorensis.*
B. Close-up of a twig showing adult eastern pine weevil feeding injury.
C. Sections of bark showing larval galleries of the eastern pine weevil.
D. A pupal chamber in bark. Close-up of G.
E. A larva of the eastern pine weevil. It is difficult to distinguish between this larva and the larvae of other species mentioned in the text.
F. An adult eastern pine weevil. Note how it blends into the background. About 6–8 mm long.
G, H. Root collar showing pupal chambers of the eastern pine weevil.
I. An adult pales weevil, *Hylobius pales,* and typical damage it causes. The weevil is about 9 mm long.

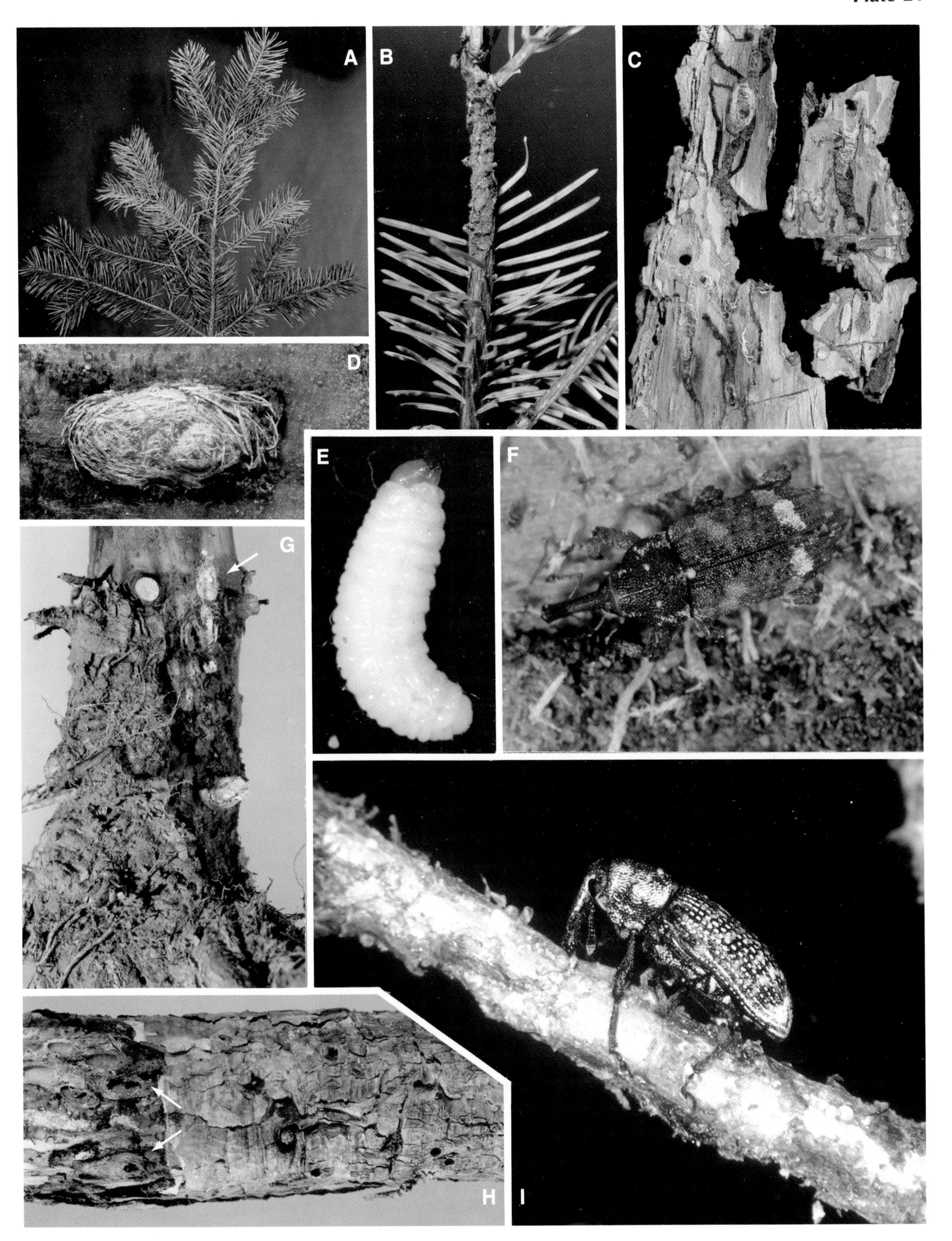
A
B
C
D
E
F
G
H
I

Pine Needle Weevils and Juniper Twig Girdler (Plate 22)

A number of weevils of the genus *Scythropus* feed on the needles of pines in western states. Their damage is distinctive in that the needles are notched intermittently along their length (B). Usually the needle dies beyond the point of damage. Under severe conditions the tree may assume a generally browned appearance.

The best known of these weevils is *S. californicus* Horn, which attacks Monterey and Bishop pines and occurs along the Pacific coast. In coastal central California the adults (D) appear in late winter and begin feeding on needles produced the previous year. By May or June the adults have completed their activities and are not seen until the following year. Eggs are laid on year-old growth in an egg chamber. To construct it, the female draws together three adjacent needles and cements them together along their length (C). Upon hatching, the young larvae drop to the ground and tunnel into the soil. They sustain themselves throughout this stage on the rootlets of pine. Pupation also occurs in the soil. The life cycle is believed to require 2 years.

S. californicus attacks only foliage that is at least 2 years old. It always leaves part of the needle uneaten. Its damage can thus be distinguished from that caused by other insect pests such as sawflies.

Although the juniper twig girdler, *Periploca nigra* Hodges, has been known in the eastern United States for more than 50 years, it has never been reported as a pest there and is not known to cause damage outside California. In the mid-1950s it caused dieback of juniper twigs in California.

Virtually all species and cultivars of ornamental juniper are attacked. Serious damage seems to be restricted to the junipers with slender stems, such as Tam (Tamariscifolia).

The adult is a tiny black moth. In the San Francisco Bay area moths are active in May, June, and July, whereas in southern California moths fly during March, April, and May. Eggs laid on the woody stems of juniper hatch into tiny larvae, which tunnel into the stems. In 8 or 9 months the larval stage is completed. After a short pupal period spent in a tunnel beneath the bark, the moths emerge from twigs to begin the cycle again.

Damage is caused by the larva, which mines beneath the bark and girdles the stems of juniper (I). The first sign of damage is the appearance of scattered yellow shoots, which later die and turn brown. Plants may be infested for 2 or 3 years before damage becomes apparent. Entire plants are never killed.

To determine if the insects are present, remove the bark from the lower part of affected juniper twigs and look for the characteristic girdling tunnels (H). Symptoms of twig-girdler activity are sometimes confused with those caused by mice, which find the thick cover of the juniper planting an ideal place to live. Mice chew bark from the twigs, which then die as a result of this girdling injury. The juniper twig girdler leaves the bark intact and girdles beneath the bark.

Once the twig-girdler larva is beneath the bark, there are no practical means of killing it. Sprays for control therefore must be applied only during the brief period when the moths are active and laying eggs. Clipping and removal of yellowed or dead twigs will improve the appearance of the juniper planting but will not control the twig girdler.

References: 87, 217, 406, 434

A. A terminal of Monterey pine showing the dead needle symptom caused by the weevil *Scythropus californicus*.
B. The notches chewed in the brown needle are another symptom of damage caused by *S. californicus*.
C. *S. californicus* lays its eggs between needles. These two needles were forced apart to show the eggs.
D. An adult *S. californicus*.
E. A view of the head of *S. californicus* showing the long sharp cusps it uses to dig its way out of its earthen pupal cell.
F. Dead or dying twigs in a mass planting of *Juniperus sabina* cv. Tamariscifolia (Tam juniper) caused by the juniper twig girdler, *Periploca nigra*.
G. Twig-girdler moth exit holes in a juniper twig.
H. Twigs of *Juniperus sabina* girdled by the juniper twig girdler. Arrow 1: loose bark not removed; arrow 2: bark removed to show the depth of the channel.
I. A portion of the bark removed to show the larva of the juniper twig girdler.

A
B
C
D
E
F
G
H
1
2
I

Cypress Bark Moth and Douglas-fir Twig Weevil (Plate 23)

Several members of the plant families Cupressaceae and Taxodiaceae are attacked by the cypress bark moth, *Laspeyresia cupressana* (Kearfott). This insect is known to occur only in coastal California from Mendocino to San Diego counties. Monterey cypress, *Cupressus macrocarpa,* is a favored host, and its cones, branch nodes, trunk, and a damaged area of any woody part of the tree are subject to infestation by the larvae.

Laspeyresia passes through two complete generations each year, one in spring and summer, the other in fall and winter. The adults, which have gray and white mottled wings with a spread of about 18 mm, lay eggs on suitable plant parts. After hatching, the larvae tunnel into the bark and produce large quantities of reddish frass, which is expelled during feeding beneath the bark (F) or within cones (C). Extensive resin flow often accompanies attack on the trunk or limbs (A). The larvae tend to feed locally, seldom moving far from the initial site of invasion. The death of plant parts—excepting some loss in germination of seeds from *Laspeyresia*-infested cones—seldom results directly from the feeding of this insect. The mature larva is about 13 mm long and creamy white with a brown head. It pupates at its feeding site and later emerges as a moth. The cast "skins" of pupae frequently protrude from the injured site (B) after the moths have escaped.

Laspeyresia commonly attacks sites on trees infected by cypress canker, a fungal disease caused by *Seiridium* (=*Coryneum*) *cardinale*. This disease is a known killer of branches and entire cypress trees. Experimentally, cypress bark moths have been shown to be capable of serving as carriers of the fungus spores, and they may aid in the spread of the disease. The actual role of the moth as a disease-spreading agent in nature has not been clearly demonstrated, however. (See Sinclair et al., *Diseases of Trees and Shrubs*.) Therefore, although it is no coincidence that the cypress bark moth and cypress canker often attack the same tree, it is the disease—not the insect—that causes the death of branches and sometimes entire trees (A).

Several tiny weevils of the genus *Cylindrocopturus* cause twig death in western conifers. Among them are the *C. eatoni* Buchanan (Figure 24), sometimes known as the pine reproduction weevil, which attacks principally ponderosa and Jeffrey pines from California into southern Oregon, and the Douglas-fir twig weevil, *C. furnissi* Buchanan (G), which occurs from British Columbia south into California, and infests Douglas-fir.

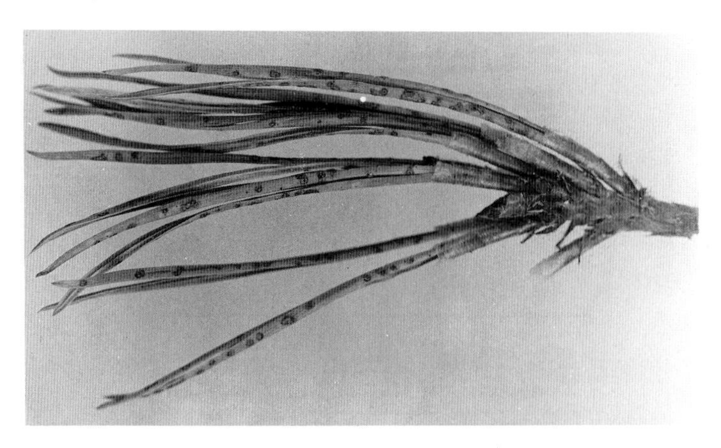

Figure 24. Feeding punctures made by the adult weevil *Cylindrocopturus eatoni* in needles of Jeffrey pine. (Courtesy U.S. Forest Service.)

Because *C. furnissi* damages young trees more than larger ones, the insect is of concern to Christmas-tree growers of Douglas-fir. Damage is said to be greatest on marginal soils during dry years.

The insect overwinters as a larva or adult. Adults lay eggs during the summer in small punctures they make in twigs. The young larvae feed by mining just beneath the bark, but mature larvae (E) often tunnel into the central pith. Pupation occurs at the end of the larval gallery (E) in the spring. As a result of larval feeding, twigs and small branches fade, turn red, and die (D). There is believed to be one complete generation and a partial second generation each year.

A tiny (2 mm long) scolytid twig beetle larva, *Pityophthorus orarius* Bright, mines the pith of Douglas-fir shoots, killing them. Larvae of the Douglas-fir weevil can easily be confused with those of *P. orarius*.

References: 87, 260–262, 274, 275, 808

A. Site of a fungus canker caused by *Seiridium* (=*Coryneum*) *cardinale* on the trunk of young Monterey cypress. Such sites are commonly infested by larvae of the cypress bark moth, *Laspeyresia cupressana*.
B. Site of infestation by the cypress bark moth. Note at arrow an extruded pupal case of a moth that has already emerged.
C. Monterey cypress cones infested by cypress bark moth larvae, showing copious amount of frass expelled during larval feeding.
D. A Douglas-fir twig killed by an adult Douglas-fir twig weevil, *Cylindrocopturus furnissi*.
E. A mature larva (*upper*) and pupa (*lower*) of the Douglas-fir twig weevil in the central portion of a twig.
F. The trunk of Monterey cypress showing frass expelled by larvae of the cypress bark moth.
G. An adult of the Douglas-fir twig weevil; only 3 mm long.

A
B
C
D
E
F
G

Conifer Bark Beetles (Plate 24)

Over 600 species of beetles are commonly referred to as bark beetles. Only the most serious pest species are dealt with here and in the following four plates. Bark beetles have an enormous effect on forests and on trees associated with recreational campsites, golf courses, and suburban residential areas carved out of conifer stands. Human activity increases the potential for tree abuse, creating environmental conditions that encourage a wide range of bark-feeding insects. The most serious injury results when vast numbers of larvae feed in the phloem of the bark, where their chewing interferes with the flow of sap and also destroys the cambium and scars the face of the sapwood. All of this damage is largely hidden from human eyes. Bark beetles also introduce a complex of parasitic and decay fungi that may contribute to tree mortality.

The black turpentine beetle, *Dendroctonus terebrans* (Olivier) (D), is primarily a central Atlantic and southern species, but it may occur as far north as New Hampshire. It is known to attack all species of southern pine and red spruce, but loblolly and slash pines are most seriously injured. It is the largest of the North American bark beetles.

In its most northern range the black turpentine beetle overwinters as an adult in corky bark. In the deep South all stages are present throughout the year. Beetles usually confine their egg laying and feeding to the basal 2 meters of the trunk (Plate 25). They especially like the stumps and buttress roots of freshly cut trees. Mated pairs of beetles work together to excavate large cavelike galleries. Such galleries may be up to 25 cm wide and 30 cm long and deeply engraved on the face of the sapwood. As many as 200 eggs are laid, aligned along one side of the gallery. Upon hatching, the larvae feed gregariously on the inner bark. By the time the brood has completed larval development, the gallery may be as much as 30 cm across. Fully grown larvae are legless, about 12 mm long, and creamy white. A life cycle may last 10–16 weeks, depending on temperature and other environmental conditions. Three generations per year may occur in the deep South.

The symptoms produced are illustrated in panels A and B. The masses of pitch on the trunk contain a small hole made by adult beetles. The pitch masses with a hole in the center are called pitch tubes and are considerably larger than those of the southern pine beetle. The hole is the only entrance to the beetles' chambers and is continually used to dispose of borings. The pitch hardens and turns white to reddish. The pitch mass is irregularly shaped and roughly 4 cm in diameter. Trees weakened by drought, by other insects, or any stressful event are most susceptible to attack. Occasionally, perfectly healthy trees become infested. Young trees the size of Christmas trees are seldom attacked.

The red turpentine beetle, *D. valens* LeConte, and the black turpentine beetle have similar habits, cause similar symptoms, and are essentially the same size, but the former, true to its common name, is a distinct reddish brown. It occurs throughout southern Canada and the United States except for the South Atlantic and Gulf Coast states, and it attacks all species of pine, occasionally spruce, fir, Douglas-fir, and larch.

In forests the eastern larch beetle, *D. simplex* LeConte, is considered to be a secondary pest of sick or weakened trees. Larches are now fairly common in northern urban landscapes, and urban stresses may make them vulnerable to injury by this beetle. Adults begin their breeding cycle in May and June, and in some locations there may be as many as three generations annually. The beetles are dark brown and between 3.4 and 5 mm long. Like all *Dendroctonus* beetles, the larvae mine in the bark phloem and lightly etch the wood surface. The eastern larch beetle can be found in Alaska, throughout Canada, and east of Minnesota to the coast and south to West Virginia.

There are about 25 species of *Ips* bark beetles in North America, and all feed on conifers, principally pine and spruce. Some species are highly host specific. They are readily distinguished from species of *Dendroctonus* by their elytral declivity and associated spines (E), their clean egg galleries, and the fact that they engrave sapwood. Behavior, particularly in their social arrangement, is distinctive. *Ips* beetles are usually polygamous. The male constructs a mating chamber in the bark phloem and constructs egg galleries. At this point, females are attracted to the male and his "staked-out" homestead. Two to seven females take up residence with the male, each with her own egg gallery radiating off of the mating chamber. The combined action of several social groups, for example, adults and their broods, may result in the girdling of trees.

The pine engraver beetle, *Ips pini* (F, G), is another native species with transcontinental distribution. It occurs from South Carolina and northern Georgia to Quebec, west to the Rocky Mountains and Pacific Coast states, plus Alaska and British Columbia. Its host trees include all of the western and eastern pines and several species of spruce. Because of its geographic distribution and wide host range, it is considered one of the most common bark beetles and at times is an aggressive pest.

The first symptom of attack is the presence of reddish boring dust that was pushed out of small holes by adult beetles and collects on rough bark. Needles then turn yellow and red and finally drop. *I. pini* attacks young as well as mature trees. In the West it commonly top-kills older trees.

The engraver beetle overwinters in duff and litter, occasionally under bark. Eggs are laid in galleries excavated by adults. They hatch into white legless larvae and feed on phloem, cambial, and sapwood tissues. One to five generations may occur each year, depending on the length of the growing season.

Pityogenes hopkinsi Swaine is a chestnut brown bark beetle, 1.5–2 mm long, that resembles *Ips* bark beetles in shape and social behavior. It is usually considered a noneconomic species, because under forest conditions it attacks weak trees and shaded branches of pines. However, these insects frequently injure newly transplanted pines, as well as ball and burlapped stock in holding areas. Injury is first noted in the spring with the appearance of splotches of sap-soaked bark associated with a tiny hole. In the vicinity of the hole there are reddish, resin-soaked borings (C). As the symptoms progress, sap creeps down the bark, becomes encrusted, and turns white. The beetles feed and breed only in thin pine bark, most noticeably in eastern white pine. Weak young trees may be killed. Adult beetle activity in upstate New York begins in late April and early May, when the beetle makes the hole described above. Two generations, and occasionally a partial third, occur each year. The beetles are capable of overwintering in the larval, pupal, and adult stages as long as they remain in bark.

None of the several western species of *Pityogenes* are known to injure ornamental conifers.

References: 63, 78, 120, 130, 193, 297, 350, 657, 690, 862

A. Rusty brown and brittle needles are symptoms of pine trees injured and dying from black turpentine beetle feeding. [Scolytidae]

B. The white pitch masses, called pitch tubes, are the results of a breeding pair of black turpentine beetles. These symptoms are always found from ground level to about 2 meters high on the trunk.

C. Tiny holes surrounded by resin-soaked bark and covered, or partially so, with fine resin-soaked "sawdust" are symptoms of injury caused by *Pityogenes hopkinsi* on eastern white pine. [Scolytidae]

D. A black turpentine beetle. The color may vary among individuals, from dark black to mahogany red; 5–10 mm long.

E. The rear end of a pine engraver beetle showing the concave declivity and spines located at the end of the wing covers. The appearance of this part of the body region is most useful in the identification of *Ips* bark beetles. Arrows point to spines.

F, G. A dorsal view (F) and lateral view of the pine engraver beetle; 3.5–4.5 mm long.

A
B
C
D
E
F
G

Eastern Pine Bark and Shoot Beetles (Plate 25)

Pine trees in the eastern United States are attacked by at least six major species of bark beetles, five of which are illustrated here. In the drawings, species of *Ips* are given common names that refer to the number of spines that border a declivity on the back of the wing covers (elytra). These spines are useful in identification but may lead to the wrong conclusions. For example, *Ips avulsus* (Eichhoff), a southeastern species, *I. pini* (Say), a transcontinental species, and *I. tridens* (Mannerheim), a western species, all have four spines. By using the known geographic distribution of a species as well as its size and the number of elytral spines, one can make a better tentative identification. The elytral declivity is a major physical feature that separates *Ips* from the other two major genera of bark beetles. If the declivity is absent, the beetle is likely to be a species of *Dendroctonus* or *Scolytus*; if present, it is likely to be *Ips*. The concave declivity functions as a scoop to keep tunnels clear of frass and boring dust. The underside of the abdomen of *Scolytus* species ascends abruptly or is concave.

In the illustration emphasis is placed on spatial distribution in the trunk. This is an appropriate generalization but an oversimplification. Of the three *Ips* species illustrated, *I. avulsus*, properly called the small southern pine engraver, has the most southern distribution (Pennsylvania to Florida and west to Texas) and is the smallest, about 2.5 mm long. It prefers the small upper limbs and top of pines, where its activities are little noticed. In the deep South its life cycle may be completed in 18–25 days, and 10 or more annual generations may occur. The larval galleries are characterized by short mines that terminate in relatively large, round pupal chambers.

I. grandicollis (Eichhoff), the sixspined ips, occurs from eastern Canada to Florida and west to Saskatchewan, Minnesota, and Texas. It prefers to colonize the upper portion of the trunk and is frequently found in fallen trees. The adult is reddish brown to black and 3–4.5 mm long. In the South six or more annual generations may occur. The galleries it forms are typical of the genus.

I. calligraphus (Germar), the eastern fivespined ips, occurs throughout the eastern United States, Nova Scotia, Ontario, Alberta, and California. It attacks the lower portion of the trunk of most species of pine. The adult is reddish brown to black and 3.5–6.5 mm long. The life cycle may be completed in about 25 days, with four to five generations per year in the Southeast and California.

The southern pine beetle, *Dendroctonus frontalis* Zimmermann, has received more research attention in the South than any other forest insect pest. Epidemics of this bark beetle have been recorded since 1882 in eastern Texas, and it has been monitored and studied almost continuously since that time. It is an aggressive killer of all southern pines. Even healthy, vigorous trees may be overcome by an epidemic population. Pines used for ornamental and shade purposes are as susceptible as forest-grown trees.

Three to nine generations of the southern pine beetle may occur each year, depending on the length of the growing season. Normally, the generations overlap, masking the actual number of generations. Selection and colonization of a host tree begins with the female beetle. After penetrating the bark of a suitable host, she constructs an egg gallery in the inner bark and is promptly joined by a male. (*Dendroctonus* beetles are monogamous.) Together they release volatile chemicals produced in their bodies called pheromones. In the presence of host odors the pheromone attracts other individuals in the vicinity to its source. The chemical released by these beetles, described as an aggregation pheromone, draws large numbers of beetles to a tree suitable for colonizing. During a southern pine beetle attack a healthy tree responds by bleeding pitch from the many attack holes (Plate 26D). Beetles make a hole in the center of each pitch mass, and the result is called a pitch tube. Hundreds of these white pitch tubes may be found on a single tree trunk. If attacking beetles overcome the tree's defenses, the females lay eggs in niches along S-shaped egg galleries. Each female is capable of laying about 160 eggs in her lifetime. The duration from egg to adult may be as little as 26 days. The adult is dark brown and is 2.5–4 mm long.

Winter is spent in the bark in any of the life stages. The first spring generation emergence has been correlated with flower opening of the flame azalea, *Rhododendron calendulaceum*. Geographic distribution of the southern pine beetle extends south from a line through southern New Jersey west through the Ohio Valley, southeastern Missouri, eastern Oklahoma and Texas, and central Arizona.

Characteristic injury symptoms include white to reddish pitch tubes, reddish boring dust clinging to rough bark, or occasionally spiderwebs at the base of the tree; the fading of green foliage to yellow or red; and finally an abundance of beetle exit holes.

Some of the pheromones produced by bark beetles and used as a means of communication between and among bark beetle species and their insect enemies have been identified and synthetically produced. These synthetic pheromones show promise for use in both control strategies and monitoring.

In 1992 a pine shoot beetle, *Tomicus piniperda* (Linnaeus), native to Europe was discovered in a Christmas tree plantation in Ohio. Since 1992 it has been found in Pennsylvania, New York, Michigan, Indiana, Illinois, and Ontario. Given its latitudinal range in Europe, its distribution in North America could reach to northern Florida.

The adult takes the same form as the native elm bark beetle (Plate 116), however *T. piniperda* is larger (3.5–4.8 mm) and darker. In Europe it overwinters as an adult in hollowed twigs or in bark galleries at the base of the trunk. In southern Europe there are two annual generations. It attacks both growing shoots and trunks, especially those of Scots pine, *Pinus sylvestris*. Other hosts include spruce, larch, several other two- and three-needle pines, and eastern white pine. In the United States the worst damage results from tip mining by the beetle. The growing points are destroyed, causing several forms of crown abnormalities such as dieback of shoots, yellowing of needles, and dead, bored-out shoots that may litter the ground under infested trees. A single beetle may kill up to four shoots during the course of its life.

References: 62, 78, 130, 193, 542, 657, 779, 781a, 853

Life cycles, shapes of galleries, and characteristics of southern pine bark beetles. [Scolytidae] (Courtesy Boyce Thompson Institute for Plant Research at Cornell University, Ithaca, NY.)

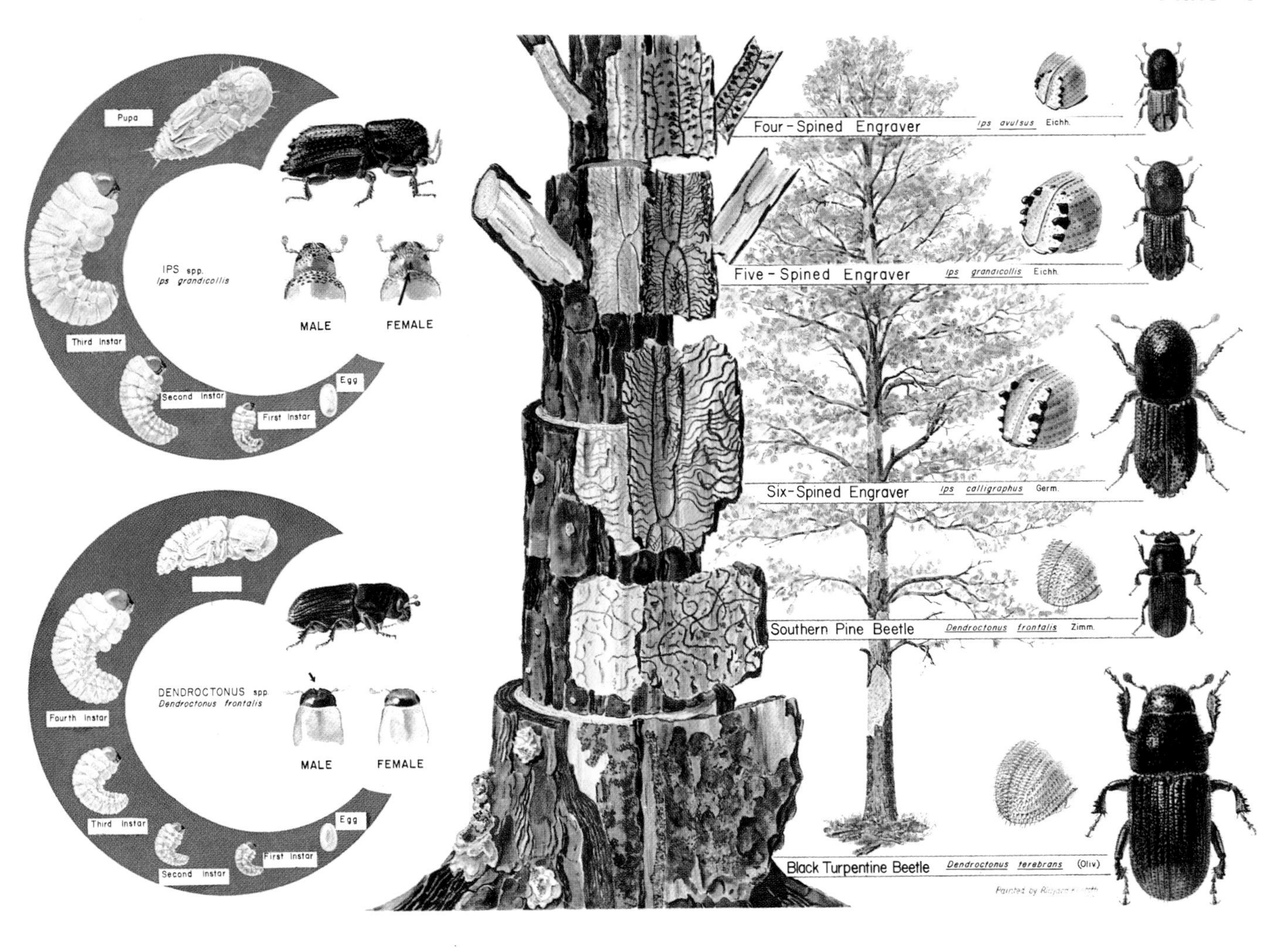
Pupa
IPS spp.
Ips grandicollis
MALE
FEMALE
Third instar
Second instar
First instar
Egg
DENDROCTONUS spp.
Dendroctonus frontalis
Fourth instar
Third instar
Second instar
Four-Spined Engraver
Ips avulsus Eichh.
Five-Spined Engraver
Ips grandicollis Eichh.
Six-Spined Engraver
Ips calligraphus Germ.
Southern Pine Beetle
Dendroctonus frontalis Zimm.
Black Turpentine Beetle
Dendroctonus terebrans (Oliv.)

Western Conifer Bark Beetles (Plate 26)

Bark beetles are important enemies of older mature conifers, especially pines. They are more likely to be a problem in areas such as the West and South, where suburbs intermingle with large tracts of mature and overmature conifers. The beetles may be attracted to yard trees, especially those weakened by the excavation and filling associated with home building.

Certain similarities exist in the life cycle and habits of bark beetles. The beetles usually overwinter beneath the bark of the host tree as larvae (C, F), pupae, or adults (I), depending on the species involved. In the spring the adult beetles tunnel their way to the bark surface, emerge, and fly in search of other trees to attack. Some beetles are attracted to the resinous odors of injured or weakened host trees. The adult (in some species the female, in others the male) initiates the egg-laying gallery after first tunneling through the bark. This gallery lies mostly in the phloem—the spongy inner white bark—though it sometimes involves the xylem or wood. The first beetles to attack a host produce and emit powerful volatile chemicals called pheromones that attract other beetles to the host. Sometimes by sheer numbers alone the beetles are able to overwhelm and kill healthy trees.

The mating of male and female usually occurs in the gallery. Eggs are laid on the sides of the gallery in characteristic patterns for each species. When the eggs hatch, the larvae form individual feeding galleries that may radiate from the egg-laying gallery (G). Other species, *Dendroctonus brevicomis* LeConte and *D. frontalis* Zimmerman in particular, form sinuous tunnels that appear to meander aimlessly in the phloem (B).

When feeding is completed, the larvae pupate in the inner or outer bark. Emergence from the tree and initiation of a new generation follows. There may be one to six or more generations each year, depending on species and geography.

As a general rule, bark beetles attack trees that are dying or in a state of decline due to a variety of stress factors such as drought, mechanical injury, compaction of soil in the root zone, smog, alteration of the water table by cuts or fills, root rots, and the like. Some species apparently are able to attack healthy trees, especially if the healthy tree is near an unhealthy one. Healthy pines located near one that has been struck by lightning are often mass attacked by beetles that emerge from the nearby injured tree. Lightning-struck trees should be removed from the yard as soon as possible, especially in the South, where few of them would survive even if not removed.

The mountain pine beetle, *D. ponderosae* Hopkins, is considered by some forest entomologists to be the most serious and destructive of the conifer bark beetles in the West. It has essentially the same body form as other *Dendroctonus* species, is black, and may be 4–7.5 mm long. Egg galleries are vertical and may be as long as 70 cm. Eggs are deposited in niches a few millimeters apart along both sides of the egg gallery (C). The developing larvae make short feeding tunnels (3–5 cm) at right angles to the egg gallery. Pitch tubes (Plate 24B) on the trunk and red boring dust lodged in bark crevices and lying on the ground at the base of the trunk indicate the presence or previous activities of adult beetles.

These insects especially become a problem in overmature forests. They will attack all thick-barked western pines; however, lodgepole is considered their favorite host.

Attacks by the various bark beetles characteristically occur in one particular part of the tree (Plate 25). The red turpentine beetle, *Dendroctonus valens* LeConte, and its close relative the black turpentine beetle, *D. terebrans* (Olivier), restrict their activities to the lower bole (stem) (Plate 24). Attack sites are conspicuous in that reddish tubes of pitch and dust protrude 2.5 cm or more from the infested tree. The western pine beetle, *D. brevicomis,* attacks the upper midbole of the tree, though it may subsequently infect the bole above and below that site. Other bark beetles, among them species of *Ips, Pityogenes* (Plate 24), and *Pityophthorus,* attack the tops and twigs. *Hylastes* species infest the root collar of newly transplanted or weakened young pines. Members of the bark beetle group *Phloeosinus* attack cypress and junipers, where they feed for a short time as adults on the small twigs. See also Plates 17 and 18.

References: 87, 193, 275

A. Pines killed by bark beetles.
B. Wandering galleries made by the western pine beetle.
C. Eggs and young larvae of the Douglas-fir beetle, *Dendroctonus pseudotsugae* Hopkins.
D. A southern pine beetle "pitched out" by a healthy pine.
E. A western pine beetle.
F. A typical bark beetle larva.
G. A typical gallery of the Douglas-fir beetle. Note the egg-laying gallery at the arrow. Larvae are at the right, at the ends of the feeding galleries.
H. A healed-over Douglas-fir beetle gallery that eventually will lead to deterioration of the stem.
I. Adult *Ips* beetles in galleries.

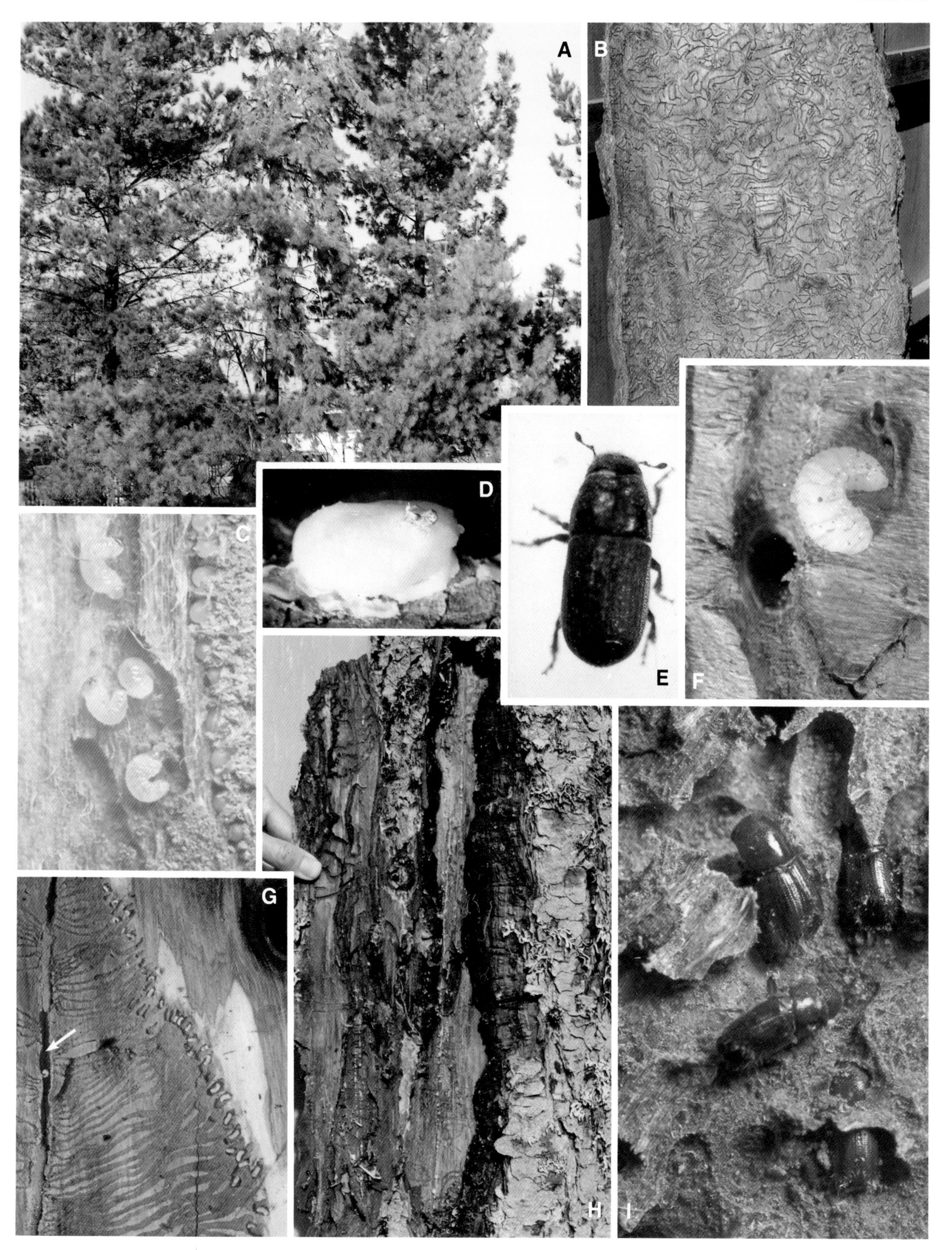
A
B
C
D
E
F
G
H
I

Cedar Bark Beetles (Plate 27)

Northern white cedar, commonly called arborvitae, is subject to injury by a small bark beetle, *Phloeosinus canadensis* Swaine, also sometimes called the northern cedar bark beetle. *P. canadensis* occurs in the northeastern part of the United States and throughout the eastern half of Canada. This species is not usually considered a major pest because it attacks mainly twigs, branches, and the bark of weakened trees. After hurricane Hazel in 1954, when there was a great deal of windthrown and damaged timber in woods and parks, and during the years 1963 through 1967, when the Northeast was plagued by drought during the growing seasons, *P. canadensis* invaded weakened arborvitae trees. Although many tree deaths could have been prevented by sound cultural practices such as the cabling, bracing, and watering of injured trees, and the pruning of twisted limbs, a high mortality rate resulted.

Both the larva and adult (C) northern cedar bark beetle injure trees. Each horizontal tunnel (D) is the work of a single larva. There were at least 72 larvae in this brood gallery. Another brood gallery on the opposite side of this trunk would have been adequate to kill the tree. Injury to *Juniperus virginiana* resulting from the feeding of adult beetles is limited primarily to the bark (C, E), and, occasionally, the foliage. The beetle, shown at the arrow point in panel C, chews holes at twig crotches to produce the flagged (or dead) twig as shown in panel B. Panel E shows a small quantity of sap that has collected and solidified in a wound, an example of a protective mechanism found in many coniferous trees.

The female beetle lays her eggs in a groove cut deeply into the sapwood. The brood develops through the autumn and resumes feeding in the spring. For more details on bark beetles, see Plates 24–26.

Several other species are found in the East. *P. dentatus* (Say) feeds principally on eastern redcedar. Like most other *Phloeosinus* species, it alone does not cause critical or life-threatening injury. But if a root-rotting fungus, such as *Fomes annosus*, is also present, the damage caused by both pests will kill the tree. *P. taxodii* Blackman is known only in the southern states and feeds on baldcypress, *Taxodium distichum*. Western *Phloeosinus* bark beetles are shown in Plate 28.

Woodpeckers often work in trees infested by bark beetles. In panel A two vertical rows of holes that were punched in the bark in a previous season by a woodpecker are visible. Note that the bird has followed one side of the gallery pattern.

References: 8, 193, 862

A. Pitlike scars caused by woodpeckers in pursuit of larvae of a *Phloeosinus* species in the bark.
B. Dead twigs caused by adult northern cedar bark beetles, *Phloeosinus canadensis,* feeding at the twig crotch in *Juniperus virginiana*.
C. A beetle feeding at a twig crotch.
D. Adult and larval galleries of the northern cedar bark beetle. The adult gallery is vertical. [Scolytidae]
E. The reddish material at the base of the twig is sap that has collected and solidified at the wound site.

A
B
C
5 mm
D
E

Cypress Bark Beetles (Plate 28)

Several species of bark beetles, all belonging to the genus *Phloeosinus,* attack trees of the families Cupressaceae and Taxodiaceae. These two families include cypress, juniper, redwoods, and certain other cedarlike trees. In fact, bark beetles found tunneling beneath the bark of limbs or trunks of such trees almost certainly are *Phloeosinus* species, for these families of trees have virtually no other bark beetle enemies.

The biology and life history of these pests are quite similar to those of the pine bark beetles (Plate 26). Breeding occurs in dead, weakened, or dying trees. Their egg and larval galleries scar the face of the sapwood (E). The larval brood is found in the trunk or larger branches. Most brood trees have been injured or weakened by other natural causes. Many newly emerged cypress bark beetles have the curious habit of feeding on the twigs of cypresses and certain other plants of the two plant families mentioned above. Twigs are usually attacked at a location 15–30 cm from their tip. During feeding, the adult hollows out, or deeply grooves, slender twigs, which then break easily in the wind (B). The dead tips, called flags, often remain on the trees for long periods of time, making them unsightly (A). Flagging is very common on Monterey cypress, *Cupressus macrocarpa,* in coastal California.

Flagging, if caused by cypress bark beetles, is not an indication that the tree is declining in health. It does indicate that somewhere in the area a dead or dying cedarlike tree is serving as a breeding place for the beetles. Flagging may continue for many months each year. In the vicinity of San Francisco Bay, however, it occurs mostly during the spring.

Important eastern species of *Phloeosinus* include *Phloeosinus canadensis* Swaine (about 2 mm long), known as the northern cedar bark beetle, which attacks arborvitae (*Thuja*) in eastern Canada and the northeastern United States (Plate 27); *P. dentatus* (Say), called the eastern juniper bark beetle, which infests eastern redcedar, *Juniperus virginiana,* and several related trees from Massachusetts to Florida and west to Texas. *P. taxodii* Blackman, the southern cypress bark beetle, attacks baldcypress, *Taxodium distichum,* in the southern United States. In the West the cypress bark beetles *P. cupressi* Hopkins and *P. cristatus* (LeConte) feed on Monterey and other cypresses in California, other Pacific coast states, and Arizona. Both western species may transmit the tree-killing fungus *Coryneum cardinale* (see Sinclair et al., *Diseases of Trees and Shrubs*).

References: 8, 87, 193, 275, 862

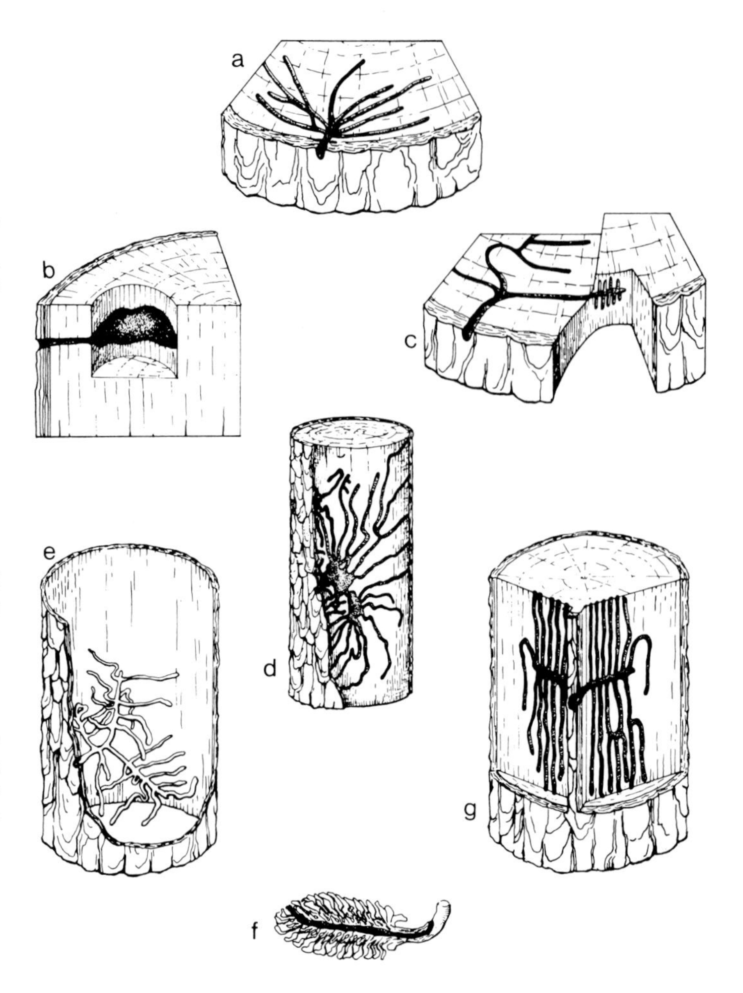

Figure 25. Different types of ambrosia and bark beetle galleries. *a,* branch—ambrosia beetles; *b,* cave—ambrosia beetles; *c,* compound—ambrosia beetles; *d,* cave—linear *Chaetophloeus* bark beetles; *e,* radiate—*Ips* bark beetles; *f,* cone—*Conophthorus*; *g,* forked—*Scolytus* bark beetles. (From J. B. Mitton and K. B. Sturgeon, *Bark Beetles in North American Conifers: A System for the Study of Evolutionary Biology,* copyright © 1982 by University of Texas Press.)

A. A Monterey cypress with many flagged twigs caused by *Phloeosinus cristatus.* [Scolytidae]

B. Three flagged twigs of Monterey cypress. One twig has not yet turned brown.

C. An adult *P. cristatus,* 3.5 mm long, on a Monterey cypress twig.

D, E. Twigs showing the nature of adult injury—bark chewed away by the beetle.

F. Gallery systems of a *Phloeosinus* species beneath the bark of Monterey cypress. Vertical channels were made by the adults. Each lateral channel was made by a single larva.

A
B
C
D
E
F

Pitch Moths (Plate 29)

The larvae of several moths of the genus *Synanthedon* attack pines and certain other ornamental conifers in many parts of the United States. The adults are clear-winged moths with yellow and black bodies; they resemble yellowjacket wasps.

Synanthedon sequoiae (Hy. Edwards), also known as the sequoia pitch moth, is a common pest on pines from California to British Columbia, and also in Idaho and Montana. The relative susceptibilities of native and introduced pines growing in an arboretum at Placerville, California, were evaluated over 2 consecutive years (Table 5).

***Table 5.* Susceptibility of selected pines to the sequoia pitch moth (Placerville, CA, 1943–44)**

Pinus species	No. trees evaluated	Avg. no. pitch masses per tree
pinaster	40	7.74
sylvestris	69	4.08
sabiniana	17	3.70
pungens	18	1.92
halepensis	34	1.82
nigra	63	1.13
torreyana	27	1.13
banksiana	22	0.95
jeffreyi	268	0.55
contorta	25	0.52
attenuata	30	0.52
densiflora	37	0.50
muricata	30	0.45
thunbergiana	10	0.40
resinosa	18	0.28
coulteri	31	0.22
ponderosa	746	0.19
virginiana	3	0.17
rigida	32	0.12
patula	35	0.11
ponderosa var. *scopulorum*	298	0.11
echinata	108	0.05
pinea	24	0.04
montezumae	61	0.04
mugo	44	0.03
radiata	279	0.03
palustris	81	0.006

Eggs are laid singly on the bark of the trunk or limbs of pines. After hatching, the larvae tunnel into the inner bark and cambium region, where they excavate a cavity and establish a feeding site. Large accumulations of pitch and frass form on the outside of the tree at the point of attack. The mature larva is about 25 mm long and has a reddish brown head and light yellowish body. New attacks often occur at the sites of old ones. Pupation takes place within the pitch mass, and adults emerge during the summer months. About half the population requires 1 year to complete a life cycle; the other half requires 2 years. Moths always emerge in the summer.

Although healthy pines and Douglas-fir are sometimes attacked by *S. sequoiae,* trees whose trunks or limbs have been mechanically injured are especially susceptible to infestation. Wounds inflicted by pruning, support wires, automobiles, and lawnmowers are often hosts to this pitch moth. It has been demonstrated that trees wounded in the spring, just before the adult moths fly, and summer, when moths are emerging, are far more likely to be attacked than trees wounded in the fall or early winter. Although accidental tree wounds cannot be avoided, those made by pruning can be controlled. Pruning should be confined to the time of year when it is unlikely to contribute to increased pitch moth attack.

S. sequoiae is not likely to threaten the life of a pine. The larvae feed locally just beneath their pitch masses. The masses, and the streaming of pitch down the trunks of trees, however, result in unsightly trees. Some limb breakage has been reported when infestations occur near the slender top of trees or on limbs. Redwood, *Sequoia sempervirens,* is not known to be a host.

The pitch mass borer, *S. pini* (Kellicott), causes pitch masses similar to those made by *S. sequoiae.* Its coniferous hosts are found along the Atlantic coast, in New England, the Midwest, the Appalachian region and southward, and eastern Canada. Hosts include Austrian pine, *Pinus nigra*; eastern white pine; Scots pine; jack pine; white spruce, *Picea glauca*; Norway spruce; and Colorado blue spruce. *S. pini* requires 2 or 3 years to complete its life cycle. Adult moths appear only during the summer months.

S. novaroensis (Hy. Edwards), sometimes called the Douglas-fir pitch moth, attacks spruce, pine, and Douglas-fir in the western states and causes a pitch mass similar to that described above.

Petrova albicapitana (Busck), known as the northern pitch twig moth, is widely distributed and attacks young jack pine and lodgepole pine and its varieties throughout the natural ranges of these trees in North America. In Alberta, Canada, they are considered one of the most important insects attacking twigs and terminals of lodgepole pine. The young larva forms a blisterlike nodule on the terminal or lateral growth of the host (D). After spending one winter under a nodule, the larva moves to a new site, usually a crotch, and forms a second nodule. Here the larva completes its feeding stage and transforms into the pupal stage. Two years are required to complete the life cycle, and adult moths appear only during the early summer. Several other *Petrova* species cause similar damage to pine, spruce, and fir in widely scattered regions of North America. Synthetic sex pheromones may be available for use in traps to assess population density and provide timing information for pest management.

References: 46, 87, 193, 194a, 275, 436, 791, 826, 869

A. The bark of Monterey pine showing several pitch masses made by the sequoia pitch moth, *Synanthedon sequoiae.* A few pitch tubes made by the bark beetle *Dendroctonus valens* are also evident.
B. A mechanical wound on Monterey pine and adjacent pitch masses (*arrow points*) of the sequoia pitch moth.
C. Pitch mass of the pitch twig moth, *Petrova comstockiana* (Fernald), and larva that has been exposed. [Torticidae]
D. A pitch nodule caused by a species of *Petrova.* Pupation occurs in the nodule. Note the pupal case.
E, F. Pitch masses on the trunk of blue spruce caused by the pitch mass borer, *S. pini.* [Sesiidae]
G. Pitch mass opened to show the larva of the pitch mass borer.

A
B
C
D
E
F
G

Balsam Woolly Adelgid (Plate 30)

The balsam woolly adelgid, *Adelges piceae* (Ratzeburg) is a member of a family closely related to aphids. This pest of true firs (*Abies* species) was introduced into the United States from Europe or Asia around the turn of the century. It has spread throughout the United States and Canada and has killed millions of feet of the finest fir timber on the continent. In some areas it has become a pest of yard trees and other ornamental firs. It does not attack Douglas-fir, *Pseudotsuga menziesii,* a tree that is not a true fir, and it does minimal damage to noble fir, *Abies procera.*

The balsam woolly adelgid has a strange life history. In the United States the entire population consists of females. The adults are less than 1 mm long and are purplish to black when the dirty white wax is removed. They are wingless, and the legs are visible only under high magnification. They remain attached to the tree, by means of deeply penetrating mouthparts, throughout their adult lives.

The crawler is the only mobile stage. Eggs hatch into active crawlers that move around on the host tree until they find a suitable place to settle, or until they are blown by the wind to another host. Because the crawlers are so small, they are capable of being carried great distances by the wind. A very small percentage of windblown crawlers reach suitable hosts; however, this passive movement is thought to be the principal means of dispersion from one host tree to another. Once the crawler has found a suitable site, it inserts its long sucking mouthparts, which are several times longer than its body, into the outer bark and sucks juices from the tree. White wax ribbons are secreted from glands along the sides and back of the crawler after it has been feeding for a while. Soon the entire body becomes covered with a white woolly material (Figure 26). There are two to four generations per year in the Pacific Northwest and two generations in the Northeast. Eggs are attached to the bark behind the female's body. All stages may be found throughout the growing season, whether in the East or West.

An early sign of damage on some trees is a swelling, or "gouting," (C) of the ends of twigs. The swelling is apparently caused by a growth-promoting substance that the adelgid injects into the tree. When the infestation is heavy, the whole trunk of the tree becomes covered by the white "wool" of the adelgids (D). New growth virtually ceases on heavily infested firs. More susceptible firs such as balsam, *Abies balsamea,* or subalpine fir, *A. lasiocarpa,* may be killed before the terminal swelling occurs.

This insect is a serious pest on both forest and landscape trees. Those responsible for the care of ornamental trees should be aware that large and older trees are usually attacked first. Many predators, both native and imported, such as the lady beetle *Aphidecta obliterata,* an "aphid" fly (*Leucopis obscura*) native to Europe, and several syrphid fly larvae are recognized as important in the natural control of these adelgids.

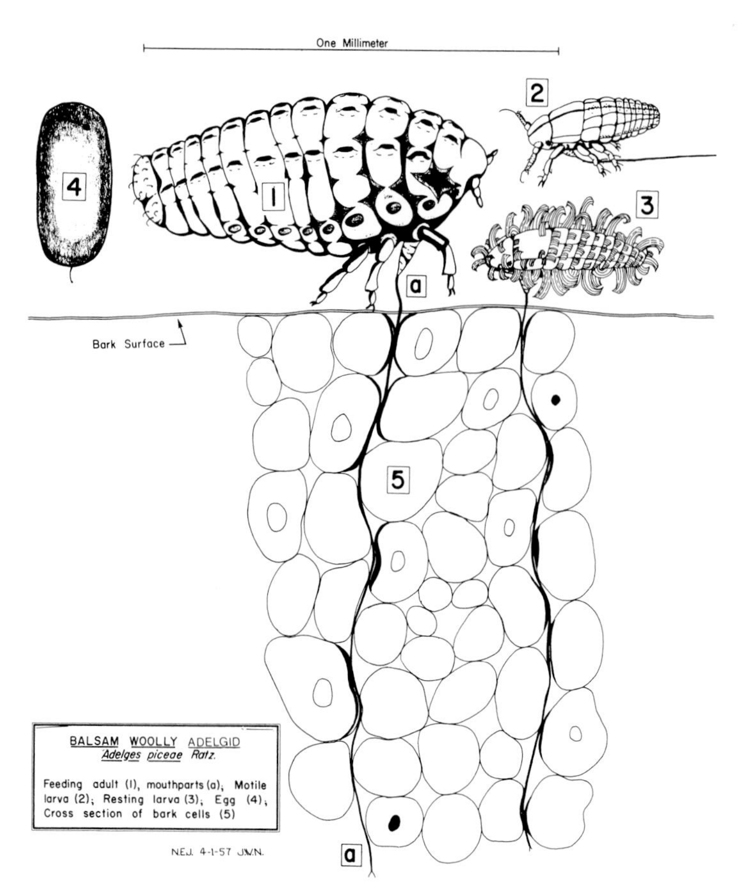

Figure 26.

References: 34, 89, 230, 275, 541

A. An uninfested silver fir on the right and one whose height growth has virtually ceased owing to infestation by the balsam woolly adelgid.
B. Adelgids on twig of Pacific silver fir.
C. "Gout" produced by feeding of woolly adelgids.
D. Woolly secretions on the trunk of infested tree.
E. An adult (*arrow*) and eggs. Highly magnified. Eggs later darken to purplish red.
F. Close-up of woolly secretions on fir twig.
G. Newly settled crawlers (*arrows*) on twig of silver fir.

A
B
C
D
E
F
G

Adelgids of Pine and Spruce (Plate 31)

The insects illustrated in this and the following plate have improperly been called aphids in older literature. They are closely related to the woolly aphids and phylloxerans but belong to the family Adelgidae (=Chermidae) and occur on conifers. Feeding sites of the various species include foliage, twigs, limbs, trunks, and gall tissue. Several species alternate hosts, between spruce and another conifer. In such cases, most produce galls on the primary host (that host where the overwintering stage occurs). (See Plates 49–50.) Some species not only alternate hosts, they also alternate generations, that is, the first generation reproduces sexually and their offspring are all asexual females who, in turn, give birth parthenogenically to males and females that start the cycle over again. Such biological complexity makes it difficult to understand their life cycle and develop control strategies.

Fourteen adelgid species are associated with conifers in the United States and Canada (Table 6). They are difficult to identify in the field. By knowing the host plant, noting the time of year the adelgid is present, and recognizing characteristic features of the gall, if present, one can make a tentative identification. A common feature of all adelgids is the production of white flocculent epidermal secretions in the form of powder, filaments, or ribbons.

Panels A, B, D, and E of Plate 31 show the pine bark adelgid, *Pineus strobi* (Hartig), at several feeding sites on two species of pine. Confusion exists about the biology and life history of most of the conifer-feeding adelgids. This species apparently utilizes spruce as an alternate host but does not complete the cycle here. On pine it repeatedly reproduces itself, and all stages feed from the bark surface, where their stylets are inserted into phloem tissue. The immature forms that overwinter begin feeding during the first days of warm weather and secrete relatively large amounts of wax, which solidify into woolly tufts over their bodies. At maturity (late April in Minnesota) egg laying begins. The eggs develop into the crawler stage, which moves to other parts of the same tree. Windblown crawlers account for distribution from tree to tree or stand to stand. When the crawler settles, it remains in place until its next molt. This insect is most abundant on the trunk of older trees (10 years+). They become so abundant that the trunk gives the appearance of being covered with snow (A, B). On seedlings and Christmas trees they are found mainly on new shoots and at dense needle clusters at the base of each year's growth (D). When large white pine stock is transplanted, an appropriate insecticide should be applied to the bark before the trunk is wrapped, otherwise a heavy infestation of pine bark adelgids will likely develop.

The pine bark adelgid occurs throughout the native range of eastern white pine and occurs also on Scots and Austrian pine. When the "wool" is pulled away from the mature adelgid, a black teardrop-shaped insect with short legs is revealed.

References: 11, 35, 137–139, 156, 193, 469, 478, 507, 643, 901, 902

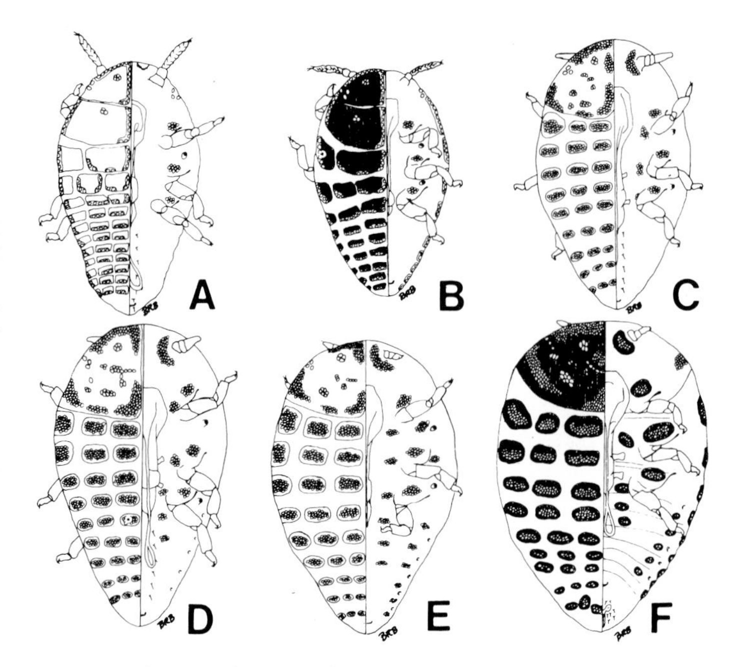

Figure 28. Life stages of sistens and progrediens of the hemlock woolly adelgid. *A,* first-instar crawler; *B–E,* settled first-, second-, third-, and fourth-instar nymphs, respectively; *F,* adult. (From McClure, ref. 902, courtesy *Annals of the Entomological Society of America.*)

Table 6. **Important species of conifer-feeding adelgids**

Name	Hosts	Distribution	Trees with galls
Adelges laricis	Larch; red, black spruce	MD to MA, west to Alb.	
A. oregonensis	Larch	Northwest	
Balsam woolly adelgid, *A. piceae*	Balsam, Fraser fir	NC to Que., B.C., WA, OR	Balsam fir
Cooley spruce gall adelgid, *A. cooleyi*	Colorado blue, Sitka, Engelmann, white spruce; Douglas-fir	Intercontinental	Spruce
Eastern spruce gall adelgid, *A. abietis*	Norway, white spruce	Eastern U.S., B.C.	Spruce
Hemlock woolly adelgid, *A. tsugae*	Hemlock	NC north to CT	
Pine bark adelgid, *Pineus strobi*	White, Scots, Austrian pine	Eastern U.S., CA	
Pine leaf adelgid, *P. pinifoliae*	Red, black spruce; white pine	Transcontinental	Spruce
P. borneri	Red pine	New England, CA	
P. coloradensis	Red, pitch pine	Western U.S., VA, MD, New England	
P. floccus	White pine; red, black spruce	Eastern U.S.	Red spruce
P. similis	Norway, white, red, Colorado blue spruce	Transcontinental	Spruce
P. sylvestris	Scots pine	Eastern U.S.	
Spruce gall adelgid, *A. lariciatus*	Larch; Norway spruce	Eastern U.S.	Spruce

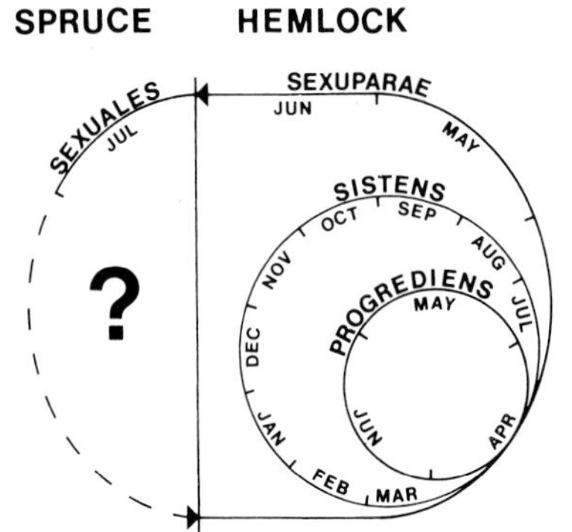

Figure 27. Polymorphic life cycle of the hemlock woolly adelgid, which alternates between hemlock and spruce. (From McClure, ref. 902, courtesy *Annals of the Entomological Society of America.*)

A, B. A heavy infestation of the pine bark adelgid, *Pineus strobi,* on trunks of eastern white pine.
C. Woolly remains of the pine leaf adelgid, *P. pinifoliae* (Fitch), on *Pinus* sp.
D. Woolly masses at the base of eastern white pine needles contain adult female pine bark adelgids.
E. Woolly masses at the base of Austrian pine needles contain eggs and adult female pine bark adelgids.

A
B
C
D
E

Adelgids of Larch, Spruce, and Hemlock (Plate 32)

Two species of adelgids feed alternately on larch and spruce. On larch they are foliage or bud and cone feeders; on spruce they cause unique galls to develop from meristematic bud (needle) tissue. These galls are similar to those of the eastern spruce gall adelgid (Plate 50).

Adelges lariciatus (Patch) goes through a complicated alternation of generations cycle in six generations; three on spruce and three on larch. The cycle with the names of the life stages in each generation is illustrated in Figure 29. The adelgid form, named *fundatrix* in the figure, overwinters on spruce as a nymph. For convenience, we call the fundatrix the first generation. At maturity they are parthenogenic wingless females and lay their eggs at the base of needles that are just beginning to develop. Upon hatching, the nymphs establish a feeding site, and gall formation begins. Gall tissue quickly grows over their bodies, and the spruce needle that should have formed begins to look like part of the epidermis of a miniature pineapple. There are 8–46 gall units (cells) per gall. Each cell contains a cavity that provides space for the second generation, called the gallicola migrans. Continued development of this adelgid and the gall parallels that of the eastern spruce gall adelgid (Plate 50). The gall caused by *A. lariciatus* usually covers only one side of the shoot. Nymphs of the first generation (sistens) on larch settle among bud scales, most commonly on old buds, and overwinter. Members of the second generation on larch may be found on cones, at the tip of new buds, at the base of buds, or outside the buds among the scales. They secrete white cottony material that partially covers their bodies.

A. lariciatus is a native species and may be found where spruce and larch grow near one another.

A. laricis Vallot (=*Adelges strobilobius*) has a life cycle similar to *A. lariciatus*. The late spring generation on larch can be spectacular. These adelgids may become so abundant that the tree may appear to be dusted with snow (A, B). Severe injury to larch has not been observed, but needles become distorted, there is reduced growth, and much honeydew (a sugary waste product) is produced. There is evidence that these adelgids can remain on larch for an indefinite period without alternating to their primary host, spruce. On larch they are able to overwinter as nymphs around buds. In early April when buds swell, nymphs migrate to the base of opening buds and commence feeding. After they mature, females secrete a powdery wax over their bodies. By mid to late April eggs are being laid, and after 2 weeks each parthenogenic female may have laid several hundred eggs. When these eggs hatch, the nymphs seek new leaves, after which they produce the signs illustrated in panel A. A nymphal stage of the stem mother overwinters on spruce. Her offspring produce galls on the newly developing shoots. These galls look much like those produced by the eastern spruce gall adelgid, but they are smaller. The gall has a distinct pink color and may engulf the entire tip of the shoot.

A. laricis may be found from New England and Ontario south to Maryland and west to Alberta.

The pine leaf adelgid, *Pineus pinifoliae* (Fitch), is transcontinental and has two secondary hosts, eastern or western white pine. Its primary host is spruce. It is regarded as one of the more serious pests of eastern white pine in New England and eastern Canada. Damage symptoms include stunted needles, drooped and dead needles and shoots, sparse foliage, crooked branches within the crown, and alternating short and normal internodes of twigs and branches. The shoots droop in mid-summer and turn red. On spruce this adelgid forms loose, terminal cone-shaped galls resembling those of Cooley spruce gall adelgids except that the latter are more elongate and that in *P. pinifoliae* galls the poorly formed chambers are intercommunicating and contain only one or two young in each chamber. When fully developed, the galls look somewhat like small hairy cones, and in that respect also they differ from Cooley spruce galls. The life history is complicated; three generations occur on spruce and two on pine. When the galls on spruce begin to dry in early summer, the chambers crack open, allowing the winged individuals of the third generation on spruce, called the gallicola migrans, to migrate to the needles of white pine. Here the parthenogenic female lays eggs and dies. Her body covers the eggs. When these eggs hatch, nymphs crawl to the new shoots, insert their mouth parts, and remain dormant until the next spring. At this stage they are dark purplish brown with a fringe of flocculant material around their bodies. From a distance, if the insects are abundant, they look like a grayish white mold on the bark and may be visible throughout the greater part of the year. When periods of warm weather arrive in the spring, they begin feeding. It is the feeding injury, or more likely the salivary fluids, that causes the chief damage by killing or weakening the twig and causing it to droop. This adelgid causes galls on spruce one year and twig droop on pine the next year in a continuing alternating cycle. These symptoms have been observed as far south as western North Carolina.

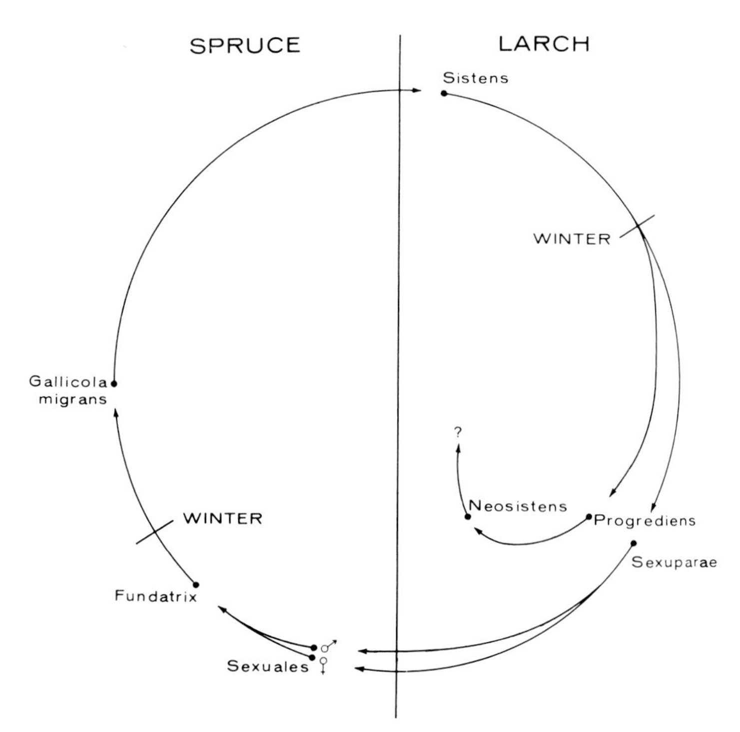

Figure 29. The life cycle of *Adelges lariciatus*, the spruce gall adelgid. There are six or more adult forms; three on spruce and three or more on larch. (From Cummings, ref. 139, courtesy *Canadian Entomologist*.)

The hemlock woolly adelgid, *A. tsugae* Annand, has been in the United States since 1927. It destroys the ornamental value of hemlock because of the presence of dirty white globular masses of cottony puffs attached to twigs or the bases of needles (E). The cottony masses, 3 mm or more in diameter, cover the adult female or her dead body and her eggs. *A. tsugae* damages the tree by sucking sap from bark phloem tissues; the tree loses vigor and prematurely drops needles, to the point of defoliation, which may lead to death.

There are four forms—progrediens, sistens, sexuparae, and sexuales—and each form goes through six life stages (egg, four nymphal instars, and adult) (see Figures 27 and 28, accompanying Plate 31). A cool weather species, the hemlock woolly adelgid completes most of its development from October through June. During the warm summer months it estivates as a first-instar nymph on young branches at the bases of hemlock needles. This, the sistens form, present from July through April, is the overwintering generation. Sistens females produce two kinds of eggs during February and March. Approximately half of these eggs develop into a winged, migratory, parthenogenic generation called the sexuparae. Upon reaching the adult stage these

---

A, B. A heavy infestation of *Adelges laricis* on larch, *Larix decidua*. Photos taken in early June in central New York. This is generation 2.

C. An adelgid nymph, *Pineus pinifoliae*, the pine leaf adelgid, on a larch needle. Note discoloration at site of stylet penetration and bending of needle.

D. Adult pine bark adelgids entwined in filamentous wax on a pine needle. (Also see Plate 31.)

E. Hemlock woolly adelgids, *A. tsugae*, and their eggs may be found under the matted white "cotton" on the twigs of hemlock. (Courtesy U.S. Forest Service.)

A
B
C
D
E

adelgids migrate to spruce, where sexuales develop. The other eggs of sistens females become progrediens. The progrediens develop through six stages without delays, ending as adults. These adults remain on the tree during June and July and produce eggs. Sistens crawlers hatch from these eggs, settle on young hemlock branches, feed for a short time, then enter into an estivation period that lasts until early October. On hemlock the settled crawlers of all forms produce white filamentous wax that covers their bodies and remains in place throughout their lifetimes.

The hemlock woolly adelgid occurs in California and the Pacific Northwest, although western hemlock is somewhat resistant. In the East it occurs from the Smoky Mountains north to the lower Hudson River Valley, Long Island, and Connecticut. Hemlock is the primary host.

References: 11, 106, 139, 299, 379

Balsam Twig Aphid (Plate 33)

The balsam twig aphid, *Mindarus abietinus* Koch, has a complex life history (Figure 30). Three distinct forms of aphids occur during the single generation this insect has each year. In the spring the overwintering eggs (E, F) hatch. After three molts the mature wingless stem mother (fundatrix) is ready to produce living young. The offspring gather in a colony around the bluish gray parent (A), where they feed on old needles as well as on buds.

The progeny of the stem mother constitute the intermediate form and are called sexuparae. Each of these individuals undergoes three molts, producing after each molt, a white waxy covering over its body (C). Aphids of this form feed in colonies, primarily on new flushes of growth. Of the three aphid forms, the sexuparae injure the tree the most. When mature, these aphids have wings and produce the final form, called sexuales. Because sexuparae have wings, they are capable of migrating by flight from the tree on which they were produced. In eastern Washington State nearly 80% of the winged balsam twig aphids in pan traps have been caught from mid-June to mid-July.

The sexuales are either males or egg-laying females. They, too, feed on needles. After mating, each mature female produces several large eggs that she deposits in bark crevices (E, F). The brown eggs, conspicuously covered with tiny rods of white wax, serve as a useful index of the amount of injury likely to occur the next year.

Balsam twig aphids produce copious amounts of honeydew. Droplets often are found scattered over the surface of needles (G) or clinging to the powdery, waxy material always associated with these aphids. If the aphid population is heavy, shoots become saturated and needles may adhere to one another; they are very sticky to the touch. Eventually, rain washes off the honeydew.

The aphids' feeding twists and curls the needles, killing some and leaving roughened twigs. Such needles remain deformed as long as they persist on the tree, although injury occurs primarily during May and June. Heavy aphid infestations on specimen ornamental trees mar their appearance. In plantations, curled, twisted needles markedly reduce the value of Christmas trees, and 2 or 3 years of recuperation may be required before damaged trees can be sold—providing the aphids are controlled during that period.

The balsam twig aphid has been recorded from alpine, balsam, Fraser, grand, and white fir; white spruce, including cultivar Conica, and Colorado spruce; and juniper. It is distributed from the New England and Appalachian states to the Pacific Coast of the United States and Canada.

References: 8, 193, 569, 673, 675

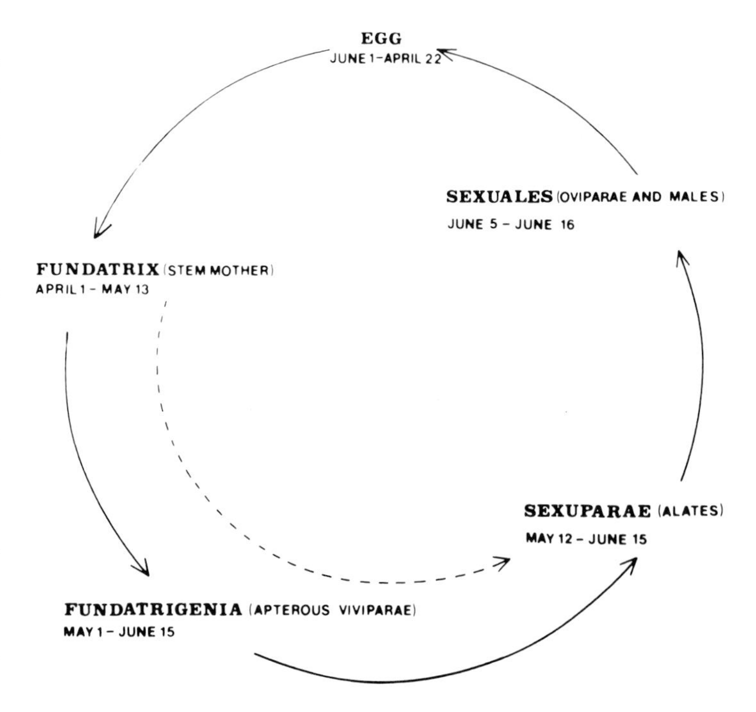

Figure 30. The life cycle of the balsam twig aphid, *Mindarus abietinus*, on Fraser fir in North Carolina. *Fundatrix*—a wingless aphid form found in spring on the primary host; hatched from egg produced by sexual union. *Fundatrigenia*—aphid form produced parthenogenically by the fundatrix. These are wingless and give birth to living young. *Sexuparae*—this form has wings and is produced by the fundatrigenia. The offspring are males and egg-laying females. *Sexuales*—forms that mate; the female produces the overwintering egg. (From Nettleton and Hain, ref. 569, courtesy *Canadian Entomologist*.)

A. Three stem mother balsam twig aphids, bluish gray, with their offspring feeding on balsam fir buds.
B. New twigs that appear wet with honeydew.
C. Immature second-stage aphids. A large quantity of a white waxy substance adheres to their bodies and to needles.
D. A cluster of very young second-stage aphids at the tip of an opening bud.
E, F. Overwintering eggs, laid in late June, covered with tiny rods of white wax.
G. Several immature third-stage sexual forms that will later produce eggs. Note droplets of honeydew.
H. Typical needle injury versus healthy needles of balsam fir.
I. A mature second-stage winged form.
J. A young stem mother just hatched from the overwintering egg.

A
B
C
D
E
F
G
H
I
J

Cinara Aphids and Other Aphids of Conifers (Plates 34–35)

Many aphids feed on conifers, causing unsightly discoloration and deformation of needles, premature needle fall, reduced shoot growth, sticky excretions with subsequent blackening of foliage from sooty mold, and even the death of the tree. A few of these aphids are illustrated here and in the following plate.

The spruce aphid, *Elatobium abietinum* (Walker), is a destructive pest of ornamental spruce along the entire western coast of North America. It has also been reported from North Carolina. This aphid is green and only 1–1.5 mm long when fully grown. By colonizing and sucking sap from older needles, it causes them to turn yellow, then drop prematurely. When infestations are severe, only those needles on the new flush of growth remain on the tree. Affected blue spruce trees sometimes take 5 or more years to recover their normal complement of luxurious foliage—and then only if the spruce aphid is controlled during that recovery period.

A light-intolerant insect, *E. abietinum* does most of its damage to the lower, shaded portion of the tree. Under outbreak conditions, however, the entire tree may be affected. The pest is most active in late winter and very early spring. This, together with the aphid's small size and its color, which allows it to blend with the foliage, makes it possible for serious damage to occur before the tree owner realizes that a problem exists. Natural enemies of the spruce aphid include spiders, lady beetle larvae and adults, and syrphid fly larvae.

Cinara fornacula Hottes, unfortunately, has been given the common name green spruce aphid. It should never be confused with *E. abietinum,* however, because it is quite large by comparison and not at all green (E, F). *C. fornacula,* like other members of this genus (Plate 35), feeds principally on the twigs, not on needles, as does the spruce aphid. Hosts of the green spruce aphid include Alberta, Colorado, Engelmann, and black spruce. Although considered a western insect, it is also found in New York, Pennsylvania, Delaware, Iowa, Minnesota, and from Ontario west to British Columbia.

C. curvipes (Patch), sometimes called the bowlegged fir aphid, feeds on the bark of true firs and *Cedrus atlanticus* and *Cedrus deodara,* both of which are introduced trees. It occurs in California, Oregon, Washington, British Columbia, Colorado, and Utah, and is also known in the Northeast, eastern Canada, and North Carolina. Large colonies of aphids often develop on the twigs or branches of host trees in the winter along the Pacific Coast, and in the spring inland. Vast quantities of honeydew rain down from heavily infested trees. High populations normally occur for only a short time on any given tree. If an infested branch is jarred, the aphids will drop.

References: 87, 275, 381, 802

A. Blue spruce severely defoliated by the spruce aphid. (Courtesy Sharon J. Collman, Cooperative Extension, Washington State Univ.)
B. Twigs of blue spruce from which all older needles have been lost to feeding by the spruce aphid. (Courtesy Gray Collection, Oregon State Univ.)
C. Cluster of wingless spruce aphids on a needle of spruce. (Courtesy Gray Collection, Oregon State Univ.)
D. A winged adult of the spruce aphid. (Courtesy Gray Collection, Oregon State Univ.)
E. A winged adult of the green spruce aphid. (Courtesy Gray Collection, Oregon State Univ.)
F. A colony of green spruce aphids on a twig of spruce. (Courtesy Gray Collection, Oregon State Univ.)
G. A colony of bowlegged fir aphids.

A
B
C
D
E
F
G

Several species of *Cinara* attack conifers throughout the United States. These aphids, many of which are quite large, long legged, and almost spiderlike, feed on the bark of smaller twigs and the main stem of small trees. A complex of three species causes serious damage to young Douglas-fir. One of these, *C. pseudotsugae* (Williams), is illustrated in panel A. *C. sabinae* (Gillette & Palmer), an important pest of juniper in the Rocky Mountain region, causes the foliage of infested plants to turn brown prematurely and fall. A closely related species, *C. tujafilina* (Del Guercio), a brown aphid with a touch of white bloom, causes foliage to turn brown and often kills branches of redcedar, arborvitae (*Thuja* spp.), and cypress in both the East and the West. On the East Coast this aphid occurs from Maine to Florida. A complex of five or more species, the most important of which is probably *C. atlantica* (Wilson), attacks southern pines. The white pine aphid, *C. strobi* (Fitch), is large; winged forms can be up to 4 mm long. Long legs enable the insects to move rapidly, especially when disturbed. Their distinct color pattern (shiny dark brown to black with a white longitudinal stripe down the middle of the back and rows of white spots along their sides) and their living in dense clusters on twigs of eastern white pine are characteristics that help identify them.

During the fall winged females of the white pine aphid lay five to six eggs in a row on needles. As the eggs age, they become shiny black (E) and conspicuous even to the casual observer. Hatch occurs in the spring, and several wingless generations develop during the growing season. The aphids damage trees by disrupting phloem and removing sap they use for food. Small trees are unlikely to tolerate large populations. Their twig growth may be seriously reduced or killed. The white pine aphid occurs in eastern Canada, New England, the Great Lakes States, and southward into the Great Smoky Mountains of North Carolina.

Two other *Cinara* species capable of serious damage to conifers are *C. cronarti* Tissot & Pepper and *C. piticornis* (Hartig). *C. cronarti*, sometimes called the black pine aphid, is native to the eastern United States from Florida north along the coast and west to Michigan. In the southern states its normal feeding sites are rust cankers and galls caused by the fungus *Cronartium quercuum* f. sp. *fusiforme* (see Sinclair et al., *Diseases of Trees and Shrubs*), leading to the suspicion that these aphids may be a vector for the fungus. In spring when the rust sporulates, aphids may be found among the loose spores beneath the crusty surfaces of the cankers and galls. The aphid colonies are seldom visible on the surface of the host, so one must remove flakes of bark or other covering to see them. Both winged and wingless individuals are dark brown with dark appendages; body length ranges from 2.5 to 4 mm. Elsewhere in their geographic range black pine aphids feed most commonly on the bark of shoots, twigs, and small branches, where they are fully exposed. They are also reported to be root feeders. When the aphids are abundant, the terminals die, owing to the loss of sap and perhaps mechanical injury. Known hosts are sand, slash, loblolly, jack, pond, and Mexican pines.

C. piticornis feeds on spruce, fir, and occasionally pines. Although records in the literature may imply a spotty occurrence, *C. piticornis*, an immigrant from Europe, is likely to be found on conifers from coast to coast in both the United States and Canada. In the eastern United States and Canada it can be an economically important pest in Christmas tree plantations. Like other *Cinara* species, these aphids live in colonies and feed on the bark of twigs and shoots. Individuals are grayish brown, plump, and spiderlike and often have a dusting of whitish powder over their bodies. Where trees support many colonies, terminal needles turn brown; on Douglas-fir the presence of *C. piticornis* is associated with slow or reduced growth. These aphids are active only in the spring. In Pennsylvania the population peaks in mid-May and is gone by mid-June. There the most common hosts are Colorado, Norway, and white spruce.

The commonly encountered North American species of *Cinara* occur in large dense colonies on the twigs of conifers; few feed on foliage. They sometimes are accompanied by ants (A), which feed on honeydew the aphids produce. This association affords the aphids protection from predators. Often, affected trees are blackened by a sooty mold that grows on honeydew. In addition, the dripping of honeydew onto cars, sidewalks, picnic tables, and other outdoor furniture is a problem when affected trees are located near homes, in parks, or along city streets.

Eulachnus rileyi (Williams) feeds on the needles of Scots, Virginia, Austrian, mugo, red, eastern white, and certain other pines. The insect is powdery gray, which accounts for its occasionally used common name, powdery pine needle aphid. Like the spruce aphid (Plate 34), it causes needles to fall prematurely. However, unlike the spruce aphid, *E. rileyi* may be found throughout the growing season. Christmas tree plantations often suffer severe damage. This aphid is widely distributed in the eastern United States and is recorded from Arizona, California, and Washington.

E. agilis (Kaltenbach), often called the spotted pine aphid (G) because of numerous black spots on its back, feeds on second- and third-year needles of Scots and red pine. It is particularly damaging under plantation conditions. The initial plant symptom is chlorosis, followed by browning and premature loss of needles. The aphid is very active and difficult to capture. Its population cycle has two peaks, the first occurring in May and the second in late summer. *E. agilis* is common in the Appalachian region as far north as New Brunswick and Maine, and in Ohio.

Schizolachnus piniradiatae (Davidson), sometimes called the woolly pine needle aphid, has been recorded from Digger, jack, lodgepole, Monterey, ponderosa, and red pines. It occurs from coast to coast across southern Canada and the northern United States. There are reports of its occurrence in North and South Carolina and Florida. *S. piniradiatae* is a powdery gray to dark green long-legged aphid that lives in colonies on host needles. Its characteristic behavior of lining up in single file also aids in field identification. In Ontario at least seven generations occur each year. The final generation of females lays overwintering eggs. These are black and beadlike with a faint whitish bloom; they are found in cluster rows on needles. Natural enemies of this species, as for many other aphid pests, include lacewings, syrphid fly larvae, lady beetle larvae and adults, and parasitic wasps. The fungi *Erynia neoaphidis* and *Zoophthora aphidis* have been reported as important natural control agents.

The balsam twig aphid, *Mindarus abietinus* Koch, because of its importance in the Northeast, the Rocky Mountain region, and the Pacific Coast states, is treated separately in Plate 33.

References: 83, 87, 193, 275, 309, 411, 910, 912

A. *Cinara pseudotsugae* aphids on Douglas-fir bark being tended by ants.
B. A winged adult *Cinara* sp. on Douglas-fir needle. [Aphididae]
C. Parasitized *Cinara* sp. The larvae of tiny wasps are inside the aphid skeletons (which have been opened), as shown at arrow points. (See also Figure 100, accompanying Plate 147.)
D. *Mindarus* sp. on western fir. (See also Plate 33.)
E. Eggs of the white pine aphid on needles of *Pinus resinosa*.
F. A cluster of *C. curvipes* on the bark of *Abies concolor*.
G. *Eulachnus agilis*, sometimes called the spotted pine aphid. (Courtesy Dept. of Entomology, Pennsylvania State Univ.)

A
B
C
D
E
F
G

Conifer Spittlebugs (Plate 36)

A number of species of spittlebugs feed on both broad-leaved and coniferous ornamental plants throughout the United States. Certainly the most striking characteristic of these insects is the frothy mass of "spittle" that clings to foliage and twigs. It is produced by the immature spittlebug to keep it moist and protect it from many natural enemies.

The most significant spittlebugs found on ornamentals overwinter in the egg stage on the host plant or on weeds or other vegetation close by. In the spring the eggs hatch, and the tiny nymphs insert their sucking mouthparts into living vegetation to feed on plant sap. While feeding, the insect produces droplets of clear fluid that surrounds its body. Bubbles of air are incorporated into the fluid by the nymph, giving rise to the characteristic spittle mass. The size of the mass depends upon the species, the size of the nymph, or the number of nymphs congregating at the same feeding site. Sometimes the frothy material encompasses a cluster of pine cones. Rains, if heavy, wash away the spittle, but it quickly reappears when dry weather returns.

Some spittlebug species complete their development on the host on which the eggs were laid. Others initially feed on low-growing herbaceous vegetation, then crawl to taller plants, and finally crawl onto woody ornamentals.

Most adult spittlebugs (F, G) are blunt, wedge-shaped insects. They are gray to brown, and measure 6–12 mm long, depending on the species. Adults are active; they can walk, hop, or fly. Many spittlebug species undergo a single generation each year. Two generations of some spittlebugs occur annually, which explains the appearance of new nymphs as late as October in some areas.

A few spittlebug species cause internodes to shorten or deform foliage on ornamental plants. Most of them, however, do not cause noticeable injury to plants, although their spittle masses are considered unsightly.

The most important conifer-feeding spittlebugs are listed in Table 7. The most damaging species is *Aphrophora saratogensis* (Fitch), also known as the Saratoga spittlebug. The most widely distributed species are *A. cribata* (=*parallela*) (Walker), the pine spittlebug, and *Philaenus spumarius* (Linnaeus), commonly known as the meadow spittlebug.

Because of its economic importance, we have chosen to give a detailed account of the Saratoga spittlebug. There is one generation per year. The egg (2 mm long), deposited under the scales of buds, in the needle sheaths, or under loose pine bark, is the overwintering stage. Upon hatching, the nymphs of this species seek and feed on low-growing herbs or shrubs, for example, sweetfern, blackberry, and hawkweed. Adults, on the other hand, feed upon several species of pine, but red and jack pine are the favored hosts. This is an obligate two-host species. Adults are about 10 mm long and tan to brown. They may be found from late June until a killing frost occurs in the fall. Adults feed by inserting their mouthparts into the cortical part of young pine twigs and then sucking out plant fluids. This action results in necrotic, resin-filled pockets in the phloem (E) that appear later as pitchy defects in the xylem. The initial symptom is a reduction in terminal and lateral twig elongation, followed by slight yellowing of the needles, twig mortality, and finally branch and tree mortality. It requires 3 or more years of heavy feeding to kill a branch or tree. The peak of adult feeding occurs in a 2-week period with a midpoint about July 30.

Table 7. **Spittlebugs that feed on conifers and other evergreens**

Species	Hosts	Distribution
Aphrophora canadensis	Pine	Pacific Coast states and B.C.
A. cribata	Pine	Eastern Canada to Great Lakes states and MO and AR
A. permutata	Pine, Douglas-fir, hemlock, spruce	Western states and provinces
A. saratogensis	Red and jack pine	Transcontinental
Clastoptera arborina	Juniper	Rocky Mt. states
C. arizonana	Acacia	Southwestern U.S. (Plate 202)
C. juniperina	Juniper	Rocky Mt. states and West
C. undulata	*Casuarina equisetifolia*	Subtropical U.S.
Philaenus spumarius	Pine, deciduous saplings	Transcontinental

The Saratoga spittlebug can be expected to occur periodically in epidemic numbers throughout the northern tier of states and adjacent parts of Canada. It is unlikely to occur in well-managed ornamental plantings. The destruction of nymphal host plants will control this pest.

The pine spittlebug, *A. cribata,* occurs throughout the eastern half of Canada and the United States, from the states bordering the Mississippi River and western Ontario east to the Atlantic Ocean. It is a serious pest of plantation-grown Scots pine, and it readily attacks jack, pitch, Virginia, slash, loblolly, white, and Japanese black and other oriental pines as well as the native spruce, balsam fir, Douglas-fir, and hemlock.

Eggs are deposited *in* the bark of twigs and occasionally in dead woody tissue on the tree. In northern parts of this spittlebug's range, the egg is the overwintering stage. Upon hatching in the spring, the nymphs seek twigs upon which to feed. Inserting their tubelike mouthparts, they pierce the phloem and withdraw sap. The excreted, partially digested, and concentrated sap is whipped into a spittlelike foam (B). Several nymphs may be found in a single spittle mass.

Adults, present during mid and late summer, also feed from the twigs but do not make spittle. They produce honeydew, which drops like fine mist to foliage and twigs, where sooty mold fungi grow. Heavy infestations of this spittlebug may cause twig, branch, and occasionally tree mortality. This insect utilizes conifer sap as its food in all of its feeding stages. The fungus *Entomophthora aphrophora* acts as a natural control agent.

References: 200, 223, 523, 657, 745

A. Spittle masses produced by the meadow spittlebug, *Philaenus spumarius,* on ponderosa pine.
B. Spittle removed to disclose nymphs of the pine spittlebug, *Aphrophora cribata,* on Scots pine. Size: 2–8 mm.
C, D. A western spittlebug, *Clastoptera juniperina,* on *Juniperus chinensis.* Note nymph at arrow.
E. Injury in the cortical tissue of red pine caused by the Saratoga spittlebug, *A. saratogensis.* The brown circular spots are dead tissue, flooded and filled with pitch. This injury blocks the vascular system.
F, G. Adults of the pine spittlebug. [Cercopidae]

A
B
C
D
E
F
G

Mealybugs of Yew and Pine, and the Woolly Pine Scale (Plate 37)

There are at least three species of mealybugs that feed and reproduce on *Taxus* (yew).

The taxus mealybug, *Dysmicoccus wistariae* (Green) (=*Pseudococcus cuspidatae* Rau), is known to occur in Massachusetts, New York, Connecticut, New Jersey, Pennsylvania, and Ohio and is believed to occur in other northeastern and central states. Severely infested plants may be sparsely foliated and the remaining needles and twigs caked with honeydew and black sooty mold. The taxus mealybugs feed in the inner bark tissue of the trunk and branches (B). When present in moderate numbers, they cluster in the crotches of twigs and branches. Adult females are present from June to August and begin giving birth to living young in early summer. In New Jersey the taxus mealybug spends the winter as a first-stage nymph in bark crevices or beneath the old, dirty white ovisac. Probably two or more generations occur each year in New Jersey and farther south. Schread (679) reported a single generation in Connecticut. This species has been collected from other woody plants such as *Prunus* species, rhododendron, dogwood, and maple, but large populations have not been reported except on dogwood. (Also see Plate 155.)

The grape mealybug, *Pseudococcus maritimus* (Ehrhorn), feeds on woody as well as other perennial plants, including tubers and corms. It has been a troublesome and persistent pest of *Taxus* in Ohio and other north central states. Biological studies of this insect's association with grape have been done, mostly in central and western states, where there are two generations per year. There are also two generations produced each year on *Taxus* in Ohio. The grape mealybug overwinters as an egg, or first instar crawler. The crawler ranges from yellow to brown. After the period of winter dormancy it is quite active. Adults are about 6 mm long, are dark purple, and are covered with a uniformly distributed white waxy powder. Filaments of wax extend from the margin of the body and from the posterior end. Although they are smaller than the taxus mealybug, the adults look much like those in panel B. The grape mealybug is described as being polyphagous; some of its host plants include English ivy, ginkgo, laburnum, pear, Japanese quince, Persian walnut, honeylocust, and species of *Grevillea, Citrus, Arbutus, Catalpa, Ceanothus, Celtis,* and *Juglans*. *P. affinis* (Essig) (=*obscurus*), commonly known as the obscure mealybug, is believed to be the most common *Pseudococcus* species on *Taxus*. However, because the two *Pseudococcus* species are difficult to distinguish from one another, host and distribution records are confused. See Plates 154–156 for other mealybugs of deciduous trees.

The woolly pine scale, *Pseudophilippia quaintancii* Cockerell, is one of the soft scales (Plate 163) but, in appearance, may be confused with mealybugs. It is covered with a white fleecy-looking secretion (C). The body of the mature female is greenish brown, hemispherical, and 2–2.5 mm long (D-1). The biology and habits of this scale are unknown. First instar nymphs, or crawlers, (E-2) have been found in mid-June near Richmond, Virginia. Much honeydew is produced and colonized by sooty mold fungi. The mold may blacken the pine needles as well as clusters of scales.

Hosts are limited to pine species, including mugo, loblolly, longleaf, pitch, Scots, and table mountain pines. The woolly pine scale is known to occur from New York to Florida and west to Louisiana, and it may occur in mixed populations with the mealybug *Oracella acuta* (Lobdell). *O. acuta* is presumed to be a North American species first described from pine in Mississippi and known to occur in the southeastern states from the Atlantic Coast west to Texas. It has also been reported on shortleaf pine from Pennsylvania. Other hosts recorded for *O. acuta* include loblolly and longleaf pines.

O. acuta is 2–3 mm long and is found enclosed in a whitish resinous cell, open at one end, from which the abdomen protrudes. Cells are attached to twigs near the bases of needles. Both young and old trees sustain direct injury resulting from insect feeding as well as indirect injury resulting from the interference of sooty mold. Photosynthesis, hence growth, is reduced when anything interrupts or reduces the quality of light reaching the needles.

The mealybug is distinguished from the cottony pine scale by its crystalline, resinous covering.

References: 319, 325, 518, 568, 679, 824

A. Typical appearance of mealybug-infested yew.

B. Female taxus mealybugs, *Dysmicoccus wistariae,* feeding on a twig. Fully grown mealybugs are about 8 mm long. [Pseudococcidae]

C. Fleecy appearance of the woolly pine scale, *Pseudophilippia quaintancii,* attached to the twig. [Coccidae]

D, E. Enlarged view of fleece-covered adult females of the woolly pine scale: (1) adult female with fleece removed, (2) crawlers or first-instar nymphs, (3) eggs, (4) a coccid insect of uncertain identity—possibly the mealybug *Oracella acuta.*

A
B
C
D
E
1
2
3
4

Golden Mealybug (Plate 38)

Plants of the genus *Araucaria* come from South America, New Zealand, the islands of the South Pacific, and Australia. They are tropical or subtropical evergreen trees, popular because of their unusual form. When *Araucaria bidwellii* was brought from New Zealand to North America, it brought with it the golden mealybug, *Nipaecoccus aurilanatus* (Maskell). This species has now become a serious pest in California. It causes needle discoloration and reduces the growth rate of *Agathis* species, *Araucaria bidwellii, Araucaria heterophylla,* and *Diplacus longiflorus.*

Eggs are laid in a spherical mass enclosed in a thin white cottony web (B). The young nymphs are dark with white furry patches (C). The adult has a characteristic longitudinal band of gold "felt" on its upper surface and small patches around its edges. Adult females range up to 3 mm long. Several generations occur each year.

Nipaecoccus nipae (Maskell), the coconut mealybug, attacks palms (particularly queen palm, *Arecastrum romanzoffianum*), *Morus* species, and *Persea* species in Florida, Louisiana, and California. An introduced predator—the lady beetle *Cryptolaemus montrouzieri* Mulsant—is an effective natural enemy of *N. aurilanatus* and other species of mealybugs. The larvae of this lady beetle, shown in panels A and F, are clothed in white fluffy wax. The adult beetle (H) is about 4 mm long. This beetle, a predator in both adult and larval stages, has been reared in large numbers and introduced into several states. It has become established in California, Florida, and perhaps other states along the Gulf Coast, but since it cannot tolerate low temperatures, it must be reintroduced each growing season into regions that experience cold weather.

Eriococcus araucariae Maskell is a scale that resembles a mealybug (see Plate 160 for related species). It was originally described in New Zealand and is now believed to occur wherever its hosts, *Araucaria* species, are grown. The specimens illustrated in this plate came from California and are commonly found on Norfolk Island pine.

So far as is known, they feed exclusively in the axils of leaflets (E). Several generations occur every year, depending on temperature and other climatic conditions. The adult female makes an egg sac in which hundreds of yellow-orange eggs are deposited (G).

References: 87, 518

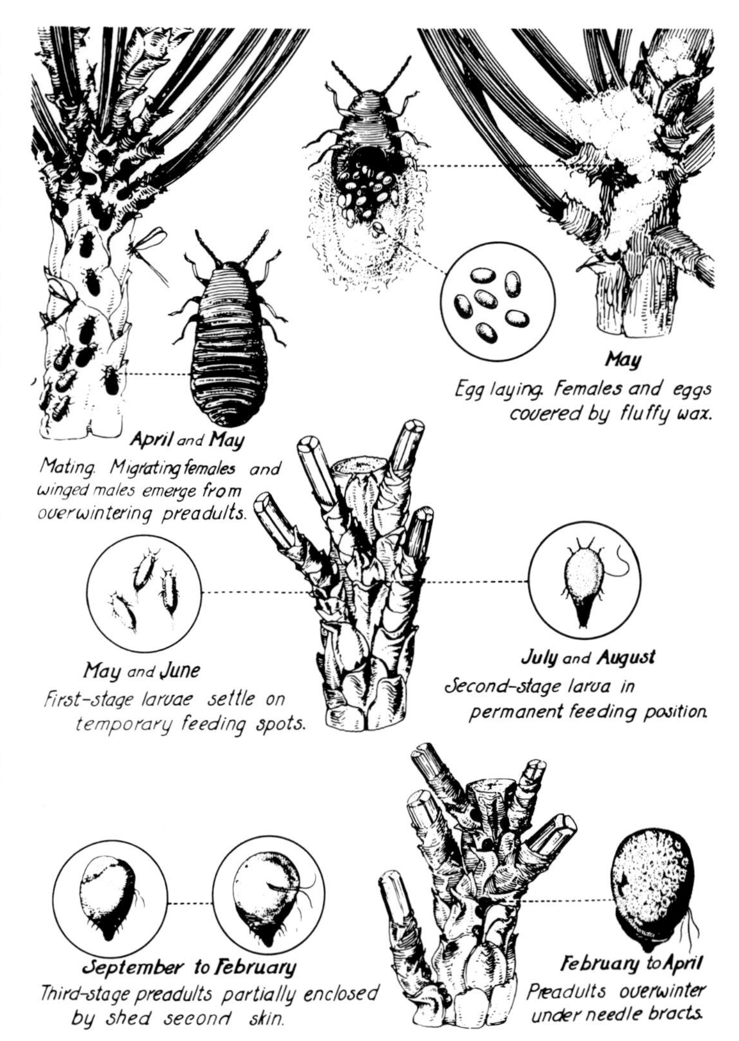

Figure 31. Seasonal history of *Matsucoccus vexillorum,* the so-called Prescott scale. (See text accompanying Plate 39 for a discussion of *Matsucoccus* scales.) (Courtesy U.S. Forest Service.)

A. Adult golden mealybugs, *Nipaecoccus aurilanatus,* and their eggs. (*At arrow*) the larva of a lady beetle, *Cryptolaemus montrouzieri.*
B. Adult mealybugs and an egg mass (*arrow*). Recently hatched young (*circle*) are black with white spots on their backs.
C. Mealybug nymphs.
D. A heavy infestation of mealybugs on *Araucaria* twig.
E. A scale insect *Eriococcus araucariae,* on *Araucaria excelsa.*
F. A magnified view of the larva of the lady beetle *C. montrouzieri.*
G. Egg sac of *E. araucariae* opened to show yellow-orange eggs. [Eriococcidae]
H. An adult *C. montrouzieri,* a lady beetle.

A
B
C
D
E
F
G
H

Matsucoccus Scales (Plate 39)

Fifteen species of scale insects in the genus *Matsucoccus* occur in North America. They belong to the family Margarodidae, and all but one attacks various species of *Pinus*. *Matsucoccus macrocicatrices* Richards, regionally known as the Canadian pine scale, has been reported on *Pinus strobus* from Ontario, Quebec, New Brunswick, Nova Scotia, and New Hampshire. In the Northeast the red pine scale, *M. resinosae* Bean & Godwin, has destroyed the majority of red pine trees planted south of their natural range, since its discovery in Easton, Connecticut, in 1946. Before then, only *M. gallicolus* Morrison, the so-called pine twig gall scale, was known from the Northeast, primarily on pitch pine and scrub pine. It also attacks other pines in the East from New England west to Missouri and south to Georgia and Florida. *M. alabamae* Morrison, sometimes called the Alabama pine scale, occurs on the bark of pines in Alabama and perhaps other Gulf states but has not caused serious destruction. The remaining 10 species occur in the West and Southwest.

The red pine scale is very difficult to detect until large populations cause severe stress symptoms in the trees. Initially, some branches show the damage, followed by involvement of the entire crown; eventually the tree dies. Severe infestations cause a decrease in new growth and needle discoloration. The needles become light green, then light yellow, before finally turning brick red. The primary host is red pine (*P. resinosa*), growing south of its natural range, but the scale also occurs on Japanese red pine, Japanese black pine, and Chinese pine. In the United States the red pine scale has been confined to southern Connecticut, southeastern New York and Long Island, northern New Jersey, and eastern Pennsylvania. Recent studies in China by McClure et al. (509) suggest that *M. resinosae* and the pine bast scale, *M. matsamurae* Kuwana, are the same species. The pine bast scale, recently a serious pest in China, is native to Japan. It may be the origin of the red pine scale, which has long been suspected to have been introduced into the United States on plants imported for use at the 1937 World's Fair.

Common predators of this scale include a lady beetle, *Cleis picta* Rand, an anthocorid bug, *Xenotracheliella inimica* Drake & Harris, and certain species of lacewings, *Chrysopa*. Parasites and diseases have not been reported. To date, natural enemies have not controlled infestations nor prevented tree mortality.

M. gallicolus, sometimes known as the pine twig gall scale, is particularly destructive to pitch pine, *P. rigida*, producing gall-like swellings on the new growth. In the eastern United States, other hosts include loblolly, shortleaf, Virginia, marsh (=pond), spruce, and ponderosa pines wherever they are grown as ornamentals. Severe infestations cause serious weakening of trees, dying branches, and eventually the death of entire trees. Normally, the initial sign of an infestation is the browning of needles near the tips of branches (A). Twigs killed by scales show many black spots (F) if examined closely. Cutting into the bark tissue reveals a pit that contains the scale or its cast "skin" (J).

Adult males are not known for this species. A single generation occurs each year. Eggs overwinter under bark scales and hatch when foliage growth is 2.5–7.5 cm long. The yellowish first-instar crawlers move to the new growth and insert their stylets into the shoot. A pit develops at the feeding site. After several weeks the plant produces a swelling that surrounds the cystlike second instar, eventually enclosing it except for a small pore. Late in July the second molt occurs and the adult female exits by squeezing through the hole in the gall tissue. After moving to the trunk or main branches, the female resides beneath bark scales, where she lays approximately 300 eggs in an ovisac.

M. macrocicatrices is mutualistically associated with the fungus *Septobasidium pinicola;* it overwinters in the cyst stage in fungal mats. After eggs are laid in bark cracks in spring, crawlers migrate to the edges of fungal mats, where they feed for nearly 2 years before reaching maturity as females early in the third spring. Males have been found only rarely. This scale does not cause serious damage.

Information on western species of *Matsucoccus* has been sum-

Table 8. **Hosts and distribution of western species of *Matsucoccus***

Species by habit group	Hosts	Distribution
On needles:		
M. acalyptus, pinyon needle scale	*Pinus aristata, P. edulis, P. balfouriana, P. lambertiana, P. monophylla*	CA, AZ, NM, UT, CO, NV, ID
Within needle sheath, at base of needle bundle:		
M. degeneratus	*P. ponderosa*	AZ
M. fasciculensis, needle fascicle scale	*P. jeffreyi, P. ponderosa, P. sabiniana*	CA, OR
M. secretus	*P. ponderosa*	CA, NV, AZ, NM, CO
On bark at base of needle bundles; in axils of twigs and branches; and on small branches in bark crevices:		
M. eduli	*P. edulis*	AZ
M. monophyllae	*P. edulis, P. monophylla*	CA
M. paucicatrices, sugar pine scale	*P. flexilis, P. lambertiana, P. monticola*	CA, OR, MT, WY
M. vexillorum, Prescott scale	*P. ponderosa*	CA, NV, AZ, NM, CO
On twigs in bark crevices and under thick bark:		
M. bisetosus, ponderosa pine twig scale	*P. contorta, P. jeffreyi, P. ponderosa, P. radiata, P. sabiniana*	CA, OR, CO
M. californicus	*P. jeffreyi, P. ponderosa*	CA, AZ

Source: Modified from Furniss and Carolin, ref. 275.

Red pine scale is manifest most readily in two ways: (1) usually as individuals or semicircular rings of settled first-instar nymphs underneath bark scales of the branches (C), and (2) masses of male cocoons on undersides of branches at the bases of needles or in branch axils. Saclike intermediate stages may also be evident, wedged in the crevices formed by bark scales.

First-stage settled crawlers of the red pine scale overwinter under bark scales. After they resume feeding in early spring, the intermediate stage, or cyst, develops during April and May. Preadults are present in May and June, and adult males and ovipositing females are present from mid-May to early June. During that time crawlers seek new feeding sites; eggs hatch about 15 days after oviposition. They mature by early August, and adults may be found until late October. During this 3-month period, eggs are laid and hatch, and crawlers of the second generation search for new feeding sites (E). Settled crawlers are present from the end of August until the following spring, when their seasonal development resumes. Although adult males are winged, they do not fly. Mating of the females is essential for reproduction.

A. Gross injury symptoms of needle and twig dieback on pitch pine caused by *Matsucoccus gallicolus*. [Margarodidae]

B. Gross injury symptoms caused by *M. vexillorum* on ponderosa pine. (Courtesy W. F. McCambridge, U.S. Forest Service.)

C. A cluster of sedentary nymphs of the red pine scale, *M. resinosae*, on red pine bark. A white waxy fringe forms along the body margin.

D. Needles may be stripped from a twig with little effort where *M. gallicolus* scales are abundant. This is a field technique to determine the degree of injury from the scale insect.

E. A red pine scale ovisac attached to the bark at the base of a needle fascicle. Several newly hatched nymphs are visible. (Courtesy C. E. Palm, Jr., College of Environmental Science and Forestry, Syracuse, NY.)

F, G. The external part of *M. gallicolus* appears as a tiny black scab. The insect is buried in the bark tissue.

H–J. Immature female *M. gallicolus*. Females are visible after the bark is removed, at the interface between bark and xylem. Panel H shows the saclike body of *M. gallicolus* excised from the bark.

A
B
C
D
E
F
G
H
I
J

marized by Furniss and Carolin (275). The accompanying table lists site preferences, hosts, and distribution for those species.

M. acalyptus Herbert, the so-called pinyon needle scale, is native to the Southwest, although it was first described from Idaho. Feeding by the scales kills older needles and weakens trees enough to make them susceptible to fatal attack by the bark beetle, *Ips confusus* (LeConte). The scale occurs widely in Idaho, Colorado, Arizona, and New Mexico, and is also found on foxtail pine in Utah and California. Trees with long-lasting severe infestations have very few needles, which are short and usually clustered at the branch tips.

This scale spends the winter as second-instar nymphs on the needles. In late April in Colorado wingless females emerge from the waxy covering and are immediately mated by the males. Females may crawl to the root collar, crotches of large branches, or fissures in the rough bark on the undersides of branches or the trunk to lay eggs. After 5–6 weeks yellow crawlers appear and migrate to the ends of branches on needles that grew the previous year. Once they insert their stylets and begin feeding, the body becomes covered with wax, then turns black. During molting the body wall ruptures down the back, and the second instar remains in place to feed for the balance of the season. Prepupal males emerge in the fall and crawl to the ground, where they spend the winter as pupae in silken cocoons. In the spring they emerge and fly to the trees in time to mate with emerging females.

In California *M. bisetosus* Morrison, locally known as the ponderosa pine twig scale, is the most damaging of the western species. The needles of infested trees may be shortened, pale, and sparse as a result of its feeding in twig axils, on the bark of branches, and on the trunk. Severely infested trees are highly susceptible to bark beetle attacks. Adult males and females emerge in early spring and mate. The females lay eggs under bark scales, and in midsummer the settled crawlers molt. They spend winter as pre-adults under bark scales.

In the 1930s *M. vexillorum* Morrison, often called the Prescott scale, caused an epidemic of twig blight on ponderosa pine in Arizona and New Mexico (see Figure 31, accompanying Plate 38). No further outbreaks have been recorded; this scale now is mostly a local problem on seedlings and young trees. Its life history and overwintering habits are similar to that of *M. bisetosus*.

References: 6, 47, 75, 275, 339a, 503, 509

Irregular Pine Scale and Monterey Pine Scale (Plate 40)

The irregular pine scale, *Toumeyella pinicola* Ferris, is one of the most common and probably the most destructive scale insects attacking pine in California. It infests Monterey pine, *Pinus radiata*; Bishop pine, *P. muricata*; Canary Island pine, *P. canariensis*; dwarf mugo pine, *P. mugo*; Aleppo pine, *P. halepensis*; Italian stone pine, *P. pinea*; and knobcone pine, *P. attenuata*. Monterey pine is attacked far more frequently than the others.

Scale insects reduce the vigor of trees by sucking plant juices; severely infested trees suffer retarded growth. Old needles turn yellow and die (B). Young trees may die if exceptionally heavy scale infestations are left uncontrolled for a period of years.

As the scales feed, they produce vast quantities of sticky honeydew, which collects on the needles, branches, and trunk. Sooty mold fungi colonize the honeydew and blacken the tree (C). Ants and yellow jackets seek the honeydew as food and create a nuisance by their presence.

The insect overwinters on the twigs of the pine tree. All the scale insects present during the winter are females. They are approximately 6 mm in diameter, robust and dimpled, and more or less circular when viewed from above. They are mottled and resemble chips of marble (E). Immature male scales (H), which resemble grains of rice, may occur in large numbers on the needles.

The irregular pine scale is occasionally found on the same tree with the Monterey pine scale, *Physokermes insignicola* (Craw), although the latter species is not common. The mature female insects (F) are about the same size as the irregular pine scale, but they are dark, shiny, and beadlike. Males are found on the needles. *P. insignicola* is similar to *P. hemicryphus* (Dalman) (=*piceae*), found on spruce in the eastern United States (Plate 41G, H).

In southern California the young crawlers of irregular pine scale begin to emerge from beneath the body of the adult females as early as mid-February. In the San Francisco Bay area crawler emergence begins in late April. During its life each adult female produces up to 2000 young. These crawlers are orange-yellow, oval, and distinctly flattened (G). They are easily visible as they crawl over, and settle on, the shoots and needles.

After settling, the young scales enlarge as they suck juices from the tree. In late summer tiny winged adult males emerge, mate with the still immature females, and then die. The females remain fixed on the shoots and continue to enlarge until spring, when they produce crawlers. Only one generation of the insect occurs each year.

References: 87, 418, 553

A. A young Monterey pine tree severely infested by the irregular pine scale, *Toumeyella pinicola*. This tree subsequently died.
B. The reduced needle length and the yellowing and browning of needles was caused by the irregular pine scale.
C. Blackened foliage results from sooty mold fungi; the fungus feeds on honeydew produced by the irregular pine scale.
D. Irregular pine scale females encrusted on a twig; this is their typical late-winter appearance.
E. A single female irregular pine scale.
F. Monterey pine scale, *Physokermes insignicola,* females on twig of Monterey pine. [Coccidae]
G. Newly hatched crawlers of the irregular pine scale on a needle.
H. Male pupal cases of the irregular pine scale on needles.
I. Settled crawlers of male irregular pine scales on a needle.
J, K. Encrustations of irregular pine scale females on twigs of Monterey pine, showing typical appearance in the spring.

A
B
C
D
E
F
G
H
I
J
K

Pine Tortoise Scale, Spruce Bud Scale, and Virginia Pine Scale (Plate 41)

Toumeyella parvicornis (Cockerell) (=*numismaticum*), known as the pine tortoise scale, is a common and sometimes serious pest of several species of pine. Its known range extends from the Dakotas and Nebraska eastward to New York and New Jersey and southward into Florida. In the North preferred hosts are Scots pine, *Pinus sylvestris*, and jack pine, *P. banksiana*. Also susceptible to infestation, but to a lesser degree than Scots or jack pine, are Austrian pine, *P. nigra*, and red pine, *P. resinosa*. Susceptible southern pines include slash pine, *P. elliottii*, and spruce pine, *P. glabra*.

The immature female scales, which are brown, wrinkled, and more or less circular when viewed from above, overwinter on the twigs of the tree. In the spring development resumes. The females reach maturity in June in the more northerly parts of this scale's range. When mature, the females are about 6 mm in diameter (C). Eggs are laid beneath the body of the adult female and the young scales, or crawlers, begin to appear in late June and early July. The young scales crawl away from the body of the parent, insert their long, slender mouthparts into the plant tissue, and begin to feed. Very soon thereafter the young scale's legs become functionless, and the insect remains fixed at the feeding site. Crawlers are often dispersed from the tree on which they were born by the wind, or they are carried away by birds upon whose feet they happen sometimes to crawl. A high proportion of crawlers never reach suitable host plants. Yet because the parent scale is able to produce about 500 young, it is not necessary for more than a few of these to survive to maintain high populations.

The male scales develop differently than the females. After feeding on needles for about a month, the males enter a brief pupal, or resting, stage (Figure 32). They then emerge as tiny winged insects, which fly about in search of females. The adult male has nonfunctional mouthparts and hence cannot feed; after mating, which must occur within a day or two, the male dies. The mated, but still immature, female remains immobile on a twig of the host pine and enters a state of hibernation with the onset of cool weather in the fall. There is never more than a single generation each year.

Young trees are more heavily attacked by the pine tortoise scale than are older trees. Generally, the lower foliage supports greater numbers of these insects than foliage higher in the tree. Needle yellowing, short needle growth, branch mortality, or the death of the entire tree can result from infestations of the pine tortoise scale. The degree of injury is related to the intensity of the infestation. The scales excrete large quantities of honeydew, which is colonized by sooty mold fungi. The infested tree, or parts of it, appears as though it has been dusted by coal soot.

Toumeyella pini King, the striped pine scale, is often confused with the pine tortoise scale, but it differs from the latter in having a white or cream-colored median dorsal stripe and black pits scattered over the dorsum. Crawlers, active in May in Virginia, choose new shoots as feeding sites. Hosts include lodgepole, red, pitch, Scots, shortleaf, and Virginia pines. The known range extends from Michigan to New York and south to Florida.

The spruce bud scale, *Physokermes hemicryphus* (Dalman) (=*piceae*), is a pest of various species of spruce, including Alberta and especially Norway spruce, *Picea abies*. It occurs throughout the northeastern states and Ontario, Canada, west to Minnesota and south to Maryland. On the Pacific Coast, it is known in Washington, Oregon, California, and likely British Columbia.

The mature scales are globular, reddish brown, 3 mm in diameter, and are typically found in clusters of three to eight at the base of the new twig growth (G). Because they so closely resemble the buds on the twig, their presence is often overlooked during casual inspection of the host (F). The lower limbs of the spruce are far more likely to be infested, and in greater numbers, than the upper branches. When populations are high, low branches are often killed. Weakened trees are said to support higher numbers of bud scales than healthy trees.

One generation of this insect occurs annually. According to Fenton (235), the young overwinter on the undersides of spruce needles, remaining dormant until late March. At that time in Wisconsin, when temperatures are high enough, they move about; in April the females move to the twigs and complete their development. They retain eggs within the body cavity. The young crawlers begin to appear about June 1. Upon settling on the new growth, the scales begin to feed and to excrete honeydew, which supports the growth of sooty mold fungi. The mold is black and gives the tree an unthrifty appearance. Tree growth is reportedly suppressed under conditions of severe attack.

Figure 32. White pupal "skins" of male pine tortoise scales attached to the bark of twigs and needles.

The Virginia pine scale, *T. virginiana* Williams and Kosztarab, was described in 1972 from specimens collected in Yorktown, Virginia, on *Pinus virginiana*. It is now known also from Alabama, Florida, Georgia, and Maryland, where it occurs on *P. clausa*, *P. elliottii*, *P. glabra*, *P. palustris*, and *P. taeda*. Not found in large numbers, it is of minor importance. It occurs under scales of the thicker bark or on twigs. Those under the bark are larger and less convex. Investigators noted that in all collections the ant *Crematogaster clara* Mayer had built protective coverings over the scale insects that consisted of soil, sand, and other debris cemented together. The scale is bisexual and mates in early September. The ovoviviparous females give birth to living young during October. There are two overlapping generations each year (854). The parasites *Tetrastichus* species and *Coccophagus fraternus* Howard (family Eulophidae) have been reared from the immature stages in Virginia.

References: 193, 235, 484, 642, 854, 896

A. A row of Scots pine attacked by the pine tortoise scale. The tree at right of center has been killed by the scale infestation.

B, C. Close-ups of pine tortoise scale insect clusters. Panel C illustrates why the insect is called a tortoise scale.

D. Twig on right shows the sooty appearance of pine foliage resulting from the growth of sooty mold fungi on the honeydew.

E, F. Mature female tortoise scale insects on needles of Scots pine.

G, H. Spruce bud scales at different perspectives. Bark flakes partially cover the scale insect, making it appear as part of the plant. [Coccidae]

A
B
C
D
E
F
G
H

Fletcher Scale (Plate 42)

The Fletcher scale, *Parthenolecanium fletcheri* (Cockerell), is also sometimes known as the arborvitae soft scale. Unlike armored scale insects, it produces no separate wax cover. Wax secretions occur as a thin transparent film that does not conceal the insect itself. Dead, dried-up females often appear as scale covers, enclosing the mass of eggs beneath (A, B). Before they produce eggs, females are hemispherical and distended with body fluid (D).

P. fletcheri is a common pest in the more northern parts of the eastern and midwestern United States, and in Canada. It occurs on arborvitae (*Thuja* spp.) and yews (*Taxus* spp.). There have also been reports of its occurrence on *Pachysandra* and *Juniperus* species. Infestations seldom cause visible injury to arborvitae. *P. fletcheri* is found primarily on the foliage of arborvitae. A serious pest of yew, it weakens the plants, causes foliage drop, and results in a heavy crust of sooty mold on twigs and needles. Twigs and stems are the predominant sites of infestation on yews, and large populations of *P. fletcheri* can frequently be seen on them.

There is great similarity between the various species of lecanium scales. The European fruit lecanium, *P. corni* Bouché, is much like Fletcher scale but has a very wide host range. However, host preference can be a general guide to identification, particularly with *P. fletcheri,* which is restricted in common occurrence to *Thuja* and *Taxus*.

One generation of Fletcher scale occurs per growing season. The second-instar nymph overwinters and matures in May. Eggs begin hatching as early as June 11 in Connecticut and as late as the first week of July in Minnesota. Oval, flat, yellowish crawlers migrate short distances on a branch or frond in search of a feeding site. As they do not crawl very far, populations tend to concentrate on certain branches of a particular plant. During the rest of the season very little growth or development occurs in the scale insect. Small, slightly convex nymphs, amber to reddish brown, remain on the plants throughout the fall and winter (E).

In the spring the juvenile scale grows quickly, and plant damage becomes obvious. Copious amounts of honeydew are produced, especially on yew, resulting in a dense black growth of the sooty mold fungi. Eggs are laid in May and hatching begins in June. Population increases often are abrupt because adult females are prolific. An average of 500–600 eggs are produced by a single female, but she may produce more than 1000. All of the eggs hatch within a short period, thus making the motile crawlers easy to control with one application of an insecticide.

References: 325, 568

A. Dead female Fletcher scales on yew. Each contains hundreds of viable eggs.
B. Dead females and developing settled crawlers (*arrows*) on arborvitae. Stippling of foliage is caused by feeding of the spruce spider mite (Plate 52).
C. Close-up of dead Fletcher scales. Two scales have been turned over to show eggs and crawlers. Both eggs and crawlers are inside circle.
D. An adult female just before egg laying; about 3 mm long.
E. Second-instar nymphs shortly after the crawler stage. Chlorotic areas are caused by the spruce spider mite.

A
B
C
D
E

Black Pineleaf Scale (Plate 43)

Known in virtually all areas of the United States and Canada where susceptible hosts occur, the black pineleaf scale, *Nuculaspis californica* (Coleman), has only occasionally inflicted sufficient damage to be considered an important pest, and even then, it was primarily in western North America. Its hosts include the following trees: *Pinus cembriodes,* three-needle pine; *P. contorta,* shore pine; *P. echinata,* shortleaf pine; *P. jeffreyi,* Jeffrey pine; *P. lambertiana,* sugar pine; *P. ponderosa,* ponderosa pine; *P. radiata,* Monterey pine; *P. rigida,* pitch pine; *P. sabiniana,* Digger pine; *P. resinosa,* red pine; *P. mugo*; *P. banksiana,* jack pine; *P. palustris,* longleaf pine; *Tsuga heterophylla,* western hemlock; and *Pseudotsuga menziesii,* Douglas-fir.

This insect is commonly found in association with the pine needle scale, *Chionaspis pinifoliae* (Fitch) (Plate 47). Because it is gray to black, it is easily distinguished from the snow white *Chionaspis.*

Infestations of the black pineleaf scale occur only on the needles. It causes injury by withdrawing plant juices from the tree, and possibly by introducing toxic substances into the needles as they feed. Infested needles tend to become spotted or blotched with yellow patches. Heavy infestations may cause premature needle drop and in extreme cases death of the tree.

The sequence of symptom expression begins with thin crowns, followed by reddish discoloration, chlorosis, and finally necrosis of needles. Severely infested trees exhibit extreme shortening of needles, the degree of shortening being a function of the degree of scale infestation and the number of years of continuous infestation. Scale infestation seriously reduces the number of years of needle retention. Historically, severe forest damage occurred in Oregon (1948), Washington (1956), Wisconsin (1959), and Great Lakes states (1943).

Trees under stress from soil moisture deficit, soil compaction, root damage, smog, or other causes tend to be especially susceptible to black pineleaf scale attack. Those along dusty roads often sustain higher infestations than those farther from such roads. This phenomenon may be due to the deleterious effect of dust on the natural enemies of the scale or on photosynthesis (which may favor scale survival) or on both.

One to three generations of this scale insect occur each year, depending on geographic location. Overwintering usually occurs in an immature stage. Eggs or live crawlers are deposited singly over an extended period. The crawlers, amber colored, disperse to other parts of the same needle, or to new, adjacent spring needles, where they settle, insert their slender mouthparts, and begin to feed. They usually insert stylets through a stoma into phloem tissue.

Many crawlers are dispersed by the wind, to be carried for considerable distances. The male scale insects emerge as tiny winged individuals and mate with immature females by mid-June. About a month later the females reach maturity and begin laying their eggs.

A wasp parasitoid, *Prospaltella* sp., normally keeps scale populations at low densities in the northwest. *Physcus varicornis* Howard is an important parasite in Wisconsin.

References: 203, 275, 810

A. An infestation of the black pineleaf scale on needles of red pine.
B. Yellowing of foliage of pinyon pine resulting from earlier attack by the black pineleaf scale.
C. Close-up of A, showing degree of infestation.
D. Close-up of the scale cover of a mature black pineleaf scale; 2 mm.
E. Mature and immature black pineleaf scales. [Diaspididae]

A
B
C
D
E

Hemlock Scale (Plate 44)

The hemlock scale, *Abgrallaspis ithacae* (Ferris) (=*Aspidiotus ithacae* = *Aspidiotus abietis*), is a native armored scale insect found on eastern hemlock, *Tsuga canadensis*. It has also been reported on spruce, *Picea* species. It feeds on the needles by piercing the tissue and sucking out cell fluids. A small yellow spot seen from the upper side of the needle is the initial symptom. If four to six hemlock scales take up residence on a single needle, the needle will turn yellow and fall from the tree. Mature scales are always found on the undersides of needles. Like all armored scales, their protective cover is not physically attached to the insect. The cover, called a test, is made of wax secreted by the insect and of its old molted "skins." The test of the female hemlock scale is dark brownish gray, round, and about 2 mm in diameter.

There have been few detailed studies on the life history, biology, and control of the hemlock scale. Stoetzel and Davidson (767) found that two generations occur each year in Maryland. Eggs produced by the first-generation females hatch in June, mature in late July, and produce second-generation eggs in August and September. The second-generation crawlers settle on the needles and spend the winter in the second instar. (See Figure 33.)

Wasp parasites are important means of natural control. They emerge, in Maryland, four times during the growing season: in early April, early June, mid-July, and early August. Insecticides for control of scales should not be used when parasites are emerging.

The hemlock scale is known to occur in Connecticut, New York, Ohio, Indiana, Pennsylvania, and Maryland. It is probably present in the other eastern states where hemlock grows.

Nuculaspis tsugae (Marlatt) is an introduced armored scale that arrived in New Jersey from Japan on the Oriental hemlocks *T. diversifolia* and *T. sieboldii* in 1910. Little is known about its present distribution except that it occurs in Connecticut and New York on the native hemlocks *T. canadensis* and *T. caroliniana,* on the two Oriental hemlocks already mentioned, and on *Abies balsamea* and *A. veitchii.* It has been taken on about 35 other conifers but found in low populations.

N. tsugae may be confused with another circular test scale, the hemlock scale (*Abgrallaspis ithacae*), also illustrated on this plate. Both attack the lower surface of needles. *N. tsugae* produces two generations a year. It overwinters as a second-instar nymph and develops to an adult in April. In Connecticut the first eggs may be found under the female's test in early May. Second-generation scales can develop through all of their stages in about 83 days and produce eggs by early August. Their eggs hatch and the nymph develops to the second instar before diapause in early December. There are five developmental stages for males and three for females.

Although *N. tsugae* has been shown to be a serious pest of native hemlocks, the Oriental species of hemlock always develop higher populations of scales. The wasp *Aspidiotiphagus citrinus* (Craw) is an important parasite of this and other armored scales, although they rarely, if ever, develop in sufficient numbers to allow for control on ornamental trees.

References: 146, 505, 507a, 508, 679, 767

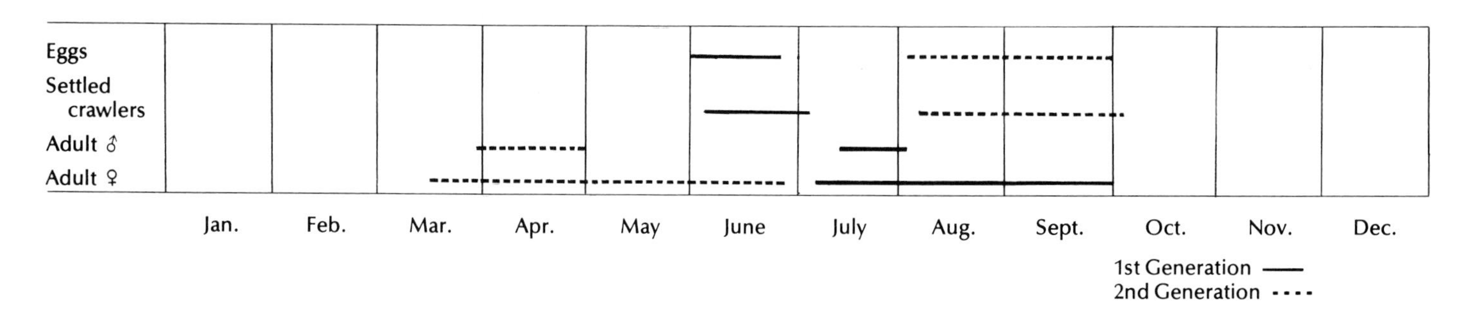

Figure 33. Seasonal history of the hemlock scale in Maryland. (After Stoetzel and Davidson, ref. 767.)

A. Typical injury to hemlock caused by the hemlock scale. View is of the upper surface of the needles.
B, C. A severe infestation by partially grown female scale insects, which locate themselves on the undersides of the needles.
D. Upper surface symptoms of *Nuculaspis tsugae* on eastern hemlock.
E. Three adult female scale covers and one immature female cover of the hemlock scale. Test diameter is about 2 mm.
F. Close-up of *N. tsugae* on the undersurface of hemlock needles.

A
B
C
D
E
F

Elongate Hemlock Scale (Plate 45)

The elongate hemlock scale, *Fiorinia externa* (Ferris), is an armored scale insect. It is primarily a pest of eastern hemlock, *Tsuga canadensis*; Carolina hemlock, *T. caroliniana*; and Japanese hemlock, *T. diversifolia*. Although it also occurs on yew, spruce, and Douglas-fir, its populations on these trees do not commonly reach damaging proportions. Such secondary hosts, if infested, are usually found near heavily infested hemlocks.

The elongate hemlock scale was first discovered in Queens, New York, in 1908. It was described as a new species in 1942. Since the 1950s severe infestations have been reported in Massachusetts, Connecticut, Pennsylvania, New Jersey, Ohio, and Maryland. A species closely related to the elongate hemlock scale has been found in the Richmond, Virginia, area. It is called *F. japonica* Kuwana. *F. japonica* is also known as a pest of pine and *Podocarpus* in California. The elongate hemlock scale is found chiefly in the northeastern United States. Very few stands of native hemlocks have been found to be infested.

The elongate scale is a destructive pest of established landscape trees and hedges. Panel A depicts the yellowing of infested needles. Plant growth is slowed and needles drop prematurely, resulting in thin, weakened plants. The elongate hemlock scale feeds by inserting its threadlike mouthparts into the needles. The flexible stylets then bend and are pushed extensively through the leaf tissue parallel to the surface. These mouthparts are as much as three to four times the length of the body (Figures 34–35).

Mature females, which may live for a year or longer, deposit eggs under a waxy cover. Egg laying is done over an extended period and is interrupted only by cold weather. Eggs hatch in about a month. First-instar nymphs (crawlers) emerge and migrate to the new needles on the same plant. There they settle on the undersides of needles, insert their mouthparts, and begin to feed. After 3–4 weeks, the crawlers molt. About 1 month later females are mature and males emerge as extremely small, delicate winged insects. Eggs are produced 6–8 weeks after mating. Dispersion to other trees occurs primarily by windblown crawlers. Winds may carry the insect 100 meters or more from point of origin.

The seasonal history is complicated by the great overlap of various developmental stages. There is a preponderance of crawlers during late May. However, all stages may be present during the entire growing season, making control of this pest very difficult. Systemic insecticides that translocate to the part of the plant where the crawlers have settled have been effective.

Parasites and predators provide some measure of natural control of elongate hemlock scale populations. A small wasp, *Aspidiotiphagus citrinus* (Craw), is most common in New York and is believed to be an effective parasite (Figure 36). Other hymenopterous parasites, along with the lady beetle *Chilocorus stigma* (Say), feed on this scale (see Figure 112, accompanying Plate 167) but apparently have little effect in controlling populations. Overpopulation can itself be a controlling factor. Drastic reductions in tree growth and in the amount of healthy needle tissue will sometimes reduce the nutritive level of the host to the point where the scales die of starvation and yet the tree recovers.

References: 144, 506, 813

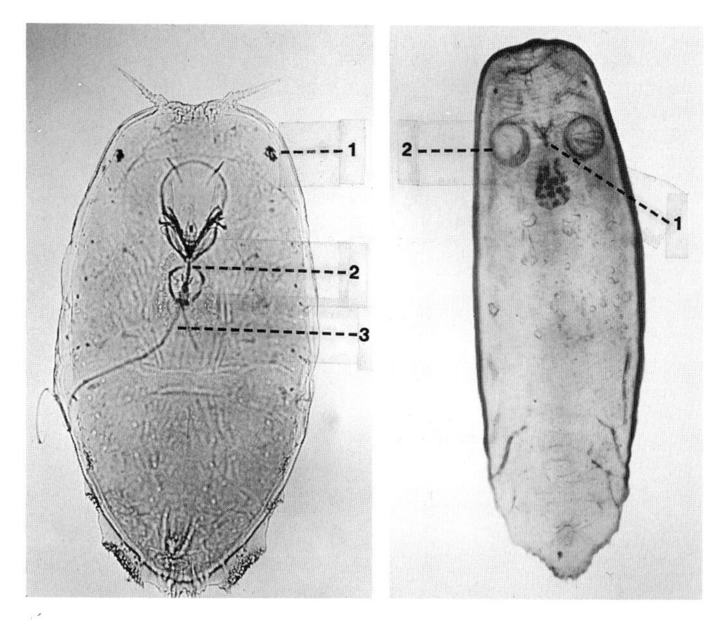

Figure 34 (left). A crawler of the elongate hemlock scale. *1*, eye spot; *2*, mouth region; *3*, part of the flagellatelike stylets. About 1.5 mm long. (Courtesy W. E. Wallner.)

Figure 35 (right). A second-instar elongate hemlock scale immediately after its first molt. *1*, mouth opening; *2*, stylets coiled inside mouth. When extended, the stylets are much longer than the insect's body. (Courtesy W. E. Wallner.)

Figure 36. Dorsal view of *Aspidiotiphagus citrinus* immediately after it emerged from a female third-instar elongate hemlock scale. Note the round emergence hole through which the parasite exited. × 57. (From Wallner, ref. 813, courtesy of W. E. Wallner.)

A. Chlorotic appearance of upper needle surfaces where elongate hemlock scale populations are high.
B. Undersides of yellowed needles showing typical damage caused by female and immature scales.
C. Immature females developing on a new cone.
D. Young and mature females on the undersurface of a hemlock needle. [Diaspididae]
E. The scale cover of a male (*arrow 1*) is white and small compared with a mature female (*arrow 2*), which is nearly finished laying her eggs.

A
B
C
D
E
1
2

Juniper Scale (Plate 46)

One can safely guess that juniper trees with leaves that have turned yellow or brown are hosts of either a spider mite (Plate 52) or the juniper scale, *Carulaspis juniperi* Bouché (=*Diaspis carueli*), particularly if the tree is located in the eastern United States. However, it takes close and preferably microscopic examination to detect these pests. The first indication of damage to the host is loss of the normal lustrous color of a healthy plant. The infested foliage fails to develop new growth, and the tree or shrub looks off-color. As the infestation progresses, the foliage of individual branches will yellow and then die (C, D). Entire plants are known to succumb as a result of scale infestation.

The female cover (test) is circular, white, and nearly 1.5 mm in diameter. This scale has one generation a year over most of its range, but some entomologists believe there may be more in the warmest regions of the United States. Adult females overwinter filled with eggs. Crawlers appear in March or April in southern California; in late April in Delaware; in late May or early June in Connecticut, New Jersey, and Ohio; and in June in western New York. Egg laying and crawler emergence continues over 30–40 days.

The juniper scale, of European origin, is now distributed from coast to coast and is known to attack various junipers and cypresses (*Cupressus*, but not *Taxodium*), *Chamaecyparis*, and incense cedar. Of the ornamental junipers, redcedars, Irish, Savin, and Pfitzer are most commonly attacked.

Carulaspis minima (Targioni-Tozzetti), a scale insect closely related to the juniper scale, attacks arborvitae (E), *Juniperus*, *Picea*, and probably all the species attacked by the juniper scale. Close observation of infested foliage will reveal many white to brownish scale covers, each of which is formed by the scales living underneath. Female scale covers are 1.5 mm in diameter and have a dark center (A). Males are elongated and are whiter than the females (B).

The scale overwinters as a mature female. The female then lays up to 40 eggs, starting in May in the Northeast but as early as March in parts of the South and West. The eggs hatch in about 2 weeks, and the yellowish crawlers seek a new site on the same host plant. They, like other scale and adelgid crawlers, are light enough to be blown by the wind to other hosts. After settling, the crawlers develop rapidly. The females molt three times and the males five. The tiny winged males are active in late summer or early fall, when they seek out and mate with females. They then disappear. In most parts of the United States a single generation occurs annually, but at least one authority suggests that two generations may occur in California.

The armored scales of juniper are difficult if not impossible to differentiate under field conditions. *C. minima* probably survives best in the southern half of the United States. Because the two species have long been confused in the literature, distribution records are not always reliable.

Figure 37. A wasp parasite, *Comperiella bifasciata* Howard, ovipositing on an armored scale. Arrow indicates ovipositor, the egg-laying organ. (Courtesy Harold Compere, Div. of Biological Control, Dept. of Entomology, Univ. of California, Riverside.)

References: 87, 359, 442, 568

A. Mature female juniper scales on juniper foliage. [Diaspididae]
B. Male juniper scales on juniper foliage.
C, D. Dead and dying foliage on *Juniperus communis* caused by a moderate juniper scale infestation.
E. *Carulaspis minima* on arborvitae found on the West Coast.

A
B
C
D
E

Pine Needle Scale (Plate 47)

The pine needle scale, *Chionaspis pinifoliae* (Fitch), is one of the most serious pests of ornamental pines in the United States. It was once referred to as the "white malady" because heavy infestations whitened the foliage of pines and some spruces.

In the eastern United States the most frequently damaged hosts are mugo and Scots pine. Austrian and red pines are also attacked at times. In the prairie provinces of Canada, spruces and pines are seriously infested. Mugo, Monterey, and ponderosa pines are favored hosts in the West. Douglas-fir and cedar are also hosts in the West. In general, infestations in the West are far less severe than in the East.

Ornamental nurseries, Christmas tree plantations, ornamental plantings, and trees planted along dusty roads are more likely to be attacked than forest trees. Light attacks generally go unnoticed and cause little damage. As the populations increase, the needles become covered with the white scale insects (B), which suck juices from the needle mesophyll and cause the needles to turn yellowish and then brown (D). Whole branchlets or branches may be killed. Continued infestations can transform beautiful ornamentals into sickly looking trees with sparse, offcolor foliage. Heavy infestations can kill the tree.

Ornamental pines should be inspected at least twice a year for evidence of infestation. The adult scales (E) are easily recognized even when only a few are present. The insects overwinter as reddish eggs beneath the female scale cover. Each female lays up to 100 eggs. The eggs hatch in May or June. The reddish nymphs (C) crawl from beneath the scale covering and migrate to a new site on the same host, or they may be blown by the wind to more distant hosts. (See Figure 38.) One or two generations occur annually, depending on geographic location. Where two generations occur, the second lot of crawlers begins to appear in late July. Whereas the female scales are wingless, the male scale has wings and is capable of flight.

It has now been determined that there are two species of scale that were unknowingly called the pine needle scale. Because the two scales cannot be confidently separated in the field and their biologies are similar, we acknowledge two scale species but treat them as one. The second species is *C. heterophyllae* Cooley. The common name pine scale has been suggested by Davidson (personal communication) and others for this species. Its hosts are southern pines and *Pinus halepensis*.

In California and Oregon anomalies have been reported in the biology of *C. pinifoliae*. For instance, unisexual populations occur. Also, in the Lake Tahoe district of northern California *C. pinifoliae* overwinters as adult females if they have been feeding on Jeffrey pine. If they have been feeding on lodgepole pine, they overwinter as eggs, which is normal for the species.

The black pineleaf scale, *Nuculaspis californica* Coleman, often attacks the same tree the pine needle scale does. The dark, more circular scale covering easily distinguishes the black pineleaf scale. (See Plate 43.)

References: 87, 136, 275, 612

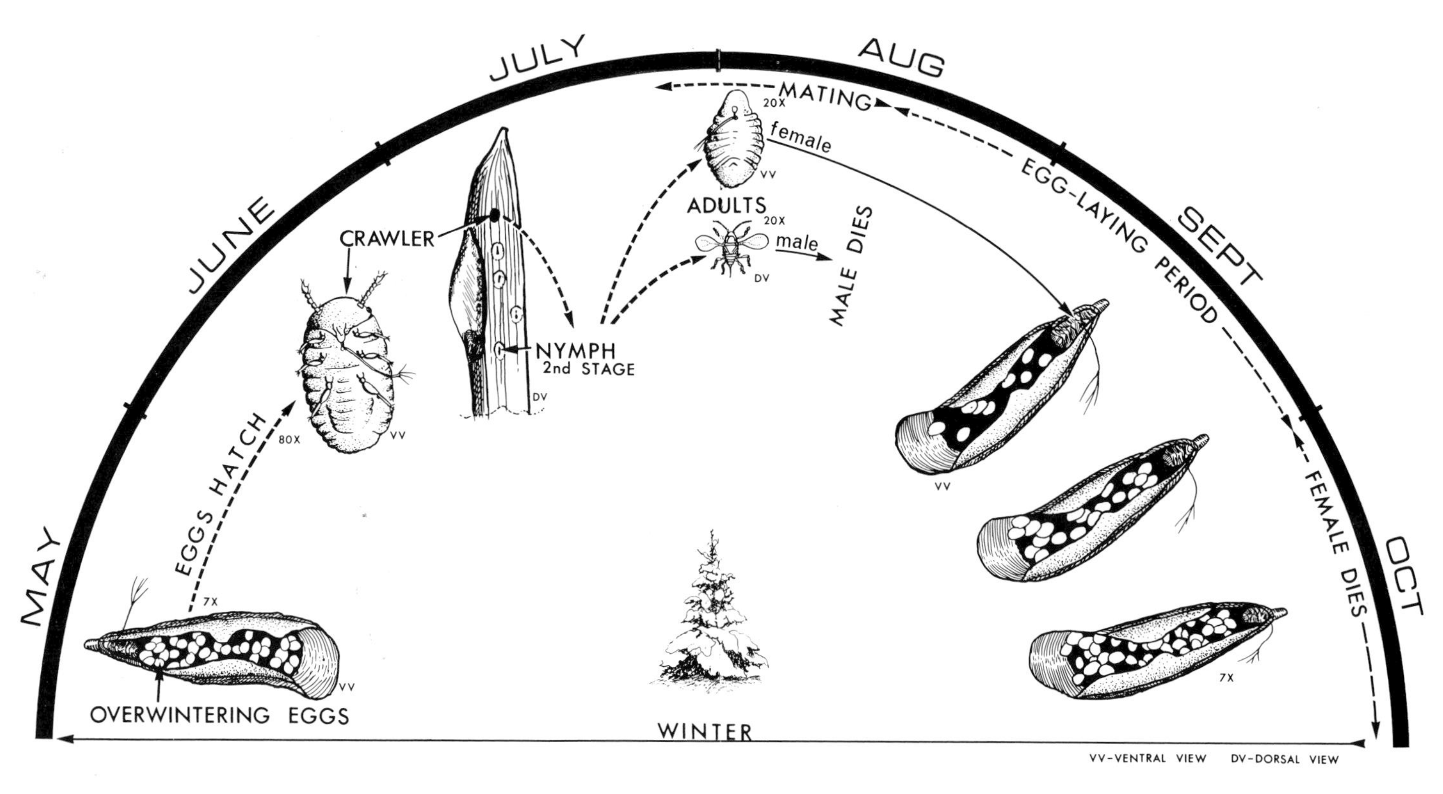

Figure 38. Life stages of the pine needle scale with one annual generation. *VV,* ventral view; *DV,* dorsal view. (Courtesy R. F. DeBoo, Manitoba Forest Research Laboratory.)

A, D. A heavy infestation of pine needle scale, *Chionaspis pinifoliae,* on mugo pine. Note the color of the infested twigs compared with the uninfested ones.

B. Moderate to heavy scale infestation on mugo pine.

C. Female scales (about 3 mm long) and newly hatched reddish crawlers.

E. Close-up of needles. The left one is covered mostly with the narrower male scale coverings, and the right one mostly with females. The circular holes (*arrows*) indicate the places at which the tiny parasitic wasps emerged after killing the female scale insects.

A
B
C
D
E

Armored Scales of Conifers (Plate 48)

The scale insects shown are poorly known, although they are serious pests of numerous ornamental trees and shrubs.

The Asiatic red scale, *Aonidiella taxus* Leonardi, is a tropical species (C). In the United States it is known only in Florida, Alabama, Maryland, and Louisiana, although it sometimes is found in greenhouses elsewhere. A severe pest of *Podocarpus macrophyllus, Taxus baccata,* and the plum yew, *Cephalotaxus* species, it feeds primarily on the undersides of leaves and causes yellow spots (A) to appear on the upper surfaces of the leaves. The scale may also be found on seed covers. Angular blotches (B) and pits may develop in the newly formed leaves. The adult female is about 2 mm in diameter. Its biology has not been studied.

Parlatoria pittospori Maskell is another introduced scale, possibly from Australia (D, E). In the United States its distribution is now believed to be limited to California. Its hosts include *Cedrus deodara, Callistemon* species, *Cotoneaster microphyllus,* mimosa, olive, *Pinus halepensis, Pinus radiata, Pittosporum tobira, Podocarpus elongatus,* rose, and *Viburnum tinus.* In its native home of South Africa it is considered to be polyphagous. This scale insect occurs only on leaves and young stems. The females are about 1.5 mm long. Its biology has not been studied.

A relatively new arrival on American soil is the armored scale *Aspidiotus cryptomeriae* Kuwana. It is currently found in New York, Connecticut, the middle Atlantic states, and Indiana. Virtually all species of conifers will act as hosts, but in Pennsylvania it is seen most often on fir (*Abies* spp.) and hemlock (*Tsuga* spp.). The female scale, not illustrated, is circular, flat, drab grayish green or tan, and 1–1.5 mm in diameter. In Pennsylvania it probably has one generation each year. Oviposition is believed to occur in June. Subsequently, the eggs hatch and crawlers settle and feed on the undersides of flat foliage. Heavy populations cause chlorosis of leaves similar to that illustrated in Plate 45A. When young needles are attacked, they become distorted, dwarfed, and mottled. These symptoms are very apparent on Fraser fir. The biology of this insect has not been studied.

References: 167, 517, 761a

A. A *Podocarpus macrophylla* terminal twig showing the yellow spots and notched leaves caused by the Asiatic red scale, *Aonidiella taxus.*

B. Upper (*left*) and lower sides of *Podocarpus* leaves showing symptoms and Asiatic red scale.

C. Both mature and immature (*arrow*) female Asiatic red scales. [Diaspididae]

D. *Parlatoria pittospori* on needles of *Cedrus deodara* collected in October.

E. Close-up of the female scale covers of *P. pittospori.*

A
B
C
D
E

Cooley Spruce Gall Adelgid (Plate 49)

The Cooley spruce gall adelgid, *Adelges cooleyi* (Gillette), has a complicated life cycle. It has at least five different morphological forms and requires 2 years and two hosts to complete its normal life cycle. This adelgid produces galls on Engelmann, Sitka, Oriental (*Picea orientalis*), and Colorado blue spruce and causes needle injury to Douglas-fir. These trees may be attacked throughout the northern part of the United States and southern Canada.

In the normal cycle of development (Figure 39), the Cooley adelgid overwinters as an immature female on spruce near twig terminals. The female matures early in the spring to become a stem mother and then lays several hundred eggs on lateral terminals. When the eggs hatch, the nymphs migrate to the new spring growth, where they feed at the base of the growing needles. Their feeding and saliva induces the development of succulent cell-like compartments (C), each enveloping an individual adelgid. Together the several compartments form a unique gall (A), which may be up to 63 mm long. By midsummer the gall becomes woody, and an opening develops at the base of each needle (compartment) on the gall (G). The new generation of adelgids escaping from inside migrate to the tips of the needles and transform into females with wings. These winged females then migrate to Douglas-fir or another spruce. On Douglas-fir they lay eggs on the needles, from which a generation of woolly adelgids (D) is produced. In upstate New York eggs on Douglas-fir hatch about mid-June. (See Plate 50C for a similar egg mass.) Douglas-fir is sometimes so heavily attacked that the foliage appears to have been sprinkled with snow. The insect produces no gall on Douglas-fir, but its feeding produces prominent yellowish spots (E, H) and bent or distorted needles. The adelgid, once settled, remains stationary. A pit is formed on the needle at the point where its stylet enters the tissue. Two or more parthenogenic generations may occur each year on the needles of Douglas-fir.

A similar woolly adelgid, *A. tsugae* Annand, occurs on the bark and needles of mountain hemlock, *Tsuga mertensiana,* and more recently it has been found on Canada hemlock, *T. canadensis,* in Virginia and Pennsylvania (Plate 32).

There are other adelgid species in the genus *Pineus* and *Adelges* that also cause galls on spruce (see "Key to Adelgid Galls on Spruce" below, and Plate 31). Some of the galls are mistaken for those caused by the Cooley spruce gall adelgid.

Key to Adelgid Galls on Spruce

1. Gall pineapple-shaped with gall needles often short and thick but never modified to thin scales 2
   Gall elongate, conelike, or appearing as a scraggly deformed twig .. 4
2. Gall terminal on twig, small, pinkish or pale green; usually on *Picea mariana* .... *Adelges laricis*
   Gall rarely terminal; foliage green except closed mouths of cells red-brown to purple 3
3. Gall needles variable but usually not less than ⅓ normal length; nymphs pale yellow with fine white pulverulency .... *A. abietis*
   Gall needles less than ⅓ normal length; nymphs reddish with abundant flocculance .... *A. lariciatus*
4. Gall conelike; needles modified to thin scales .... *Pineus pinifoliae*
   Gall elongate or appearing as a scraggly twig; gall needles not scale-like 5
5. Gall needles radiating to expose gall surface with lateral joining of tissue between swollen needle bases; gall terminal on twig 6
   Gall needles more or less obscuring gall surface and with no lateral joining of tissue between swollen bases of needles; mature galls often elongate and appearing as scraggly twigs .... *P. similis*
6. On *Picea mariana,* rarely others .... *P. floccus*
   On *Picea pungens* .... *A. cooleyi*

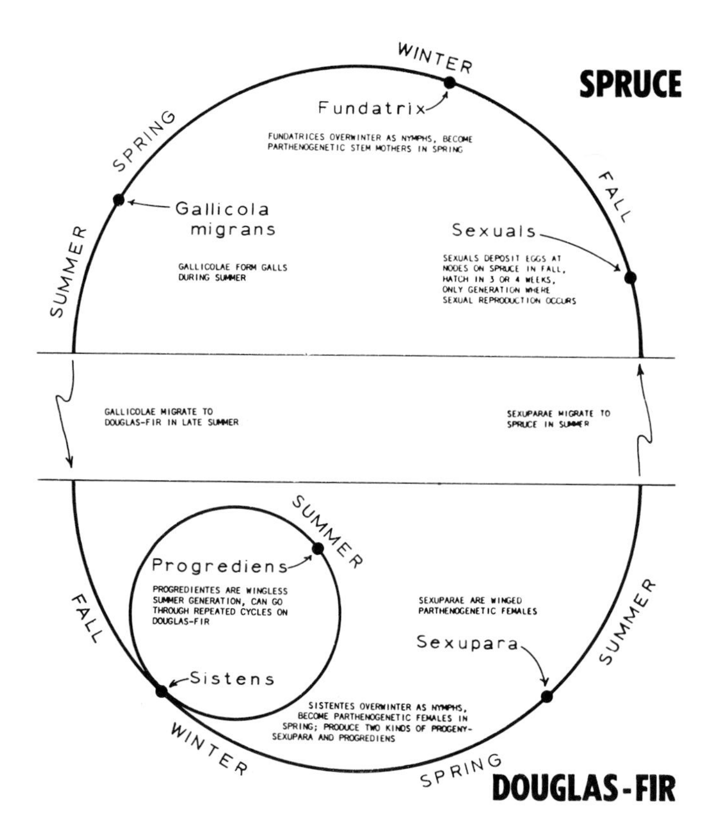

Figure 39. Life cycle, hosts, and developmental stages of the Cooley spruce gall adelgid. (After Annand, ref. 11.)

This key is modified slightly from one that appeared in volume 102 (1971) of the Proceedings of the Entomological Society of Ontario, in a paper titled "The Adelgidae (Homoptera) on Forest Trees in Ontario with a Key to Galls on Spruce." Permission to reprint has been granted by the author, O. H. Lindquist, and the Great Lakes Forest Research Centre, Sault Ste. Marie, Ontario.

References: 137–139, 193, 275, 469, 682

A. Fully formed but unopened galls on Colorado blue spruce. Note that the galls are at the tips of twigs.
B. A Cooley spruce gall after the adelgids have left and at the time when most homeowners notice the damage.
C. A gall sliced open to show the small cells in which the adelgids develop.
D, E, H. Wool-covered adelgids and damage to Douglas-fir needles.
F. A third-instar nymph on a Douglas-fir needle.
G. Recently opened cells in a Cooley spruce gall. A winged adelgid may be seen at the arrow point.

A
B
C
D
E
F
G
H

Eastern Spruce Gall Adelgid (Plate 50)

The eastern spruce gall adelgid, *Adelges abietis* (Linnaeus), is a primary pest of Norway spruce. Occasionally it damages Colorado blue, white, and red spruce. Introduced from Europe before 1900, it has spread throughout the northeastern United States and southern Canada. This insect causes galls to develop at the basal portion of shoots (A), thereby weakening the stems so that they will more readily break from weight of snow or other physical stresses. They cannot be removed by shearing without disfiguring the tree. Galls detract from the beauty and symmetry of trees. If abundant, they decrease a tree's vitality.

Old literature calls the eastern spruce gall adelgid an aphid, which may cause confusion.

This adelgid overwinters as a partially grown female, sometimes called a stem mother, and matures in early spring. The female lays her eggs about the time new buds are ready to break. From 100 to 200 eggs are surrounded by a coat of woolly wax (B, C). They hatch within 10 days, and the nymphs begin feeding on new needles. In a few days they move to the bases of the needles. Their continued feeding causes abnormal twig growth and the subsequent development of the gall. The gall tissue grows around the young insects, and until the galls crack open, during mid to late summer, the nymphs remain entirely protected from parasites, predators, and unfavorable weather. As sections of the gall open (E), the nearly mature nymphs crawl out, settle on a needle, cast their nymphal "skins," and transform to winged, egg-laying females. They lay their eggs in unprotected masses, usually near the tips of needles. When the eggs hatch, the young nymphs attach themselves to a terminal twig near or at a dormant bud to spend the winter.

Many horticulturalists and forestry scientists have observed what appears to be spruce tree resistance to the gall maker. A 5-year plantation study provided some evidence that the ratio of galled to gall-free trees approaches 1:1. It is likely that future research will result in the marketing of resistant clones of Norway and white spruce.

Other adelgids cause galls on Norway, red, black, and white spruce, and some may be mistaken for those caused by the eastern spruce gall adelgid (see "Key to Adelgid Galls on Spruce," accompanying Plate 49). Another shoot gall maker on spruce is the gall midge *Mayetiola piceae* Felt (Figure 40).

References: 193, 360, 469, 620, 781, 851, 862

Figure 40. Shoot gall on *Picea glauca* caused by the midge *Mayetiola piceae*. These galls are sometimes confused with adelgid galls on spruce. Midges emerge from the gall in late May. (Drawing by Carl Whittaker.)

A. An old eastern spruce gall and a mass of eggs at the base of a bud (*arrow*).
B. Egg masses of the eastern spruce adelgid, *Adelges abietis,* at the base of an expanding Norway spruce bud.
C. The "cotton" pulled apart to show the black eggs, now about ready to hatch.
D. A Norway spruce twig. Note the several newly formed pineapple-shaped galls at the bases of the new twigs; these were caused by the eastern spruce gall adelgid.
E. An eastern spruce gall on Norway spruce. The gall is about ready to open and release the young adelgids.
F. A new gall cut open to show the cells and the young adelgids inside. [Adelgidae]

A
B
C
D
E
F

Conifer Gall Midges (Plate 51)

There are many troublesome species of gall midges, the maggot or larval stages of which cause galls on needles, cones, buds, and shoots. Some midges live in needles and shoots or become imbedded in the bark of twigs (Plate 15) or cause resinous nodules to form. Their feeding and presence produce abnormalities that may threaten the life of the host, kill shoots and twigs, stunt the host or parts of it, or cause insignificant cosmetic injury. Most of them are very narrow in their host range, but those with broad host ranges get most of the attention.

Shortened needles, usually with greatly swollen bases, are characteristic of attack by the Monterey pine midge, *Thecodiplosis piniradiatae* (Snow & Mills). Among the pines occasionally infested by this pest, which has a rather restricted distribution in coastal central California, are the bigcone, *Pinus coulteri*; Digger, *P. sabiniana*; Bishop, *P. muricata;* and knobcone, *P. attenuata*. As the common name of this pest suggests, the favored host of the midge is Monterey pine, *P. radiata*.

The adult insects are tiny, delicate flies that deposit their eggs on the terminal buds of the pine tree during January, February, and March. The larvae, upon hatching, migrate to the bases of the newly forming needles, where they burrow into the needle tissue. Their feeding causes a clamlike swelling to develop. If, during the spring and summer months, the basal portion of an infested needle is carefully dissected, tiny white larvae may be found (B). The mature larvae are reported to leave the gall in November and December and to drop to the soil, where the pupal stage is spent. A single generation occurs each year.

The midge attacks only the current season's needles. The following year such needles turn yellow and then brown, and drop prematurely from the tree. Severe infestations have been reported. Trees may be nearly denuded by the premature loss of older infested needles, and the current season's needles may be much shorter than normal. Typically—and for reasons yet unknown—single, widely scattered trees are attacked by the Monterey pine midge year after year.

Contarinia coloradensis Felt, not illustrated, causes a bulbous swelling at the needle base. Affected needles are short and drop after the first year. This species limits activities to ponderosa pine. Its damage resembles that shown in panel C.

T. resinosae Kearby is found on red pine embedded in the slightly swollen bases of the current season's needles. The needles usually appear normal but turn reddish brown and drop prematurely. *Pinyonia edulicola* Gagné causes spindle-shaped galls at the base of pinyon pine needles. This midge is primarily a problem in urban landscapes, and if abundant may cause defoliation. Complete defoliation will kill the tree. *T. pinirigidae* (Packard) larvae live in pockets between two of the three needles of *Pinus rigida,* causing an uneven enlargement at the base of the needles. Needles are short and sometimes recurved (D). *Janetiella coloradensis* Felt is found only on pinyon pine and causes the needle base to enlarge. The fascicle also enlarges to encase the gall. As many as three larvae may be found in the gall, each in a separate capsule. *Mayetiola piceae* Felt causes galls that may be confused with those of eastern spruce gall adelgid (Plate 50). Eggs are laid at the base of newly developing needles in late May. Plant tissue grows around the young larva and in a short time completely encloses it. Needles drop from the gall tissue. If galls completely surround the shoot, the terminal portion ceases to grow and dies (Figure 40, accompanying Plate 50). Hosts include white and red spruce, with questionable reports of galls found on Norway spruce.

The balsam gall midge, *Paradiplosis tumifex* Gagné, is responsible for the development of galls on the needles of balsam and Fraser fir (Figure 41). Eggs are laid early in the spring on elongating fir buds; when eggs hatch, each larva crawls to an immature needle, where it settles and begins to feed. Within a week, and with continuous exposure to gall-inciting chemicals from the maggot, gall tissue grows around the insect in a characteristic form (E). The vascular bundle is not affected by gall formation, but the cells lining the resin ducts are altered. The larva vacates the gall in late summer, drops to the ground, and burrows into the duff where it pupates and spends the winter. After the larva has vacated the gall, the leaf dies and drops to the ground. Growers of Christmas trees must be especially diligent to keep this pest out of their plantations.

Balsam gall midges are likely to occur for 2 or 3 consecutive years and then be absent or nearly so for several years. They can be found where balsam grows from the Canadian Maritime Provinces west to Wisconsin and south along the Appalachians to North Carolina. Associated with this gall maker is an inquiline, *Dasineura balsamicola* (Lintner), thought for many years to be the causal organism. *D. balsamicola* lays its egg(s) near the larva of *P. tumifex* before the gall tissue covers the insects. After the gall is nearly closed, the inquiline egg(s) hatches and the larva crawls into the gall opening, where it grows much faster than the gall maker. The gall maker eventually dies, and the inquiline emerges and drops to the soil, where it completes its

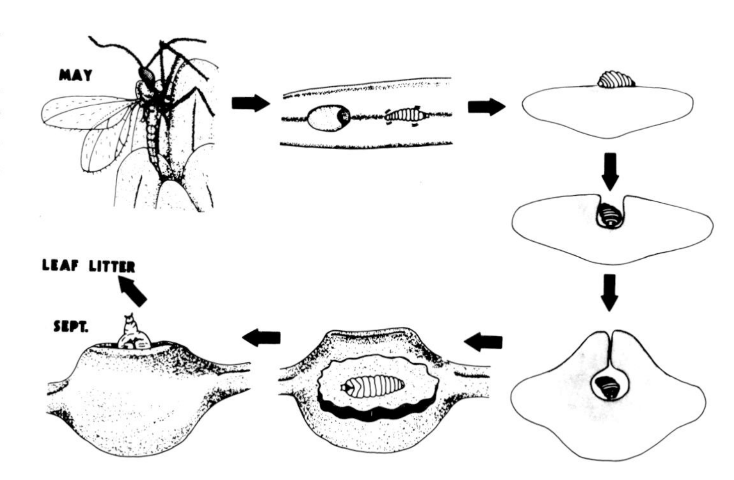

Figure 41. Life cycle of the balsam gall midge. The shaded regions in the needle cross section (*right*) indicate the zones of maximum influence by the larva. (From West and Shorthouse, ref. 834, courtesy J. D. Shorthouse.)

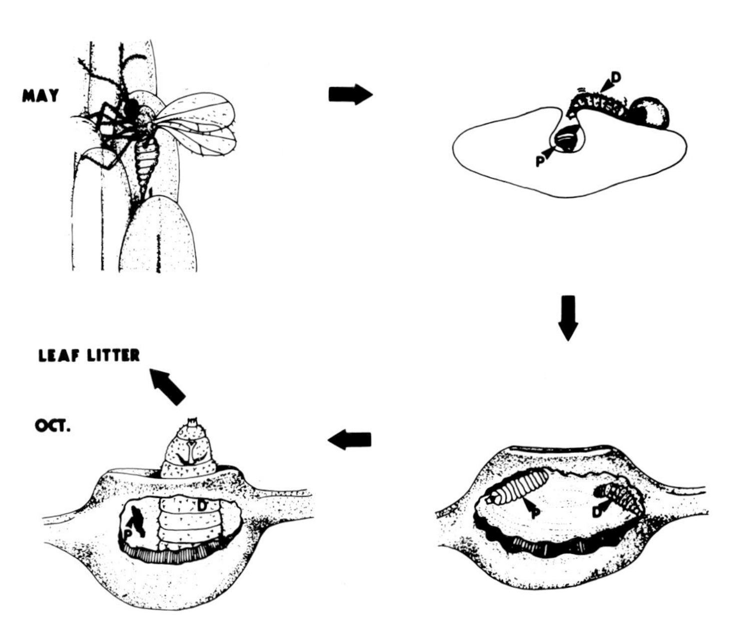

Figure 42. Life cycle of *Dasineura balsamicola,* an inquiline of the balsam gall midge, *Paradiplosis tumifex,* as it develops in the gall caused by *P. tumifex. D, Dasineura; P, Paradiplosis.* (From Shorthouse and West, ref. 695, courtesy J. D. Shorthouse and *Proceedings of the Entomological Society of Ontario.*)

A. Shortened needles with bulbous bases are characteristic of attack by the Monterey pine midge, *Thecodiplosis piniradiatae.*
B. Swollen base of a needle cut open to expose larvae of the Monterey pine midge. The gall midge is about 1.5 mm long.
C. A single fascicle of injured Monterey pine needles.
D. A pitch pine injured by *T. pinirigidae.* [Cecidomyiidae]
E. Close-up of a galled balsam fir needle.
F. A balsam fir twig with needle galls caused by the balsam gall midge, *Paradiplosis tumifex.*

A
B
C
D
E
F

development (Figure 42). The two larvae are readily distinguished. *D. balsamicola* has spicules and setae on the abdomen; *P. tumifex* is smooth. There are two species of parasitic wasps in the genus *Platygaster* that have the potential for significantly reducing balsam gall midge populations.

The cypress twig gall midge, *Taxodiomyia cupressiananassa* (Osten Sacken), causes heavy spongy galls of varying sizes from leaf bud tissue. When numerous, the galls may cause baldcypress and pondcypress branches to droop under their weight. Typical galls are oval, green, and about 20 mm long. Each gall may contain up to 15 yellow-orange maggots in individual cells. The larvae overwinter in the gall and emerge as flies for about a month beginning in mid-May. Female flies lay their eggs on newly developing leaves. At the end of the growing season the galls turn brown. During autumn galls drop to the ground with the leaves. Some control may be obtained by raking and destroying the fallen leaves and galls. There are two generations per year.

This insect is found in the midsection of the United States from northern Illinois south to Louisiana and Florida. For other gall midges, see Plate 15.

References: 77, 114, 391, 488, 590, 695, 706, 834

Spider Mites of Conifers (Plates 52–53)

Spider mites attack conifers as readily as they do the broad-leaved plants. Many conifer-feeding mites build up to damaging numbers in the spring and early summer and sometimes again in the fall. Hot, dry summer weather often causes a marked cessation in their activities.

As they feed, spider mites destroy the chlorophyll-bearing cells at the surface of the needle or scale of the conifer. This results in a flecking, stippling (B), or bleaching of the affected foliage; some of the leaves turn brown and later drop as a result of the mite injury. These symptoms may be confused with air pollution injury. Webbing may or may not be found, depending on the mite species involved.

To determine whether a plant is infested with spider mites, hold a sheet of white paper beneath some foliage and jar the foliage sharply. If the tree is infested, mites will soon begin to crawl around on the paper. Although very small, less than the size of ground pepper, the crawling mite can be easily seen against the white background. The presence of a few mites is of no concern. However, if dozens appear on the paper, take some measures to reduce the mite population before serious damage occurs to the plant.

Large and destructive populations of mites sometimes result from improper use of insecticides on conifers. Some pesticides kill the natural enemies of the spider mites without eliminating the mites themselves. Consult a pest control specialist for advice about the use of miticides.

Probably the most destructive conifer-feeding spider mite in Canada and the United States is the spruce spider mite, *Oligonychus ununguis* (Jacobi) (Plate 53F). This species attacks all spruce, arborvitae, juniper, hemlock, pine, Douglas-fir, Siberian larch, and sometimes other conifers. Tiny strands of webbing are often found among the needles on which this mite feeds (Plate 53). Damaged needles on spruce turn reddish brown.

Of greatest concern in northern climates, the spruce spider mite is usually a serious problem in landscape conifers, but it also occurs in natural forests. Overwintering eggs are placed under bud scales, in the axils of needles, or under webbing on the stem or branches. Development of larvae and nymphs (Figures 43–44) requires about 3 and 6 days, respectively. The larva has only three pairs of legs. After its first molt there are four pairs of legs. The adult, about 0.5 mm long, varies from dark green to dark brown. Three or more generations develop each year, with successive generations produced at intervals of 2–3 weeks. All active forms feed on needles, preferring old over new needles. This mite is readily dispersed by wind. It is primarily a problem in the spring and fall.

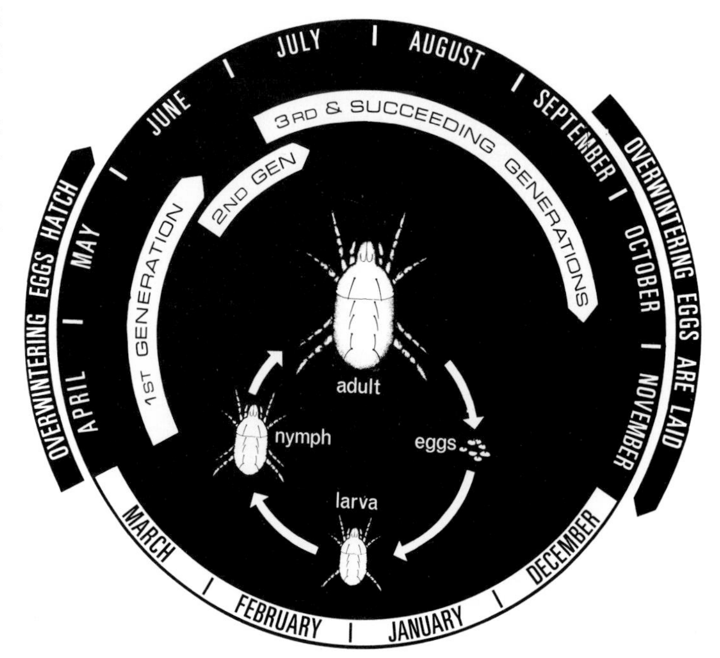

Figure 43. Life cycle of the spruce spider mite. Note only three pairs of legs on the larval stage. (From Peterson and Hildahl, ref. 613, courtesy Northern Forest Research Centre.)

In the western states several other species of *Oligonychus* are destructive to conifers. *O. subnudus* (McGregor) is a serious pest of young Monterey pine, *Pinus radiata,* especially when it is grown for Christmas trees (D). Populations of *O. subnudus* tend to be highest in the spring.

Another species, *O. milleri* (McGregor), makes its appearance in the summer and fall on the same trees. Neither *O. subnudus* nor *O. milleri* is a web-forming mite.

For additional information on spider mites, see Plate 228.

References: 282, 407, 568, 613, 633

A. Unaffected foliage (*top*) in contrast to the bleached, bronzed foliage of a pine infested with *Oligonychus* spider mites.
B. Stippling of pine foliage, the result of feeding by *Oligonychus subnudus*.
C. Infestation of *O. subnudus* on needles of Monterey pine. The tiny pinkish objects are the mites.
D. Close-up of *O. subnudus.* [Tetranychidae]

A
B
C
D

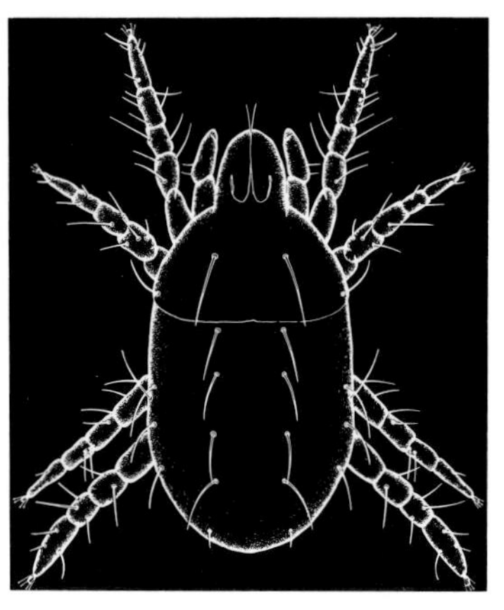

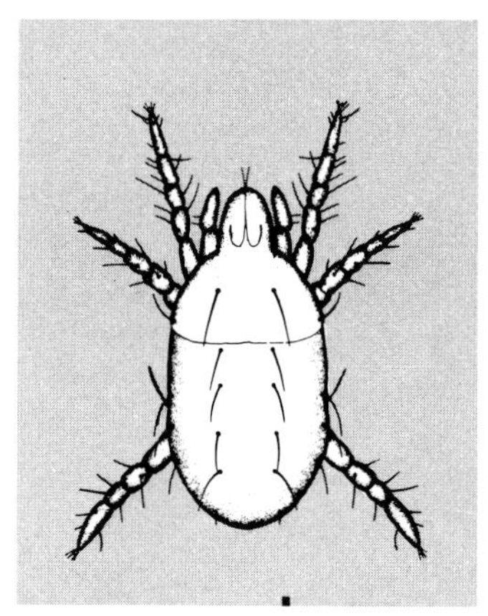
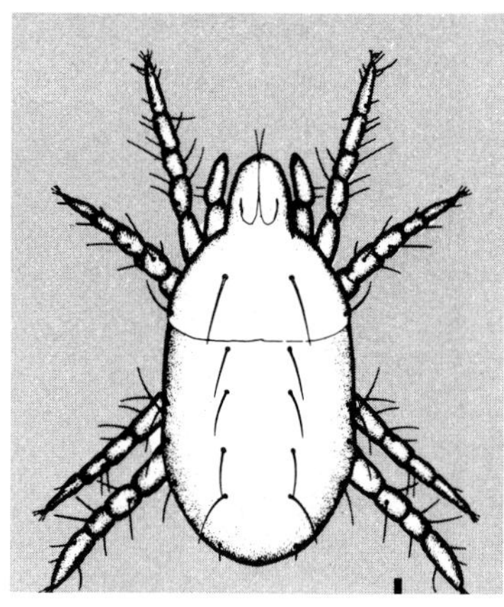

Figure 44. Life stages of the spruce spider mite. (From Peterson and Hildahl, ref. 613, courtesy Northern Forest Research Centre.)

A. Damage to *Picea glauca albertiana* by the spruce spider mite, *Oligonychus ununguis*.
B. Off-color foliage of *P. abies* caused by the spruce spider mite. Note the webs characteristic of this mite.
C. Eggs of the spruce spider mite.
D. Characteristic damage by spider mites to hemlock. Mites feed on the undersides of these leaves.
E. Mite-caused damage of varying intensity to needles of Norway spruce.
F. A spruce spider mite nymph on spruce needle. White spots are pores called stomata. (See also Plate 228.)

A
B
C
D
E
F

Eriophyid Mites of Conifers (Plate 54)

Figure 45 (left). A rosette-type witches'-broom on Scots pine symptomatic of damage caused by the eriophyid mite *Trisetacus gemmavitians.* (Courtesy D. G. Nielsen, Ohio State Univ., OARDC, Wooster.)

Figure 46 (right). Galls on *Juniperus* sp. caused by the eriophyid mite *Trisetacus quadrisetus.* (Courtesy H. A. Denmark, Florida Dept. of Agriculture, Div. Plant Industry, #680121-8.)

There are probably more unknown and undescribed species of eriophyid mites than any other group of arthropods. They are very small (their sizes are measured in micrometers), wormlike, and slow moving. On conifers, injury is expressed by chlorotic needles, by dwarf, distorted, or short needles, by rosetted bud/needle clusters similar to a witches'-broom, by galls, and by partial defoliation of old as well as the current season's needles. Eriophyid mites are to be found in or on the buds and foliage of all North American conifer species. Some mite injury is similar to the injury caused by air pollution and acid rain, therefore some abiotic causes must be brought into the diagnostic equation. Because mites are not visible without magnification, they often go undetected until it is too late to prevent damage to the tree.

Trisetacus campnodus Keifer & Saunders is an eriophyid mite that attacks the needles of Scots pine, *Pinus sylvestris,* in the Northwest. These mites typically occur at the needle base beneath the sheath. In the initial phase of attack, mites are found at the flat interface where two needles come together, but as the mite population increases, the entire needle base covered by the sheath may be invaded. Eventually the epidermis of the entire needle base is destroyed. These conditions are accompanied by chlorosis, stunting, and often twisting of the needle distal to the sheath. Needle length may be reduced as much as 70%.

In the East *Platyphytoptus sabinianae* Keifer, *Setoptus jonesi* Keifer, and *T. ehmanni* Keifer cause the short needle syndrome of Scots pine, which resembles that of *T. campnodus* in the West. *T. gemmavitians* Nielsen & Balderson causes a proliferation of buds, which in trying to elongate, form a mass of short twisted needles in rosette form (Figure 45). The rosettes are conspicuous and at times are even attractive. This mite occurs only on Scots pine both in the East and West.

One juniper gall mite, *T. quadrisetus* (Thomas), is widely distributed in the United States. Its presence causes galls (Figure 46) on *Juniperus communis, J. silicicola, J. virginiana,* other *Juniperus* species, and *Cedrus.* This mite feeds on the terminal needles, enlarging their bases and forming gall-like terminals. Little is known about its biology and life cycle.

Other closely related species feed on the newly developing juniper berries, transforming them into galls.

The growing tips of *Juniperus* species, particularly the Andora types, and northern white cedar are feeding sites for *T. juniperinus* (Nalepa) and *T. thujivagrans* Smith, respectively. *T. chamaecypari* occupies a similar niche on oriental arborvitae and Alaska cedar. All three inhabit the growing tips of the host and cause foliar and shoot distortion—shoots may twist or become sharply angled—as well as a variety of symptoms that often end in necrosis of all or part of the tip. The overall appearance of the afflicted trees or bushes ranges from a loss of gloss or luster to rusty brown from an abundance of brown tips. The feeding injury results in very slow terminal growth or complete cessation of growth, prompting the common name tip dwarf mites.

Typically these mites occur in large numbers at or near the bases of newly developing scalelike foliage. To find them one must pull back a newly formed leaf that shows early symptoms and examine the basal area with a hand lens or a dissecting microscope (see panels B and C for relative size of mites to leaf).

All three species overwinter as adults in the tips. They migrate to new sites in the spring and fall but only under very humid conditions. Eventually every shoot may harbor sizable colonies. The mites are difficult to control because they are protected inside the tips. Care should be taken not to confuse juniper tip midge injury (page 46) with that caused by tip dwarf mites or fungal disease. Tip dwarf mites are most likely to be found on ornamentals rather than on trees and shrubs in forests. They are distributed throughout the Northeast, the Great Lake states and eastern Canada.

The hemlock rust mite, *Nalepella tsugifoliae* Keifer, feeds openly on the needles, whereas those mites affecting pine are usually found between the needles within the needle sheath. The hemlock mite sucks juices from the hemlock needle. When mite populations are high, the normally dark green foliage of the hemlock turns blue and later yellowish before dropping (A). This mite has been a pest of hemlock in nurseries on Long Island and in Virginia. It causes most of its damage in the spring. By midsummer, when the damage is most evident, the mite population diminishes to a very low level. This mite has been collected from a number of other conifers, including fir, spruce, and species of *Taxus, Pseudolarix,* and *Torreya.* Another species of *Nalepella* causes distinct galls on the tips of Douglas-fir branches.

The taxus bud mite is treated in Plate 230.

References: 171, 425, 426, 674, 770

A. An uninfested hemlock twig (*left*) and one that has been heavily fed on by *Nalepella tsugifoliae,* the hemlock rust mite (*right*).
B. A highly magnified section of one needle showing numerous mites and eggs.
C. A magnified section with a single female mite and a group of three eggs that are about the same color as the body of the mite.

A
B
C

INSECTS THAT FEED ON BROAD-LEAVED TREES AND SHRUBS

Dogwood Sawfly (Plate 55)

This and the following five plates demonstrate some of the general features and behavioral characteristics of the sawflies such as egg-laying habits, larval postures, prolegs, and overwintering forms.

The three insects shown in this plate are instars of *Macremphytus tarsatus* (Say), a sawfly. Each has a different color pattern from the others, and in fact, if the growth of these larvae were not followed, one might think that they were of different species.

M. tarsatus is found in the Northeast and in the Great Lakes states. It is a common defoliator of dogwood, primarily the gray dogwood, *Cornus racemosa*. One generation occurs each year. The larvae overwinter inside cells prepared in soft or decaying wood and occasionally in structural timbers. Adult sawflies emerge over a rather long period from late May through July. They lay eggs on the undersides of leaves, depositing well over 100 on a single leaf. When the eggs hatch, the larvae feed together and skeletonize the leaf (G). As the larvae mature, they eat all of the leaf except the midvein. After the second molt, the bodies of the larvae become covered with a white powder-like material (C) that can be rubbed off. At their final molt they completely change their color pattern (E). At this stage they wander about, apparently in search of an ideal location for hibernation. Fully grown larvae, which are about 25 mm long, become quite a nuisance and may cause some damage. They prefer to build their overwintering home in rotting wood found lying on the ground, but they will burrow into composition wood-fiber wallboard used in modern homes and will also burrow into clapboard siding. Redwood siding or furniture is just as acceptable to them. Woodpeckers can detect the presence of larvae in the siding and may cause additional damage in their efforts to get a meal.

The color changes of the larva have given entomologists reason to speculate that the white color of the feeding stage mimics bird droppings and thus helps to protect them from enemies. Likewise, the spotted pattern of the last-instar larva camouflages it as it crawls over duff and litter beneath the bushes.

A closely related species, *M. varianus* (Norton), which also feeds on dogwood, lays its eggs in an interesting manner. The female cuts a slit in the leaf epidermis and inserts her egg in the soft parenchyma. This causes a slight elevated bulge, easily seen from the underside of the leaf.

Reference: 193

A. Several larvae of the dogwood sawfly, *Macremphytus tarsatus,* feeding on *Cornus stolonifera.*
B. Typical posture of a resting dogwood sawfly larva.
C. Dogwood sawfly larvae feeding on the edge of a leaf. Note that the legs are not covered by the white powdery material.
D. A fully grown dogwood sawfly larva in its final color form. Note the number of abdominal prolegs.
E. A larva in the act of molting.
F. Recently hatched dogwood sawfly larvae on the underside of a leaf. The leaf shows the feeding injury typical at this stage of the larva's development.
G. Dogwood leaves: (*left to right*) lower leaf surface with a young colony of sawfly larvae; upper leaf surface showing injury; a leaf completely skeletonized.

A
B
C
D
E
F
G

Mountain-ash Sawfly and Dusky Birch Sawfly (Plate 56)

The mountain-ash sawfly, *Pristiphora geniculata* (Hartig), is found throughout the northeastern United States and Canada south to West Virginia and as far west as Michigan. It survives only on the foliage of *Sorbus americana, S. aucuparia, S. decora,* and *Sorbaronia hybrida.* The larvae devour all of the leaflet except the midvein, a symptom that identifies the pest (A). This insect occurs in both the Old and New World, but authorities differ as to whether it is of European origin.

Adults that produce the first generation emerge from the soil over a 6-week period beginning in late May, continuing into early July. Second-generation adults emerge about mid-August. They lay their eggs in slits cut in the epidermis with the ovipositor, near the edge of the leaflet. Adults die in about a week. Eggs of the first generation begin to hatch in early June, and some larvae may be found in the trees until early August. Second-generation larvae may be found in September. In either generation, the larvae require about 3 weeks to complete this stage of development. The second generation is small and often inconsequential. The prepupa is the overwintering, diapause state. Some individuals may remain in diapause for several years.

The larvae feed in colonies, devouring all the foliage on one branch before moving to another. Mature larvae drop to the ground, burrow into the soil, and prepare an earthen cell where they spin a paperlike cocoon similar to those shown in panel C.

Though it is a defoliator, this insect seems to cause little lasting harm to the tree. It has been postulated that it does not seriously deplete the food reserves of the tree because it feeds primarily during the mid-growing season.

There is little information about the effects of parasites and pathogens on the mountain-ash sawfly.

The dusky birch sawfly, *Croesus latitarsus* Norton, is widely distributed from Newfoundland to Florida, across Canada to Alaska and south to Utah. Its preferred host is gray birch, but black, red, paper, and yellow birch are also sometimes attacked. One to two generations occur annually. It spends the winter as a prepupa in a cocoon in the soil, transforming to a pupa in the spring and emerging as an adult during May. Adults of the second generation may emerge from the middle of July to the middle of September.

The fully grown larva has a shiny black head and yellowish green body (F). In the earlier stages its markings are not as distinct (E). Each mature larva assumes a curious S-shaped posture around the edge of the leaf when alarmed or not feeding.

The birch sawfly, *Arge pectoralis* (Leach), not illustrated, is an important defoliator of all birch species and occasionally alder and willow. It occurs throughout transcontinental Canada and to a lesser extent the northeastern United States. Adults appear in June and July from their overwintering pupal sites in the ground under host trees. Larvae are present on foliage from mid-June to mid-September. They have a pale yellowish green body with several lineal rows of black spots, and the head is dull orange. When fully grown, they are 20–26 mm long. Their form and feeding behavior resemble those of the dusky birch sawfly.

References: 193, 199, 254, 660

A. Young mountain-ash sawfly larvae, *Pristiphora geniculata,* on leaves of mountain ash. [Tenthredinidae]
B. Sawfly larvae on a single leaflet.
C. Typical sawfly cocoons.
D. A mature mountain-ash sawfly larva; about 18 mm long.
E. A mature larva (*top*) and younger larvae of the dusky birch sawfly, *Croesus latitarsus,* on birch leaf.
F. A mature larva of the dusky birch sawfly on birch leaf. The brown spots on the leaves in E and F are caused by the birch leafminer, *Fenusa pusilla* (Lepeletier) (see also Plate 84).

A
B
C
D
E
F

"Slug" Sawflies (Plate 57)

There is a group of sawfly larvae commonly called "slugs" because they superficially resemble true slugs. With the exception of the bristly roseslug, they have a slimy, nonsegmented appearance. Careful observation, however, reveals their characteristics as sawfly larvae: the mouthparts are tucked under the thorax and are not visible from the dorsal surface, the thorax is enlarged, and the abdomen tapers toward the end.

The pear sawfly, *Caliroa cerasi* (Linnaeus), was introduced from Europe and is now distributed from the North Atlantic states to California. In Canada it is known from Ontario to British Columbia. It feeds on the foliage of a wide variety of trees and shrubs, including hawthorn, mountain ash, shadbush, crabapple, cotoneaster, and all of the flowering fruit trees with the exception of Bradford pear.

The pear sawfly passes the winter in the ground as a fully grown larva, changing to a pupa in early spring. Adult sawflies emerge in late May and June. They deposit eggs singly, on the underside of the leaf, forcing each into a pocket between the two epidermal layers. Upon hatching, the larva feeds as a skeletonizer on the upper side of the leaf (A). The injured area quickly turns brown, and if enough leaf surface has been consumed, the leaf drops prematurely. It is not uncommon for unattended trees to be completely defoliated. The larvae feed on the leaves for about 4 weeks. In their early stages they are greenish black and slimy (D). Fully grown larvae are yellow and about 13 mm long. They drop to the ground to form pupal cells in the soil and transform to adults in time for a second generation by early August.

Caliroa quercuscoccineae (Dyar), the scarlet oak sawfly, is one of seven species in the genus that feed on oak foliage. All are found in the eastern half of North America and all are potentially important defoliators. Damage by *C. quercuscoccineae* occurs primarily in the larval stage. Larvae are skeletonizers and may feed from either surface of a leaf, removing most of the interveinal tissue except for one epidermal layer (F). Adult sawflies damage leaves as they lay eggs.

There are two and occasionally three generations per year in Kentucky. Adults emerge in the spring and lay their eggs in full-sized oak leaves. The female selects the upper leaf surface and uses her sawlike ovipositor (Figures 47–48) to cut the upper epidermis for a length sufficient to accommodate a disk-shaped egg. When she has finished depositing eggs, they form a continuous chain along major leaf veins (E). Upon hatching, the larvae usually crawl to the underside of the leaf and begin to feed. These larvae are gregarious, mostly black to dark green, and covered with slime. The coating helps them adhere to the feeding surface. The last-instar larva (the female has six instars) takes no food and is a wandering stage (B). During the wandering period it sheds the slime covering, changes color, and loses about 20% of its weight. Because it cannot remain attached to the leaves without the slime, it drops to the ground. There, it burrows into the duff and loose soil, makes a cocoon, and goes into diapause. It takes about 38 days to complete all growth stages in each of the two generations.

The egg-laying female, who makes the feeding choice for her offspring, shows a distinct preference for pin oak, *Quercus palustris,* and scarlet oak, *Q. coccinea. Q. velutina,* the eastern black oak, is also an acceptable host. *C. quercuscoccineae* is found in most of New England south to North Carolina and west to Louisiana and Wisconsin. Many parasites and pathogens usually keep this insect under control. The wasp *Trichogramma minutum* Riley is a parasite of the egg, and a pathogenic single-cell flagellate protozoan, *Herptomonas* sp., kills late-stage larvae.

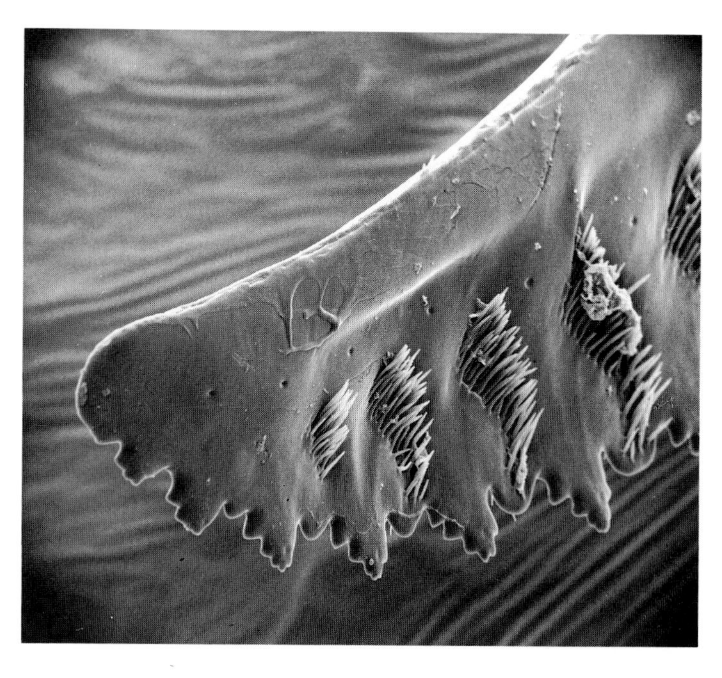

Figure 48. Scanning electron micrograph of the tip (apical portion) of one surface of the ovipositor of the scarlet oak sawfly showing the serration. About × 750. (Courtesy G. L. Nordin, Univ. of Kentucky.)

Figure 47. Scanning electron micrograph of the two-part ovipositor, or "saw," of the scarlet oak sawfly, *Caliroa quercuscoccineae.* About × 90. (Courtesy G. L. Nordin, Univ. of Kentucky.)

References: 193, 221, 410, 470, 536, 580, 712, 836

A. Leaves of the *Crataegus stricta* severely injured by pear sawfly larvae. Note larva at arrow point.
B. Larvae of the scarlet oak sawfly, *Caliroa quercuscoccineae,* on a pin oak leaf showing the skeletonized injury caused by larval feeding. Earlier instars are greenish black. (Courtesy G. L. Nordin and E. L. Johnson, Univ. of Kentucky.)
C. Mountain ash leaves injured by pear sawfly larvae. [Tenthredinidae]
D, G. A larva of the pear sawfly and skeletonized injury to a hawthorn leaf. The full-grown larva is about 13 mm long.
E. Chains of scarlet oak sawfly eggs, embedded in pin oak leaf parenchyma. (Courtesy G. L. Nordin and E. L. Johnson, Univ. of Kentucky.)
F, H. A pin oak defoliated by scarlet oak sawflies. Photos taken in early September in Ithaca, NY. [Tenthredinidae]

A
B
C
D
E
F
G

Rose Sawflies (Plate 58)

There are three injurious sluglike sawflies of rose: *Cladius difformis* (Panzer), commonly known as the bristly roseslug; *Endelomyia aethiops* (Fabricius), the roseslug; and *Allantus cinctus* Linnaeus, the curled rose sawfly.

The roseslug (B) can be found from British Columbia east to the Atlantic Coast. *Rosa* species are their only food plants. The female sawfly deposits her eggs singly in pockets along the edges of leaves; using the sawlike ovipositor (see Figures 47–48, accompanying Plate 57), she cuts a slit in the leaf and gouges out a pocket between the leaf surfaces for the egg. After hatching, the larvae skeletonize the upper surface of the leaf. When fully grown, about 13 mm long, the larva enters the ground and constructs its overwintering cell. In central New York all larvae are in the soil by early July and remain there until they pupate the following spring. There is but one generation each year.

The bristly roseslug is found over a large geographic range, though not in warm climates. When fully grown, the pale green larva measures about 16 mm and has many bristlelike hairs over its body. The young larva begins feeding as a skeletonizer from the underside of foliage but later chews large holes from leaves. There may be five to six generations each year.

The curled rose sawfly larva is 19 mm long when mature and is the largest of the three species. Though it rarely occurs in great numbers, even a few can seriously damage a rose garden. The larva injures plants by devouring leaves and boring into pruned twigs. It begins by skeletonizing leaves; later it devours the entire leaf except for the main vein. When ready to pupate, the larva bores into the pith of twigs, killing a portion of the stem and opening it to fungal infections. The larva is pastel green dorsally and is marked on the thorax and abdomen with white dots. The head is yellowish with black eye spots. In Maryland, Kentucky, and Ohio two generations are believed to occur each year.

Little is known about the natural enemies of the various sawfly slugs. With the exception of the pear sawfly slug, they rarely occur in large numbers, a fact that suggests that natural forces hold them in check. Slugs can easily be washed from trees and shrubs with a strong jet of water from a garden hose. They are not capable of crawling back onto the tree or bush.

Certain sawfly larvae that are borers feed in stems or tender shoots of trees and shrubs. Some of these are closely related to the pigeon tremex (Plate 238), as is the case of *Hartigia trimaculata* Say. There are few biological data about *H. trimaculata,* but it is assumed to produce one generation each year. The adult appears in early June and lays its eggs in punctures made in current-season rose or blackberry canes. Upon hatching, the larva may girdle the twig (D), causing the tip to wilt (E). As the larva matures, it enters the pith and feeds downward. There is usually one borer per cane. *H. trimaculata* may be found along the Atlantic Coast from Florida to Quebec and west to Louisiana and the Rocky Mountains and Mexico.

A related species, *Hartigia cressoni* (Kirby), occurs in California, Nevada, Oregon, and Montana. The adult horn-tailed wasp appears in April and May. The female inserts a single white egg into canes of rose or bramble fruits. Eggs hatch into very small larvae, which then spirally girdle the tips, causing them to wilt and die. Individuals of this species attain a length of 22–25 mm. *H. cressoni* also bores into the pith of the canes and sometimes gets into the larger roots.

References: 109, 193, 217, 536, 836

A. A rose slug near full development; about 13 mm long.
B. Rose leaves damaged by the roseslug, *Endelomyia aethiops.* [Tenthredinidae]
C. Curled rose sawfly larvae in resting and locomotion posture; about 19 mm long.
D. A nearly mature *Hartigia trimaculata* larva; about 21 mm when fully grown.
E. A wilted rose shoot girdled by the sawfly *H. trimaculata.*
F. A rose shoot dissected to show the girdling tunnel caused by an *H. trimaculata* larva.
G. An early-instar *H. trimaculata* excised from its tunnel.

A
B
C
D
E
F
G

Sawflies (Plates 59–60)

Selected cultivars of the black locust, *Robinia pseudoacacia*, a common tree in farm wood lots in the eastern half of the United States, are now offered in the trade as ornamental shade trees. Other small *Robinia* trees or multiple-stem shrubs such as *Robinia hispida*, a native to the southeastern United States, have also been given a place in ornamental plantings. *Robinia* is subject to several defoliators, leaf miners, and webbers (Plates 80, 87, 114). The rose acacia, *R. hispida*, and black locust occasionally harbor sawfly larvae. The sawfly *Nematus tibialis* Newman, for example, is associated with both of the named *Robinia* species. *N. hispidae* Smith was first described in 1984 and is associated only with the rose acacia.

Adult *N. tibialis* appear in early May in western New York and, like many other sawflies that feed on deciduous woody plants, insert their eggs into the interveinal leaf parenchyma. The newly hatched larva is light green and chews holes in the interior of leaflets and later devours the entire leaf except for the midvein. As larvae mature, they become edge feeders (C). When fully grown, they are about 15 mm long. The larvae are believed to overwinter in the soil. Comstock reported two and possibly three generations per year; we have recognized only one. No males have been found. Their geographic range is believed to extend from New York to South Carolina.

N. hispidae and *N. tibialis* feed on rose acacia at about the same time of the year. Eggs of *N. hispidae* are deposited in the leaflet midrib. Newly hatched larvae are black, and larvae mature about 2 weeks later than those of *N. tibialis*. Little is known about the biology and behavior of either species.

Three sawflies are known to feed on azaleas in the eastern United States, two in the family Tenthredinidae and one in the Argidae. *Amauronematus azaleae* (Marlatt), a tenthredinid, was first reported in New Hampshire in May 1895, feeding on the edges of young azalea leaves. Some of them were collected as larvae and reared; adults emerged the following March. Larvae have since been collected from New York, Pennsylvania, and Washington, DC, and are likely to be found elsewhere in New England, parts of eastern Canada, and the central Atlantic states. Adult females are predominantly black, with whitish orbits and face below the antennae. The larvae are green, like an azalea leaf, with an amber head, black eyespot, and black tubercles laterally below the abdominal spiracles (D). They are about 10 mm long. The annulets on abdominal segments 2 and 3 bear setae. Larvae feed predominantly on mollis hybrid azaleas. Larval development is complete by the end of June. There is one generation per year.

N. lipovskyi Smith was described in 1974 from adults reared from larvae collected in 1923 on swamp azalea (*Rhododendron viscosum*) in Massachusetts. Adult females, 4.5–5.5 mm long, are mostly pale orange with black antennae and numerous black spots or areas on the thoracic and abdominal segments. Larvae, about 10 mm long, are green with setae on annulets of abdominal segments 1–8. Early-instar larvae are gregarious and feed on the edges of foliage. Adults have been collected from Maine to Virginia, and in Alabama. Adults appear in April in Virginia and May in New England. Larval feeding is predominantly in late April and May in Virginia and June in New England. One generation occurs each year. Deciduous flame azalea (*R. calendulaceum*), including such cultivars as Exbury, is attacked. Expect this sawfly larva to feed on most mollis hybrid azaleas. Other harmful species of *Nematus* may be found on willows and poplars, particularly in prairie states.

Arge clavicornis (Fabricius) has been reported from Connecticut and Massachusetts. The adult females are black, except that each tibia and part of each tarsus is whitish. The larvae are green with an amber head, black eyespot, and a longitudinal black stripe on the head above the clypeus. A lobe is conspicuous next to each tarsal claw. Abdominal segments 1–8 each has three annulets. Annulets 2 and 3 have black tubercles bearing setae. Adults of this species are active in July; larvae are present during August and September. Eggs are inserted in the leaf in a row around the leaf edge.

The larva of the blackheaded ash sawfly, *Tethida cordigera* (Beauvois), is an occasional defoliator of red and white ash (F), particularly of shade trees. It is distributed throughout much of New England, south to Florida, and west to Texas and California. Larvae are without wax or hair. In central New York they are found in late June. Fully grown larvae are about 18 mm long.

A related species, the brownheaded ash sawfly, *Tomostethus multicinctus* (Rohwer), is found in the same geographic area and causes the same kind of injury. It looks much like the blackheaded ash sawfly but has a brown head.

References: 141, 193, 217, 682, 713

A. A partially defoliated branch of rose acacia damaged by *Nematus tibialis*.

B. A second-instar *N. hispidae* on the hairy rachis of a rose acacia leaf.

C. A pair of *N. tibialis* larvae feeding on the edge of a rose acacia leaflet.

D. A larva of the sawfly *Amauronematus azaleae* Marlatt, an occasional pest of azalea in the Northeast.

E. The green midveins of azalea leaves are positive symptoms of injury caused by the sawfly larva *A. azaleae*.

F. A nearly grown blackheaded ash sawfly larva, *Tethida cordigera*. [Tenthredinidae]

A
B
C
D
E
F

Macrophya punctumalbum (Linnaeus), a European sawfly, was first found in the United States in 1979 but has been known in part of Ontario since 1932 and in British Columbia since 1934. It is now probably more widespread than records show. *M. punctumalbum* feeds on privet, *Ligustrum ovalifolium, L. vulgare,* and species of *Fraxinus* and *Syringa*. Our experience with this insect is limited to California privet and green ash, *Fraxinus pennsylvanica*.

In New York adults (predominantly females; males are normally scarce) appear on privet foliage about mid-May. They feed on the foliage, producing characteristic rasping marks and small rectangular holes (C). They begin to lay eggs in early June and place them either singly or in chains of two to seven under the upper epidermis of leaves. The embedded eggs are usually located at the apex or along the periphery of the leaf. First-instar larvae appear in early June. Females apparently oviposit over an extended period, because first-instar larvae can be found until late June. Most larvae mature by late July and drop to the soil, where they presumably burrow to make earthen cells and prepare cocoons for diapause. One generation is produced each year.

Early-instar larvae excavate circular holes 1–3 mm in diameter in the interior of the leaves. Young larvae move into a curled (head-to-tail) posture to feed. Mature larvae are capable of consuming the entire leaf but avoid the midvein (B).

Some control can be attained on privet by trimming it late in the spring before the larvae can mature. Trimming or pruning dislodges larvae from their feeding sites. If they drop to the ground, they cannot crawl over a long vertical distance to get back onto foliage.

Oak foliage supports several species of sawflies in the genus *Periclista*. The sawfly illustrated in panel E is believed to be *Periclista media* (Norton). The larvae are solitary feeders on white oak and devour all or portions of the leaves. Because they are rarely, if ever, found in large numbers, they are not of major economic importance. One generation occurs each year. Larvae are found in May in the vicinity of Washington, DC, and in June in central New York. Fully grown larvae are about 26 mm long. Little is known about the biology of *P. media*.

Pamphilius phyllisae Middlekauff is a web-spinning sawfly (not illustrated) that confines its feeding to northern red oak, *Quercus rubra,* and is therefore believed to be distributed throughout the natural range of this species. It, too, is a defoliator. A serious outbreak occurred in Pennsylvania forests in 1964. Its life cycle requires 1, 2, or more years, depending on unknown ecological conditions. Adult emergence begins in Pennsylvania about mid-May and continues to early June. The eggs, laid fully exposed along the oak midrib on the underside of the leaf, are light colored and sausage shaped. Newly hatched larvae crawl to the edge of the leaf while laying down a bed of silk. Each aligns itself with the edge and, using silk, rolls the edge of the leaf under, covering itself in the process. It is a solitary insect and feeds within the leaf roll. After each molt, it makes another roll, usually on the same leaf. Because one leaf contains enough tissue to feed four or five larvae, the form of the rolls becomes less distinctive as the feeding area enlarges. Fully grown larvae are about 25 mm long; they have a pale green body and dark, heavily pigmented head. They drop to the ground after an arboreal larval stage of about 19 days and burrow into the soil, where they diapause.

The striped alder sawfly, *Hemichroa crocea* (Geoffroy), is another European insect. It defoliates various species of alder (F). In the northwest it feeds on filbert. It now is transcontinental across southern Canada and the northern United States.

Adult striped alder sawflies emerge from the soil in mid to late May. In early June they lay eggs in rows along the midrib on the underside of alder leaves. The female cuts small slits in the epidermis of the midrib and forces an egg into each slit, pushing it under the epidermis. Upon hatching, larvae feed communally on the leaf until only the midrib remains. First-instar larvae are white, but as they grow through their several instars, they become more heavily pigmented. Larvae are fully grown in about 20 days (F). First-generation larvae enter the soil for a short diapause about mid-July. Second-generation adults begin to lay eggs in late July, and by mid-September the larvae enter the soil, make an earthen cell, and cover themselves with a brown papery cocoon. They overwinter in this state. Males are rare; parthenogenesis is the most common method of reproduction. *Bessa selecta* (Meigen), a tachinid fly, is an effective parasite.

The butternut woollyworm, *Eriocampa juglandis* (Fitch), is a defoliator of black walnut, butternut, and hickory. As is the case of most free-living sawflies, the overwintering stage occurs in the soil in a papery cocoon. Adults emerge in the spring and begin to lay eggs in early June (Alabama). Eggs are laid in the midrib of leaflets. The larvae, which feed gregariously, are unusual because of a white woolly substance that covers their bodies (G). Under the "wool," which rubs off easily, is a smooth green caterpillarlike body with a white head (see also Plate 55). They pupate in the soil in cocoons covered with grainy soil particles that are attached to the cocoon surface. There is one generation per year.

The butternut woollyworm is presumed to occur throughout the range of *Juglans nigra,* black walnut.

Eriocampa ovata (Linnaeus) (not illustrated) is sometimes called the alder woolly sawfly and is closely akin to and looks much like the butternut woollyworm. It is another European species now widespread in eastern Canada, the northeastern United States, and the Pacific Northwest. A defoliator of *Alnus* species, it feeds on the soft leaf tissue between the lateral veins. The leaf skeleton of major veins is its trademark. It also characteristically ignores the foliage on the leader of its host trees.

In British Columbia adult sawflies appear in early May and begin to lay eggs by mid-May. They insert their eggs into the midrib in rows on the upper side of the leaf. Eggs hatch in a few days and the larvae feed on the interveinal tissue. All the larval stages are covered by a white woollike material secreted from a gland in the integument. The white molt "skins," attached to twigs and branches, look much like bird droppings. Larval development requires about 18 days.

There are two generations of *E. ovata* each year. Overwintering takes place in the soil, where the larva spins a brown silk cocoon for protection. Diapause occurs in the prepupal stage.

The elm sawfly, *Cimbex americana* Leach, (not illustrated) is the largest of the North American sawflies and occurs throughout the United States. The larvae feed primarily on the foliage of elm and willow but are occasionally found on alder, basswood, birch, maple, and poplar. The full-grown larva is about 45 mm long, wrinkled, and yellowish to greenish white, with a median black stripe and many whitish spots that make its "skin" appear grainy. When at rest it is usually coiled on a leaf. Adults may kill shoots by gnawing on the bark especially in the tops of trees. There is one generation each year.

References: 72, 199, 208, 374, 383, 400, 750

A. Nearly full-grown *Macrophya punctumalbum* larvae on California privet; about 16 mm long. Note nature of injury by older larvae.

B. California privet injury caused by early-instar larvae of *M. punctumalbum*.

C. Injury to privet foliage by adult *M. punctumalbum*. The black varnishlike spots are fecal material.

D. An adult female *M. punctumalbum* on a privet leaf. [Tenthredinidae]

E. A fully grown sawfly larva, probably of the genus *Periclista,* on white oak. Note how the tail wraps around the leaf. This is a typical position for many free-living sawfly larvae.

F. The gregarious striped alder sawfly is shown as an edge feeder on an alder leaf. Full-grown larvae are about 20 mm long.

G. The butternut woollyworm on walnut. This specimen was collected in July on Long Island.

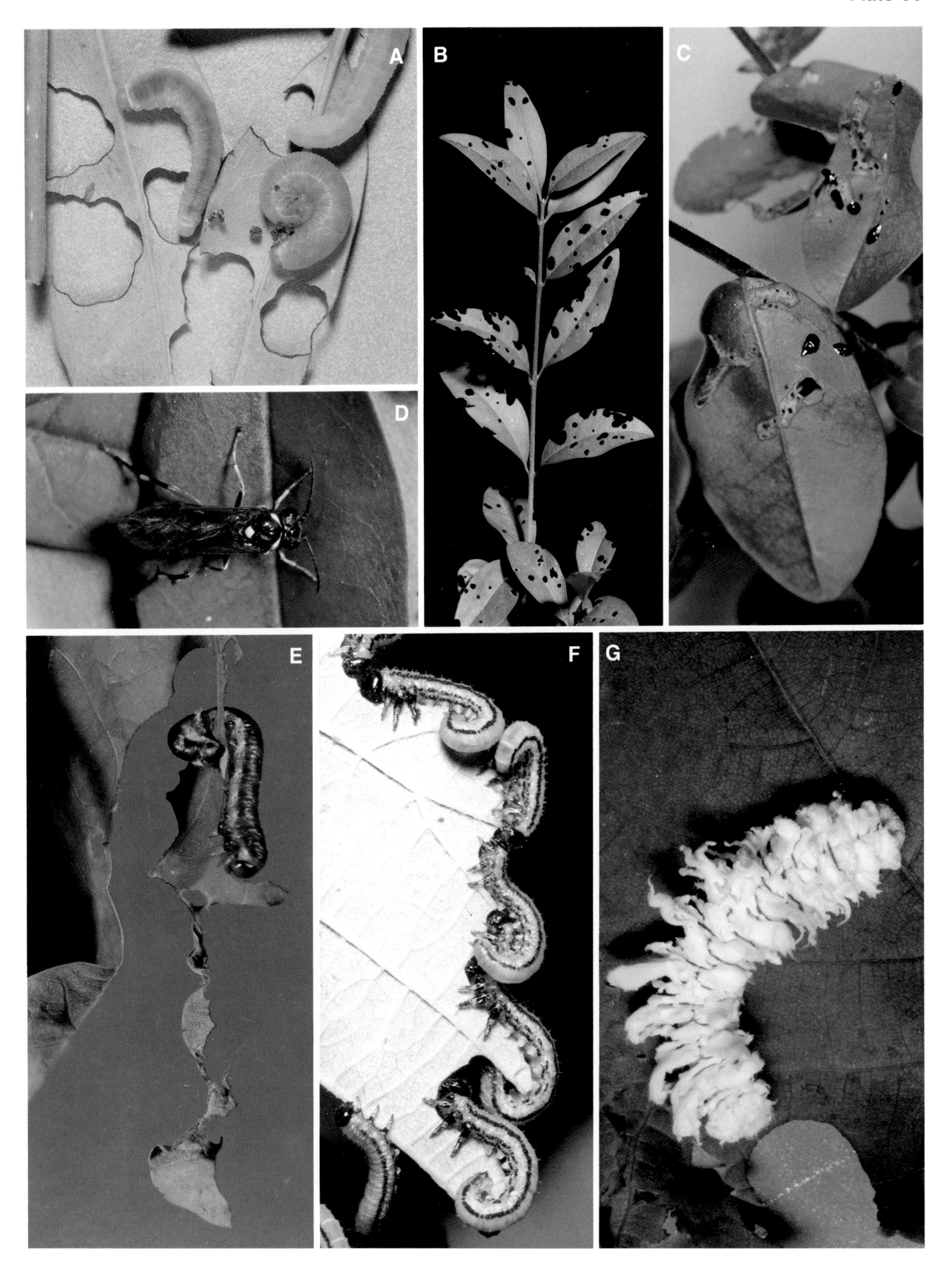
A
B
C
D
E
F
G

Gypsy Moth (Plates 61–62)

Gypsy moth, *Lymantria dispar* (Linnaeus), is a familiar name, especially to northeasterners. Probably no pest of trees has received more publicity or cost more to control.

In 1870 this announcement was made by a well-known entomologist: "Only a year ago the larvae of a certain owlet moth, *Porthetria* [*Lymantria*] *dispar,* were accidentally introduced by a Massachusetts entomologist into New England, where it is spreading with great rapidity. It happened this way. Mr. Trouvelot, then living at 27 Myrtle Street in Medford, Massachusetts, was in search of a silk moth that would survive in America. He brought eggs of the gypsy moth to his home where some larvae or possibly adult moths escaped" (256). The gypsy moth was known to be a serious pest of forest and shade trees in Europe, and Trouvelot apparently knew this because he informed local authorities of the moths' escape.

Nothing was done, and within 12 years the insect became a serious nuisance to those living on Myrtle Street. The local residents assumed it to be a native pest. The personal testimonies of residents of Myrtle Street in the early 1880s were a preview of things to come. One woman stated: "I went to the front door and sure enough the street was black with them [caterpillars]." Another resident who was out of town for three days in June of 1889 related: "When I went away the trees in our yard were in splendid condition and there was not a sign of insect devastation on them. When I returned there was scarcely a leaf upon the trees." Another resident testified to having collected 4 quarts of caterpillars from one branch of his apple tree. Another said that the caterpillars covered one side of his house so thickly that it was impossible to tell the color of the house. Many people disliked going outdoors because the caterpillars dropped from the trees onto them. Streets and sidewalks were slippery in places because of the crushed caterpillars.

Placed in an environment devoid of its natural enemies, the gypsy moth multiplied and spread so rapidly that today it is a pest of trees in over 518,000 sq km (200,000 sq miles) of northeastern forestland. Despite strict quarantine regulations and chemical eradication and control programs, which have undoubtedly helped control the gypsy moth's activities, it continues to occupy new territory. (Automobiles and camping trailers often transport the gypsy moth's eggs, and sometimes other stages of the insect.) With the present ban on the use of persistent insecticides such as DDT, the insect appears to be spreading even more rapidly. It has been found with increasing frequency in the Carolinas, Kentucky, Michigan, Wisconsin, Colorado, Idaho, California, British Columbia, and Ontario. In Oregon (1984) gypsy moths defoliated Douglas-fir. In 1991 moths of the Asian race were discovered in Tacoma, Washington, and Portland, Oregon. Whether the species will spread throughout all of the United States remains to be seen.

Adult gypsy moths are rather large. The wingspan of the female is about 5 cm. The male is dark brown (Plate 62E), and the female nearly white, with wavy, blackish bands across the forewings (Plate 62D, E). Eggs are deposited on tree trunks, branches, fences, buildings, or other suitable places in masses of 100–600 or more, and are covered with a dense mass of tan or buff-colored hairs (D). The individual eggs are pelletlike and range from brown to black.

The larvae hatch from early April to late May, the peak hatching period coinciding with the flowering of shadbush. The tiny larva often remains on the egg mass for several days before climbing the tree to feed. It then spins a silken thread, suspends itself from a leaf, and is swayed back and forth by light breezes. If the wind velocity is great enough, it may become airborne. In a wooded area the wind may carry a larva several hundred yards. In open terrain larvae are reported to be transported several miles. The larva at this stage is hairy and basically dark. When fully grown (Plate 62G), it may be up to 5.5 cm long.

As the caterpillar (larva) matures, its feeding habits change. It feeds at night and descends from the tree to take refuge in shady places. On a heavily infested tree it may continue to feed throughout the daylight hours.

Larvae feed on a number of hosts. Preferred are leaves of apple, alder, basswood, hawthorn, oaks, some poplars, and willows. Less preferred hosts include elm, black gum, hickories, maples, and sassafras. A few larvae from a massive population may occasionally feed on beech, hemlock, white cedar, pines, and spruce. Ignored or only rarely fed on are ash, balsam fir, butternut, black walnut, catalpa, redcedar, dogwood, holly, locust, sycamore, and tuliptree.

Oaks suffer from the gypsy moth's attack more than other species. Most deciduous trees can withstand one or two consecutive years of defoliation before severe decline or death occurs. Conifers will die after one complete defoliation.

The larval stage lasts about 7 weeks. After it has completed feeding, the larva finds a sheltered place and pupates in a brownish black pupal case. Pupal cases are often found on the bole of the host tree in clusters accompanied by molted "skins" of the last caterpillar instar.

Moths begin to emerge about the middle of July, males appearing several days earlier than the females. The female does not fly; she crawls to an elevated place and emits a liquid substance called a sex attractant (pheromone), which volatilizes and is carried in the air. With the proper wind conditions, the odor will be detectable and attractive to a male moth for a distance of about 1.6 km (1 mile). The sex attractant has been synthesized, and the synthetic pheromone is used as a survey tool to determine the presence of male moths. After mating and depositing eggs, the adults soon die without feeding.

Control has been effected by human manipulations as well as through a number of natural factors. A bacterial pathogen called *Bacillus thuringiensis* kills larvae. Now prepared commercially, it is effective against *L. dispar* and certain other caterpillar species. Important parasites and predators from Europe and Asia have been reared and released, with some success. The major parasites are *Ooencyrtus kuvanae* Howard, an egg parasite; *Blepharipa pratensis* (Meigen), a fly parasite of the caterpillar; and *Calosoma sycophanta* Linnaeus, a large and colorful predatory beetle. The wasp shown in Figure 60 (accompanying Plate 83) is a parasite of gypsy moth larvae.

The fungus *Entomophaga aulicae,* a control agent for gypsy moth larvae in the Orient, has good potential for use as a microbial insecticide in North America. Another fungal insect pathogen, *E. maimaiga,* has been a factor in gypsy moth population crashes in the Northeast. In general, transmission of fungal pathogens to insects and subsequent population control require long periods of moist weather. About 10% of natural populations of gypsy moths harbor a latent virus that becomes lethal when activated. Infected gypsy moths pass the virus on to their offspring. If the virus could be activated through human manipulation, injurious populations could be controlled.

In outbreak areas there have been numerous reports of human allergic respiratory and skin reactions from an abundance of airborne wing scales and other tiny fragments from gypsy moth larvae, adults, and molted larval skeletons.

References: 181, 193, 256, 573

A. A hillside in the Hudson Valley of New York. The trees at the top of photograph were defoliated by the gypsy moth, *Lymantria dispar.*

B. Egg masses on the underside of an oak limb.

C. Gypsy moth larvae on partly devoured oak leaves.

D. Close-up of egg masses. The small pinholes mark exit holes of egg parasites. [Lymantriidae]

E. A pupa, an empty pupal case, and an egg mass.

A
B
C
D
E

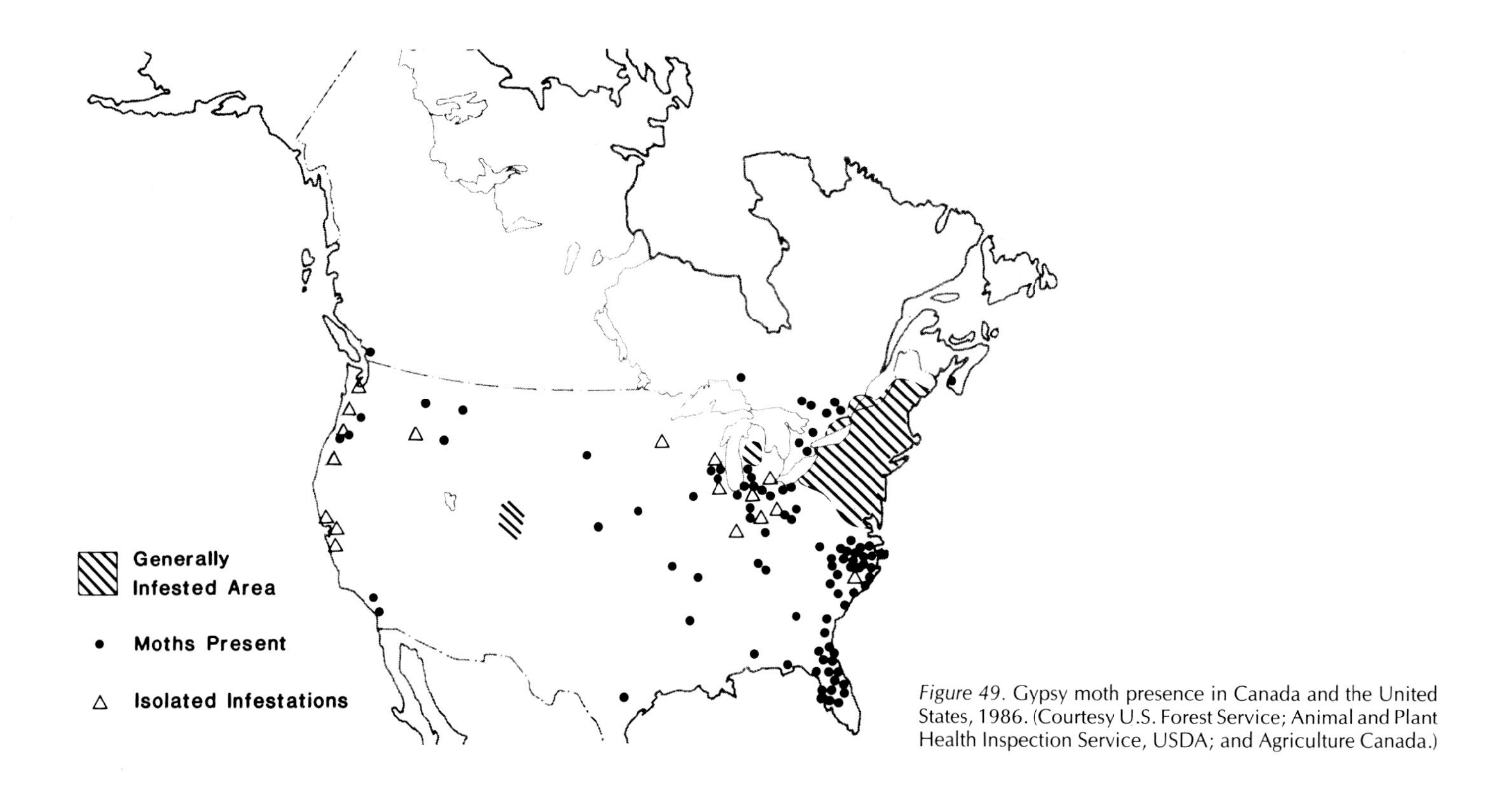

Figure 49. Gypsy moth presence in Canada and the United States, 1986. (Courtesy U.S. Forest Service; Animal and Plant Health Inspection Service, USDA; and Agriculture Canada.)

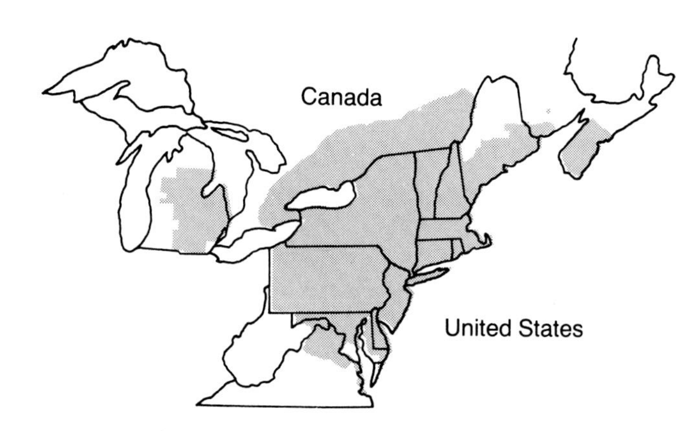

Figure 50. Areas of general infestation of gypsy moth, 1986. (Courtesy Animal and Plant Health Inspection Service, USDA; and Agriculture Canada.)

A. Pupae and egg masses of the gypsy moth webbed together on a hemlock twig.
B. A mass of pupae. One dead larva hangs alone, killed by a viral disease called wilt.
C. A twig showing several larvae killed by the wilt disease.
D. Female gypsy moths laying eggs.
E. Male (dark) and female gypsy moth.
F. A roadway lined with oak trees completely defoliated by gypsy moth larvae.
G. A fully grown gypsy moth larva.
H. Front view of a gypsy moth larva.

A
B
C
D
E
F
G
H

Cankerworms (Plate 63)

The fall cankerworm, *Alsophila pometaria* (Harris), and the spring cankerworm, *Paleacrita vernata* (Peck), are two of the more common and injurious of the many species of loopers, spanworms, and inchworms that attack ornamental trees. The common names of these two insects indicate the respective seasons during which their eggs are laid. The fall cankerworm lays its eggs in late November and early December, and the spring cankerworm in late February and early March. The females of both species are wingless. Both are indigenous to North America and occur from Nova Scotia and southern Canada to the Carolinas, Texas, Colorado, New Mexico, California, Montana, and Manitoba. They have a wide host range but are important pests primarily of elm, apple, oak, linden, and beech.

Eggs of both species hatch as soon as buds begin to open in the spring, and they occur together in mixed populations. Young larvae feed on buds and unfolding leaves. There are both green and dark larvae of each species. The larvae often drop from the leaves on silk strands of their own making, from which they are often detached by the wind and blown considerable distances. Larvae devour all but the midrib of the leaf and often defoliate entire trees (B). Four to 5 weeks after egg hatch larvae enter the soil to pupate, usually in early June.

The fall cankerworm larva can be distinguished from the spring cankerworm larva because the former has three pairs of prolegs (including a small pair on the fifth abdominal segment) whereas the spring cankerworm has two pairs. For detailed information about the spring cankerworm, see Plate 64.

The eggs of the fall cankerworm are laid in carefully aligned masses of about 100 on small twigs (F, G). The spring cankerworm lays its spindle-shaped eggs in clusters of about 100 in the crevices of rough bark on larger limbs and trunk.

Because the wingless females (E) must crawl from the ground into the trees to lay their eggs, attempts have been made to control their populations by banding tree trunks with a sticky material such as tanglefoot. The surest control is achieved by spraying the tree with a residual contact insecticide soon after the larvae first become active and after the leaves have expanded. Chemicals sprayed on rapidly growing leaves will not provide sufficient coverage and adequate control. Bacterial and viral diseases help to regulate infestations naturally. *Bacillus thuringiensis* is a naturally occurring spore-forming bacterium that causes a fatal disease in cankerworms. Commercial formulations of the bacterium, under several trade names, are available from garden stores and have proved effective in the control of cankerworms.

See Plates 5 and 64 for illustrations of other loopers.

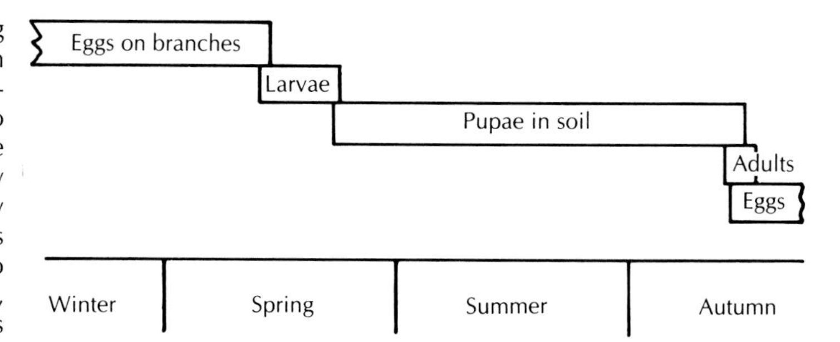

Figure 51. Seasonal development of the fall cankerworm in Illinois. (Courtesy J E. Appleby, Illinois Natural History Survey.)

References: 193, 359, 364, 415

A, C, D. Color variations of the fall cankerworm, *Alsophila pometaria*. [Geometridae]

B. The fall cankerworms seen here have nearly defoliated a hickory twig.

E. Adult female cankerworms; note the absence of wings.

F, G. Fall cankerworm eggs; in F some have hatched.

A
B
C
D
E
F
G

Loopers (Plate 64)

These insects are placed in the family Geometridae, which means "earth measurer." Those shown here represent a large group of leaf-feeding caterpillars, some of which are among our most destructive defoliators. Some entomologists consider this family as the most destructive of the foliage-feeding insects. The larvae are distinguished by a reduced number of abdominal prolegs and by their peculiar method of locomotion. They loop the body, then stretch forward either by crawling with the thoracic legs or by raising the front legs and straightening out the body (C, H). This characteristic has prompted a number of common names such as geometers, inchworms, loopers, measuring worms, and spanworms. Some 1200 species are found in the United States and Canada. The economically important species appear in cycles. They are present for 2 or 3 years in large numbers; then their numbers suddenly drop, and they seemingly disappear for 5–8 years, only to reappear.

Mahonia has a tough prickly leaf that would seem to be unpalatable to any animal, but for the moth *Coryphista meadii* (Packard), which in its larval stage feeds exclusively on *Mahonia* and *Berberis* cultivars, tough fibrous leaves apparently are no obstacle. Where larvae are abundant, the cultivars of these genera are defoliated by midsummer.

There are two races of *C. meadii*. The dark form is dull brownish gray and has several narrow, wavy black bands across its wings. The other form is lighter.

Both have a wingspan of about 38 mm, and both overwinter as pupae. Adults emerge in the spring—about mid-May in New York. Eggs are usually laid singly on either side of the leaf. The female produces an average of 125 eggs; these she lays at night. Eggs hatch in 3–4 days, and the larvae immediately start to feed on the leaves. At first they are skeletonizers and leave behind an exposed network of veins (B). Later they devour entire leaves. These caterpillars move in a looping motion and have a double pair of prolegs (E). They feed for about 14 days, and when fully grown, about 24 mm long, they crawl to the soil and burrow in a short distance. There, they pupate and overwinter in this condition. Three generations occur each year in New York, as well as in Idaho. In Ohio the caterpillars are active until October. The *C. meadii* larvae are sometimes called barberry loopers.

C. meadii occurs sporadically throughout much of the United States, eastern Canada, and British Columbia.

Several fly and wasp parasites undoubtedly are major factors in holding *C. meadii* in check. Birds devour many of the pupae by digging them from the ground.

The linden looper (C), *Erannis tiliaria* (Harris), periodically defoliates many deciduous trees, including linden, apple, birch, elm, hickory, maple, and oak. It is a native insect found in the northern half of the United States from the Atlantic Coast to the Rocky Mountains and in Quebec, Ontario, and the Northwest Territories of Canada. The linden looper frequently occupies the same kinds of trees as the cankerworm (Plate 63).

Its eggs are laid three or four to a cluster under loose bark on the trunk and larger limbs. They hatch at about the time buds open in the spring. The caterpillars feed on the foliage for about 1 month. Although the color patterns of the individual linden loopers are essentially the same, the intensity of their coloring varies. They possess two pairs of prolegs and move with a looping motion. When fully grown they measure about 37 mm long and are quite active. They crawl to the ground, where they tunnel an inch or more into the soil to pupate. All this activity occurs before July. Moths emerge from the soil during October to December, at which time they lay eggs. The wingless female can be identified by the two rows of large black spots on the back; they measure about 13 mm long. They must crawl up the tree to lay eggs. One generation occurs each year.

The spring cankerworm, *Paleacrita vernata* (Peck) is another spring-feeding caterpillar (F) with a geographic distribution almost identical to that of the elm spanworm. They get their name because females lay their eggs in the spring, whereas the fall cankerworm (Plate 63) female lays her eggs in the fall. Host trees include elm, hickory, maple, ash, beech, basswood, oak, and apple.

Young larvae feed by making holes, often described as shot holes, from the underside of leaves. As the larvae grow, they enlarge the small hole; finally they devour all the interveinal tissues, avoiding the major leaf veins. The tattered appearance of leaves illustrated in Plate 63B is similar to the injury caused by the spring cankerworm.

The insect overwinters as a larva in an earthen cell under a host tree; it forms no cocoon. Adults emerge shortly after frost leaves the ground. Female moths are wingless, resembling those illustrated in Plate 63E but much more hairy. They crawl up the trunk of a tree to deposit their eggs in loose clusters of about 100, under loose bark scales or in deep crevices. Sometimes they lay eggs on large branches. The eggs hatch to coincide with the opening of leaf buds. It takes about 30 days to complete the larval stage (F). When fully grown, larvae drop to the ground, enter the soil, and prepare for hibernation. There is one generation per year.

As is the case with many species of cankerworms, dispersion takes place when the larvae are in the first or second instar. They attach a silk thread to a leaf or twig and drop, suspended by the thread. Wind currents can blow them for many meters.

The elm spanworm, *Ennomos subsignarius* (Hübner), occurs throughout the eastern half of the United States and Canada and in spots west of the Mississippi River. Its common name is misleading, for it is a general feeder on a wide range of deciduous trees. It is found most often on hickory, oak, ash, red maple (*Acer rubrum*), elm, basswood, beech, and *Aesculus* species.

Injury by the elm spanworm takes the form of perforated leaves, often described as shot holes, made by young larvae as they feed from the lower surface of leaves. Young larvae require juvenile leaves. As they mature, larvae eat most of the leaf but avoid the midrib and

---

A. Ragged leaves on a *Mahonia* bush. Damage was caused by *Coryphista meadii,* sometimes called the barberry looper.
B. Two larval stages of *C. meadii.*
C. Two color forms of the linden looper, *Erannis tiliaria*; 37 mm long. [Geometridae]
D, E. Fully grown larvae of *C. meadii.* Note the looping posture and the double pair of prolegs.
F. Spring cankerworm; about 22 mm long. (Courtesy New York State Agricultural Experiment Station, Geneva.)
G. Chainspotted geometer; about 50 mm long. (Courtesy New York State Agricultural Experiment Station, Geneva.)
H. Fall cankerworm caterpillar, for comparison with the spring cankerworm. (Also see Plate 63.)

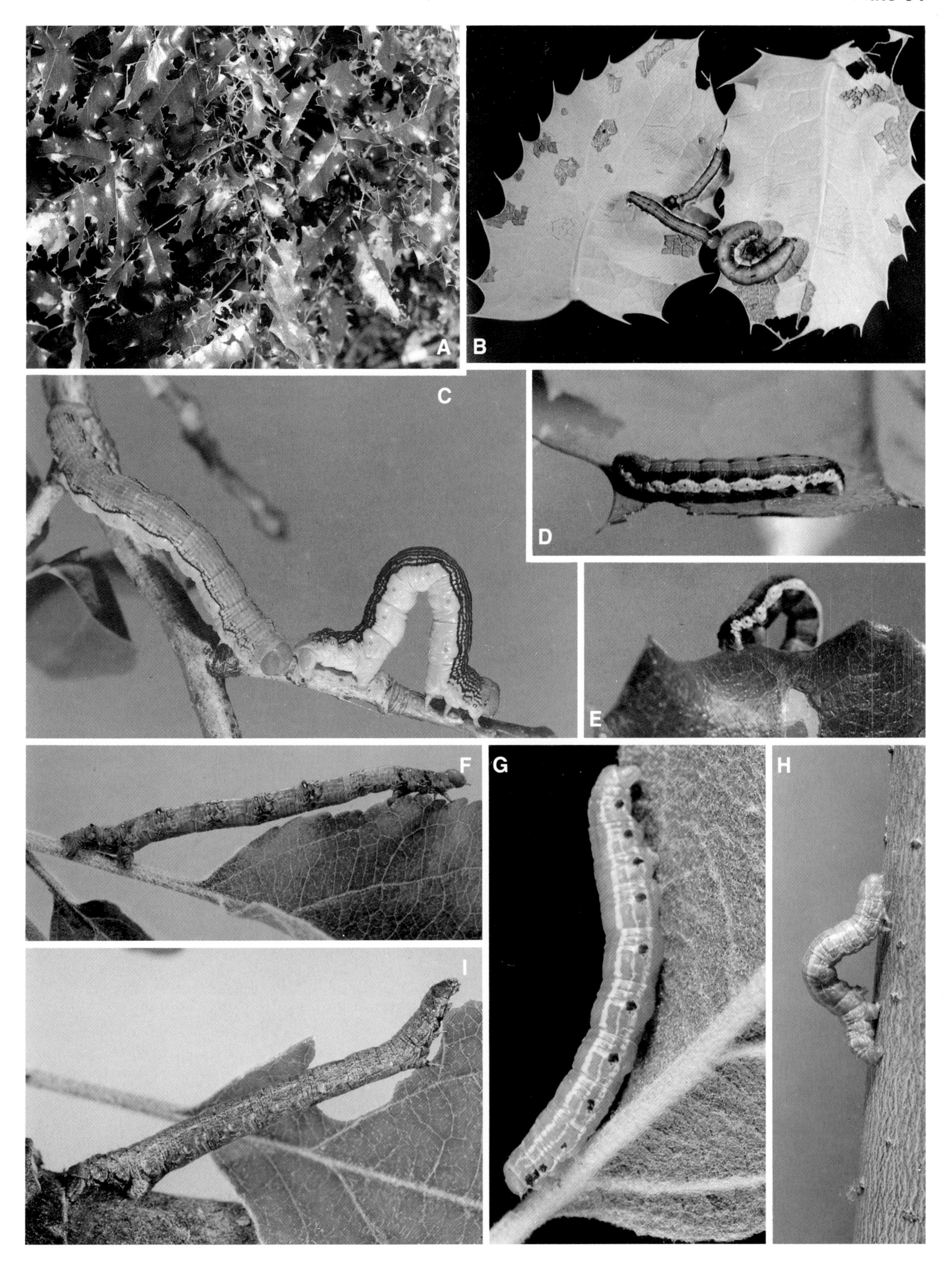
A
B
C
D
E
F
G
H
I

petiole. Larval development requires about 29 days, followed by pupation in netlike cocoons attached to partly eaten leaves. Snow white moths appear in August and lay their eggs in clutches on the underside of twigs. The eggs overwinter.

Two important egg parasites, *Ooencyrtus* and *Telenomus*, appear in great numbers when moths are abundant.

The chainspotted geometer, *Cingilia catenaria* (Drury), is found from the Maritime Provinces south to Pennsylvania, and is recorded in spots all the way to Colorado and Alberta. They are solitary feeders as larvae (G) and occur during midsummer. They feed on both deciduous and coniferous trees. In the United States they are more likely to be found on deciduous trees such as birch, willow, poplar, oak, cherry, and blueberry; in Canada they are often associated with conifers such as spruce, balsam fir, and larch. They overwinter as eggs. A multicapsid nuclear polyhedrosis virus (MNPV) has been reported to naturally infect populations of the chainspotted geometer in Quebec and Ontario.

When American holly leaves show deep marginal notching, the larva of the holly looper, *Thysanopyga intractata* (Walker), is the most likely cause. The larva has anal prolegs and one pair of abdominal prolegs. When fully grown, it is about 2 cm long and light green. It pupates in the soil under the host plant. The moths are migratory and move north from the Gulf states in the spring. The holly looper may be found anywhere within the natural range of American holly. Damage may occur every year, although complete defoliation is rare. Host plants of record also include Chinese and English holly and yaupon (*Ilex vomitoria*).

The spear-marked black moth, *Rheumaptera hastata* (Linneaus), is a northern species that flies during the day. The moths have black wings that span about 25 mm and are interspersed with many white areas. When abundant, they are conspicuous as they flit around roadside drainage ditches, culverts, and puddles of water near their favored host, paper birch. The larvae fold leaves and feed by skeletonizing the leaves within the protection of the folds. They may be found from late June through August. The full-grown caterpillars are about 25 mm long and dark brown to black, hence the common name black looper. Although considered a northern species, the spear-marked black moth can be found in the Appalachian and Rocky mountains as far south as North Carolina and New Mexico.

See Table 9 for information about other economically important geometrids.

***Table 9.* Common loopers of North America**

Name	Hosts	Distribution
Bruce spanworm, *Operophtera bruceata*	Maple, beech, aspen	Across Canada, WI eastward
Cherry scallop shell moth, *Hydria prunivorata*	Cherry	Maritime Provinces to NC west to WI
Half-wing geometer, *Phigalia titea*	Various hardwoods, e.g., oak, hickory, maple, blueberry	Southern Canada to Appalachians
H. undulata	Azaleas, spirea, willow, poplar	Northern states and adjacent Canada
Spear-marked black moth, *Rheumaptera hastata*	Birch, alder, willow, rose, hazelnut, poplar	Range of paper birch and quaking aspen
Winter moth, *O. brumata*	Oak, apple	Eastern provinces, B.C., northeast and northwest U.S.

References: 9, 81, 94, 98, 193, 207, 355, 364, 422, 510, 567, 595, 634, 660, 683

California Oakworm (Plate 65)

The California oakworm (or oak moth), *Phryganidia californica* Packard, is not known to occur outside California. It is, however, the most important insect affecting oak in that state. The cyclic nature of its population levels causes it to be very common in some years and virtually absent in others. Trees are seldom, if ever, killed as a result of defoliation by the oakworm, but the landscape is marred and the shade-producing value of affected trees is markedly reduced. Falling excrement, or frass, produced by the feeding larvae and by the young larvae as they hang on silken threads are additional nuisances where pools and patios are situated beneath oaks. Nearly all species of western oaks are subject to attack by this insect.

The insect passes through two complete generations each year in all but the southern parts of its range, where a third annual generation is reported. In the San Francisco Bay area it overwinters as an egg or young larva. Larval development proceeds rapidly with the coming of warm temperatures in the spring. When young, the larvae feed by removing only one surface from the leaf (C). The injured parts of these leaves then turn brown. When the larvae are about half grown, they begin to feed completely through the leaves, excising large sections from them and collectively defoliating the tree.

At maturity the larva is about 31 mm long. Pupation occurs on leaves, limbs, or trunks of oak trees, other trees nearby, or on other objects such as fences and houses. Moths emerge from pupae in May, June, or July. Adults fly in greatest numbers at dusk. The cycle is then repeated, and the subsequent generation of moths occurs in October and November. These moths lay eggs on leaves, branches, and limbs.

When infestations are heavy, either of the two generations is capable of severely defoliating oaks over a wide area. Because most of the eggs are deposited on leaves and because the leaves of deciduous oaks drop late in the fall, carrying the eggs or newly hatched larvae away with them, the deciduous oaks normally are spared spring infestations. Evergreen oaks, on the other hand, commonly are infested by both generations of the insect.

The California oakworm is attacked by a number of parasites and predators. Also, disease-causing microorganisms sometimes cause the death of large numbers of the feeding caterpillars. The bacterium *Bacillus thuringiensis* Berliner is an effective control agent when sprayed to coincide with the presence of early third-instar larvae.

References: 85, 93, 217, 340

A. An adult female moth of the California oakworm, *Phryganidia californica*. [Dioptidae]
B. Eggs on underside of an oak leaf. Newly laid eggs do not have the reddened areas shown here.
C. A young larva at rest on underside of live oak leaf. Typical feeding on only one leaf surface is shown beneath the larva.
D. A pupa of oakworm attached to the underside of a leaf.
E. Young larvae of the California oakworm. Note what appears to be an oversized head; this is typical of young larvae.
F. Mature larvae of California oakworm. Note that large larvae feed completely through the leaf tissue.
G. A valley oak, *Quercus lobata*, defoliated by the California oakworm.

A
B
C
D
E
F
G

Fruitworms and Other Caterpillars (Plate 66)

Fruitworms are most often associated with the pome fruit–growing industry. There are at least 10 species that form a distinct pest complex, most members of which are basically green, though in a variety of color patterns. The caterpillars also have similar habits; they feed during the same period, cause the same type of feeding injury (they are gross feeders), and produce one generation annually. Native insect species, their original food plants are thought to be the foliage and fruit of various indigenous deciduous trees and shrubs.

The insect that bears the common name green fruitworm, *Lithophane antennata* (Walker), is illustrated in panel A. Moths appear from September to November and overwinter in sheltered areas such as the loose bark on living trees or in leaf litter. They become active again in the spring when the maximum daily temperature reaches 15.6°C and remains above 7.2°C for an hour or two after dark. Mating and egg laying occur under these conditions. Eggs may be found singly or in masses. Upon hatching, the tiny larva usually crawls to an opening bud, enters it, and begins to feed. The caterpillars and their host leaves develop simultaneously, and each larva devours several leaves. The larvae go through six instars and pupate in the soil. Caterpillars feed in the trees until the end of June. There is one generation each year.

A wide range of woody plants serve as hosts, including azalea, rhododendron, mountain laurel, forsythia, rose, and trees such as oak, hickory, maple, ash, elm, willow, redbud, dogwood, beech, linden, cherry, plum, pear, and crabapple. The distribution of green fruitworms is transcontinental, but they are most common in the northeast quadrant of the United States and adjacent Canada. Male moths are attracted to baited sex pheromone traps and may be collected throughout April in New York.

Other destructive species of the so-called green fruitworms include *Orthosia hibisci* (Guenée), sometimes called the speckled green fruitworm; *Amphipyra pyramidoides* (Guenée), sometimes called the humped green fruitworm; and *Himella intractata* Morrison, sometimes called the fourlined green fruitworm. *O. hibisci* and *A. pyramidoides* are common defoliators of deciduous trees in Pacific Coast states and British Columbia.

Urodus parvula (Hy. Edwards), an ermine moth, is found from Washington, DC, to Florida and west to the Mississippi River. A heavy outbreak was reported on red bay, *Persea borbonia,* in northern Florida in 1980. The larvae (B) are nibblers, and unless they occur under outbreak conditions, they rarely devour an entire leaf. Pupation takes place in a unique netlike transparent cocoon that dangles at one end of a leaf. The moth has a smoky black body and wings. Its wingspread is about 15 mm. Known hosts include hibiscus, *Cinnamomum camphora, Quercus* species, *Persea* species, and *Citrus* species.

The silverspotted skipper, *Epargyreus clarus* (Cramer), is a butterfly valued by collectors because of its beauty. Adults feed on nectar and fly during daylight hours. There are two generations per year in the South, only one in the North. The pupa overwinters in a loose chrysalis that is usually tied around or among leaves on the ground. The skipper emerges in the spring and lays eggs on foliage. The larvae usually feed in concealment by webbing leaves together. Heavy defoliation occasionally occurs but never over large geographic areas. The larvae are yellow to leaf green, with a dull red head; when fully grown, they are about 50 mm long. They are solitary and feed at night. Silverspotted skipper larvae most frequently feed on black locust, wisteria, honeylocust, and in Florida on *Amorpha* and *Cassia* species. In the central Atlantic states larvae occasionally damage alfalfa, requiring emergency control measures. They may be found throughout United States and southern Canada (G, H).

References: 49, 111, 193, 255, 367, 653, 654

A. The green fruitworm, *Lithophane antennata,* feeding on apple foliage. The larva illustrated is about 3.2 cm long and nearly full grown. [Noctuidae] (Courtesy New York State Agricultural Experiment Station, Geneva.)

B. The larva of an ermine moth, *Urodus parvula*; about 8 mm long. [Yponomeutidae] (Courtesy W. Dixon, Florida State Dept. of Agriculture.)

C, D. The larva of *Urodus parvula* inside a net-type cocoon on the undersurface of a redbay leaf. (Courtesy W. Dixon, Florida State Dept. of Agriculture.)

E. One of the brown fruitworms, *Lithophane innominata.* (Courtesy New York State Agricultural Experiment Station, Geneva.)

F. A nearly mature larva of the silverspotted skipper. The full-grown larva is about 50 mm long. Note yellow spot on head. (Courtesy Gray Collection, Oregon State Univ.)

G, H. The silverspotted skipper. G, upper body and wing surface; H, lower body and wing surface, from whence its name. Wingspan about 5 cm. [Hesperiidae]

A
B
C
D
E
F
G
H

Walnut Caterpillar (Plate 67)

The walnut caterpillar, *Datana integerrima* Grote & Robinson, feeds on the foliage of black walnut, butternut, Japanese walnut, Persian walnut, pecan, and several species of hickory. Other reported hosts include birch, oak, willow, honeylocust, and apple. Walnut caterpillars always feed in clusters. They devour all the leaves on one branch before they move to another. Isolated trees usually suffer more than forest or orchard trees. Defoliation leads to sunscald, which further weakens the trees and leaves them vulnerable to wood-boring insects.

The pupa overwinters just beneath the surface of the ground, usually under the host tree. The adult, a heavy-bodied moth with a wingspan of 40 mm, emerges from its pupal case late in the spring. The buff-colored forewings have narrow, dark lines across them. Eggs are laid on the undersides of leaves (A) in masses of 300 or more. First-stage larvae (B) are skeletonizers. During the second stage the larvae devour all of the leaf except for the midvein. As they near full size, they devour the entire leaf including the petiole. Third- and fourth-stage larvae are brick red (C, F). They feed a month or more before reaching maturity. The mature larvae are about 50 mm long with grayish black bodies clothed with long, grayish white hairs (F).

The larvae have several interesting habits. When disturbed, they arch their heads and tails as though to fight off a predator. When molting, they crowd together on a branch or trunk and all molt at the same time, leaving an ugly patch of furlike hair and "skin."

The walnut caterpillar is distributed throughout most of the eastern United States and as far west as Minnesota, Kansas, and Texas. In the northeastern states and west to Michigan one generation occurs each year. Two generations are common from North Carolina to Arkansas and south.

Eggs are subject to attack by several wasp parasites. Larvae are often heavily parasitized by the fly *Archytas metallicas* (Robineau-Desvoidy).

Other common caterpillar pests of walnut are *Schizura concinna* (J. E. Smith), the redhumped caterpillar (Plate 70), and *Datana angusii* Grote & Robinson (Plate 69).

See Plate 69 for additional *Datana* caterpillars.

References: 228, 369, 449

A. Egg masses on the undersides of leaflets.
B. Eggs and newly hatched larvae.
C. A cluster of walnut caterpillars in their red stage.
D. Second-stage walnut caterpillar larvae feeding on leaves of *Juglans cinerea*. The brown leaves have been skeletonized by the first-stage larvae.
E. Walnut caterpillars massed on a trunk, about ready to molt. [Notodontidae]
F. The upper specimen is a fully grown walnut caterpillar. The lower specimen will molt again.

A
B
C
D
E
F

Nymphalid Caterpillars (Plate 68)

Many insects are endowed with beautiful color and form, as is the butterfly *Nymphalis antiopa* (Linnaeus). Even the caterpillar is attractively colored. Nonetheless, this insect can be a serious defoliator of shade trees such as elm, willow, poplar, birch, hackberry, and linden. Elm and willow (A) are the preferred food plants.

At the latitude of New York City two generations occur each year. The butterflies hibernate during the winter in secluded places where they can avoid rain and wind. As a result, they are among the first butterflies to appear in the spring after the buds of food trees open. They lay eggs in masses of 300–450 on twigs of newly burgeoning tree buds. The ribbed eggs are yellow at first but later turn black. The larvae, commonly called spiny elm caterpillars, are gregarious; they devour leaves, one branch at a time. When fully grown, they are 50 mm long. Their spines, though formidable in appearance, are harmless. The abdominal legs as well as the single row of spots on the back are red. The black areas are abundantly sprinkled with small white dots. After the larva has completed its growth, it suspends itself from a twig and hangs downward to form a chrysalis (C). The butterfly emerges within a week, and a second generation begins in early August in the southern part of the insect's range. This beautiful butterfly, commonly called the mourningcloak (E), is found throughout the United States and in Canada as far north as the Arctic Circle. It is both a woodland and city-dwelling species, and at times it is abundant enough to be classified as a serious pest. It is easily controlled by the bacterial insecticide *Bacillus thuringiensis* (Figure 52). It is also attacked by numerous wasp parasites. Black- and yellow-billed cuckoos, among other birds, sometimes devour the larvae.

Related to the mourningcloak butterfly is the California tortoiseshell, *Nymphalis californica* (Boisduval), found throughout western United States and British Columbia. Its larvae are similar to those shown in panel A but are marked dorsally with a row of bright yellow spines mounted on small blue humps (tubercles). Larvae, which are present from late June to mid-July, are 30 mm long when fully grown. Adults appear in late July and swarm in large numbers. The butterfly stage overwinters and lays eggs in the spring. One generation occurs each year.

Varnishleaf ceanothus, *Ceanothus velutinus,* and snowbush, *C. cordulatus,* are principal hosts. Larvae also feed on species of *Amelanchier* and *Salix,* on manzanita, and on other plants.

Hackberry, including sugarberry (*Celtis laevigata*), is subject to defoliation by the caterpillars of two butterfly species, *Asterocampa celtis* Boisduval & LeConte and *A. clyton* Boisduval & LeConte. Outbreaks often leave trees totally denuded. The former is regionally

Figure 52. Mummified larva of the mourningcloak butterfly, killed by the bacterial pathogen *Bacillus thuringiensis.*

called the hackberry butterfly; the latter is sometimes known as the tawny emperor.

These insects are likely to be found wherever native species of *Celtis* grow, including the eastern two-thirds of the United States and the Canadian Niagara peninsula. They seem to feed only on *Celtis.* In southern states adults (B) and larvae both have the ability to hibernate. Three annual generations may occur in the South, but in northern Ohio there is only one.

Eggs are usually laid in clusters on the underside of hackberry leaves, one spherical egg on top of another. Cluster size commonly exceeds 300 eggs, and all eggs in a cluster hatch about the same time. Initially, caterpillars are whitish with a black head (D), but after they feed, the body becomes greenish. Fully grown larvae are devoid of hairs and may reach a length of 37 mm. Larvae migrate downward to the ground when they have finished feeding, then randomly select other vegetation, fence posts, or buildings to climb in preparation for pupation in a chrysalis. Adults are found in early July. Circumstantial evidence indicates that in northern regions overwintering occurs in the larval stage.

References: 273, 454, 688, 741

A. Caterpillars of the mourningcloak butterfly feeding on willow leaves. [Nymphalidae]

B. The hackberry butterfly, *Asterocampa celtis.* Its wingspread is about 65 mm. (Courtesy F. L. Oliveria, U.S. Forest Service.)

C. Chrysalises of the mourningcloak butterfly hanging from a defoliated willow.

D. A hackberry butterfly egg mass with a cluster of first-instar larvae feeding on a sugarberry leaf. (Courtesy F. L. Oliveria, U.S. Forest Service.)

E. The mourningcloak butterfly. Color patterns vary somewhat with individual butterflies. The center portions of the wings may range from dark maroon to nearly black. The wingspan ranges up to 63 mm.

A
B
C
D
E

Notodontid Caterpillars (Plates 69–70)

The orangehumped mapleworm, *Symmerista leucitys* Franclemont (A), is a northern species not known to occur south of Pennsylvania. Outbreaks have been recorded from Maine to Minnesota and in all of the adjacent Canadian provinces. In northern New York moths appear in early June and lay eggs from mid-June to mid-July. They lay the cup-shaped eggs in single-layered clusters on the underside of leaves; each female usually lays more than 200 eggs. Upon hatching, larvae crawl to the upper surface of the leaf and consume the epidermis and mesophyll. Initially they feed gregariously, often touching one another, but they become less gregarious as they grow and finally become solitary, feeding from the leaf margins. Larvae are present from late June until frost. The insect overwinters on the ground in cocoons camouflaged with leaves. Eggs are often heavily parasitized by a wasp belonging to the genus *Telenomus*.

The orangehumped mapleworm feeds predominantly on sugar maple, but it also feeds on American beech, basswood, white birch, American elm, red maple (*Acer rubrum*), northern red oak, chestnut, and ironwood (*Ostrya virginiana*).

The saddled prominent, *Heterocampa guttivitta* (Walker), is a native moth found from eastern Canada south to Florida and Texas and west to Colorado and Nebraska. Its status as a severe defoliator is probably limited to northeastern states and eastern Canada.

The saddled prominent has one generation per year and overwinters as a pupa in the organic litter under host trees. In New York and western New England a majority of the adults emerge over a 2-week period beginning about the last week of May. The moths are strong flyers during twilight hours. They normally rest during the day on the trunks of trees, and their protective coloration keeps them hidden from all but the most discriminating searcher. They lay eggs over a 4- to 5-week period, beginning about the second week of June. They are globose, pale green, about 1 mm in diameter, and are deposited individually on the underside of host leaves. Females have the capacity to produce about 500 eggs. Upon hatching, the larva has a bizarre but no doubt protective appearance. It is dark reddish with a pair of black antlerlike horns located immediately behind the head. These horns are lost after the first molt. The first three instars are skeletonizers and feed from the undersides of leaves. The resulting irregular skeletonized patches on the leaves transmit light; these patches are the most obvious early indications of a developing infestation. The last two instars feed from the leaf margin and are gross defoliators. Fully grown larvae are about 38 mm long, smooth "skinned," and variable in color (C, D). The saddled prominent is a late-season feeder, typically stripping foliage in late July and August. Sugar maple, beech, yellow birch, and apple are the preferred hosts. These caterpillars are not often found as defoliators of shade trees.

Datana major Grote & Robinson, known commonly as the azalea caterpillar, is a serious defoliator of azalea. It is largely limited to the southeastern quadrant of the United States, frequently posing a problem in and south of Maryland. Eggs are laid in clusters of 80–100 in late spring on the underside of leaves. The developing larvae (F) begin feeding in colonies as skeletonizers on the underside of leaves. They remain gregarious but soon become gross feeders, devouring the entire leaf. As they mature, the larvae become highly colored (B). The fully grown caterpillar, 50 mm long, has a red head and legs, and a black body with longitudinal rows of yellow spots and sparse white hairs. The caterpillar seems to prefer indica azaleas but has been reported on blueberry in Delaware, red oak in Maryland, and andromeda and apple in the mid-Atlantic states. Little has been published about this animal, but it is presumed to overwinter as a pupa. In some parts of the South there is a partial second generation, but one generation is the rule. Most of the damage occurs in August and September; in Florida damage continues through October.

The yellownecked caterpillar moth, *D. ministra* (Drury), is difficult to distinguish from the walnut caterpillar moth (Plate 67), *D. integerrima* Grote & Robinson. The yellownecked caterpillar (E, G, H) has a rather broad range of host plants including crabapple, flowering peach, cherry, almond, quince, maple, elm, butternut, walnut, oak, hickory, chestnut, beech, linden, witch-hazel, birch, locust, sumac, azalea, and boxwood. Widely distributed throughout the United States, it is most common in the Appalachian Mountains and foothills. It is known in British Columbia and probably occurs in other Canadian provinces. *D. ministra* is common in the Appalachian Mountains and foothills.

As with other *Datana* caterpillars, *D. ministra* spends its winters as a pupa in an oblong cell 20–25 mm underground. Moths appear in June and July and lay their eggs in clusters of 25–100 or more on the undersides of leaves. The eggs of a single cluster usually hatch at about the same time, and the larvae feed gregariously for 4–6 weeks. Newly hatched caterpillars skeletonize the lower leaf surface; older ones devour all but the petiole. When disturbed, larvae lift their head and tail portions high above the rest of their bodies as though to strike at a potential enemy. The yellownecked caterpillar gets its name from the bright orange-yellow segments behind its head. When fully grown, it crawls down the trunk and burrows into the soil. Male moths are attracted in moderate numbers to lights.

Several natural enemies attack this species. Birds such as robins and bluejays feed on the yellownecked caterpillar, and it is often attacked by predaceous bugs and parasitic flies.

D. angusii Grote & Robinson is a common pest of walnut. Fully grown larvae are black with pale lines and are clothed with white hairs. Their feeding behavior and life cycle are similar to those of the walnut caterpillar (Plate 67).

References: 3, 31, 161, 306, 596

A. Orangehumped mapleworm on a sugar maple leaf. This caterpillar is nearing full growth; 40 mm long. (Courtesy J. G. Franclemont, Cornell Univ.)

B. A nearly fully grown azalea caterpillar; about 50 mm long. (Courtesy J. Baker, North Carolina State Univ.)

C, D. Two color phases of the saddled prominent; 38 mm long when fully grown. (Courtesy D. C. Allen, New York State College of Environmental Science and Forestry.)

E. A cluster of yellownecked caterpillars on oak.

F. A cluster of newly hatched azalea caterpillars. (Courtesy J. Baker, North Carolina State Univ.)

G, H. Yellownecked caterpillars. G, yellow-red color phase; H, black-yellow phase.

A
B
C
D
E
F
G
H

The caterpillars of all the insects illustrated in this plate are large, gluttonous feeders. When abundant, they can quickly strip a tree of its foliage.

The redhumped caterpillar, *Schizura concinna* (J. E. Smith), is a pest of many ornamental, fruit, and nut trees throughout the continental United States. The moth is grayish brown with a wing expanse of slightly over 25 mm. Its spherical white eggs are laid in groups of 25–100 on the undersides of host leaves. Newly hatched larvae feed gregariously at first but tend to disperse as the caterpillars mature. The mature larva, about 38 mm long, is yellow with a red head and is lined longitudinally with orangish, black, and white stripes (B). Several prominent black tubercles arise from the back, and there is a large reddish hump on top of the first abdominal segment that gives the insect its common name. After completing growth, the caterpillar migrates down the tree trunk to pupate and overwinter inside a silken cocoon constructed in the soil or among plant refuse on the soil surface. One to five generations occur each year, depending on climatic conditions.

Larvae consume all the leaf tissue except the major veins. Damage is generally confined to scattered branches in the tree, though defoliation of entire trees occurs periodically over a widespread area. In California the late summer and early fall generations of *S. concinna* are more important on fruit trees, such as walnut, than are earlier generations. Reduced foliage results in sunburned fruit.

The redhumped caterpillar feeds on the following woody plants: birch; hickory; *Cercis canadensis* and western redbud; dogwood; hawthorn; walnut; butternut; black locust; sweetgum; apple; persimmon; poplar; aspen; cottonwood; *Prunus* species; pear; rose; willow; elm; and *Vaccinium* species.

Several related species of *Schizura* that may be noticed from time to time include the unicorn caterpillar, *S. unicornis* (J. E. Smith), and *S. ipomoeae* Doubleday. The latter species has been reported from the eastern states and from Idaho, Utah, and California. It feeds on maple, birch, *Ceanothus* species, honeylocust, apple, plum, oak, *Rubus* species, elm, and *Vaccinium* species. Neither *S. unicornis* nor *S. opomoeae* resembles *S. concinna* while in the larval stage. Larvae of both species are protectively colored in greens and browns so that their appearance is much like bits of leaves with brown or reddish edges. Both species have a prominent dorsal projection.

The orangestriped oakworm, *Anisota senatoria* (J. E. Smith), is one of four closely related moths that are often abundant in the eastern United States. The other three species are the spiny oakworm, *A. stigma* (Fabricius); the pinkstriped oakworm, *A. virginiensis* (Drury); and the greenstriped mapleworm, *Dryocampa rubicunda* (Fabricius). These four species share many habits and have similar physical features. *A. senatoria* and *A. virginiensis* moths both have a single white dot on their forewings. The larva of each has two recurved and flexible horns on the second thoracic segment, and the anal plate has three to four short terminal spines. All of these larvae are marked with conspicuous stripes and have either short or long spines along the abdomen. The *D. rubicunda* larva has a reddish head and seven dark green stripes running the entire length of its body. Two prominent, slender, but flexible horns attached to the second thoracic segment provide distinctive identification characters. When fully grown, the larva is about 36 mm long. Two annual generations are produced in the South and are timed so that a maple or oak tree may be defoliated twice in the same season. Outbreaks of the greenstriped mapleworm have occurred on sugar and red maple (*Acer rubrum*) in eastern Canada and the northeastern states. In 1981 it was reported to have defoliated silver maples in Missouri. It is often associated with maples and oak in Nebraska. Larvae are commonly found in trees during late July and early August.

The orangestriped oakworm (C) is about 55 mm long, and when fully grown it crawls from trees and burrows 2.5–10 cm into the soil. There it builds an earthen cell, where it pupates and passes the winter. The moths first appear in early summer and are attracted to artificial light at night. Over the period of a month the female will deposit from 1 to more than 500 eggs in a single cluster on the underside of an oak leaf, usually on lower branches. The newly hatched larvae are greenish yellow but have the same characteristic hornlike projection as the mature larvae. The early instars are gregarious and skeletonize the leaf, consuming all except a network of veins. Older caterpillars eat all but the main vein, usually defoliating one branch before going on to another. One generation occurs each year in the insect's more northerly range.

The orangestriped oakworm moth has a rather wide distribution. It has been reported from Wisconsin and Michigan east to New England and eastern Canada. It is also known in Georgia and Kansas, although it is more abundant in the northern United States than in southern areas. According to Herrick (359), the orangestriped oakworm moth seems to prefer white and scrub oak. It definitely prefers oak foliage but also feeds on maple, hickory, birch, and hazelnut. Destructive outbreaks of this insect have been reported in New York, Michigan, New Jersey, and other states. Usually forest trees suffer most, but trees in parks and along city streets are, on occasion, severely attacked.

The Nevada buck moth, *Hemileuca nevadensis* Stretch, and its eastern kin the buck moth, *H. maia* (Drury), belong to the family Saturniidae, the giant silkworm moths (Plate 74). The pandora moth (Plate 6) is also a closely related species.

The eastern buck moth passes the winter as an egg in a cluster encircling oak twigs, or occasionally on willow, cherry, or hazel. Eggs hatch in late April or May, and the larvae are present in trees from May to August. Larvae are gregarious and pass through five instars, pupating in the ground. Fully grown larvae are about 60 mm long. In autumn when deciduous tree leaves begin to fall, the day-flying moths appear and deposit eggs. If people allow the larvae or their cast skeletons to come in contact with skin, a rash often develops.

It is said that the name buck moth, given at the turn of the century, arose because the moth appears at the beginning of the hunting season. The larvae feed on a wide range of deciduous oaks. The insects' range extends from Nova Scotia and Maine to Georgia, west to Louisiana and Oklahoma and to the western edge of the Great Plains.

The Nevada buck moth also overwinters in egg clusters on twigs of its host plants, willow and poplar. The spiny black caterpillar (E) of this defoliator feeds gregariously when young but disperses in its final instar. This western species is reported as far east as Oklahoma and Nebraska. Its life history is similar to the buck moth of the East.

The common names redhumped caterpillar and redhumped oakworm may be confusing, but they represent distinct, related species in the family Notodonidae. The color patterns of the redhumped oakworm, *Symmerista canicosta* Franclemont, and the orangehumped mapleworm, *S. leucitys* Franclemont (Plate 69), may be similar. However, these caterpillars are readily separated by host, as is implied by their common names.

The redhumped oakworm overwinters as a pupa in the duff under host trees. In Michigan moths begin to emerge by early June and individuals continue to emerge until mid-August. A few pupae remain quiescent through the summer and spend two winters dormant. The female moth deposits about 300 eggs in several single-layered circular masses on the underside of oak leaves. The larvae (F) feed on the upper surface of the leaf and rest on the undersurface. They are gregarious, both resting and feeding in colonies. They feed on all the leaf tissue except the midrib. If larvae are disturbed or dislodged, they drop and dangle from silken threads, which may cause an additional annoyance to people who live near infested trees. Larvae may be found in trees through September.

The redhumped oakworm is a native whose numbers are largely controlled by natural means. However, cyclic outbreaks occur at 10- to 15-year intervals and last 1–3 years. White oak appears to be the

---

A, B. Feeding larvae of the redhumped caterpillar.

C, D. Feeding larvae of the orangestriped oakworm; 55 mm long. Note the short spines on the abdomen, particularly those on the terminal segment. [Saturniidae]

E. A larva of the Nevada buck moth; about 44 mm long. [Saturniidae] (Courtesy D. Leatherman, Colorado Forest Service.)

F. A nearly grown orangehumped maple worm; about 44 mm long. (Courtesy J. G. Franclemont, Cornell Univ.)

G, H. Poplar tentmaker caterpillars in typical nests; about 40 mm long when fully grown. (Courtesy J. D. Solomon, U.S. Forest Service.)

A
B
C
D
E
F
G
H

choice of food, but all other oak species found in the Northeast, the northcentral states, and adjacent Canada are attacked.

Ichthyura inclusa (Hübner) has been given the common name poplar tentmaker. As it feeds, this defoliator folds or ties leaves together and lives in the protection of the tentlike structure that is lined with silk (G, H); 10–20 larvae may huddle together in the protection of one tent. The poplar tentmaker does not consume the leaves it uses for habitation. It habitually avoids the major leaf veins, leaving skeletal remnants as a symptom. Abandoned nests often remain in the trees throughout the winter. Moths appear in the spring and lay their eggs in clusters on the undersides of leaves. In the South, where two generations occur each year, larvae may be in trees from May to October. Pupae overwinter on the ground in loose cocoons.

The poplar tentmaker occurs from southern Canada to Georgia and Texas and west to Colorado. It is an important defoliator primarily in regions where two generations occur each year.

References: 193, 359, 380, 490, 539, 547, 660

Tussock Moth Caterpillars (Plates 71–72)

Tussock moths are uniquely hairy in their caterpillar stage and have been placed in two families, the Arctiidae and Lymantriidae. They are often described as general feeders, taking the foliage of conifers and/or deciduous trees. (See Plates 6 and 10 for other tussock moths.)

The satin moth, *Leucoma salicis* (Linnaeus), was given its common name because of the moth's characteristic white satin sheen (C). It is an imported pest, native to Europe and Asia. In 1920 it was found almost simultaneously in Massachusetts and British Columbia. In the intervening years it has spread throughout New England, New York, and the Canadian Maritime Provinces as well as Washington, Oregon, and northern California.

This insect overwinters as a small larva in a silken cocoonlike bag attached to the trunk or branch of a host tree. On the West Coast the cocoons may be camouflaged with moss and lichens. The larva comes out of hibernation to feed after the leaves have formed. Fully grown caterpillars (A) may be 50 mm long. The pupa (B) is glossy black, and after about 10 days in this stage, the adult emerges. Moths, which are on the wing through much of July, are strong fliers and are attracted to lights. Females lay eggs during July, often on leaves, but eggs may be found anywhere on the tree, where they are deposited in masses of up to 400. Upon hatching, the young larva feeds on the epidermis of leaves. After feeding for 5 or 6 days, the larva molts inside a small, flat web. After molting, it continues to skeletonize leaves until early fall or the first frost, when it prepares for hibernation. Shade and ornamental poplars and willows are most severely attacked. Occasionally the larvae feed on oak and aspen.

Many of the native and imported parasites of gypsy moth also attack the satin moth. The two most important parasites in the East are a fly, *Compsilura concinnata* Meigen, and a wasp, *Eupteromalus peregrinus* Graham. On the West Coast serious outbreaks of the satin moth are kept in check by parasites such as the native tachinid fly, *Tachinomyia similis* (Williston).

American dagger moth is the common name given to *Acronicta americana* (Harris). The photograph in panel F indicates why this insect was so named. The fully grown larva is about 50 mm long, with a shiny black head. It occurs throughout the eastern part of the United States and Canada. Although the dagger moth is quite common, it rarely occurs in large or damaging numbers. Its food plants include apple, basswood, boxelder, maple, oak, and willow.

Orgyia leucostigma (J. E. Smith), the whitemarked tussock moth, is most widely distributed in the eastern half of North America but occurs from Florida to British Columbia. The species has been subdivided into at least four distinct geographical populations. This insect overwinters in the egg stage. Each female deposits about 300 eggs in a conspicuous frothy white mass on her spent cocoon (D). Between April and June, depending on location, the eggs hatch and the larvae migrate to the leaves. Young larvae skeletonize leaves, but older larvae chew the entire leaf except for the main veins and petiole. The fully grown larva is about 31 mm long with a reddish orange head and yellowish body tufted with distinctive hair. The mature larva pupates in a grayish cocoon, which contains many body hairs and is constructed on twigs, branches, or trunk bark. One to three generations of the whitemarked tussock moth occur each year.

The male moth is ash gray with a wingspan of about 26–30 mm. The female is dirty white, hairy, wingless, and about 13 mm long. Records show at least 60 host species, the more common of which are maple, horsechestnut, birch, apple, sycamore, poplar, linden, elm, rose, fir, and larch. The southern subspecies commonly feeds on live oak, redbud, pyracantha, and mimosa.

The pale tussock moth, *Halysidota tessellaris* (J. E. Smith) (E), is a general feeder but limits its feeding to the foliage of deciduous trees and shrubs. It is primarily an eastern species but has been reported as far west as New Mexico and Colorado. It is also found in Ontario and Quebec.

Moths appear in June and July in the middle Atlantic states and Ohio Valley. Eggs are laid in masses on the underside of host tree leaves. The olive green, hairy larvae feed interveinally from mid-July until the end of September and attain a length of 30–35 mm. Winter is spent in a hairy, brownish gray, ball-like cocoon that may be attached to almost any such substrate as a tree trunk, a fence, or under stones. One generation occurs each year. The biology of this moth is poorly known.

Host plants include sycamore, American elm, Siberian elm, ash, boxelder, chokecherry, oak, beech, apple, and others. The tiny hairs from larvae and pupae may irritate human skin.

The rusty tussock moth, *Orgyia antiqua* (Linnaeus), gets its common name from the color of the male moth. It is the most northern and most widely distributed member of the genus, found across North America south to the middle Atlantic states, in northern California, and in the Pacific Northwest.

The larva is predominantly blackish with orange or yellow spots. The pencil hairs are black, and the four dorsal abdominal tufts may be light brown or whitish. The head is nearly black; the legs and underbody are light yellowish or green.

It too has a wide host range that includes deciduous trees and shrubs, nearly all conifers except *Juniperus* species, and broad-leaved evergreens such as rhododendron. The life cycle parallels that of the whitemarked tussock moth, but it produces only one generation per year. Synthetic sex pheromones developed for several of the economically important tussock moths have the potential to control these species by disrupting mating.

Many other species of tussock moths occur on various ornamental plants in different parts of the United States. All appear sporadically and are quite numerous one year and virtually absent the next. A large number of parasites and predators, along with microorganisms that cause disease and subsequent death of larvae, are believed to be responsible for the irregular appearance of these insects as pests.

References: 8, 20, 46, 193, 217, 238, 301, 390, 403, 533, 591, 749

A. Larvae of the satin moth, *Leucoma salicis*, on *Populus deltoides*. A fully grown larva is 35–50 mm long.
B. A pupa of the satin moth.
C. An adult satin moth. [Lymantriidae]
D. A cluster of tussock moth eggs on a juniper twig. The eggs were laid on the female's old cocoon.
E. Three larval stages of the pale tussock moth. Fully grown larvae are 30–35 mm long. [Arctiidae]
F. The American dagger moth, *Acronicta americana*, on *Acer platanoides*. [Noctuidae]
G. A fully grown larva of the whitemarked tussock moth on *Cotoneaster horizontalis*. [Lymantriidae]

A
B
C
D
E
F
G

(Tussock Moth Caterpillars, continued, Plate 72)

Orgyia vetusta (Boisduval), the western tussock moth, is commonly found on the Pacific Coast from British Columbia to southern California. It feeds on willow, hawthorn, manzanita, oak, walnut, crabapple, pyracantha, California holly, coffeeberry, and other plants. In southern California two generations may occur each year; in other areas there is but one. In northern California this insect lays eggs in late summer in feltlike masses upon old cocoons. The eggs overwinter and hatch early in the spring. The young larvae skeletonize the leaves (C); mature larvae devour the entire leaf. Mature larvae are 13–22 mm long and can be a nuisance when they wander to find suitable places for pupation (A). They pupate on leaves, bark, and twigs of shrubs and trees, and on fences, eaves of buildings, and other constructed objects (D). The larva is attractively colored with spots of bright red and yellow. It has four round brushlike tufts of hair on its back, and hornlike tufts of black hair both posteriorly and anteriorly. It is attacked by a number of parasites, mainly parasitic wasps and tachinid flies, but these are usually inadequate for control.

The hickory tussock moth, *Lophocampa caryae* (Harris), is also known as the hickory tiger moth. This name comes from the markings on its wings (J). It is found from Missouri and Minnesota east to the Atlantic Ocean. It also occurs in southern Canada.

This insect produces one generation each year and overwinters as a pupa inside a densely hairy oval cocoon (F). Moth emergence begins in May and continues into July. Eggs are laid in clusters of 100 or more on the undersides of leaves. Hickory tussock moth females have functional wings, but because of their comparatively heavy bodies, they are weak fliers. Young larvae feed gregariously; later they scatter to feed. The caterpillars require about 3 months to complete their development. During this time color patterns change (G, H), but the caterpillar is basically white, with a black head. When fully grown it is about 37 mm long. The caterpillars build their cocoons on or near the ground; in New York state this occurs in late September. Host trees include hickory, walnut, butternut, linden, apple, basswood, birch, elm, black locust, and aspen.

Several wasps parasitize the hickory tussock moth but rarely hold it in check.

Halysidota harrisii Walsh, the sycamore tussock moth, attacks American sycamore and London plane throughout the range of these hosts. Outbreaks of this insect have occurred in New York City, where its feeding has removed half or more of the foliage of street and park trees. On Long Island larvae are found in early July. Feeding is most active at dusk. The larva chews from the edge of the leaf and eats all of it except major veins. The life cycle of this species parallels that of other East coast tussock moths (Plates 6, 70).

References: 85, 147, 193, 217, 275, 682, 805

Key to Fuzzy Caterpillars Occurring on Trees and Shrubs East of the Rocky Mountains from Late July to September

1. Caterpillar lightly hairy (body visible through the hair) and/or with hair in tufts 2
   Caterpillar densely hairy with or without hair in tufts or pencils 5
2. Caterpillar with at least two long black hair pencils in front and one behind, and four dense whitish tufts on the back; colorfully marked ... 3
   Caterpillar without four dense whitish dorsal tufts 4
3. Caterpillar with coral red head; body distinctly marked with one white or yellow longitudinal line on each side of the back; only three black hair pencils on each side .... whitemarked tussock moth (Plate 71G)
   Caterpillar with head dark chestnut to black; body without light dorsal lines; one or two additional black hair pencils on each side (four or five total) .... rusty tussock moth
4. Caterpillar grayish yellow with off-white hairs; in loose tents ... fall webworm (Plate 75)
   Caterpillar strikingly white with crisp black spots and a row of dorsal black hair pencils .... hickory tussock moth (Plate 72H)
5. Caterpillar distinctly marked and banded 6
   Caterpillar with indistinct marks and bands 7
6. Caterpillar bright yellow (occasionally cream-colored) with black on both ends and a series of black tufts along back .... spotted tussock moth
   Caterpillar black on both ends with an orange midband and no dorsal tufts .... banded woolybear
7. Caterpillar with drab olive to gray hair that darkens along the back forming a vague line; no dorsal hair pencils ..... pale tussock moth (Plate 71E)
   Caterpillar variously colored but without vague middorsal line; three or more long black hair pencils 8
8. Caterpillar with gray, yellow, or white hair and five long black hair pencils (two paired and a single one on the posterior end) .... American dagger moth (Plate 71F)
   Caterpillar ranging from white to gray with brownish dorsal hairs; three unpaired black hair pencils along the back .... alder dagger moth (*Acronicta dactylina*)

This key is modified from one that appeared in *Forest and Shade Tree—Insect and Disease Conditions for Maine,* August 1990, a newsletter published by the Maine Dept. of Conservation. Permission to reprint has been granted by the author of the key, R. G. Dearborn.

A. A fully grown larva of the western tussock moth.
B, D. Cocoons of the western tussock moth on a leaf and a fence.
C. Young western tussock moth larvae on skeletonized willow leaves.
E. An adult male moth.
F. A cocoon of the hickory tussock moth, *Lophocampa caryae.*
G, H. Color forms of the larvae of the hickory tussock moth.
I. A fully grown sycamore tussock moth larva, *Halysidota harrisii.*
J. A hickory tussock moth. [Arctiidae]

A
B
C
D
E
F
G
H
I
J

Subtropical Caterpillars (Plate 73)

This plate and text describes four subtropical caterpillars common in Florida and some areas of the Gulf Coast states.

Syntomeida epilais jucundissima (Dyar), often called the oleander caterpillar, feeds only on oleander, a plant that is poisonous to most animals. Three generations occur each year; larvae are present in March, December, and July. It is likely that generations overlap because young larvae (B), fully grown larvae (C), and adults have been observed together in early December.

The moth is purplish with white dots on greenish black wings. Female moths lay oval, light yellow eggs in clusters of 25–75 on the undersides of oleander leaves. The eggs are laid close together but never touching. Upon hatching, the young larvae first eat their own eggshells, then begin feeding gregariously on the underside of the leaf (B). They first skeletonize the leaves and later completely devour them (A). Fully grown larvae are about 30 mm long. Cocoons are loosely woven from silk threads produced by their silk glands. The life cycle requires about 60 days for completion; adults live about 9 days.

The larva has developed a defense against poisonous latex alkaloids that flow to wound sites through the oleander leaf's vascular system. They first skeletonize leaves from the underside, a type of feeding that does not trigger the production of latex. When the larvae reach a length of about 18 mm, they change feeding habits to become solitary feeders, devouring all of the leaf blade except the midvein. At this stage the insects must deal with the poisonous latex in the sap. To avoid most of the toxic alkaloids, they bite repeatedly into leaf veins at the petiole as well as the blade. Where petiole and major veins are wounded by manibular bites, a viscous yellow latex forms, thereby concentrating the poison. The larvae thus effectively deactivate the latex system, rendering the distal tissues essentially free of the alkaloid and safe for them to consume.

The yellow oleander, *Thevetia peruviana,* is used as an ornamental plant in Florida and California. In Florida a caterpillar, *Palpita flegia* (Cramer), feeds on yellow oleander as a leaf skeletonizer. It ties the leaf remains together with silk. The mature caterpillar, which may be up to 37 mm long, is a dingy white ornamented with black and yellow spots.

The palm leafskeletonizer, *Homaledra sabalella* (Chambers), is the most important pest of palms in Florida. It skeletonizes the fronds, usually on the lower side, leaving the leaf folds packed with frass. Although palms are seldom killed, the upper surfaces of their leaves turn brown, destroying their ornamental value (D). A single female may deposit up to 70 eggs on the underside of a young leaf. After the eggs hatch, larvae feed under the eggshells for a few days and collectively spin a tentlike mat or web that protects the colony as it feeds. The frass is mixed with a continuously growing web (E). The larvae live in colonies of 35–100 individuals. When fully grown, the larva is about 15 mm long. It pupates under cover of the web. Generations occur continuously throughout the year, and moths may be collected at almost any time. The moths, which are frequently attracted to light at night, do not feed. They live only 3–10 days. *H. sabalella* has been found on the following palms and palmettos: butia, coconut, latania, and date palms; and dwarf, cabbage, and saw palmetto. *Plochius amandus* Newman, a carabid beetle, is an active predator of *H. sabalella* that helps keep the moth in check. The beetle larvae live under the web within the colony of caterpillars.

In Florida, California, and states bordering the Gulf of Mexico the cabbage palmetto, *Sabal palmetto,* is used as a native ornamental. In some areas it is an important honey plant; bees produce a choice table honey from its nectar. *Litoprosopus futilis* (Grote & Robinson), sometimes called the cabbage palm caterpillar, feeds on the flowers and is occasionally abundant enough to remove all traces of bloom. The mature larva, about 37 mm long, is pinkish, punctuated with numerous shiny black spots called tubercles. The head and thoracic shield are also black.

References: 74, 135, 162, 165

A. The oleander caterpillar, *Syntomeida epilais jucundissima,* feeding on oleander leaves.

B. A colony of young oleander caterpillars skeletonizing the underside of a leaf.

C. Fully grown oleander caterpillars.

D. A palm frond showing dead areas of leaves. The damage was caused by the palm leafskeletonizer, *Homaledra sabalella.*

E. Frass moved aside to expose a larva.

F. Close-up of a palmetto palm showing the webbing filled with frass under which is a colony of palm leafskeletonizers. [Coleophoridae]

A
B
C
D
E
F

Armed Caterpillars (Plate 74)

The caterpillars shown here are widely varying in appearance. They have distinctive transverse rows of hairs and beautiful tubercles; some have poisonous spines and all are unusually colorful.

The stinging rose caterpillar, *Parasa indetermina* Boisduval (Limacodidae), not illustrated, is of no economic importance and is mostly known for causing urticaria in people who come in contact with it. The larva, about 20 mm long when fully grown, is strangely colorful and bears urticating hairs on lateral fleshy yellow lobes. Four dark lines run the length of the dorsum. The head is concealed, and unless the animal is moving it is difficult to tell one end from the other. The larvae are solitary feeders and are never abundant. Food plants include apple, oak, redbud, sycamore, and many woody shrubs. In North Carolina larvae are found in July. They range from New York to Illinois and south.

Harrisina americana (Guérin), shown in panels A–D, is commonly known as the grapeleaf skeletonizer. It is found from New England to Florida and west to Missouri. This species becomes abundant periodically, devouring the leaves of Virginia creeper, *Parthenocissus quinquefolia,* as well as those of wild and cultivated grapes. These destructive caterpillars can be found in central Florida in early September, and in Connecticut from late June to mid-August. In Florida adults are on the wing in March and September and lay their lemon-colored eggs in clusters on the undersides of host leaves. Two generations occur annually in the South and one in the North.

The eggs hatch at about the same time on any given leaf, and the larvae line up side by side, sometimes two rows deep, and skeletonize the foliage (B). As they approach maturity, they become less and less gregarious and finally live separately. Fully grown larvae, about 11 mm, attach themselves to the woody stems and spin thin white cocoons (D). The moth is black (F) with a wingspan of about 25 mm.

In California and the Southwest *Harrisina brillians* Barnes & McDunnough, sometimes called the western grapeleaf skeletonizer, is at times a serious pest of grape. It has been controlled largely by the use of insect parasites, particularly the wasp *Apanteles harrisinae* Muesebeck.

Sphinx caterpillars that attack woody plants are not usually pests; rather, they are a curiosity because they are large and ornate. Moths are often sought by collectors because of their beauty and phenomenal ability to hover like hummingbirds.

The cherry sphinx caterpillar, *Sphinx drupiferarum* J. E. Smith, is a typical example (E). It may be found from Nova Scotia south to Georgia, west through Mississippi, Arkansas, Kansas, Colorado, Utah, and California, and north into British Columbia. Its food plants include apple, plum and other *Prunus* species, hackberry, and lilac (*Syringa vulgaris*). Little is known about its life history.

Automeris io (Fabricius), commonly called the io moth, is a favorite of moth collectors, partly because of the large eyelike spots on its hind wings. People should learn to recognize the larva (I) if for no other reason than to avoid touching it. It is covered with spines, each of which is connected to a poison gland. People who come into contact with these spines often react as if stung. The itching response may be mild or severe, depending on one's sensitivity to allergens.

The larvae feed on various trees and shrubs such as rose, willow, oak, sycamore, elm, beech, poplar, maple, birch, ash, and linden. Io moths are common in the East but are rarely found in large numbers. One generation occurs each year. The moths are attracted to lights.

The cecropia moth, *Hyalophora cecropia* Linnaeus, is the largest of our native silk moths, with a wingspan of 140–165 mm. It is common in the insect collections of both amateur and professional entomologists. The larva (G, H) feeds on at least 50 species of plants, including linden, maple, boxelder, elm, birch, willow, hawthorn, and poplar. When fully grown, it measures 75–100 mm. It forms a large double-walled overwintering cocoon that is often attached to a twig of its host plant. Artificial control is often unnecessary because the larvae are usually heavily parasitized by the tachinid fly, *Lespesia ciliata* (Macquart). One generation occurs each year.

References: 217, 370

A. A fourth-stage larva of the grapeleaf skeletonizer, *Harrisina americana.*

B. A grape leaf partly skeletonized by the larvae of the grapeleaf skeletonizer. Note the larva at the juncture of the leaf blade and petiole.

C. A fully grown larva of the grapeleaf skeletonizer; about 11 mm long. [Zygaenidae]

D. Cocoons of the grapeleaf skeletonizer.

E. A fully grown cherry sphinx caterpillar, *Sphinx drupiferarum.* (Courtesy D. A. Leatherman, Colorado State Forest Service.)

F. An adult female moth of the grapeleaf skeletonizer.

G. A fully grown larva of the cecropia moth, *Hyalophora cecropia.*

H. The anterior end of a cecropia caterpillar with white eggs of a parasite attached to its exoskeleton.

I. The fully grown io moth caterpillar, *Automeris io.* [Saturniidae]

A
B
C
D
E
F
G
H
I

Webworms (Plate 75)

The fall webworm, *Hyphantria cunea* (Drury), is a native of North America and Mexico. In 1940 it was recorded in Hungary and has since become an important pest of trees throughout Europe and parts of Asia. Many of the important forest and shade tree pests of the United States were accidentally introduced from Europe and Asia; the fall webworm is one of a much smaller number of insect pests that went the other way.

It feeds on almost all shade, fruit, and ornamental trees except conifers. In the United States the fall webworm attacks at least 88 species of trees, and in Europe, 230 species of trees, shrubs, ornamentals, and annual plants. In Japan its diet includes 317 plant hosts.

In the United States its feeding habits vary from region to region, and preferences range from pecan or black walnut in some areas, to persimmon or sweetgum in others. In western Appalachia and the Ohio Valley, American elm, maples, or hickory are preferred; in the western United States the fall webworm is common on alder, willow, cottonwood, madrone, and fruit trees.

This defoliator is distributed throughout most of the United States and Canada. At one time it was thought that there were two species, *H. cunea* and *H. textor.* Now there are known to be two races of a single species—one blackheaded (F) and the other redheaded. These races differ in the markings of both the larvae and adults, as well as in their food habits and biology. For example, in the southern United States adults of the blackheaded race appear one month earlier than those of the redheaded race. The females of the blackheaded race deposit their eggs on the undersides of leaves in single-layer masses during May to July, whereas the redheaded form deposits most of its eggs in double layers in mid-April. After laying her eggs, the female covers them with white hairs from her abdomen. When the larvae hatch, the blackheaded form is yellowish green to pale yellow, with two rows of dark tubercles along the back. Its head is black, and its body covered with fine hair. When fully grown, the blackheaded form is yellowish or greenish with a broad, dark stripe along the back. The redheaded race is tawny or yellowish tan with orange to reddish tubercles.

Larvae of the fall webworm pass through as many as 11 stages of development. In each stage feeding occurs within a distinctive web made of silk produced by the larvae. The blackheaded race forms a flimsy web; that of the redheaded race is larger and more compact. When alarmed, all the larvae in a "nest," whether of the black- or redheaded race, make jerky movements in perfect rhythm, possibly as a defensive mechanism. Depending on climate, one to four generations of the fall webworm occur per year; in Louisiana, four are common.

More than 50 species of parasites and 36 species of predators of the fall webworm are known in America. The egg parasite *Telenomus bifidus* Riley is probably the most effective in controlling fall webworms. *Apanteles hyphantriae* Riley and *Meteorus hyphantriae* Riley are important parasites of the caterpillar. A similar number of natural enemies is reported in Europe.

On small trees nests of the fall webworm may be cut out and destroyed. The insect is detrimental mainly to the beauty of the host and is thus more a nuisance than a threat to the health of the tree. Nests always occur terminally on the branches of the host.

Uresiphita (=*Tholeria*) *reversalis* (Guenée), the genista caterpillar, is a pyralid moth that is likely to be distributed transcontinentally in the southern states. An infested host is likely to attract attention because of damaged or lost foliage, webs around clusters of leaves, fecal pellets tangled in the webbing, and the bright contrasting colors of the larvae.

Very little is known about the biology and behavior of this species. There are probably two generations per year in its southern range (Florida, Georgia, Texas, California). In Texas second-generation larvae may be found in August. A single generation occurs in New York. Eggs are laid in overlapping clusters on the undersurface of leaves. The larva's body is pale yellow contrasted with black spots, fine white hairs, and shiny black head (D). If the larvae are disturbed while feeding, they drop to the ground. They pupate in thin white cocoons that fully reveal the pupa. The cocoon may be attached to buildings as well as to vegetation. The moth is brown; the hind wing is yellow with a brownish apex. *Ephialtes sanguineipes* Cushman is the only known parasite.

Its hosts are known to include a crape myrtle, honeysuckle, *Cytisus* species and other brooms, and species of *Buddleia, Laburnum,* and *Sophora.* A closely related species feeds on pine foliage (Plate 4C, D).

The boxwood webworm, *Galasa nigrinodis* Zeller, is a pyralid moth found primarily on English boxwood. In Virginia one instance of the larvae on American boxwood was recorded, and a severe infestation that entailed considerable webbing and leaf damage was observed on Stokes holly (*Ilex crenata* cv. Stokes) in Richmond; the larvae have also been recorded elsewhere in the state. Lepidopterists have reported finding the adults in New York, Pennsylvania, and North Carolina. Boxwood webworms and their damage are very inconspicuous and easily overlooked. The larvae feed on foliage of inner branches and are found only by spreading the plant open to look for partially chewed leaves and frass among webbing along the branches. In one case, branches were found girdled within the webbing. Adults are rusty orange and nearly 1.25 cm long, with broad "shouldered" wings that are rectangular and laid flat dorsally over the body when at rest. They are present in June and July. Young larvae can be found beginning in July and August. By late May they reach a length of 1.25 cm or more and are dark gray to nearly black when mature. They overwinter in webbing along the branches.

References: 193, 218, 362, 516, 533, 548, 584, 816

A. A nest of young fall webworms, *Hyphantria cunea.* [Arctiidae]

B, C, F. The color of the fall webworm caterpillar varies; the black dots are always distinctive.

D. Larvae of the genista caterpillar, *Uresiphita reversalis,* about 30 mm long. [Pyralidae]

E. A large web on a black cherry tree constructed by the fall webworm.

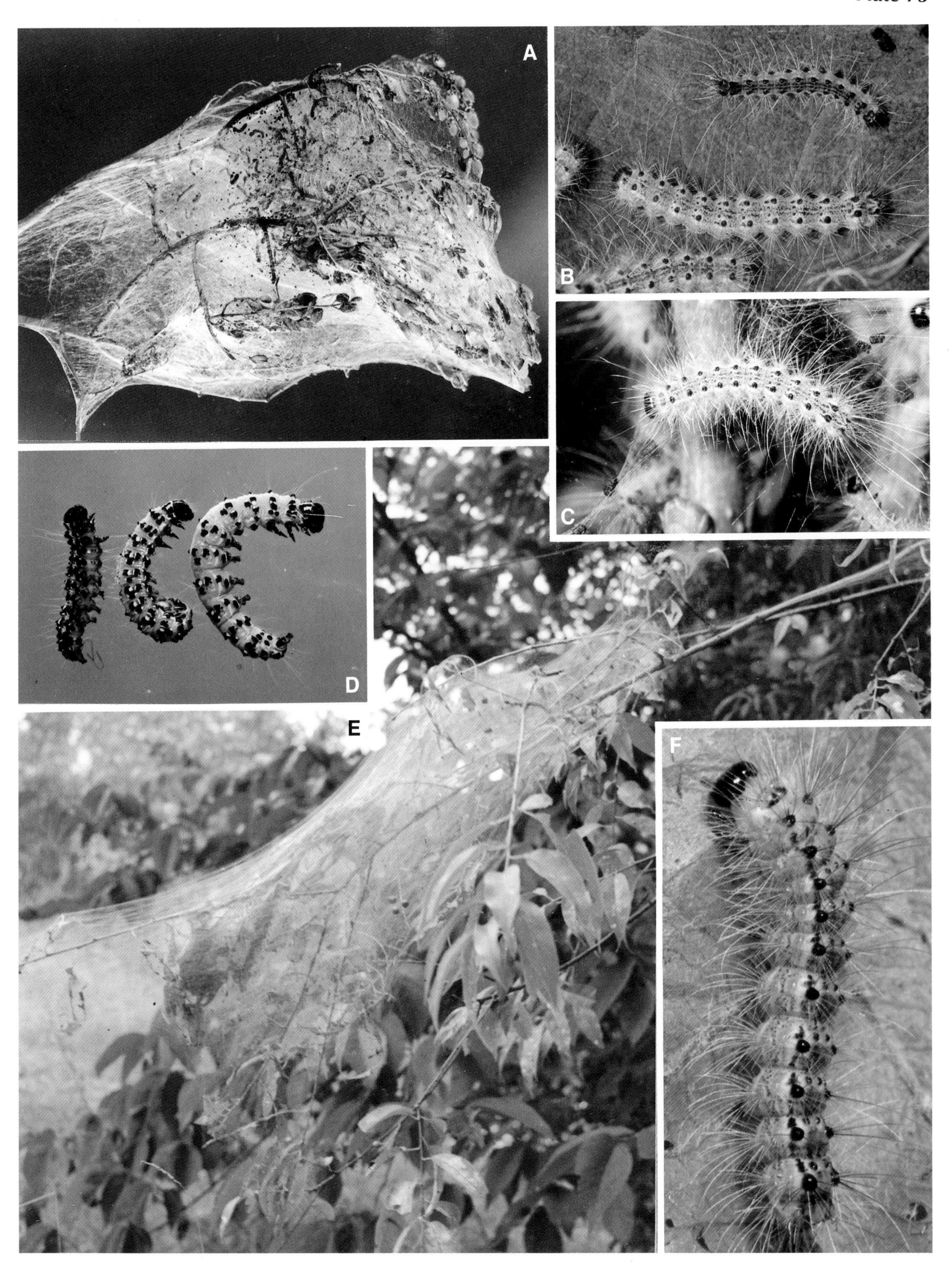
A
B
C
D
E
F

Eastern Tent Caterpillar and Forest Tent Caterpillar (Plate 76)

The eastern tent caterpillar, *Malacosoma americanum* (Fabricius), is a native defoliator whose presence in the United States was reported as early as 1646. Large numbers of this caterpillar occur at intervals of approximately 10 years. Before the advent of huge outbreaks of the gypsy moth, *Lymantria dispar* (Linnaeus), in the 1970s and 1980s, some authorities considered the eastern tent caterpillar to be the most widespread defoliator of deciduous shade trees in the eastern United States. At times the entire countryside seems festooned with its silken tents.

Its name notwithstanding, the eastern tent caterpillar occurs as far west as the Rocky Mountains. Its favorite hosts are wild cherry, apple, and crabapple, but occasionally it feeds on deciduous forest and ornamental trees such as ash, birch, blackgum, redgum, willow, witch-hazel, maple, oak, poplar, cherry, peach, and plum.

This insect overwinters as an egg, which is laid in distinctive masses that encircle the smaller twigs of the host plant (D). These masses are as long as 19 mm, contain 150–350 eggs, and appear varnished.

Larvae hatch from eggs in the spring about the time wild cherry leaves begin to unfold (E). The young caterpillars quickly gather at a major branch fork or crotch and begin to build a web from which they go forth to feed on newly opened leaves. Larvae spin a fine strand of silk wherever they go. As the caterpillars grow, so does the size of the tent. When populations are large, whole trees become covered with webbing and all leaves are devoured.

The fully grown caterpillars are generally black with a white stripe down the back and a series of bright blue spots between longitudinal yellow lines (A). When the larvae reach this stage, they leave the host tree and search for a place to spin white cocoons, on fences, tree trunks, or other natural or constructed objects. Within the cocoon they transform into reddish brown moths with two whitish stripes running obliquely across each forewing. The adults emerge in late June or early July. One generation occurs each year.

Damage can be reduced on small trees by getting rid of the egg masses during the winter or by clipping and destroying the tents and their occupants on rainy, cool days when they are still small. Larvae do not venture out of their tents to feed during inclement weather.

The forest tent caterpillar, *Malacosoma disstria* Hübner, does not form a tent in the usual sense. The larvae construct a silken mat on the trunk or a branch of the host tree. From there they forage in all directions but begin by concentrating on one branch at a time. The egg masses, like those of *M. americanum,* encircle twigs and are coated with a varnishlike substance called spumaline, but they are square at the ends, in contrast to the spindle-shaped masses of *M. americanum.* Overwintering takes place as a fully developed embryo inside the egg shell. When they hatch, the larvae tend to migrate high in the tree, where they feed on expanding flower and leaf buds. After buds open, larvae feed on foliage. They are gregarious during early instars. At molting time they form the silken mats, usually on the tree trunk. The fully grown larva of *M. disstria* (C) is easily distinguished from that of *M. americanum* by the series of keyhole-shaped spots along its back (*M. americanum* has a single solid stripe).

Pupation occurs in pale yellow cocoons often spun in folded leaves and attached to any nearby vegetation and occasionally to buildings. The adult is a tan moth, about 40 mm long with two dark brown oblique stripes on each forewing. Flight and mating activities begin in late afternoon and continue through most of the night. Moths are attracted to lights at night. There is one generation each year.

The forest tent caterpillar is a major defoliator of deciduous trees from Louisiana and Georgia to the Canadian provinces. In Louisiana it is a severe pest of water tupelo and other species of *Nyssa.* In Canada its principal host is trembling aspen. Other hosts include sweetgum, various species of oak, birch, ash, maple, elm, and basswood.

Parasites are often abundant in populations of *M. disstria,* which probably accounts for the cyclic nature of outbreaks. Parasites include a fly, *Sarcophaga aldrichi* Parker, and several parasitic wasps. *S. aldrichi* can be so abundant that it becomes almost as much of a nuisance as the caterpillar. Superficially the fly resembles a large house fly. A polyhedrosis viral disease also occurs in nature and may play a major role in reducing populations of *M. disstria. Bacillus thuringiensis,* a commercially available bacterium, is a good microbial means of caterpillar control. Several fungi in the genus *Entomophthora* are known pathogens of the forest tent caterpillar. Some of these are good candidates for commercial production in future biological control programs.

See Plate 77 for tent caterpillars that occur principally in the western United States.

References: 49, 193, 483, 660

A. A mature larva of the eastern tent caterpillar, *Malacosoma americanum*; about 5 cm long. [Lasiocampidae]

B. *Prunus serotina* defoliated by the eastern tent caterpillar. Note tent at branch fork (*center*).

C. A mature larva of the forest tent caterpillar, *Malacosoma disstria*; about 4 cm long.

D. An egg mass of the eastern tent caterpillar.

E. Young larvae of the eastern tent caterpillar, recently hatched from egg mass (*center*).

A
B
C
D
E

Western Tent Caterpillars (Plate 77)

Several species of tent caterpillars of the genus *Malacosoma* are important pests of woody ornamental plants in the western United States. Their names, principal hosts, and known distributions are given in Table 10.

Although some authors report the absence of larval tents in localized populations of *M. californicum* and *M. constrictum,* generally the life cycles and habits of tent caterpillars in the West are similar to those species found in the East (Plate 76). Their immature stages are similarly attacked by a variety of natural control agents. For the western tent caterpillar, the most important parasitic wasps are *Tetrastichus malacosomae* Girault, an egg parasite, and *Bracon xanthonotus* Ashmead, a parasite of mature larvae.

Tent caterpillars that commonly occur in the East are shown in Plate 76.

References: 87, 275, 751, 752

Table 10. **Tent caterpillars of the western United States**

Name	Principal hosts	Distribution
Forest tent caterpillar, *Malacosoma disstria*	Poplar, birch, alder, oak, willow, cherry, peach, plum, prune, pear, apple, quince, hawthorn, rose	Continental U.S. (except AK), southern Canada
Pacific tent caterpillar, *M. constrictum*	Oak	CA, OR, WA
Sonoran tent caterpillar, *M. tigris*	Oak	Southern Great Plains, southern Rocky Mountain states, Mexico
Southwestern tent caterpillar, *M. incurvum*	Poplar, willow, *Prunus* spp.	CO, UT, NV, AZ, central Mexico
Western tent caterpillar, *M. californicum pluviale*	Oak, willow, poplar, birch, alder, madrone, ceanothus, redbud, hazel, ash, California holly, apple, almond, apricot, cherry, prune, plum, California coffeeberry, currant, antelope brush	Rocky Mountain states and westward to Pacific Ocean, NY

A. Egg masses of the western tent caterpillar, *Malacosoma californicum pluviale,* on birch seed pods. [Lasiocampidae]
B. Aggregation of larvae of the western tent caterpillar on coast live oak.
C. Western tent caterpillar larvae, as found on apple in western New York.
D. Mature larvae of the western tent caterpillar.
E. A tent of the western tent caterpillar, with larvae crawling over its surface. Tents are enlarged as larvae grow. Note the fecal droppings of caterpillars in the web.
F. Defoliation of a branch of coast live oak by a colony of the western tent caterpillar.
G. A cocoon of the western tent caterpillar. (Courtesy D. A. Leatherman, Colorado State Forest Service.)

A
B
C
D
E
F
G

Uglynest Caterpillar, Oak Webworm, Rollers, and Tiers (Plate 78)

Some insects do a tidy job of chewing up their host trees, but not the uglynest caterpillar, *Archips cerasivorana* (Fitch), or its near relative, the oak webworm, *A. fervidana* (Clemens). These caterpillars spin a dense web around the feeding site that becomes filled with their excrement and bits and pieces of leaves. The larvae are gregarious and nearly 20 mm long when fully grown. They are yellowish green with shiny black heads. The larvae pupate in the mess they create. The moth of the uglynest caterpillar is dull orange with a wingspan of 18–25 mm. Emergence of the adult occurs from July to September, depending on climate. The eggs are laid in the stem or trunk in masses of 25–200 in late summer and early fall and remain attached to the host plant all winter. The individual egg masses are oval, brown, and covered with a semitransparent rubbery secretion. Eggs hatch in May and June, and the larvae begin building a web nest, where they remain until they emerge as moths.

The uglynest caterpillar is generally distributed through the northern states and Canada, where its favorite hosts are roses, hawthorn, and cherry, although it is found on other species. It seldom does any lasting damage to its host, but its ugly nests are detrimental to the beauty of ornamental plants.

The adult oak webworm is a brownish moth with a slightly shorter wingspan than the adult uglynest caterpillar. The hind wing is gray, whereas that of its close relative is orange. The larvae (C) are generally greener than the uglynest caterpillar (D). The oak webworm feeds on scrub, red, black, and scarlet oaks. Its life cycle and habits are similar to that of the uglynest caterpillar. A sex pheromone has been identified for the oak webworm.

The leaf tiers *Croesia albicomana* (Clemens) and *C. semipurpurana* (Kearfott), the oak leaftier, are often associated with *A. semiferana*. Their larvae are about 12 mm long when fully grown and are a dirty white to light green. They have a pale brown head and brown to black thoracic legs. Newly hatched larvae are found in Connecticut from mid-April to early May. Larvae that hatch early enter unopened buds and feed on the newly forming leaves. Large populations can destroy nearly all the buds on a tree. Those buds that survive frequently produce leaves with a series of round holes, symptoms of earlier injury. Older larvae feed more openly, since they are somewhat protected by their webbing and folded leaves. Mature larvae drop by means of silken threads to the ground, where they pupate. Moths emerge in June and July and lay eggs on twigs with rough bark. One generation occurs each year, but damaging populations usually appear in cycles of several years.

C. albicomana and *C. semipurpurana* are found from Ontario and Quebec south to Texas and California. Injury occurs in late April and early May in Connecticut and Pennsylvania. Both have a preference for red and scarlet oaks in the Northeast, and both cause spring defoliation, which results in greater injury than does midsummer defoliation.

The fruittree leafroller, *A. argyrospila* (Walker), is also a pest of oak, hawthorn, white birch, elm, maple, hickory, black and Persian walnut, California buckeye, poplar, cherry, pear, and rose. It is a common pest of apple and crabapple. Its larvae feed on leaves and fruit in the spring. One generation occurs each year. It may be found wherever host plants grow.

A. negundana (Dyar) occurs in widely scattered locations—from Florida to Washington, and in Canada. Its principal host is boxelder, although it has also been observed on honeysuckle, American elm, alder, and paper birch.

The behavior of the larvae is much like that of other solitary *Archips* species. The larvae roll the leaves and feed from the inside. Larvae reach 20 mm long. Damage occurs in the spring, and there is one generation each year.

See Plates 99–101 for other leaf tiers and rollers.

References: 37, 50, 193, 574, 592, 627

A. Black cherry twigs, webbed together by the uglynest caterpillar, *Archips cerasivorana*. [Tortricidae]
B. Nest of the uglynest caterpillar, opened up to show the young larvae.
C. A larva of the oak webworm, *A. fervidana*, on an oak leaf.
D. A larva of the uglynest caterpillar on cherry.
E. The anterior half of a larva of *A. semiferana* (see also Plate 99). (Drawing by Joseph A. Keplinger; from Chapman and Lienk, ref. 110, courtesy New York State Agricultural Experiment Station, Geneva.)
F. A nest of uglynest caterpillars on rose.
G. A nest of uglynest caterpillars on *Ilex glabra*.
H. Web and injury of the oak webworm. Note larva at arrow.

A
B
C
D
E
F
G
H

Euonymus Caterpillar (Plate 79)

Yponomeuta cagnagella (Hübner), the euonymus caterpillar, a European species widespread on the continent as well as the British Isles, was first reported in Ontario in 1967. The larvae feed in colonies that envelop the foliage in large silken webs (A). They are defoliators primarily of *Euonymus europaea* (the tree form); *E. kiautschovicus, E. alatus,* and in Europe they are also reported on *E. japonicus*. In North America it is currently known in Ontario, Michigan, and in several New York localities. On this side of the Atlantic it must be considered as a potential pest of at least the euonymus species named here.

The female moth lays eggs in mid to late July in New York, usually on twigs, branches, and in the axils of buds on *E. europaea*. During oviposition the eggs are covered with a gummy secretion that hardens to a scalelike form that makes them difficult to see. The eggs hatch by mid-August, and larvae immediately prepare to overwinter under their eggshells. There is no further activity until early in the following year, when larvae begin to make small webs and feed gregariously on new leaves. As larvae mature, the size of the web increases and soon envelops large branches. Fully grown larvae are about 20 mm long. By late June cocoon formation begins (B, C). The cocoons are constructed side by side and hang with the longer axis in a vertical position. The open end is up. Before pupating, the larva positions itself within the cocoon so that its head faces toward the opening of the cocoon. The adult emerges in late June in Ontario. There is one generation per year. A pheromone has been used to trap females and is a potential means of pest management.

The moth's forewings have attractive black spots on a white background. Wingspread is about 24 mm (Figure 53). The larva and moth of *Y. cagnagella* look much like those of the ermine moth *Y. padella* (Linnaeus) and *Y. malinella* (Zeller). Moths of the three species cannot be separated on the basis of morphological characters. *Y. padella* larvae feed on apple and hawthorn, and *Y. malinella* appears to limit its feeding to apple foliage. *Y. malinella* is thought to occur only in the Pacific Northwest and British Columbia. Young larvae overwinter, resulting in much early defoliation the following spring, first from leaf mining by solitary larvae, later from gross feeding by colonies of caterpillars in tents.

The ailanthus webworm, *Atteva punctella* (Cramer), while often of litte consequence, can occasionally defoliate tree of heaven, *Ailanthus altissima*. This insect is an ermine moth, and like other ermine moth caterpillars, the larvae cluster together in a loose web. The larvae each have five white longitudinal lines on an olive-brown background. They may be found in late summer.

Reference: 375, 883

Figure 53. Yponomeuta cagnagella moths showing their distinctive black spots. All three *Yponomeuta* species mentioned here have the same color pattern. (Courtesy Peter Scorrar, Univ. of Guelph.)

A. A nest of *Yponomeuta cagnagella* caterpillars feeding on the leaves of *Euonymus alatus*.
B. A group of cocoons. Note their vertical orientation.
C. Close-up of cocoons. Some of the larvae have not pupated.
D. Mature caterpillars.
E. Close-up of a cocoon being constructed by the larva.

A
B
C
D
E

Bagworms (Plates 80–81)

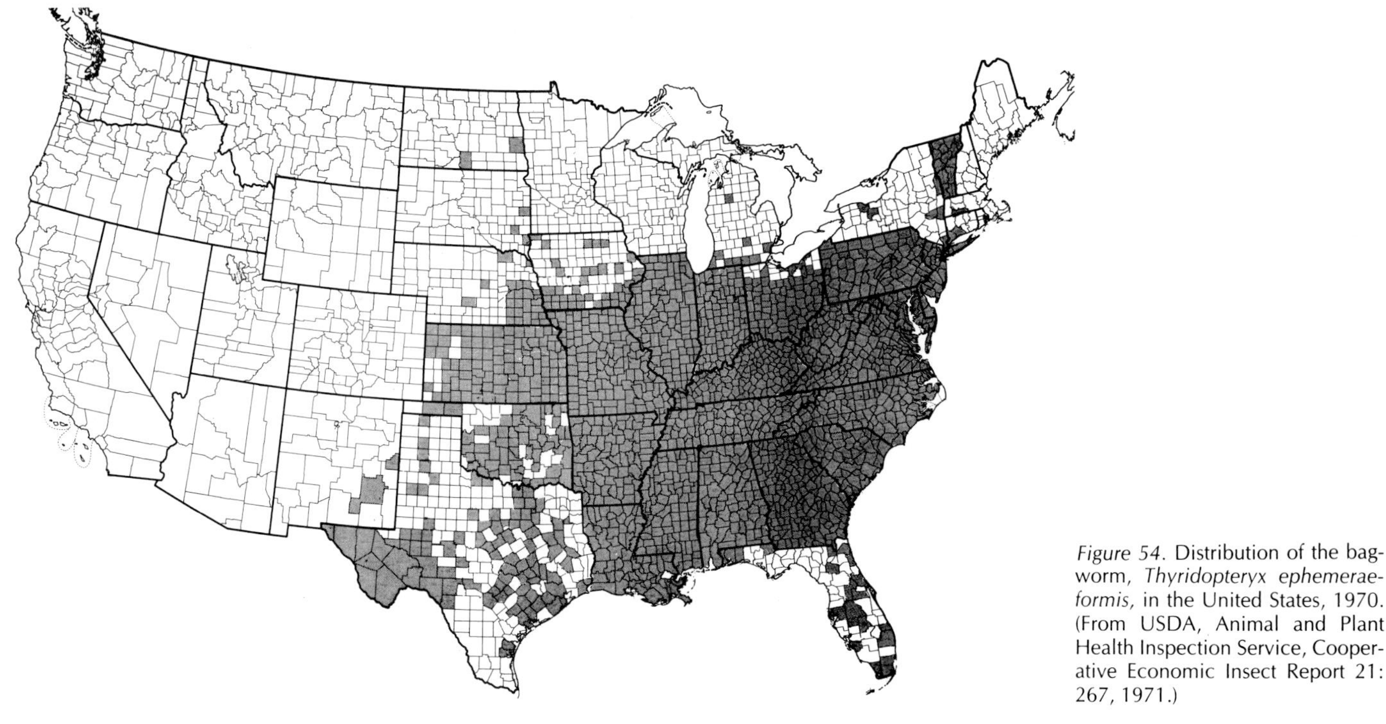

Figure 54. Distribution of the bagworm, *Thyridopteryx ephemeraeformis,* in the United States, 1970. (From USDA, Animal and Plant Health Inspection Service, Cooperative Economic Insect Report 21: 267, 1971.)

There are 20 species of bagworms in the United States, but only 2 or 3 cause enough damage to plants to be of economic importance. This plate and the next show the bagworm, *Thyridopteryx ephemeraeformis* (Haworth). Indigenous to America, it has a wide range of host plants as well as an extensive geographic distribution (Figure 54). According to Davis (149), the bagworm feeds on a total of 128 plant species. It is a common pest east of the Rocky Mountains but is less common in latitudes north of Massachusetts. It can be devastating in the South as a pest of woody ornamentals.

The female bagworm never looks like a moth. She is eyeless and lacks wings, legs, antennae, and functional mouthparts. Her body is soft, yellowish white, and almost devoid of hairs; she never leaves the bag that she made as a larva. The male moth emerges from his bag and flies to the female, mates, and dies in a few days. Males are black and have almost clear wings that span about 25 mm.

From the pairing, the female produces 500–1000 eggs in a single mass, all within the bag. The egg is the overwintering stage. In Florida, however, there is no dormant period. Eggs hatch in the mid-Atlantic states in late May and early June, and larvae immediately begin to feed and construct their protective cases.

The bag is made of silk and bits of twigs or leaves interwoven to disguise and strengthen the case. When the larva is small, it feeds on the epidermis on the upper side of the leaf with the bag pointed upward (Plate 81B). Epidermal feeding results in a brown spot on the leaf. Later, if feeding on a broad-leaved host, the larva moves to the lower surface and eats all the leaf except for the larger veins. As the larva grows, it enlarges its bag. When the larva is fully grown, its bag may be 30–50 mm long. It pupates inside the bag, a structure homologous to a cocoon. The change from pupa to adult requires 7–10 days, depending on temperature.

Dispersal of this insect from one plant to another depends on movement of the caterpillar. Because of the bagworm's feeding behavior and limited movement, a lone host plant may harbor a huge population in a single season.

The principal harm done by this insect is the destruction of foliage by the caterpillars. In some parts of the country bagworms are found predominantly on arborvitae and juniper. In Ohio and Pennsylvania they are commonly found on hardwoods such as black locust, maple, and sycamore. Other hosts include elm, buckeye, and boxelder. In Maryland and Virginia bagworms can be found on arborvitae, 13 types of juniper, eastern redcedar, Chinese elm, honeylocust, and eastern white pine, loblolly and Virginia pines, hemlock, Norway maple, deodar cedar, and spruce. In Texas they flourish on cedar, cypress, and willow. In Florida they may be found feeding on the new leaves of cabbage palmetto.

In northern Florida fully grown larvae, as well as first-stage larvae, may be seen during the winter months. This indicates that continuous generations probably occur each year (no hibernation or overwintering period). In southern California another species, *Oiketicus townsendi* Cockrell, commonly feeds on ash, pear, sycamore, willow, and locust. In Florida and the Gulf Coast states another bagworm species, *O. abbotii* Grote, feeds on several subtropical trees and shrubs, including citrus.

There are several parasitic insects that are important in reducing numbers of bagworms. The ichneumon wasp parasites *Itoplectis conquisitor* (Say) and *Chirotica thyridopteryx* (Riley) are perhaps the most important. Unfortunately, populations of the bagworm usually increase to seriously damaging proportions before parasites are effective. *Bacillus thuringiensis,* a bacterial insecticide, is effective in controlling bagworms.

A sex pheromone has been identified that, when used in traps to lure the moths, has successfully interfered with the male moth's mating behavior, resulting in a nonpolluting and effective means of control. Unfertilized eggs do not hatch.

References: 149, 193, 231, 393

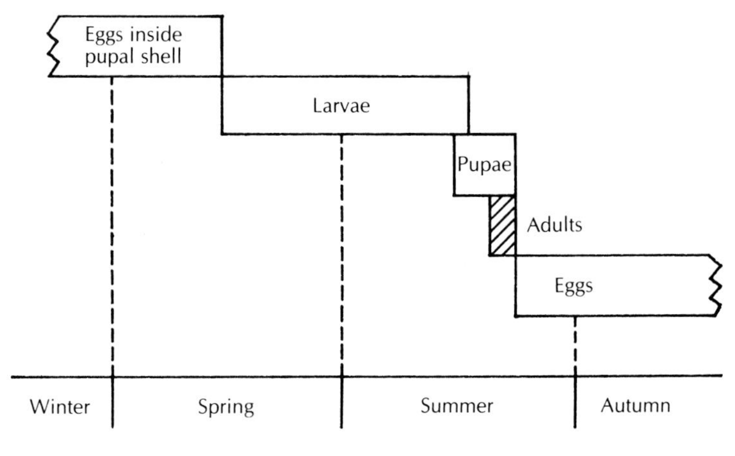

Figure 55. Seasonal development of the bagworm in Illinois. (After J. E. Appleby, Illinois Natural History Survey.)

A. A black locust, *Robinia pseudoacacia,* partially defoliated by the bagworm, *Thyridopteryx ephemeraeformis.*

B. A bag with a larva partially exposed. The larva is feeding on a leaf stem.

C. Bagworms on eastern redcedar, *Juniperus virginiana.* [Psychidae]

A
B
C

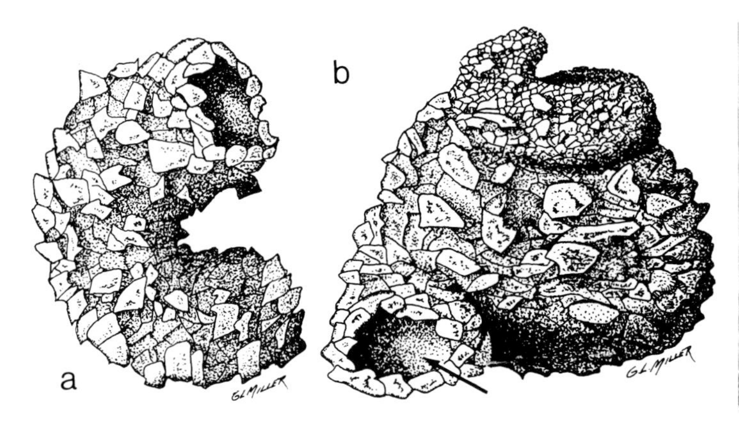

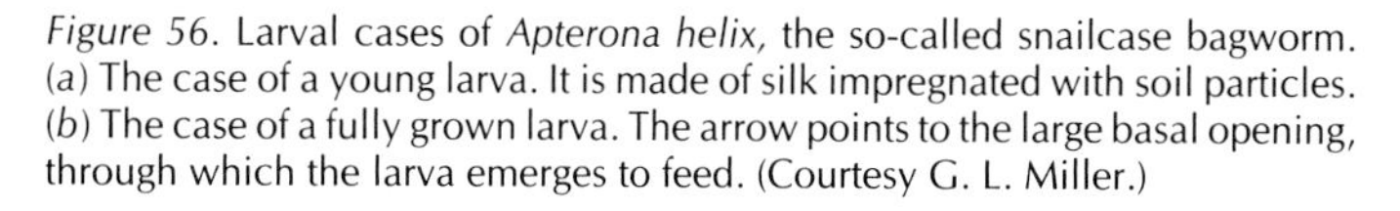

Figure 56. Larval cases of *Apterona helix,* the so-called snailcase bagworm. (*a*) The case of a young larva. It is made of silk impregnated with soil particles. (*b*) The case of a fully grown larva. The arrow points to the large basal opening, through which the larva emerges to feed. (Courtesy G. L. Miller.)

Figure 57. A larva of *Apterona helix* protruding its head and thorax from its case. (From Suomi and Akre, ref. 906, courtesy D. Suomi and R. D. Akre, Washington State Univ.)

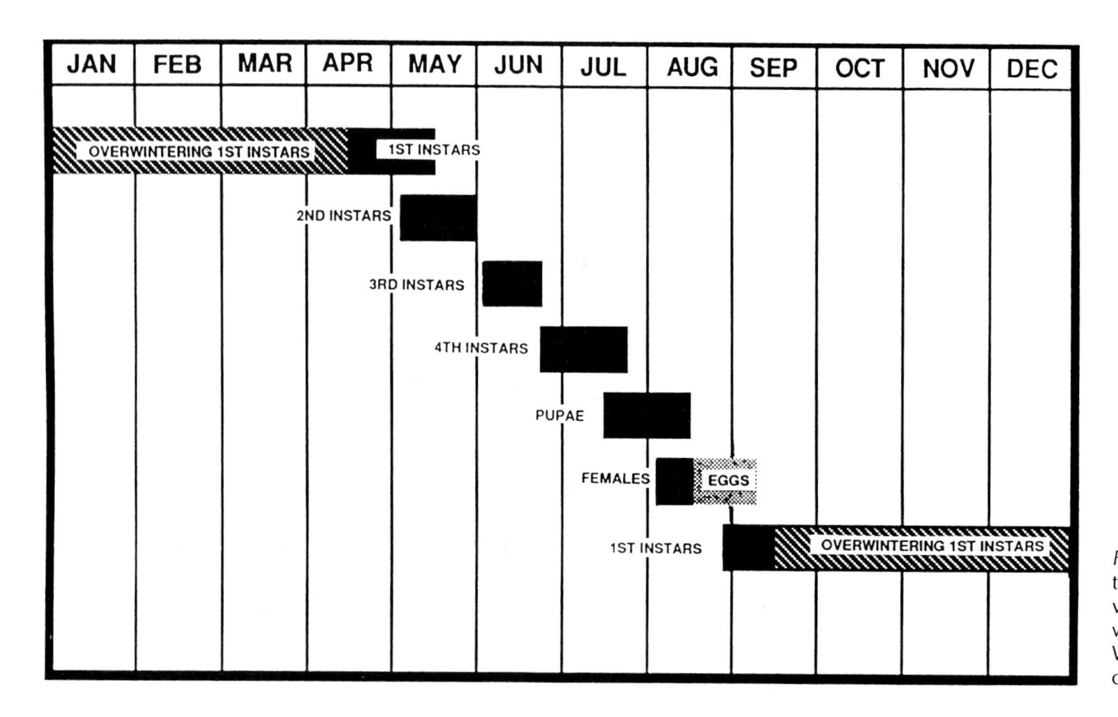

Figure 58. Inferred seasonal history of *Apterona helix* in Pennsylvania. The first-instar larvae overwinter in the female case. (From Wheeler and Hoebeke, ref. 913, courtesy A. G. Wheeler.)

Apterona helix (Siebold), regionally known as the garden bagworm or the snailcase bagworm, is a European species that was first recorded in the United States in Nevada County, California, in 1940. Since then it has extended its range to include all states west of the Rocky Mountains. In the Northeast it was first recorded in New York in 1962 and now occurs in several states in the region including Michigan and Pennsylvania.

This highly unusual insect is believed to be exclusively parthenogenetic in the United States, for no males have ever been found. In Europe, however, equal numbers of males and females occur in some populations, although parthenogenesis is known to occur there also. The most characteristic attribute of this insect is its larval case (Figures 56–57), which alone enables recognition.

The larvae feed at night by protruding their heads from their snaillike cases (Figure 57). In California all feeding takes place from mid-April through early July, a pattern essentially followed in Pennsylvania as well (Figure 58). Pupation takes place in the case, with transformation to adult occurring throughout July. The wingless, somewhat wormlike females each deposit several dozen eggs inside the case. The young larvae overwinter within the case, which is typically attached high off the ground to a tree trunk, branch, or building wall. On emergence from the parent case in the spring, the larvae are naked, but they soon begin to construct their own cases, in which they remain until they die. At first the case consists of a few dozen grains of soil that adhere to the body. As the larva grows, it adds more soil particles, so that at maturity the case is helical and 3–5 mm in diameter and 5 mm in depth. The color of the case varies depending on the soil from which it is built.

A. helix feeds on a wide variety of vegetation. First-instar larvae are leaf miners and cause circular clear or yellowish imperfections in the leaf blade. This feeding is similar to the leaf mining done by casebearers (*Coleophora*) (see page 186). As the larvae age, they will skeletonize as well as bite through and ingest all tissue layers of the leaf. Woody ornamental food plants include rose, apple, pear, cherry, *Ceanothus,* grand fir, Douglas-fir, and ponderosa pine. Its importance as a nuisance on the external walls of buildings, outdoor furniture, and the like may exceed its importance as a plant feeder.

References: 149, 904, 906, 913

A–C. *Feijoa sellowiana* foliage showing feeding injury by first- and second-stage larvae of the bagworm, *Thyridopteryx ephemeraeformis.* Note that the first-stage larva is on the upper surface of the leaf in B.

D. This bag contains a pupa.

E. A fully grown larva removed from its bag.

F. A bag opened to show a pupa.

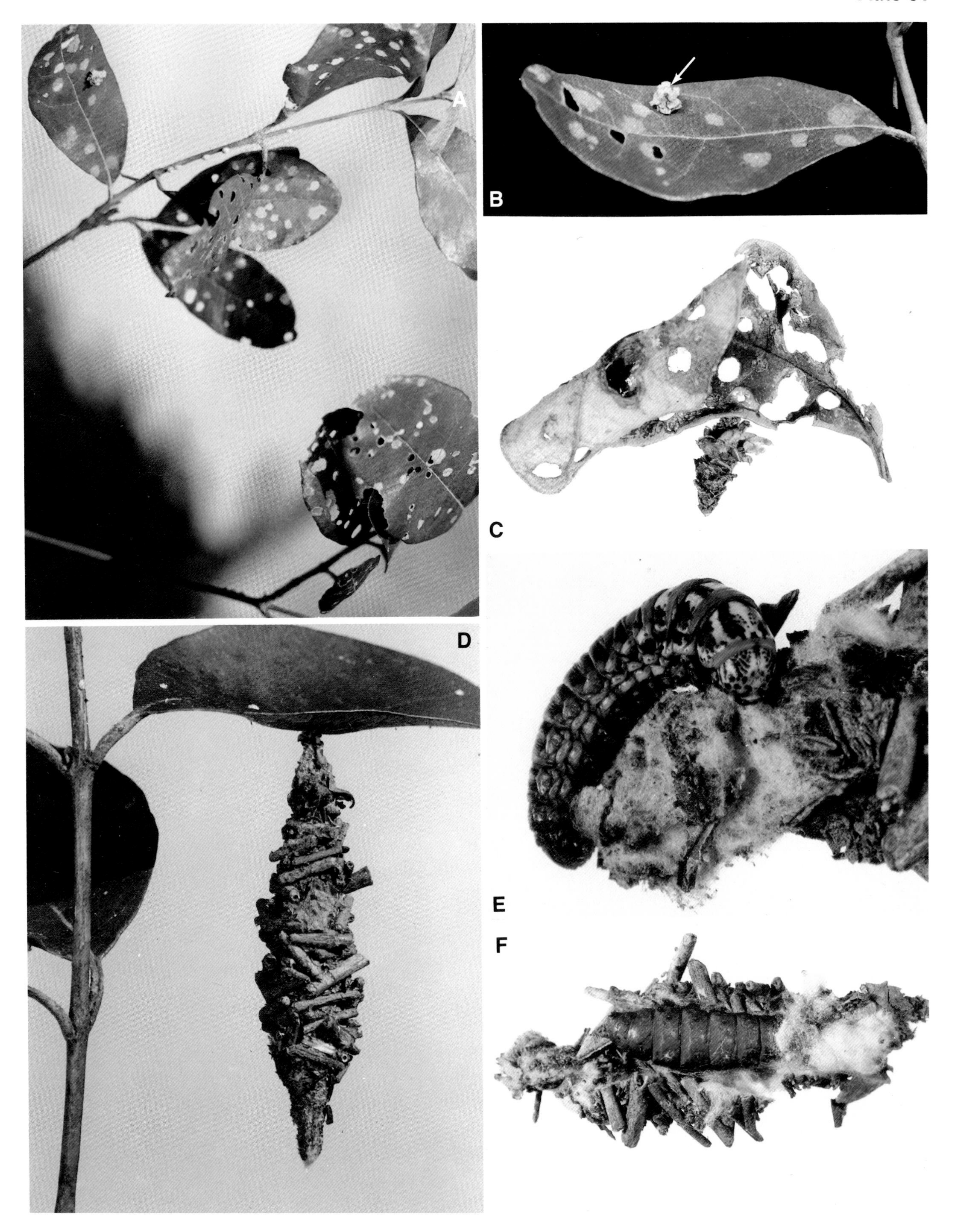
A
B
C
D
E
F

Mimosa Webworm (Plate 82)

The mimosa webworm, *Homadaula anisocentra* Meyrick (=*albizziae* Clarke), came to the United States from China. It was first detected in the United States in 1940 on mimosa, *Albizia* (=*Albizzia*) *julibrissin*. It is from this plant that it gets its common name. In the United States, however, the honeylocust, *Gleditsia triacanthos*, including the thornless cultivars, is a more important host.

Although the mimosa webworm is primarily an eastern and midwestern pest, several serious infestations were detected in California in the 1960s. In the South mimosa is heavily attacked by the webworm. It is reasonable to believe that eventually the insect will be found wherever its host plants are grown in the United States.

Observations made in Indiana indicate that the thornless honeylocust (cultivar Sunburst) is highly susceptible to attack by the mimosa webworm, whereas the cultivars Moraine, Shademaster, and Imperial are less susceptible. Since all were subject to attack, however, it is recommended that none of the thornless cultivars be planted without provision for annual inspection and control.

The adult is a silvery gray moth with wings that are stippled with black dots. The wingspan is about 13 mm. When fully grown, the larvae are about 16 mm long and vary from gray to brown. On the body of the larva are five longitudinal white stripes.

The insect overwinters in the pupal stage in cocoons located under scales of bark on the trunk of the host tree, or in plant refuse underneath the tree. In the Midwest the moths appear in early to mid-June and lay their pearly gray eggs on tree leaves. The larvae web together several leaflets and feed on the foliage within the protection of the web. Commonly, a number of larvae feed together and construct a large, unsightly nest of webbing (E). When alarmed, the larvae can move rapidly and may drop from the nest on a thread of silk. Upon reaching maturity, the larvae pupate, a new generation of moths appears in August, and the cycle is repeated. It is the pupa of this second generation that overwinters. In western New York this second generation emerges in early September, but offspring generally do not survive the winter. In Florida, southern Georgia, and Alabama three or more broods emerge each year between May and September.

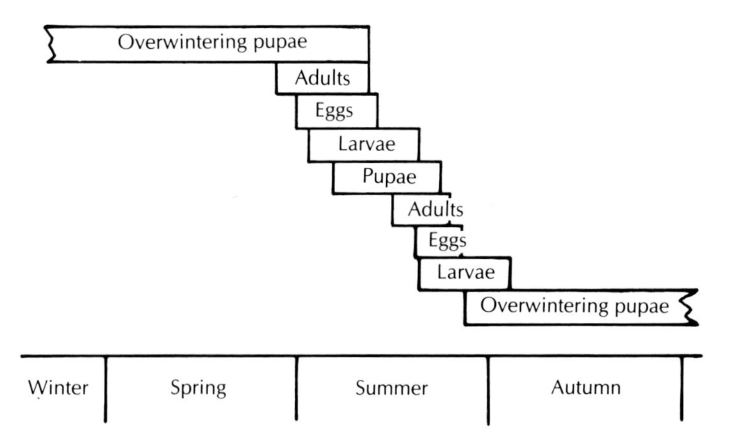

Figure 59. Seasonal development of the mimosa webworm in Illinois. (Courtesy J. E. Appleby, Illinois Natural History Survey.)

Often, an entire tree may appear covered by an unsightly tent of webbing (A). Although foliage is lost as a result of larval feeding, it is difficult to assess the extent of damage to the leaves, since this injury is often hidden by the webbing.

References: 8, 16, 210, 568, 822

A. Field view of mimosa damaged by the mimosa webworm.
B. Brown foliage and tents of webbing are characteristic of attack by the mimosa webworm. [Plutellidae]
C. Close-up of the webbing and brown leaflets caused by the feeding of young larvae.
D. Several mature larvae and their protective tent.
E. Webbing made by larvae of intermediate size. Note the larva at arrow point.

A
B
C
D
E

A Bougainvillea Caterpillar (Plate 83)

Bougainvillea is a popular subtropical flowering shrub. It can be found in Florida, in the Gulf Coast area, and in California. Damage caused by the feeding of the caterpillar of a pyralid moth, *Asciodes gordalis* Guenée, limits its use as an ornamental shrub and destroys its beauty. Although the insect's biology has not been studied, it has been observed in Florida every month of the year except June, an indication that there are several generations each year. After folding a leaf, the larva feeds on it from inside the fold. When the partially devoured leaf ceases to afford some protection, the larva moves on to another leaf, folds it, and continues to feed. As a consequence of this process, the foliage of the infested plant generally appears ragged.

The larvae are about 24 mm long when fully grown and are about the same green color as the foliage (C).

Figure 60. A wasp parasite, *Meteorus pulchricornis* (Wesmael), laying an egg on a gypsy moth caterpillar. Related species attack the bougainvillea caterpillar. (Courtesy Ronald Weseloh, Connecticut Agricultural Experiment Station.)

A, B, D, F, G. These photographs show the extent and kind of damage caused by *Asciodes gordalis*.

C. The bougainvillea caterpillar. Note the protective coloration.

E. An adult female *A. gordalis*. The wingspan is approximately 22 mm. [Pyralidae]

A
B
C
D
E
F
G

Birch Leaf-mining Sawflies (Plate 84)

There are numerous species of insects that live and feed inside leaves. Such specialized feeders may be collectively described as leaf miners or gall makers. The leaf miners feed between the two epidermal layers and devour the spongy mesophyll, the palisade cells, or both. The mine symptoms take the form of a blotch or a winding gallery called a lineal mine. Leaf miners are among the smallest of leaf-eating arthropods. Their habitat is distinguished from that of all other foliage-feeding insects, restricting their movement but providing some protection from predators. The upper and lower epidermis of the leaf remains intact except in the case of infestations of a group called the shield bearers (Plate 96).

Leaf miners must accommodate to cramped living quarters. The thinner the leaf, the flatter or tinier the larva. The larval form may change as the leaf miner develops and as space is provided in the mined-out area. Leaf miners often have specialized mouth parts. The head and mouthparts must remain flattened if larvae are to continue to feed on the mesophyll layer within the leaf. Some have hooks that perform like a scythe. The larvae lacerate cell walls and suck the fluid contents into the mouth. Others have mandibles that may face forward or downward. If they face forward, the larvae must feed forward and in one plane. Their food is made up of both solids and liquids. Those leaf miners whose mandibles face downward are often capable of making several mines, either in the same leaf or an adjacent one. They are usually voracious feeders and eat almost without interruption.

The next 12 plates illustrate leaf miners in the wasp, fly, moth, and beetle families.

A sawfly called the birch leafminer, *Fenusa pusilla* (Lepeletier), is a common pest of gray birch, *Betula populifolia,* and paper birch, *B. papyrifera.* It rarely feeds on black, yellow, European white, or river birch. Entomologists in Connecticut first recorded the sawfly's appearance in the United States in 1923. It came from Europe and has rapidly spread throughout northeastern North America.

The adult sawfly is black and is about 3 mm long (E). It appears in the spring when the first birch leaves are half grown. The adults may be visible when they hover about birch trees or crawl over the surface of leaves. Eggs are laid singly in the new leaves and may be easily seen if the leaf is held up to the light. They are positioned between the palisade and the spongy mesophyll layers. Eggs soon hatch, and the larvae begin to mine the leaf, feeding on the tissue between the leaf surfaces. At first, the mines are separate and small, but soon several of them coalesce to form a single large, hollowed-out blotch on the leaf (C).

The larva is flat. When fully grown it is about 6 mm long. The period of larval development varies from 10 to 15 days. When mature, the larva cuts a hole through the leaf and drops to the ground. There, it builds a cell in which pupation takes place; 2–3 weeks are required for transformation into the adult stage.

The female sawfly deposits her eggs only on newly developing leaves, never on older leaves. In the spring every leaf of a tree may be mined, giving it a brown color. A new crop of leaves soon develops, but by this time a second generation of egg-laying sawflies are again in the trees. Two to four generations of this sawfly occur per year, depending on the length of the growing season. In the Catskill region of New York three generations commonly occur.

Vigorous, growing trees are the most attractive to ovipositing adults. Such trees are able to withstand the attack of these insects for several years before showing symptoms of decline.

Investigations on the biological control of this insect in the United States have produced very little practical information. Work by Fiori and others (246, 247) has produced evidence that *Betula davurica* is resistant to *F. pusilla* oviposition and that there is reduced egg laying in *B. maximowiczana, B. schmitii,* and *B. costata* by 95–100%, as compared with *B. populifolia.* Use of these birches would likely eliminate *F. pusilla* as a pest in ornamental plantings.

Three additional leaf-mining sawflies on birch have become increasingly numerous and are found from the Canadian maritimes and New England to the prairie provinces and Minnesota. They, too, are of European origin. *Profenusa thomsoni* (Konow) and *Heterarthrus nemoratus* (Fallén), also known as the birch-leafmining sawfly, are both late-season feeders. In contrast to *F. pusilla,* they attack mature leaves more within the crown and produce a single generation each year. *P. thomsoni* makes its blotch mine away from the leaf margins. *H. nemoratus* makes small blotch mines that usually begin at leaf margins. *P. thomsoni* females lay their eggs singly in minute slits cut in the upper epidermis of leaves. As the larvae develop, they are found in groups that feed in irregularly shaped mines. Fully grown larvae are about 7 mm long. Prior to leaf drop the larva cuts a hole in the epidermis, drops to the ground, and burrows into the soil, where it constructs an overwintering cocoon. *H. nemoratus* females lay their eggs in small pockets in a leaf tooth on the leaf margin. The larvae make brown blisterlike blotches on the upper surface of the leaf. When fully grown, they are about 8 mm long. At maturity larvae wrap themselves in a silk cell in the center of the mine and overwinter inside the fallen leaf on the ground.

Messa nana (Klug) is less common than and not as widespread as the birch leafminer, perhaps because it has been in North America for the shortest time. The two leafminers coexist on paper birch and oviposit in the spring at about the same time but apparently do not compete. *M. nana* lays its eggs at the ends of the leaf's teeth and on fully expanded leaves situated proximally on the branch. Injury, though similar to that caused by the birch leafminer, is more likely to be at or near the tip of the leaf. Males are unknown. Females appear in late May (southern Ontario). There is one generation each year.

Two exotic ichneumoid parasites, *Lathrolestes nigricollis* (Thompson) and *Grypocentrus albipes* Ruthe, have been established in several northeastern states. Over time these biological control entities should reduce birch leafminer populations.

References: 49, 158, 246, 247, 660, 776

A. Birch leaves after the first brood of leaf miners have completed their development.

B. Larvae of the birch leafminer, *Fenusa pusilla,* inside a leaf.

C. Leaves of *Betula populifolia* showing an early stage of damage caused by the birch leafminer.

D. A severely infested birch tree with completely brown leaves, the result of damage done by birch leafminers.

E. Adult sawflies, *Fenusa pusilla.* [Tenthredinidae]

Sawfly Leaf Miners and Case Bearers (Plates 85–86)

A small, legless larva—that of the elm leafminer, *Fenusa ulmi* Sundevall—is responsible for the injury shown in panels A and B. *F. ulmi* is a European species that entered this country before 1898. In the United States and Canada it attacks Scotch and Camperdown elms (cultivars of *Ulmus glabra*); English elm, *U. procera*; and American elm.

The adult insect is a black sawfly, about 3 mm long. In New England it may be found flying about its food plant in May. It lays eggs in the leaf tissues through slits the insect cuts with its sawlike ovipositer in the leaf's upper epidermis. In about 1 week the eggs hatch, and the young larvae begin to make their blotchlike mines in the leaves. The mines first appear as tiny whitish spots. The larvae confine their feeding to the area between the leaf's lateral veins and eat the green tissue between the two layers of epidermis. Several mines may coalesce to form large blotches.

The larva completes its development in late spring, cuts a hole through the epidermis, drops to the ground, and burrows into the soil to a depth of nearly 25 mm. It then prepares a papery cocoon within which it remains as a pupa through the summer, fall, and winter. One generation occurs each year.

Damaged leaves may remain on the tree throughout the growing season (A). Injured areas in the leaves turn brown and eventually drop out, leaving irregular holes. Such injury spoils the appearance of ornamental trees and reduces the tree's vigor and growth.

The insect is well established in the Northeast and in the Great Lakes region, including southeastern Canada. It has been observed as far west as North Dakota, Minnesota, and Wisconsin.

The elm casebearer, *Coleophora ulmifoliella* McDunnough, is a native North American moth. It is buff colored with gray markings and has a wingspan of about 13 mm. It occurs throughout the Northeast and as far west as Michigan. Older literature erroneously identifies it as *C. limnosipennella*, a European species.

After mating in late July, the female lays her eggs on the leaves of elm. Upon hatching, the larva enters a leaf and makes a small mine between the epidermal layers. Soon the larva emerges from the mine and makes a case, which is enlarged as the larva grows. This tubelike case, which the insect carries throughout the balance of the larval stage, serves as a protective covering. The cases are characteristic of the genus *Coleophora*.

Mines made by the elm casebearer take the form of small rectangular spots. When cool weather approaches in the autumn, the insect migrates to twigs, where it fastens its case and spends the winter in an immature larval stage. In the spring, it migrates to the new leaves and resumes feeding as a miner, but never does it allow its entire body to leave the case. The case rarely exceeds 6 mm in length (C). Pupation occurs within the case and the adult moths emerge in July. A single generation occurs each year.

The amount of damage caused by a single caterpillar is small, but considerable injury can result from heavy infestations.

There are several related species of *Coleophora*. Each makes a case and causes similar injury on other trees and shrubs. The group includes the birch casebearer (E), *Coleophora serratella* (Linnaeus), considered by some Canadian entomologists to be the most important defoliator of white birch in Quebec and the Atlantic provinces. It is another introduced pest from Europe now found in all of the northeastern states and southern Ontario. In addition to feeding on white birch, these insects are also reported on other birches, alder, elm, hazel, apple, and boxelder.

Early birch casebearer injury looks somewhat rectangular and is always located between the veins. Extensive mining results in functional, if not complete, defoliation. There is one generation a year. Birch casebearers overwinter as larvae, each in its own brown case attached to crotches of twigs, buds, or crevices in bark. Cases are normally found in clusters. In the spring larvae resume feeding on buds and newly developing leaves. On leaves, they continue to feed on the soft parenchyma between the two epidermal layers, mining as far as they can without leaving their cases. It is during the spring that the heaviest damage is done. Grossly mined leaves turn brown, and from a distance a tree may appear to have been scorched by fire. Twigs and limbs die back with complete defoliation.

Birch casebearers pupate inside their cases, and in the Northeast moths emerge in July. They deposit eggs on the undersides of leaves among leaf hairs along the midrib and large veins. They hatch in early August, and the young bore through the epidermis and feed as does any leaf miner. It is during the second instar that the larva modifies its feeding behavior and makes an oval case of leaf epidermis to cover its body. Larvae live in these cases but extend their legs out to crawl over the leaf surface and to feed for short distances on the parenchyma between the leaf surface layers. Before the leaves drop in autumn, larvae crawl to hibernation sites on the host tree.

The larch casebearer (Plate 11) feeds on all larch species. *C. sacramenta* (Heinrich), sometimes called the California casebearer, feeds primarily on willow, but also on almond, cherry, and peach. The pecan cigar casebearer, *C. laticornella* Clemens, is a minor pest of pecan, hickory, and black walnut. Its range extends from northern Florida and Texas to New Hampshire.

References: 81, 119, 289, 359, 566

A, B. Leaves of *Ulmus glabra* injured by the larva of the elm leafminer, *Fenusa ulmi*. Note the larva inside the mines in B. [Tenthredinidae]

C. The elm casebearer, *Coleophora ulmifoliella*. Larvae are inside the brown cases. Note the leaf injury as seen from the underside of the leaf. [Coleophoridae]

D. Leaf epidermis removed to expose the elm leafminer larva.

E. Birch casebearer cases attached to the leaf contain full-grown larvae about 5 mm long. The case is about 6 mm long.

A
B
C
D
E

The larva of a sawfly known as the European alder leafminer, *Fenusa dohrnii* (Tischbein), mines the leaves of many species of alder. Blotches develop (A) and turn brown; such discoloration serves as a means of identifying the presence of the larvae.

The European alder leafminer was introduced into the United States sometime before 1891. It can probably be found wherever alders grow, but mainly in the eastern quadrant of the United States, and in southern Canada.

The eggs of the European alder leafminer are laid in the leaves. Several larvae often develop in a single leaf. The larval form resembles that shown in panel C. After completing their development, larvae drop to the ground to pupate. The larva overwinters in a cocoon on the ground. The adult generally emerges sometime in May. It is a dark-bodied wasplike insect and is about 6 mm long.

As many as three generations of this insect occur each year. This, of course, means that it may be found throughout the growing season. Little is known about its parasites and predators because complete biological studies have not yet been made.

The various cultivars of the European alder, *Alnus cordata,* are most susceptible to attack by this insect. Some investigators have found that *A. rugosa, A. macrophylla,* and *A. viridis* are immune.

Panel D shows *Crataegus crus-galli* with mine injury caused by *Profenusa canadensis* (Marlatt), sometimes called the hawthorn leafmining sawfly. It is a wasplike sawfly that was first described by entomologists as a pest of sour cherries, *Prunus cerasus.* The insect, however, is a primary pest of *C. crus-galli.* At the beginning of the growing season it is in the ground in its pupal stage. Adults appear when the host tree's first leaf clusters begin to unfold and the blossom buds begin to open. Females lay eggs singly in the upper epidermis at the base of the leaf. Upon hatching, the larva feeds on the parenchyma between the leaf surfaces. Several larvae may feed on a single leaf. They mine toward the distal end, generally keeping close to the margin until large blotches, which often cover half the leaf, become apparent. The larvae have flattened bodies with three pairs of legs; they average about 7 mm long when fully grown. They abandon the foliage about the middle of June by dropping to the ground. There, they make earthen cells in preparation for the winter.

This insect can be a serious pest of certain species and cultivars of *Crataegus. C. persimillis* and *C. erecta* may be severely damaged, whereas the leaves of *C. mollis* are rarely mined. From a distance heavily infested trees have a brownish cast as though they have been singed by fire. Injury to the tree is not long-lasting, especially in the years that new leaves develop rapidly.

P. canadensis is indigenous to North America and is found in the Northeast west to Iowa, Missouri, and Arkansas. Two wasp parasites, *Aptesis segnis* (Provancher) and *Trichogramma minutum* Riley, often control its populations.

In Pacific Coast states foliage of the western sycamore, *Platanus racemosa,* may be damaged to near functional defoliation by *P. platanae* Burks, also known as the sycamore leafmining sawfly. It initiates, as a first-instar larva, a short lineal mine that is later usually encompassed into a blotch mine. It consumes both palisade and spongy mesophyll cells, leaving the two epidermal layers to turn brown. The center of the blotch is always filled with dark fecal refuse. The blotch mine is most impressive from the upper surface of the leaf. Because 15–20 mines are commonly found in a single leaf, however, some of these mines may coalesce. Like many sawfly species, *P. platanae* overwinters in soil either as a larva or pupa. The adult female emerges by mid-February in southern California and deposits her eggs in newly expanding leaves. A second generation develops in late April and a third in late July. There may be three to five overlapping generations per year. The larvae are similar to that illustrated in Plate 85D.

References: 49, 86, 594, 832

A. Typical leaf mines caused by the European alder leafminer on *Alnus glutinosa.* [Tenthredinidae]

B, C. Leaf mines of the hawthorn leafmining sawfly. Note the larva at the arrow point in C.

D. *Profenusa canadensis* mines on *Crataegus crus-galli* foliage.

A
B
C
D

Leaf-mining Beetles (Plate 87)

One of the serious insect pests of the black locust, *Robinia pseudoacacia*, is *Odontota dorsalis* (Thunberg), which is commonly known as the locust leafminer. In its adult stage this insect is a beetle 6 mm long (A). The beetle hibernates wherever winter protection can be found, often in litter under its host tree. In the spring beetles emerge and begin feeding on the black locust's developing foliage. After a short time they deposit flat, oval eggs on the undersides of leaves. The eggs soon hatch, and the larvae eat into the inner layer of leaf tissue, forming a mine. As the larvae feed and grow, the mines enlarge. The terminal portion of the leaflet is the preferred feeding site (Figure 61). The mine, from its beginning, takes the form of an irregular blotch. When fully grown, the larvae are flattened and yellowish white; they have black legs, a black head, and an anal shield. They pupate in the mine. Upon emerging from their pupal cases, the beetles skeletonize the undersurface of leaves. In southern Ohio and adjacent areas, a second annual brood occurs.

If a tree grows two sets of leaves in one growing season, the combined feeding of larvae and adults may destroy both sets. When this happens in successive years, the tree will die. In early summer affected black locusts may look as though they have been swept by fire. Larvae feed on black locust, false indigo (*Amorpha fructicosa*), bristly locust (*Robinia hispida*), *Sophora japonica*, and golden chain tree; adult beetles feed on dogwood, elm, oak, beech, cherry, wisteria, and hawthorn as well. They also feed on several herbaceous plants.

The locust leafminer occurs with regularity in Pennsylvania, West Virginia, and Ohio and all along the highland areas near the Ohio River. The species is also reported frequently in Alabama and Georgia, and in the area along the eastern seaboard extending north into Canada. It is known as far west as Kansas and Manitoba and south to Mississippi.

There are several wasp parasites of the locust leafminer, including *Trichogramma odontotae* Howard and *Spilochalcis odontotae* Howard. The wheel bug, *Arilus cristatus* (Linnaeus), is a common predator in southern states and feeds on leaf miners while they are still inside the leaf tissue. A eulophid pupal parasite, *Closterocerus tricinctus* (Ashmead), is a highly effective parasite of locust leafminer pupae in West Virginia. *Lopidea robinae* Knight is one of its mirid predators.

Sumitrosis rosea (Weber) is also a locust leaf miner. It is closely related in distribution, biology, behavior, host, range, and mine characteristics. It differs in size (about 3.6 mm), color (yellowish brown), and form (its elytra are coarsely punctate with the apex truncated). It is rarely as abundant as *Odontota dorsalis* but often occurs in the same tree.

Both larvae and adults of *Baliosus ruber* (Weber), the basswood leafminer, can cause extensive foliar injury. Females deposit eggs on host leaves about mid-June, and the larvae feed as leaf miners, forming large blotch mines. The mines are fully developed by mid-July. Full-grown larvae are about 6 mm long. Adults appear in August and feed on the upper surface of leaves, skeletonizing irregular spots throughout the leaf. When adults are abundant and feeding is intense, the entire leaf may appear skeletonized. Adults are dark reddish yellow and have wedge-shaped, truncated wing covers with distinct longitudinal ridges. If disturbed when feeding or resting, they drop to the ground or take flight while falling. They are strong fliers. Adults overwinter in leaf and twig litter beneath host trees. Hibernation ends in the spring about the time leaf buds open. Leaf feeding resumes, again resulting in skeletonized leaves. Distributed throughout the native range of American basswood, the basswood leafminer attacks all *Tilia* species, which appear to be the primary hosts; other hosts include apple, birch, cherry, hophornbeam, oak, and maple.

The willow flea weevil, *Rhynchaenus rufipes* (LeConte), is a small blackish snout beetle with reddish yellow legs (D). It is found in the cooler regions of North America, including the Appalachians, the Rocky Mountains, Canada, and Alaska, and it is also known to occur in Massachusetts, California, and Oregon. It overwinters under loose bark, or under stones and ground surface debris, reappearing in the spring to feed on the buds, and later on the tips of new shoots and foliage. The weevil chews tiny circular pits in these plant parts, and sometimes the injury is sufficient to kill new shoots. The pit-type injury to leaves becomes a hole after the remaining epidermis turns brown and drops out. These holes may be less than 1 mm in diameter. The adults feed primarily on willow, but it also uses other hardwood trees such as elm, aspen, oak, birch, cherry, and apple for food. Larvae appear in early summer, penetrate the leaf, and make small blotch mines in willow leaves (C). Later they pupate in the mine. Adults in the new generation appear in August and feed on leaves, causing more tiny holes. With the approach of cool weather adult beetles often congregate in the sides of buildings. A related species, *R. ephippiatus* (Say), feeds on cottonwood or willow foliage both as an adult and a larva. When ready to lay eggs, the female chews a tiny round hole in the midrib, then deposits an egg in the puncture. When the egg hatches, the larva tunnels along the injured midrib for a short distance, then enters the blade's parenchyma. In cottonwood the mine destroys about one-third of the leaf blade. The adult beetle, about 2 mm long, is a mottled brown with a white scutellum. These beetles occur from Quebec to California and south to Mississippi.

Figure 61. Mines in black locust leaflets caused by larvae of the locust leafminer. (After Weaver, ref. 820.)

Brachys aeruginosus Gory is a buprestid beetle that, as a larva, mines the leaves of *Fagus* species. The insect might well be called a beech leafminer. Very little is known about its biology and life history, perhaps because it is of little economic importance. Its eggs are deposited on the lower surface of the leaf, then covered with a transparent secretion that glistens conspicuously long after the larvae have matured and left the leaf. The larvae, when fully grown, are 6 mm long and whitish with a slight tint of green. They are legless and flat with deep-cut abdominal segmentation. The adult is shown in panel F.

Related species include *B. ovatus* Weber, which mines in oak leaves, and *B. aerosus* Melsheimer, which mines the leaves of elm and oak. These three species occur throughout much of the eastern quadrant of the United States.

References: 40, 49, 101, 193, 269, 343, 373, 546, 566, 819a, 820, 844

A. An adult locust leafminer. [Chrysomelidae]

B. A digitate mine always associated with the midrib vein on black locust leaves; it is believed to be caused by the caterpillar of *Parectopa robiniella*, a moth.

C, D. An adult and mine of the willow flea weevil, *Rhynchaenus rufipes*, on *Salix nigra*. Length about 2 mm. The two dark spots in the mines (C) are pupae. [Curculionidae]

E. A broad serpentine leaf mine found on *Populus trichocarpa* in southern California; the insect that made the mine is probably one of the leaf-mining caterpillars.

F. A *Brachys aeruginosus* beetle; 3–4.5 mm long. It mines beech leaves as a larva.

G. Leaf mines caused by *B. aeruginosus*.

A
B
C
D
E
F
G

Oak Leaf Miners (Plate 88)

Across the United States the larvae of many species of moths mine oak leaves. There are several common species in the eastern states. One of these is known as the solitary oak leafminer, *Cameraria hamadryadella* (Clemens). It attacks scrub, post, black, blackjack, red, and white oaks (B). As the common name suggests, there is only one larva per mine, but there may be many miners per leaf. Fully grown larvae are about 4 mm long. Another common species is the gregarious oak leafminer, *C. cincinnatiella* (Chambers), which attacks primarily white oaks. As many as a dozen larvae of this leaf miner may occur in a single large mine.

Light to moderate attacks of either species on oaks leave splotched foliage. Heavy attacks may completely kill the leaves. In 1971 a heavy attack on white oak in Arkansas killed 80–100% of the foliage by August, making the landscape look as if autumn had arrived early. Although normally leaf miners cause no life-threatening injury to their hosts, they may reduce the ornamental value of a yard tree because of the mined foliage.

Depending on the location and species of the oak leafminers, there can be two to five generations per year. Because both species pupate and overwinter inside dried leaves on the ground, they can be controlled if the leaves are raked and destroyed in the fall. However, if the trees in need of protection are near other oaks or near a wood lot, additional measures must be taken. Fortunately, the populations of oak leafminers vary greatly from year to year and seldom reach such high numbers that they need to be controlled by insecticides. At least 14 species of parasitoid wasps keep these leaf miners in check.

References: 193, 368, 566, 681

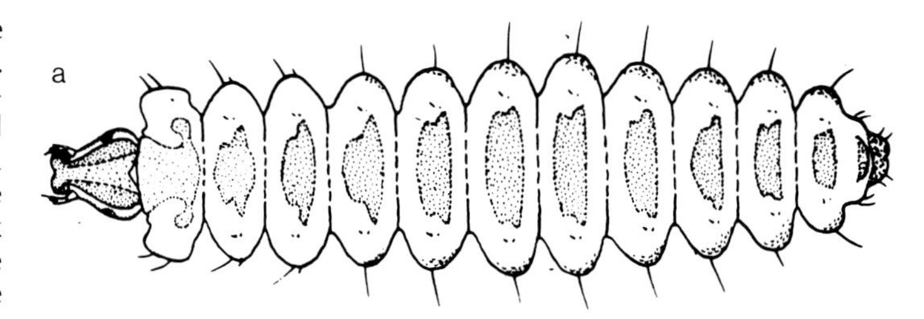

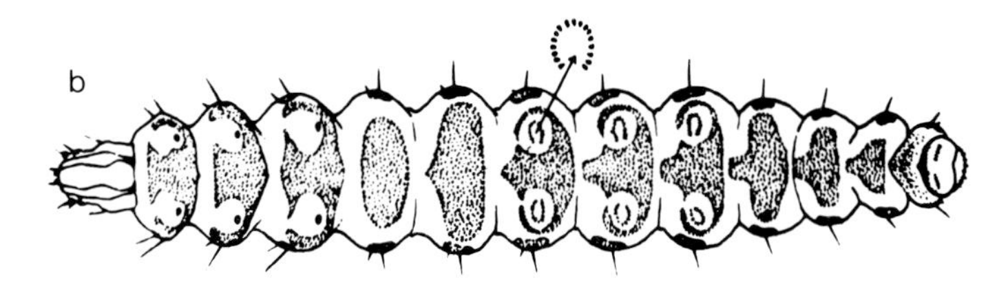

Figure 62. The larval forms of two *Cameraria* leafminers that feed on oak foliage. *a, C. ostryarella,* 5 mm long, dorsal view; *b, Cameraria* sp., 5.5 mm long. Note the crochets on the abdominal prolegs. (From Peterson, ref. 609, courtesy H. N. Peterson.)

A, B. Splotches on leaves of white oak were caused by the solitary oak leafminer, *Cameraria hamadryadella.* [Gracillariidae]

C. Damage caused by *Cameraria* sp. to *Quercus virginiana.*

D. Blotch mines on leaves of *Q. robur,* collected at Long Island, caused by the solitary oak leafminer.

A
B
C
D

Blotch and Serpentine Leaf Miners (Plates 89–90)

Insects that occur on woody ornamentals that are seldom "domesticated" frequently are not studied in great detail. Such is the case with most of the insects illustrated in this plate.

Madrone, *Arbutus menziesii,* also called arbutus, occurs on the West Coast from British Columbia to Mexico. This tree is difficult to transplant and does not survive except on well-drained sites. As a consequence, it has very minor status in the nursery trade, although naturally occurring trees are prized by their owners and admired by Pacific Coast travelers.

Larvae of the moth *Gelechia panella* Busck feed on manzanita (*Arctostaphylos*) as well as on madrone. Blotch-type mines (A), which turn brown, as well as external leaf etching (B) by larvae are symptomatic of attack. The fully grown larva is about 16 mm long with a yellow-brown head and a yellowish green body. Larvae occur on plants at about the time the new flush of leaves appears. Moths are believed to be the overwintering stage. This insect is known only from California.

The larva of another moth, *Marmara arbutiella* Busck, makes a sinuous leaf mine (F) in madrone throughout the range of this plant. The insect also has the fascinating habit of mining not only in the leaves but tunneling through the petiole and into the cortex of twigs. Sometimes it moves back into another leaf. Many species of miners in the genus *Marmara* alternate between leaf mining and epidermal stem, twig, and shoot mining. Such mines are readily visible through the epidermis of stems of young eastern white pine trees (see Figure 63, accompanying Plate 91). Another species, *Coptodisca arbutiella* Busck, a shield bearer, cuts out elliptical holes (C) in the leaves of madrone and manzanita (also see Plate 96).

The larva of a fly, *Paraphytomyza cornigera* Griffiths, makes a white serpentine mine (E), normally following the leaf margins of snowberry, *Symphoricarpos albus,* and also of twinberry, *Lonicera involucrata.* There are probably other *Symphoricarpos* and *Lonicera* hosts as well. This insect is known to occur in New York, California, British Columbia, Saskatchewan, and Ontario. A single generation develops early in the summer.

The aspen leafminer, *Phyllocnistis populiella* (Chambers), a caterpillar, occurs on trembling aspen, *Populus tremuloides,* and its hybrids throughout northern United States and southern Canada. Related *Phyllocnistis* are said to cause similar mines in the leaves of other *Populus* species.

The moth is the overwintering stage, and all evidence suggests that its overwintering site is below ground. It lays eggs singly near the tips of the young leaves in the spring. After hatching, the larva makes a characteristic sinuous mine (D). The entire generation requires about two months at the latitude of British Columbia, and one generation occurs each year.

Mined leaves dry, shrivel, and drop prematurely—often by midsummer. Severe infestations year after year are reported to reduce aspen tree growth, and occasionally to kill trees. Periodically this insect develops to outbreak numbers in Wyoming, Idaho, Alberta, and British Columbia.

References: 125, 266, 270, 275, 353, 395, 427, 746, 747

A. Blotch mines of *Gelechia panella* in leaves of madrone.

B. The older larvae of *G. panella* skeletonize leaves from the underside. The larvae feed from black frass tubes. (Also see Plate 98.) [Gelechiidae]

C. Damage to a madrone leaf by *Coptodisca arbutiella* [Heliozelidae], whose larvae make elliptical holes. Note also the twisting mine made by *Marmara arbutiella.*

D. A leaf of *Populus* mined by *Phyllocnistis* sp. [Gracillariidae]

E. Mines of *Paraphytomyza cornigera* in leaves of snowberry.

F. Lineal mines of *M. arbutiella* in madrone leaves. [Gracillariidae]

A
B
C
D
E
F

This plate shows examples of blotch leaf miners placed taxonomically into one of three moth families: Gracillariidae, Tischeriidae, and Nepticulidae. The family Gracillariidae has the greatest number of species that feed on woody plants. The moths in these families are all very small, some with a wingspan of less than 5 mm. Tables 11 and 12 provide facts about some of the more important species found in America north of Mexico.

Table 11. Common *Tischeria* species found on trees and shrubs

Species	Hosts	Nature of mine	Distribution
admirabilis	Rose	Blotch; leaflet folds cover mine	Central U.S.
badiiella	*Quercus alba*	Blotch; one margin borders vein	Eastern U.S.
ceanothi	*Ceanothus* spp.	Long blotch may cross major veins	Western U.S.
citrinipennella	Oaks	Elongate, marginal blotch; leaf curls	Central and eastern U.S., adjacent Canada
discreta	Evergreen oaks	Blotch	Western U.S.
immaculata	*Ceanothus griseus*, others	Long blotch	Western U.S.
mediostriata	Oaks	Blotch usually marginal	Western U.S.

Tischeria malifoliella Clemens, given the common name appleleaf trumpet miner, has long attracted the attention of naturalists and horticulturalists because of its colorful mine (A, B). This moth belongs to the family Tischeriidae, which contains numerous species that injure the leaves of ornamental trees and shrubs.

The appleleaf trumpet miner, a native species, limits its feeding to the foliage of apple, crabapple, and members of the genus *Crataegus*. Its presence often results in premature leaf drop and general weakening of the tree. The adult emerges from its pupa in the spring about the time leaves are fully formed. It deposits eggs singly on the upper surface of the leaf. When the egg hatches, the larva enters the leaf at the spot where the egg was attached so that the larva never becomes exposed to the external environment. The larva feeds on the palisade layer, making the mine more visible from the upper surface of the leaf. One caterpillar inhabits each mine. The larva is flat with distinct body segments but no thoracic legs. Two to four annual generations may occur—two in upstate New York and New England, and four in the Washington, DC, area. About 30 days are required to complete the summer generation(s). It overwinters on the ground as a larva.

The mine is described as a trumpet blotch mine, one that has a narrow serpentine beginning that gradually broadens to a blotch. The mine of this species is so distinct that it alone can be used to identify its maker. The insect's distribution is largely limited to the eastern quadrant of the United States and adjacent Canada.

Table 12. Common leaf blotch miners

Species	Hosts	Leaf surface	Distribution
Camereria hamameliella	Witch-hazel	Upper	East
Phyllocnistis liriodendronella	Tuliptree, magnolia	Lower	South
Phyllonorycter alnicolella	*Alnus* spp.	Upper	Western U.S.
P. crataegella	Apple, cherry, quince	Lower	East
P. felinella	*Platanus* spp.	Lower	Western U.S.
P. incanella	*Alnus* spp.	Both	Western U.S.
P. lucetiella	Linden, basswood	?	Atlantic states
P. salicifoliella	Willow and poplar	Lower	trans-America
P. trinotella	Red and Norway maple	Lower	Central U.S. and East

Ectoedemia platanella, which lacks a common name, belongs to the moth family Nepticulidae. The biology and behavior of this leaf miner resemble that of the appleleaf trumpet miner. The greenish larva is only slightly flattened, and when not feeding it retracts its head deeply into the thorax. Four larval instars are apparent; the fully grown larva is about 4 mm long. When finished feeding, the larva cuts a hole in the leaf epidermis and drops to the ground, where it pupates; it is presumed to overwinter as a larva or pupa. The number of annual generations has not been determined, but we expect three or four. Presumably, hosts are limited to the genus *Platanus*, but it has been found at least once in the foliage of *Quercus rubra*. The mine (C) is a distinct blotch with the center containing much frass. All of the palisade mesophyll is eaten. The mine begins as a very slender track (note arrow point). We do not know the limits of this moth's distribution, but it has been seen in much of the Northeast and in Ontario. We presume that it occurs over the natural range of the American sycamore. Several species in this genus mine in the bark and cortex of twigs.

The aspen blotch miner, *Phyllonorycter tremuloidiella* (Braun), and *Cameraria hamadryadella* (Clemens) (Plate 88) provide examples of the many species in these genera. From these species we generalize biological statements that apply to all members of the family Gracillariidae. They all make blotch mines (E) that may be more visible from one side of the leaf than the other. The visibility of the mine and other of its features are species-dependent. Most of these species spin silken threads across the loosened epidermis. As the silk dries, it shrinks, causing the roof of the mine to form a tentiform ridge (566). Currently, any leaf miner that causes the ridge formation is described as a tentiform leaf miner, for example, the spotted tentiform leafminer, *P. blancardella*.

In their early instars these larvae are flat, with thoracic segments that are often considerably wider than the abdominal segments. During the early instars (1–3) the larvae are sap feeders. Their mandibles cut the cells, and the insect sucks the fluids into its digestive system. Older larvae (4–5 instars) are tissue feeders and devour cells and membranes. With the difference in food source, that is, palisade tissue, such mines vary in their ability to transmit light when the leaf is held to a light. The head and mouthparts are also flattened, with the mandibles in the most forward position. Larvae never leave the mine and pupate there. Depending on the species, there may be more than one larva per mine and more than one generation per year.

Members of this genus are commonly found in great abundance in oak, hickory, beech, chestnut, witch-hazel, linden, birch, apple, pear, and maple. They are distributed throughout North America where their host plants normally grow.

Lilac and privet are closely related plants. Both are subject to injury by the lilac leafminer, *Caloptilia* (=*Gracillaria*) *syringella* (Fabricius), which, in its caterpillar stage, causes a blotch-type leaf mine. This species originated in Europe and is now found throughout the northeastern states, in parts of eastern Canada, and in the Pacific Northwest. It occurs occasionally on *Fraxinus, Deutzia,* and *Euonymus* species.

Eggs are laid along the midrib and other veins on the undersides of the leaves. The newly hatched larva enters the leaf directly under the shell and forms a linear mine that cannot be seen from the topside of the leaf. The second-instar larva enlarges the mine to form a blotch (F). Mines may coalesce, resulting in several larvae feeding in a single mine. The last-instar larva usually crawls out of the mine, folds the leaf, and makes a thin white cocoon within the protective fold. The lilac leafminer is capable of spending the winter in either its larval or pupal stage. In the Northwest it is common to see young larvae hibernating in the mine. In Alberta overwintering occurs at the soil surface under leaf litter. Several generations occur each year, and moths may be found throughout the summer months. This leaf miner has been reported attacking the leaves of mountain ash in Canada.

A related moth, *Caloptilia fraxinella* (Ely), feeds on privet and ash as a leafminer. The whitish caterpillars are about 23 mm long when fully grown. Their mines resemble those of the lilac leafminer.

References: 84, 193, 229, 270, 275, 353, 493, 560, 566, 591, 852

A, B. Leaf mines caused by the appleleaf trumpet miner. Note the exposed larva at the arrow point in B. The fully grown larva is about 5 mm long.

C. The mine of *Ectoedemia platanella* on the leaf blade of American sycamore. Note the serpentine portion of the mine.

D, F, G. Leaf mines caused by the lilac leafminer. D and F, mines in lilac; G, mines in privet leaves. [Gracillariidae]

E. Leaf mines in a cottonwood leaf believed to be caused by the aspen blotch miner.

A
B
C
D
E
F
G

Cherry Leaf Miner and Cambium Miners (Plate 91)

Hollyleaf cherry, *Prunus ilicifolia,* is a popular ornamental tree on the West Coast. The leaf-mining larva of a moth, *Paraleucoptera heinrichi* Jones, is perhaps its most severe pest. Little is known about its biology or distribution. The mine occurs on young leaves, and the injury becomes noticeable by mid-July (A, C). The mine ordinarily follows the edge of the leaf. When fully grown, the larva, about 4 mm long, leaves the mine and crawls to the upper surface of the leaf to pupate. Before it makes its cocoon, it constructs an H-shaped web tent (B, E). Under the tent it makes a disk-shaped cocoon. In the vicinity of Oakland, California, the adults (D) begin to emerge about mid-August. One generation is believed to occur each year. Hollyleaf cherry is also attacked by whiteflies (Plate 152).

Cambium miners make linear or serpentine mines in the thin bark of saplings, twigs, and small branches. Their name is a misnomer, because the mining actually occurs in the cortex and phloem, immediately under the epidermis. The mines are readily visible and may be alarming to the homeowner who has never seen them before. Young, thin-barked trees like maple, shadbush, cherry, ash, birch, holly, white pine (Figure 63), Douglas-fir, white fir, and shrubs like rose are most susceptible. *Phytobia amelanchieris* (Greene), a fly, mines the cambium of *Amelanchier canadensis.* The mine may begin on a twig and continue down to the root. The larval stages of a few tiny beetles and moths also produce cambium mines. These mines cause little damage to shade trees and shrubs.

Larvae of *Phytobia pruinosa* (Coquillett), a fly, are cambium borers on birch. When fully grown, the larva may reach 30 mm in length and a little over 1 mm in diameter (Figure 64). One generation occurs each year. *P. setosa* (Loew) mines the bark of red (*Acer rubrum*) and sugar maples. Cambium borers do not measurably interfere with the growth and vigor of the tree. If a host tree is cut into lumber, the boards have dark flecks, marks of old mines that were produced years before.

Other cambium borers in the genus *Marmara* are illustrated in Figure 65.

References: 249, 303, 330, 489, 746

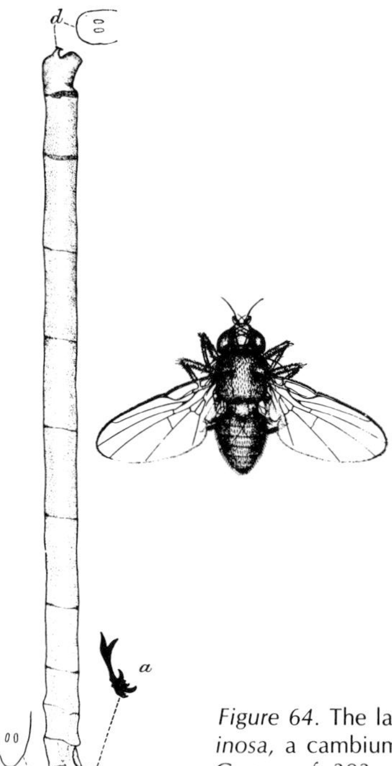

Figure 64. The larva and adult of *Phytobia pruinosa,* a cambium miner found in birch. (From Green, ref. 303, courtesy *Journal of Agricultural Research.*)

Figure 63. A cambium mine on eastern white pine caused by the caterpillar of *Marmara fasciella.*

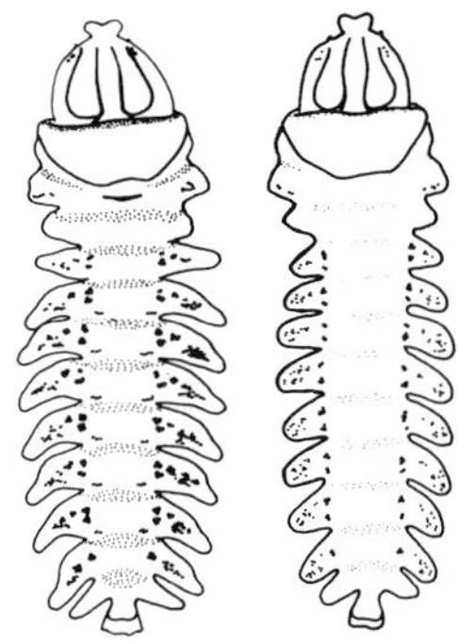

Figure 65. These larvae mine the bark of ash and belong to the genus *Marmara.* They are approximately 5 mm long and are very flat. (From Fitzgerald, ref. 249, courtesy *Annals of the Entomological Society of America.*)

A, C. Leaves of hollyleaf cherry, *Prunus ilicifolia,* mined by the caterpillar of *Paraleucoptera heinrichi.*
B. A larva in the process of making its "tent" and cocoon.
D. The adult moth of *Paraleucoptera heinrichi.* The wingspan is about 14 mm. [Lyonetiidae]
E. Close-up of the "tent." × 5.
F. Several cocoons and tents may occur on a single leaf.

A
B
C
D
E
F

Maple Petiole Borer, and Shoot and Twig Borers (Plate 92)

The maple petiole borer, *Caulocampus acericaulis* (MacGillivray), a sawfly with peculiar habits, was introduced into the United States from Europe. Larvae mine in the petioles of maples, causing the petioles to break a short distance from the leaf blade (A–D). Damaged leaves fall from the tree in May and June, causing considerable concern to the homeowner. However, seldom is the leaf drop serious enough to require control measures. Several maple species are subject to attack by this insect.

A single generation is produced each year. The adults emerge in May and lay their eggs near the base of the petioles of maple leaves (F). The newly hatched larva makes a tunnel and eats practically all the inner tissues of the petiole. In about 1 month the larva reaches maturity. The fully grown larva is only 8 mm long and resembles a weevil larva (E). The larva remains in the petiole stub attached to the twig; later it drops to the ground, where it pupates 5–8 cm below the soil surface. Raking and disposing of the fallen leaves will not reduce the insect population.

The full range of this insect is uncertain, but it has been seen in Connecticut, Massachusetts, New York, and New Jersey south to Alabama, west to Michigan and Illinois.

Boxelder, *Acer negundo,* and red maple, *A. rubrum,* appear to be the only hosts of the boxelder twig borer, *Proteoteras willingana* (Kearfott). The boxelder twig borer destroys the dormant buds of its host in the fall and early spring, kills the succulent new shoots during May and June, and causes swollen shoots and the subsequent development of adventitious growth. When the eggs hatch in mid-July, the tiny caterpillars feed by skeletonizing a small portion of the leaf. The larva during its leaf-feeding stage covers the feeding area with silk webs mixed with frass. In autumn the larva bores into the petiole and dormant leaf buds. There, it spends the winter. In the spring the larva burrows into the new shoot, where it completes its development. The plant responds to this intrusion by producing what some authors term a gall, but it is little more than a swollen twig. Frass accumulates at the point of entrance, and this serves as an excellent sign of the larva's presence (Figures 66–67). The fully grown larva is whitish yellow with a brown or nearly black head and is up to 12 mm long. In Canada the female moth emerges in early July and shortly thereafter lays her eggs.

Affected trees become bushy after several years and undesirable as shade trees. The pest generally occurs wherever boxelder grows; from the east coast to North Dakota, and in Canada.

Figure 67. A swollen boxelder twig, split open to show the tunnel and twig borer. [Tortricidae]

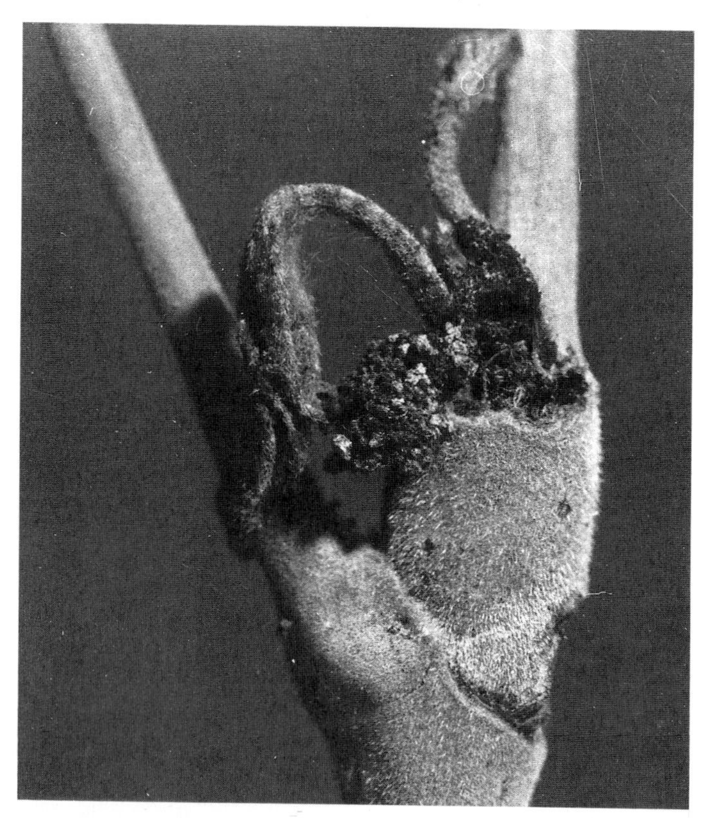

Figure 66. Damage by the boxelder twig borer, *Proteoteras willingana.* Note the accumulation of frass near the entrance hole.

P. aesculana Riley and *P. moffatiana* Fernald larvae bore into the buds and terminal shoots of most native and exotic maples, occasionally causing economic damage in nurseries. *P. aesculana* also bores into the buds and shoots of horsechestnuts and buckeyes. The larvae of both species usually expel frass through a hole at the base of a bud or leaf petiole. The frass accumulates there, held together by silk, providing a sign of the larva's presence. Terminal buds are hollowed out, and the bud shells (bud scales) soon drop from the tree. In New York the moths of *P. aesculana* appear in June and early September. They are brownish gray and have a wingspan of about 14 mm. Found in the northeast quarter of the United States and the border provinces of Canada, *P. aesculana* occurs as far south as the mountains of North Carolina and Tennessee. *P. moffatiana* is a more northern species. Its moths, present in July and August, have a slightly wider wingspan (16–18 mm), and the wing scales have a distinct greenish tinge. The larvae overwinter in the shoots.

The cottonwood twig borer, *Gypsonoma haimbachiana* (Kearfott), feeds in the pith and buds of rapidly growing twigs of several *Populus* species. The eastern cottonwood, *Populus deltoides,* seems to be a favorite host. The larva or caterpillar of the cottonwood twig borer is about 15–17 mm long when fully grown. The feeding injury stunts growth and kills twigs; repeated attacks result in a bushy tree. The twig borer is found throughout the natural range of eastern cottonwood.

References: 49, 538a, 610, 868

A. Healthy leaves of *Acer saccharum* in contrast to one damaged by the feeding of the maple petiole borer. Note that some leaves have already broken off their stems.
B. Before dropping off, infested leaves shrivel and change color (*arrow*).
C. Black and shrunken petioles resulting from larval feeding.
D. The end of a broken leaf stem.
E. A larva and borings. [Tenthredinidae]
F. Close-up of a puncture where the egg has been laid. The petiole becomes discolored from larval activity.

A
B
C
D
E
F

Azalea Leaf Miners (Plate 93)

The azalea leafminer, *Caloptilia azaleella* (Brants), is a leaf miner for only the first half of its larval life. Later, it feeds externally as a leaf roller or tier, rolling a leaf or tying several together to form a self-protective canopy. The larvae rarely destroy entire leaves but greatly disfigure them. They feed singly and, if disturbed, move rapidly either forward or backward. The azalea seems to be the only food plant of this insect.

The eggs of the azalea leafminer are white and are laid singly on the underside of a leaf along a midrib or vein. Upon hatching, the young larva enters the leaf directly beneath its eggshell and feeds initially as a leaf miner (B), causing a blisterlike blotch mine. The mined area turns brown. It then crawls to the upper surface of the leaf and, by means of silk, pulls the leaf over its body and proceeds to feed by chewing holes. The larva also may tie the newly expanding leaves together at the tip of a shoot and feed there in the same manner. When mature, the larva often selects an undamaged leaf, rolls it up, and pupates inside. A small moth emerges after about 1 week, mates, and begins the cycle again.

The golden yellow moth is about 10 mm long and has a wingspan of 12 mm (F). Adults are very secretive, and most of this stage is spent hidden among the leaves of their host. On Long Island there are at least two generations of the azalea leafminer each year. In Georgia and Alabama three or four generations occur annually, and in Florida breeding is continuous. In Oregon, where there are three generations per year, the winter is spent as a last-instar larva or pupa in a rolled leaf, or occasionally as a tiny miner in a leaf.

This insect has become a destructive pest throughout the range of evergreen azaleas. It is found from Florida to Texas and north to Long Island, West Virginia, and the Ohio Valley. It is also found in northern California and the Pacific Northwest. At one time it was considered primarily a pest of greenhouse-grown azaleas. Now, container-grown and field-grown azalea nursery stock and plants in the landscape are commonly attacked. If the insect is controlled early, the plants will quickly outgrow the injury it causes.

Identifying the insects that attack azaleas can be difficult, because several leaf rollers, tiers, and miners feed on azalea foliage. The fruittree leafroller (Plate 99) feeds on azalea foliage. *Olethreutes cespitana* Hübner is an occasional leaf roller of azaleas, but in its northern range it is predominantly a solitary leaf roller of aspen. *Ptycholoma peritana* Clemens mines leaves; in central New York there are two generations per year. It is a rather general feeder that prefers low-growing plants.

References: 31, 164, 682

A, B. An azalea plant damaged by the leaf-tier stage of the azalea leafminer.

C. Azalea leaves damaged by first-instar larvae of the azalea leafminer.

D. Blotch and lineal mines caused by the azalea leafminer.

E. A full-grown azalea leafminer larva; about 12 mm long. (Courtesy Gray Collection, Oregon State Univ.)

F. An adult azalea leafminer in a resting position. (Courtesy Gray Collection, Oregon State Univ.)

A
B
C
D
E
F

Boxwood Leafminer (Plate 94)

Boxwood has been used for hundreds of years in England and Europe in both public and private formal gardens. Hundreds of these plants were brought into the United States in the 1800s. Many of them found their way to the large estates of Long Island, New Jersey, Rhode Island, Delaware, eastern Pennsylvania, Virginia, and the Chesapeake Bay area. Many of these old original plants of *Buxus sempervirens* still exist.

With the introduction of other species and cultivars of the boxwood came the boxwood leafminer, *Monarthropalpus flavus* (=*buxi*) Schrank, but it was not reported in the United States until 1910. It is now considered by many to be the most serious insect enemy of boxwood and is found from the Atlantic to the Pacific wherever boxwood grows.

The boxwood leafminer has a single generation each year. It passes the winter as a partly grown larva in the leaves. During the first warm days of spring it becomes active and grows rapidly. It transforms into orange-colored pupae late in April and emerges as a fly when weigela begins to bloom. When it is time for emergence, the pupa forces its way partly out of the mine, where the pupal case may cling for several days after the fly has emerged. The adults are tiny (2–3 mm long), fragile, yellowish orange gnatlike flies. Females soon begin laying eggs in the upper side of the current season's leaves and insert each egg deep into the tissues. The female lays an average of 29 eggs and dies hours after her last eggs are laid. Eggs begin to hatch about 3 weeks after being laid. Larvae grow slowly during the summer. Many larvae may inhabit the same mine.

Injury is caused when larvae feed in the soft parenchyma tissue. Mined or blistered leaves are evident from midsummer until the leaves are shed from the plant. After new growth develops, mines are not evident for several weeks. Egg punctures are conspicuous, however, on the undersides of the leaves. Infested leaves are spotted yellow (B) and may drop prematurely. Infested plants grow poorly and lack the dense foliage characteristic of the healthy plant. Continuous infestations result in dead twigs and a weakened plant subject to disease, and winterkill in the colder areas.

Eleven cultivars of *B. sempervirens,* together with *B. microphylla* and *B. harlandii,* are subject to heavy infestations. However, English boxwood varieties *B. sempervirens suffrutiosa, pendula,* and *argenteo-varigata* are seldom damaged. There are few known natural enemies of the boxwood leafminer.

References: 318, 718

A, C. *Buxus microphylla* cv. Japonica with leaf-miner symptoms, as seen from both upper and lower leaf surfaces.
B. *B. harlandii* with leaf-miner symptoms.
D, F. Upper and lower surfaces of infested *B. harlandii* leaves.
E. The lower epidermis of *B. harlandii* removed to show the young larvae.
G. A pupa of the boxwood leafminer. [Cecidomyiidae]

A
B
C
D
E
F
G

Leaf-mining Maggots and Holly Budmoth (Plate 95)

Holes in leaves do not ordinarily provide clues adequate to identify the causal agent. Any number of insects, bacterial and fungal pathogens (see Sinclair et al., *Diseases of Trees and Shrubs*), and even cold temperatures at a critical time during bud development can cause holes with smooth or irregular margins. Figure 68 illustrates holes made by the larva (maggot) and adult of a leaf miner, *Agromyza viridula,* on red oak. These holes result from feeding injury by the adult female fly and her developing larvae in juvenile leaves. The leaves at the time of injury are never more than 4.5 cm long. Because the mouthparts of the female are unable to penetrate leaf tissue, she uses her sharp yet flexible ovipositor to pierce the epidermis. She often macerates the palisade parenchyma and spongy mesophyll by moving her ovipositor in side-to-side and circular motions. Then she withdraws the ovipositor and moves to lap up the fluids that flow from the puncture. Occasionally, she lays an egg into the tissue instead of macerating it.

The initial injury is inconspicuous and measures about 0.7 mm at its widest point. The injured leaf area soon turns brown and dries in the form of a disk. As the leaf continues to grow, the necrotic disk containing the larva separates from the living tissue and drops to the ground. The resulting hole continues to expand in direct proportion to enlargement of the leaf (Figure 68) until it is 2.3–10 mm in diameter. Occasionally, leaves may be so riddled with holes that the holes coalesce to form tattered edges.

Very little is known about this and most other agromyzid leaf miners on woody plants. *A. viridula* makes its first appearance when red and white oak buds open, and it is present until the leaves are about 5 cm long. In Maine mining by the larvae was observed during the second week of June. Fully grown larvae emerge from their mines and drop to the ground to pupate. Adults (Figure 69) may also appear in the middle of the growing season if a second flush of growth takes place or water sprouts develop. Geographic distribution of *A. viridula* includes the East Coast states from Maine to Georgia and probably the entire eastern half of the United States.

Figure 68. Oak leaf shot-hole damage caused by the fly *Agromyza viridula.* Injury begins when the leaf is in the juvenile stage. Arrow points to a necrotic disk. The hole in the mature leaf is about six times the size of the original injury. (Drawing by J. R. Baker, North Carolina State Univ. Cooperative Extension.)

Figure 69. A female fly, *Agromyza viridula*; about 3 mm long. (Drawing by J. R. Baker, North Carolina State Univ. Cooperative Extension.)

Seven species of leaf miners feed on holly (Table 13). One of them, *Phytomyza ilicicola* Loew, a fly, has been given the common name native holly leafminer. It feeds on *Ilex opaca, I. crenata,* and related cultivars but lays its eggs only in American holly. Kulp (451), however, believes that *P. ilicicola* will lay its eggs in *I. aquifolium* but that the larvae are unable to complete development on this host.

***Table 13.* Leaf miners of holly**

Species	Hosts	Distribution
Phytomyza ditmani	*Ilex longipes, I. serrata*	MD, VA
P. glabricola	*I. glabra*	MA south to VA, OH
P. ilicicola	*I. opaca, I. aquifolium, I. cumicola*	MA and NY south to SC and FL
P. ilicis	*I. aquifolium*	CA, OR, WA, mid-Atlantic states
P. opacae	Same as *P. ilicicola*	NJ south to VA
P. verticillatae	*I. verticillata*	MD, VA
P. vomitoriae	*I. vomitoria*	GA, FL, CA

The native holly leafminer is found throughout the native range of its host. One generation occurs each year. It overwinters as a larva in the leaf mine, pupating in March or April. Adult flies emerge over a period of about 6 weeks in the spring, starting after a few new leaves have formed. At about 10 days old, females begin to lay eggs in the leaf. They seek the underside of a newly developing leaf, pierce the epidermis with the ovipositor, and deposit an egg in the mesophyll. A tiny green blister appears on the leaf as the first injury symptom.

Upon hatching, the larva begins to make a narrow mine that appears dark brown from the upper leaf surface (G). It broadens the mine in late fall and completes the long blotch in late winter. Larvae are yellow maggots (E) and, when fully grown, will be about 1.5 mm long. The

---

A, B. American holly infested by the native holly leafminer, *Phytomyza ilicicola.*

C, D. Japanese holly, *Ilex crenata,* leaves showing feeding punctures caused by one of the holly leaf miners.

E. A holly leafminer larva, *P. ilicis,* excised from an English holly leaf; about 1.5 mm long. Its mouth is near the arrow point. [Agromyzidae]

F. An adult native holly leafminer; about 1.5 mm long.

G. An American holly leaf showing early mines and feeding punctures made by the ovipositor of the native holly leafminer.

H. An English holly leaf with mines produced by the native holly leafminer. This species cannot complete its development in English holly.

I. Larva and pulled-apart web of *Rhopobota naevana,* the holly bud moth; about 12 mm long. [Tortricidae]

J. The holly bud moth. Its wingspan is about 15 mm.

A
B
C
D
E
F
G
H
I
J

mouthparts feature dark sickle-shaped hooks that lacerate the palisade and spongy parenchyma cells. Because the larva feeds slowly, the current year's mines are easily overlooked. In heavy infestations, every leaf of a tree can be mined. Then the tree may drop many of its leaves and remain thin-crowned and unthrifty until new growth begins the next spring.

Injury to the host tree is caused not only by the mining activities of larvae but also by the feeding of the adults. Only the adult female can injure the leaf from without, by jabbing and puncturing the leaf with her sharp ovipositor. From this minute wound flows a few drops of sap imbibed by the female and nearby males. These wounds leave tiny round, deep scars visible from both sides of the leaf (D, G). Leaf distortion will result from numerous feeding punctures.

P. ilicis Curtis (E), commonly called the holly leafminer, was introduced from Europe and feeds only on *I. aquifolium*. This leaf miner is found on both the East and West Coasts. In the Pacific Northwest it is a pest in the commercial orchards of "Christmas" holly. The biology of this species and the damage it causes resemble those of *P. ilicicola*. However, eggs of *P. ilicis* are laid in the midvein of the leaf, and young larvae tunnel in the vein until late fall. The linear portion of its mine is not as evident as that of *P. ilicicola*.

P. opacae Kulp is a pest of *I. opaca*. The mines of this species are quite linear and may traverse the length of the leaf two or three times. Mines have a yellowish orange color.

P. verticillatae Kulp mines the leaves of *I. vomitoria*. It forms linear, irregular mines. Little is known about its biology.

P. glabricola Kulp mines the leaves of inkberry, *I. glabra*, forming blotch mines. It is common in New England and occurs south to Virginia and west to Ohio. Two generations occur each year.

In the Pacific Northwest *Rhopobota naevana* (=*R. unipunctana*) is known as the holly budmoth (I, J), but its official common name is blackheaded fireworm. Although the larva begins to feed on opening buds, it must be considered primarily a leaf roller. As such, it produces a large amount of silk that not only holds the leaf in a rolled position but also protects the larva. *R. naevana* produces one or two generations per year. The overwintering eggs start hatching about mid-May and continue through the month. Mature larvae drop from foliage and pupate in the duff. Females lay their eggs singly on the undersides of leaves, and second-generation larvae, where they occur, appear by the end of July.

Some entomologists consider *R. naevana* a generalist because it feeds on apple, cherry, blueberry, holly, and other ornamental plants, although it is not now recognized as a pest of ornamentals in the East. It is presumed to be of European origin and is now known from North Carolina to Maine and in the Pacific Northwest, including British Columbia. Its biology has been studied almost exclusively on cranberry and holly in the Pacific Northwest, although its life history in Massachusetts is known to be similar.

References: 268, 290, 329, 339, 451, 510, 623, 681, 797, 827, 836

Madrone Shield Bearer and Other Shield Bearers (Plate 96)

The madrone shield bearer, *Coptodisca arbutiella* Busck, attacks the leaves of madrone, *Arbutus menziesii*, and is also believed to be responsible for identical injury to the foliage of manzanita, *Arctostaphylos* species, and the strawberry tree, *Arbutus unedo*. Damaged leaves appear as though they have been perforated by a paper punch. The madrone shield bearer is found only on the Pacific Coast from California to British Columbia.

The adult is a tiny silvery gray moth with a wingspan of only 5 mm. The females lay eggs in the spring on leaves that are about two-thirds expanded. It is not clear whether the eggs hatch after a normal incubation period of a few days or are delayed in hatching until fall, but there is no evidence of mining by the larvae for about 6 months after the eggs are deposited. It is possible that after normal hatching occurs, larvae undergo a long period of inactivity at the site of egg deposition. Where eggs have been deposited, a tiny discolored dot appears.

The first evidence of larval activity is visible in the autumn. Sinuous mines that are evident on the leaves soon become blotch mines. In late winter the larva (Figure 70) completes development and cuts an elliptical section from both the upper and lower leaf surfaces. The insect later pupates within this section. This elliptical section, also called a case or a shield, drops from the foliage or is carried off and fastened to the bark by the mature larva inside. One or more perforations may appear in each affected leaf, depending on the intensity of attack. Moths appear in March and April in the San Francisco area.

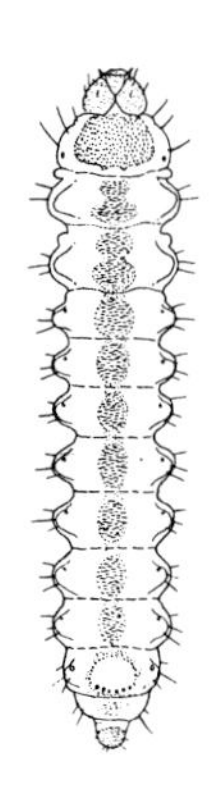

Figure 70. A tupelo leafminer; 5 mm long. Dorsal view.

Other shield bearers belonging to the genus *Coptodisca* occur on a number of plants such as myrtle, oak, grape, poplar, apple, pecan, and cranberry. The biology of these species often departs appreciably from that of the madrone shield bearer. For example, the resplendent shield bearer, *C. splendoriferella* (Clemens), a species associated with apple, has two generations each year. Several others do not have the long time lag between egg deposition and the appearance of the larval mine. All species, however, have the peculiar habit of cutting out elliptical leaf sections, and it is on this basis that the common name *shield bearers* was assigned to this insect family.

A common and sometimes devastating shield bearer, the tupelo leafminer, *Antispila nysaefoliella* Clemens, is a pest of black tupelo, *Nyssa sylvatica*, also called sourgum. The insect may be found wherever its host grows. Injury occurs first as a small linear mine, later as an oval blotch mine that may be as wide as 25 mm. When the larva is ready to pupate, it cuts a 10-mm hole in the mined area of the leaf, attaching the two epidermal leaf disks to its body (similar to D). With the disks attached, it drops to the ground and pupates. There may be several generations each year. The fully grown larva is about 5 mm long and is flattened, with distinct segments. It has a pale green body and a yellowish head. There are no legs.

References: 217, 353, 602, 682

A. A manzanita twig with most of its leaves destroyed by the madrone shield bearer, *Coptodisca arbutiella*. [Heliozelidae]
B. Close-up of an injured leaf.
C. Larvae and pupae inside the leaf disks. Some have dropped out, showing the characteristic elliptical holes.
D. Leaves of black tupelo injured by the tupelo leafminer. Larvae are visible by transmitted light.

A
B
C
D

Yellow Poplar Weevil (Plate 97)

Odontopus (=*Prionomerus*) *calceatus* (Say), the yellow poplar weevil, is also known as the sassafras weevil, the magnolia leafminer, and the tuliptree leafminer (A). These common names were derived from names of its hosts. It also feeds on *Laurus nobilis*. The yellow poplar weevil is found from Massachusetts to Florida and west to Louisiana, Iowa, and Michigan. It is most abundant in the central Appalachian area.

Both larvae and adults cause injury to their hosts, the adult by feeding on buds and leaves and the larvae by mining the leaves. The beetles overwinter in duff and leaf litter found beneath their host trees. During warm days in early spring they fly to trees and feed on buds and new leaves. In May they lay their eggs in the midrib on the undersides of leaves. After the eggs hatch, the larvae mine the leaves and cause blotch-type mines. On tuliptree and *Magnolia grandiflora* leaves (D), the blotch starts near the apex. The larvae are white, legless, and less than 2 mm long. In tuliptree there may be up to nine larvae in a single blotch mine. Most of the larval activity occurs in late May and June. Larvae pupate in the mine. Adults of the new generation emerge and for a short time feed on leaves in the same manner as their parents. After July adults greatly reduce their activities and are difficult to find. However, they have been observed in magnolia trees as late as August in Mississippi. Adults may invade houses in search of suitable overwintering sites. They may be found nearly every year on fence row sassafras on Long Island and occasionally on deciduous magnolias in the same area.

Damage done by this insect is not of great economic importance. In the deep South some damage occurs on magnolia leaves, but the damage is rarely very serious. In 1967 and 1968 there was a severe infestation on tuliptree in parts of West Virginia, Ohio, and eastern Kentucky, where severe leaf injury occurred.

References: 96, 348, 449

A. An adult yellow poplar weevil, *Odontopus calceatus*; about 2 mm long.
B. *Sassafras albidium* leaves and buds injured largely by adult beetles. The injury by larvae usually occurs at or near the tip of the leaf.
C. A *Magnolia grandiflora* leaf showing adult injury by the yellow poplar weevil. (Courtesy L. C. Kuitert, Univ. of Florida.)
D. Leaves of *M. grandiflora* showing larval injury. Injury occurs at the tip of the leaf. (Courtesy L. C. Kuitert, Univ. of Florida.)
E, F. Deciduous *M. soulangeana* leaves with adult injury.

A
B
C
D
E
F

Maple and Other Trumpet Skeletonizers, and Maple Leafcutter (Plate 98)

The maple trumpet skeletonizer, *Epinotia aceriella* (Clemens), feeds mainly on the foliage of sugar maple and red maple, *Acer rubrum*. Sugar maple is its principal host, but there are also records of its feeding on red, black, white, and chestnut oaks as well as on hawthorn and beech. The moth is found from Ontario to New Brunswick south to North Carolina and west to Michigan.

Moths appear in the spring as early as April. They are gray, with a wingspan of about 15 mm, and are attracted to lights at night. The female lays its eggs singly in a random pattern on the undersurfaces of leaves. In central New York the eggs are found in late June through mid-July.

Damage is caused by the larva, which eats the tissue between the larger veins on the underside of the leaf, leaving the leaf's thin upper epidermal layer intact. It spins a silken web on the underside and then folds the leaf. Inside this fold it constructs a trumpet-shaped tube (B) made of frass and silk. The length of the tube may exceed 5 cm. The larva feeds from within the protection of this tube and skeletonizes the area beneath the web. The remains of the web can be seen in panel B on the green area of the leaf.

The larvae occur on the leaves from early July through early October. The fully grown larva, 13 mm long, has a pale yellowish green body and a yellowish head. It drops to the ground to construct its cocoon between two fallen leaves. Although activity is obvious, the injury it causes is considered slight even when the insects are abundant because the damage occurs late in the season. Minor infestations of maple trumpet skeletonizers can be found every year; only occasionally do outbreaks occur. Infestations can be reduced by raking the leaves from under infested trees, then composting or otherwise destroying them.

Other leaf-tying or leaf-folding insects are common on many trees and shrubs (Plates 78, 99–101). *Episimus tyrius* Heinrich, an olethreutid (moth) leaf tier, is occasionally found in early summer on red and sugar maples. When fully grown, larvae are about 10 mm long.

Several caterpillars in the genus *Acrobasis* tie, fold, and skeletonize leaves, and to protect their feeding area and themselves, they make trumpetlike tubes of frass and silk. The tubes also serve as resting places and are enlarged as the larvae grow. Because the injury and frass tubes resemble those of the maple trumpet skeletonizer (B) and could be mistaken for them, we briefly describe five species of trumpet skeletonizers belonging to the genus *Acrobasis*.

A. minimella Ragonot could appropriately be called the oak tubemaker. It occurs only on *Quercus* species and appears to prefer *Q. marilandica* and *Q. alba*. The fully grown larva has a yellowish white head and purplish brown body and is up to 15 mm long. The frass tube is brown, up to 50 mm long, and attached to a leaf vein; it is never moved. Larvae feed on the undersurface of leaves, consuming only the tissue near the opening of the frass tube. They skeletonize leaves, removing about 15 square centimeters of surface from each leaf. *A. minimella* can be found in the eastern coastal states from New Jersey to Florida and west to Texas. In eastern North Carolina there are two generations each year.

Fully grown larvae of *A. exsulella* (Zeller) are about 14 mm long and have markings similar to those of *A. minimella*. They feed only on the foliage of *Carya* species. Throughout its distribution—Maryland south to Florida and west to Texas and Arkansas—*A. exsulella* has two generations per year.

The larvae of *A. juglandis* (LeBaron), the pecan leaf casebearer, skeletonize the foliage of *Carya* and *Juglans* species. When fully grown they are up to 17 mm long and have a dark reddish brown head and olive green body. The finished frass tube is 13–22 mm long and brownish. There is a single generation each year throughout its range, which extends from Ontario south to Florida and west to the Mississippi River.

A. betulella Hulst, also called the birch tubemaker, feeds exclusively on *Betula* species. The larvae are 17–23 mm long when fully grown and have a reddish brown head and dark purplish body. In the spring larvae that have overwintered bore into unfolding birch buds, but as the foliage grows they tie leaves together with silk and begin constructing a frass tube. Each larva skeletonizes large portions of two to three leaves. The species is widespread in Canada, the northern United States, Colorado, and northern California. Although larvae in various stages of development are found throughout the growing season, there is only one generation per year.

A. betulivorella Neunzig occurs on river birch in the southeastern United States.

The maple leafcutter, *Paraclemensia acerifoliella* (Fitch), is widely distributed over the northeastern United States and southern Canada. Although its preferred host plants are the sugar and red maples, the insect has also been found on beech and birch that grow adjacent to maple trees.

Premature browning of leaves is a symptom of the damage caused by this species. Close observation of an infested tree will reveal numerous small blotch mines made by the feeding of young larvae. This symptom occurs in June and is often overlooked. The most characteristic signs of attack are circular holes in the leaves (D), which are usually not apparent until late July or August.

Ordinarily, the insect is unimportant; occasionally, however, populations become abundant in limited areas and injury to foliage may be severe. If injury is severe for several consecutive years, older trees may be weakened or killed. Damage to scattered trees, as in landscape plantings, is seldom serious.

The small, steel blue moth emerges in late May. It lays its eggs in tiny slits on the undersides of the leaves. The newly hatched larva mines the leaves for about 2 weeks. After this, it cuts two oval pieces from the leaf and sandwiches itself between them to form a case in which it lives (E). The larva reaches from its case to eat the surface of the leaf, skeletonizing a circular area around the case. When it has consumed all of the leaf within its reach, the larva drags the case away to another feeding site. An oval green area is left, surrounded by the circular dead area (D).

The larva is not more than 6 mm long. It is dull white, with a brownish head and thorax. The fully grown larva crawls or drops to the ground in September to pupate.

References: 128, 193, 357, 570, 630, 660, 664

A. Folded maple leaves damaged by the maple trumpet skeletonizer, *Epinotia aceriella*. Eriophyid spindlegalls are also present (see Plate 232).

B. One of the folded leaves opened to show the extent and type of injury. Note the trumpet tube of the larva and the white web (*arrow*) used to fold the leaf together.

C. The work of two trumpet skeletonizers in a single sugar maple leaf. [Tortricidae]

D. The oval disks removed from this leaf have become the case in which a maple leafcutter caterpillar resides. The round skeletonized areas with green centers (*arrow*) are feeding sites of the nearly mature larva. [Incurvariidae]

E. *Left:* an intact case of the maple leafcutter. *Right:* the disk-shaped case opened to show the larva.

F. Close-up of a feeding site of the maple leafcutter larva. The case of a maple leafcutter (*arrow*) is on the leaf surface.

A
B
C
D
E
F

Leaf Rollers, Tiers, and Webbers (Plates 99–101)

This is the first of three plates illustrating a few common defoliators that are mostly solitary and that roll, fold, or tie leaves together with silk. Larvae feed inside these shelters, either as skeletonizers or gross feeders. Most of these larvae become very active when disturbed and wiggle vigorously, often backward, and either drop to the ground or hang suspended on a thread of silk. Approximately 200 species of these moths occasionally become numerous enough to injure ornamentals. The most abundant species are those that choose plants, like cottonwood, that produce new leaves throughout the growing season. Second- and third-generation larvae are thus provided with tender, succulent, and nutritious food. When nurseries force other plant species, causing them to produce additional foliage following fertilization, the populations of leaf rollers and tiers are only encouraged. Most of these insects cannot be tolerated under nursery conditions. A few are serious forest defoliators. There are at least four families of moths that contain species with these feeding habits, but the family Tortricidae contains the greatest number of pests. A summary of the leaf rollers and tiers common to woody ornamentals is given in Table 14, accompanying Plate 101.

In panel A we illustrate leaf tying by the large aspen tortrix (also see text accompanying Plate 100), leaf folding by the fruittree leafroller, and leaf rolling by *Pseudosciaphila duplex* (Walsingham), sometimes called the aspen leaftier. These insects are all transcontinental in distribution. The illustrations alone may be deceiving, because they represent very early stages of symptom formation and do not illustrate foliage feeding. The first feeding symptom, skeletonizing, is observable only by unrolling or pulling apart the silk-tied foliage.

The caterpillar of *Pyramidobela angelarum* Keifer is a leaf tier. This insect was first recorded in southern California in 1934 and is found in the northern part of the state as well. It probably represents an introduction from a more tropical region. Breeding apparently occurs year round on its specific host, *Buddleia calbillea. P. angelarum* webs, skeletonizes, and tatters leaves and feeds also on the terminal buds (B, C). The larva is yellow to green and measures 12–14 mm when fully grown. The moth is ashen, with a wingspan of 16–21 mm.

The oak leafroller, *Archips semiferana* (Walker), is one of a complex group of leaf rollers, tiers, and shredders that feed on oaks, particularly the northern oaks. Others in this group include *A. fervidana* (Clemens), known as the oak webworm, which is a gregarious species with nesting habits similar to a web nest caterpillar, *Argyrotaenia quercifoliana* (Fitch). When larvae of these species are placed side by side, species identification is easy; however, when they are examined individually in the field, identification is often difficult.

Young larvae of the oak leafroller feed by webbing together the newly emerged leaf clusters, concealing and protecting themselves in the folds of the leaves. They skeletonize small patches of the upper leaf surface within the webbed area. Later, as the larva rolls and webs new leaves, it removes most of the interveinal tissue, thereby creating a shabby mass of webbed and shredded foliage. Several young larvae usually occupy the same leaf cluster. As the larvae grow, they may become solitary and roll a single leaf lengthwise or crosswise, or roll only a single lobe of a leaf. These leaves become tattered, with only the major veins intact. In outbreak populations this insect is a serious defoliator. In Pennsylvania massive outbreaks occurred in 1970, 1971, and 1972. During these epidemics it became apparent that the larvae were carried from one place to another by the wind. The oak leafroller is widely distributed in North America but is largely limited to the range of deciduous oaks. Other hosts include hazel, apple, and *Pyrus* species.

The moths, active in July, lay their keg-shaped eggs in an oval mass, usually on the side of or just below a twig crotch. In contrast to other tortricids, they place their eggs side by side and on end rather than in a shinglelike pattern. The female covers them with scales from her abdomen, which offers some protection. Eggs overwinter and hatch in close synchrony with bud break of *Quercus rubra* and *Q. coccinea*. The larvae go through five instars. In northern Michigan all injury is done by the end of June. Pupae may be found in rolled green leaves in the tree crown, or attached to trunk bark or understory vegetation. There is one generation each year. Eggs are often parasitized by *Trichogrammatomyia* species. Two tiny wasps, *Itoplectis conquisitor* (Say) and *Phaeogenes gilvilabris* Allen, are important pupal parasites of the oak leafroller.

Redbanded leafroller, *Argyrotaenia velutinana* (Walker), larvae feed on a wide variety of plants, including spruce, balsam fir, larch, and hemlock. Chapman and Lienk (110) believe that plants belonging to the Rosaceae such as crabapple, hawthorn, as well as birch, alder, trembling aspen, basswood, and willow, are primary hosts. Larvae feed by skeletonizing leaves, and when abundant, they are capable of defoliating trees. The first-instar larva usually begins feeding on the underside of a leaf along the midrib or larger veins. The feeding area is protected with white silk webbing, the larva working between the web and leaf surface. Larvae are usually light green with a greenish yellow head (F), but their color may vary according to the food they eat.

The redbanded leafroller is distributed eastward from Ontario to the Atlantic provinces and in the United States from Missouri east to Virginia.

Overwintering occurs as a pupa within a folded leaf on the ground. Moths emerge about the time apple buds begin to show green tips. They lay whitish eggs in a shinglelike pattern in an oval egg mass that may be as large as 5 mm in diameter (Plate 100G shows an egg mass of similar form). First-generation eggs are laid on the bark of a trunk or the undersides of scaffold limbs. Second-generation eggs are usually laid on green leaves.

First-generation larvae, in upstate New York, complete their development by mid-July. There are two generations each year except in the South, where three and sometimes four generations may occur. The continuous presence of brood allows feeding to extend from early spring throughout the growing season; the result, often, is large populations and severe defoliation.

Eggs are often parasitized by the wasp *Trichogramma minutum* Riley. Synthetically developed sex pheromones can be used in traps as a monitoring tool.

Entomologists specializing in fruit insects give *Amorbia humerosana* Clemens the common name whiteline leafroller because of a distinguishing line low on the head (G). Based on host plant records, we conclude that this species is a rather general feeder of both gymnosperms and angiosperms. Among host plants most frequently named are apple, crabapple, hawthorn, blueberry, sumac, alder, chestnut, pine, spruce, balsam fir, and tamarack. This insect is found from Nova Scotia to Florida and west in scattered locations of Arkansas, Iowa, Wisconsin, in all of the bordering Canadian provinces, and the Maritimes. Adults of this solitary leaf roller and tier appear in June, and larvae are found from mid-June through September.

The whiteline leafroller overwinters in a silken cell formed within a fallen leaf or in conifer needle litter. The moth appears about the time of apple bloom. It lays eggs on the upper surfaces of leaves in masses that are covered with a milky substance. Larvae grow slowly and become fully grown in late summer. There is one generation each year.

See Plate 78 for other leaf rollers.

References: 50, 110, 118, 193, 263, 471, 538a, 556, 574, 592, 627, 630, 634, 649

A. Aspen leaves *tied together* by a young larva of the large aspen tortrix, *folded* by the fruittree leafroller, and *rolled* by the larva of *Pseudosciaphila duplex.* [Olethreutidae] (Courtesy D. A. Leatherman, Colorado State Forest Service.)

B, C. Damaged *Buddleia calbillea* leaves caused by the rolling, webbing, and feeding of *Pyramidobela angelarum.*

D, E, H. Egg, adult, and larval stages of the oak leafroller, *Archips semiferana.* Note moth's body scales on the egg mass. The full-grown larva is about 20 mm long. [Tortricidae] (All photos courtesy New York State Agricultural Experiment Station, Geneva.)

F. Redbanded leafroller larva; about 15 mm long.

G. *Amorbia humerosana,* sometimes called the whiteline leafroller, on apple; about 26–32 mm long. [Tortricidae] (Courtesy New York State Agricultural Experiment Station, Geneva.)

A
B
C
D
E
F
G
H

The term *tortrix* is used as a part of the common name for several leaf-rolling caterpillars. It means "the habit of twisting or rolling leaves to make a nest." Such is the case with the orange tortrix, *Argyrotaenia citrana* (Fernald), a western species known in the three Pacific states and coastal British Columbia. It is a pest in woody ornamental nurseries and in residential areas. As its name implies, it is also a pest of citrus, as well as many other fruit crops. Hosts include both gymnosperms and angiosperms such as pine, acacia, aralia, eucalyptus, euonymus, walnut, oleander, penstemon, rose, pittosporum, pepper tree, oak, and willow.

A. citrana is a subtropical species without highly distinct generational patterns. In some parts of its range there are two distinct generations. It lays eggs in clusters on either leaf surface or on smooth green twigs or shoots. The flattened and oval eggs are arranged so that they overlap one another, like those illustrated in panel G.

The newly hatched larva is 1.5 mm long and grows to 12–14 mm (B). Larvae, like others in the family Tortricidae, are very active; when disturbed, they wiggle sideways or backward, and in doing so, they may drop to the ground or hang suspended on a silken thread. They are solitary feeders, primarily leaf rollers, although they may locate in shoot tips and hide among the unfolding leaves, which they web lightly together for a temporary nest (A).

Pupation occurs in the last larval nest, and the pupa is covered by a thin transparent cocoon. Adults have the form illustrated in panel D and in Plate 99E. The orange tortrix may be parasitized by many wasp species, such as those in the genera *Apanteles* and *Hormius*.

The genus *Sparganothis* contains a large number of leaf tiers, leaf rollers, and skeletonizers that are considered more a nuisance than a threat to woody ornamental plants. None have approved common names. A few have narrow host ranges, such as *Sparganothis directana* Walker, which utilizes chokecherry as its primary food source. It ranges throughout the northeast quadrant of North America, where this tree is common. *S. sulfureana* (Clemens) is a more general feeder found in the eastern half of Canada and in the United States as far west as Utah and Washington and as far south as Georgia. It feeds on tips and shoots of both soft and hard pines, such as eastern white and loblolly pine; on tamarack; and on the foliage of apple, crabapple, hawthorn, blueberry, honeylocust, elm, and cherry. It overwinters as a first-instar larva in a tiny cocoonlike sack called a hibernaculum, attached to leaf debris lying on the ground. It resumes feeding in the spring but not necessarily on the same plant it used in the fall. First-generation larvae (C) complete their development by late spring. Adults (D) are on the wing from early June to early July. They lay 40–50 eggs in clusters on foliage. Their life and seasonal history is very similar to that of the obliquebanded leafroller. Feeding injury by *S. sulfureana* on deciduous plants is fairly distinctive. It consists of a rather crumpled mixture of dead or dying leaf fragments held together with silk and skeletonized leaves. On conifers these insects feed on shoots and needles. The life cycle in New York and New Jersey suggests that this species may have more than one generation per year in its southern range. *S. acerivorana* (MacKay) occurs in the Great Lakes states and east to Maine, as well as in Canada along the border with these states. Full-grown larvae are yellowish green and about 21 mm long. They feed primarily on sugar and red maple foliage but, strangely, also feed occasionally on the needles of Scots and red pine.

The apple-and-thorn skeletonizer, *Choreutis pariana* (Clerk), occurs from Virginia to northern California and north into parts of Canada. It is a pest of apple, birch, crabapple, cherry, hawthorn, willow, and mountain ash. Like the other larvae illustrated in this plate, they roll and skeletonize leaves, which in time turn brown (H) and drop prematurely. The moth overwinters (except in the Pacific Northwest, where the pupa is the overwintering stage) and after mating in the spring lays tiny green eggs on the undersurface of leaves near the midrib. After hatching, the young larva feeds on the lower epidermis under a loose covering of silk. Later, it moves to the upper surface, where it ties the edges of the leaf together at the base, forming a cone-shaped protective covering. Within the cones larvae skeletonize the upper epidermis between the veins. When disturbed, they become very active and wiggle backward out of the rolled leaf and hang suspended on a silken thread. This insect develops rapidly and in some parts of its range may produce four generations per year. It is related to the mimosa webworm (Plate 82).

The obliquebanded leafroller, *Choristoneura rosaceana* (Harris), is another transcontinental species, but it does not occur in the more arid portions of the Southwest. It is a general feeder and has 79 known host species. Chapman and Lienk (110) believe that most of these are chance infestations, established principally by airborne first-instar larvae. The primary hosts belong to the rose (Rosaceae) family, for example, hawthorn, rose, *Prunus* species, cotoneaster, and apple. Nonetheless, they have caused notable injury to a wide variety of ornamental trees and shrubs including ash, basswood, birch, dogwood, hazel, horsechestnut, lilac, linden, maple, mountain ash, oak, poplar, sycamore, willow, pyracantha, rhododendron, spirea, and viburnum. There are accounts of their feeding on conifers such as pine, spruce, and hemlock, seriously damaging seedlings in nurseries.

Like other leaf rollers, they utilize silk to roll the sides of the leaves upward or downward, sometimes into a rather tight cylinder, and feed inside the shelter of the rolled leaf. They usually choose leaves on rapidly growing terminals; sometimes they tie and roll several leaves together. Gross damage occurs after several hours of feeding, and the tied leaves look very ragged (see Plate 99A, B for similar injury). They also damage buds. Larvae overwinter in the second- or third-instar stage. When dormancy is broken in the spring, they bore into leaf buds, notably those of apple, maple, and rhododendron, often disbudding the twig, which eventually results in a forked terminal.

The larva overwinters in a well-camouflaged hibernaculum located under old bud scales or roughened areas in twig crotches. In New York the overwintering larvae (F) complete development by the end of May. In June females lay eggs in silvery oblong egg masses (G) on the upper surface of leaves. Upon hatching, larvae disperse by lowering themselves on strands of silk and become airborne on gusts of wind. In the northern portion of its range there is but one generation per year, but *C. rosaceana* occurs in the deep South as well, and we suspect that several generations develop there.

The large aspen tortrix, *Choristoneura conflictana* (Walker), is a leaf tier (Plate 99A). With the exception of the forest tent caterpillar, no insect is more widespread or consumes more aspen leaves than this species. It is primarily a northern species but occurs throughout the range of trembling aspen in Canada and the United States. It overwinters as a second-instar larva in a silken web on rough bark at the base of a tree. Before buds break in the spring, the larva crawls up the tree, chews a hole in a swollen bud, and enters it, causing round holes in the tightly rolled, newly developing leaf. After buds open and form leaf clusters, the larva webs three or four leaves together and feeds inside. Feeding continues until early June, when the larva reaches maturity. By this time, it devours all parts of the leaf. Mature larvae are dark olive green or nearly black, with a black head; they are 20–25 mm long. They pupate in a silken cocoon attached to several leaves. Adults appear in late June, mate, and the females lay eggs. They lay their pale green eggs in flat clusters on the upper leaf surface, and the individual eggs overlap like shingles, much as do those of the spruce budworm (Plate 7E). Upon hatching, tiny larvae web the flat surfaces of three or four leaves together to form a shelter in which they feed by skeletonizing the leaf surfaces. Many first-instar larvae hang by a silken thread and are blown by wind, which facilitates dispersion. In early August larvae congregate at the base of a tree in search of an overwintering site. There is one generation per year.

The greatest potential impact from defoliation by the large aspen tortrix is on valuable trees in urban and residential areas. This pest is

---

A. Damage to the growing tips of leaves caused by the orange tortrix. [Tortricidae] Photo made in Oregon. (Courtesy Gray Collection, Oregon State Univ.)

B. An orange tortrix larva; about 14 mm long. Those feeding on pine are green. (Courtesy Gray Collection, Oregon State Univ.)

C. A larva of *Sparganothis sulfureana*; about 14–16 mm long.

D. The moth stage of *S. sulfureana*; wingspan 15–17 mm.

E, H. A larva of the apple-and-thorn skeletonizer on an apple leaf; about 12 mm long. A thin transparent layer of silk covers the larva and its feeding area. H. Margins of skeletonized leaves roll upward and turn brown from feeding injury. [Choreutidae] (Courtesy Gray Collection, Oregon State Univ.)

F. A larva of the obliquebanded leafroller; about 20–24 mm long.

G. An egg mass of the obliquebanded leafroller. (Panels C, D, F, and G courtesy New York State Agricultural Experiment Station, Geneva.)

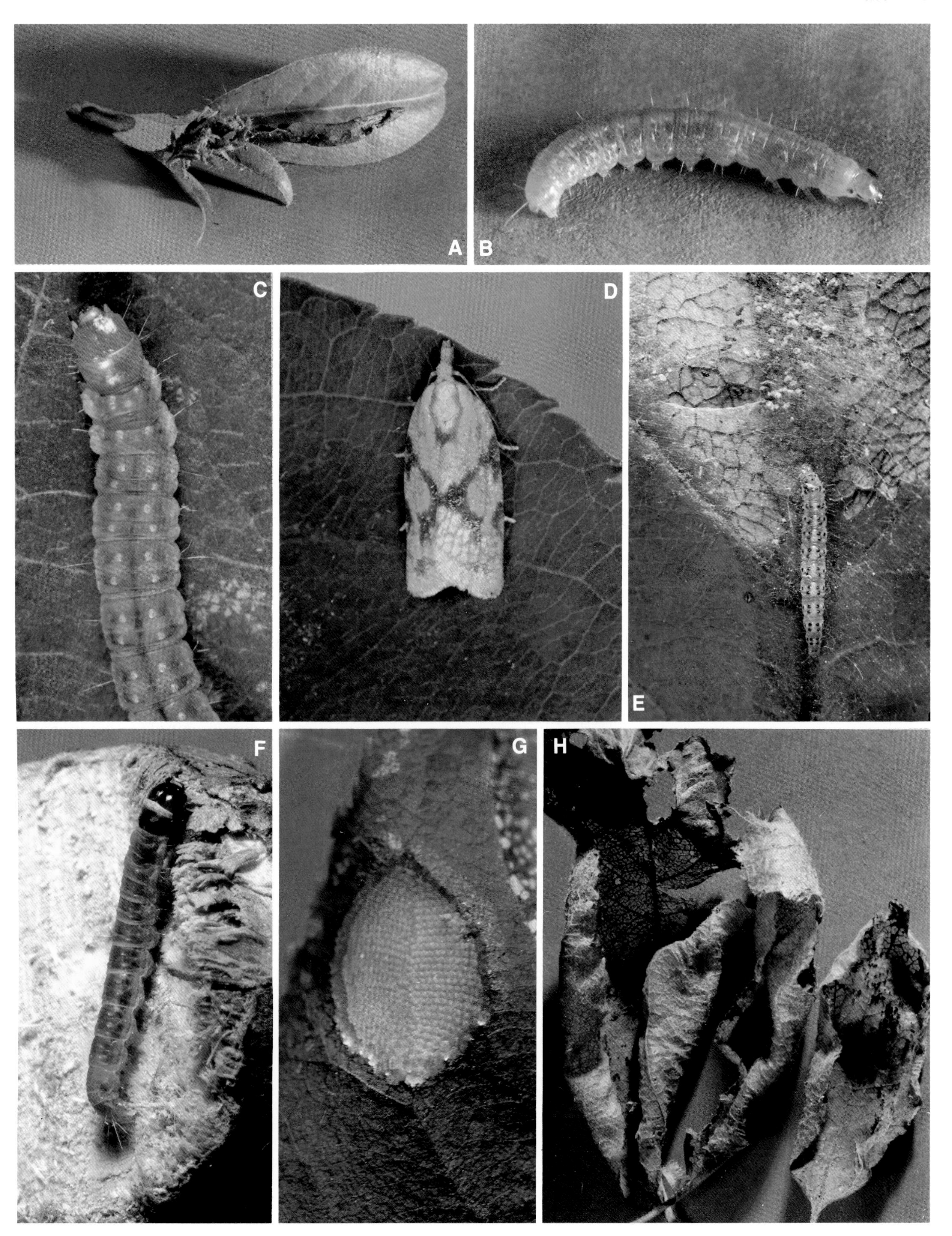
A
B
C
D
E
F
G
H

subject to numerous insect parasites, among which is a tachinid fly, *Omotoma fumiferanae* (Tothill). A commercially available biological insecticide (*Bacillus thuringiensis*) provides significant foliage protection.

References: 44, 110, 275, 387, 452, 538a, 660, 699, 796

***Table 14.* Leaf rollers and tiers common to shade trees and other woody ornamentals**

Species	Hosts	Distribution
Ancylis discigerana	Birch, aspen	Eastern Canada, northeastern U.S.
A. plantanana	Sycamore	Sycamore distribution
Apotomis spp.	Birch, alder, *Corlyus* spp., Salix spp.	Northern transcontinental
Archips argyrospila	Apple, oak, quince, rose, etc.	Transcontinental
A. griseus	Oak, hickory, apple	Eastern half of U.S.
A. negundana	Boxelder, *Lonicera* spp., *Alnus* spp.	Northern transcontinental
A. semiferana	Oaks	Northeast quadrant of U.S. and Ont.
Argyrotaenia franciscana	Lupines, California Christmas berry, dwarf chaparral broom	Pacific states
A. juglandana	Hickory, viburnum, *Prunus* spp.	WI and Ont. east to NY and FL
A. quercifoliana	Oak, *Rhamnus* spp., *Hamamelis* spp.	Que. to Man., south to TX and FL
Choristoneura conflictana	*Populus tremuloides,* other broad-leaved trees	Transcontinental
Croesia semipurpurana	Northern oaks	Eastern half of U.S. and Canada; TX
Epinotia solandriana	Birch, aspen, *Alnus* spp.	Transcontinental Canada, Northeast
Pandemis limitata	Oak, *Alnus* spp., sassafras, birch, rose, hazel	Transcontinental, GA north
P. pyrusana	*Cornus stolonifera,* apple, cherry, plum	Western species: B.C. and CA
Psilocorsis cryptolechiella	Oak, beech, birch, maple, pecan	Carolinas north to Nova Scotia and Ontario
Spilonota ocellana	Apple, *Prunus* spp., oak, laurel	Transcontinental

Archips rosanus (Linnaeus) does not have an approved common name, but it has several colloquial names; for example, in Oregon it is known as the filbert leafroller. The larvae are easily confused under field conditions with those of *A. argyrospila,* the fruittree leafroller (see Table 14. This pest of European origin is considered by some entomologists to be a general feeder, but we believe the primary hosts are privet and lilac. Other hosts include alder, apple, almond, basswood, hawthorn, hickory, laurel, pear, poplar, rose, oak, sycamore, and willow.

A. rosanus occurs in Oregon, Washington, and British Columbia and from the Great Lakes states east to the Atlantic, including those Canadian provinces that border the eastern and central states. It has the potential for a much more southern distribution.

In western New York this species lays eggs in late June or early July. Masses may be 3–7 mm in diameter, and the eggs overlap like fish scales. Each female lays two to four egg masses on the bark of trunk and scaffold branches, often within 26 cm of the ground. Eggs overwinter and hatch about the second week of May in New York and in mid-April in Victoria, British Columbia. Larvae feed on all stages of foliar growth; they live and feed within webbed or rolled leaves (C), usually confining these functions to shoot terminals.

Populations can be reduced by spraying foliage with the bacterial insecticide *Bacillus thuringiensis* when the larvae are about half grown.

Athrips rancidella (Herrich-Schaeffer) (=*Cremona cotonestri* Busk) is a European species now found in all of the Pacific Coast states and British Columbia. Its potential for movement east by way of nursery stock is great.

A. rancidella is a webworm that spins dense silken webs among foliage and twigs. Larva (B) forage from these webs and skeletonize leaf surfaces. Within 3 weeks the entire plant may be a mass of silk, frass, and brown leaves (D). Larvae overwinter attached to twigs or branches in a silken hibernaculum. Moths emerge over a period of about 30 days between mid-May and mid-June in Berkeley, California. There is one generation per year. The major host plants in the west are *Cotoneaster horizontalis* and *C. congestus.* In Europe this insect is also reported to feed on *Prunus* species.

The leaf crumpler, *Acrobasis indigenella* (Zeller), was a well-known pest in apple and cherry orchards before the development of synthetic organic insecticides. It is now more commonly a pest in ornamental nurseries and of landscape plants. The feeding and protective habits of the larvae are what injure plants. First-instar larvae are skeletonizers, but older larvae consume all the leaf tissues except the midrib. Larval protective habits make host plants ugly; the larvae construct cases made of dead leaf fragments, silk webbing, and frass pellets, attaching this material to twigs and branches (F). The inside of the case is lined with silk. The larva forages away from the case and carries back leaves, which it consumes in front of its case. If the leaf is not fully consumed, the larva incorporates the fragments into the case with silk. It later uses the case for hibernation and pupation. Sometimes two or more larvae can be found within a case. During the last larval stage the case may be as long as 40 mm; it may remain attached to a twig for several months, long after the moth has emerged.

In Illinois moths begin to emerge in late June. Peak emergence occurs about mid-July. Females lay eggs for a period of about 30 days from early July to early August; however, the longevity of a single moth female averages only about 10 days. Eggs are laid both singly and in masses along the veins on the underside of the leaf. Larval development is interrupted in the fall by cold weather. Feeding resumes in mid-April, about 2 weeks after *Cotoneaster* leaves appear. Foliage consumption is greatest in June and early July. The mature larva is about 16 mm long (E). There is one generation per year in its northern range. According to Neunzig (570), there are two generations per year in North Carolina and the southeastern United States.

Host plants include a wide range of *Cotoneaster* species and cultivars, as well as *Malus, Pyrus, Crataegus, Pyracantha,* and *Cydonia* species. The leaf crumpler occurs in the eastern half of the United States and Canada, Texas, the states bordering the Mississippi River, and all of the Pacific Coast states. There is at least one important internal parasite of the larva, the fly *Nemorilla pyste* (Walker).

References: 2, 113, 122, 285, 371, 482, 538a, 570, 616, 628

A. The caterpillar stage of *Archips rosanus*; about 18 mm long. [Tortricidae]
B. The caterpillar stage of *Athrips rancidella*; about 13 mm long. [Gelechiidae] (Courtesy Gray Collection, Oregon State Univ.)
C. Flowering almond leaves rolled and tied by the larva of *Archips rosanus.* (Courtesy Gray Collection, Oregon State Univ.)
D. Webs and damage to cotoneaster caused by *Athrips rancidella.* (Courtesy Gray Collection, Oregon State Univ.)
E. Larvae of the leaf crumpler and their open case attached to a twig of cotoneaster. Larvae are nearly full grown; 16 mm. (Courtesy J. E. Appleby, Illinois Natural History Survey.)
F. An opened case with the overwintering larva of the leaf crumpler.
G. Leaf crumpler cases in the overwintering condition attached to a cotoneaster twig.

A
B
C
D
E
F
G

Skeletonizers of Oak, Birch, and Apple (Plate 102)

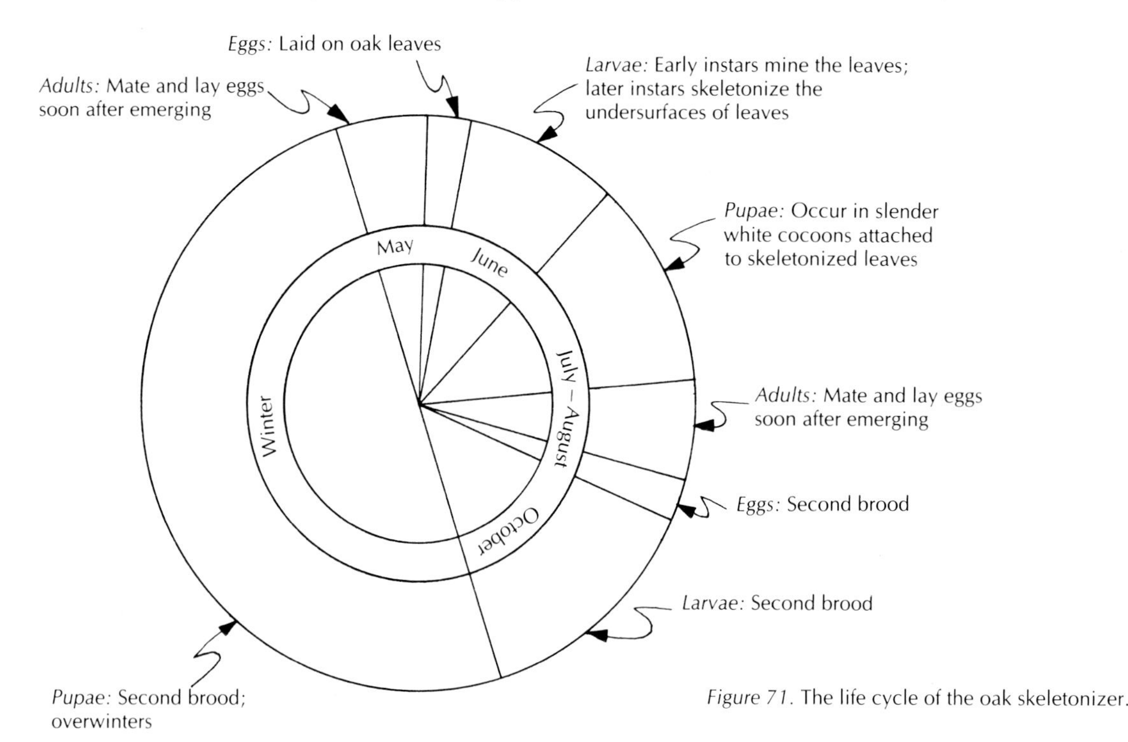

Figure 71. The life cycle of the oak skeletonizer.

The oak skeletonizer, *Bucculatrix ainsliella* Murtfeldt, is a native species that occurs throughout the Great Lakes states east to New England and south to North Carolina and Mississippi. It also occurs in southern Quebec and Ontario. Its hosts are various species of deciduous oaks and chestnut. The moths initially appear in late May and lay eggs on the undersides of fully grown leaves. Upon hatching, the larvae begin to feed on the epidermis and soft parenchyma layers of the leaves, causing one side of the epidermis to resemble a translucent, intact vein matrix (A). Fully grown larvae are about 7 mm long (E) and are pale yellow, sometimes tinted with green. If larvae are disturbed, they drop a few feet on a silken thread, often seeming to hang in midair. These suspended larvae become a nuisance when the tree is located in a yard or over a sidewalk.

When preparing for its several molts, the larva spins small, flat webs (B) within which it transforms to the next instar. Before the final molt, it spins a characteristic white ribbed cocoon (E) on a leaf, branch, trunk, or nearby object.

There is some uncertainty as to how this insect passes the winter. It most likely hibernates as a pupa in its cocoon. Just prior to emergence, the pupa, through the propelling action of its abdomen, protrudes through the cocoon (C). The moth's wingspread is about 7 mm.

In parts of New England and New York two generations of the oak skeletonizer occur each year. In 1971 a heavy infestation occurred in a city park in Binghamton, New York. The leaves of affected trees were badly damaged by July 8, at which time the insects were nearing the end of their first generation. The green leaf color was almost gone by early September near the end of the second generation. This kind of defoliation affects tree growth in the following year and tends to make the tree more vulnerable to attack by wood-boring insects. Defoliation has been reported as the primary cause of top dieback in oaks.

Populations of the oak skeletonizer fluctuate widely from year to year, indicating that natural factors such as parasites and predators affect their numbers in cycles. Little is known about these natural enemies.

The birch skeletonizer, *B. canadensisella* Chambers, is a pest of native as well as exotic birches. Like the oak skeletonizer, its larvae feed on the leaves and produce the type of damage shown in panel D. Its biology is similar to that of the oak skeletonizer except that there is only one generation each year and the moths are active from late June to late July. The birch skeletonizer seems to occur in outbreak numbers on a 10-year cycle. It is distributed throughout the eastern range of birch as far west as Minnesota and throughout southern Canada to the Pacific Ocean. This insect is also known in Colorado.

B. pomifoliella Clemens may occasionally be a problem on apple, flowering crabapple, cherry, shadbush, and hawthorn. It is about the same size as the oak skeletonizer and has two generations each year. The cocoons vary from white on crabapple to dark brown on hawthorn. Cocoons on apple and quince look like the oak skeletonizer cocoon (C). This species is distributed from the Rocky Mountains to the Atlantic Coast and from Texas throughout much of Canada.

On the West Coast the oak ribbed casemaker, *B. albertiella* Busck, lives on *Quercus agrifolia* and *Q. lobata*. Its biology and injury to foliage is essentially the same as for *B. ainsliella,* and its cocoons are identical.

References: 49, 193, 217, 267, 284

A. A view of upper surface of a *Quercus rubra* leaf. Injury was caused by the oak skeletonizer.
B. The underside of a red oak leaf. Larvae molt beneath the white webs.
C. Cocoons of the oak skeletonizer with empty pupal skins protruding toward the bottom of the photograph.
D. Birch skeletonizer injury as seen on the upper leaf surface. Note cocoons and molting webs.
E. Larva and cocoons of the oak skeletonizer. [Lyonetiidae]

A
B
C
D
E

Elm Leaf Beetles (Plate 103)

The elm leaf beetle, *Xanthogaleruca (=Pyrrhalta) luteola* (Muller), was accidentally introduced into the eastern United States early in the nineteenth century. It now occurs throughout the country wherever elms, *Ulmus* species, are important shade trees. Essentially all species of elm are attacked, but local preferences for one kind of elm or another are noticeable where a number of species are available to the insect. The elm leaf beetle readily feeds on the leaves of Japanese zelkova, *Zelkova serrata*.

The insect overwinters in the adult stage in houses, sheds, and in protected places outdoors (under the loose bark of trees or house shingles, for example). Elm leaf beetles often become a nuisance inside homes during autumn and spring when they begin to enter or exit overwintering quarters. In the spring the adults fly back to elm trees and eat small, roughly circular holes in the expanding leaves. Eggs are laid by the female beetle in clusters on the leaves (Figure 72). A female may produce as many as 600–800 eggs during her lifetime. Upon hatching from the eggs, the tiny black grublike larvae begin to feed on the undersurfaces of the leaves. As the larvae grow, they become green to yellow with lateral black stripes (E). Larval feeding results in a skeletonization of the foliage. The upper surface of the leaves and the veins are left intact. Badly affected leaves soon turn brown and drop from the tree prematurely. Trees that lose their leaves as a result of elm leaf beetle damage commonly put out a new flush of growth, which may be consumed by the resident insects as quickly as it appears. When conditions are severe and trees lose leaves for several consecutive years, limbs or the entire tree may die.

At the end of the feeding period larvae migrate to the lower parts of the elm tree and pupate openly on the ground at the base of the tree, or in cracks, crevices, or crotches on the trunk or larger limbs. Transformation to the adult stage requires a week or two. At this time the new adults return to the foliage of the same or adjacent elms and lay eggs. In the late summer and fall the adult beetles leave the host tree and seek a suitable site for overwintering.

Elms differ in their suitability as a food for the elm leaf beetle. For example, English elm (*Ulmus procera*), Siberian elm (*U. pumila*), and the "urban elm" [*U. pumila* × (*U. hollandica* cv. Vegita × *U. carpinifolia*)] are better food sources than American elm (*U. americana*), Chinese elm (*U. parvifolia*), and *U. wilsoniana*. Siberian and urban elms are strongly preferred by beetles under laboratory conditions. It appears that the Chinese elm may be a good substitute for American elm under some circumstances. Eulophid wasp parasites *Tetrastichus brevistigma* Graham and *T. gallerucae* (Fonscolombe) are pupa and egg parasites, respectively, of the elm leaf beetle. These parasites, if properly protected and manipulated, may provide adequate pest management for the elm leaf beetle in many areas of the United States.

In California a tachinid parasite, *Erynniopsis antennae* Rodani, is used to control the larval stage.

Figure 72. A cluster of elm leaf beetle eggs, highly magnified to show the characteristic sculpture of the chorion. × 10.

Two generations occur each year throughout the United States, although a third and sometimes a fourth generation are reported to occur in the warmer parts of the West.

The larger elm leaf beetle, *Monocesta coryli* (Say), feeds on all native and oriental elms and is recorded as a feeder on birch, hawthorn, hazel, and pecan. Adults emerge from the ground in the spring by the time native elm leaves have reached full size. They feed in the top of trees and lay their eggs in clusters on the underside of leaves. The metallic reddish brown and gregarious larvae feed interveinally, often leaving a nearly perfect vein skeleton as a diagnostic symptom. Full-grown larvae are about 13 mm long. These insects occur from southern Pennsylvania south to Georgia and west to the prairie states, and they are very common in the piedmont region of the Carolinas and Virginia. Adults are dull yellowish brown and have a dull green band at the end of the wing covers. They are much larger than the more common elm leaf beetle.

References: 10, 15, 49, 230a, 312, 315, 479

A. An elm tree damaged by the elm leaf beetle.
B. A branch of elm showing the nature of leaf injury by larvae of the elm leaf beetle.
C. A leaf of elm showing cluster of eggs (*circle*) and larvae feeding in vicinity of midvein.
D. An adult beetle and two egg masses of the elm leaf beetle. Adult length approximately 5 mm.
E. Close-up of a leaf section, showing two feeding larvae and the nature of feeding injury.
F. Mature larvae and pupae of the elm leaf beetle on the soil surface beneath an elm tree.

A
B
C
D
5mm
E
F

Leaf Beetles (Plate 104)

The leaf beetle *Pyrrhalta viburni* (Paykull), a European import, was first reported in North America from the Niagara peninsula in Ontario in 1947. It apparently did not survive at this site but appeared again in 1978 in Montreal, Quebec City, and Ottawa. As far as known, *P. viburni* limits its feeding to *Viburnum* species, but it appears to have preferences within the genus. Both larvae and adults devour the leaves and give them a tattered appearance at first, but larvae in dense populations eat all of the leaf except the major veins (A).

The egg is the overwintering stage. The female chews square holes (1 × 1 mm) in the bark of twigs and deposits several eggs in each hole. She covers these holes with a cap of black excrement and fragments of chewed bark and wood (D, E). The cap soon hardens, providing protection from predators and perhaps providing insulation. For the first few weeks there is a sharp color contrast between the excrement and wood fragments. A female may produce about 500 eggs. In May of the following year eggs hatch and the larvae commence feeding on the young, readily accessible foliage. Larvae feed gregariously and complete their development in 8–10 weeks. Adults begin to appear about mid-July and may be seen until the first frost.

Known hosts include *Viburnum opulus* and cultivars, *V. dentatum*, and *V. rafinesquianum*. The adults feed on and lay eggs in *V. lentago*, *V. acerifolium*, and *V. trilobum*. It is not known whether larvae will survive on these last three species. With a high reproductive potential, a favorable overwintering location, and an abundance of host plants both native and ornamental, this beetle is considered a potentially serious threat. Easy movement to other geographic areas through nursery commerce and by amateur horticulturalists is likely to bring about rapid dispersion.

P. decora decora (Say), the gray willow leaf beetle, is a native that feeds solely on willow. It is transcontinental, occurring across the northern states and adjacent Canada, but it is more of a problem on shelterbelt trees in the prairie states and provinces. It produces one generation per year. Like other species in the genus, both larvae and adults eat leaves. The beetle is 4–6 mm long, has light to dark brown wing covers (elytra), and has three black irregular spots on the dorsum of the thorax.

An alder leaf beetle, *Chrysomela mainensis* Bechyne, is found in a broad band from Newfoundland and the northeastern states to Colorado, the northern Rockies, Washington, and the Canadian Northwest Territories. Both larvae and adults eat leaves. Larvae feed gregariously on the tissue between the veins (B). Only the midrib and lateral veins remain, which provides a good diagnostic symptom. There is one generation per year. Adults overwinter and lay their eggs on the leaves of alder in early June (Maine).

Colaspis pseudofavosa Riley is a bright metallic green leaf beetle ranging in size from 4.6 to 5.8 mm. A southeastern species, it feeds primarily on the foliage of azalea and *Photinia* species. Adults may be found year-round in Florida. As is true of most species of *Colaspis* the larvae are root feeders.

Neochlamisus cribripennis (LeConte), commonly called the blueberry case beetle, skeletonizes foliage as a larva and feeds on the bark of blueberry stems as an adult. Blueberry appears to be its only host. Its distinctive larval case makes identification easy. One writer described it as looking like "the burned end of a wooden match stick." In Ontario the adult overwinters and begins to lay eggs in May. The tiny pink eggs attached to the stem of the host plant are each covered with a dark bell-shaped mass of excreta. There may be several rows of excreta-covered eggs along the length of the stem. Soon after hatching, the larva begins to construct its case. It feeds on the foliage for about 4 weeks. See page 400 for a related species. *N. platani*.

With the increased use of black locust and its cultivars in ornamental plantings has come concern about the tiny black beetle (2–2.5 mm long) *Apion nigrum* Herbst. It could also be called the black locust weevil. The adults breed in the seeds of black locust and chew small, usually round holes in the leaflets, often riddling them. This weevil can be found from the Ozarks, Kansas, and Iowa east to New England and south to the southernmost parts of the Appalachians. Adults may be present from late April through June. See Plate 87D for a weevil of similar size and color.

References: 48, 660

A. Leaf injury to *Viburnum opulus* caused by *Pyrrhalta viburni*.
B. Larvae, nearly fully grown, of *Chrysomela mainensis* feeding on alder; up to 9 mm long. (Courtesy Sharon J. Collman, Cooperative Extension, Washington State Univ.)
C. An adult *P. viburni*; 4.5–6.5 mm long. [Chrysomelidae]
D. Defoliated twigs of *V. opulus* with egg sites of *P. viburni* capped by black fecal matter and wood chips.
E. Close-up of the fecal egg caps.
F. Close-up of egg caps (shown in D) removed.

A
B
C
D
E
F

Cottonwood Leaf Beetles and Related Species (Plate 105)

The beetles shown on this plate, collectively called leaf beetles, feed on the leaves of poplars, including aspen, willows, and alders. Each of them, while in the larval stage, feeds by skeletonizing the leaf (A, B).

The cottonwood leaf beetle (G), *Chrysomela scripta* Fabricius, is found throughout the United States. It feeds on the leaves of eastern cottonwood (*Populus deltoides*), other poplars, and several willows. It also feeds on the succulent bark of seedling cottonwoods and willows, and kills the tree or retards its growth.

This species hibernates under bark, litter, and forest debris. Beetles may be collected in large numbers under or near cottonwood or willow trees in the winter. In the spring, after leaf growth begins, they become active and fly to host trees to feed on the leaves and twigs. In a few days the female beetles begin to lay their lemon yellow eggs in clusters of 25 or more on the undersides of leaves. The eggs hatch in 2 weeks or less, and the young larvae immediately begin to feed in groups on the underside of the foliage. After a few days of aggregated feeding they separate and feed alone. The very young larva is black.

The larva reaches full size in less than 2 weeks and then develops into a pupa. The pupa attaches itself to leaves and hangs head downward. After a second 2-week period the adult emerges and repeats the cycle. Four or more generations occur annually in the South and West. In the North two and, occasionally, a partial third generation occur.

These beetles are serious skeletonizers, particularly in the southern and western states. Immature stages are attacked by a variety of natural enemies, including ants, lady beetles, stink and assassin bugs, lacewings, spiders, parasitic flies, and wasps.

C. interrupta (F, H) is closely related to *C. scripta,* and the two are often confused. They have a similar geographic distribution, although *C. interrupta* is believed to be more northern. Most collection records tend to substantiate this, although in April 1972 *C. interrupta* was recorded as far south as Gainesville, Florida. Records show it to be present all through the mountainous regions of the West and also in Alaska. *C. interrupta* prefers willow but is frequently found on poplar and alder. Both *C. interrupta* and *C. scripta* have great variation in color markings within their species. The insects shown in panels F and H are both *C. interrupta,* even though the wing markings shown in panel F are subdued black spots and those shown in panel H are jet black.

The larvae of both species have a remarkable protective mechanism. In the black tubercles along the thorax and abdomen are glands that produce a foul-smelling milky fluid. When the larva is disturbed, it emits a droplet of this fluid. Predators rarely attack these larvae, and it is assumed that they are repelled by the repugnant odor. When the danger ends or the disturbance ceases, the larva retracts this droplet, giving us an example of nature's conservancy. The larvae of these two species, according to some writers, cannot be distinguished.

Other leaf beetle species are commonly found in the West and North. *C. tremulae* Fabricius feeds on aspen and poplars in the Pacific Northwest, and *C. californica* (Rogers) and *C. aenicollis* (Schaeffer) feed on willow in California. *C. crotchi* Brown is frequently encountered in Canada, where it feeds on both willow and poplar. It bears the common name aspen leaf beetle. *Calligrapha bigsbyana* (Kirby) (D) is found in the northeastern and north central states, and it, too, is a defoliator of willows. A similarly marked *Calligrapha* beetle feeds on alder in Maine and northeastern Canada.

The elm calligrapha, *Calligrapha scalaris* (LeConte), feeds on the leaves of elm, linden, alder, and willow in both its larval and its adult stage. It is likely to occur in scattered locations in Canada, the Northeast, and the Midwest. Severe infestations on elm have also occurred in Kansas and other plains states. The elm calligrapha looks somewhat like *C. bigsbyana* (D) both in color and size. The beetles hibernate during winter and begin to lay eggs in the spring when leaves are beginning to develop. A new generation of adults appears about midsummer and feeds on the leaves until autumn.

References: 49, 95, 346, 359, 514, 705, 783

A. A cottonwood shoot with foliage damaged by *Chrysomela interrupta.* Note beetle at arrow.
B. *Salix caprea,* the pussywillow, showing the typical skeletonized injury caused by leaf beetles. When the skeletonized leaf dries, it tatters in the wind, breaking off and leaving only the main vein.
C. A larva of *Chrysomela interrupta.*
D. An adult *Calligrapha bigsbyana* and the damage typically caused by its feeding; about 8 mm long.
E. A larva of *Chrysomela scripta,* the cottonwood leaf beetle.
F, H. Both beetles are *Chrysomela interrupta*; there is great variation in the color pattern.
G. The cottonwood leaf beetle.

A
B
C
D
E
F
G
H

Imported Willow Leaf Beetle and Flea Beetles (Plate 106)

In the United States' commerce with other nations, it is inevitable that some insects of foreign origin are brought accidentally onto American soil. This is the case with *Plagiodera versicolora* (Laicharting), commonly known as the imported willow leaf beetle. Since it was first found in the United States in 1911, this beetle has become an important pest of willows and poplars.

The adults (B, E) vary from nearly black to greenish blue. They hibernate under loose bark or in other protected places on the tree trunk. In the spring, shortly after the leaf buds open, they emerge to feed on foliage and lay eggs. Their glossy, pale yellow eggs (F) are laid in irregular masses on the undersides of leaves. These eggs hatch in a few days and the blackish larvae begin feeding on the undersides of leaves. Later they may be found on the leaf's upper or lower surfaces. In 3–4 weeks the larvae transform into pupae, which are about 10 mm long, are yellowish brown, and have dark markings. They remain as pupae for a short while, then transform into adults.

The larva causes the major damage to the plant by chewing the leaf so as to expose a network of leaf veins. The adult chews holes or notches in the foliage as seen in (B). When abundant, this insect can turn a beautiful ornamental willow into an unsightly plant.

There may be four generations of *P. versicolora* during a single growing season in Virginia or North Carolina but probably only two or three in other parts of its range. It is often abundant in New England, the Atlantic Coast states, and eastern Canada. Recent records report it as far west as Michigan. Most species of willow are acceptable as food, although some are more frequently attacked than others. Air pollution studies suggest that leaves exposed to excessive ozone are preferred. *Salix nigra, S. lucida, S. alba,* and *S. interior* are important willow hosts. The Lombardy poplar and eastern cottonwood are also food plants of the willow leaf beetle.

Schizonotus sieboldi Ratzeburg, a chalcidoid wasp, is considered to be a highly efficient parasite of the imported willow leaf beetle pupa. At one location in New York state this insect parasitized about half of the second-generation pupae. The imported willow leaf beetle is preyed upon by lady beetles, lacewings, predacious bugs, and spiders.

The alder flea beetle, *Macrohaltica* (=*Altica*) *ambiens* (LeConte) (D), is transcontinental. It feeds on the foliage of alder; adults chew holes in leaves but larvae feed gregariously and skeletonize the upper leaf surface. Fully grown larvae are about 8 mm long; they are brown to black on the dorsum and yellowish on the venter and have a shiny black head and short legs. The cobalt blue adults overwinter and appear in the spring, to resume feeding and eventually lay eggs on foliage. There is one generation per year.

These insects usually occur in innocuous numbers, but occasionally there is a population explosion. If the larvae run out of food, which occasionally occurs when populations are high, they wander and crawl on the sides of nearby buildings, becoming a nuisance around dwellings. Adults cause similar problems in dwellings in the fall and spring as they enter and leave hibernation sites.

Another flea beetle, *Altica ignita* Illiger, is poorly known; it is a potential pest of rhododendrons and other ericaceous plants. Both adults and larvae feed on leaves, but the skeletonization caused by larvae is of most concern. Skeletonized areas soon drop out, resulting in irregular holes in leaves. Most damage occurs in early June. Adults are bronzy green, are about 4 mm long, and have the same form as *M. ambiens,* shown in panel D. Larvae are dirty yellow and have brown spots. *A. ignita* is a native species and likely to occur in regions where its hosts grow wild.

References: 49, 275, 359, 382, 646, 775, 798, 806

A. Imported willow leaf beetle larvae. Note the injury to the upper leaf surface of *Salix babylonica.* [Chrysomelidae]

B, E. Each photograph shows a willow leaf beetle. Note that the feeding injury of the adult differs from that of the larva.

C. A cluster of willow leaf beetle larvae on the underside of a leaf. Note the nature of injury.

D. An adult alder flea beetle; about 6 mm long. [Chrysomelidae] (Courtesy Sharon J. Collman, Cooperative Extension, Washington State Univ.)

F. A cluster of imported willow leaf beetle eggs on the underside of a willow leaf.

A
B
C
D
E
F

Fuller Rose Beetle (Plate 107)

The Fuller rose beetle, *Asynonychus godmani* Crotch (=*Pantomorus cervinus*), is a common pest in both adult and larval stages. It is a voracious feeder on outdoor ornamentals, on fruit trees and particularly citrus, and on vegetable crops in all of the southern states. Although it has pest status in the South, it also occurs in New York, New Jersey, Illinois, Wisconsin, and west to Oregon. It was first recorded in California in 1879.

The adult, a brown to gray snout beetle (E), lacks the ability to fly. Males of the species are unknown. It overwinters as an adult or larva in the soil and, in the case of the adult, also under plant refuse or other debris. Eggs, which may be found throughout the growing season, are yellow, cylindrical, and deposited in masses of 5–50 on the ground, in bark crevices, or sandwiched between leaves. The eggs are covered with a white sticky substance. A female may produce, parthenogenically, over 200 eggs during her reproductive period.

The larva is a whitish legless grub with a yellowish head capsule and contrasting black mandibles. After hatching from the egg, it enters the soil to feed on both feeder and suberized roots, sometimes completely girdling suberized roots. When fully grown it will be about 10–12 mm long. At this time the larva moves near the soil surface, where it pupates. The adult emerges from the pupal cell, crawls to the host plant, and begins feeding. Fuller rose beetles are long lived for insects and feed primarily at night. There are two generations per year in Florida, where peak adult populations occur in early June and again in early September.

Adults cause the greatest injury to ornamental plants. They devour leaves, buds, and blossoms, leaving behind them trails of dark fecal material on uneaten foliage (C). A wide variety of woody ornamental plants are susceptible to injury by adult weevils. Some of the more common hosts are species of *Abutilon* (flowering maple), *Acacia* (acacia, wattle), *Acer negundo* (boxelder), azalea, *Camellia, Cissus, Citrus, Deutzia, Diospyros* (persimmon), *Dracaena, Feijoa* (pineapple guava), *Fuchsia, Gardenia, Hibiscus, Malus,* palm, *Pentstemon, Plumbago, Prunus* (peach), *Pyrus, Quercus,* and rose.

References: 177, 217, 682, 875

A, B. Foliage of *Hibiscus* sp. (A) and boxelder (B) tattered by adult Fuller rose beetles. Note adult beetles (*circle*) on the boxelder.

C. Close-up of boxelder foliage damaged by the adult Fuller rose beetle. Fecal matter is deposited on the foliage and is an easily identifiable symptom of the Fuller rose beetle's presence.

D. Buds of *Hibiscus* sp. destroyed by the Fuller rose beetle. Note beetles feeding.

E, F. Adult Fuller rose beetles; 7–9 mm long. [Curculionidae]

A
B
C
D
E
F

Seagrape Borer and a Seagrape Gall Midge (Plate 108)

The seagrape borer, *Hexeris enhydris* Grote, is probably the most important pest of the seagrape, a small tree common to central and southern Florida. The larva of this pest bores into the stem, the leaf petiole, and occasionally the main vein of leaves. The weakened stem is subject to wind breakage. If the leaf petiole is attacked, the leaf dies and turns brown, providing a diagnostic symptom.

The fully grown larva is about 15 mm long, with a light brown head and a pale brown, partially translucent body. It chews one or more small holes in the stem from which it ejects small masses of dark frass. Pupation occurs in the stem. There are probably three generations each year. Adults appear in December, from March to May, and from July to September. The moth has a wingspan of 34 mm. Its body is pale brown and its forewings have numerous wavy rusty lines. The biology of this species has not been fully studied. Its only known host is the seagrape.

A minute gall midge, *Ctenodactylomyia watsoni* Felt, can also substantially reduce the ornamental value of seagrape as a result of the numerous galls produced by this pest on the seagrape's leaves. Each gall is hemispherical, succulent, and has rather thick walls. The galls are visible from both surfaces of the leaf (Figure 73). If one is cut open, a small maggot or larva will be found. If there is a hole in the gall, the adult midge has emerged (Figure 74).

The flatid planthopper, *Metcalfa pruinosa* (Say), may also be found on the leaves of seagrape (Plate 203), as can the greedy scale *Hemiberlesia rapax* (Comstock); an armored scale, *Melanaspis coccolobae* Ferris (Plate 178); and the black citrus aphid, *Toxoptera aurantii* (Fonscolombe) (Plate 143).

Reference: 526

Figure 73. The lower surface of a seagrape leaf showing galls of the midge *Ctenodactylomyia watsoni.* Two-thirds natural size. (Courtesy Div. of Plant Industry, Florida Dept. of Agriculture and Consumer Services.)

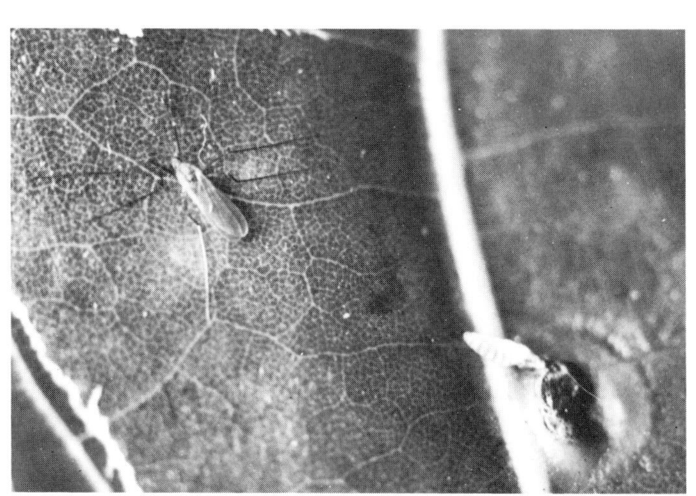

Figure 74. An adult seagrape gall midge and empty pupal case on a seagrape leaf. (Courtesy Div. of Plant Industry, Florida Dept. of Agriculture and Consumer Services.)

A, B. A seagrape, *Coccoloba uvifera,* with dead leaves, which are symptoms of the presence of the seagrape borer, *Hexeris enhydris.*
C. Leaf vein damage from borer feeding.
D. An exit hole with frass at the base of a leaf.
E. A hollowed-out stem with borer inside.

A
B
C
D
E

Walnut and Hickory Insects (Plate 109)

The walnut tree is useful for both shade and food. Because it has a dual purpose, close observation and special care are necessary to prevent insects from defoliating the tree or from ruining its fruit.

The butternut curculio, *Conotrachelus juglandis* LeConte, feeds on butternut and Siebold, Persian (English), and black walnut. It is an eastern species found from Texas, Kansas, and Wisconsin eastward to the Atlantic Coast; it is also found in south central Canada.

Injury to the tree is caused by both larvae (B) and adults (C). The adult feeds on nuts, tender twig terminals, and leaf petioles; the larva burrows into the nuts, young shoots, leaf petioles, and stems. Of greatest consequence to the health of the tree is larval injury to stems and branches. New shoots are sometimes killed back to the wood of the previous season.

C. juglandis overwinters in its adult stage, probably in ground litter. It lays its first eggs in new twig growth. Later, as nutlets begin to form, it deposits the eggs in crescent-shaped excavations that the adult curculio chews in the nutlet's husk near the blossom end.

The larvae become fully grown in 4–5 weeks. In West Virginia they begin to leave the infested nuts and twigs about the middle of July. At this time they enter the ground to pupate. The new adult emerges from the soil in late summer to feed on terminals, shoots, and leaf petioles until cold weather causes it to hibernate.

The black walnut curculio, *C. retentus* Say, closely resembles the butternut curculio in both behavior and appearance. Adults feed on young shoots and leaf petioles. They lay eggs in the immature nut. Black walnut is the preferred host. This species may be found from Pennsylvania and New Jersey south to North Carolina and west to Mississippi, Arkansas, Kansas, Missouri, and Illinois.

Other major pests of walnut fruits are husk maggots in the genus *Rhagoletis* and the codling moth, *Laspeyresia pomonella* (Linnaeus). Husk maggots, as their name implies, feed in the exocarp. They may have an economic impact when the damage they cause results in the downgrading of nut kernels. Although the codling moth is of European origin, it is a fruit pest around the world wherever apples and walnuts are grown. In the western United States it is a serious pest of Persian walnut. In the East there have been locally severe infestations, but it is not considered a major pest.

The female moth (E) lays her eggs on foliage, twigs, and nuts. The location of her eggs depends on the time of the year she is active. The young larva enters the nut from the calyx end, where two nuts come into contact (D). After this invasion, it acts as a borer and expels frass, keeping its tunnels fairly clean. The frass that collects on the outside of the nut is a sign of the larva's presence. The larva is creamy white in its early stages, but when fully grown it is pinkish white with a brown head and about 25 mm long. In the South and in California there may be two or three generations each year.

Mites also cause damage to walnut. An eriophyid gall mite, possibly *Aceria brachytarsus* (Keifer), caused the formation of the excrescence shown in panel F. The galls are formed in late spring and persist through the life of the leaf. This mite causes galls only on the upper side of the leaf. The gall is hollow and contains hundreds of tiny mites that feed on the succulent tissue (Plate 235). They are invisible to the unaided eye. In addition to the eriophyid gall mite, a number of spider mites cause injury to walnuts, particularly in the western states.

Conotrachelus aratus (Germar), sometimes called the hickory shoot curculio, feeds primarily on hickory and pecan. Eastern Massachusetts south to Florida and west to Texas and Kansas is its geographic range. The adult beetle, a weevil, overwinters on the ground and awakens from hibernation soon after the host trees begin to grow in the spring. In the deep South adults begin feeding in late March and April on shoot tips and leaf petioles, but the damage is usually negligible. The female lays her eggs singly in tender shoots and leaf petioles, but first she must chew a hole or niche in the tissue big enough to accommodate the egg. After hatching, the crescent-shaped larva feeds by tunneling within the plant part (Figure 75). Once fully grown, about 6 mm long, it vacates its tunnel, drops to the ground, and burrows into the soil, where it pupates. During August and September the adult emerges, but it does little other than feed to maintain itself. There is one generation per year. Several larval parasites are capable of significantly reducing the population.

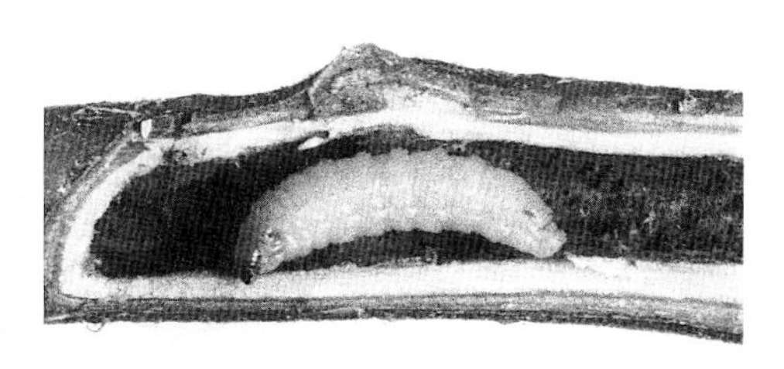

Figure 75. A larva of *Conotrachelus aratus* in the shoot of a hickory. The larva is crescent-shaped and legless. (Courtesy J. D. Solomon, U.S. Forest Service.)

Figure 76. A shoot damaged by a *Conotrachelus aratus* larva. The hole was originally the egg niche chewed out by the adult female. The larva is inside the enlarged, swollen area above the hole. (Courtesy J. D. Solomon, U.S. Forest Service.)

References: 41, 64, 250, 403, 533, 886

A. Black spots on the stem (*arrow*) were made by the adult butternut curculio, *Conotrachelus juglandis,* on *Juglans* sp.
B. A new shoot split open to show larvae of the butternut curculio.
C. An adult butternut curculio. [Curculionidae]
D. A Persian (English) walnut infested with larvae of the codling moth, *Laspeyresia pomonella*. Note the accumulation of frass between the two nuts.
E. The codling moth.
F. Leaf galls on *Juglans hindsii* caused by an eriophyid gall mite.

A
B
C
D
E
F

Japanese Beetle, European Chafer, and Rose Chafer (Plate 110)

The Japanese beetle, *Popillia japonica* Newman, was first found in the United States in 1916. New Jersey was its point of entrance, and from here it has spread through much of the eastern United States. Local infestations have occurred in Illinois, Kentucky, Tennessee, Michigan, Iowa, Missouri, and California. Some of these have been eradicated. Its southernmost limit is Georgia, and its range extends north to Ontario and Nova Scotia in Canada. It continues to spread each year. Climatological studies predict that it will reach all of the southern states along the Gulf of Mexico. Its spread to northern areas will be limited because of the cold. In New Hampshire, Vermont, and Maine some of the grubs require 2 years to mature, though a normal cycle is completed in 1 year. Its spread west into the plains states may be inhibited by lack of adequate rainfall during the late summer months and by cold winter temperatures. The coastal areas of Washington and Oregon are believed favorable for Japanese beetle development.

The Japanese beetle overwinters in the northeastern states as a partially grown grub in the soil below the frost line. In the spring the grub resumes feeding, primarily on the roots of grasses, and then pupates near the soil surface. Adults begin to emerge about the third week of May in North Carolina and the first week of July in upstate New York, southern Vermont, and New Hampshire.

The adults fly to trees, shrubs, or other food plants to begin feeding. Beetles live for 30–45 days. After mating, they lay eggs in small groups in the ground at a depth of 2.5–10 cm. Each female is capable of laying 40–60 eggs. These hatch in about 2 weeks, and the young grubs feed on fine rootlets until cold weather drives them deep into the soil.

Both beetles and larvae seriously injure plants. Grub injury may be unnoticed until the plants are badly damaged, often beyond recovery. The roots of premium grasses seem to be the choice food for grubs, but they may also consume young roots of woody ornamental nursery stock. Vegetable seed beds are also choice feeding sites for grubs. The adults feed on nearly 300 species of plants. Upon emergence, they usually feed on the foliage and flowers of low-growing plants such as roses, grapes, and shrubs; later, they feed on tree foliage. On tree leaves beetles devour the tissue between the veins, leaving a lacelike skeleton (A). At a distance severely injured trees appear to have been scorched by fire. The beetles completely consume rose petals and leaves with delicate veins. They are most active during the warmest parts of the day and prefer to feed on plants that are fully exposed to the sun.

Of the woody ornamental plants attacked, the rose seems to be the favorite host. Other plants that receive extensive damage include crape myrtle, flowering crabapple, flowering cherry, Japanese maple, Norway maple, gray birch, sycamore, pussy willow, linden (*Tilia tomentosa* is somewhat resistant), elm, Virginia creeper, certain varieties of azalea, flowering quince, weigela, wisteria, California privet, bayberry, viburnum, and others. Beetles have not been observed feeding on dogwood, forsythia, American holly, snowberry, or lilac.

The USDA and several state experiment stations along the Atlantic Coast have done extensive work on the biological control of this insect. From this work, ways have been developed to produce and disseminate a disease-causing bacterium commonly called milky disease. The milky disease organism, *Bacillus popilliae,* is commercially available for application by home gardeners. Naturally occurring microorganisms that reduce grub populations include pathogenic fungi, protozoans, and nematodes. The important parasitic insects are *Tiphia vernalis* Rohwer and *T. popilliavora* Rohwer, both of which are wasps. A parasitic fly, *Hypercteina aldrichi* Mesnil, attacks adult Japanese beetles in New England. In Massachusetts and Connecticut the rates of parasitism have reached 30%. Birds and toads eat large numbers of Japanese beetle adults as well as grubs. Moles, shrews, and skunks also eat large quantities of grubs.

The European chafer (E), *Rhizotrogus* (=*Amphimallon*) *majalis* (Razoumowsky), is another accidentally introduced foreign insect. It has a life cycle similar to that of the Japanese beetle, and the grubs of both species look similar. The larva feeds almost exclusively on the roots of grasses. The adult does little feeding and causes no noticeable damage. Spectacular mating swarms, however, attract much attention. About the middle of June the adults begin to emerge from the soil, appearing only for brief mating flights. Around sunset, on warm days, thousands of these insects swarm like bees around trees or large shrubs. They fly for about half an hour, mate, and then the females enter the soil to lay eggs. Mating flights may be repeated several times in a season. Beetles are most abundant from mid-June to mid-July.

The European chafer is currently found in New York state, Connecticut, New Jersey, western Pennsylvania, and in limited localities in Ohio. In Canada it is known only in the Niagara Falls, Ontario, area. A USDA quarantine for this insect has helped to slow its westward movement.

The rose chafer (D), *Macrodactylus subspinosus* (Fabricius), skeletonizes leaves in much the same way as the Japanese beetle. The rose chafer feeds on many kinds of plants; if deprived of one plant, it will take another. It is strongly attracted to blossoms, particularly to rose and peony flowers. It commonly feeds on the leaves of trees and shrubs.

The rose chafer overwinters as a larva. The grub enters the pupal stage in the early spring and emerges as an adult beetle in June. The adult lives for about 1 month and lays its eggs most commonly in grassland soil. Upon hatching, the larva feeds on the roots of grasses and weeds.

Rose chafer development requires light sandy soil, a factor that greatly limits its distribution. The larva is a white grub with legs, and when fully grown it is 18 mm long. The rose chafer occurs in restricted localities throughout the Northeast, in eastern Canada, and also in Colorado.

Birds are often killed by eating the adult rose chafer. Its body contains a poison that affects the heart of small, warm-blooded animals.

The western rose chafer, *Macrodactylus uniformis* Horn, greatly resembles the eastern species, and its habits are similar. It occurs primarily in Arizona and New Mexico.

References: 49, 217, 249a

A. Typical feeding injury to cherry caused by the adult Japanese beetle, *Popillia japonica.* [Scarabaeidae]

B. Severe Japanese beetle damage to a linden leaf.

C. An adult Japanese beetle; about 10 mm long.

D. An adult rose chafer, *Macrodactylus subspinosus*; about 12 mm long.

E. The European chafer, *Rhizotrogus majalis,* in all stages of development. Adult length is about 13 mm. (Courtesy G. A. Catlin, New York State Agricultural Experiment Station, Geneva.)

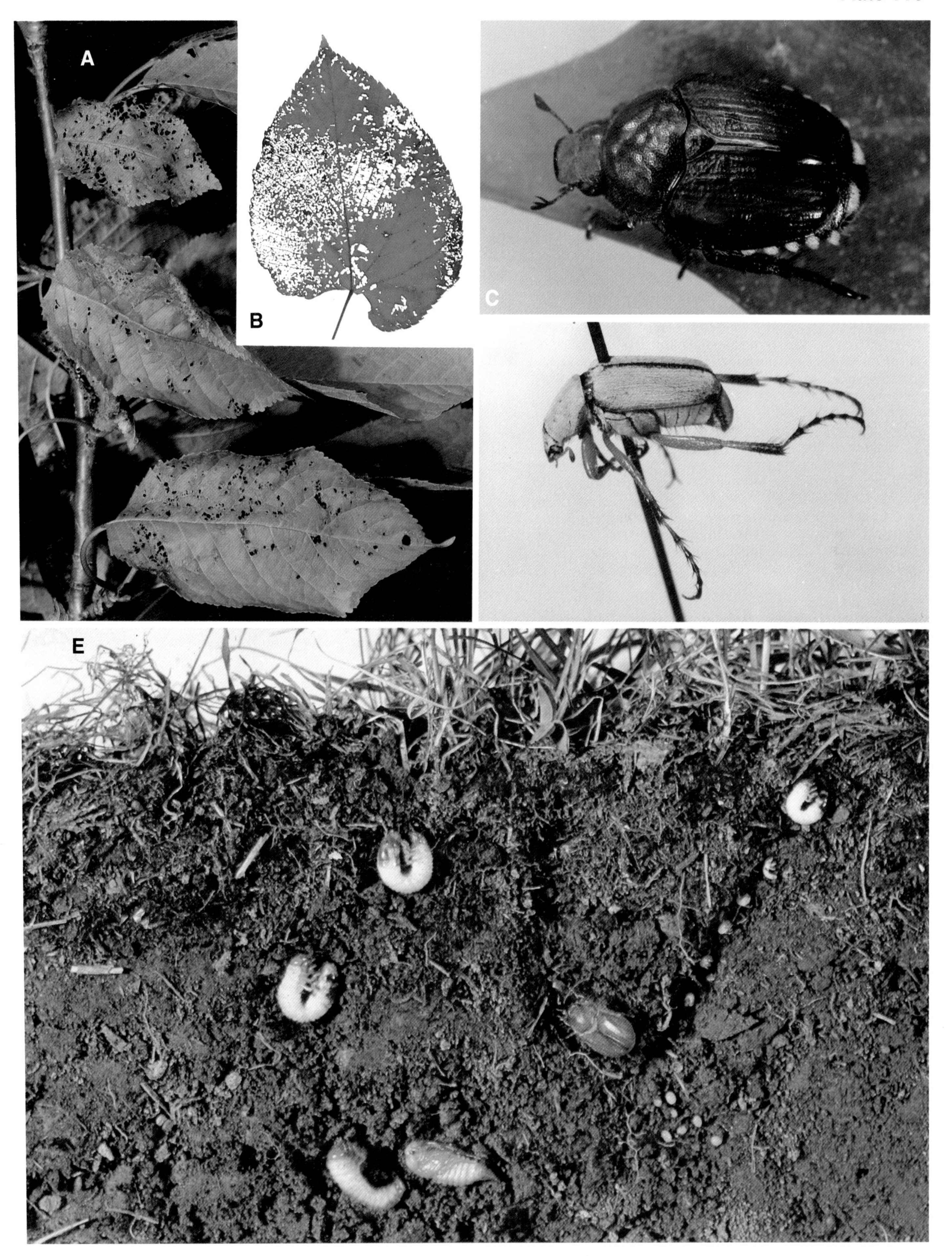
A
B
C
E

Baccharis Leaf Beetle and Rose Curculio (Plate 111)

The baccharis leaf beetle, *Trirhabda flavolimbata* (Mannerheim), is one of many related leaf beetles that feed on sagebrush (*Artemisia* species) and other wild rangeland shrubs or goldenrod (*Solidago* species) in many parts of North America. *T. flavolimbata* feeds on dwarf chaparral broom, or coyote bush (*Baccharis pilularis*) only in the San Francisco region of California. Another common western species is *Trirhabda pilosa* Blake, which feeds on sagebrush in Nevada, Wyoming, California, and in the provinces of Alberta and British Columbia. *Trirhabda nitidicollis* LeConte feeds on rabbitbrush, *Chrysothamnus nauseosus,* in areas of low rainfall in the Rocky Mountains and the Pacific slope. This plant, and others like it, are being used more and more in xeriscape plantings in urban areas subject to drought and limited water supplies. In the case of rabbitbrush, look for an increasing incidence of *T. nitidicollis* as the density of this shrub increases. Adults may be found feeding on foliage throughout the growing season. Unlike most *Trirhabda* species, which have only one generation each year, *T. nitidicollis* produces two or more generations annually. Many species of *Trirhabda* are known in the western United States but only six species are reported from the East. Here *Trirhabda bacharidis* Weber feeds on ground bush, *Baccharis halimifolia,* a plant that can tolerate coastal saltwater spray.

The life histories of most *Trirhabda* species are similar. Instead of overwintering in the adult stage, which is usual for the leaf beetle family Chrysomelidae, this species spends the winter in the egg stage, on the ground. The eggs hatch in the spring, and the young larva climbs a suitable host plant, where it feeds first on tender terminal foliage. The older larva is less discriminating in its choice of food and feeds readily on all foliage (A). At maturity, the larva, which measures about 13 mm long, returns to the ground where it forms a pupal case just below the surface of the soil. The adult baccharis leaf beetle appears in midsummer and lays eggs soon thereafter. The beetle is commonly a metallic blue or green, although some species are somber in color. It feigns death and drops to the ground quickly when its host plant is disturbed. There is only one generation each year. The larvae likewise are metallic blue-green.

Widespread serious damage to sagebrush has resulted from defoliation by *T. pilosa* in parts of western North America. Where sagebrush is considered a pest plant on rangeland, interest has been expressed in the possibility of utilizing *T. pilosa* as a means to control the plant.

The rose curculio *Merhynchites* (=*Rhynchites*) *bicolor* (Fabricius), occurs in a variety of black to red color phases. It is destructive to roses throughout the United States and is most common in the cooler northern regions. Although roses—both cultivated and wild—are its principal host, the rose curculio also feeds on wild brambles. The greatest damage to roses is caused when adults feed on flower buds. The adult punctures the floral parts contained inside buds. Later, if the flowers succeed in opening, these floral parts are riddled with holes, resulting in ragged, unsightly blossoms. If flower buds are not plentiful, the adult curculio may feed on the tips of new rose shoots, causing the death of the terminals. At other times it gouges the stems of buds, causing the bud to wilt and die.

Females lay eggs in the flower buds. After hatching, the white legless larva completes its development by feeding on the reproductive parts of the blossom. Larval development may occur in blossoms that remain on the plant or in those that drop to the ground. The fully grown larva leaves the bud and enters the soil, where it passes the winter. Pupation takes place in the soil in the spring. The new adult then emerges from the soil several weeks later, begins feeding on available host plants, and the cycle is repeated. The adult is about 6 mm long. Apparently, there is only one annual generation wherever this insect occurs.

References: 25, 36, 65, 217, 376, 445, 496, 543

A. Larvae of the baccharis leaf beetle, *Trirhabda flavolimbata,* feeding on *Baccharis* sp. Note larva at arrows.
B. Close-up of a larva of the baccharis leaf beetle.
C. A rose flower damaged by the rose curculio.
D. An adult rose curculio.

A
B
C
D

Weevils and Other Root-feeding Beetles (Plates 112–114)

Root weevils are insidious pests of ornamental shrubs. Adult feeding makes notches on the edge of leaves, but the injury is often inconspicuous unless large numbers of weevils are present. The larva feeds unseen on roots in the ground and, if undetected, may consume most of the feeder roots, causing the plant to die abruptly. Such plant losses often are attributed to other causes by those unaware of the presence of the weevils. The diagnosis of dead, wilting, or declining plants should include examination of the roots. Early detection is possible through recognition of adult feeding symptoms (Plate 113).

The most injurious weevils on broad-leaved evergreens and yew are species of *Otiorhynchus*. Most common of the several species are the black vine weevil, *Otiorhynchus sulcatus* Fabricius, and the strawberry root weevil, *O. ovatus* Linnaeus. The adult of *O. sulcatus* is 9–13 mm long, whereas that of *O. ovatus* is slightly less than 6 mm long. *O. rugosostriatus* (Goeze) in the West (Figure 77) and *O. rugifrons* (Gyllenhal) in the East are intermediate in size. The relative size and appearance of these species are illustrated in this plate. The strawberry root weevil is more prevalent as a garden pest, and the black vine weevil is more important on woody ornamental plants. Both of these species, however, attack a wide variety of hosts. Conifer nurseries are occasionally devastated by the strawberry root weevil. Larvae chew the bark from roots and root collars of small seedlings and transplants. Most damage to arborvitae is caused when adults devour new foliage and girdle stems.

The black vine weevil, also known as the taxus weevil, is a native of Europe. It is the most destructive and widespread of the root weevils throughout the northern half of the United States and in Canada (Figure 78). It is destructive, primarily in its larval form, to yew, hemlock, rhododendron, and several other broad-leaved evergreens, as well as to strawberry and some greenhouse plants such as impatiens. As an adult it has a wide host range that includes both deciduous and herbaceous plants. It is a serious pest in nurseries as well as in landscape plantings.

The adult is black and is the largest of the *Otiorhynchus* species found on ornamentals. It is active at night. During the day the adult hides in dark places on the stems of very dense plants or in ground litter and mulch. When disturbed on stems or leaves, it drops to the ground, where protective coloration makes it nearly invisible. All adults are female and cannot fly. A period of 2–3 weeks of feeding occurs before the adult begins to lay eggs. During midsummer, the weevil alternately feeds and lays eggs for a month or more.

Injury caused by adult feeding on broad-leaved evergreens and other hosts consists of marginal notching of the leaves. The damage is indistinguishable from that caused by the Japanese weevil, Fuller rose beetle, and other weevils. However, because other weevils do not feed on yew, notched and cut-off needles are positive diagnostic symptoms of black vine weevil. To detect weevil infestations early on yew, it is necessary to examine needles at the center of the plant near the main stems. If there are many notches on most of the host plant's foliage, the roots will likely be severely injured by the weevil's progeny. The larva, or grub, is highly destructive to plants. Adult feeding rarely injures plants seriously, although it might be extensive. Larval feeding takes place from midsummer, when the eggs hatch, and continues into the fall, and then again in the spring. Feeder roots and bark on larger roots are consumed by the larger larvae in the spring, which often results in the death of the plant. Root feeding may not cause noticeable injury to growing plants in the nursery, but infested plants may die after they have been transplanted. Species of root weevil grubs cannot always be distinguished from one another in the field. All are legless, C-shaped, and white with brown heads (G). They may be found at a depth of from 2–40 cm in the soil around the roots. Pupation occurs in the soil near the surface. The pupa is milky white and has conspicuous appendages.

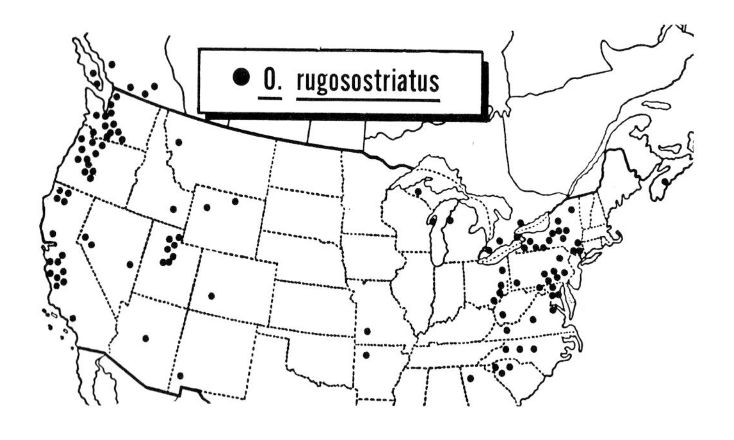

Figure 77. Generalized distribution of the rough strawberry root weevil, *Otiorhynchus rugosostriatus,* in the United States and Canada. (From Warner and Negley, ref. 911, courtesy R. E. Spilman, USDA.)

There is usually one generation each year. Partially grown larvae overwinter in the soil. Occasionally, adults may survive by living in houses during the winter. Adult emergence from the soil usually starts about mid-June in the Northeast, and egg laying begins 1–2 weeks later. In northern California adults begin feeding and laying eggs in March and continue through October. Each female lays as many as 500 eggs in the soil near the base of the plant over a period of 2–3 weeks. The eggs hatch in 10 days to 2 weeks, and the larvae tunnel through the soil to feed on roots. In the East the most extensive feeding occurs in late May and early June, just before pupation. There has been some success in California in controlling black vine weevil larvae and pupae with the insect parasitic nematode *Heterorhabditis heliothidis* Kahn, Brooks & Hirschmann and *Steinernema feltiae* Woutz, Marack, Gerdin & Bedding (=*Neoaplectana carpocapsae*), which is available from commercial producers.

In the Northwest the woods weevil (also known as the raspberry bud weevil), *Nemocestes incomptus* (Horn), and the obscure root weevil, *Sciopithes obscurus* Horn, are destructive to azalea (A), rhododendron, viburnum, camellia, rose, yew, and others (Plate 113H). *Nemocestes* larvae chew the bark of stems beneath mulch and also feed on roots deep in the soil. Development of this weevil is continuous throughout the year, although activity is reduced during the summer. The adult weevil is about 6 mm long. Because males have not been seen, the species is considered to be parthenogenic.

Many other root weevils are pests of ornamentals: the arborvitae weevil, *Phyllobius intrusus* Kono, feeds on arborvitae, retinospora, and juniper in New England; the Asiatic oak weevil, *Cyrtepistomus castaneus* (Roelofs), on oak and chestnut in the East (Plate 114); the citrus root weevil, *Pachnaeus litus* (Germar), is found on citrus in Florida; the imported longhorned weevil, *Calomycterus setarius* Roelofs, feeds on woody and herbaceous plants in the Northeast and west to Illinois; the twobanded Japanese weevil, *Callirhopalus bifasciatus* (Roelofs), is associated with a wide variety of hosts in the Northeast and the South through Virginia (Plates 113–114); and the Fuller rose beetle, *Asynonychus godmani* Crotch (=*Pantomorus cervinus*), occurs on a variety of ornamentals in the South (Plate 107).

Many of these weevils are parthenogenic; that is, males are not required for successful reproduction. The adult cannot fly, because the hard, shell-like forewings are fused together. The larva and pupa of many species look very similar, making identification difficult unless adults of the species are also present. The larvae of most species feed throughout the root zone.

References: 205, 217, 413, 460, 532, 552, 716, 757, 804

A. A larva of *Nemocestes incomptus.* Note damage to the bark of an azalea cutting.

B. *Phyxelis melanothrix,* a native root weevil in the East.

C. An adult of the strawberry root weevil, *Otiorhynchus ovatus*; 4–6 mm long.

D. An adult of the rough strawberry root weevil, *O. rugosostriatus*; 7–9 mm long. [Curculionidae]

E. An adult of the black vine weevil, *O. sulcatus*; 10 mm long.

F, H. A rhododendron stem girdled by black vine weevil larvae.

G. Typical appearance of root weevil larvae and pupae. Note that the larvae are legless.

A
B
C
D
E
F
G
9mm
H

The obscure root weevil, *Sciopithes obscurus* Horn, is a native insect and the most common root weevil of *Rhododendron* species in northern California and the Pacific Northwest. Larvae develop on roots of its host plant, and the adult feeds on foliage. The most serious injury is caused by the larvae (Plate 112G illustrates larvae of similar form) as they feed on young field-grown or container plants or rooted cuttings. Adults are present during the summer and chew notches in leaf margins (H). This weevil's life cycle is similar to that of the *Otiorhynchus* weevils (Plate 112).

It has been known for many years that some species of *Rhododendron* are never or rarely fed on by *S. obscurus*. The plants' resistance is attributed to the volatile chemicals they produce through foliar glands. One of the several chemical compounds was recently identified and was found to act as a powerful feeding inhibitor. This discovery of a self-protecting chemical produced by certain rhododendrons will likely be the basis of further research that ultimately may benefit growers and plant fanciers. The following species are known to be resistant to the obscure root weevil: *Rhododendron edgeworthii*, *R. luteum*, *R. mergeratum*, and *R. trichostomum*.

The cranberry rootworm, *Rhadopterus picipes* (Oliver), has an exceedingly broad host range but is not found west of the Rocky Mountains. Early literature emphasized its injury to cranberry and apple and its broad geographical distribution. It now attracts attention because of the damage it does to woody ornamentals in both nursery and residential areas.

Injury occurs when adults feed on foliage. The characteristic feeding injury is adequate for tentative field identification. The adult makes elongate, sometimes curved, cuts from the upper surface of leaves (G). Injury has been most pronounced in container nurseries and in home lawns located under dense pine tree canopies where moist conditions prevail. This pest is not likely to cause problems on clean, cultivated, open-grown plants. *R. picipes* is nocturnal and hides in ground litter near its host during the day.

The beetle is about 5 mm long, shiny brown with a slight greenish sheen, and looks like a typical leaf beetle (Chrysomelidae). In Mississippi most beetles emerge from late April to mid-May. There are always a few that do not follow the pattern and emerge later in the season. After feeding for about 2 weeks, the beetles mate and females move under the leaf litter to lay eggs. Larvae are thought to feed on herbaceous plant roots and have not been found in container stock. We have not seen injury caused by larvae. There is one generation per year.

Some of the ornamental hosts of this beetle are *Ilex crenata* and its cultivars, *I. cornuta* and cultivars, camellia, *Viburnum japonicum*, *Rhododendron* species, wax myrtle, cherry laurel, *Photinia* species, silver maple, golden raintree, sycamore, spirea, and rose; some native wild plants that are hosts are Virginia creeper, several oaks, *Magnolia virginiana*, sumac, and sassafras.

References: 54, 185, 337, 585, 911

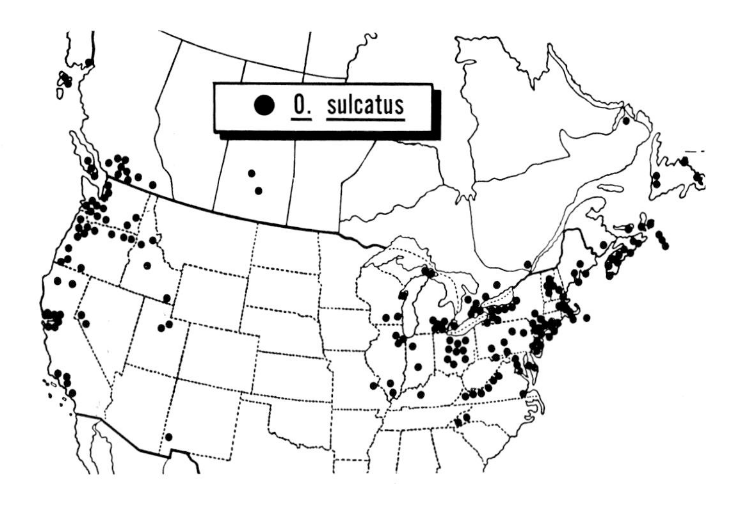

Figure 78. Generalized distribution of the black vine beetle, *Otiorhynchus sulcatus*, in the United States and Canada. (From Warner and Negley, ref. 911, courtesy R. E. Spilman, USDA.)

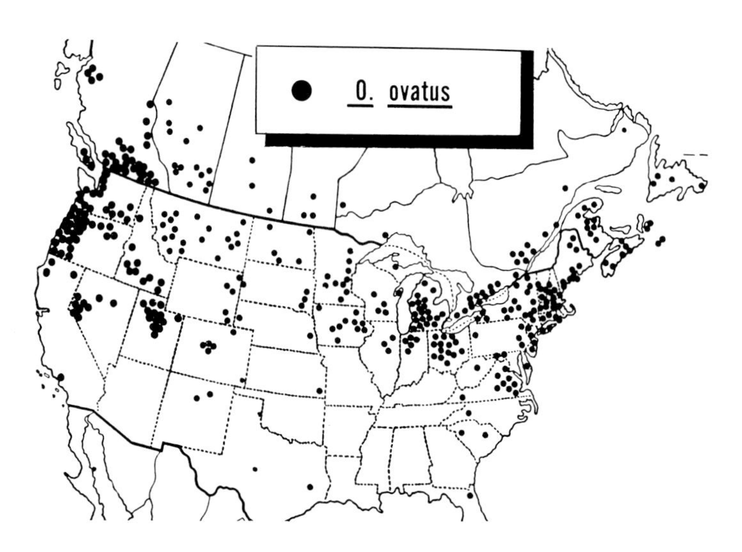

Figure 79. Generalized distribution of the strawberry root weevil, *Otiorhynchus ovatus*, in the United States and Canada. (From Warner and Negley, ref. 911, courtesy R. E. Spilman, USDA.)

A. Typical feeding injury by black vine weevils on leaves of privet.
B. Fatal injury to roots of yew resulting from attack by black vine weevil grubs. Note that the tender bark has been chewed from most of the roots.
C. Adult black vine weevil feeding injury on severely infested azalea.
D. Black vine weevils on yew.
E. Injury on euonymus caused by the feeding of adult weevils.
F. An adult black vine weevil; 9–13 mm long. [Curculionidae]
G. Foliar injury typically caused by cranberry rootworm on camellia and maple. (Courtesy A. D. Oliver, Louisiana State Univ.)
H. An adult obscure root weevil and typical foliar injury. Its legs hold its body high off the leaf as it feeds. (Courtesy Gray Collection, Oregon State Univ.)

A
B
C
D
E
F
G
H

The twobanded Japanese weevil, *Callirhopalus bifasciatus* Roelofs, was first found in the United States in 1914. It now occurs in New England, the mid-Atlantic states, Kentucky, Indiana, Illinois, and probably other states. Host plants are numerous and include privet, rhododendron, azalea, mountain laurel, camellia, hemlock, mimosa, Japanese barberry, forsythia, *Abelia* and *Koelreuteria* species, viburnum, dogwood, lilac, holly, ash, spirea, rose, *Deutzia* species, and weigela, as well as many others. Hedges may be badly injured by the adult weevil, particularly new leaves and shoots. The weevil feeds in the daytime and is easily overlooked because of its size and color. If alarmed, it drops to the ground and feigns death. In the course of feeding, it chews notches in the leaves until eventually only the petiole is left. One investigator found 265 adults on a single spirea bush. Symptoms of plant injury may be confused with those caused by *Otiorhynchus* weevils (Plate 112). Weevil larvae feed on forsythia and privet roots, but the degree of damage has not been determined. Evidence is lacking, but we believe that further research will associate larvae with a wide range of woody plants. Present knowledge leads us to believe that foliage feeding by the adult rather than the larva causes the most injury. Such feeding occurs mostly on new and inner foliage. Foliage damage becomes more evident by late summer as the growth rate of the plant slows down and the effect of continuous adult feeding accumulates. Adults are most abundant in September.

Adults, eggs, and larvae may overwinter on the ground beneath host plants in plant debris and partly decomposed organic matter. Those adults that overwinter remain inactive until the first warm days in April (Long Island), and then they crawl to a shrub and begin feeding. Damage by the overwintered adults is not noticeable because of the rapid spring growth of the plant. Egg laying begins in mid-May and lasts throughout the summer. The weevils lay tiny white eggs on the inside of a folded leaf in what Zepp (882) called an egg pod. Upon hatching, larvae drop to the ground and burrow into the soil. Fully grown larvae are white, 7.5–8.5 mm long, and legless; they have the same general features of other curculionid larvae (Plate 112G). There is one generation per year.

The adult weevil is 6 mm long. Its color varies from light to dark brown, with mottled color bands on the wing covers (C). The wing covers are fused, and because the insect cannot fly, dissemination is relatively slow. Morphologists have found all specimens are females, indicating the probability that there are no males and that females can produce viable eggs without mating. One generation occurs each year.

Calomycterus setarius Roelofs, a weevil native to Japan, was first reported in the United States in New York in 1929. It has since spread to much of New England south to Maryland and Virginia and west to Iowa.

The adult lays eggs in the ground in grassy areas, and the developing larvae feed on roots. Adult weevils begin to appear in late June and may be found in abundance during July and early August. They are grayish, 4 mm long, and darken as they mature. Major injury to the plant results when the adult weevil chews on the foliage, leaving the margins of the leaf blade scalloped. Lights at night attract the weevils, and they often become nuisance pests inside the home.

The Asiatic oak weevil, *Cyrtepistomus castaneus* (Roelofs), is an insect native to northeastern Asia (D, E). It was first reported in this country in New Jersey in 1933 and has since spread into New York, the central Atlantic states, and west to Missouri and the Ozark Mountains. The adult weevil is greenish gray and, with age, may become nearly black. It is about 6 mm long. Males of this species are unknown. It feeds on the foliage of susceptible hosts by eating the interveinal area. The lower leaves of a host plant are attacked first. *C. castaneus* feeds on 36 or more species of woody plants, including oaks, Chinese chestnut, beech, hickory, black locust, red maple (*Acer rubrum*), dogwood, willow, tuliptree, spicebush, sweetgum, sycamore, apple, redbud, persimmon, and viburnum. If *C. castaneus* occurs in large populations, it may become a household pest. It and *Calomycterus setarius* have similar life cycles.

Phyllobius intrusus Kono, also known as the arborvitae weevil, often shares its coniferous hosts with *C. setarius*, and the two species are frequently collected from the same host. *P. intrusus* feeds on foliage as an adult and on roots as a larva; its hosts are *Thuja*, *Chamaecyparis*, and *Juniperus* species.

The fully grown larva (about 6 mm long) overwinters in the soil. Adults appear in Rhode Island in mid-May and feed on newly formed leaves at the terminal ends of twigs. On arborvitae a slightly irregular, cup-shaped area about 1–2 mm in diameter is eaten, and after a few days the tissue surrounding the consumed area becomes necrotic. The beetles have a tendency to feed lightly here and there so that no large area of plant foliage is consumed at one time. They insert their eggs 2–3 mm into the soil, and the developing larvae feed on root bark. There is one generation per year. Adult weevils are black, but the elytra (wing covers) are covered with metallic green scales. There are both males and females; their length varies, 5.5–6 mm.

Phyllobius oblongus (Linnaeus) defoliates many hardwood trees but is especially damaging to sugar and mountain maples, elm, yellow birch, and serviceberry. Eggs are laid in midsummer a few centimeters below the soil surface. After hatching, the larvae feed on tree roots, mature, and overwinter in the soil. Adults are brown weevils, 4–6 mm long, and appear in host trees in early June. Feeding by day, they straddle the leaf margins and chew semicircular scallops along them. They are found throughout the Northeast into Michigan and the border provinces of Canada.

In 1959 a ligustrum weevil, *Ochyromera ligustri* Warner, was found in Raleigh, North Carolina, feeding on Japanese privet, *Ligustrum japonicum*. The Orient is believed to be the beetle's origin. Since 1959 it has been found in South Carolina, Georgia, Florida, Virginia, and North Dakota. It probably has a wide distribution in the mid-Atlantic states. Adults feed on foliage by making perforations in the interior of the leaf blade and occasionally marginal notches. The predominant symptoms are round or oblong holes about 3–4 mm long. The adult weevil, 3–4.7 mm long, is shiny brown with golden yellow hairlike scales. The insect is most abundant on host plants in late June to early July (North Carolina). When feeding or resting in bushes, it is easily alarmed and drops to the ground where it feigns death. Females lay their eggs in seed capsules or fleshy fruit as soon as the fruit is well formed. Larvae feed on the fruit and seed capsule and burrow into the seed, where they complete their development. There is one larva per seed. Adults begin to emerge from seeds by mid-May. There is one generation each year. Hosts include Japanese privet, *Ligustrum amurense*, *L. lucidum*, lilac, and grape.

References: 413, 430, 497, 716, 836, 865a, 877, 882

A. A cluster of rhododendron leaves notched along the margins by the twobanded Japanese weevil, *Callirhopalus bifasciatus*.

B. An adult twobanded Japanese weevil feeding on the edge of a leaf. The adult is about 6 mm long.

C. A pair of twobanded Japanese weevils. [Curculionidae]

D, E. An adult Asiatic oak weevil, *Cyrtepistomus castaneus*.

A
B
C
D
E

Vectors of Elm Diseases (Plate 115)

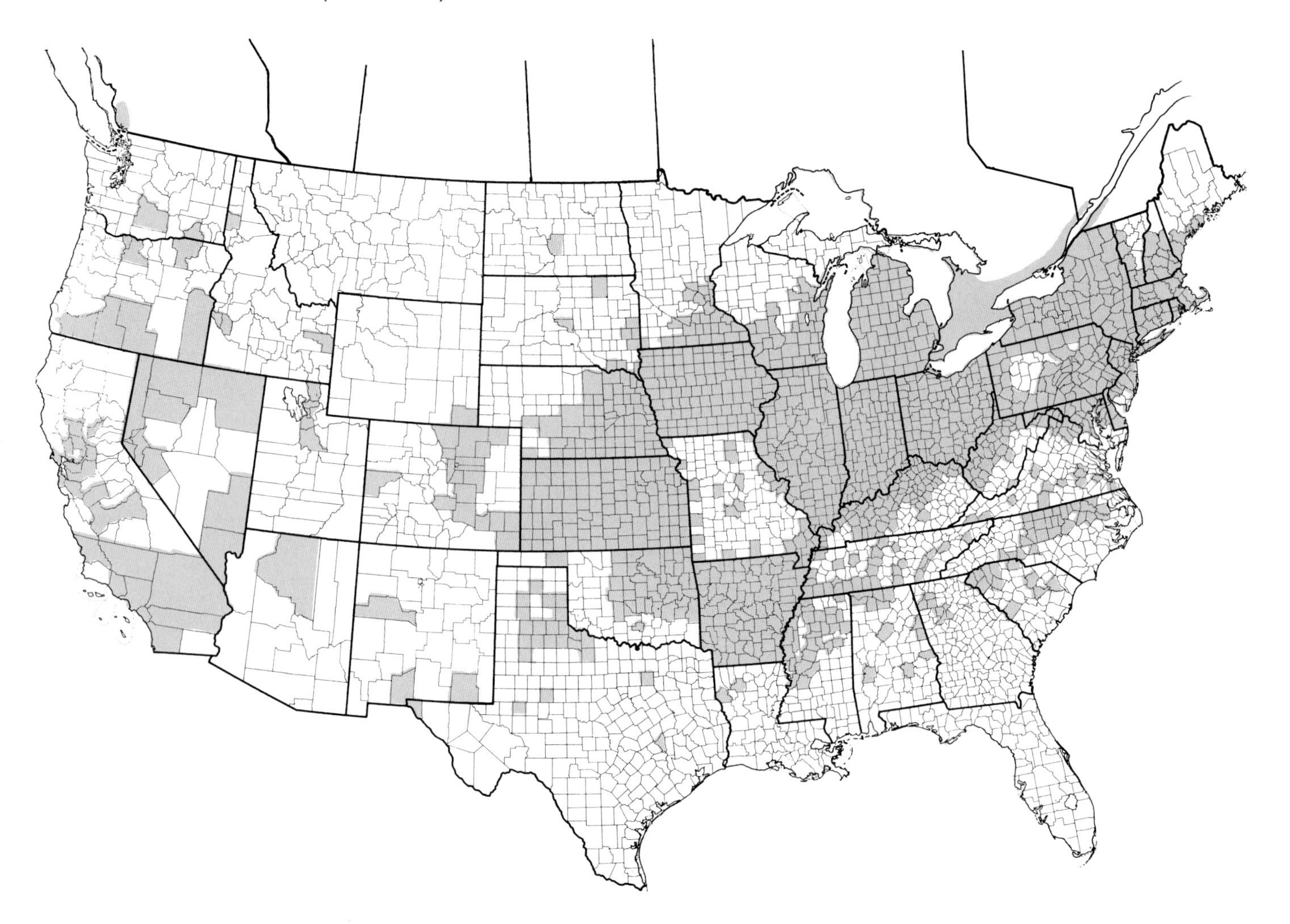

Figure 80. Distribution of the smaller European elm bark beetle, *Scolytus multistriatus*, in the United States and Canada, 1987. (Courtesy Pest Survey, Animal and Plant Health Inspection Service, USDA.)

Two major elm diseases in the United States and Canada are discussed in detail and illustrated in Sinclair et al., *Diseases of Trees and Shrubs*. Dutch elm disease, though transmitted primarily by the exotic *Scolytus multistriatus* (Marsham) and the native *Hylurgopinus rufipes* (Eichhoff) (Plate 116), may also be transmitted by *S. mali* (Bechstein), another insect vector introduced from Europe. The latter has acquired the common name larger shothole borer (Plate 117), and is a bark beetle pest of apple, cherry, plum, and slippery elm as well as American elm. Its known distribution is presently limited to the northeastern states, but there are no known ecological factors that prevent its spread to other geographic regions where *Malus* species, its presumed favored hosts, grow.

Both the adults and the larvae of the above-named beetles cause injury. *S. mali* adults feed on phloem in twig crotches and/or on the sides of twigs. Frequently, several beetles will gather at one spot and make a broad patchlike wound. They usually select weakened branches of apple or other hosts for breeding purposes, seeming to prefer fairly large branches, 15–37 cm in diameter being the usual size. Weak branches die after being attacked by larvae.

Adults are shiny black with dark reddish brown wing covers (elytra). They are 3.5–4.5 mm long. In the lower Hudson River valley beetles begin to emerge in late May; peak emergence occurs in early June. Rogue specimens may emerge as late as August and be present through the middle of September. Adults live for 2 months or more. Eggs are laid in the usual fashion of scolytids in niches along the wall of an egg gallery in the inner bark. The immatures do most of their feeding in the summer and autumn and usually overwinter as full-grown larvae. There is one generation each year.

Elm yellows is a disease caused by a mycoplasmalike organism that has many potential leafhopper and planthopper vectors. A few that have been collected from or reared on elm are *Empoasca fabae* (Harris), *Erythroneura basilaris* (Say), *E. obliqua* (Say), and *Typhlocyba ulmi* (Linnaeus), and all have the capacity for spreading this disease. *Scaphoideus luteolus* Van Duzee is a known vector. (See Plates 199–201 for more details about leafhopper biology, behavior, and injury to plants.)

References: 32, 327a, 603

A. Early symptoms of elm yellows, a disease transmitted by a leafhopper.
B. An American elm with characteristic Dutch elm disease symptoms in the crown of the tree.
C, D. The smaller European elm bark beetle feeding in twig crotches. (Courtesy Jack Clark, Pavel Svihra, and *California Agriculture*.)

A
B
C
D

Elm Bark Beetles (Plate 116)

Elm bark beetles were of little concern until Dutch elm disease arrived in America. It is through these insects that the fungus responsible for the disease is able to enter the tree and produce its devastating effect. The native elm bark beetle, *Hylurgopinus rufipes* (Eichoff), and the smaller European elm bark beetle, *Scolytus multistriatus* (Marsham), are the primary insects involved in disease transmission. Of the two species, the European elm bark beetle is the more important vector of the disease in most areas because of its breeding dominance over the native species. The adult *S. multistriatus* is shiny, reddish brown, and about 3 mm long. The slightly smaller *H. rufipes* is dull brown and has a rough body surface. The adults of the two species are easy to distinguish. However, their eggs, larvae, and pupae are so similar that in the field they cannot be separated unless their galleries (under the tree bark) are well developed and can be examined.

The native elm bark beetle (C) occurs in the eastern and central states from Maine to Virginia and west to Mississippi, Kansas, and Minnesota. It has also been found in Ontario and is likely to be found throughout the range of elm in Canada. It overwinters either as a fully grown larva in dying trees or as an adult in the bark of the trunk or of large limbs of living elms. Adults begin to appear in May and make their egg-laying galleries in dying or recently killed trees. These galleries extend across the grain of the wood (A). Two or more generations occur each year. Adults are difficult to find during the growing season. Look for them in shallow burrows at the base of young elm trees.

The European elm bark beetle (D2), as its name implies, is a native of Europe. In the United States it was first discovered in 1904 near Boston. Since then it has spread to most of the states and Canadian provinces that lie east of the Rocky Mountains and to Utah, Nevada, the Pacific Coast states, and British Columbia (see Figure 80, accompanying Plate 115). Implications from work done in Canada (655) are sufficient to assume that *S. multistriatus* will become established throughout the elm's northern range.

This species overwinters as a full-grown or nearly full-grown larva in the inner bark. Larval and pupal development is completed in the spring and adults begin to emerge about the middle of May. The emergence period extends over several weeks. The adult feeds on young bark, selecting most often the crotches of elm twigs (B) as a feeding site. When the female is ready to lay eggs, she chooses an unhealthy elm tree and bores into the bark of the trunk or larger branches to prepare the egg-laying gallery, which is oriented parallel to the grain of the wood. The gallery shown in panel F was produced by one female, which laid 62 viable eggs.

Each tunnel radiating from the central egg-laying gallery is the work of one larva. When it completes the larval stage, the insect pupates at the end of its feeding gallery. Upon emerging from the pupa, the adult chews a hole through the bark and becomes free to migrate to other elms. Two or more generations occur each year, except in this species' most northern range, where only one generation can be expected. Several major studies have shed light on bark beetle behavior. New chemicals, particularly the pheromones, can be used to alter and manipulate insect behavior in favor of population control. Already, pheromones have proven their effectiveness in attracting beetles to their death on sticky traps. The day is close at hand, if not already here, when potent, volatile, species-specific communication chemicals will be an important component in integrated management of elm insect and disease control.

References: 26, 81, 455, 655

A. Brood galleries of the native elm bark beetle. Note the horizontal alignment of the central egg gallery.
B. Adult European elm bark beetle feeding injuries in the crotches of American elm twigs.
C. An adult native elm bark beetle, *Hylurgopinus rufipes*; 2.5 mm long. [Scolytidae]
D. (1) A newly emerged native elm bark beetle. (2) The European elm bark beetle *Scolytus multistriatus*; 3 mm long.
E. Elm bark cut away to expose bark beetle larvae.
F. Brood galleries of the European elm bark beetle. Note the vertical alignment of the central egg gallery.
G. Bark beetle larvae as they appear from the inside of the bark.

A
B
C
D
1
2
E
F
G

Shothole Borer and Other Bark Beetles (Plate 117)

Trees that have been attacked by *Scolytus rugulosus* (Müller) appear as though they have been blasted with a shotgun; hence this insect's common name, shothole borer (C). The shothole borer and the peach bark beetle, *Phloeotribus liminaris* (Harris), attack unhealthy fruit and ornamental trees. Trees in neglected orchards are particularly susceptible. The shothole borer attacks our common fruit trees, wild plums, cherries, serviceberry, and, occasionally, declining shade trees such as mountain ash, hawthorn, and elm. The peach bark beetle attacks the stone fruits, mountain ash, elm, and mulberry. The two insects have similar habits and life histories. Only the shothole borer will be discussed in detail.

The beetle is attracted to weakened and dying trees. It bores through the bark, making a hole about the diameter of pencil lead and leaving reddish dustlike borings in the bark crevices. The parent gallery of the shothole borer parallels the grain of the wood, whereas that of the peach bark beetle normally runs across the grain. As do other bark beetles, the shothole borer lays eggs in small niches along the sides of the parent galleries. The larvae that hatch from these eggs feed out from these galleries, generally at right angles, gradually enlarging their side galleries as they develop. Pupation occurs under the bark, and the new adult bores out through the bark, leaving a shot-hole effect. Two generations occur annually in the northern part of the United States, and three generations are common further south.

The shothole borer was introduced into this country from Europe and now occurs throughout the United States. The peach bark beetle occurs from New York westward to Michigan, south to Tennessee, and eastward to Maryland.

Attacks by either insect can be minimized by keeping trees in good health. Proper cultivation, fertilization, and irrigation all help to prevent attacks. Dead or dying trees and prunings should be promptly removed from the orchard or home grounds and burned, chipped, or taken to the dump to destroy existing beetle populations and prevent reproduction.

The hickory bark beetle, *S. quadrispinosus* Say, is considered by some forest entomologists to be the most destructive insect attacking *Carya* species. It also attacks *Juglans cinerea*. These bark beetles feed both externally and internally on a tree, and if a significant number attack, they may damage the tree past its ability to recover before the critical symptoms are noticed. Symptoms in the usual order of appearance are wilted leaves, dead twigs and limbs, premature dropping of leaves, broken but not severed twigs in the crown, rosetted terminal growth, and quantities of fine boring dust on bark flaps and at the base of the tree. Vigorous trees are seldom attacked for breeding purposes.

In Michigan adults begin to emerge about the middle of June. They fly to the crowns of healthy host trees to feed at the base of leaf petioles and twig crotches (Plate 115). If the beetles are very abundant, partial defoliation may occur. After maturation feeding, the beetles seek weakened host trees where they construct galleries and deposit eggs. The injury they cause is chiefly by beetles and larvae as they mine in the phloem, thus killing the tree. The centipede-shaped gallery in the bark and etched on the interface of sapwood is so characteristic that the hickory bark beetle can be identified by the gallery alone (Figure 82).

Adults are short, stout, dark brown to black beetles 4–5 mm long. Larvae, when fully grown, are 5–8 mm long and look like the larva shown in panel E. There is one generation per year in the North and two in the South. This species occurs in southern Quebec and Ontario and the eastern half of the United States, including all of the Gulf Coast states.

The pitted ambrosia beetle is a common name for *Corthylus punctatissimus* Zimmerman. It occurs from the Atlantic Coast to the prairie states, in eastern Canadian provinces bordering the states, and south to Georgia. In Canada it is primarily a pest of young maples; in the United States it breeds in a variety of trees and shrubs such as dogwood, *Carpinus caroliniana*, sassafras, rhododendron, azalea, and hazel.

Adults are black and about 3.5 mm long; they overwinter in the tunnels from which they developed (Figure 81). In the spring they emerge and seek new sites to make tunnels and lay eggs. They prefer

Figure 81. A pitted ambrosia beetle, about 4 mm long and shiny black. Note the many tiny pits in the wing covers.

young plants with stems 4–12 mm in diameter. The initial boring site is near ground level. After reaching the cambium region, they produce a spiral tunnel that may girdle the stem. Slender stems are most always killed. The tunnels are covered with a black fungal stain that is a prominent diagnostic feature. Other symptoms include the "sawdust" that is pushed out of their tunnels and wilting if girdling takes place. Injury is most likely to occur in dense plantings. As is the case with all ambrosia beetles, these insects do not eat wood or bark. They excavate tunnels in living trees and shrubs and inoculate the wood with a fungus that grows on the walls of the tunnels. The fungus serves as food for both larvae and adults. The male stores ambrosia (fungal) spores in a specialized organ located in the prothorax; he carries the spores to new brood sites and releases them there. In the North there is one generation per year. (See Figure 25, accompanying Plate 28.)

There are several other ambrosia beetles that may need to be monitored, if not controlled, such as *Xyleborus saxeseni* (Ratzeburg), a polyphagous beetle that can infest most deciduous tree species and some pines. It is dark brown and varies in size from 1.6 to 2.5 mm long. *Xylosandrus crassiusculus* (Motschulsky), an aggressive introduction from East Africa, attacks most *Prunus* species, sweetgum, oak, and presumably other hardwood species. It is reddish brown and up to 3 mm long. It is not easily distinguished from other scolytids. *Xyleborus dispar* (Fabricius), a European introduction, attacks unthrifty limbs and trunks of *Malus* and *Prunus* species, walnut, *Corylus* species, and a number of shade tree species. It is dark brown to black and as long as 3.5 mm. Both *Xyleborus sayi* (Hopkins) and *Xylosandrus germanus* (Blandford) feed on the fungus *Fusarium solani*. This fungus causes cankers in tuliptree, cottonwood, oaks, and water tupelo (*Nyssa aquatica*) (see Sinclair et al., *Diseases of Trees and Shrubs*). *X. germanus* attacks healthy as well as weakened trees, and its tunnels may result in terminal dieback. In multiple-stem shrubs, stems with small diameters may be killed. In the spring (April–May) the female beetles initiate new galleries, where they lay their eggs in specially prepared brood chambers. The galleries are found in the trunk or branches from 30 cm to 1.4 m above the ground and extend into the wood as deep as 25 mm. There is a wide range of host plants, including ash, beech, birch, baldcypress, dogwood, elm, holly, honeylocust, linden, maple, pine, redcedar, rhododendron, sycamore, sweetgum, and willow.

References: 45, 217, 245, 298, 514, 532, 660, 682, 719, 740, 870a

---

A. A twig of *Prunus laurocerasus* killed by the shothole borer.
B. An adult shothole borer, *Scolytus rugulosus*; about 2 mm long. [Scolytidae]
C. Emergence holes made by adult shothole borers.
D. "Healed-over" bark, evidence of unsuccessful attack by adult shothole borers.
E. A shothole borer larva; about 2 mm long.
F. An adult (*arrow*) constructing its egg gallery.
G. An adult (*arrow*) boring into a dying twig.
H. Bark removed to reveal larval galleries.

A
B
C
D
E
F
G
H

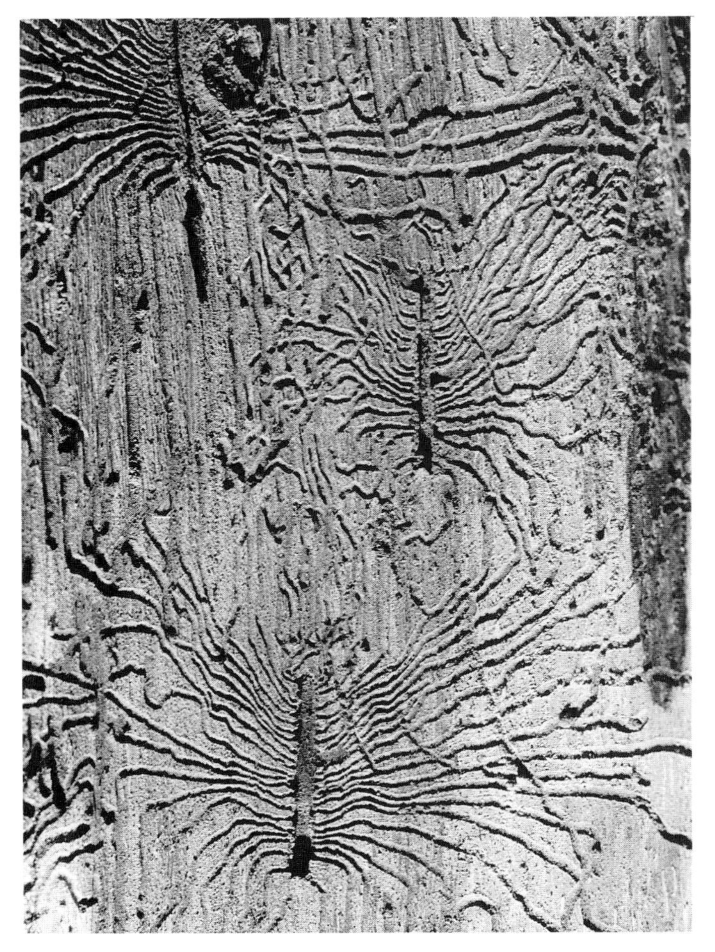

Figure 82. Galleries etched in the inner bark of hickory by a pair of hickory bark beetles and their brood. The gallery width is about 5–6 cm.

Figure 83. Bark beetle galleries of *Hylesinus* (=*Leperisinus*) sp. in ash. Transverse tunnels, made by a pair of adults, are deeply engraved on the inner bark and the summer wood. Larval galleries, barely visible in photo, are at right angles above and below the tunnels. (Courtesy D. A. Leatherman, Colorado Forest Service.)

American Plum Borer (Plate 118)

The American plum borer, *Euzophera semifuneralis* (Walker), is a minor pest of fruit trees, mainly apple and peach. In the late 1950s and 1960s many London plane trees, *Platanus acerifolia,* were planted as street trees in eastern cities. They received considerable abuse, and many were weakened or suffered bark damage. Consequently, they were subject to infestation by the American plum borer.

In the larval stage the American plum borer injures weakened trees by feeding on the inner bark and cambium. The dogwood borer occupies similar sites (Plate 123G).

It overwinters as a mature larva inside a thin, but tough white cocoon. The cocoon is usually located in the crevices of, or under, loose bark (B). The white cocoon distinguishes this borer from the dogwood borer, which has dark brown or black cocoons usually covered with frass. In Virginia the insect pupates in early April. Moths begin to appear in late April. They mate and soon lay eggs. The moth shown in panel D, a female, has a wingspan of about 25 mm. Her eggs are deposited singly or in small groups, usually in cracks or under loose bark. The female lays about 26 eggs and lives only about 8 days. The eggs hatch after a few days and each larva crawls about until it finds a broken place in the outer bark. When it has found such a place, it invades the tree. Larval tunnels or galleries are shallow and are filled with frass. They have frequent openings to the outer bark, where red frass accumulates (E). In young trees early-season infestations may be recognized by wet spots on the bark where sap oozes from the cambium in response to the presence of the feeding insect. The fully grown larva is 16–28 mm long (F) and requires 4–6 weeks to complete this stage.

The American plum borer injures mountain ash, mulberry, flowering apple, peach, cherry, persimmon, walnut, pecan, hickory, olive, linden, poplar, sweetgum, ginkgo, and sycamore. It occurs from New York to Florida and west to Arizona and Colorado.

A number of parasites and predators of the American plum borer have been recorded. They include the wasp parasites *Idechthis* species and *Mesostenus thoracicus Cresson,* and the larva of the predatory beetle *Tenebroides corticalis* Melsheimer. Woodpeckers also help keep the American plum borer in check.

References: 66, 740

A. A London plane tree damaged by the feeding activities of the American plum borer, *Euzophera semifuneralis.*
B. Large sections of bark can be easily stripped from trees heavily infested by the American plum borer. Note the white cocoons.
C. A young London plane tree. Note the numerous scars and cut places in the bark resulting from willful abuse. Larvae are active in the vicinity of the larger scars.
D. A pinned specimen of the American plum borer.
E. An older tree trunk with much loose bark and red frass accumulations, the result of feeding by the American plum borer. The fully grown larva is 25 mm long.
F. The inner surface of some loose bark with two cocoons. Note larva at arrow.
G. A sycamore with plum borer infestation starting where a guy wire was located.

A
B
C
D
E
F
G

Wood-boring Caterpillars (Plates 119–121)

Euzophera ostricolorella Hulst is a root crown borer of tuliptree and southern magnolia. The larvae feed in the phloem from about 5 cm below to 12.5 cm above the soil line of young as well as large established trees. This insect occurs from New York to Florida in the eastern United States. An infestation was reported in western Kentucky and Tennessee in 1954, and a serious infestation of 4-3" caliper trees was found in a Virginia nursery in 1982. The moths fly in the spring, although a partial emergence in the fall of 1955 was reported. Larval development is completed by late summer or early fall. Mature larvae are about 25 mm long. When they are ready to pupate, they tunnel near the bark surface. The tunnels are stained a deep black, and a blackish ooze accompanied by frass flows from the entrance holes. When numerous, the larvae can girdle a tree. (See page 252 for a related species, the American plum borer.)

The caterpillar stages of moths illustrated in Plates 119–121 feed primarily in the xylem, where they make tunnels of various lengths and diameters. Such damage causes structural weakness in a limb or trunk and the loss of lumber value; it reduces aesthetic value in the case of a shade tree; and it opens the tree to wood-rotting fungi (see Sinclair et al., *Diseases of Trees and Shrubs*). Borer-infested shade trees are particularly susceptible to wind breakage, which often results in property damage. The insects remain hidden from view, often for more than a year, but not without producing visible symptoms, such as frass, commonly protruding from their entrance hole, lodged in bark crevices, or on the ground at the base of the tree. The frass of each species is distinctive and adequate for insect identification (730).

Paranthrene simulans (Grote) (G) lacks a common name, but it belongs to a family known as clearwing moths (Sesiidae). Moths in this group have few scales on their wings, which makes the veins readily visible. They look like wasps and fly during the day rather than at dusk or at night. Being moths, they cannot sting. *P. simulans* ranges throughout much of the eastern half of Canada and the United States. They require 2 years to complete a generation. Adults are active from early April to early August, depending on geographic location. Males have been captured in pheromone-baited traps in Georgia from early May to July and in central Mississippi from early June to early August. In the northern part of its range this insect lays its eggs on the thin bark of small branches and the trunks of American chestnut and young red and white oak and elm trees. In the South (Mississippi) it chooses larger trees in the red oak group, most frequently the Nuttall oak, and selects the base of the tree, including root flanges, for egg deposition. Forbes (255) in New York and Solomon and Morse (734) in Mississippi, several decades apart, observed that the moths are much more common in odd-numbered years.

Each larva (F) makes a single unbranching tunnel similar to the one shown in Plate 120F. Entrance holes are 10–16 mm in diameter. The gallery slopes upward, and its total length may be 9–10 cm. It is narrower and shorter than a carpenterworm gallery. The gallery is kept clear by the larva, but the entrance hole may be obscured by the accumulation of frass and wood chips held together by a light silken web (A). External symptoms of injury include wet spots around the borer's entrance hole, loose clumps of frass held together by silk at the entrance hole or on the ground under the entrance hole, and, if the attack site is on small saplings or branches, gall-like swellings.

Larvae pupate in their galleries (D). The moths do not eat. They live only a few days and die soon after they mate and lay eggs.

A related species, *P. tabaniformis* (Rottenberg), attacks small branches and trunks of poplars and willows throughout much of the United States and Canada. It, too, has a 2-year life cycle throughout most of its range. The moth is about 16 mm long, shorter than *P. simulans*, which averages 18–19 mm. Its body is charcoal black with three narrow but bright yellow transverse bands. In its northern range *P. tabaniformis* is often associated with galls produced by *Saperda* borers (Plates 130–131).

Clearwing moths and their damage are more abundant than casual field observations reveal. The extent of a population may be reckoned through the use of pheromone traps.

The leopard moth, *Zeuzera pyrina* (Linnaeus) (E, I), is an introduction from Europe and North Africa and was first found in New Jersey in 1887. Since then it has continued to spread westward and southward. The moth may appear throughout the growing season from May to September. It is a heavy-bodied weak flyer, with a wingspan of 30–40 mm. The male moth is strongly attracted to light. Eggs (H) are laid in masses, or, more commonly, in small clusters in bark crevices. A single female may lay as many as 800 eggs. Upon hatching, the larva often crawls some distance before boring into the bark. The larva is a wood feeder and tunnels into the heartwood, where it does most of its feeding. If a larva bores into a branch too small for it to complete its development, it chews its way out and finds a bigger branch. A single larva may kill a limb 7–8 cm in diameter. Larvae are also common in the trunk. Development is usually completed in 2 years. The caterpillar (B) has a brownish black head. When fully grown, it is about 50 mm long. It pupates in its gallery.

An infested tree may exhibit several symptoms. An accumulation of sawdustlike frass on the bark indicates the presence of borers; however, dead branches may be the first evidence of damage. Exit holes are often found in the tree limbs, sometimes with a portion of the old pupal case remaining in the hole.

The leopard moth larva feeds on more than 125 species of deciduous trees, among which are maple, elm, beech, ash, oak, walnut, chestnut, poplar, willow, lilac, and all the flowering *Malus* and *Prunus* species. The larvae are most injurious to shade trees. Even though the moth is widespread, it is not an important pest because it is rarely found in large numbers. Woodpeckers and other birds are its major predators. Control is difficult after the insect has become established in a tree.

References: 81, 209, 359, 392, 730, 734, 738, 897

A. The trunk of a young oak, *Quercus palustris*, containing one or more *Paranthrene simulans* larvae. Note the frass and dark sap-stained area.
B. A leopard moth larva, about half grown, in its gallery. (Courtesy J. D. Solomon, U.S. Forest Service.)
C. An old entrance hole of *P. simulans* exposed to show its relative size.
D. An empty pupal case of *P. simulans* protruding from an elm branch.
E. A female leopard moth. Note the sharp ovipositor.
F. A larva of *P. simulans* exposed in its tunnel.
G. An adult female *P. simulans*; wingspan 28 mm.
H. Eggs produced by the female in E.
I. A male leopard moth, *Zeuzera pyrina*; wingspan about 38 mm. [Cossidae]

A
B
C
D
E
F
G
H
I

(Wood-boring Caterpillars, continued, Plate 120)

Wood-boring insects that feed in xylem often have a life cycle of 2 or more years. The carpenterworm, *Prionoxystus robiniae* (Peck), requires 3–4 years to complete its development in the northern states and in Canada. Certain broods in Mississippi regularly emerge in 1-year cycles. Except for one summer month of its life cycle, the insect is in the larval stage, feeding within the host. When fully grown, the larva is 50–75 mm long, which makes it one of the largest known wood-boring caterpillars. The female moth (B) has a wingspan of about 75 mm; the male moth is considerably smaller. Both male and female are endowed with a protective coloration that makes them nearly invisible when resting on the corky bark of host trees. Their forewings are light gray and mottled with black. The hind wings of males are marked with orange. The carpenterworm moth is often a prize specimen of amateur insect collectors.

In southern California the moths begin to emerge in early April. In the lower Mississippi valley emergence coincides closely with the leafing-out and flowering of pecan. Emergence begins in June in North Dakota and the Canadian prairie provinces. Wherever this insect occurs, emergence of the brood continues for about 2 months. Egg laying begins shortly after the moths emerge and lasts about 1 month. Sticky egg masses may be found on the trunks or branches of susceptible hosts, often in crevices, wounds, or in the scarred areas resulting from previous larval infestations. A single female may lay 300–600 eggs. Upon hatching, the hairy reddish pink larva bores into the sapwood. It remains there for most of the larval stage (the number of instars varies, from 8 to 31), but as it approaches maturity it extends the tunnel into heartwood (E, F). Tunnel openings in the corky bark are irregular but somewhat rectangular and often covered by frass and wood chips held together by loose silk webbing. All through its development, the larva maintains contact with the outside and expels large quantities of sawdust and frass that cling in masses to the external bark (C). During these forays to the gallery entrance, it feeds on the inner bark and keeps the entrance free of callus tissue. The mature larva may be up to 14 mm in diameter; it is greenish white except for the dark brown head and has distinct abdominal prolegs (Figure 84) and several short, stout hairs on each body segment. Its tunnels may exceed 15 cm in length and be up to 18 mm in diameter.

Over a period of time the activities of the carpenterworm larvae may prove disastrous to the host tree and to adjacent buildings. The strength of the infested wood may be greatly reduced, making the tree or its large branches subject to breakage by high winds. In addition, the bark and wood may be severely scarred, ruining the wood for saw-log purposes.

Host trees include elm, ash, birch, black locust, oak, cottonwood, maple, willow, apricot, pear, loquat, and, occasionally, ornamental shrubs. In the eastern and southern states red oaks show the heaviest damage. In the prairie states and Canadian provinces green ash, *Fraxinus pennsylvanica,* is the chief host; in the Rocky Mountain region poplars are favored; and in California, American elm and live oak, *Quercus agrifolia,* are most commonly attacked. The carpenterworm is distributed over the entire United States from Maine to California and southward to the Gulf of Mexico, and in all of the border provinces of Canada. It is reportedly the most destructive borer of hardwoods in Missouri. Little is known about its natural enemies. However, California scientists have demonstrated that single treatments of the parasitic nematode *Steinernema feltiae* Woutz, Marack, Gerdin & Budding (=*Neoaplectana carpocapsae*) prepared in a water suspension and squirted into entrance holes can suppress or potentially eradicate carpenterworms from orchards as well as shade trees. The nematode has a mutualistically associated bacterium, *Xenorhabdus nematophilus,* which occurs in its gut. When the nematode invades the body of the borer host, it releases bacterial cells and the host dies from bacterial infection 24–48 hours later. Investigators have experimentally infected numerous insect species with the nematode. Its broad host range and phenomenal reproductive capacity has stimulated much research on biocontrol.

A related species, *Prionoxystus macmurtrei* (Guérin), sometimes called the little carpenterworm, is often confused with *P. robiniae* in all stages. It is primarily a pest of oaks in the eastern half of North America. A wet spot on the bark made by oozing sap is an early symptom of the borer's presence. No part of the tree is immune from attack. Branches approximately 2.5 mm in diameter often contain tunnels of this insect. Branches high in the crown are often killed, which may ruin a tree's symmetry. In Canada, 3 years are required to complete one generation. A fully grown larva is about 62 mm long.

Paranthrene robiniae (Hy. Edwards), the western poplar clearwing, is one of the colorful clearwing moths (D) that looks much like a vespid wasp. The larva feeds in the bark phloem and xylem of *Populus trichocarpa* (G), other poplars, willow, and birch. The specimen shown in panel D was collected in Claremont, California, in mid-August. It is a western species found throughout the northern Rocky Mountains, Alberta, British Columbia, and along the Pacific Coast to southern California. (See Plates 119 and 121–123 for photographs of other clearwing moths and their wood-boring caterpillars.)

The western poplar clearwing has a 2-year life cycle. Female moths deposit eggs singly in bark crevices and wounds. Upon hatching, the larvae tunnel into phloem tissues, later moving into the xylem. Here they feed during two successive summer and fall seasons. In northern California pupation occurs in the spring, and adults are found from May through July. Its cycle in southern California is unclear. Moths have been found in July and August and from February through May. The presence of actively feeding larvae can be detected by sawdust-like frass that accumulates at or near their tunnel opening. Stressed trees such as those that are newly planted or damaged are preferred hosts. Infested trees may have deformed stems or branches and are slow growing. Larvae and pupae may be controlled biologically by the nematode *Steinernema feltiae* Woutz, Marack, Gerdin & Budding.

References: 193, 209, 345, 359, 398, 419, 467, 515, 559, 737, 739

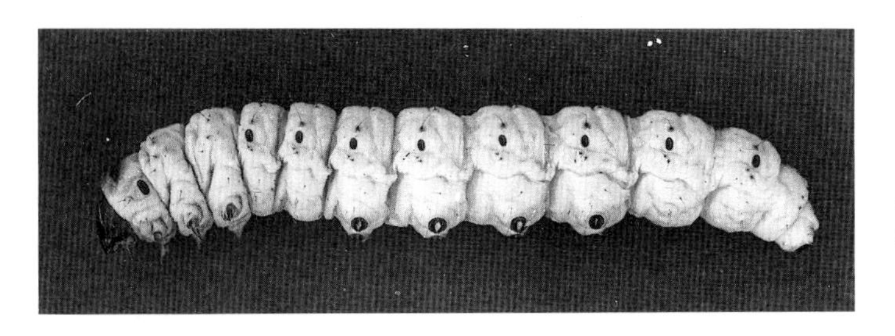

Figure 84. A fully grown larva of the carpenterworm, *Prionoxystus robinae.* Note the abdominal prolegs and the sharp, hooklike thoracic legs. (Courtesy J. D. Solomon, U.S. Forest Service.)

A. An American elm on a ranch in the foothills of the Sierra Nevada. It is heavily infested with the carpenterworm, *Prionoxystus robiniae.* The tree has probably been continuously infested for 25 years. Note size of the bark scars. [Cossidae]

B. An adult female carpenterworm moth. (Courtesy J. D. Solomon, U.S. Forest Service.)

C. Scarred bark with large quantities of frass and "sawdust" (*circle*) expelled by carpenterworm larvae.

D. An adult clearwing moth, *Parathrene robiniae.* [Sesiidae]

E. The exposed tunnel of a carpenterworm in the trunk of a Nuttall oak. (Courtesy J. D. Solomon, U.S. Forest Service.)

F. Tunnels in a branch chewed by carpenterworm larvae.

G. Injury to the trunk of *Populus trichocarpa* caused by the larvae of *Paranthrene robiniae.*

A
B
C
D
E
F
G

The rhododendron borer, *Synanthedon rhododendri* (Beutenmüller), is the smallest of the native clearwing moths. Rhododendrons are the preferred hosts; however, mountain laurel (*Kalmia latifolia*) and deciduous azaleas may be heavily infested, especially if they are located near infested rhododendrons. Injury (G, H) shows up most clearly in the fall. Early symptoms are similar to drought stress. Leaves lose their sheen; progressively, they become pale green, then olive, and finally chlorotic. When examining a plant for borer injury, look for branches that may not have attained normal growth and compare them to branches of nearby plants. If any of these symptoms are observed, search the bark for holes, particularly at limb crotches and scars, irregularities, and accumulations of sawdust not only on the scaly bark but also on the ground beneath. If the observation is made in late May and early June (in Maryland), some holes may contain pupal "skins" that extend about halfway out of the holes (E). These features are fairly reliable evidence that the injury has been caused by the rhododendron borer. Past infestations may be recognized by loose bark that covers longitudinal scars (H).

Moth emergence begins in late May in Delaware and central Maryland and ceases within the first week of July. After mating, the female seeks sites for laying eggs, most frequently old pruning scars, narrow V crotches, scars where the inflorescence was broken off while still green, and old larval feeding galleries. Eggs are tucked deeply into cracks and bark crevices. After laying her eggs, the female dies within 2 days. Upon hatching, the larva chews a hole to the inner bark, where it makes long tunnels that are eventually filled with small reddish pellets. By late fall the larva moves to sapwood and hibernates. It resumes feeding around mid-March, at which time it makes more holes to the outer bark and pushes out large amounts of frass. By early May it is fully fed and prepares for pupation by chewing its way to a site in the bark. Here it creates an oblong cell, leaving a paper-thin layer of outer bark for its exit. A cocoon is formed, which is covered with wood chips and frass. When metamorphosis is complete, the pupa wiggles out of its cocoon and, using the spines on its head, cuts an exit hole and wiggles its way to the outside until about half of its body is exposed. In this position the pupal skeleton splits and the wasplike clearwing moth emerges. There is one generation each year.

Sex pheromones have been synthetically produced and have been used successfully in traps to monitor the presence of and to estimate populations of rhododendron borers. Such pheromones are available for several other clearwing moths.

Other clearwing moths damage ornamental trees and shrubs. Throughout the United States the peachtree borer, *S. exitiosa* Say, attacks trees in the genus *Prunus*. The landscape shrub *Prunus* ×*cistena*, sometimes called the purpleleaf sand cherry, is highly susceptible. The larva of the peachtree borer feeds on the inner bark at the root crown, where it often girdles the tree. It behaves much like the larva of other clearwing moths. The adult emerges from its tunnel in early July in central New York and has a wingspan of about 28 mm. Its body has a metallic purplish black luster. The female lays eggs over a period of about 6 weeks. The newly hatched larva is capable of chewing through the external bark to reach its primary feeding site, the phloem.

S. pyri (Harris), the apple bark borer, is a fairly common clearwing moth in the eastern half of the United States and bordering Canadian provinces. Before the advent of synthetic organic insecticides, this species was common in apple orchards. It is now more likely to be found in ornamental crabapple, hawthorn, *Amelanchier* species, and mountain ash. *S. pyri* looks much like the dogwood borer (Plate 123), has a similar biology, and produces similar injury, but is 1–2 mm longer. In its northern range adults are found in late June; in the Gulf States, in early May.

In all the Pacific Coast states and as far east as New Mexico a clearwing moth, *S. resplendens* (Hy. Edwards), often becomes a pest of western sycamore, *Platanus racemosa*. The larvae are voracious feeders in the inner bark but rarely in the wood. They prefer to feed in the trunk of older trees, that is, trees on which the bark has begun to exfoliate. Recurring attacks by the larvae cause the sycamore to lose all of its pleasing bark character; it becomes dry and acquires a checked appearance (A), like dried mud in the summer sun. Large sections of bark, some as thick as 20 mm, slough off, giving the trunk an unsightly appearance. When the loose bark is scraped away, numerous meandering tunnels are exposed. This insect causes similar damage to coast live oak, *Quercus agrifolia*.

The moths emerge over a long period, from May to early August, with peak emergence in June and July. They have the typical wasplike form, are blue black, and have yellow bands around the abdomen. Eggs are ovoid, golden, and laid singly in small cracks or depressions on the trunk. Upon hatching, the larva bores into the inner bark and makes twisting tunnels over an area of about 100 square cm. The larva is pinkish white with a brown head and is 18 mm long when fully grown. In southern California it feeds through most of the winter months and pupates in the early spring. One generation occurs each year.

The lesser peachtree borer, *S. pictipes* (Grote & Robinson), is well known to peach growers. It is also a pest of flowering cherry and almond. *S. pictipes* is frequently found at or near a tree crotch, or at the site of an old trunk injury or fungal canker. Newly hatched larvae can reach their feeding sites in the phloem only by boring through injured bark such as a pruning scar, frost crack, or canker. Injury typical of that caused by this pest is shown in panel F. Such damage often results in sucker and adventitious growth. Borer injury frequently results in a deposit of amberlike gum on the outer bark, a symptom called gumosis, which has various causes. This insect is concentrated primarily in southern and eastern regions, although it also occurs as far north as eastern Canada.

The hornet moth, *Sesia apiformis* (Clerck), resembles the European hornet, *Vespa crabro* Christ (Plate 238), in size and color. It was introduced from Europe about 1880 and is primarily a pest of *Populus* species. The larvae bore into roots, trunks, and large branches, where they excavate extensive tunnels. Young trees are sometimes killed. The insect requires 2 years to complete its life cycle. The larva spends its first winter in xylem tunnels. The insect spends the second winter in an exposed root or in the wood at the base of the tree, where it is wrapped in a cocoon made of silk and wood borings. Mature larvae are white with reddish heads and are 30–50 mm long. Pupation occurs in the spring. In New York this clearwing moth emerges during May and June. It is brownish black with yellow markings on the head and thorax, and it has yellow bands on the abdomen. Its wingspan is 32–44 mm. This insect occurs in the northeastern states and in California.

Sesia tibialis (Harris), the American hornet moth but sometimes called the cottonwood crown borer, is a native clearwing moth with habits similar to the hornet moth. It is a northern species, transcontinental in distribution. It apparently feeds solely on *Populus* species.

Little is known about the natural enemies of any of these species. Woodpeckers may be important predators, but because they are forest-inhabiting birds, they are not commonly found in cities where many of these insects flourish.

References: 85, 126, 194a, 209, 482a, 563, 564, 660, 832

A. A western sycamore, *Platanus racemosa*, with bark injured by *Synanthedon resplendens*.

B. An adult male rhododendron borer, *S. rhododendri*. (Courtesy John Neal, USDA.)

C. Close-up of sycamore bark showing the dry, caked appearance typical of a site under attack by *S. resplendens*. Note the empty pupal case (*arrow*) at the opening of a larval tunnel.

D. Black knot (gall) infection on a flowering cherry branch caused by *Apiosporina morbosa* fungus. From the areas infected by fungus, six dogwood borers emerged. (See Sinclair et al., *Diseases of Trees and Shrubs*.)

E. Close-up of the pupal "skin" of *S. resplendens*.

F. Typical injury caused by the lesser peachtree borer, *S. pictipes*, on flowering cherry. In the black hemispherical knot one or more larvae may be found.

G, H. Rhododendron branches showing scars resulting from earlier rhododendron borer infestations.

A
B
C
D
E
F
G
H

Lilac Borer and Banded Ash Clearwing (Plate 122)

The lilac borer, *Podosesia syringae* (Harris), also known as the ash borer, is a day-flying clearwing moth that mimics a paper wasp (*Polistes* species) in its adult stage. Larvae (D) injure lilac, ash, privet, and occasionally other members of the olive family by making tunnels and feeding in the bark and wood of stems, trunks, and branches. Such feeding results in visible external holes and scars (A). This insect has a wide geographic distribution, from Miami, Florida, to Saskatoon, Saskatchewan, west to Salt Lake City, and in Sacramento, California. Throughout its range, it apparently produces one generation each year except in parts of Canada, where a 2-year cycle has been reported. Moths begin to appear as early as February in Florida, but throughout the rest of the South the population peaks in May. In Ohio, North Dakota, and other northern states peak moth flight occurs in June. The precise time of moth emergence can be forecast by the use of pheromone traps and by prediction based on growing degree-days. Adult activity usually ceases by August 1.

Female moths lay eggs during daylight. Their flattened, oval, tan eggs are about 0.7 mm long. They deposit them singly or in clusters in bark crevices or ridges, sometimes on smooth bark. Each female lives about 1 week and produces an average of 395 eggs over about 5 days.

After hatching, the young larvae chew into the bark and feed laterally and vertically in phloem tissue. The first symptom of injury is a slight sap flow mixed with frass at the penetration site. Later, larvae concentrate their feeding on sapwood. Light-colored "sawdust" accumulates in clumps (C) at the entrance sites as well as on the ground beneath, for the larvae characteristically keep their tunnels clear of debris. Larvae, which have been feeding since spring, overwinter in their tunnels in the final instar. Completed galleries (Figure 85) are about 7.3 cm long and 5–7 mm in diameter. Each larva that completes development produces two diagnostic holes in the trunk of its host: an irregularly shaped entrance at the bottom of the gallery and a round exit hole (4–5 mm diameter) at the top. Open-grown ash and urban street trees are highly susceptible to borer attack. Avoid pruning when moths are present. Eggs are almost always laid in or near wounds.

Figure 85. Injury evidence to green ash by the lilac borer, *Podesesia syringae*. *A,* Larval gallery extending inward, upward, and back to the surface; upper portion of gallery contains pupal chamber plug. *B,* Irregular shaped entrance below and circular exit above. *C,* The bark scars that result from the "healing" of entrance and exit wounds remain as evidence of attack. *D,* A woodpecker excavation several inches above the larval entrance. (Courtesy J. D. Solomon, U.S. Forest Service.)

***Table 15.* Clearwing moths, their known hosts, flight periods, and distribution**

Scientific Name	Common name	Hosts	Flight period in north central U.S.	Distribution
Albuna fraxini	Virginia creeper clearwing*	Virginia creeper: roots	Late July–Aug.	East to Rocky Mts.
Alcathoe caudata	Clematis borer	*Clematis virginiana*: stem, roots	Late July–Aug.	Mississippi Valley east, Canada
Paranthrene asilipennis	Oak stump borer	*Quercus*: wood, root flares	May–June	East to Rocky Mts.
P. dollii	Poplar clearwing borer	*Populus, Salix*: wood	May–June	Eastern U.S., Canada
P. robiniae	Western poplar clearwing	*Populus, Salix, Betula*: wood		West of Rocky Mts.
P. simulans	Oak borer*	*Quercus*: wood	May–June	Eastern U.S.
P. tabaniformis	Dusky clearwing*	*Salix, Populus*: root, trunk	June–July	Northern U.S., Canada
Podosesia aureocinta	Banded ash clearwing	*Fraxinus*: trunk, branches	Late Aug–Sept.	Eastern U.S., Canada
P. syringae	Lilac borer, ash borer	*Fraxinus, Syringa*: stem	May–early Aug.	U.S., Canada
Sesia tibialis	American hornet moth, cottonwood crown borer	*Populus*: roots, trunk	July–early Aug.	U.S., Canada
Synanthedon acerni	Maple callus borer	*Acer*: inner bark	Late May–July	Eastern U.S., Canada
S. acerrubri	Red maple borer*	*Acer*: bark of branches	June–July	Eastern U.S., Canada
S. decipiens		*Quercus*: galls	July	Rocky Mts. east
S. exitiosa	Peachtree borer	*Prunus*: lower trunk to roots	Late June–Aug.	U.S., Canada
S. fatifera		*Viburnum*: stems	July	Eastern U.S., Canada
S. fulvipes		*Betula*: bark, summer wood	Late May–June	Eastern U.S., Canada
S. pictipes	Lesser peachtree borer	*Prunus*: branches, crotches	Mid-July	East of plains states, Canada
S. pini	Pitch mass borer	*Pinus, Picea*: trunk	Mid-July	Northeastern U.S., Canada
S. pyri	Apple bark borer	*Malus, Pyrus, Crataegus*: bark	July	Eastern U.S., Canada
S. rubrofascia	Sour gum clearwing*	*Nyssa* (blackgum): bark, wood	Late July	Eastern U.S.
S. scitula	Dogwood borer	*Cornus, Prunus, Malus* (many others): bark	July–Sept.	Mississippi Valley east, Canada
S. viburni	Viburnum borer	*Viburnum*: bark of branches	July	Eastern U.S.

*Name not approved by the Entomological Society of America as of 1990.
Source: Modified from Taft et al., ref. 908.

Podosesia aureocincta Purrington & Nielsen is now the correct name for the insect formerly named *P. syringa fraxini*. It has been given the common name banded ash clearwing. The lilac borer and banded ash clearwing look very much alike, but there are some subtle differences. The banded ash clearwing can be differentiated from the lilac borer by the presence of a narrow gold band around the fourth abdominal segment that contrasts with a dark blue-black body. Adults are found from July through December. Larvae overwinter in the second instar and have crochets on one pair of the abdominal prolegs. More differences include the size and color of eggs.

The injuries caused by the banded ash clearwing resemble those caused by the lilac borer. Crown dieback and basal sprouts are symp-

A. An old lilac stem showing several deep scars of past lilac borer infestations.
B. Pupal skeletons remain at the exit hole where banded ash clearwing moths emerged from an ash stem. (Courtesy D. A. Leatherman, Colorado State Univ.)
C. A young lilac with fresh sawdustlike frass produced by an active borer. The entrance site was at ground level.
D. An excised lilac borer, *Podosesia syringae*. The thoracic legs are small.
E. An adult female lilac borer. Although this insect is a moth, it looks at first glance like a wasp.

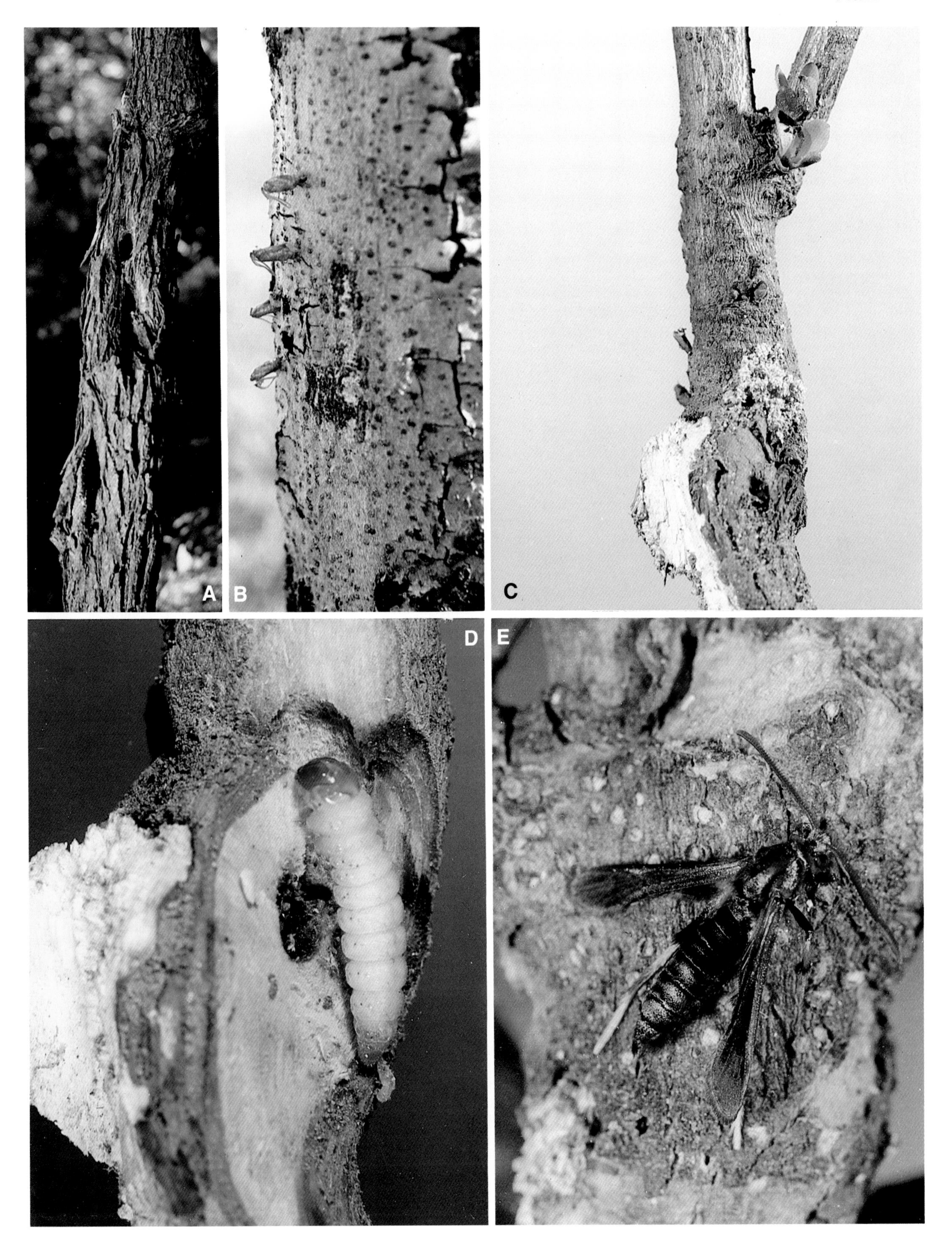
A
B
C
D
E

toms of injury, as are oozing sap and frass at the larvae's entrance holes. The latter two symptoms signify the presence of actively feeding larvae.

The host plants of *P. aureocincta* appear to be limited to various species of ash (*Fraxinus*). Its distribution is limited to New York south to Florida and west to Indiana and Oklahoma.

Because of the life cycle differences between the two species, chemical control measures must be adapted to each. Synthetic sex pheromones have been developed as monitors for determining the presence of adults. Woodpeckers are probably the most important natural enemy. Ants, especially *Crematogaster clara* Mayr, are frequent predators. Parasites include species of *Apanteles* and *Lissonota, Phorocera signata* Aldrich & Webber, and *Phaeogenes ater* Cresson. *Beauveria* is a known fungal pathogen.

References: 626, 637, 728, 731, 733

Dogwood Borers (Plate 123)

The flowering dogwood, *Cornus florida,* is one of our most beautiful flowering trees. Boring insects greatly limit its success in home and park plantings. Artificial control of these pests is frequently necessary. The most serious is *Synanthedon scitula* (Harris), commonly known as the dogwood borer (E). In the South this clearwing moth has been called the pecan borer. The adult's appearance is similar to that of a wasp. It is a swift flyer and is on the wing during the daylight hours. Although only one generation occurs annually, moths may emerge throughout the summer. In the vicinity of New York City the first moths of the season appear after mid-May; farther south, they appear earlier.

The timing of adult emergence is linked to plant phenological development. When the last of the dogwood flower petals are ready to drop and weigela begins to bloom, dogwood borers have begun to emerge in the dogwood host. The flight of moths continues until September. Potter and Timmons (625) believe that there may be at least two distinct populations, based on host preference and the fact that development may be slower in some hosts (e.g., emergence from apple in Kentucky occurs predominantly during August and September). However, peak emergence of adults from apple burr-knot tissue at root stocks occurs in mid-July in central New York. The precise time of emergence in any locality can be established through the use of sex pheromone traps.

The eggs are laid, singly or side by side, on both smooth and rough bark, but the female moth is particularly attracted to injured bark. After hatching, the light brown larva wanders about until it finds a satisfactory point of entrance to the tree's inner bark. The larva cannot chew through the rough, corky bark but must find ready access to phloem through wounds or scars. The crotches of limbs are also likely places of invasion. These borers, which are phloem and cambium feeders, remain in the inner bark throughout their developmental period. Their tunnels are indistinct, and they mine an area much the way a blotch leafminer does. The fully grown larva is 14 mm long; it is white, with a pale brown head (H).

Cultivated trees are usually more heavily infested than those that grow in woodlots. The sloughing of loose bark is an early symptom of attack. Dieback (F) and adventitious growth are advanced symptoms. Since dogwood borers commonly live in the bark, their presence may be indicated by coarse brown sawdustlike frass that has been pushed from the burrows. Old trees that are infested annually may persist in an unhealthy condition for years. The dogwood borer, which is distributed throughout the eastern half of Canada and the United States, has the broadest host range of all the clearwing moths. Its numerous hosts include flowering cherry, chestnut, apple, mountain ash, hickory, pecan, willow, birch, bayberry, oak, hazel, myrtle, loquat, and others. In apple orchards the dogwood borer is frequently found in the callus tissue between the scion and root stock. The insect is also found in twig galls caused by other insects or fungi. The Korean dogwood, *C. kousa,* is said to be resistant to dogwood borer.

The dogwood twig borer, *Oberea tripunctata* (Swerderus), is a threat to many species of trees and shrubs including viburnum, blueberry, deciduous and evergreen azaleas, laurel, *Pieris japonica, Oxydendrum arboretum,* and fruit trees in the genera *Malus* and *Prunus.* Adults (B) feed on the underside of host leaves by chewing deeply into the midrib, often along its entire length. Dark callus scar tissue grows around the vein. The leaf curls downward at the tip, making an abnormal bow, especially noticeable in rhododendrons. Damage by the larva is much more serious and can sometimes be life threatening to the host.

The female beetle prepares a stem for oviposition in late June or early July by girdling a shoot in two places about 13 mm apart. The first of the two girdles is several centimeters below the bud. The egg is deposited in the phloem between the two girdled areas. Upon hatching, the larva bores into the pith, going upward at first for a few millimeters, then turning downward, soon to feed in 2- or 3-year wood. At disjunctive intervals along an occupied twig, the larva cuts off the distal part of the twig from within. It spends the first winter below a plug of coarse wood fibers in the hollowed stem. In the spring it continues to hollow out the stem, and by fall, if in an azalea, it may have reached the root crown. In the meantime it makes frequent holes (C) to the surface to push out wood fibers and frass from its tunnel. In rhododendron it overwinters as a mature larva in a root or in the root crown; it pupates in the spring and emerges as an adult. It requires 2 years to complete its life cycle.

Only sparse distribution records exist for this insect. It is known from Ohio and Pennsylvania and probably can be found throughout the natural range of dogwood, laurel, and rhododendron.

O. bimaculata (=*ulmicola*) (Chittenden) is a closely related species. Its larva occurs in branches of hickory, dogwood, and black cherry. Adults are active in June and July.

The black twig borer, *Xylosandrus compactus* (Eichhoff), (not illustrated) is one of the few ambrosia beetles that chew galleries in the wood of healthy trees; it is frequently encountered on *Cornus florida.* It is a subtropical species introduced into the southern United States, probably from Latin America. It is not a tree killer, but it can cause loss of growth and substantial aesthetic damage. Symptoms of damage include wilted foliage and shoots, and ultimately dead twigs and small branches. Close observation of trees bearing these symptoms will often reveal round holes in the bark about 0.8 mm in diameter that are located on the sides of terminals and undersides of twigs and branches.

The adult beetle is solid black; about 1.5 mm long, it has the same general form as other ambrosia beetles (Plate 117). The female constructs an entrance tunnel into the pith or wood to a depth of 1–3 cm, where it forms a small cavity and deposits a loose cluster of eggs. Up to 20 females and their broods may be found in branches of 8–20 mm in diameter. This insect introduces an ambrosial fungus (e.g., *Fusarium solani*) (see Sinclair et al., *Diseases of Trees and Shrubs*) into the chamber. By the time eggs hatch a fungus culture is growing on the walls of the chamber. Larvae feed on a mixture of the fungus and xylem, all the while extending the length of the tunnel. The mature larvae, about 2 mm long, are creamy white and legless (Plate 117E shows a similar larva). Pupation and mating occurs inside the chamber. A new brood of adults emerges through the entrance that was made by the parent. In central Florida adults overwinter in damaged branches and brood production begins in April. Well over 200 species of woody plants are susceptible to the black twig borer. Included among the ornamental shrubs and shade trees are species of *Acer, Carya, Cassia,* and *Cercis,* camphor tree, dogwood, *Koelreuteria* species, sweetgum, species of *Persea* and *Platanus,* and water oak. The black twig borer occurs in Florida and the Gulf States to Louisiana.

See Plate 135 for more twig borers.

References: 180, 209, 572, 619, 624, 625, 652, 665, 793, 811

A. A twig of dogwood, *Cornus florida,* showing typical injury caused by the dogwood twig borer, *Oberea tripunctata.* The larva normally feeds downward, or toward the trunk.
B. An adult dogwood twig borer;́ 14 mm long.
C. Small holes in the bark where the dogwood twig borer larva expels dry frass.
D. A dogwood twig borer excised from its channel in the pith.
E. An adult dogwood borer, *Synanthedon scitula;* a clearwing moth. Body length about 15 mm.
F. A flowering dogwood showing typical crown dieback caused by severe dogwood borer injury.
G. Scars on the trunk and limbs, along with adventitious growth, are symptoms of dogwood borer injury.
H. A dogwood borer larva; about 14 mm long. (Courtesy J. M. Ogrodnick, New York State Agricultural Experiment Station, Geneva.)

A
B
C
D
E
F
G
H

Twig Pruners and Twig Girdlers (Plate 124)

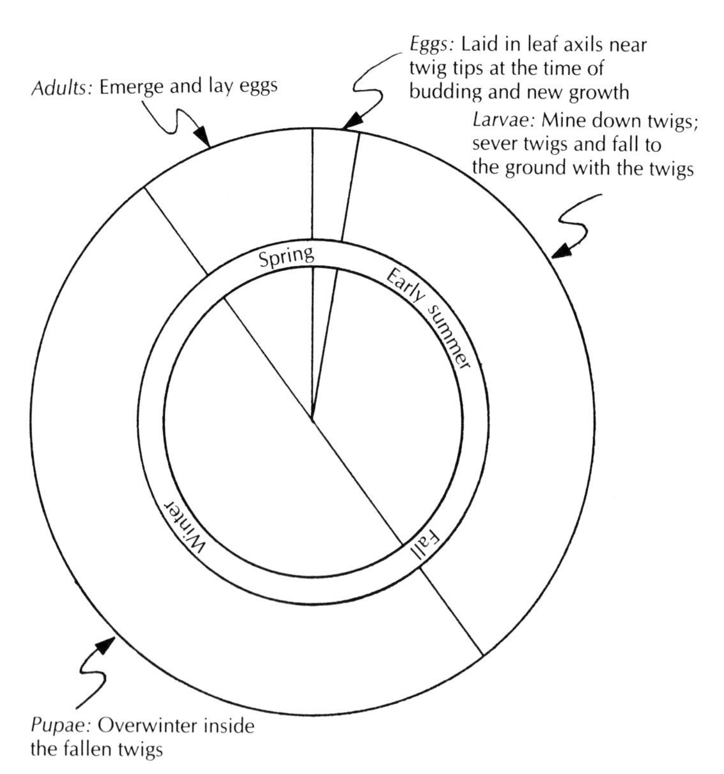

Figure 86. The presumed life cycle of the twig pruner, *Elaphidionoides villosus.*

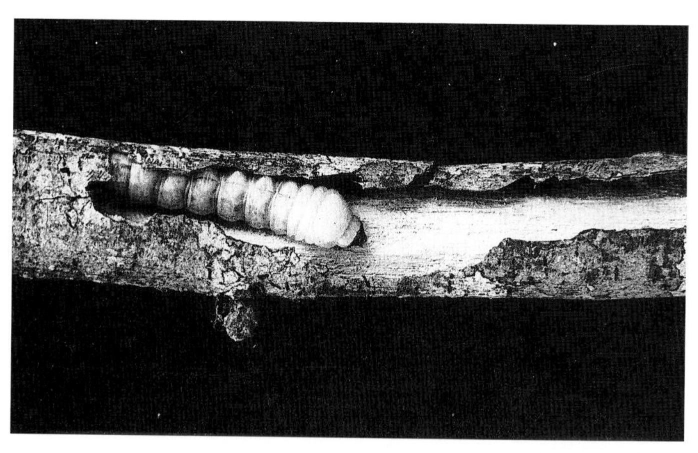

Figure 87. A larva of the twig girdler, *Oncideres cingulata,* inside a twig it has hollowed out, leaving a paper-thin shell of bark. The larva is about 22 mm long. (Courtesy J. D. Solomon, U.S. Forest Service.)

There are a few species of wood-boring beetles that tunnel in twigs during their larval development. The twigs, partially severed by the larva, break easily and drop to the ground or hang loose on the tree. Twig pruning, whether it be done by insects or humans, increases the density of a tree crown by encouraging latent buds to grow. Excessive pruning damages the tree and can greatly alter its shape. An abundance of small branches and twigs lying under a tree late in the growing season is a symptom that should invite examination. If the broken ends are clean cut, the cause has likely been the twig pruner, *Elaphidionoides villosus* (Fabricius). Split open one of those twigs, and you will find a creamy white roundheaded borer (C). The larva is immediately recognizable by its long lemon yellow hairs, especially on the prothorax. Twigs and small branches 6–50 mm in diameter are subject to attack.

Accounts about the life history of this beetle are sketchy and inconsistent. Light trap catches in North Carolina indicate that the adult twig pruner first appears in late April. From the fragmented data available, we believe that Figure 86 illustrates a fairly accurate account of its life cycle in the mid-Atlantic states.

In late spring the female chews a small niche in the bark of a twig at a leaf axil and there lays an egg. After making its entrance hole, the larva begins to feed in the center of the small branch, making a tunnel toward the base. The twig remains alive through the summer, but as autumn approaches the larva moves from the center to the sapwood by making concentric circular cuts, leaving only the thin bark to hold the branch in place. The larva retreats into its center tunnel and plugs it at the severed end with fibrous frass. A gust of wind will cause the branch to break and fall to the ground. The larva continues to feed inside the broken branch and overwinters as a pupa. In the spring the adult emerges from the hollowed-out branch. There is one generation each year in the Northeast and the middle Atlantic states.

The twig pruner has a long list of acceptable hosts: oak, chestnut, hickory, pecan, maple, linden, flowering fruit trees, redbud, sweetgum, hackberry, sassafras, persimmon, honeylocust, elm, quince, and wisteria. It occurs from New England west to Wisconsin and Saskatchewan and south to the Gulf states. Specimens have been collected in spotty locations west of the Mississippi River.

There are other species of twig pruners with similar feeding habits. In southern Ontario and the eastern states *E. parallelus* Newman attacks oak and hickory. Its life history parallels that of *E. villosus.* Note that the term *pruner* is given to those borers whose larva sever the twig from the inside; the term *girdler* applies when the adult severs the twig from the outside.

The twig girdler, *Oncideres cingulata* (Say), damages twigs as an adult (G). They appear, in the Piedmont region of North Carolina, in late August and September. The female deposits eggs, singly, beneath the bark in a slit she prepares in a terminal or lateral twig; then she chews a continuous notch around the twig, girdling it. Each twig may have 3–8 eggs. Apparently the larva is unable to develop successfully in wood that contains large volumes of sap. Girdled twigs soon die and break off; the eggs hatch in the fall, and the larvae remain dormant inside twigs on the ground. They grow rapidly in the spring, tunneling toward the severed end and feeding only in the woody portion. They make a few small circular holes in the bark to eject pellets of frass. The fully grown larva is cylindrical and 18–25 mm long when mature (Figure 87). It completes its development during September, when it emerges as an adult beetle to repeat the cycle. Females live 6–10 weeks, and during this time each will deposit 50–200 eggs. One generation usually occurs each year, although some individuals do not complete their development until the second season. Hosts of the twig girdler include elm, oak, linden, hackberry, redbud, apple, hickory, pecan, persimmon, poplar, basswood, honeylocust, dogwood, some flowering fruit trees, and others. Injury, particularly where leaders are destroyed, would be intolerable in the nursery. Landscape trees may be badly misshapen. Twigs and terminals that are killed average more than 0.5 meter long. Control can be attained by collecting the dropped twigs in the fall to burn or bury them. This beetle is generally distributed throughout the eastern United States from New England to Florida and west to Kansas, Texas, and Arizona.

The insects, or their injury, shown in panels A, E, F have been identified only as a species of the genus *Agrilus.* They are twig pruners of red oak and other shade trees in the Northeast. The injury caused by this species is particularly noticeable in early September. See Plate 125 for a description of the oak twig girdler, *Agrilus angelicus* Horn, in California.

References: 46, 116, 236, 474c, 515, 740, 755

A. A red oak tree with dead twigs—symptom of attack by *Agrilus* sp. (Also see Plate 125A for similar symptoms.)
B. A drawing of the twig pruner made by L. H. Joutel in 1900. Following his legend: (7) a larva in its tunnel; (7a) a thin shell of bark, the wood being nearly all eaten; (8) a pupa in its burrow; (9) an adult beetle.
C. A fully grown larva of the twig pruner, *Elaphidionoides villosus,* exposed in its tunnel; about 25 mm long. Its head is to the right. [Cerambycidae]
D. An egg of *Agrilus* sp. laid near a leaf scar.
E. An *Agrilus* sp. larva in a red oak twig (*arrow*).
F. A broken red oak twig where an *Agrilus* sp. larva has been feeding.
G. The twig girdler, *Oncideres cingulata*; about 15 mm long. (Author's collection; photographer unknown.)

A
B
C
D
E
F
G

Oak Twig Girdlers (Plate 125)

Patches of dead foliage scattered throughout canopies of oak trees (A) in California are characteristic symptoms of the activities of the Pacific oak twig girdler, *Agrilus angelicus* Horn. It is considered the major insect pest of oak in southern California. The California live oak, *Quercus agrifolia*, seems to be affected more than other oaks. Most complaints of damage caused by this insect arise from infestations in the California live oak. Such complaints are probably due to the dense human population in areas where this tree is native. The interior live oak, *Q. wislizenii*; the Engelmann oak, *Q. engelmannii*; and several introduced oaks are reported to be susceptible to attack as well.

The adult insect is a brownish bronze beetle about 7 mm long (C). After mating, the female lays eggs singly on the twigs of the tree's most recent growth. The eggs hatch after 2–3 weeks, and the larvae bore directly into the twig.

The larva is white and legless, and it looks like a string of flattened sausage links because of the definite constrictions between the body segments. When mature, it measures about 19 mm (E).

After hatching and boring into the twig, the larva tunnels just beneath the bark, making a slender gallery in the direction of the older twig growth. The larva then begins to girdle the twig, spiraling in the direction of older growth. At this time, the leaves distal to the girdled area begin to die; additional leaves continue to die as the girdling advances. After nearly 2 full years within the twig, the larva reverses its course and tunnels back toward the tip of the twig for a distance of up to 15 cm. There, it pupates and emerges as an adult several weeks later. Emergence of adults occurs any time between May and September. Along the southern California coast, most of the adults appear in late June and early July. Inland, peak emergence takes place from late May to early June. The entire life cycle requires approximately 2 years.

Figure 90. The D-shaped exit hole of *Agrilus acutipennis* in the bark of an overcup oak. (Courtesy J. D. Solomon, U.S. Forest Service.)

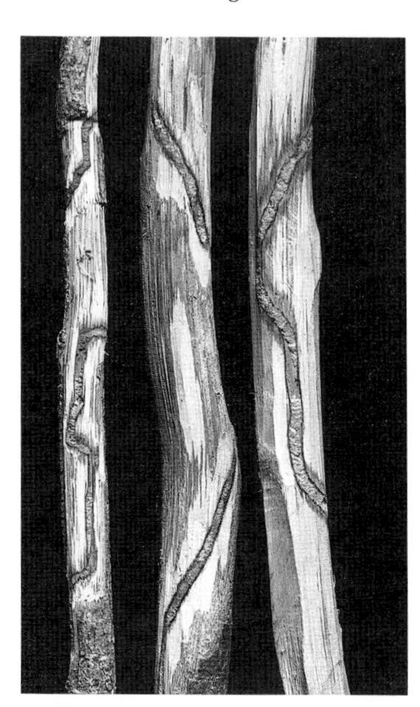

Figure 88. An oak sapling with bark removed to expose the larval galleries of *Agrilus acutipennis*. (Courtesy J. D. Solomon, U.S. Forest Service.)

Figure 89. A larva of the flat-headed borer *Agrilus acutipennis* in an oak sapling. (Courtesy of J. D. Solomon, U.S. Forest Service.)

The cause of dying patches of foliage on oak is best ascertained by pruning away several of these patches and removing the twig bark at the juncture between the living and dead foliage. If the Pacific oak twig girdler has been the cause of the damage, a gallery filled with brown powdery frass and possibly a larva will be found (D). Dieback may be initiated by other insects, such as the roundheaded oak twig borer, *Styloxus fulleri* (Horn). Instead of a spiral gallery beneath the bark, this insect makes a linear tunnel directly through the central part of the twig.

A. acutipennis Mannerheim is another pest of oak, particularly white oak and overcup oak, although other oaks in the white oak group probably serve as hosts as well. It is widely distributed from Maine south to Florida and west to Texas and Colorado, where it attacks saplings as well as mature trees. The larvae cause injuries by tunneling in the sapwood. By cutting away the bark of infested saplings, one can observe the feeding sites (Figure 88). The galleries are long and crooked—they occasionally spiral around the stem—and they are flat oval in cross section and tightly packed with frass. Adults emerge in May and June from D-shaped holes in the bark (Figure 90). A generation is completed in 2 years.

See Plate 124 for more twig-girdling insects.

References: 85, 217

A. Patches of dead foliage on California live oak caused by the oak twig girdler, *Agrilus angelicus*. [Buprestidae]
B. Close-up of an oak twig showing dead foliage beyond point (*arrow*) girdled by the oak twig girdler.
C. An adult oak twig girdler.
D. Twigs showing area of girdling.
E. A larva of the oak twig girdler removed from its gallery, tail end out.
F. A hickory twig showing larval injury by *Agrilus* sp.

A
B
C
D
E
F

Poplar-and-Willow Borer (Plate 126)

Willow trees are plagued by many insect pests. In 1882 a willow weevil native to Europe was discovered in the United States. Since that time this weevil, the poplar-and-willow borer, *Cryptorhynchus lapathi* (Linnaeus), has spread from the East to the West Coast and as far south as North Carolina. It is also established throughout southern Canada. Weevils, as a group, are known for their peculiar mouthparts that are attached to the tips of their long snouts (D).

The life cycle of the poplar-and-willow borer appears to vary depending on the geographic location. In southern coastal British Columbia the life cycle may require 2–3 years. There, most adults of a brood emerge from their pupal cells in host trees in autumn, then hibernate on the ground during winter. They are in the trees in the spring and early summer but do not lay eggs until July and August. It is not uncommon for a few adult weevils to overwinter a second time.

In the Northeast few adults overwinter; instead, the larvae in various instars overwinter in sapwood. When plant growth resumes in the spring, the larva grows rapidly and expels much frass through openings of its own making to the outside of the tree. This frass clings to the bark and serves as an external sign of the borer's presence. Sap oozing from the bark may also signal an infestation. The most pronounced injuries are old bark scars that are horizontal and deep and often have right-angle extensions (E) and exposed, stained wood. Trunks that have been repeatedly attacked are often honeycombed with tunnels. Pupation takes place in June. In New York weevils appear during the latter half of July and are most abundant in August, at which time they lay eggs. The adults are active in the evening and morning. When disturbed, they feign death by folding their legs and dropping to the ground. They seldom fly.

The adult weevils feed on the green smooth bark of shoots by chewing small holes in the bark, but the injury caused is minor. They make holes or slits for eggs at lenticels, branch bases, or on the edge of damaged bark of stems at least 25 mm in diameter. White eggs are produced over a period of 4–6 weeks; they are laid singly or in groups of two to four. Upon hatching, the young larva feeds on the inner bark. The larva has a C-shaped posture. When fully grown it is about 6 mm long. Typical injury caused by boring of larva is shown in panels B, C, and E. When ready to pupate, the larva usually bores inward and upward and prepares a pupal cell in the center of the stem.

The poplar-and-willow borer feeds on, and lays its eggs in, all *Salix* and most *Populus* species common in the United States and Canada, although *P. tremuloides* is not affected. It also feeds on *Alnus* species, *Betula pumila,* and *B. nigra*. Preference is given to young willows with smooth bark. Through most of its range there is one generation per year.

Injury caused by the larva is greatest on nursery stock and on newly set trees in landscape plantings. Infested trees may be killed or may lose their form and become bushy from coppice or adventitious growth (A). Young *P. deltoides* respond to larval injury by producing bulbous swellings. After completion of insect development, these swollen areas exhibit emergence holes and much scar tissue.

References: 338, 500, 682

A. A severely scarred pussy willow stem with several adventitious sprouts. These grow as a reaction to severe injury by the poplar-and-willow borer.
B. Tunnels in the wood where larvae have prepared pupal chambers.
C. Larvae in what will become their pupal chambers. [Curculionidae]
D. An adult poplar-and-willow borer; 8–10 mm long. Note the snout. The posterior part of the wing covers and abdomen is covered by pale yellow scales. On a dark background the adult is nearly invisible.
E. And old tunnel scar caused by a larva.

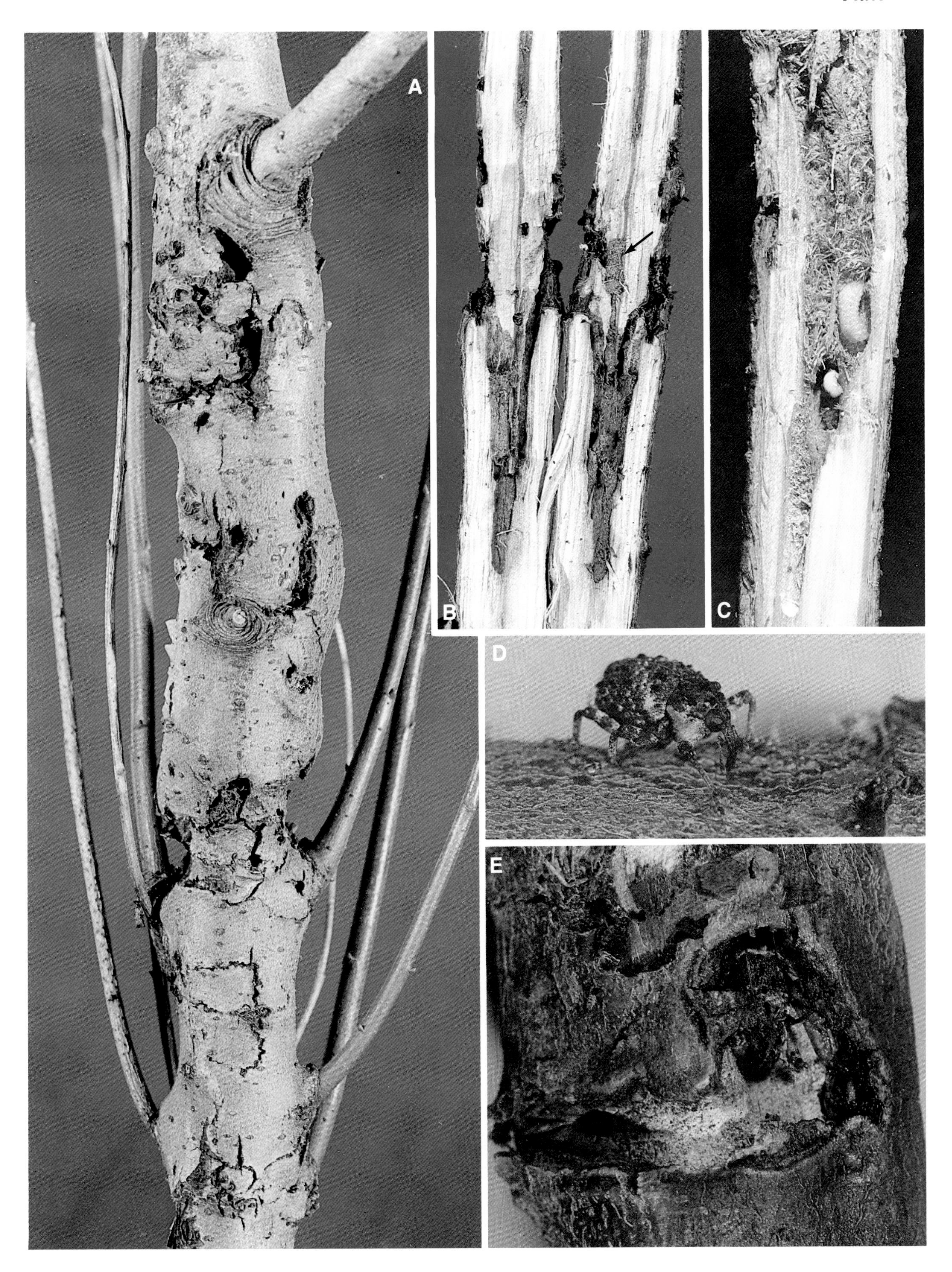
A
B
C
D
E

Flatheaded Borers (Plate 127)

The larvae of many wood-boring beetles can be classified as either flatheaded or roundheaded borers. They damage shade trees by tunneling into the conducting tissue (xylem proximate to the cambium and phloem). More dangerous to the tree's health, however, may be the fungi that enter through the tunnels made by these insects. Those illustrated in this plate are all flatheaded borers.

The flatheaded appletree borer, *Chrysobothris femorata* (Olivier) (B), is a common species that feeds on many kinds of deciduous shrubs as well as shade and fruit trees such as maple, oak, hickory, sycamore, tuliptree, willow, rose, and cotoneaster. The beetle is about 12 mm long, has a somewhat flattened appearance, and varies from a dark metallic brown to dull gray. Some adult beetles emerge from their pupae late in the spring; others continue to emerge throughout the summer and into November in the South. The female lays her eggs singly in bark crevices and produces more than 100 eggs during her reproductive period. The legless yellowish white larva enters the bark at the place where the eggs were laid. If the tree is vigorous, the borer may be killed by heavy sap flow; if the tree or shrub is in a weakened condition, or the bark has been badly damaged, the larva enters and develops rapidly while producing long tortuous tunnels in the phloem, partially filling them with powderlike frass. A single larva may girdle a young tree. Points of larval activity can usually be detected by white, frothy sap oozing from cracks in the bark. Injured areas often become depressions, and later the bark may split at the injury site. Wounds may be enlarged year after year by succeeding generations. Newly transplanted ornamental trees and shrubs are particularly susceptible to injury. When nearly mature, either in the spring or fall, the flatheaded appletree borer tunnels deep into the heartwood (C), where it pupates. Adults feed at the bases of twigs, on partially defoliated young trees.

The flatheaded appletree borer is found predominantly in the eastern and central states, but is known throughout the United States and southern Canada. One generation occurs each year. Regular fertilization and irrigation of most ornamental plants will promote vigorous growth and reduce attacks by this insect. A horticultural tree wrap, binding the trunk of a newly planted tree, is reported to be a successful protective measure against the flatheaded appletree borer.

A species closely related to the flatheaded appletree borer and called the Pacific flatheaded borer, *Chrysobothris mali* Horn, causes the same kind of injury, usually under similar conditions (H). It occurs on the West Coast and in the Rocky Mountain states and provinces. Severe damage is usually associated with such plant stress as above-normal summer temperatures and sunburned bark. This insect is considered one of the worst enemies of newly planted trees and shrubs in the Pacific Coast states and British Columbia. It attacks 70 or more woody plants, including the rose. In some higher elevations more than a year is required for a complete generation. Eggs are usually laid in cracks and crevices in the bark. The embryo, upon completion of its development within the eggshell, bites a hole in the bottom and mines directly into the bark, never exposing itself to the outside world. The larvae make tunnels in the phloem and wood and pack them with frass. Adults are sun loving and quick in flight. They may be found from April through July in California. Larvae may be found in any portion of a mature tree, but on young trees their attack is confined to the trunk. The borer has a number of natural enemies. The most common predator in Oregon is the mite *Pyemotes ventricosus* (Newport), which attacks the larva in its tunnel.

The twolined chestnut borer, *Agrilus bilineatus* (Weber), is a pest of *Quercus* and *Castanea* species with a few reports of injury to species of *Fagus*. The oaks most frequently attacked are chestnut oak, white, black, red, scarlet, and burr oaks. Chapman (112) showed that healthy as well as stressed oaks can be attacked by the twolined chestnut borer. Subsequent studies pointed out that vigorous trees are relatively immune and that weak trees support large beetle populations, which often kill the tree. Outbreaks of defoliators like gypsy moth, forest tent caterpillar, cankerworms, and oakworms are most always followed by large and seriously injurious populations of the borer. If a tree weakened from any cause supports these borers for 2–3 years in succession, gradual death occurs from the top down. Infested trees may be recognized by sparse, small, and discolored foliage and branch dieback.

The twolined chestnut borer usually produces one generation of offspring each year. When they feed on vigorous trees, 2 years may be required to complete a generation. Other unfavorable host conditions may also contribute to variations in the insects' developmental time.

Adults begin to appear in early June and may be found through mid-September in Wisconsin. They lay eggs in late June and early July in clusters of up to 10 eggs, in cracks and crevices in the trunk bark, usually in sections exposed to sunlight. Beetles are sun loving and somewhat gregarious; they feed indiscriminately on the leaves of many kinds of hardwood trees. Newly hatched larvae bore directly into the bark and establish themselves primarily in the phloem, making winding, zigzagging tunnels filled with frass that resemble those illustrated in Figure 91, accompanying Plate 128. By the time larval development is completed, the tunnel may be 30 cm or more long. The cream-colored larvae are legless; they have flat heads and an elongate, rather delicate body (A). Pupation takes place in the outer bark in the spring. Adult eclosion occurs in the pupal cell. The adult chews a D-shaped exit hole in the bark like those made by the bronze birch borer. Twolined chestnut borer attack usually begins in the tree crown and proceeds downward along the bole in each succeeding year of infestation. This insect is found in eastern Canada south to Georgia and Texas and west to the Rocky Mountains.

In the 1960s and 1970s the thornless cultivars of honeylocust became popular shade trees for urban properties. As a result of overplanting, we anticipate an increasing problem on honeylocust with the buprested borer, *Agrilus difficilis* Gory. Current distribution records are sparse and discontinuous, but the insect is present or likely to be present from New York and New Jersey south to Georgia and west to Texas, Colorado, and Utah.

The adult has the same form as other *Agrilus* species (Plate 128). Its length is 8–12 mm and its upper surface (elytra and thorax) varies from metallic greenish black to bronzish purple; it has distinct yellowish to white bands on the sides of the abdomen. Adult emergence begins in early June and continues through part of July (Ohio). In Pennsylvania adults have been collected from early June through early October. In all locations they feed on honeylocust foliage, making notches along leaflet margins. Females produce about 30 eggs, laying them on twig leaf scars and in bark crevices. Each egg is covered with a frothy substance that hardens after exposure to air. The young larvae bore into the trunk or branches, causing the tree to ooze sap at the site of penetration, which may harden into a mass of gum. The larval galleries are similar to those of the bronze birch borer and the twolined chestnut borer. Fully grown larvae are flat, strongly segmented, and up to 25 mm long; they resemble the larva shown in panel A. The adult's exit hole in the bark is D-shaped. Repeated infestations bring about a gradual decline and dieback of the twigs and branches in the crown. A single infested tree may contain 45 or more larvae per meter of trunk. *A. difficilis* colonizes large and small honeylocust trees as well as branches more than 1.5 cm in diameter. Trees under stress are most likely to be attacked.

A. burkei Fisher, a native insect, attacks several species of alder in British Columbia, Washington, Oregon, California, Nevada, Idaho, and Wyoming. In California it is a particular pest of white alder planted in landscapes. Girdling beneath the bark by the larvae causes dieback of branches, tree tops, even entire trees. Most losses have occurred where the planted trees have not had adequate irrigation throughout the year.

A. A twolined chestnut borer larva, *Agrilus bilineatus*, in oak bark; about 25 mm long. [Buprestidae] (Courtesy Connecticut Agricultural Experiment Station.)

B. A larva of *Chrysobothris femorata*, the flatheaded appletree borer, in the crotch of a mountain ash tree. Note that the tree wrap did not prevent egg laying or injury.

C. Larvae of *C. mali*, the Pacific flatheaded borer, in the heartwood of almond.

D. A flatheaded borer, *Dicerca lurida*; about 14 mm long.

E. A twolined chestnut borer adult; about 13 mm long. Note the rather prominent stripes on the wing covers.

F. Larvae of the Pacific flatheaded borer; about 13 mm long when fully grown.

G. Larvae of the flatheaded appletree borer; about 12 mm long. [Buprestidae]

H. An adult Pacific flatheaded borer; about 10 mm long.

A
B
C
D
E
F
G
H

Other *Agrilus* species attack black locust, hackberry, dogwood, hickory, maple, persimmon, redbud, and walnut. Most species are host specific.

Dicerca lurida (Fabricius) (D) is another of the many species of flatheaded borers that feed in hardwoods. Its eggs are laid on the bark of *Carya, Tilia,* and *Carpinus* species. The larvae make extensive tunnels, primarily in bark phloem. They are found throughout the region of eastern forests but are not a threat to healthy trees.

References: 1a, 45, 87, 103, 124, 145, 191, 232, 242, 299, 344, 461a, 492, 511, 651, 707a, 907

Bronze Birch Borer (Plate 128)

Agrilus anxius Gory, known as the bronze birch borer, is a beetle native to North America. It occurs throughout the range of birch, from Newfoundland to British Columbia south to New Jersey, Ohio, and Idaho. It can be a serious pest of forest and shade trees. Attempts to extend the range of white birches into milder climates, such as West Virginia, Kentucky, and Tennessee, has resulted in increased landscape maintenance costs largely because the plant is out of place. Such trees are placed under additional heat stress, making them more susceptible to the omnipresent bronze birch borer.

This insect overwinters as a larva in one of several instars. According to Loerch and Camerson (476), the fourth instar composes more than 90% of the overwintering population. When the sap begins to flow, these larvae begin to feed again. Those that wintered in the fourth instar molt again and pupate very near the bark surface. When ready to emerge, the adult chews a D-shaped hole in the bark to escape. The shape of the emergence hole is a means of positive identification. Adult emergence begins in Canada near the end of June and lasts approximately 6 weeks. Emergence peaks about the middle of July, and the period of flight and egg laying lasts until about the end of August. Eggs are laid either singly or in clutches, deposited under bark flaps, or cracks in the trunk or branches. Upon hatching, the larva bores directly into the bark and begins to make its feeding gallery, which runs in a serpentine manner at the interface of xylem and phloem (E and Figure 91). A larva requiring 2 years to complete its cycle will produce a gallery 75–125 cm long. The larvae are relatively long and flat, with a head that is a bit wider than the body. They have a unique tail segment that terminates in a pair of minute brownish black, forcepslike horns (C). The borer may have a 1- or 2-year life cycle, but the latter is more common. The length of the cycle is mainly governed by host condition and by the time of year the egg is laid and subsequently hatches.

According to Barter (42), "The bronze birch borer larvae cannot survive in healthy trees. Successful larval development is dependent upon the host being in a weakened condition from repeated unsuccessful attack or other injury such as defoliation, adverse weather conditions or old age." The latter is important because most white birch species are relatively short-lived, seldom living longer than 50–80 years. The common multiple stem gray birch is old at 30 years.

Injury is caused by larval feeding galleries that, in effect, girdle the trunk or branch. Trees that survive borer attack have conspicuous swollen areas on the trunk (B) caused by the "healing" process. Chlorotic leaves and sparse foliage are early symptoms of borer presence; these are most evident in the upper crown and are sometimes accompanied by increased adventitious growth in the lower crown (A). If such adventitious growth develops, it is accompanied by twig dieback in the upper crown. The adults feed on birch leaves but seem to prefer alder or poplar. Adult feeding injury is, however, insignificant.

The birch species most susceptible to injury by the bronze birch borer are *Betula papyrifera, B. pendula* and its cultivars, and *B. populifolia. B. maximowicziana* (the monarch birch), *B. platyphylla* var. *japonica,* as well as *B. nigra* and other brown bark species, are said to be resistant or more tolerant. Birch identification is fraught with difficulty, according to some plant specialists. Be sure of the identification, especially when selecting birch for resistance to borer attack and for low-maintenance plantings.

Woodpeckers and parasites such as *Phasganophora sulcata* Westwood (Hymenoptera) are major natural control agents of the bronze birch borer. These animals are not, however, generally effective in urban or suburban situations. Vigorously growing birch trees are less likely to be damaged by the bronze birch borer than are trees in poor condition.

A closely related species, the bronze poplar borer, *Agrilus liragus* Barter & Brown, feeds in at least five species of *Populus,* including *P. tremuloides, P. grandidentata, P. balsamifera, P. trichocarpa,* and *P. deltoides.* Its life history and the morphology of its immature stages are very similar to those of the bronze birch borer. Adults of the two species are so similar that they are often confused. Adults of the bronze poplar borer are blackish with a faint touch of metallic green and 7–12 mm long. Adults emerge about 2 weeks earlier than the bronze birch borer, but a life cycle requires 2 years.

Figure 91. Bark removed from a birch trunk to show the winding galleries of bronze birch borer larvae. (Courtesy Ontario Ministry of Natural Resources.)

The presence of the bronze poplar borer larva causes deterioration and frequently death of the host through its feeding on branches and the main stem. Any natural event that weakens a tree increases its susceptibility to the bronze poplar borer. Injury symptoms on poplar are similar to injury symptoms on birch caused by the bronze birch borer.

This insect is found from the Canadian maritimes west to British Columbia and in the states bordering Canada. It is also known in Pennsylvania, Colorado, Utah, and Arizona.

Agrilus angelicus Horn, the Pacific oak twig girdler, has the dubious honor of being the number-one pest of shade tree oaks in southern California (Plate 125).

References: 39, 42, 43, 193, 359, 476, 660, 704

A. A paper birch, *Betula papyrifera,* that is dying as a result of bronze birch borer attack.
B. Lumpy bark is an external indication of attack by the bronze birch borer.
C. A bronze birch borer larva; about 28 mm long.
D. *Left:* adult male bronze birch borer; *right:* the female. [Buprestidae]
E. Bark removed to show larval galleries.

A
B
C
D
E

Locust Borers (Plate 129)

Black locust, *Robinia pseudoacacia,* and its several cultivars are useful and important shade trees, and they are also important in the farm wood lot as a source of decay-resistant fence posts. The locust borer, *Megacyllene robiniae* (Forster), and the locust twig borer, *Ecdytolopha insiticiana* Zeller, are wood-boring pests. The locust borer attacks the trunk as well as branches as small as 45 mm in diameter. The borer prefers locust trees that are at least 4 years old, but once the trunk reaches 15 cm in diameter it is attacked less often. The adult locust borer does not harm the tree, but the larva may riddle the wood with its feeding and tunneling (Figure 92). Even so, few trees are killed outright.

The borer is found in most parts of North America where locust is grown. It is particularly damaging to host trees in the central and northeastern United States. The adult (D) may be found in early September, and until the first frost it feeds on goldenrod pollen (*Solidago* species). Because of this insect's yellow markings, the casual observer may easily mistake it for a wasp, since both frequent the flowers of goldenrod. After maturing, feeding, and mating, the female deposits one to several eggs in deeply fissured bark crevices, beneath loose bark scales, and in holes where borers of another generation emerged. Eggs are creamy white, oval, and about 3 mm long. They hatch in about a week, and larvae bore directly into the bark until they reach living bark tissue (phloem). By mid-June of the next year the larva has made a shallow burrow on the phloem/sapwood interface that may measure 25 mm in diameter (Figure 92). At the end of this cavity it bores upward and inward toward the center of the trunk, then the tunnel turns sharply and descends into the heartwood for about 8 cm. When the larva has completed its development, the gallery will be about 12 cm long and 8 mm in diameter. The gallery is kept clear of refuse, which drops in piles on the ground under the bark openings. When the larva is ready to pupate, it prepares a plug of wood chips and frass (E). The mature larva is about 25 mm long. When it chews on the wood, its mandible action is audible. The adult uses its larval gallery opening to emerge. There is one generation per year. A female sex pheromone has been identified, which has opened new avenues for monitoring and control. Few parasites have been identified, and none have been studied.

M. caryae (Gahan), the painted hickory borer (G), is nearly the same size and color as the locust borer. We have observed that this insect breeds in freshly cut hickory logs, but other investigators report its breeding in other hardwoods such as ash, butternut, black walnut, black and honeylocust, hackberry, mulberry, and oak. Adults are active only in the spring, so the two species are never found at the same time of the year. To avoid damage by the painted hickory borer, cut timber or fireplace wood in July or August so that it can dry before adults emerge the following spring. There is one generation per year.

"Locust twig borer"—the common name of *E. insiticiana,* a moth belonging to the family Tortricidae—is a misnomer because the insect attacks only black locust shoots, not twigs, causing an elongate, spindle-shaped swelling (F). Though generally considered to be an eastern species, this borer has been reported in Arizona, California, and Colorado, and in Manitoba and Ontario. Shoots are seldom killed but are disfigured, weakened, and stunted and are easily broken by wind. In nurseries weak shoots cannot be tolerated. Eggs are laid on the shoot, and upon hatching, the larva bores into the shoot, usually at the base of a thorn, and feeds in the pith. The larval entrance hole becomes the site of refuse disposal, and dark frass gradually accumulates at this point on the shoot surface. When fully grown, the larva changes from yellow to bright crimson with a reddish brown head. It leaves the safety of the mine to pupate between leaves, either in the tree or on the ground. A generation is completed in about 6 weeks. In Kentucky and nearby states two generations occur each year. The first-generation moths emerge in early May to late June; those of the second generation appear from early July to early September. Larvae of the second generation do not complete their growth, and they overwinter in the twig.

References: 59, 193, 278, 281, 336, 359, 474, 481, 690, 876, 909

Figure 92. Black locust logs with bark removed to show sapwood damage. The dark scars are points of injury caused by young locust borer larvae. (Courtesy U.S. Forest Service.)

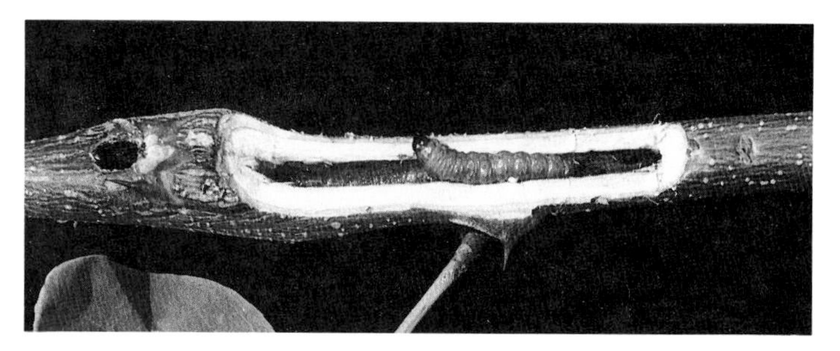

Figure 93. A locust twig borer in its tunnel in a black locust shoot. (Courtesy J. D. Solomon, U.S. Forest Service.)

A. Holes of emerged locust borers in trunk of black locust. Vigorously growing trees cover old holes with callus tissue, producing a scar (*arrows*).
B. A swollen trunk is another symptom of locust borer injury.
C. Trunk split to illustrate locust borer galleries in the heartwood.
D. An adult locust borer feeding on goldenrod flowers. [Cerambycidae]
E. Pupa and frass plug near the end of a gallery.
F. An adult painted hickory borer. Note close resemblance to D.
G. Swollen shoots of black locust caused by the locust twig borer. [Tortricidae]

A
B
C
D
E
F
G

Roundheaded Borers (Plates 130–131)

Many wood-boring beetles of importance to woody ornamentals belong to the family Cerambycidae. The adults are commonly known as longhorned beetles because they bear antennae, which are about as long as the body of the female and are up to four or five times the length of the body of the male (A). Cerambycid larvae are known as roundheaded borers because of their body structure and characteristic feeding tunnels. Most species in this family infest and feed on dead or dying trees. There are, however, a number of important pests in this group that attack living healthy trees and shrubs.

The ponderosa pine bark borer, *Canonura* (=*Acanthocinus*) *princeps* (Walker), is an example of a species that avoids healthy trees. This insect attacks only dying or dead pines and feeds in the bark (A, B). *C. princeps* is a western species that occurs on the Pacific Coast in British Columbia, Washington, Oregon, and California on *Pinus ponderosa* and *P. sabiniana. Acanthocinus spectabilis* (LeConte) occurs on *P. ponderosa* and *P. scopulorum* from Montana and South Dakota through the Rocky Mountain states into Mexico. Other similar species associated with pine include *A. nodosus* (Fabricius), which occurs from Pennsylvania south to Florida and west to Texas, and *A. obsoletus* (Olivier), which occurs in eastern North America west to Minnesota and Texas.

The genus *Saperda* contains several economically important species that injure living trees. *Saperda vestita* Say, the linden borer, attacks both native (basswood) and European lindens, especially the cultivar Greenspire, and poplars. The larva, about 25 mm long when fully grown, feeds primarily in the wood of large branches, trunk, and root crown; it attacks both weak and healthy, nursery and established shade trees. The life history of these borers is similar to that of the roundheaded appletree borer (Plate 131). In central New York adults begin to emerge in mid-May and may be collected throughout the summer. Beetles feed on leaf petioles, large veins on the undersides of linden-basswood leaves, and the bark of new shoots. They are about 18 mm long, heavy bodied, yellowish gray, and usually embellished with three irregular black spots on each elytrum. The linden borer is endemic to the eastern half of North America but may be found wherever *Tilia* species grow.

S. inornata Say, and *S. fayi* Bland, are gall-making saperdas. Their habits and development are similar. *S. inornata* is most frequently found in sapling stems of trembling aspen, *Populus tremuloides*. In Michigan adults appear in late May to early June and lay eggs in the bark of twigs and saplings. Unique in their egg-laying habit is the construction of a horseshoe or shield-shaped egg niche chewed in the bark. Each niche is about 3 mm at its widest point, and the open end is up. A single egg is forced between the bark and xylem at the curved part of the "horseshoe." One to five egg niches are made in a ring around the twig. Twigs most frequently selected for egg laying are 5–15 mm in diameter. The egg niches remain visible and identifiable for several years as dark brown half-moon scars (Plate 131A).

Upon hatching, the larva feeds in the phloem, and the tree reacts by producing callus tissue. As this tissue accumulates, a globose gall forms (Plate 131A). One and sometimes two adults complete their development in each gall. The life cycle may be completed in 1 year, but eggs laid late in the growing season may require 2 years to complete their cycle. Adults feed on leaves from the host tree.

Severe damage is usually limited to young saplings. Larval feeding weakens the twig and makes it subject to breakage by gusts of wind. Wounds made by the female beetle in preparation for egg laying provide an entry for a devastating canker fungus pathogen, *Hypoxylon mammatum* (see Sinclair et al., *Diseases of Trees and Shrubs*).

S. inornata occurs from Minnesota and Saskatchewan east to the Maritime Provinces and south to Pennsylvania, Illinois, and southern Colorado. *S. fayi* is often called the thornlimb borer. It causes spindle-shaped galls on twigs of hawthorn. This beetle is found in the northeast and midwest. Its life history and egg-laying behavior are similar to those of the other twig-feeding saperdas.

The following species of *Saperda* attack only weakened, dying, or recently killed trees: *Eutetrapha* (=*Saperda*) *tridentata* Olivier attacks elm; *S. discoidea* (Fabricius), hickory; *S. lateralis* (Fabricius), hickory; and *S. puncticollis* Say, Virginia creeper.

The cottonwood borer, *Plectrodera scalator* (Fabricius), feeds on poplars and willows, and it probably occurs throughout the natural geographic range of the cottonwood, *Populus deltoides,* considered its major host. Both adults and larvae cause injury. All larval damage occurs at the root crown or in buttress roots of host trees of any age or size. Young trees may be girdled, but more often they are structurally weakened and easily broken by wind. Adults feed on petioles and the tender bark of shoots. They may remove large chunks of bark, causing the shoot to break, though it remains hinged to the proximal part of the shoot. The distal part of the shoot becomes a blackened flag and a symptom of adult injury. When beetles are numerous, their feeding on petioles may defoliate portions of trees.

Adults (C) begin to emerge from their larval galleries in late May. Emergence continues in Mississippi into early July but may be delayed by a month or more farther north. After a period of maturation feeding (before eggs develop), the female begins to lay eggs. To do this, she must dig into the soil to reach the proper site on the tree. Using her mandibles and legs, she loosens the soil beside the root crown or buttress root and burrows into the ground. She then shreds the bark with her mandibles to make a niche; then with her ovipositor she forces a single egg into the shredded bark. This procedure is repeated 20 or 25 times, until she lays her complement of eggs. The eggs are white to yellowish, elliptical, and about 3 mm long.

When eggs hatch, the larvae mine downward in the inner bark and gradually move into the xylem. Their galleries vary in form and length, dependent upon the size of the root. In small 1- and 2-year-old trees the gallery may be 10–20 cm long (Plate 131D). In the tap or large buttress roots of older trees the larva hollows out an area that is roughly oval, with a diameter of approximately 50–75 mm. Portions of galleries are packed with frass, especially if the larvae are feeding below the soil surface.

The larvae are legless, as is the case with many other cerambycids, and when fully grown are 32–38 mm long. Pupation occurs within the gallery. The life cycle is completed in 1 or 2 years. A 1-year cycle is common in the roots of young cottonwoods in the South.

Because most larvae feed below the soil line, they are well protected from both predators and parasites. Woodpeckers often capture those few that are exposed above the ground line. A fungal disease of the larva has been found, but it is not well dispersed in nature.

The sugar maple borer (E), *Glycobius speciosus* (Say), is a pest of living sugar maple throughout the native geographic range of *Acer saccharum*. This borer does not normally kill trees, but it weakens them by opening the wood to other borers and decay organisms. Injury is most often recognized in old trees along village streets and road sides. The female beetle usually chooses an open-grown tree whose trunk is 11–40 cm in diameter (at breast height) on which to lay her eggs. The most susceptible points of attack on a tree occur where the diameter is 7–27 cm. Active borer damage is easily missed by those unfamiliar with the symptoms. After adult emergence, several years may pass before bark sloughs off, exposing the old gallery engraved in dead sapwood. Old scars, called cat-faces, may encompass as much as one-third of the trunk diameter. After several years of reinfestation a tree loses vigor; leaves may be small and discolored in portions of the crown, and they tend to suffer premature senescence.

Adults begin to emerge in late spring and may be seen around maple

References: 231, 275, 307, 377, 515, 547, 571, 576, 577, 660, 736

A. An adult male of the ponderosa pine bark borer, *Canonura princeps*; about 30 mm long.

B. A larva and typical tunnel of the ponderosa pine bark borer in pine.

C. An adult cottonwood borer, *Plectrodera scalator*; about 32 mm long.

D. A swollen stem or gall of poplar with larvae of *Saperda populnea* and tunnels. (Also see Plate 131B, C.)

E. A pinned specimen of the sugar maple borer.

F. A drawing by L. H. Joutel of the life stages of the sugar maple borer, *Glycobius speciosus* (Say): (1) place where egg was laid; (1a) another with more larval activity, showing borings thrown out by the larva; (2) borer as it would appear in September from an egg laid in the same season; (3) nearly fully grown borers; (4) female beetle; (5) hole through which the beetle emerged; (6) sawdust or borings packed in a burrow.

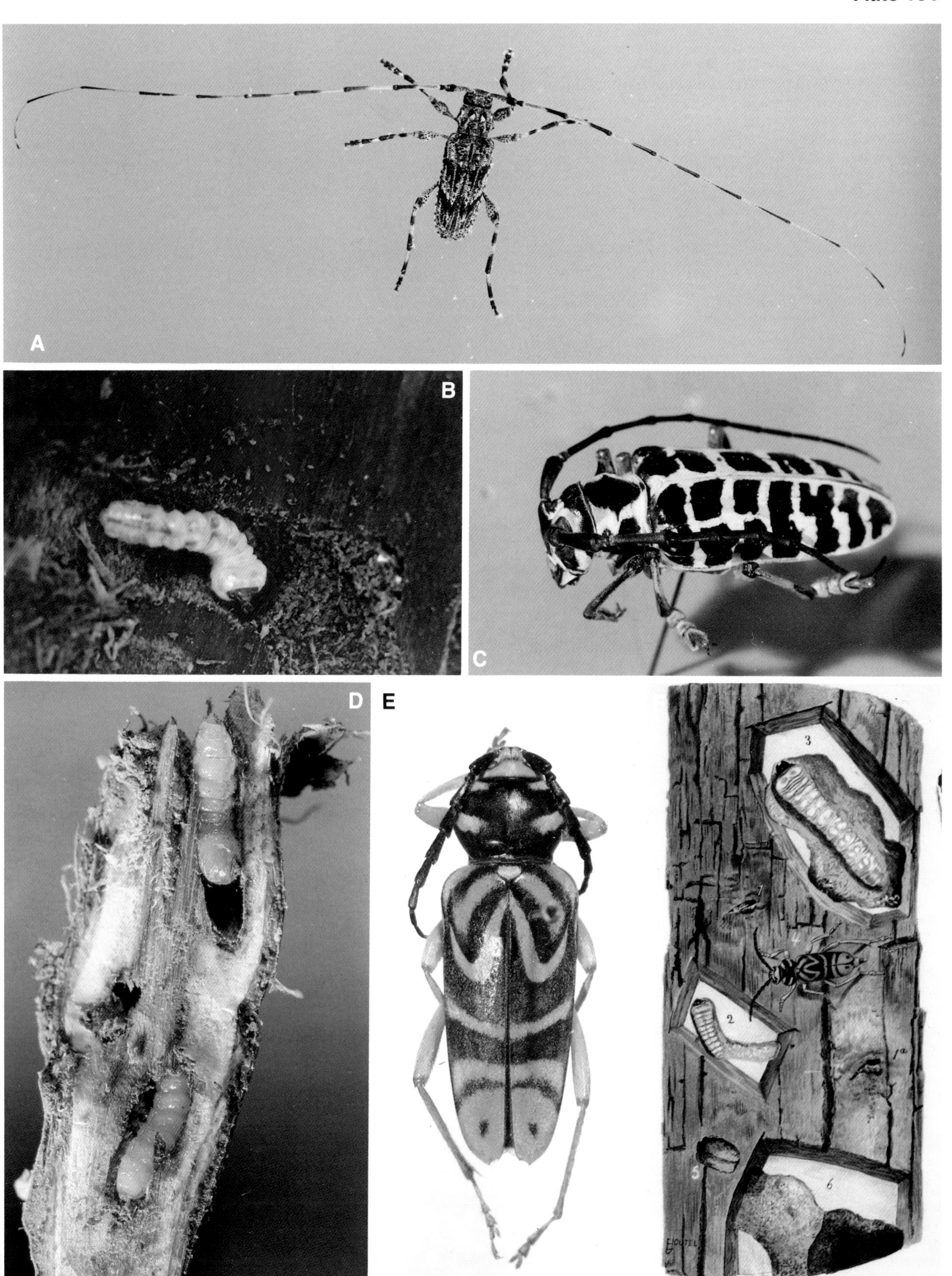
A
B
C
D
E
3
4
2
1a
5
6

trees for most of the summer. Egg deposition does not begin until late July and early August. The female cuts a small slit in the bark, where she lays an egg. When the egg hatches and the larva bores into the phloem, some "bleeding" takes place, and the bark is stained for a short distance around the hole. The larva generally makes its tunnel horizontally the first year in the interface between sapwood and phloem. As it nears full growth, the larva turns inward and upward toward the heartwood, pupating near the end of the gallery. The fully grown larva may be 40–50 mm long. It requires 2 years to complete its cycle. Note that the adult sugar maple borer, the locust borer (Plate 129), and the painted hickory borer (Plate 129) have similar markings.

Saperda populnea (Linnaeus) is another example of a gall-making saperda. It develops in the twigs of poplar or willow. Eggs are laid in a notch cut in the bark and pushed between the phloem and xylem. The resulting galls are often so numerous that on a given tree nearly all twigs with diameters of about 15 mm may have one or more spindle-shaped galls (B). The gall, with its tunnel(s), weakens the twig and makes it subject to wind breakage. An infestation in a young tree will greatly modify its normal growth pattern. Full-grown larvae are about 25 mm long and 8 mm in diameter; they are legless (C). Very little has been written about the biology of this species in the United States. This insect, a native of Europe, is currently thought to be limited in distribution to the states west of the Rocky Mountains.

Of the species belonging to the genus *Saperda,* the roundheaded appletree borer, *Saperda candida* Fabricius, does the most damage to tree crops. In the mid-1880s it was a serious problem for apple producers in the northeastern United States. Next to the codling moth, it was the worst enemy of the apple tree. With current pest management programs, however, it is now of little concern to fruit growers. Even so, the insect remains a major pest of several ornamental trees and shrubs, including hawthorn, mountain ash, quince, shadbush, cotoneaster, and flowering crabapple. The larva does the greatest damage, but adults feed on the fruit, bark, and foliage of host trees.

This insect is found throughout the northeastern quadrant of the United States as well as in the hilly areas of Georgia, Alabama, and South Carolina. It is also known from Texas north into southern Canada.

It takes 2–3 years for a roundheaded appletree borer (E, F) to complete its life cycle. The adults emerge from infested trees in May in the latitude of Washington, DC, and in June in central New York state. The female beetle lives about 40 days, devoting much of this period to egg laying. She is secretive, depositing her eggs at night, normally in young healthy trees. She makes a longitudinal slit, about 13 mm long, in the bark of a tree trunk, usually at or near the ground, and then inserts a single egg between bark and xylem. The egg soon hatches, and the larva bores into the sapwood. Newly hatched larvae are 3–4 mm long. The larva moves upward or downward in the trunk (depending on the time of the year), all the while enlarging the tunnels and compounding the damage. In the early spring of the third year it changes to a pupa and soon emerges as an adult.

The golden and downy woodpeckers are the only natural enemies that seem to have any effect on populations of the roundheaded appletree borer. There are very few insect parasites of *S. candida.* One investigator wrote: "Possibly no other economic insect of equal importance has had so few natural enemies recorded definitely and specifically as has the roundheaded appletree borer" (234).

The redheaded ash borer (G), *Neoclytus acuminatus* (Fabricius), is a pest of ash, elm, hickory, hackberry, oak, linden, redbud, walnut, birch, beech, maple, dogwood, holly, black locust, honeylocust, sassafras, lilac, sweetgum, and crabapple. Forest entomologists consider it a pest of cut logs. However, the female lays her eggs on weakened or newly set trees. The larva works in the inner bark and summer wood, cutting off the flow of sap. In young trees, burrows may extend both horizontally and vertically through the trunk, making it subject to breakage during high winds. This borer is a severe pest of young nursery stock in the north central states. The full-grown larva is short, cylindrical, and densely clothed with fine, silky lemon-white hairs. Initial attacks are difficult to diagnose because there are no open entrance holes. It is one of the few cerambycid larvae that attack live trees and do not maintain contact with the outside. The tunnels are packed with frass. Young larvae feed on the inner bark, later boring deep into the xylem. The redheaded ash borer overwinters in the trunk. Adults begin to emerge in April about the time that red maple is in bloom. In Illinois one generation occurs most years; occasionally, a partial second generation appears. This insect occurs in New England, the Appalachian Mountain area, the north central states, the Ozark Mountains, and west from Missouri to Arizona. Specimen ornamental trees can be protected from this pest by good cultural practices such as fertilization and watering during periods of drought.

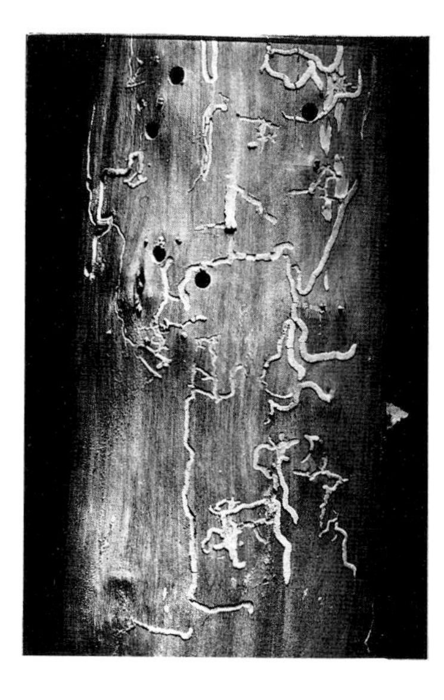

Figure 94. Bark removed to expose the frass-packed galleries and exit holes made by several redheaded ash borers. (Courtesy J. D. Solomon, U.S. Forest Service.)

A native of Australia, the eucalyptus longhorned borer, *Phoracantha semipunctata* (Fabricius), was discovered in Orange County, California, in 1984. It has since spread throughout most of southern California and into several counties in the northern part of the state. Eucalyptus of many different species is attacked and killed by this insect, particularly trees that are stressed by drought or are declining for other reasons. The adult beetles fly at night. Fertilized females deposit spindle-shaped eggs in clusters in bark crevices or beneath loose bark flakes or plates. Newly hatched larvae score the bark surface, then tunnel through the bark and mine in the cambium region. Mature larvae are about 33 mm long and are similar in color and shape to those shown in Plate 130F. They penetrate the xylem to pupate; new adults then chew their way through the bark to the surface. It is the larvae's girdling of the tree beneath the bark that kills the tree. Several generations of this borer occur each year.

References: 193, 217, 234, 236, 361, 905

A. A gall in an aspen twig caused by the larva of *Saperda inornata.* The gall is 2 years old; note the brown egg niche and round hole. The hole was made by a woodpecker.

B, C. Twig galls in western cottonwood, *Populus fremontii.* In C, a gall has been cut open, exposing the larva of *S. populnea.*

D. Larval injury by the cottonwood borer, *Plectrodera scalator,* to the tap root of a cottonwood sapling. (Courtesy J. D. Solomon, U.S. Forest Service.)

E. An adult roundheaded appletree borer, *S. candida*; about 24 mm long.

F. A larva of a roundheaded appletree borer in cotoneaster; about 24 mm long.

G. The redheaded ash borer, *Neoclytus acuminatus.* [Cerambycidae]

H. A newly emerged *S. populnea* adult.

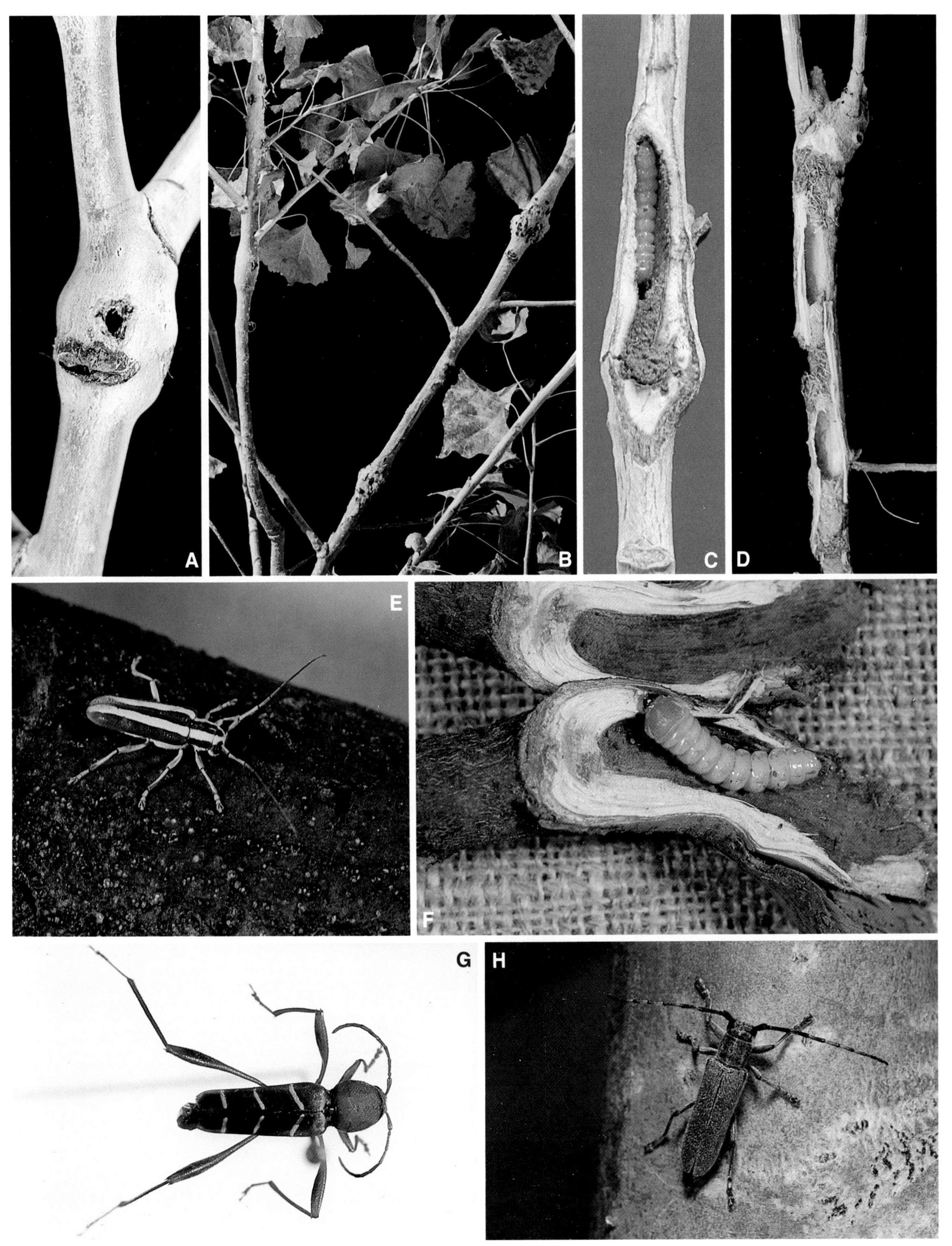
A
B
C
D
E
F
G
H

Root-feeding Beetles (Plate 132)

There are relatively few insects that feed on tree roots. Some, like the cicada (Plate 236), cause insignificant root injury. Root borers, however, may slowly bring about the demise of what was once a healthy tree. The prionid root borers are capable of killing trees and other woody plants. They are also among the largest of the North American beetles. Three economically important species in the United States and Canada have similar life cycles and look much alike. They normally select mature specimen trees or shrubs or those growing in open stands or in heavily used recreation areas. The most severe damage and greatest frequency of injury occurs to host trees growing in light or sandy soils. The first symptoms of decline resemble nutrient deficiency: leaves may be smaller and lighter in color. Symptoms progress to a thin leaf crown and twig dieback, and finally to the death of the entire tree. All injury is caused by the larvae; adults do not feed. Only by excavating and examining the roots can the cause of visible injury be confirmed.

Prionus imbricornis (Linnaeus) has been given the common name tilehorned prionus. Larvae feed mostly in buttress roots and other large roots close to the root crown. Their tunnels may be up to 15 mm in diameter. Several larvae may be found feeding independently in a small section of a large buttress root. Their feeding period lasts 3–5 years, during which time roots may be so riddled that individual tunnels are obscure or nonexistent. The area where they have been feeding is filled with dark, decomposing frass and wood chips. Any solid wood quickly decomposes.

In Georgia and North Carolina adults (D) emerge in June and early July. The female presses her eggs into the soil, usually in clusters about the base of a tree. She is capable of producing several hundred eggs, most of which will be laid near the roots of a single tree. Such a concentration of eggs may make the selected tree a candidate for destruction. The nocturnal adults are attracted to incandescent light. Eggs hatch by early August and the larvae make their way to one of the primary roots. Young larvae first feed on bark and later enter the xylem. If they have entered a small root, it is soon girdled or severed and often completely devoured. They then move through the soil in search of another root. Some of the galleries in large roots are filled with wood chips and frass. The walls of root tunnels quickly turn dark brown to black, probably because of the action of decay organisms (B). Fully grown larvae may be 7–9 cm long (C). This larva differs from other cerambycids that feed on living trees because of their large size and the presence of tiny thoracic legs. Pupation occurs in earthen cells within 10 cm of the soil surface.

The tilehorned prionus attacks the roots of a wide range of hardwood trees, particularly species of *Malus, Carya, Prunus, Castanea, Maclura, Populus, Tilia,* and *Pyrus*. It is found from the prairie states east to the Atlantic states and from Ontario south to Florida and Texas.

P. laticollis (Drury) has been given the common name broadnecked root borer. Its biology, behavior, and the injury it causes resemble those of the tilehorned prionus, but they have a greater biotic potential. Some females have been known to lay more than 1200 eggs. They lay them singly but deposit several at the same general site in the soil or on partially exposed roots. The eggs are oval and light yellow; in the longest dimension they measure about 3.5 mm. It has been estimated that larval development requires 3–4 years. Adult males fly and are attracted to light at night; females have not been observed to fly. Adults probably do not feed. A fly, *Helicobia rapax* (Walker), is a common larval parasite.

The broadnecked prionus is found in southeastern Canada south to Georgia and west to the Mississippi River, with spotty records in Minnesota, Arkansas, and Oklahoma. Hosts include many trees and shrubs such as high bush blueberry and species of *Malus, Tilia, Quercus, Populus, Prunus, Cornus, Acer,* and *Rhododendron*.

The California prionus, *P. californicus* Motschulsky, is distributed in the Southwest, in all of the Pacific states, and in British Columbia. The larvae feed on a wide range of hardwood trees and shrubs such as species of *Malus, Eucalyptus, Prunus, Juglans, Cydonia, Chaenomeles, Quercus, Populus,* and *Arbutus*. Eggs about 4 mm long are deposited 25–35 mm below the soil surface. Larvae tunnel into live roots, sometimes making tunnels in the xylem; at other times they make tunnels in a spiral shape in both bark and xylem, which results in a girdle that brings about a more rapid demise of the tree or shrub. Pupation tunnels, made in the soil, may be 1 meter long. Pupation takes place in May; in New Mexico beetles emerge in June or early July. Larvae, when fully grown, measure up to 7.5 cm long and 15 mm in diameter. The Indians of the Southwest consider them to be food delicacies. Adult females are 38–55 mm long and may lay 200–300 eggs. One generation requires 3–5 years.

The whitefringed beetle is a name given to three weevil species in the genus *Graphognathus*. They were accidentally imported from South America about 1935. At present, they are a serious agricultural pest in the Gulf states, and their distribution is largely limited to the southeast quadrant of the United States. They can be expected to survive as far north as New York City and in a line west to southern Iowa. Natural advance into noninfested areas will be slow because the adult weevils are unable to fly. The adult, a snout beetle, is about 12 mm long, primarily gray with a light-colored band along the outer margin of each wing cover—the feature responsible for its common name. There are no known males; each individual is a female capable of producing viable eggs without mating. The whitefringed beetle has one generation each year and overwinters as a larva in the soil. Pupae are most numerous in early July, and adult emergence is at its peak in mid-July. Eggs are laid in masses of up to 60 eggs, usually placed on objects in contact with the soil. The larva is legless and C-shaped; when full grown it is about 12 mm long.

These beetles have a wide host range and have gained greatest notoriety from feeding on vegetables, soybeans, alfalfa, peanuts, and tobacco. They can cause serious injury in an ornamental nursery and on specimen trees and shrubs. Adults and larvae both cause injury. Adults notch foliage and feed at night. Larval injury to roots is of most concern, largely because it often goes undetected until too late to take preventative measures. On woody plants larvae feed on the soft suberized bark of roots (F). Damage is also similar to that shown in Plate 113. Some of the most common woody plants attacked are species of *Carya, Prunus, Salix,* and *Platanus*. Expect them to feed on most of our common woody ornamental plants. The best-known natural enemies of whitefringed beetles are nematodes in the genus *Neoaplectana*.

References: 46, 57, 193, 224, 227, 278, 474b, 537, 601, 740, 881

A. A pecan tree sparsely foliated because of the feeding injury to roots caused by tilehorned prionus larvae. Normal trees in background. (Courtesy J. A. Payne, USDA.)

B, C. Larvae of the tilehorned prionus: B, a larva in its root tunnel; C, a larva excised from a root; length about 7 cm. Note the thoracic legs. (Courtesy J. A. Payne, USDA.)

D. Adult tilehorned prionus beetles. The female (*right*) is about 40 mm long. (Courtesy J. A. Payne, USDA.)

E. A cottonwood borer in the root of a sapling cottonwood tree. (See also Plates 130–131.) (Courtesy J. D. Solomon, U.S. Forest Service.)

F. A sycamore root damaged by the whitefringed beetle. Sometimes all of the bark is removed from roots. (Courtesy J. D. Solomon, U.S. Forest Service.)

A
B
C
D
E
F

Trunk Borers (Plate 133)

The cerambycid borer *Goes pulverulentus* (Haldemann) is colloquially called the living beech borer. Its larvae are trunk-boring pests of such hardwoods as oaks, elm, sycamore, and wild cherry. They occur in the entire eastern half of the United States and Canada and are found in prairie states where hardwood trees have been brought in for shade, shelterbelt, and landscape purposes.

The primary injury is caused by larvae as they feed in sap and heartwood, creating large open galleries with an elongated entrance hole on the bark surface. Attacks are most common in trunks 8–12 cm in diameter, but they also occur in larger trunks and branches. A tree that has been attacked is likely to be reinfested year after year. Repeated attacks structurally weaken the tree, and the scarred bark is open to attack by other borer species such as carpenterworms (Plate 120) and wood decay fungi (see Sinclair et al., *Diseases of Trees and Shrubs*).

The adult (A) has a mottled brownish rusty color, although the mottling is not discernible in the photograph. Adults emerge in the spring over a 2-month period, beginning in late April in Mississippi and early June in northern states and Ontario. The adult feeds on the leaf midrib and the bark of twigs of hardwood trees, occasionally killing the apical end of injured twigs. Beetles live for about 4 weeks, a relatively long life for adult insects. They are reportedly attracted to light traps that utilize low-wattage incandescent bulbs. When the female is ready to lay eggs, she chews irregular oval, craterlike niches in the bark (B) about 5–7 mm in diameter and deposits a single egg near the center of each depression, normally in contact with the sapwood. After depositing the egg, the female covers it and the adjacent area with a brownish pastelike secretion that hardens. Typically there is a cluster of egg niches on the trunk that have been produced by more than one female. New egg niche clusters are produced every year by a new brood of females.

When the egg (see Plate 134D for an egg of similar appearance) hatches, the larva bores directly into the sapwood, which usually produces leakage around and below the niches, staining the bark. As the larva grows, the entrance hole is enlarged and the niche symptom is lost. As the larva nears maturity, it packs the entrance hole with masses of excelsiorlike wood fibers (D). Larvae are white to cream-colored, distinctly segmented and nearly cylindrical, with the anterior end slightly larger (a similar larva is illustrated in Plate 131F). Because they are legless, their movement is restricted to their galleries. Upon hatching, the larva is about 4 mm long and may reach up to 30 mm when development is complete. The life cycle generally requires 3 years, but a few individuals may require 4–5 years.

Population control is carried out by several biological factors. Larvae are frequently cannibalistic. When their galleries cross, a battle may develop, with only one combatant surviving. Carpenter ants (Plate 239) attracted to fermenting sap frequently chew away the bark around individual egg niches, killing the egg or young larva or both. Woodpeckers are primary predators. There are no known parasites. The shape of the tunnel makes it easy for a person to kill the larva by forcing a soft wire into the gallery, impaling it.

The red oak borer, *Enaphalodes rufulus* (Haldeman) (not illustrated), is a related species that occurs in the same geographic region as *G. pulverulentus*. Its host range is limited to the genus *Quercus*, especially *Q. rubra*, *Q. velutina* (black oak), and *Q. coccinea* (scarlet oak).

Although *E. rufulus* resembles *G. pulverulentus* in color and size of larva and adult, here we stress differences. The *E. rufulus* female places her eggs in bark cracks or under lichen patches. Newly hatched larvae bore into the corky bark and feed in the phloem and sapwood the first year, expelling finely textured frass that accumulates at the entrance hole and on adjacent bark. In the second year, when the larva approaches maturity, the tunnel may reach 15–25 cm in length and 12 mm in diameter. The entrance hole usually doubles as the exit hole. Adults are attracted to traps baited with tupelo honey or molasses. The life cycle is completed in 2 years. Adults emerge from mid-June through August but only in odd-numbered years. Studies by Galford (279a) show that nitidulids (sapbeetles), ants, and woodpeckers are important natural predators.

G. debilis LeConte has acquired the common name oak branch borer. Its distribution includes all of the eastern United States and Canada where deciduous oaks grow. The borers' damage is limited to small branches and terminals; they cause gall-like swellings and cankerlike lesions in trees of all ages. Much of the injury is overlooked in urban areas as well as the forest.

The beetle looks much like the white oak borer, but it is smaller, ranging from 10 to 16 mm long, and more reddish. Mature larvae are 14–18 mm long and resemble those of other *Goes* species. The female chews widely separated egg niches in the bark. Larval galleries are not uniform. Some look like those illustrated in Plate 134C; others extend in both directions from the entrance. Gallery entrances are almost always kept tightly plugged with yellowish frass. The size of completed galleries varies, from 30 to 75 mm long and 4 to 6 mm in diameter. In Mississippi part of a brood completes its development in 3 years and the remainder in 4 years; development may require more than 4 years in the insects' northern range.

The hickory borer, *G. pulcher* (Haldemann), may be the most important wood borer of the genus *Carya*. Larvae tunnel in the sapwood and heartwood of young trees and cause serious damage along the entire trunk. Their attacks in large trees are restricted mostly to the branches and upper bole. The biology and injury caused by the hickory borer closely parallel that of *G. pulverulentus*. Adults of the two species are about the same size, but the hickory borer is more colorful because of a dark reddish brown band in a V-shape across the wing covers. Color characteristics in this beetle and others of the genus fade, making color differentiation between species unreliable. Expect to find the hickory borer wherever hickory and pecan grow.

The white oak borer, *G. tigrinus* (De Geer) (E), is considered to be among the most important pests of living white oaks. It is found in the eastern United States, including the states bordering the Mississippi River. Its hosts are limited to members of the white oak group, including overcup, burr, chestnut, and post oak. Young open-grown trees less than 15 cm in diameter are favored for attack. *G. tigrinus* may also feed in small-diameter limbs but is mostly in the stem within 8 meters of the ground. Once infested, a tree is likely to be reinfested.

Most of the injury is internal within the xylem core. Many external symptoms resemble those of other wood-boring beetles, for example, an oval egg niche about 5–8 mm in diameter on the exterior bark, ribbons of yellowish frass and excelsiorlike wood fibers at the entrance hole and in piles around the infested tree, and bark around the

---

A. A male (*left*) and female *Goes pulverulentus*; 18–28 mm long.

B. A cluster of egg niches chewed in the bark of a young oak tree by female *G. pulverulentus*. Each niche (*arrows*) is 5–7 mm in diameter and contains one egg. (Specimen courtesy J. D. Solomon, U.S. Forest Service.)

C. A longitudinal section of a young red oak tree cut to show larval galleries made by *G. pulverulentus*. The outlined gallery (*arrows*) shows the usual configuration of the work of a single larva. The upper hole cut through the bark is the adult exit hole; the lower hole is the site of larval entrance into the tree. Galleries are 11–17 cm long and up to 1.5 cm in diameter. (Specimen courtesy J. D. Solomon, U.S. Forest Service.)

D. The excelsiorlike mass of wood fibers mark the plugged entrance holes of *G. pulverulentus* larvae. Such plugs are put in place by the larva as it nears maturity. The open hole at the top is an adult exit hole.

E. An adult white oak borer, *G. tigrinus*. The base color is silver to tan, with two brown bands across the wing covers. Length is about 35 mm. [Cerambycidae]

F. Injury 2 and 3 years old to a red oak trunk caused by *G. pulverulentus*. Egg niches are less distinct than the enlarged larval entrance holes. (Specimen courtesy J. D. Solomon, U.S. Forest Service.)

G. The gallery entrance of the white oak borer *G. tigrinus* on a 7.5-cm diameter trunk of an overcup oak, *Quercus lyrata*. The two bark scars (*arrows*) above and to the right of the entrance hole mark borer holes made in previous years. (Specimen courtesy J. D. Solomon, U.S. Forest Service.)

H. An elongate entrance hole to the gallery of a white oak borer in an overcup oak. (Specimen courtesy J. D. Solomon, U.S. Forest Service.)

A
B
C
D
E
F
G
H

entrance hole discolored and wet from oozing sap. If the insect has developed successfully, there will be two holes in the bark, one above the other but separated by a distance of 12–18 cm. Callus tissue covers these holes soon after the adult emerges, and the scar appears as seen in panel G at the arrow points.

In the South adults emerge from late April through May; emergence is later in its northern range. In Michigan it is present into early August. Adults feed on twig bark and leaf petioles without producing significant damage. They are rarely observed in nature because of their cryptic and nocturnal habits. They deposit eggs singly in niches, like those described above, that are usually widely separated on the trunk or limbs. Newly hatched larvae bore directly into sapwood, later into heartwood. Larvae mature in 3–5 years and produce a gallery (Plate 134C) that may be up to 23 cm long and 1.5 cm in diameter. When mature, larvae are 23–32 mm long, whitish, legless, and deeply segmented (Figure 95). The gallery is kept clear except for a frass plug placed there by the larva before pupation. There is never more than one larva per gallery.

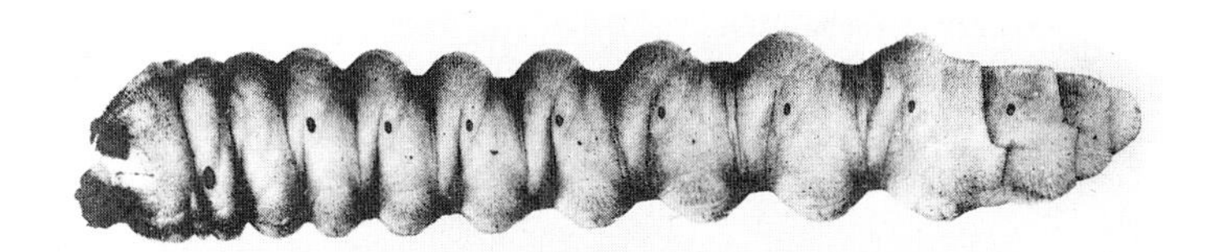

Figure 95. A full-grown larva of the white oak borer, *Goes tigrinus*. Lateral view. (From Craighead, ref. 131a.)

Woodpeckers are probably the only predators of this borer, and they actively seek these larvae as food in the dormant season. There are no reported parasites, although parasites are known in other wood borers.

References: 184, 279, 279a, 344, 433, 727, 729, 730, 739, 742

Wood-boring Beetles (Plate 134)

The poplar borer, *Saperda calcarata* Say, is a pest of all species of live trees in the genus *Populus*. The insect occurs wherever poplars grow, including trees planted in the prairie states and the Canadian provinces. Some forest entomologists consider it to be the most destructive insect enemy of poplar in the Rocky Mountain region. It is probably of equal importance in the lower Mississippi valley and delta.

Injury is caused by both the adult and the larva, but adult feeding and egg-laying injury is inconsequential. The feeding and boring of the larva in wood (G) weakens the tree structurally and makes large limbs and the trunk vulnerable to wind breakage (B) and attack by other borers and decayed fungi (see Sinclair et al., *Diseases of Trees and Shrubs*). Specimen trees or any trees growing in open areas are more heavily infested than those in dense stands. Trees less than 5 cm in diameter are seldom attacked.

The most discernible injury symptoms are black scars on the trunk, entrance/exit holes in the bark, wet areas caused by sap that exudes around the holes, and light-colored fibrous frass lodged in bark crevices and piled at the base of the tree below the holes. Small trees may be killed when young larvae feed in the phloem-sapwood interface. All other injury is hidden under the bark. The adult is a robust grayish to greenish beetle with antennae as long or longer than its body. Emergence of the adult varies with the geographic region, but usually occurs from early June through August. Adults feed on the bark of twigs. Egg laying presumably occurs throughout the summer. Usually a single egg is laid in an oblong hole (15 mm) gnawed in the bark of the trunk or occasionally in exposed roots at the collar. Females seem to concentrate their egg laying in a small area of bark rather than randomly over the bole, resulting in clusters of larvae in a small area (G).

Figure 96 (left). A full-grown larva of the poplar borer, *Saperda calcarata*, in its tunnel. (Courtesy J. D. Solomon, U.S. Forest Service.)

Figure 97 (right). A poplar limb honeycombed with tunnels made by the poplar borer. The pupal chamber is at the end of the tunnel above the coarse fiber plug. (Drawing by L. H. Joutel, in Felt, ref. 231.)

A. An adult poplar borer, *Saperda calcarata*; 21–30 mm long. The basic color may vary from gray to greenish. [Cerambycidae]

B. Poplar borer damage to a cottonwood tree; 75 mm in diameter. A longitudinal section exposes the galleries made by several borers. The broken stem is an example of wind breakage. (Specimen courtesy J. D. Solomon, U.S. Forest Service.)

C. A complete gallery in longitudinal section made by the white oak borer in overcup oak, *Quercus lyrata*. The top opening is the exit hole through which the adult emerges. The arrow points to a wood fiber plug in the gallery. (Specimen courtesy J. D. Solomon, U.S. Forest Service.)

D. Eggs of the poplar borer; about 3 mm long. (Specimen courtesy J. D. Solomon, U.S. Forest Service.)

E. A Columbian timber beetle gallery in red maple, *Acer rubrum*. The short horizontal chambers where larvae develop are called cradles. They are about 10 mm long. (Photo courtesy J. D. Solomon, U.S. Forest Service.)

F. Holes in the bark of a cottonwood tree that mark the sites of actively feeding poplar borers. (Specimen courtesy J. D. Solomon, U.S. Forest Service.)

G, H. A small cottonwood tree with a "window" cut in the stem to show how wood may become riddled with tunnels made by the poplar borer. Arrows point to excelsiorlike plugs made by larvae before pupation. H, typical wood and bark damage by woodpeckers, the major borer predators. (Specimen courtesy J. D. Solomon, U.S. Forest Service.)

A
B
C
D
E
F
G
H

Peterson (611) in Canada found that females preferred healthy trees for oviposition. Reinfestation of trees damaged by the borer is common.

After hatching, the first-instar larva feeds for several weeks in the phloem and sapwood, greatly expanding the entrance area immediately under the bark. Fully grown larvae are 30–35 mm long (Figure 96); their small, dorsoventrally flattened heads are distinctive. After 2–4 years of larval feeding in sap and heartwood (the time depends on climatic conditions), the tunnel may attain a length of 20–30 cm. The gallery is open and widest in the sapwood adjacent to the bark; it angles upward into the heartwood but does not branch. As the larva nears maturity, it prepares a pupal cell and blocks the tunnel with a plug of shredded wood fibers (Figure 97). At the end of the pupal stage the adult chews through the plug and emerges through the same hole that it used to enter the tree and to dispose of wood chips and frass while it was a larva. The life cycle may be completed in as few as 2 years in the South (Mississippi) and as many as 4 years in the Canadian prairie provinces.

Parasites, predators, and disease help keep infestations of *S. calcarata* in check, but they have not been studied. Some control can be obtained in specimen trees by pushing a wire into the active gallery to crush the larva.

The Columbian timber beetle, *Corthylus columbianus* Hopkins, is a native ambrosia beetle also described as a pin hole beetle because of the size of the hole cut in the bark by the adult beetle. It occurs throughout much of eastern United States. (See Figure 81, accompanying Plate 117, for an illustration of a beetle similar in size and shape.) Breeding occurs in vigorously growing deciduous trees such as maple, linden, birch, tuliptree, sycamore, and oak. All tunnel construction (E) is done by adult pairs. The male inoculates the tunnel with a yeast that grows on the walls. The yeast is food for both larva and adult. This beetle is rarely abundant enough to become a threat to shade trees. Two or three generations occur each year.

References: 1, 193, 236, 377, 547, 611, 660

Shoot and Twig Borers (Plate 135)

The most commonly encountered sawflies are those that feed on foliage; examples are pine sawflies and the mountain-ash sawfly. *Euura atra* (Jurine) is one of a small group of sawflies that either are shoot borers or cause galls on buds, leaves, or shoots of various species of willow. *E. atra* is a borer that attacks young stems and lateral shoots. The infested shoot usually dies before the end of the growing season. The following season some of the stem tissue dies, and if the fungus *Cytospora* sp. (see Sinclair et al., *Diseases of Trees and Shrubs*) is introduced, a young tree is not likely to survive.

This insect overwinters as a larva, usually in a transparent cocoon-like sack constructed in the pith of a shoot. Pupation occurs in the early spring, and adults emerge in late May or early June (in the Canadian prairie provinces and northern New York). By means of a sharp ovipositor, the female inserts white eggs into the pith of new shoots. The eggs hatch in mid-June, and the larva feeds in the pith (A). In the process of development, the larva tunnels through several centimeters of the pith and pushes frass and molted "skins" to one end of the tunnel. As it nears maturity, it prepares an exit hole (D) and then plugs it with plant fragments, frass, and webbing. By late September its preparation for winter is complete. Each gallery contains one larva. There is one generation each year.

Host plants are limited to willow, but some species are favored more than others: *Salix acutifolia, S. alba* var. *vitellina, S. babylonica, S. viminalis, S. purpurea, S. lapponum,* and *S. cinerea*. This sawfly is found from the Canadian Maritimes to Alberta, the bordering states, and New England. It is primarily a pest under nursery conditions. The downy woodpecker, *Picoides pubescens* (Linnaeus), is a major predator, but it frequently causes as much or more damage to the host as does the insect.

The willow shoot sawfly, *Janus abbreviatus* (Say), is a shoot-boring pest of willows and poplars such as *Salix nigra, S. babylonica, S. alba, Populus tremuloides, P. balsamifera,* and *P. deltoides*. The female sawfly causes injury during the process of egg laying; the larva, as it utilizes the plant as food. The female uses her ovipositor to girdle shoots with a ring of punctures. The shoot may begin to wilt in response to these punctures in 30 minutes, and within a day or two it turns black. By girdling the shoot, the female interrupts the plant's vascular system and gives her progeny a better chance for survival. After making the punctures, the female lays one or occasionally two eggs in the pith of the shoot below the girdle. Upon hatching, the larva tunnels and feeds toward the girdle for a few millimeters, then turns downward and tunnels for 7 cm, often the entire length of the shoot. As larvae feed, they must dispose of the undigestible plant tissue by packing the undigested material, called frass, in the older part of the tunnel. Each tunnel, or gallery, is occupied by a single larva. When fully grown, the larva chews a short passage into the wall of the tunnel to the outer bark, leaving intact a thin layer of shoot epidermis to hide the opening. This passage will later be the adult sawfly's exit hole. The larva retreats a short distance into its tunnel and constructs a thin cocoon in preparation for metamorphosis. It overwinters in this condition. In the insect's northern range there is but one generation per year; in Mississippi there are three. The first generation is completed by the end of June; the second is completed by the end of August; and the third-generation larva overwinters.

The European corn borer, *Ostrinia nubilalis* (Hübner), is an occasional pest of woody plants. The larvae bore into rapidly growing shoots and tunnel in the pithy center, often causing the death of the terminals. The corn borer can be a problem in production nurseries especially if corn or potatoes are grown nearby. In urban settings dahlias and chrysanthemums are prime hosts for corn borers, and second-generation borers from such flowers may attack almost any woody plant that produces a second flush of growth. Succulent shoots of linden, honeylocust, hibiscus, and catalpa, among others, harbor European corn borer larvae. Full-grown larvae are about 25 mm long; the body is flesh-colored and most body segments are marked wtih four dark spots. The larvae overwinter in the shoots.

The willow shoot sawfly is found in Canada from the maritimes to Manitoba south to Mississippi and Arkansas and in spotty locations in shelterbelt plantings of the prairie states.

Twig girdlers are longhorned beetles that were named for a peculiar habit of the female as she prepares the plant to receive her egg. To provide young larvae with a suitable habitat, she chews an encircling groove through both bark and outer woody tissue of a living twig or small branch. This action, of course, girdles the twig, but before the twig dies she lays an egg at its distal end. The egg hatches and the larva bores into inner bark, where it feeds for a short time; then it moves to the center of the twig, feeding toward the girdle. While this is going on, the twig dies and the dead leaves appear like a flag in the crown of the tree. In time, a gust of wind will cause the girdle-weakened twig to

---

A. A full-grown larva of the sawfly *Euura atra* in the pith of *Salix alba* var. *vitellina*. The larva is about 8 mm long.

B. The S-shaped larva of the willow shoot sawfly, *Janus abbreviatus,* in black willow; about 10 mm long. (Specimen courtesy J. D. Solomon, U.S. Forest Service.)

C. A dead *S. discolor* twig killed by the willow shoot sawfly. [Cephidae]

D. An emergence hole for the adult sawfly *E. atra* on a willow twig.

E. A young tepeguaje tree girdled by the huisache girdler *Oncideres pustulatus* in Kingsville, Texas. [Cerambycidae] (Courtesy J. E. Gillaspy, Texas A&I Univ.)

F. A mortally damaged hawthorn leader attacked by an *Oberea* stem borer. The center of the stem is hollowed out by the larva. The small holes are used as ports for ejecting frass.

G. An adult huisache girdler, *Oncideres pustulatus.*

H. A stem borer, *Oberea delongi*; about 12 mm long. Variable in color. The shape and form of this beetle is typical for most of the cerambycid stem borers.

A
B
C
D
E
F
G
H

break and fall to the ground. Most of the larval development occurs while the twig is on the ground.

Oncideres pustulatus (LeConte), the huisache girdler (see Plate 124 for another twig girdler), attacks mimosa, *Acacia* species, and several native trees, for example, huisache in the Southwest. In southeastern Texas *O. pustulatus* is a limiting factor in the use of mimosa as a landscape tree. The beetle (G) is slow moving and more likely to drop to the ground when disturbed than to fly. This insect confines its attacks to small branches and stems 20–40 mm in diameter.

Longhorned beetles in the genus *Oberea* are borers of twigs and small branches that also feed in small diameter stems. *Oberea delongi* Knull and *O. schaumii* LeConte are examples of borers that feed on twigs and branches of various species of poplar. *O. delongi* is more likely to attack small branches and terminals, which results in a swollen, sometimes gall-like appearance. Infested terminals occasionally become crooked or break at the point of the insect's entrance. Adult emergence, which begins in April in Mississippi, is accomplished by the insect's cutting a typical circular emergence hole through the bark. Beetles feed extensively on the midrib and lateral veins of leaves, causing conspicuous damage. They lay their eggs in notches they cut in the bark of branches or stems that have a diameter of 5–12 mm. The larvae tunnel down the center of the stem. There is one generation per year. The range of *O. delongi* is not well identified but probably occurs from Ohio southward to the Gulf states.

O. schaumii LeConte also feeds on a wide range of *Populus* species and is found throughout a greater geographic range than *O. delongi,* approximately the eastern half of North America. The adult is about 15 mm long; its color varies, but the thorax is generally orange to yellow and the wing covers are dark, usually black. Oviposition begins in late June in Wisconsin and late April in Mississippi. The female selects small stems or lateral branches and deposits a single egg in a roughened, rectangular niche. The newly hatched larva bores directly into the center of the branch. Before it has become fully grown (about 12–15 mm long), it chews a tunnel about 15 cm long. One or more small holes are cut through the bark from the main tunnel and are used as ports for ejecting frass. Its life cycle requires 1–2 years in the South and as many as 4 years in the most northern part of its range.

O. myops Haldeman, sometimes called the rhododendron stem borer or the azalea stem borer, feeds on the leaves, lower twigs, stems, and roots of rhododendron, azalea, laurel, blueberry, and *Oxydendrum arboreum*. Adults feed on the underside of leaves on the midvein, causing the earliest damage to hosts. Callus tissue forms around the injured vein, and the leaf curls abnormally. In late June to early July female beetles lay their eggs in new shoots several inches below the bud. Upon hatching, the larva bores down the twig and is likely to overwinter in the stem. The following year it continues to bore downward and overwinters in the roots. Along the larva's path to the root one may find clusters of small holes in the bark where frass has been expelled. Sometimes the frass will accumulate in bark crevices. During the larva's second year quite a bit of "sawdust" may be extruded, accumulating on the soil at the base of the host plant. Signs of larval injury are frass and "sawdust" and broken or dead terminal shoots and twigs. Full-grown larvae are creamy white and about 25 mm long.

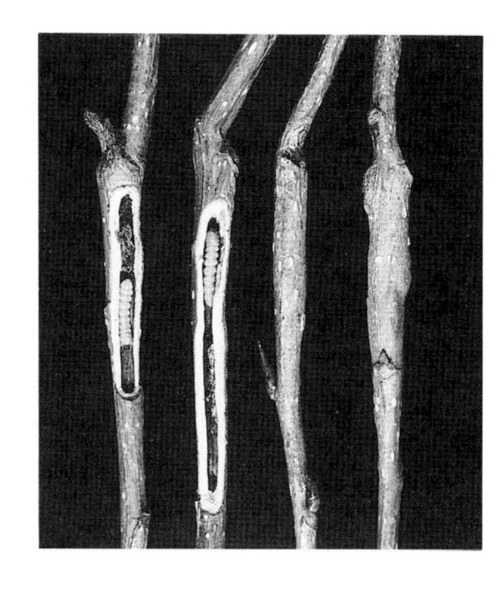

Figure 98. Poplar twigs showing larvae of the twig girdler *Oberea delongi* and the injury they cause. (Courtesy J. D. Solomon, U.S. Forest Service.)

O. myops is a native species. It is likely to be found throughout the natural range of its host plants. The beetles are slender and elongate and look much like adult dogwood twig borers (see Plate 123B).

See Plate 92 for the boxelder twig borer.

References: 433, 461, 473, 578, 660, 735, 870

Acacia Psyllid (Plate 136)

The acacia psyllid, *Acizzia* (=*Psylla*) *uncatoides* (Ferris & Klyver), was discovered in California in 1954, apparently after its accidental introduction from Australia. It now occurs in Hawaii and Arizona as well, where it causes a chlorosis and dieback of the branch tips on susceptible species of *Acacia* and *Albizia*. It is also a nuisance pest, for the adults are attracted to lights at night. When infested landscape plants extend over patios or other outdoor living areas, the white pellets of solidified honeydew excreted by the immature psyllids create a nuisance.

The adult psyllid (E) is a small leafhopperlike insect that lays its yellow eggs (D) in clusters on the tips of its host plants. The nymphs (F) feed by sucking plant juices. On especially susceptible plants, thousands of nymphs, adults, and eggs have been recorded from a single branch tip.

Reproduction takes place throughout the year, but populations in California are highest in May. At least eight generations of this insect occur each year in the coastal central part of the state.

Evaluation of 31 species of *Acacia* for their susceptibility to the psyllid in northern California revealed the following:

Immune or highly resistant to infestation include *A. albida, A. armata, A. aspera, A. baileyana, A. cardiophylla, A. collettiodes, A. craspedocarpa, A. cultiformis, A. cunninghami, A. cyanophylla, A. dealbata, A. decurrens, A. gerardii, A. giraffae, A. iteaphylla, A. karoo, A. mearnsii, A. obtusata, A. parvissima, A. pendula, A. podalyriifolia, A. robusta, A. saligna, A. spectobilis,* and *A. triptera*.

Moderately resistant to infestation are *A. cyclops, A. implexa, A. longifolia,* and *A. penninervis*.

A. melanoxylon and *A. retinodes* are highly susceptible.

The psyllid problem in California is much less critical today than it was soon after the introduction of the insect. The change may be explained, in part, by the success of a biological control effort consisting of the introduction from Australia and subsequent colonization and release of a tiny black lady beetle, *Diomus pumilio* Weise. Another lady beetle, *Harmonia conformis* (Boisduval), was introduced in California but did not become established. It did become established in Hawaii, however, where it reportedly has brought the psyllid under control on koa, *Acacia koa,* trees.

References: 435, 438, 558, 618

A. An infestation of the acacia psyllid, *Acizzia uncatoides,* on leaves, stem, and flowers.
B. An infestation of the acacia psyllid on *Acacia longifolia*. Note the dead tip (*arrow*) killed by psyllids.
C. A heavy nymphal infestation on a stem of a very susceptible acacia.
D. Clusters of eggs of the acacia psyllid.
E. An adult acacia psyllid.
F. A nymph of the acacia psyllid.

A
B
C
D
E
F

Psyllids (Plate 137)

Insects in the family Psyllidae, like other homopterans, have piercing-sucking mouthparts and wings that are held in an inverted V shape over the abdomen when the insect is at rest. Most psyllids are host specific, if not to the species level then to the genus. This fact makes identification easier. Those who need to have a psyllid identified by a taxonomist should provide the specialist with the specific name of the host plant. Also, whenever possible, one should submit both adults and nymphs. Some psyllids, such as the pear psylla, *Cacopsylla* (=*Psylla*) *pyricola* Foerster, feed on leaves; others, such as boxwood psyllid, *Cacopsylla* (=*Psylla*) *buxi* (Linnaeus), feed on developing buds and new leaves. The hackberry psyllids, as well as a number of others, cause galls on leaves and in buds (see Plates 217–219).

The pear psylla, *C. pyricola,* is the most serious pest of fruit pears in the United States and may be found wherever fruit pear trees grow. These insects feed by ingesting sap from the leaf phloem. When populations are high, leaves may be covered with honeydew excrement and subsequently blackened by the growth of sooty mold fungi on the honeydew. There may be three to five generations of the psylla each growing season. Some of the literature states that all varieties and species of *Pyrus* are attacked, but we question the validity of the statement with regard to ornamental and flowering pears. For example, we have never seen pear psylla on Callery pear. Both the psylla and all species of pear currently used for fruit and amenity purposes are of European and Asian origin. Though there is some uncertainty about whether ornamental pears host psyllids, the historical background should alert horticulturalists to the potential problem.

The boxwood psyllid is very common in the more temperate parts of the United States. It is highly conspicuous, because of the leaf-cupping effect it produces on host bushes, but it is not as destructive as some pests. Leaf cupping results from injury done to tissue in rapidly growing foliage.

Boxwood psyllids overwinter as very small orange spindle-shaped eggs. The eggs are inserted between the bud scales of the host bush by females during early summer. Only the tips of the eggs protrude past the edge of the bud scale. As soon as buds begin to open in early spring, the eggs hatch and nymphs emerge to begin feeding. Studies have shown that embryonic development is complete before winter and that nymphs overwinter inside the eggshell.

As nymphs develop on the expanding foliage, leaves become cupped, enclosing several nymphs in a pocket of leaves. The nymph produces waxy filamentous secretions that partly cover its body and provide it with additional protection. By late May and June adults appear. The adults are greenish jumping plant lice. They spring from the leaves to become airborne and sustain flight with their wings.

Adults may occasionally bite humans, but the bites are not serious. In severe infestations thousands of adults may occur on relatively few plants at the time of oviposition. They lay their eggs under the bud scales. Only one generation occurs annually. The American boxwood is most frequently and severely infested. The English boxwood is less severely attacked.

There are a number of other psyllids that feed on ornamental trees and shrubs. *Calophya flavida* Schwarz occurs in the eastern United States on smooth sumac. *C. flavida* nymphs spend the winter on terminal twigs of this tree (G). The nymphs are small, flat, and black, although they have a white fringe. *C. nigripennis* Riley, another eastern species, may be found from Pennsylvania south to Alabama. *C. californica* Schwarz, known in California (H), is also a pest of sumac, *Rhus ovata*. *C. triozomima* Schwarz occurs on *R. trilobata* in Arizona, Colorado, and California, and on *R. aromatica* in Missouri. This insect has a yellow abdomen.

Willow serves as host to more species in the genus *Psylla* than any other plant group in North America. The records, however, are primarily from the western United States. Among these species are *P. alba* Crawford and *P. americana* Crawford. Both species are generally whitish to light green. Adults overwinter and lay their eggs on foliage or catkins in the spring. *P. annulata* Fitch is an eastern species found on the twigs and leaves of sugar maple.

Crawford (134) published a monograph on the Psyllidae in 1914, listing some 29 genera and about 161 species of psyllids in the Americas. Many, if not most, of these insects are known to feed on ornamental trees and shrubs, including pine, spruce, arborvitae, sumac, willow, mimosa, boxelder, redbay, ceanothus, currant, raspberry, alder, birch, pear, hazel, laurel, maple, apple, and plum.

References: 134, 217, 404, 405, 682

A. Typical cupping of leaves results from feeding by the boxwood psyllid.
B. Boxwood psyllid nymphs and their white, waxy, and cottony filaments in cupped leaves.
C. A newly eclosed adult boxwood psyllid. Note the old nymphal "skin." (Courtesy Donald Specker, Cornell Univ.)
D. The typical appearance of psyllid nymphs and their wax secretions.
E. Psyllid nymphs and their waxy deposits on a dogwood leaf.
F. A psyllid nymph on dogwood. The pest shown is color adapted to the leaf surface.
G. Overwintering nymphs of *Calophya flavida* on sumac.
H. The psyllid *Calophya californica* on *Rhus ovata*.
I. A psyllid nymph outside its protective waxy web.
J. The dark, flattened appearance of overwintering *Calophya flavida* nymphs. Note the fringe of white filaments.

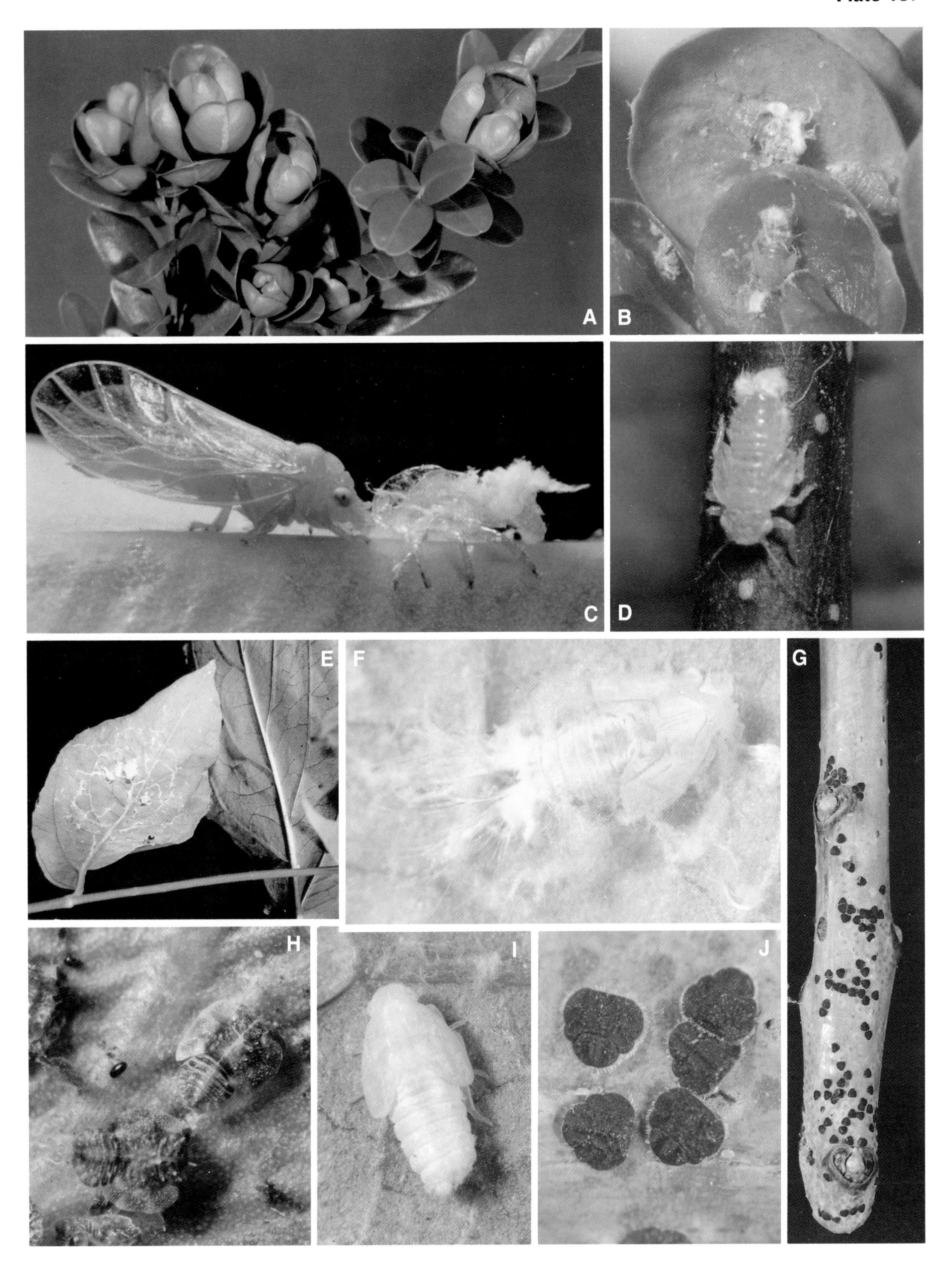
A
B
C
D
E
F
G
H
I
J

Tuliptree Aphid and Other Foliage-feeding Aphids (Plate 138)

Free-living aphids that occur in colonies on the foliage of shade trees and other ornamental plants are common, persistent, and troublesome pests. They vary greatly in color from yellow, green, pink, or red to purple, brown, or black. There is also considerable variation in size between species. In small numbers their effect may be innocuous; in some cases, even large numbers may not be seriously injurious to plants. Some aphids are important, not because of their direct feeding on the plant, but because they act as vectors, or carriers, of viral diseases. The winged adult aphid has considerable mobility, and the aphid family has the highest reproductive potential of all insects. The adult's excreted honeydew droplets are annoying when they cover cars and furniture in outdoor living areas. The subsequent growth of sooty mold fungi on the honeydew is unsightly, and sometimes injurious, whether on leaves or on people's possessions.

The types and habits of various species of aphids differ widely. Some infest only one kind of host, others have alternate hosts, and still others have a wide variety of hosts. A list of aphid species that attack shade trees alone would be extensive and would include at least one species for each of most of our common trees.

Aphids may occur throughout the growing season, but most species are very seasonal. The tuliptree aphid, *Illinoia* (=*Macrosiphum*) *liriodendri* (Monell), occurs wherever the tuliptree, *Liriodendron tulipifera,* grows (A). In Illinois this aphid is found from mid-June through most of October. Large populations often develop by late summer. Some leaves, particularly those in the outer canopy, turn brown and drop prematurely. Tuliptree damage by these aphids resembles that shown in panel A in Plate 173. If tuliptrees used as street trees become infested, a sticky mess from honeydew droplets will foul sidewalks, and the cars parked under the trees will be spotted. On hawthorn and flowering crabapple, the apple grain aphid, *Rhopalosiphum fitchii* (Sanderson), is present only in the very early spring. It then spends the remainder of the season on grain crops. The apple aphid, *Aphis pomi* De Geer, does not migrate but continues to feed on apple from early spring until new growth ceases in midsummer. The leaves of willow tend to be infested with aphids from spring to late summer.

In most cases the piercing-sucking feeding activities of aphids produce inconspicuous signs of damage. Leaves may, however, be slowly depleted of plant food, which gradually weakens the tree.

Injury to a host by the walnut aphid, *Chromaphis juglandicola* (Kaltenbach), is an example of the type of conspicuous injury aphids can cause. Like many free-living aphids, walnut aphids overwinter as eggs on host twigs (E) and buds. The black eggs usually hatch in early spring when buds begin to open. Feeding, which occurs primarily on the undersides of leaves, results in stunted leaves, nuts, and twigs. This species is an occasional pest of Persian (English) walnuts in California, where the wasp parasite *Trioxys* sp. usually keeps the population low. This aphid is likely to occur wherever Persian walnuts are grown; in the East its damage is minimal.

The walnut aphid was introduced from Europe in 1909. It is a small yellow insect, about 1.5 mm long at maturity, and lives in colonies. Both males and females are produced as fall approaches. After mating, females lay eggs that overwinter. Egg-laying females are wingless and can be distinguished from other aphids by two dark bands across their backs, one much broader than the other.

Injury by the black pecan aphid, *Melanocallis caryaefoliae* (Davis), is shown in panels C and D. The adult aphids have a series of large black tubercles on the back and sides; they are not found in crowded colonies. This species occurs wherever hickory or pecan are grown. A related southern species, *Tinocallis* (=*Sarucallis*) *kahawaluokalani* Kirkaldy, occurs in large colonies on the twigs and leaves of crape myrtle.

In hickory wood lots of the North as well as in the pecan orchards of the South it is common to see extensive patches of foliage marred by large yellow angular blotches (D). This is the first symptom of black pecan aphid injury. The yellow areas eventually turn brown as the season advances. As aphids increase in number, leaf damage increases, resulting in premature leaf drop. Some entomologists speculate that discoloration of the leaf tissue is caused by the aphids after they inject toxic salivary fluids into the leaf at the feeding site. The aphids feed on both sides of the leaves and seem to prefer the shaded inner parts of the tree. In the South as many as 15 generations occur during a growing season. In the fall adult females lay their eggs on twigs and bark, and the aphids overwinter as eggs.

Calaphis juglandis (Goeze), sometimes called the duskyveined aphid, is another European species presently limited to Persian walnut in the Pacific Coast states. It lives in closely packed colonies attached to the midvein on the underside of leaves (F). Several generations are produced each year. Injury to the plant is similar to that by the walnut aphid.

References: 388, 403, 463, 464, 533

A. The underside of a tuliptree leaf with a developing colony of *Illinoia liriodendri,* the tuliptree aphid.
B. Tuliptree aphid nymphs and wingless adult females.
C. Moderate injury to hickory leaves caused by the black pecan aphid.
D. An early stage of hickory leaf injury caused by the black pecan aphid.
E. Black eggs of the walnut aphid, *Chromaphis juglandicola,* matted on the bark of Persian walnut twigs.
F. A colony of wingless aphids, *Calaphis juglandis,* feeding from the midvein of a Persian walnut leaf. (Courtesy Helmut Riedl, Mid-Columbia Experiment Station, Hood River, Oregon.)

A
B
C
D
E
F

California Laurel Aphid and Podocarpus Aphid (Plate 139)

The California laurel aphid (A) is sometimes mistaken for an immature whitefly. Careful examination is required to avoid this error. In the western hemisphere *Euthoracaphis umbellulariae* (Essig) is known only from California. It occurs primarily on California laurel, *Umbellularia californica,* but has been found in small colonies on common, or American, sassafras, *Sassafras variifolium*; camphor tree, *Cinnamomum camphora*; and avocado, *Persea americana*.

In all its life forms this insect is blunt and robust. The wingless (egg-laying form) females are covered with white waxy material. The winged aphids are jet black. This insect occurs in all its stages on the undersides of the older leaves of the host tree, often in neat rows (B).

The wingless female looks much like a juvenile whitefly. It is oval and appears to have a fringe of waxy filaments or plates (Plate 152). Apparently, these females do not move once they have settled and begin to feed. Individuals become fixed to the leaf surface and produce white cottony wax, which envelops their bodies.

The California laurel aphid excretes large quantities of honeydew, on which black sooty mold grows. Often, the honeydew drops to leaves below, and these are also blackened. No direct damage to the foliage of California laurel has been noticed as a result of the feeding of *E. umbellulariae*.

In central California migration from tree to tree by the winged form of this aphid occurs from spring to fall. Wingless forms may be found on the leaves throughout the year, evidence of numerous overlapping developmental stages. The number of generations that occur each year is unknown.

Podocarpus species are evergreen trees native to Australia and Asia. In the United States one of the insects that feed on podocarpus is an aphid, *Neophyllaphis podocarpi* Takahashi. This aphid probably was introduced into the United States when the first podocarpus plants were imported. It is presently found in Hawaii, California, Texas, Florida, Mississippi, and Louisiana. In the future the aphid very likely will be found wherever podocarpus is grown. At least twelve *Podocarpus* species are grown as ornamentals in the United States; all are susceptible to this aphid.

The aphid causes a stunting and curling of the new terminal leaves. On older leaves a hard resinlike substance may appear at feeding sites (E). The aphids feed primarily on the undersides of leaves, on young twigs, and on fruit stems (D). They produce honeydew that gives the podocarpus fruit and leaves a bluish white bloom. In time, sooty mold fungi grow in the honeydew, which imparts a dusty bluish appearance to affected plant parts.

The biology of this aphid under conditions in the United States is unknown. Several generations are produced each year. The adult aphid is about 1.3 mm long, and under its waxy bloom the body is reddish purple.

References: 164, 215, 365, 668

A. Light infestation of the California laurel aphid, *Euthoracaphis umbellulariae,* on the undersides of leaves of the California laurel. [Aphididae]

B. Parasitized California laurel aphids. They are characteristically lined up in rows on either side of a vein.

C. California laurel aphids on a camphor tree leaf.

D. Clusters of podocarpus aphids, *Neophyllaphis podocarpi,* on shoots and fruit stems of *Podocarpus macrophylla*.

E. Resinous substance on leaves at the feeding sites of the podocarpus aphid.

A
B
C
D
E

Aphids of Beech, Birch, and Apple (Plate 140)

Phyllaphis fagi (Linnaeus) is an aphid that appears to have but a single host: beech. It has no generally accepted common name, although some authors have called it the woolly beech leaf aphid because of the quantity of waxy woollike filaments that extrude from its body. The specimens shown in panels A and B are on *Fagus sylvatica* and were collected during July. These aphids are gregarious and are found almost exclusively on the undersides of leaves. Large numbers of cast "skins" may be found attached to leaf hairs, adding to the whitish appearance of the leaf.

P. fagi is widely distributed in the East, including eastern Canada, and it also occurs in Kansas, Utah, all of the Pacific Coast states, and British Columbia. In general, it is able to survive wherever beech is grown. Beech trees in the United States seem capable of maintaining huge populations of this aphid year after year with little observable tree injury. In Europe, however, where this aphid originated, a great deal of harm to beech has been reported. On specimen shade trees, a heavy infestation could probably not be tolerated, because of the tremendous quantity of honeydew produced by the aphids. Clear droplets of fresh honeydew are shown in panel B. The fluid is very sweet and sticky; it is the same substance that, when dried, was called manna in the time of Moses. Honeydew is an attractive food for bees, wasps, and ants, and it also serves as food for certain fungi, commonly called sooty molds. These molds are black and are unsightly when found on ornamental plants (Plate 142).

Another woolly aphid common to beech is the beech blight aphid, *Grylloprociphilus imbricator* (Fitch), which feeds primarily on the bark of twigs and small branches. The beech blight aphid looks much like the woolly alder aphid shown in Plate 144, although greater quantities of snow white down are attached to *G. imbricator*. It occurs from Maine to Florida and west to Kansas, and alternates between beech and bald cypress.

Hamamelistes spinosus Shimer also has no accepted common name, but it might well be called the spiny witch-hazel gall aphid. This aphid has a complex and most interesting life history; it alternates between two host plants of ornamental interest: witch-hazel (*Hamamelis* spp.) and birch (*Betula* spp.).

H. spinosus may survive the winter in two ways: as an egg, laid on witch-hazel twigs, or as a hibernating female, sometimes called a "pupa," on a birch tree. In the spring, after winter dormancy, the egg hatches. The resulting aphid develops into what is called a stem mother. Its feeding results in the formation of a bud gall on the host (Figure 99). Plant buds thus affected become green to reddish oblong galls that are about 18 mm long and that are covered with long, sharp spines. Each gall is hollow and contains numerous reddish young aphids, the offspring of the stem mother. An exit hole at the base of the gall allows the mature aphids to emerge and fly to their secondary host, birch.

While spring development occurs on witch-hazel, changes take place in the overwintering aphid on birch. The hibernating female, or "pupa," becomes active at about the time birch leaf buds are opening. It moves from the bark to the new leaves, where it gives birth to young aphids. The insect's growth and reproduction is extremely rapid, and the leaves of an infested host soon acquire symptomatic corrugations. The undersides of leaves within the corrugated folds fill with aphids and white granular material (E, F). Winged migrants develop on the leaves and seek witch-hazel on which to lay their eggs and complete the life cycle. All of these activities occur before the end of June. Photographs shown here were taken in south central New York state in early June. The host is *Betula populifolia*.

H. spinosus is transcontinental, but primarily a northern species. A related species, *Atarsaphis* (=*Hamamelistes*) *agrifoliae* (Ferris), feeds on the foliage of coast live oak in California. Injury by *H. spinosus* ranges from light to severe, that is, it ranges from premature leaf drop to dead twigs and branches. There is keen competition in some areas between the birch leafminer, *Fenusa pusilla* (Plate 84), and this aphid. Where both occur, the birch leafminer dominates and thereby reduces the aphid population.

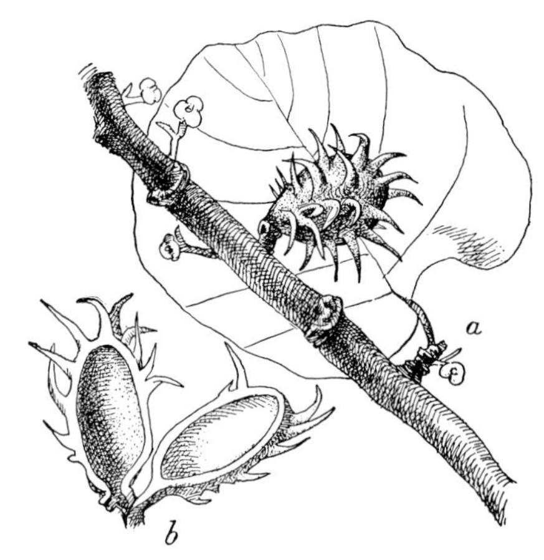

Figure 99. A bud gall on witch-hazel caused by the aphid *Hamamelistes spinosus*: *a*, the mature gall; *b*, section of the empty gall. (Drawing by Theodore Pergande, ref. 607.)

In the aphid genera *Betulaphis, Calaphis, Callipterinella,* and *Euceraphis* there are numerous species that feed on the underside of birch foliage. Most are European introductions that now are broadly spread over Canada and the United States. Because birch trees do not flush or senesce their leaves simultaneously, these aphids find fresh foliage throughout most of the growing season, which allows for numerous generations. When there is a choice, the European aphids in California feed predominantly on the introduced European birches such as *Betula pubescens* and *B. pendula,* and they probably follow the same feeding preference in the East. *Calaphis betulaecolens* (Fitch)—a common large green aphid—often develops into large populations that produce quantities of honeydew that smuts much of the foliage. During the dormant season dark-colored egg clusters, usually located at branch crotches, may be found on trees heavily infested during the previous growing season. All common species of birch are suitable hosts for this pest, which may be found throughout North America wherever birch is grown. It is not a serious pest of forest birch but is frequently a problem on shade and specimen trees. In the spring and early summer *Callipterinella calliptera* (Hartig), *Betulaphis brevipilosa* Borner, and *Euceraphis betulae* (Koch) utilize catkins as well as foliage as a food source. *E. betulae* is a large, active yellowish to greenish aphid. It utilizes secondary and main veins as feeding sites, and in autumn it prefers senescing leaves. If disturbed, it readily drops from its feeding site.

The rosy apple aphid, *Dysaphis plantaginea* (Passerini), is a two-host aphid that is obligated to an alternation of hosts. Its primary hosts are *Malus* species; the egg overwinters on the bark (D). Newly hatched nymphs move to newly developing foliage and often establish massive colonies of pink to purplish aphids on the undersides of leaves. Infested leaves eventually wrinkle, curl, and twist inward, sometimes becoming grossly distorted. This aphid feeds only on the most tender foliage. Several generations are completed during the spring. In early summer winged aphids are produced, and they fly to their summer host, plantain (*Plantago lanceolata*), a common weed. The rosy apple aphid is not found on apple in the summer. In autumn winged aphids are again produced, and they migrate back to their primary host, where, eventually, eggs are laid. This aphid may be found wherever *Malus* species grow.

References: 217, 231, 232, 314, 359, 463, 711

A. Several beech leaves with *Phyllaphis fagi* aphids and cast "skins." [Aphididae]
B. The underside of a beech leaf showing aphids, cast "skins," and large droplets of honeydew.
C. *Betula populifolia* leaves showing the corrugation symptoms caused by *Hamamelistes spinosus*. Normal leaves are on the right.
D. Black overwintering eggs of the rosy apple aphid on the bark of apple.
E. Close-up of a birch leaf showing the corrugations and the chlorotic secondary veins on the upper side caused by *H. spinosus*.
F. Close-up of *H. spinosus* aphids on the underside of a birch leaf.

A
B
C
D
E
F

Spirea Aphid and Aphids of Manzanita (Plate 141)

Aphis citricola Van der Goot has been given the common name spirea aphid. It is believed to be native to North America and is distributed throughout the United States. Spirea seems to be its basic ornamental host plant, yet in Florida it has been recorded from 11 other plants. It is often found in large colonies. Leaf curl (B, C) and stunted foliage are typical of the injury caused by this aphid, which feeds on the undersides of leaves or on stems.

Over most of the range of the spirea aphid, its sexual form produces eggs in the fall, apparently laying them on any species of *Spiraea*. In central and southern Florida, as well as in southern California, this aphid appears to have no sexual stage and hence lays no eggs. Except for the sexual form, all mature aphids are parthenogenic and give birth to living young. During the summer both winged and wingless forms are produced. Winged forms appear when aphids become crowded in the colony and when suitable food becomes scarce.

When the food supply becomes limited on one plant, winged aphids fly to more suitable or succulent plants to start another colony. Each female may produce up to 70 offspring. As many as 44 generations in a single year have been recorded in Florida.

Meteorological conditions and natural enemies have a tremendous effect on the spirea aphid as well as on other aphid species. Heavy rain and strong winds knock them to the ground, where they die. They have numerous parasites and predators. Occasionally, the fungus *Empusa fresenii* helps control aphid populations, especially in the South.

Another *Aphis* species, *A. ceanothi* Clarke, is reddish brown and black and can severely injure ceanothus, particularly *Ceanothus intergerrimus*. This species occurs primarily on the undersides of leaves, but heavy populations may force them to feed on the young tender stems as well. By August in northern California, a population may increase to the point where the leaves and twigs of the host plant die. *A. ceanothi* is known to occur in the Pacific Coast states, British Columbia, Colorado, and Idaho. Biological studies have not been conducted on this species.

Wahlgreniella nervata (Gillette) is a pale green aphid (D) that occurs in the West on wild and cultivated roses (*Rosa* spp.), and on manzanita (*Arctostaphylos* spp.), California Christmas berry (*Heteromeles arbutifolia*), madrone (*Arbutus menziesii*), and strawberry tree (*A. unedo*). Both terminals and leaves are colonized, but little leaf curling results. This aphid produces large quantities of honeydew, which makes the foliage of its host plant sticky. Cast "skins" (E) remain attached to leaves, stuck there by honeydew. On manzanita severe infestations kill leaves and twigs.

Manzanita is the only known host of *Tamalia coweni* (Cockerell), referred to by some authors as the manzanita leafgall aphid. Records show it to be present on Long Island and in Connecticut, Illinois, Minnesota, Nevada, Colorado, and the Pacific seaboard states, as well as in British Columbia. This aphid causes conspicuous galls on leaf margins, or the gall may encompass the entire leaf. The galls are bright red when newly formed but turn brown as they age (F). The aphid colony lives within the protection of the galls. Syrphid fly larvae are often found feeding on the aphids within the galls.

References: 217, 291, 293, 388, 463

A. Winged adult spirea aphids, *Aphis citricola*.
B. Young shoots of citrus. The curling of leaves was caused by the spirea aphid.
C. Injury to spirea caused by the spirea aphid.
D. A winged adult *Wahlgreniella nervata*.
E. Manzanita showing some of the symptoms of infestation by *W. nervata*.
F. Injury to the leaves of manzanita caused by the gall-making aphid *Tamalia coweni*. New galls are bright red.

A
B
C
D
E
F

Snowball Aphid and Other Foliage-feeding Aphids (Plate 142)

The plant injuries shown here were caused by at least five species of aphids. Panels A and D show injury caused by the snowball aphid, *Neoceruraphis* (=*Aphis*) *viburnicola* (Gillette), on two cultivars of viburnum. Although this aphid has secondary host plants, their identity is unknown. Viburnum is the host on which it overwinters, because it is on the twigs and buds of this plant that the aphid lays its eggs in the fall. The eggs hatch at the time the buds open in the spring. These buds serve as the feeding site for the developing aphids. At maturity, the insects are plump and about 2.5 mm long. They are bluish white, and their bodies seem to be dusted with white powder. As stem mothers, they produce large numbers of young aphids without mating.

Within 3 weeks after bud break the leaves of the viburnum are grossly misshapen; sometimes even the leaf stem is bent. Severe leaf curling damage to *Viburnum opulus*, *V. opulus* 'Sterilis', *V. acerifolium*, and *V. prunifolium* occurs, but other species of viburnum, such as *V. tomentosum*, are immune. Twisted foliage is a primary symptom. Two months after the eggs hatch, all the progeny leave the viburnum, but it is not clear just where they go. In September migrant forms return to the viburnum and give birth to the sexual aphid forms that produce the overwintering eggs. *N. viburnicola* is a transcontinental species, but it is primarily found in northern states and in Canada.

A second species of aphid, *Aphis viburniphila* Patch, known as the viburnum aphid, seems to have a year-round association with this plant. Unlike *N. viburnicola*, this species does not cause leaf deformation. In the Northwest the bean aphid *Aphis fabae* Scopoli causes foliage injury to viburnum. It has a black or dark olive green body and overwinters as an egg on many shade trees and shrubs such as species of *Euonymus*, *Viburnum*, *Philadelphus* and *Amelanchier*, and on *Fatsia japonica*, English ivy, tuliptree, and manzanita. A European study produced evidence that *V. opulus* is significantly more infested by *A. fabae* in the vicinity of heavily traveled highways, which suggests that auto-produced air pollutants are beneficial to the development and reproduction of certain aphids and that such plants will require more maintenance when placed near roads and parking lots.

The aphid injury to mountain ash shown in panel B was probably caused by the apple aphid, *A. pomi* De Geer. This is a European species believed to have been introduced into this country prior to 1851. It is now considered to be worldwide in distribution.

The apple aphid, a very destructive insect, has been the subject of many intensive entomological investigations. Its life history is simple compared with many aphid species; it requires but a single host. Like most aphids, it spends the winter as a dormant egg. In winter clusters of these black eggs may be seen on twigs of the previous summer (see Plate 138E for similar eggs). The eggs hatch while the new leaves are very young, and the summer forms emerge. These may be winged or wingless. In Virginia there may be as many as 17 generations of summer aphids on apple. In the fall, when leaves begin to turn color, sexual forms are produced. These lay eggs to carry the species through the winter.

The apple aphid may take up residence on numerous host plants, but apple seems to be its favorite. All the hosts of the apple aphid are in the rose family; they include species of *Crataegus*, *Prunus*, *Pyrus*, *Malus* (flowering crabapple) and *Chaenomeles* (flowering quince) as well as *Sorbus americana*, the plant shown in panel B. On all host plants the injury caused by aphids appears to be essentially the same: the leaf blades and petioles become curled and stunted. Early in the year new shoots may be so heavily infested that all growth stops. Under ornamental nursery conditions, this aphid cannot be tolerated; likewise, if many leaves were curled on a specimen tree (as shown in B), the tree would certainly not be accepted as ornamental. The common aphid predators such as lady beetles, lacewings, nabid bugs, and syrphid fly larvae, as well as parasitic wasps, help keep this pest in check.

Myzus persicae (Sulzer), also known as the green peach aphid (C), is a universally distributed insect. Its biology and behavior are well known. Leonard (463) called it "our commonest aphid" when describing the aphids of New York. Essig (217) describes it as one of the most common aphids in North America. In the northern two-thirds of the United States the green peach aphid overwinters as a shiny black egg attached to the bark of peach, plum, cherry, and other trees of the genus *Prunus*, including the flowering cultivars.

The aphids' chronological development is not complicated. The overwintering hosts are considered to be primary hosts, for it is on them that the sexual forms are produced. Two or three generations of green peach aphids develop asexually on *Prunus* species in the spring. The aphids then produce winged forms that migrate to any one of 100 or more plant species for continued summer development. Most of the aphids' summer hosts are herbaceous plants, but some are woody. The woody ornamental summer hosts include *Hedera helix* and species of *Clematis*, *Citrus*, *Juglans*, *Bougainvillea*, and *Vinca*. In the milder sections of the country, the green peach aphid is abundant on the summer hosts throughout the year. It apparently never goes through the sexually produced egg stage.

The green peach aphids' reproductive rates have been carefully studied. Under controlled conditions a single parthenogenic female may give birth to three to six fully formed young per day for several weeks. The rate of reproduction of all aphid species is dependent on quality of food, host plant species, and temperature. The food quality of a host plant can to some degree be controlled by the application of fertilizer and other chemicals.

The summer form of this aphid, the form most likely to be seen, is a pale yellowish green. It has red eyes. Close observation will reveal three rather distinct lines of green extending from the back of the head to the end of the body.

In addition to injury it causes by direct feeding, the green peach aphid is capable of transmitting more than 100 different plant viruses. These viruses are best known on vegetable crops, where they seriously reduce production.

In addition to the common aphid predators, which include lady beetles, lacewings, syrphid flies, and nabid bugs, fungal diseases aid in natural control. A number of parasitic wasps also provide a degree of natural control of this species.

Potato aphid is the common name given to *Macrosiphum euphorbiae* (Thomas). It is about 3 mm long and has green and pink color forms. Though not a major pest of woody ornamentals, it is a severe pest of many vegetable crops. The potato aphid overwinters in the colder regions of the United States as an egg on rose, *Cotoneaster* species, and possibly other rosaceous plants. In the spring the hatched migrant forms fly to other plants to produce the summer generations. In addition to vegetable crops, this species infests crabapple, species of *Citrus*, *Fuchsia*, *Cornus*, *Ribes*, *Lonicera*, and *Euonymus alatus*. In warm regions it is found the year around, with the egg stage omitted altogether. The potato aphid is commonly found throughout the United States and Canada.

Panel E shows severe smutting caused by the feeding of aphids on citrus fruit and foliage. The honeydew becomes so thick that the leaf is completely covered. The black color comes from one of several fungi, commonly called sooty molds, that grow on the honeydew. As the leaf continues to grow, and if the aphids do not persist, the thin coat of smut ruptures and peels away from the leaf. Injury is caused not only by the presence of the feeding aphids, which reduces the amount of sap in the tree, but also by the presence of the sooty mold covering, which interferes with the photosynthetic process carried out by the leaves.

References: 71, 217, 292, 388, 463, 464

A, D. Two cultivars of viburnum with leaf injury caused by the snowball aphid, *Neoceruraphis viburnicola*.

B. The apple aphid, *Aphis pomi*, caused these leaves on *Sorbus americana*, mountain ash, to become deformed.

C. The green peach aphid and the potato aphid in a mixed colony on daphne.

E. Black sooty mold mixed with honeydew on a citrus leaf. Pressure from the expanding leaf causes the black film to break and peel.

A
B
C
D
E

Black Citrus Aphid and Other Foliage-feeding Aphids (Plate 143)

The black citrus aphid, *Toxoptera aurantii* (Fonscolombe), is an important and common pest of camellias in many parts of both the United States and Europe. As its common name implies, this species attacks species of citrus, but it also occurs on *Pelea, Straussia,* and *Coffea* species in Hawaii, and species of *Ixora* and *Clusia,* seagrape, *Acer saccharinum, Pittosporum* species, crape myrtle, *Eleagnus pungens, Cassytha filiformis, Ficus* species, and *Ilex aquifolium* in the eastern and southern United States.

The black citrus aphid is 2 mm long and usually shiny black but may vary from dull black to the color of mahogany. Although most abundant in the early spring, it persists all year in citrus orchards. Colonies are formed at the buds or on the leaves of the host. On camellias the aphids often congregate in large numbers on the buds of the flowers (A). Leaf cupping, curling, or twisting are symptoms associated with its damage. Other symptoms include the inevitable blackening of plant parts resulting from the production of honeydew by the insects, and, on camellia, flower malformation. It may be found from New York to Florida and west to California and Oregon. It is notably missing from the north central and prairie states.

The black citrus aphid has many natural enemies. These include lady beetle larvae and adults, lacewing larvae, and syrphid fly larvae. The most dramatic of these natural enemies are tiny parasitic wasps that develop inside the bodies of the aphids (see Figure 100, accompanying Plate 147). At maturity, the parasite cuts a tiny hole in the back of the dead aphid, through which it escapes, leaving a brown mummified shell as evidence of its activities. These "mummies" persist for some time on the leaves of the host plant.

Eucallipterus (=*Myzocallis*) *tiliae* (Linnaeus), sometimes called the linden aphid, probably occurs wherever basswood, littleleaf European linden, or other linden shade tree species are grown. It feeds on the undersides of leaves in colonies. Street trees are frequently infested with them, and the honeydew from these aphids soils cars parked under the trees. These aphids are green with black lateral stripes. Winged forms have clouded wings. There is continuous development of new progeny throughout the host's growing season. Several hymenopterous parasitoids attack the linden aphid, but rarely in sufficient numbers to provide pest management. It is common, however, to find many brown to bronze aphid mummies attached to leaves.

Maple trees are frequently attacked by several leaf-feeding species of aphids belonging to two genera, *Drepanaphis* and *Periphyllus.* There are at least 18 species of *Drepanaphis,* and all feed on the foliage of various species of maple. One or more species may be found wherever maple grows. The genus *Periphyllus* contains eight species in North America, and all but one species feed on maple foliage. Large populations frequently develop in the absence of natural control organisms. When trees are heavily infested, copious amounts of honeydew collect on objects beneath them. Summer leaf drop may occur, and it is not uncommon for silver and Norway maple to nearly defoliate.

Periphyllus lyropictus (Kessler), the Norway maple aphid, most frequently occurs on Norway maple. It clusters along the veins on the underside of leaves, feeding in the phloem sieve tubes. The species may be further identified by its red eyes and somewhat hairy and yellowish green body with brown markings. Several generations develop during the growing season.

Drepanaphis and *Periphyllus* species are found throughout the United States and in parts of Canada. They may also infest *Aesculus* species and sycamore. All stages occur on a single host.

References: 202, 211, 414, 710, 836

A. A camellia bud infested with the black citrus aphid.
B. Close-up of black citrus aphids and cast "skins" on a camellia bud. [Aphididae]
C. Black citrus aphids on a camellia leaf.
D. Skeletons of black citrus aphids (on a camellia leaf) that have been parasitized by a minute wasp. Note the exit hole of the wasp parasite.
E. A colony of black citrus aphids on the leaves and stem of camellia.

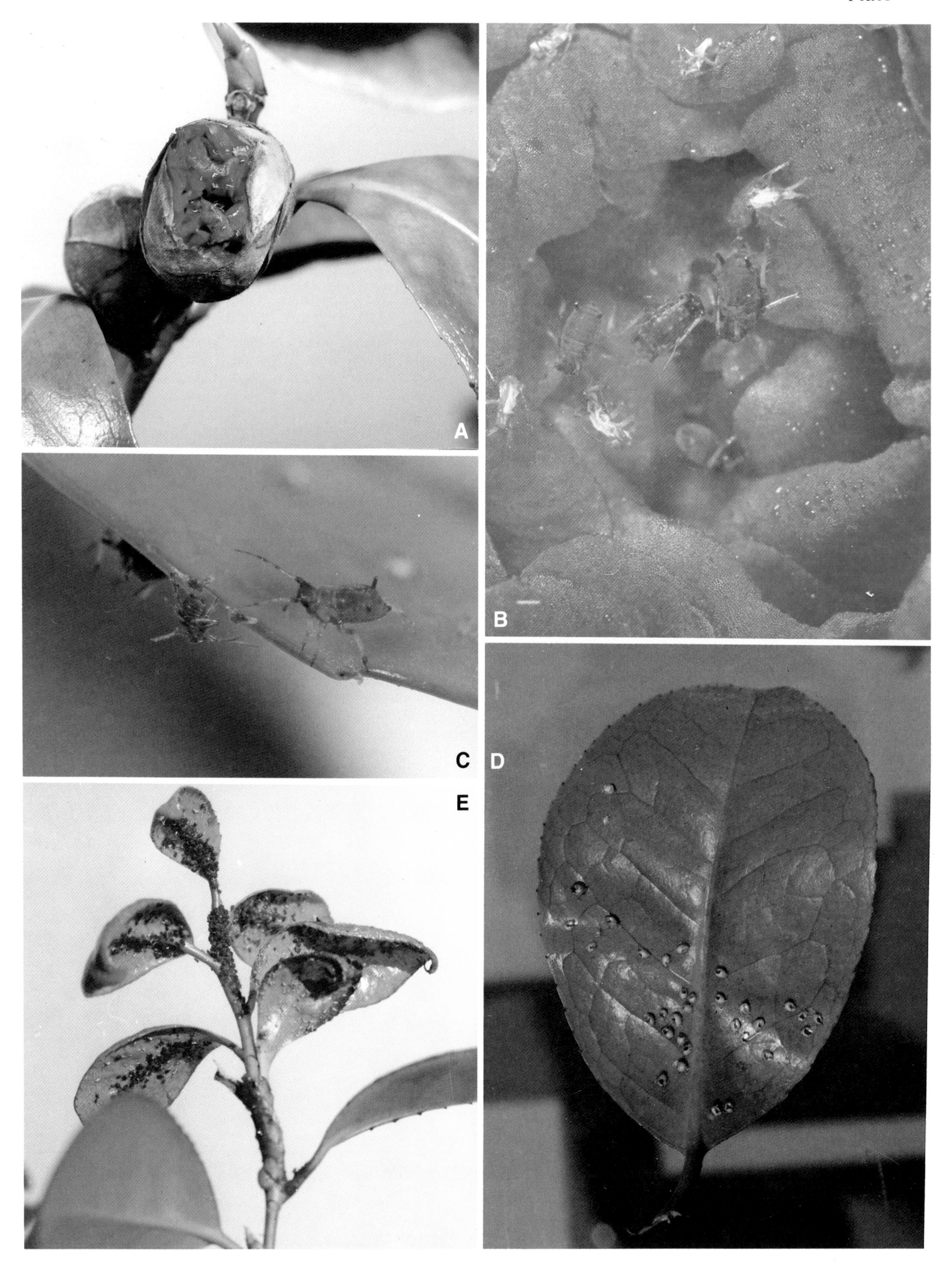
A
B
C
D
E

Woolly Aphids (Plates 144–145)

Those aphids shown in panels A and B are believed to be *Stegophylla quercicola* (Monell) on red oak, *Quercus rubra*. An appropriate common name would be the oak woolly aphid. These aphids spend part of their lives in colonies on either side of oak leaves. Several aphid forms may be found concurrently in these colonies, including the parthenogenic females, which appear as plump milky-white individuals, and winged males.

A large colony of these aphids may cause the host leaf to curl, the only noticeable injury to the host. Heavily infested shade trees are made conspicuous by the white patches of flocculent wax strands, produced by the aphid colony. The colony becomes covered with the wax mixed with cast molt "skins." As the colony ages, the white patches begin to look like dirty snow, giving the tree an unsightly appearance. *S. quercicola* is known to occur in the eastern states to North Carolina, the north central states, and Kansas.

There are five species of *Stegophylla* known in the United States, and three of these are also known to occur in Canada. All of them feed on various species of oak and are limited to oak. Several generations occur each year, but very little is known about their biology and developmental cycle. *S. quercifolia* (Gillette) feeds on scrub oak in California, but it occurs on other oaks in the Midwest and eastern states. *S. essigi* Hille Ris Lambers is limited in its host range to *Q. agrifolia* and is known only on the West Coast.

The woolly alder aphid is the common name given to *Paraprociphilus tesselatus* (Fitch). The prolific production of white waxy filaments is the most distinctive sign of its presence on an infested plant. Its biology is of special interest to the naturalist. Two hosts are required: alder (Plate 145A–C, G) and silver maple.

Eggs are deposited in the fall on the bark of maple trees. The eggs hatch in the spring as soon as the new leaves appear. Immediately, the immature insect seeks out the underside of a new leaf and settles at the midvein (C). Young aphids are all females (stem mothers) capable of reproducing asexually. They produce large colonies that collectively withdraw great quantities of sap from the host, causing its leaves to curl inward to make a protective cover for the enclosed insects. The young of the stem mothers develop into migrant forms that, by the end of July, fly to alder trees (E). This winged migrant is quite large, measuring 10 mm from wing tip to wing tip. Like the stem mother, it reproduces asexually. The young aphids, now on alder, collect on the twigs to start a new life cycle for the species. A number of generations develop on alder (Plate 145A–C), and produce large amounts of white, fluffy wax. Honeydew is also produced, and foliage on infested trees may appear wet from its abundance. Each aphid is covered with white down in a checkerboard pattern. A maturing colony may be completely concealed beneath a dense covering of the white waxy material. The aphids beneath are plump and gray and measure about 2 mm long.

In the vicinity of Washington, DC, migrants mature on alder and fly back to the trunk and branches of maple trees around the first of October. These are sexual forms, both male and female. After mating, the female produces but one egg, which she deposits on loose bark and covers with white woolly down.

Not all the aphids on alder mature and fly back to maple; some remain throughout the winter in closely packed colonies that are so crowded that the aphids literally stand on their mouthparts, which are embedded in the bark. It would appear that the species can exist indefinitely on alder, reproducing without fertilization. The aphids shown in Plate 145 were photographed in late August in south central New York state, indicating that late development of the leaf form may occur in the North. The alder forms have also been seen with some frequency in the Adirondack Mountains of New York state in early August.

The numerous predators of the woolly alder aphid include lacewings and lady beetles. Ants invariably tend woolly alder aphids, particularly those on alder, to obtain honeydew.

The woolly alder aphid is not considered particularly injurious but may become quite annoying around homes because of the amount of white, woolly threads that accumulate on the ground under heavily infested trees.

The reported range of the woolly alder aphid is from Manitoba to New Brunswick south to Florida and west to the Mississippi River.

A related genus, *Prociphilus,* represents another group of woolly aphids that normally have unlike alternate hosts; one is an aerial host on bark or foliage, the other subterranean on the bark of roots. The subterranean host for members of *Prociphilus* species is usually a conifer. *P. caryae* (Fitch) (=*P. alnifoliae* Williams) alternates between *Amelanchier alnifolia* and/or *Heteromeles arbutifolia,* the primary hosts, and pine; *P. americanus* (Walker) alternates between *Fraxinus* and *Abies* species. *Fagiphagus* (=*Prociphilus*) *imbricator* (Fitch), the beech blight aphid, is a white to yellowish aphid that covers its body with white waxy filaments similar to those shown in panel D. It is thought to feed only on beech, most frequently on small branches and twigs. Clusters of these aphids attract attention in August and September. Winged forms have black markings on the thorax. This species overwinters as eggs on twigs and buds. There is little biological information about any of these aphids. The extent of injury to their coniferous hosts has not been studied.

References: 463, 464, 608, 708, 710

A. A red oak leaf with a colony of aphids believed to be *Stegophylla quercicola*.

B. Close-up of the colony, showing the mature wingless form, the immature form, and the winged form of the aphid.

C. A colony of woolly alder aphids, *Paraprociphilus tesselatus,* on silver maple.

D. Close-up of two woolly alder aphids covered with wax strands.

E. The winged adult of the woolly alder aphid.

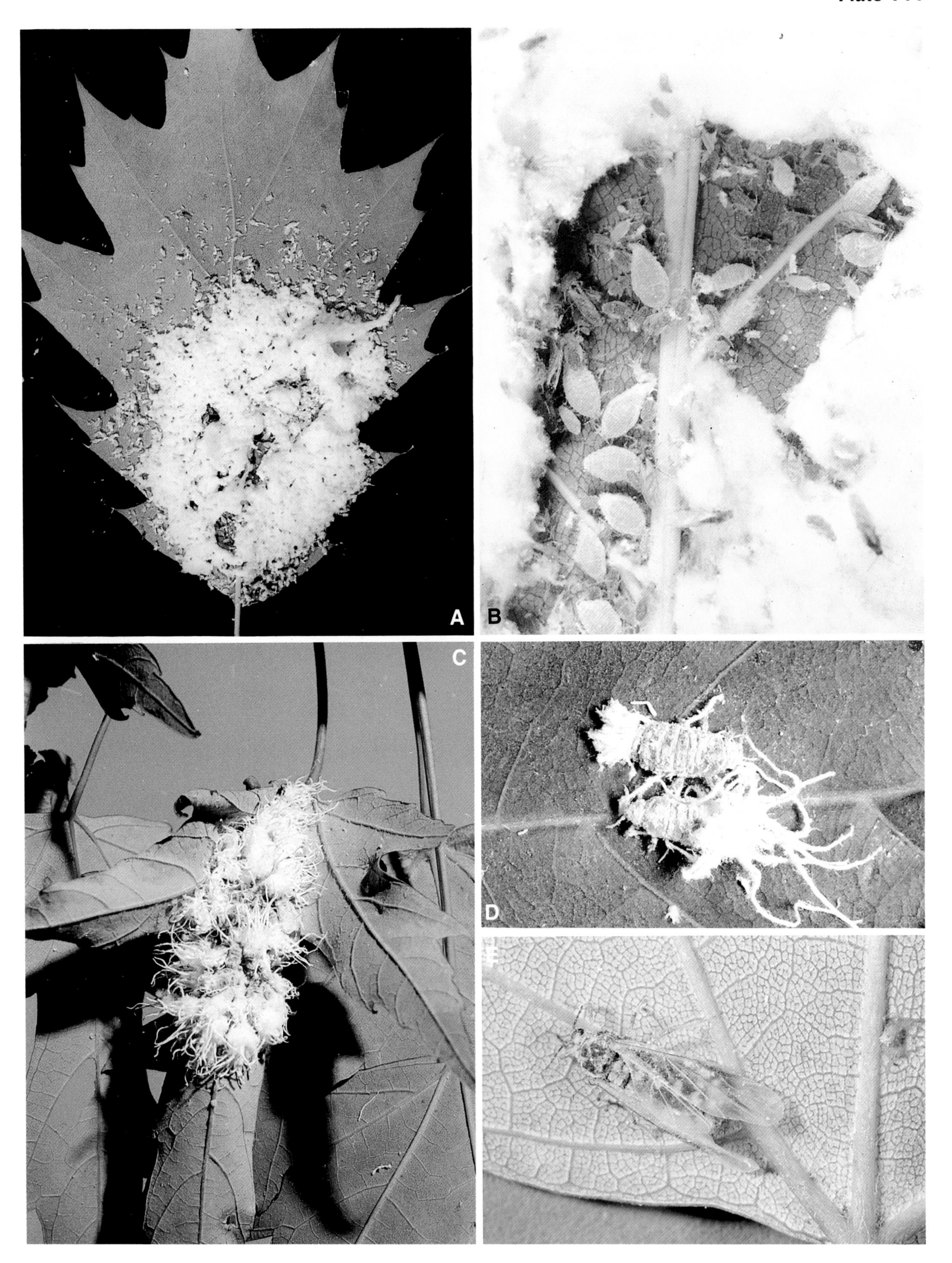
A
B
C
D
E

(Woolly Aphids, continued, Plate 145)

Under natural circumstances neither animals nor plants cover themselves with clothing or a parasol of white unless it aids in protection or survival. White quickly catches the eyes of foes. The woolly aphids, the woolly adelgids, and a number of closely related sucking insects clothe themselves in white waxy ribbons or threads. Apparently, this habit is beneficial to the insect, but exactly how no one is quite sure. The white color certainly does not aid in camouflaging the aphid from its enemies.

Some of these woolly aphids are curiosities of nature rather than serious pests, but sometimes they may become abundant enough to cause plant owners concern. The aphids shown in panels A–C are the woolly alder aphid, *Paraprociphilus tessellatus* (Fitch), and are described in Plate 144.

It is claimed that all members of the genus *Eriosoma* have two hosts, one of which is a species of *Ulmus*. This is probably true if elm is available, but it is certainly not an obligate arrangement with all these aphids. *Eriosoma lanigerum* (Hausmann) in the western United States survives very well on its summer host and overwinters as subterranean aphids on the roots of summer hosts (Plate 150). *E. crataegi* (Oestlund) alternates annually between elm and hawthorn and is a transcontinental species. In the summer it is a common pest of English hawthorn. Clusters of aphids covered with white flocculent material may be found on stems and branches. *E. rileyi* (Thomas) apparently remains on elm the year around, causing knots and gall-like growths on twigs and limbs. Large woolly masses of aphids are signs of their presence. It, likewise, is a transcontinental species.

The woolly elm aphid, *E. americanum* (Riley), alternates annually between elm and *Amelanchier* species. Eggs are laid in the fall by females whose sole function is to produce a single egg; the female is not even endowed with functional mouthparts, and when her egg is laid, she dies. The egg is placed in a bark crevice where it is able to withstand the rigors of winter. In the spring, when the elm leaves are unfolding, the egg hatches into a young wingless female that seeks out the underside of a leaf and proceeds to feed on its sap. At maturity she is capable, without mating, of giving birth to young aphids. Within a few days she can produce as many as 200 young, all females. Soon, the infested leaf becomes curled, and it serves to protect the colony from wind, rain, and other elements of nature. By the end of June the colony has become so crowded and full of cast "skins" that individuals begin to spill over onto other leaves. About this time a winged generation matures. These migrants, again all females, desert the elm leaves and instinctively seek *Amelanchier,* which is variously known as shadbush or serviceberry. Upon finding a tree and establishing residence, they begin to give birth to young, again all females. The young crawl down the twigs and branches to the underground parts of the plant. This is the summer destination of *E. americanum.* Several generations are produced underground, and the roots of *Amelanchier* are injured. In the fall winged females are produced underground; they migrate back to the elm to perpetuate the species and start the cycle again.

A tiny wasp, *Aphelinus mali* Haldeman, is a native parasite of *Eriosoma* species and of aphids in other genera. Predators common to aphid species such as lacewings, lady beetles, and syrphid fly larvae are often present. However, these predators can do little to reduce leaf damage. By the time the predator population has reached an effective density, the migratory aphids have often vacated the curled elm leaves and have flown to their secondary host.

Two mirid (plantbug) predators are often found in the curled leaves of elm. *Deraeocoris aphidiphagus* Knight feeds on *E. americanum* and *D. nitenatus* Knight feeds on *E. lanigerum.* The nymphs of both bugs are covered with a white waxlike material similar to that which covers aphids.

References: 354, 463, 464, 544, 597, 598, 608

A–C. Mature woolly alder aphids on black alder, *Alnus glutinosa.* [Aphididae]

D–F. Aphids believed to be woolly apple aphids on hawthorn (*Crataegus*) species.

G. Woolly alder aphids under and beside long strands of waxy filaments.

H. The woolly elm aphid on elm, *Ulmus americana,* mostly cast "skins."

I. A syrphid fly larva feeding on an aphid colony.

J. The woolly psyllid, *Psylla floccosa,* on *Alnus* sp. Insects other than aphids also produce flocculent masses of wax.

A
B
C
D
E
F
G
H
I
J

Leaf and Stem Aphids (Plates 146–147)

Aphids have so many and such varied ways of existence that no generalized account can adequately describe them. Their behavior is governed largely by food preference and feeding site. Some feed underground on roots, others on bark, twigs, leaves, flowers, and fruit. The facing plate shows some examples of aphids that feed on the stems, flowers, and leaves (blade and petiole) of their hosts. Panel A shows all stages of the cotton, or melon, aphid, *Aphis gossypii* Glover, on the leaf and flower of *Hibiscus*. The dark forms are mature females, which may vary in color from yellowish green to dark mottled green in their wingless stage. In the Northeast the cotton aphid overwinters as an egg on catalpa and rose of sharon. During the growing season it has been collected on at least 40 plant species in New York state, but it is most frequently found on herbaceous plants. Some woody ornamental tree and shrub hosts include *Bougainvillea* species, camellia, *Euonymus japonica,* crape myrtle, and species of *Pittosporum, Platanus, Prunus, Salix, Eucalyptus, Persea,* and *Malva.* The cotton aphid is a cosmopolitan species occurring throughout the United States, and in parts of Mexico and Canada. Where it is abundant, it frequently is heavily parasitized by tiny parasitic wasps. The arrows point to brown "mummies" that often result when aphids are killed by such parasites.

Heavy aphid populations result in generally unthrifty plants, distorted leaves and flowers, and wilt symptoms during times of water stress. Several viral diseases, which are particularly severe on vegetable crops, are transmitted by this aphid.

The rose aphid (B), *Macrosiphum rosae* (Linnaeus), has both pink and green forms; it feeds on the stem and developing bud of its host. Its entire life history is presumed to be confined to rose. Single plants can tolerate fairly high aphid populations; however, the quality and quantity of flowers are reduced. Rose connoisseurs, plant nursery owners, and greenhouse producers find high aphid populations on roses particularly intolerable. The rose aphid is common throughout the United States and Canada wherever roses are grown.

Illinoia (=*Macrosiphum*) *rhododendri* (Wilson) looks much like the rose aphid but is found only in California and the Northwest on rhododendron. *A. hederae* Kaltenbach, commonly called the ivy aphid, is shown in panels C and D. This species usually occurs in black clusters, which seem to stand one on top of the other. The primary host plant is English ivy, *Hedera helix,* of which there are more than 15 commercial cultivars. Growing tips and young leaves are the preferred feeding sites. When the aphids are abundant, they may cause severe stunting of the growing tips and leaf malformation. The ivy aphid produces large quantities of honeydew. Ants are attracted to the honeydew and often tend these aphids in a symbiotic relationship.

In southern California ivy aphids may be abundant all year. Official collection records indicate that this species is likely to be found wherever English ivy grows. The ivy aphid has also been reported on species of *Aralia, Euonymus,* and *Viburnum.*

A. craccivora Koch (E) has been given the common name cowpea aphid because of its affinity for members of the legume family, which includes many ornamental trees and shrubs. The cowpea aphid's body is shiny and black. Its legs and the basal half of its antennae are either white or pale yellow. Immature forms vary from dark green to brown to black. The aphids shown in panel E are feeding on leaf petioles and stem of *Wisteria floribunda.* Injury to this host was severe. Leaf blades were severely scorched because of the quantity of fluid extracted by the feeding aphids. Tree hosts include species of *Laburnum, Robinia, Gleditsia, Pyrus, Malus, Citrus, Eucalyptus, Bougainvillea,* and *Cassia.* Feeding occurs primarily on young twigs.

Records of this aphid's geographic distribution show it to occur all along the Atlantic Coast, in all the southwestern states (including Texas), and in Colorado.

References: 217, 388, 463, 464

A, F. *Aphis gossypii,* the cotton, or melon, aphid, on *Hibiscus* sp. Note aphid mummies at arrows in A.

B. The rose aphid, *Macrosiphum rosae,* on *Rosa* sp. Note the two color forms. [Aphididae]

C. A badly deformed ivy leaf with ivy aphids, *Aphis hederae,* attached to the petiole.

D. A tightly packed ivy aphid cluster on stem, leaf, petiole, and blade of ivy.

E. *Aphis craccivora,* the cowpea aphid, on the petioles and stem of *Wisteria floribunda.*

G. Winged and wingless adult females of the cowpea aphid.

A
B
C
D
E
F
G

The foxglove aphid, *Acyrthosiphon* (=*Aulacorthum*) *solani* (Kaltenbach), is probably best known as a pest of potato. In California it often ruins *Aucuba japonica* as an ornamental plant. New leaf growth is stunted and curled, and if the infestation is severe, leaves may die (A). Much sticky honeydew is formed and many generations develop, making it a continuous problem over the growing season. Dust adheres to the honeydew, adding to the plant's unsightly appearance. Similar symptoms may occur from scale insect infestations. Other ornamental hosts include *Myrica* species, *Paulownia tomentosa, Ilex aquifolium, Vinca major, Viburnum* species, *Philadelphus* species, and *Hibiscus calyphyllus*.

Pterocomma (=*Clavigerus*) *smithiae* (Monell) is a large aphid, about 3.5 mm long, with bright orange cornicles (see arrows in B). It feeds gregariously on the bark of willows and many species of poplar and has been reported on silver maple. Periodically it appears in damaging numbers. These aphids have a rather simple life cycle and are content to remain on the same host through many generations. In the autumn eggs are laid on twigs. These eggs are yellow at first and later turn black. Adults have been observed from May through October. Those illustrated in panel B were collected in June in central New York state, although this species is usually most abundant in August and September. *P. smithiae* occurs from New Brunswick to Florida and west to Texas, Colorado, and California to British Columbia.

Longistigma caryae (Harris) is commonly called the giant bark aphid. The length of its body is about 6 mm. Because of its long legs, however, it appears much larger. Winged migratory forms of this aphid are also produced (E). Their color pattern is distinctive, and their short, black cornicles make field identification easy. An egg-laying female is depicted in panel D. (In C, the arrow points to one of many *Asterolecanium* scale insects; see Plate 168.) Eggs overwinter and are deposited late in the growing season—in October in northern Illinois. These eggs are laid in bark crevices as well as on the smooth bark of smaller branches. The aphids shown in these pictures were on *Quercus rubra* and *Q. palustris*. Other hosts include beech, sycamore, linden, birch, chestnut, hickory, pecan, walnut, and willow.

The giant bark aphid may be found throughout the growing season, but it becomes most numerous during the late summer. It is capable of causing severe injury to its host. Twigs or branches may die owing to its feeding activities. When the aphid population is large, a twig may be completely covered with eggs. The eggs are black and look much like those shown in panel E of Plate 138. The giant bark aphid is found throughout most of the United States except the Pacific Northwest and a few of the Rocky Mountain states.

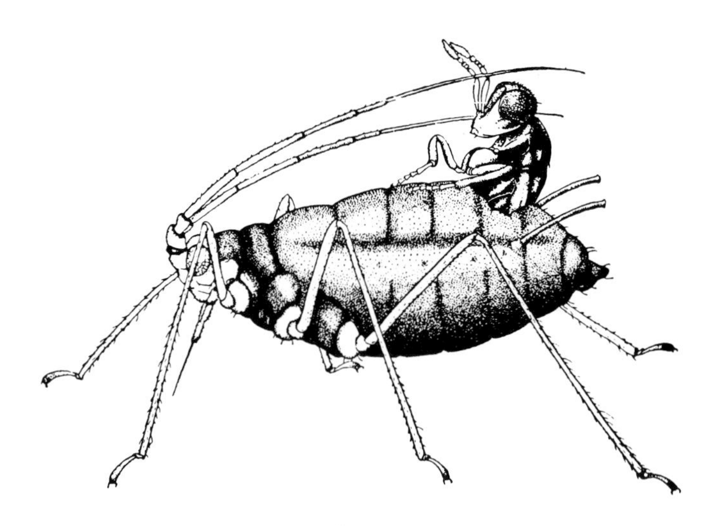

Figure 100. A wasp parasite emerging from an aphid. (Drawing by Grace H. Griswold, Cornell University Press.)

Tuberolachnus (=*Lachnus*) *salignus* (Gmelin), sometimes called the giant willow aphid, is a large (4–5 mm) gray to black twig-feeding insect. Willow is its only host. The amount of damage inflicted by this insect is probably minor, but its presence is likely to alarm homeowners. It may be found wherever willow grows.

References: 253, 388, 463, 651

A. Variegated *Aucuba japonica* injured by *Acyrthosiphon solani,* the foxglove aphid.
B. A cluster of black willow aphids, *Pterocomma smithiae,* on the bark of weeping willow. Arrows point to the orange cornicles.
C. Wingless female giant bark aphids, *Longistigma caryae,* on the bark of red oak. Note that the aphid blends in with its background. Arrow points to a pit scale, *Asterolecanium* sp.
D. A giant bark aphid on a contrasting background.
E. The winged form of the giant bark aphid together with three immature females.

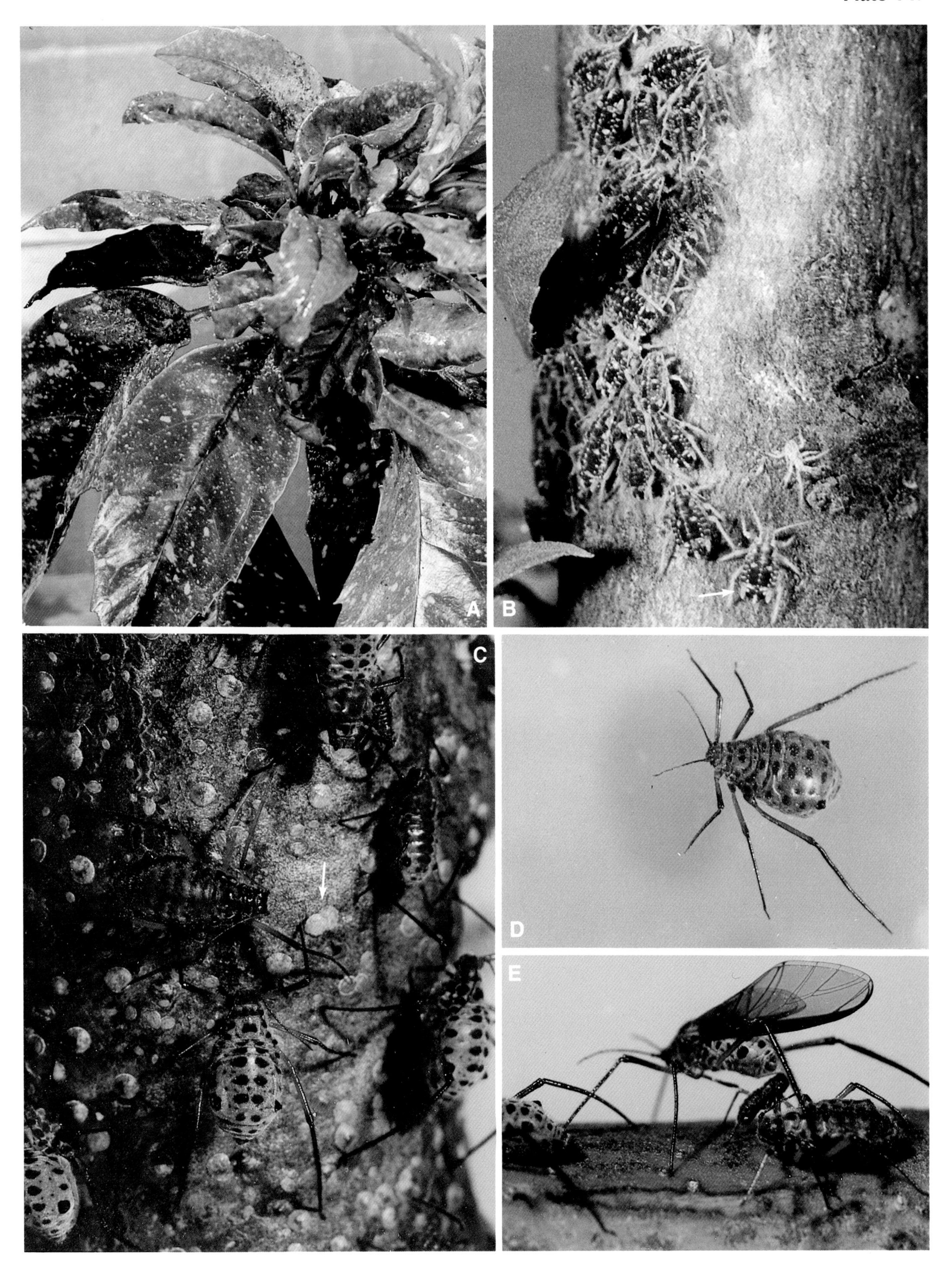
A
B
C
D
E

Aphids of Oleander and Bamboo (Plate 148)

The aphid *Aphis nerii* Fonscolombe is primarily an urban pest. It occurs in large populations where its host plant oleander, *Nerium oleander,* is regularly pruned, watered, and fertilized, for this aphid prefers to feed on actively growing terminal shoots.

A. nerii is a European species that gained access to this country some years ago. It is commonly referred to as the oleander aphid; milkweed is a secondary host. In the North oleander is used as a house plant and placed outside during the summer. Oleander aphids may be attracted to such plants and may require treatment with a pesticide before being brought indoors. The insect is bright yellow with black markings. It appears in the early spring and colonizes the young shoots and buds of oleander. Huge colonies often develop and persist into early summer, when warm temperatures and natural enemies contribute to their decline. Oleander is such a vigorous plant, however, that very little damage results from this pest. The aphid produces honeydew in copious amounts, reducing the plant's aesthetic appeal. The insect is now believed to have transcontinental distribution. A native braconid wasp, *Lysiphlebus testaceipes* (Cresson), is an important parasite of the oleander aphid in California.

Takecallis arundinariae Essig and *T. arundicolens* Clarke, both known as bamboo aphids, are common in California wherever bamboo is grown. *T. arundinariae* has been found in Oregon and British Columbia on the West Coast, and in New York, New Jersey, Maryland, and North Carolina. These aphids are pale yellow with black markings. The exact markings vary with the particular species. Both species infest the undersides of leaves. Here, large colonies are often formed, causing the foliage to blacken as a result of the sooty mold fungi that colonize the insects' honeydew. Weakening of the plant may also occur. Virtually nothing is known of the biology and seasonal activities of the bamboo aphid. Both species are believed to be of Asiatic origin.

References: 212, 217, 316

A. A colony of oleander aphids, *Aphis nerii,* on the foliage of oleander.
B. A colony of oleander aphids (adults and nymphs) on an oleander shoot.
C. Most of the oleander aphids shown here have been killed by a species of parasitic wasps.
D. Off-color foliage and dead leaves are symptoms of aphid injury to bamboo.
E. A colony of *Takecallis arundinariae* (adult females, nymphs, and winged aphids) on bamboo.

A
B
C
D
E

Leaf, Shoot, and Twig Aphids (Plate 149)

Utamphorophora (=*Amphorophora*) *crataegi* (Monell) is sometimes called the fourspotted hawthorn aphid (A, B). It feeds on twigs, shoots, and leaves of *Crataegus* species, causing leaves to curl.

The egg (Plate 150F) is the overwintering stage, and in Connecticut eggs hatch in early April. The wingless individuals that hatch from eggs, called stem mothers, give birth to living young. Both stem mother and her progeny feed on foliage. The progeny, in turn, give birth to another generation, and these aphids are presumed to remain on *Crataegus* hosts. Other species of *Utamphorophora* are known to have an alternate summer host, but none are known for the fourspotted hawthorn aphid. There may be several generations of the summer form. In late summer winged females develop, and if migration is to occur, they will do so at this time. Winged females, in mid to late September, give birth to wingless males and females. Once mature, these mate, and the female produces the overwintering egg, which she lays in bark crevices on the host tree.

This aphid is considered as a common but not a severe pest throughout its geographic range. There are, however, many records of populations that feed, in legion numbers, on twigs and shoots (A). In the fall they congregate in huge masses on twigs and branches. After a few days they line up into columns or rows and move, in endless succession, both horizontally and vertically, with columns of aphids going in both directions. At present this behavior is unexplained. The species is known from Nova Scotia to Florida and Louisiana, the north central and Rocky Mountain states, and the Pacific Northwest including British Columbia. They are likely to occur on exotic as well as native hawthorn species.

Nearctaphis (=*Anuraphis*) *crataegifoliae* (Fitch) is an example of an aphid with two obligate hosts. Its primary host(s)—the host(s) that receives the overwintering eggs and upon which the first progeny develop in the spring—are hawthorn, quince, or pyracantha. In the fall they lay eggs on twigs. The eggs hatch after buds open in the spring, and the nymphs feed on the newly developing leaves. At maturity, all individuals are female and give birth to active nymphs. These nymphs mature in a short time and give birth, asexually, to another generation of nymphs that, upon maturing, have wings and take flight to search for their summer host, any of several herbaceous legumes such as clover. Several wingless generations develop on clover during the summer. Winged individuals develop again in early September and begin their migratory flight back to the overwintering host. When they arrive back on their primary host, they give birth to young that develop into males or females. These individuals mate and produce the overwintering eggs.

N. crataegifoliae injures its woody host only in the spring during the months of May and June (in Maine) by withdrawing sap from developing leaves. The leaves respond by twisting into swollen dark purple curls. This aphid has been collected in 40 of the contiguous states and all of the Canadian provinces bordering the states.

Most aphids are difficult to identify under field conditions. By associating this aphid with its host, taking note of the time of the year when it is present, and observing the animal for unique body features, fairly good field accuracy can be attained. The nymph that becomes the spring migrant is pink to red. The winged migrant has a long beak, and its body is black and reddish yellow.

The honeysuckle aphid, *Hyadaphis tataricae* (Aizenberg), is common to eastern Europe and was first recognized on the North American continent in Quebec in 1976 and in Illinois in 1981. The aphid, along with its injury, is now found from the maritimes and New England west to Idaho and Alberta, and it appears to be spreading rapidly. It can now be expected in any location where bush honeysuckle grows. It is probably the most damaging arthropod pest of honeysuckle.

Injury to honeysuckle is the result of the plant's response to toxins or growth regulator substances in the aphid's saliva. The response is expressed in very short shoot elongation, the growth of dormant buds into a witches'-broom, the folding upward of dwarf leaves from the midvein axis, a fading to light green in the leaves, and the frost kill of all of the broom components (C, D).

Honeysuckle aphids, like most aphids found in the north temperate zone, overwinter as eggs. These are laid in the fall on shoots, twigs, and small stems, and they hatch in the spring at the time of bud break. The aphid that develops from the egg is described as a stem mother. She gives birth, asexually, to vast numbers of living young. She and her offspring feed on new shoots on the undersides of honeysuckle leaves. The second and all additional summer generations of aphids feed on the upper surface of leaves, causing them to fold upward. It is in the folded leaf that aphids may be found (E). Summer aphids, about 2 mm long, are pale green to cream colored, and they have a fine powdery wax over their bodies. In September winged males and wingless females may be found.

Studies currently indicate that there are some resistant and tolerant species and cultivars. However, we are reluctant to give any names because of the confusion in the taxonomy of honeysuckles. Species of *Lonicera* known to produce witches'-brooms include *L. tatarica, L. korolkowii,* and *L. microphylla*.

References: 23, 70, 305, 710, 803

A. A colony of *Utamphorophora crataegi,* the fourspotted hawthorn aphid, on a large twig of the hawthorn *Crataegus crus-galli*. Picture made in Ithaca, NY, in early September.

B. A cluster of fourspotted hawthorn aphids feeding on the underside of a hawthorn leaf. [Aphididae]

C. A witches'-broom on a terminal of tatarian honeysuckle caused by the honeysuckle aphid *Hyadaphis tataricae*. This symptom is highly visible in early August.

D. Witches'-brooms on honeysuckle remain highly visible in the winter. The tiny brown leaves are often retained throughout the winter.

E. A deformed honeysuckle leaf, unfolded to reveal the honeysuckle aphid.

A
B
C
D
E

Woolly Root and Twig Aphids (Plate 150)

The woolly apple aphid, *Eriosoma lanigerum* (Hausmann), causes woody galls on twigs, branches, and roots of apple, hawthorn, mountain ash, and pyracantha. Additional injury includes loss of vigor and loss of aesthetic value through the production of honeydew. On its primary host, American elm (*Ulmus americana*), it causes clusters of rolled, twisted, dwarfed, or otherwise malformed leaves at the growing tip of a shoot. These abnormal leaf clusters are called rosettes.

Conflicting results in research and natural history observations by the various investigators makes the life history of this insect difficult to summarize. All of the truth is not known, and what is known may have been faultily analyzed.

There are two overwintering forms: eggs on American elm, and nymphs (occasionally adults) on the roots of rosaceous hosts. It is generally accepted that American elm is the primary host and that if eggs are laid on this tree, they are produced by the progeny of migrants that fly to elm from their summer hosts in the fall. In the spring eggs hatch, and the nymphs and future offspring cause the rosettes described above. In late spring winged aphids are produced. The entire population vacates elm and flies to apple or another rosaceous host. Here, woolly clusters of wingless aphids amass on the bark around the base of a petiole or on scar tissue around a wound. Galls may form around these feeding sites (A, C). While this has been going on above ground, aphids on the roots have been feeding and reproducing into larger colonies. Roots respond by developing more and larger nodules, or galls (Figure 102). By the middle of the growing season there is a slow but continuous walking migration of aphids from aerial colonies to the roots and a corresponding migration from root colonies in the opposite direction.

In early autumn winged aphids appear in both aerial and root colonies and fly to elm, where sexual (male and female) aphids develop. They mate, and each female lays a single egg. In the western United States and British Columbia, and probably many other places, migration back to elm does not occur. Aerial colonies are usually killed by frost; the root colonies are protected, allowing for overwintering and a continuous infestation on the rosaceous host.

The woolly apple aphid has been recorded in 33 states and in Canada from Nova Scotia to British Columbia.

References: 131, 366, 394, 599

Figure 101. A woolly apple aphid, *Eriosoma lanigerum,* on an apple root. (Courtesy P. L. Lehman, Florida Dept. of Agriculture and Consumer Services, Div. of Plant Industry. #702701-12.)

Figure 102. Galls on the roots of an apple seedling caused by the feeding of woolly apple aphid colonies. (Courtesy Florida Dept. of Agriculture and Consumer Services, Div. of Plant Industry. #702702-15.)

A. Woody galls caused by the woolly apple aphid and a colony of aphids on a small branch of an apple. The aphids are underneath the patches of "wool." (Courtesy New York State Agricultural Experiment Station, Geneva.)

B. Colonies of woolly apple aphids as seen through dense pyracantha foliage.

C, D. A pyracantha twig showing galls caused by the woolly apple aphid. The aphids had vacated this host by the time the photograph was made.

E. A colony of black aphids feeding on the succulent shoot and twig of the currant *Ribes sanguineum.* These aphids probably belong to the genus *Aphis.*

F. A mass of eggs of the fourspotted hawthorn aphid (see also Plate 149).

A
B
C
D
E
F

Whiteflies of Rhododendron and Azalea (Plate 151)

The rhododendron whitefly, *Dialeurodes chittendeni* Laing, is an important pest of rhododendron in the United States. It arrived in this country about 1932 on rhododendrons imported from England, where the pest had been known for many years. Entomologists believe, however, that the insect is native to the Himalayan region of Asia.

The rhododendron whitefly prefers to feed on the tender terminal leaves of its host plant. The adults lay eggs on the undersides of leaves, and when the nymphs hatch, they affix themselves there to feed for the duration of their immature lives. Heavily infested foliage takes on a yellow mottled appearance on its upper surface. On some varieties a curling of the leaf margins occurs as a result of the whitefly's feeding. Honeydew produced by the nymphs drops to the leaves below, giving them a varnished appearance. Sooty mold fungi grows on the honeydew-covered leaves, changing the varnished appearance to that of a leaf dusted with coal soot. This type of injury is both harmful and ugly, and on a plant such as rhododendron, which retains its leaves for about 3 years, the sooty appearance will likely remain until the leaf is shed.

So far as is known, the rhododendron whitefly attacks only rhododendron. Research has shown a great deal of difference between the susceptibility of plant varieties. Those plants that have a thick and leathery epidermis often escape infestation.

In the state of Washington a single generation of the rhododendron whitefly occurs annually. The insect overwinters in the nymphal stage. During the growing season there is much variation among the developmental stages found at a single location, and one or another of the immature stages may be found at any time. Adults, however, occur only from mid-May until early August.

The azalea whitefly, *Pealius azaleae* (Baker & Moles), is widespread in the eastern and southern United States and in California, wherever the hairy-leaved *Azalea ledifolia alba* (=*mucronatum*) is grown. This evergreen azalea is hardy as far north as Washington, DC, and is widely grown in southern gardens. It evidently has been used as a parent by many hybridizers, because named varieties now on the market have hairy leaves and are susceptible to whitefly attack. In the immature stages the azalea whitefly is pale yellow to orange and lacks the distinctive waxy fringe characteristic of many other juvenile whiteflies.

For additional information on whiteflies that are destructive to woody ornamental plants, see Plates 152–153.

References: 217, 458, 836

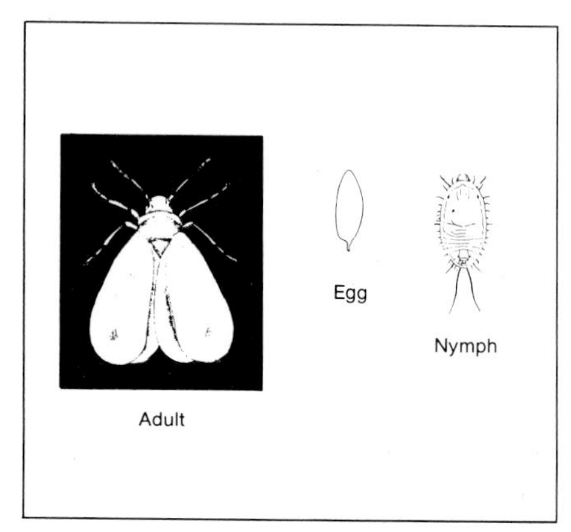

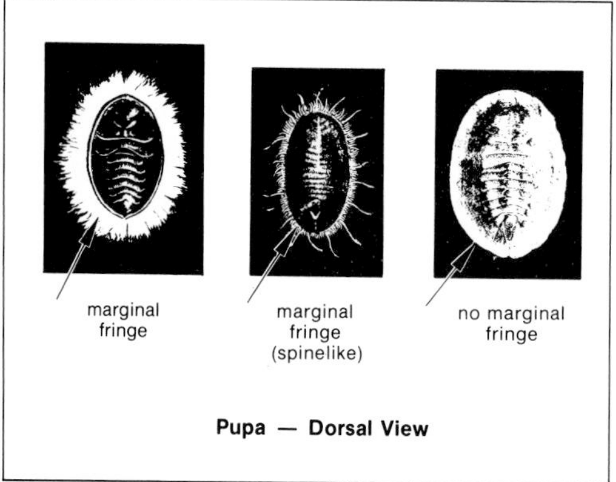

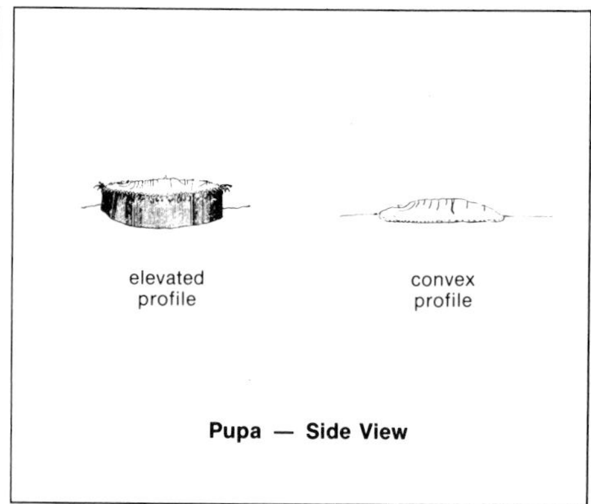

Figure 103. Whitefly life stages. (Courtesy R. J. Gill, in "Color-Photo and Host Keys to California Whiteflies," California Dept. of Food and Agriculture, Sacramento.)

A. The yellowing, cupping, and rolling of terminal leaves of rhododendron are all characteristic symptoms of injury by the rhododendron whitefly.
B. The undersurfaces of rhododendron leaves showing numerous whitefly adults.
C. Adults and eggs of the rhododendron whitefly. [Aleyrodidae]
D. Immature azalea whiteflies. Note the black discoloration caused by sooty mold fungi.
E. Close-up of the underside of an azalea leaf showing immature whiteflies.

A
B
C
D
E

Greenhouse Whitefly and Relatives (Plates 152–153)

In the United States whiteflies are far more important economically on *Citrus* fruit crops and on greenhouse-grown ornamentals than on ornamental plants grown out-of-doors. By and large, where outdoor plants are concerned, whiteflies are much more abundant in the southern portions of the country than in northern regions.

The wings of the whitefly are commonly covered with a powdery substance. Adults vary in size from 1 to 3 mm and are noticeable to the casual observer only when present in large numbers. Usually, eggs are laid by the adult females on the lower leaf surface (D, G). The eggs are initially light yellow, but those of many species turn gray as incubation proceeds. The oval eggs are attached to the plant by a stalk. They require 1–3 weeks to hatch, depending on the season. The young nymphs, or immature whiteflies, crawl about on the lower leaf surface for several hours and then settle to feed. After settling, the juveniles remain fixed in the same position until they grow to maturity. The nymph has a flattened, scaly form, and the nymphal period requires 3–4 weeks. A most distinctive feature of the nymph is the fringe of waxy material that projects radially outward from its body. This fringe may consist of fine strands or thick wavy plates, depending on the species of whitefly. Its characteristic appearance is helpful in making field identification of species. At the end of the nymphal stage the pupa appears. The insect spends this stage within a case that is fixed to the leaf. The case is somewhat more bulbous and segmented than the covering of the nymph. The adult whitefly emerges from this case through a T-shaped slit on the dorsal surface of the case.

On any infested leaf whiteflies may be found in all stages of development. There are several generations each year, the exact number depending on the species and its location. The insect overwinters in either the nymphal or pupal stage. Damage to plants by whiteflies is primarily caused by the removal of sap by the nymphs. Heavy feeding results in yellowing and drying of the foliage. Fully as important as direct feeding injury is that caused by the honeydew produced by the nymphs. The honeydew covers the leaves and imparts a blackened appearance as a result of sooty mold fungi that colonize the honeydew.

The greenhouse whitefly, *Trialeurodes vaporariorum* (Westwood), is present throughout the United States on many greenhouse plants but persists out-of-doors in the southern parts of the country. Some of its preferred woody ornamental hosts are ash; dogwood; sycamore; sweetgum; *Gleditsia triacanthos*—honeylocust; *Robinia pseudoacacia*—black locust; *Solanum pseudocapsicum*—Jerusalem cherry; and species of *Berberis*—barberry; *Cercis*—redbud; *Fuchsia*; *Hibiscus*; *Lantana*; *Rhamnus*—coffeeberry; and *Rosa*—rose.

The wasp parasite, *Encarsia formosa* Gahan, is an effective biological control agent of the greenhouse whitefly under greenhouse conditions.

The Stanford whitefly, *Tetraleurodes stanfordi* (Bemis), and *Aleuroplatus coronatus* (Quaintance) the crown whitefly, are well known in coastal California, where they attack the coast live oak, *Quercus agrifolia,* and, to a lesser extent, tanbark oak, *Lithocarpus* species, and coffeeberry, *Rhamnus* species. *T. acacia* (Quaintance), sometimes called the acacia whitefly, feeds on the orchid tree, *Bauhinia purpurea; Cassia brachiata; Rhamnus californica;* as well as *Acacia* species.

The deer brush whitefly, *Aleurothrixus interrogatonis* (Bemis), commonly occurs on *Ceanothus* species in California (B). The nymph of this species is pale yellow to tan and is characteristically surrounded by a transparent jellylike ring. Other forms of whitefly pupae are depicted in Plate 154.

References: 79, 107, 202, 217, 538

A. Fuchsia leaves infested by the greenhouse whitefly.

B. Ceanothus leaves supporting a high population of deer brush whiteflies, *Aleurothrixus interrogatonis*. These may be confused with ceanothus lace bugs (see Plate 206).

C. *Robinia pseudoacacia* leaflet with Stanford whitefly pupae, *Tetraleurodes stanfordi*. The pupa is about 0.8 mm long. The fringe of white wax filaments is characteristic of several whitefly species.

D. Eggs, presumably those of the Stanford whitefly.

E. Mahonia leaves heavily infested with deer brush whiteflies.

F. Infestation by a species of *Tetraleurodes* on the lower surface of leaves of the California bay.

G. Eggs of the greenhouse whitefly are about 0.5 mm long. They are laid on end and are attached to the leaf by a short stalk.

H. Nymphs of the greenhouse whitefly measure about 1 mm. The number of hairlike wax filaments is less than that of the Stanford whitefly in C. (See Plate 154E, F for photographs of other whitefly pupae.)

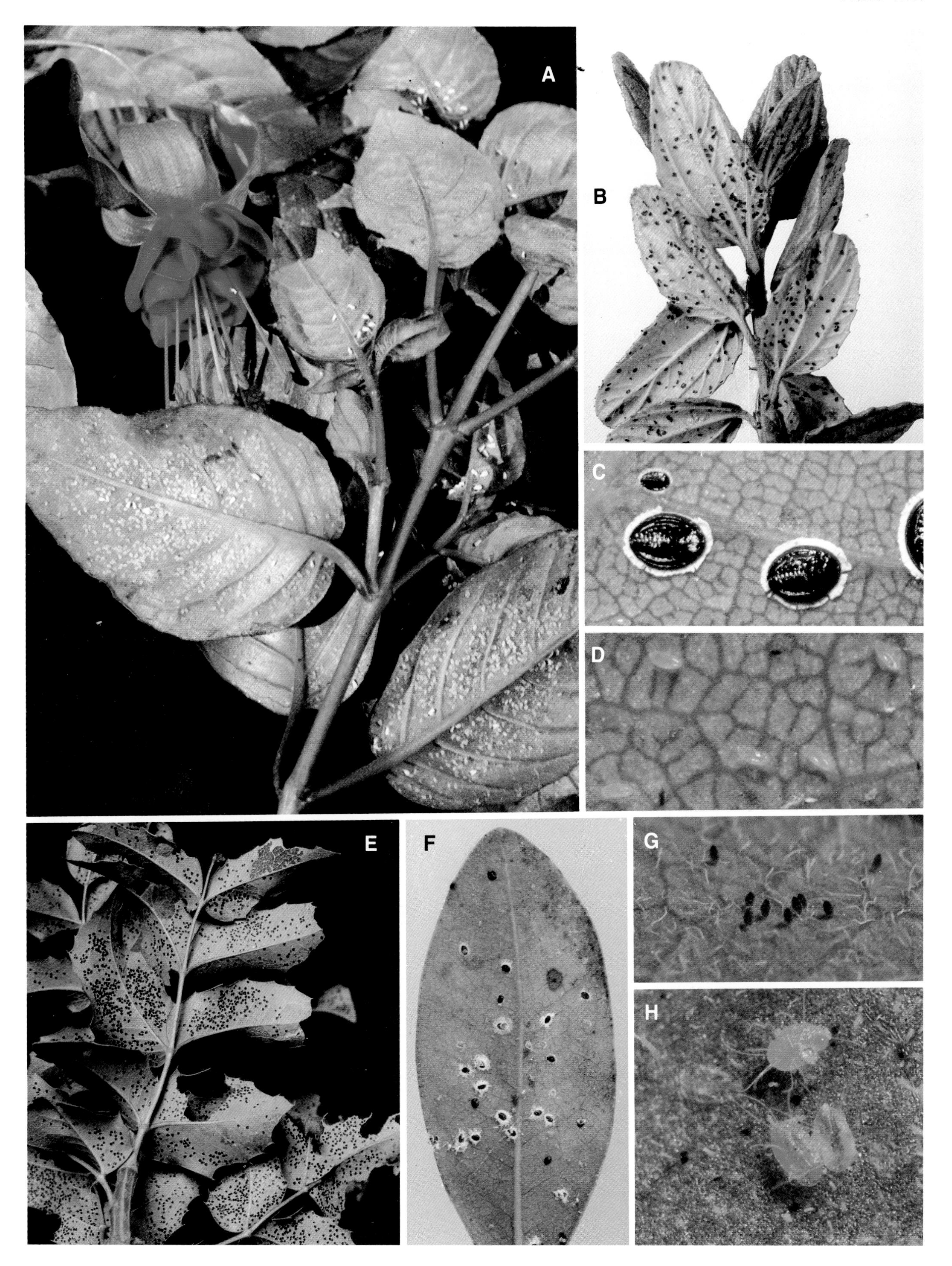
A
B
C
D
E
F
G
H

The woolly whitefly, *Aleurothrixus floccosus* (Maskell), was first observed in the United States in 1909 and is believed to have been accidentally brought into this country from Cuba. Its common name comes from the fluffy covering of long, curly, waxen filaments that cover the quiescent stage, often called the pupa (D, E).

In Florida four generations occur each year. The adults are active during December and January, and again in May, July, and October. They are much like other whitefly species in shape and size (A) but are distinctly more yellow, though the color difference is not apparent in the photograph. They differ from other whiteflies in their sluggish habits; they are so sluggish that the adult frequently lays its eggs on the leaf on which it developed.

Eggs are laid in circles (Figure 104) on the undersides of leaves. The first-stage larva is light green. Soon after hatching it inserts its sucking mouthparts and does not move again until it becomes an adult. The third-stage larva develops a woolly coat, which remains on it throughout the pupal stage (E).

The injury caused by the woolly whitefly is threefold: it withdraws great quantities of sap from the trees; excretes copious amounts of honeydew, on which sooty mold grows and subsequently interferes with the proper functioning of the leaves; and creates conditions favorable for severe infestations by purple scale, *Lepidosaphes beckii* (Newman). The honeydew given off by this species is quite syrupy and is eagerly sought by ants and wasps.

Table 16. **Other whiteflies that infest ornamental plants**

Scientific name	Common name	Ornamental hosts
Aleurocanthus woglumi	Citrus blackfly	*Mangifera indica*, citrus, *Rhaphiolepsis indica*, *Citrofortunella mitis*
Aleuropleurocelus nigrans	Black aleyrodid	Manzanita, madrone, *Ceanothus* spp.
Aleuroplatus berbericolus	Barberry whitefly	Oregon grape, loquat, *Malus* spp., *Prunus* spp., persimmon
Aleyrodes amnicola	Willow whitefly	Willow, gooseberry
Bemisia tabaci	Sweetpotato whitefly	*Hibiscus rosasinensis*
Dialeurodes citrifolii	Cloudywinged whitefly	Citrus, *Ficus* sp., gardenia
Parabemisia myricae	Japanese bayberry whitefly	Citrus, gardenia, avocado, camellia, mulberry, others
Tetraleurodes acaciae	Acacia whitefly	*Acacia* spp., *Calliandra* spp., redbud, orchid tree, *Bauhinia* spp., mimosa
Trialeurodes merlini	Cottony manzanita whitefly	Manzanita
T. vittata	Grape whitefly	Coffeeberry

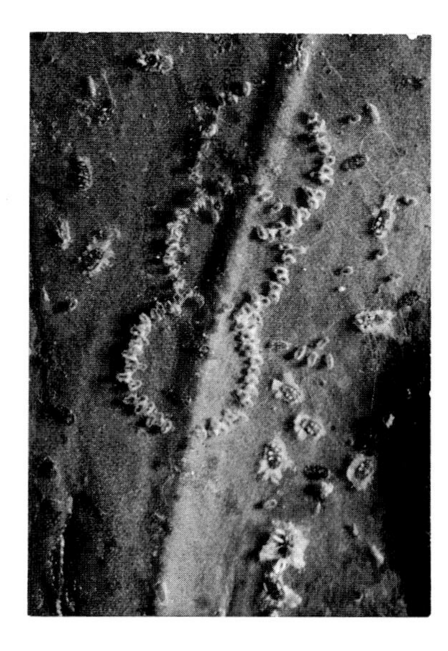

Figure 104. Woolly whitefly eggs laid in a circular pattern on the underside of leaves. (Courtesy A. Hamon, Florida Dept. of Agriculture and Consumer Services, Div. of Plant Industry.)

The woolly whitefly, in the United States, is a pest of many woody and herbaceous subtropical plants. It is rather common in Florida and the coastal regions of the Gulf states. Woody plants subject to attack include both ornamental and orchard plants, for example, species of *Citrus* and *Calliandra, Eugenia uniflora, Ficus* species, *Rhododendron indicum*, and seagrape. The insect is kept in check with little help from humans by the minute wasp parasite *Eretmocerus haldemani* Howard and by *Encarsia braziliensis* (Hempel). *E. haldemani* also attacks other species of whiteflies.

A striking red fungus, *Aschersonia aleyrodis* (F), is an effective natural control organism that attacks the nymphal stage of whiteflies. Spore solutions of the fungus have been prepared and sprayed on citrus orchards in Florida, resulting in successful control of the pest.

References: 304, 323, 449, 818

A. Young hibiscus leaves with many adult greenhouse whiteflies, *Trialeurodes vaporariorum*. [Aleyrodidae]
B. *Catalpa japonica* badly damaged by *Tetraleurodes* sp. in California. Leaves are not only mottled from feeding injury but are disfigured by the growth of sooty mold.
C. A citrus leaf with a colony of woolly whiteflies.
D. Close-up of the cottony filaments formed by the woolly whitefly. Note pupa at arrow.
E. A magnified view of woolly whitefly pupae. They are approximately 2 mm long.
F. A fungus, *Aschersonia aleyrodis*, which is pathogenic to immature whiteflies. The red fungus covers the juvenile whiteflies.

A
B
C
D
E
F

Mealybugs and Whiteflies (Plate 154)

Mealybugs derive their common name from the white waxy, mealy secretions that cover their bodies. Mealybugs infest all plant parts: feeder roots, root crowns, stems, twigs, leaves, flowers, and fruits. The longtailed mealybug, *Pseudococcus longispinus* (Tarioni-Tozzetti), is a plant pest in Florida, Texas, and California and occurs as a greenhouse pest in most states. This species feeds on new shoots and leaves and has a wide range of hosts, including gardenia (A), pittosporum, *Pinus pinea, Cedrus deodora, Ceanothus* species, citrus, persimmon, *Hedera helix,* hibiscus, *Ficus* and *Jasminum* species, apple, oleander, avocado, eucalyptus, yew, and rhododendron.

The major feature for field identification of the longtailed mealybug is its long posterior wax filaments (D). These filaments may be as long as the insect's body. They break off readily, but in any large cluster of mealybugs there normally are several individuals that have retained their long tails. Adult females are about 3 mm long. They multiply rapidly, and all stages usually occur on a host plant at the same time. It is believed that in warm climates they never produce eggs but instead give birth to living young called crawlers. Ants sometimes aid in distributing mealybugs from plant to plant. Ants are attracted to the mealybugs' honeydew excretions.

Injury to the host plant is caused by loss of sap, which results in discolored and wilted foliage, deformed leaves (A), and the eventual death of the affected parts. Mealybugs often share their host with other species of homopterous insects such as soft scales (C) or whiteflies (E). In Florida, mealybug populations are greatly reduced by *Entomophthora fumosa,* a parasitic fungus.

The apple mealybug, *Phenacoccus aceris* (Signoret), is another insect of European origin that has a wide host range. Some of its woody ornamental hosts include pyracantha, cotoneaster, laurel, blueberry, honeysuckle, maple, cherry, plum, elm, linden, chestnut, dogwood, filbert, persimmon, mulberry, oak, camphor, magnolia, apple, hawthorn, and species of *Cercidiphyllum, Sorbus,* and *Myrica.*

Injury, as with most phloem-feeding insects, occurs because of the vast amount of sap removed from shoots, twigs, and foliage. The excess sap is excreted as honeydew, which is sticky. Dust adheres to the honeydew and sooty mold fungus grows on it, making an infested plant look dirty and unthrifty. This mealybug overwinters as a second-instar nymph on twigs and branches. The adult female can be distinguished by a longitudinal ridge of white wax on top of the body. If placed in alcohol, the body appears green. When sexually mature, the female produces a dirty white ovisac at the end of her abdomen within which she deposits eggs.

The wasp parasite *Allotropa utilis* (Meusebech) is important in biological control of the apple mealybug.

P. aceris occurs in the northeastern states, eastern Canada, Idaho, all of the Pacific Coast states, and British Columbia.

The obscure mealybug, *Pseudococcus affinis* (Maskell), is a slow-moving oval insect that is lightly covered with white powdery wax that does not completely hide its light gray body color. Fully grown females are 3–5 mm long. It is considered to be the most common and widespread mealybug in California and feeds on herbaceous as well as woody plants. This is one of several mealybugs that feed on above-ground plant parts (twigs and shoots) as well as on roots below the soil surface.

Woody host plants include redbud, quince, *Citrus* species, *Baccharis* species, *Magnolia* species, coral tree, *Malus* species, *Pittosporum* species, *Quercus agrifolia, Platanus racemosa,* and *Salix* species. In northern states it shares the greenhouse with the longtailed mealybug. In warm areas, namely the Gulf Coast states, it occurs as an outdoor pest.

The mulberry whitefly, *Tetraleurodes mori* (Quaintance), was once thought to be subtropical in distribution. The literature recorded it as occurring in most of the southern states including California. In 1990 it was found in New York in the foothills of the Catskill Mountains, where it was causing severe damage and mortality to understory *Kalmia latifolia*. It is now known to occur in Quebec and the Maritime Provinces. Mulberry is probably one of its primary hosts, but it also feeds on Norway and red maple (*Acer rubrum*), boxelder, wax myrtle, American holly, *Cornus florida, Eugenia* species, sweetgum, *Citrus* species, *Kalmia latifolia, Itea virginica,* and less frequently on *Persea* species, *Magnolia macrophylla,* and *Pittosporum tobira.*

All stages of this foliage feeder are always found on the underside of host leaves. The pupa is distinctive, shiny black with a cottony white fringe of wax filaments along its outer margin (see Plate 152C for a similar pupa). Its length is about 7 mm. Adults are bright yellow. The first pair of wings have red and brownish black spots.

Injury to host plants occurs from sap removal, resulting in premature leaf drop and an unthrifty appearance. Vast amounts of honeydew are excreted and are quickly colonized by sooty mold fungi. In California this whitefly is frequently parasitized by a species of *Eretmocera* and preyed upon by chrysopids.

Aleurochiton forbesii (Ashmead) is one of the few northern species of whitefly found from Georgia to Canada and west to Missouri and Wisconsin (E). It occurs exclusively on maples and feeds on the undersides of leaves. It rarely causes measurable damage to its host trees. Little is known about its biology and behavior.

Siphoninus phillyreae (Halliday), known locally as the ash whitefly, was first reported from the Western Hemisphere in Los Angeles County, California, in 1988. Populations have reached extraordinary levels in southern California, and several counties in northern California reported the insect late in 1989. On the basis of its occurrence in the Old World, from Poland and Ireland to the north to Sudan and Ethiopia to the south, infestations in most of the southern two-thirds of the United States seem possible in the years ahead. The ash whitefly infests many ornamental hosts including ash, flowering pear, toyon, western redbud, crape myrtle, tuliptree, privet, lilac, pomegranate, serviceberry, flowering quince, crabapple, and pyracantha.

The ash whitefly is a cause of concern for several reasons. In heavy infestations the clouds of adult insects milling about make life unpleasant outdoors. The honeydew that the nymphs excrete gives foliage a sooty appearance. These excretions also foul cars, outdoor furniture, sidewalks, and anything else beneath infested trees. Ants, wasps, and other insects attracted to the honeydew add to the nuisance problem. Premature loss of leaves results from the sucking of sap by large populations of nymphs.

The ash whitefly adults look much like other whiteflies. The late-instar nymphs can be separated from most other whitefly nymphs by the broad central stripe of wax on the dorsal surface.

In the Old World the ash whitefly is not reported to be a common pest except when it has been introduced into areas lacking effective natural enemies or when populations of natural enemies have been disrupted by the use of pesticides. The prognosis for successful biological control of this insect in the United States, therefore, seems bright.

References: 202, 239, 324, 449, 518, 553a, 639, 644, 758, 884

A. Colonies of the longtailed mealybug, *Pseudococcus longispinus,* on a shoot and leaves of *Gardenia augusta.*

B. A shoot colony of longtailed mealybugs. × 4.

C. Mealybugs of the genus *Pseudococcus. At arrow:* a recently settled soft scale crawler. The insects are on *Viburnum opulus nanum.*

D. Longtailed mealybugs on *Ficus benjamina.* [Pseudococcidae]

E. Pupae of the whitefly *Aleurochiton forbesii* on *Acer rubrum.* Length of the pupa is 1.5 mm. [Aleyrodidae]

F. A whitefly pupa (probably the greenhouse whitefly) belonging to the genus *Trialeurodes.* Length of the pupa is 0.7 mm. (See also Plate 152.)

A
B
C
D
E
F

Comstock Mealybug (Plate 155)

The Comstock mealybug, *Pseudococcus comstocki* (Kuwana), is believed to have originated in Asia. It was first reported in the United States in 1918 at two distant points, California and New York. It has since spread to all the coastal states and into the Ohio and Mississippi river valleys. Its hosts include mulberry, maple, pine, *Catalpa bungei,* peach, apple, pear, Boston ivy, holly, boxwood, avocado, California privet, regal privet, citrus, Dutchman's pipe, elm, *Elaeagnus* species, *Euonymus alatus, Aesculus* species, *Hibiscus* species, olive, *Paulownia tomentosa,* persimmon, *Photinia villosa,* poplar, viburnum, wiegela, wisteria, and yew.

The biology of this insect was studied in Virginia on the umbrella catalpa, *Catalpa bungei.* The Comstock mealybug overwinters in the egg stage. The eggs hatch in May, and the young mealybugs feed throughout the spring on the undersides of the host's leaves. Usually the young nymphs are unnoticed until sooty mold grows on their deposits of honeydew, making the plant unsightly. As the nymphs approach the adult stage, there is a tendency for them to migrate from the leaves and to cluster on the older branches at a pruning scar, or a node, or at the base of a young branch. When a number of individuals begin to feed at the same location, a knotlike gall forms.

Oviposition occurs on the bark in early summer, and the generation hatched from these eggs in turn produces eggs in the fall. Eggs are deposited into a saclike structure attached to the female's abdomen. The sac is covered with a grayish white wax. When the female lays all of her eggs, she dies, and her dead body adheres to the ovisac.

Nymphs and adults do all their feeding in phloem tissue, from leaves (leaf petiole and veins), from callus tissue on the bark of the trunk and large branches, and from the thin bark around the base of small branches. They are especially attracted to trees that have been fertilized and are growing vigorously. They injure plants by removing sap and by inducing abnormal plant growth in the form of knots and adventitious growth (B). When leaves are covered with honeydew, dust and dirt collect. When sooty mold develops, the affected leaves may stop functioning and drop prematurely.

Sex pheromone traps have been used successfully to monitor mealybug populations and provide data for timing pest management techniques. There are several wasp parasites of the Comstock mealybug. Among them are *Allotropa burrelli* Muesebeck, *A. convexifrons* Muesebeck, and *Clausenia purpurea* Ishii. Predators of this pest include the lacewings *Chrysopa rufilabris* Burmeister and *Hemerobius stigmaterus* Fitch.

References: 295, 389

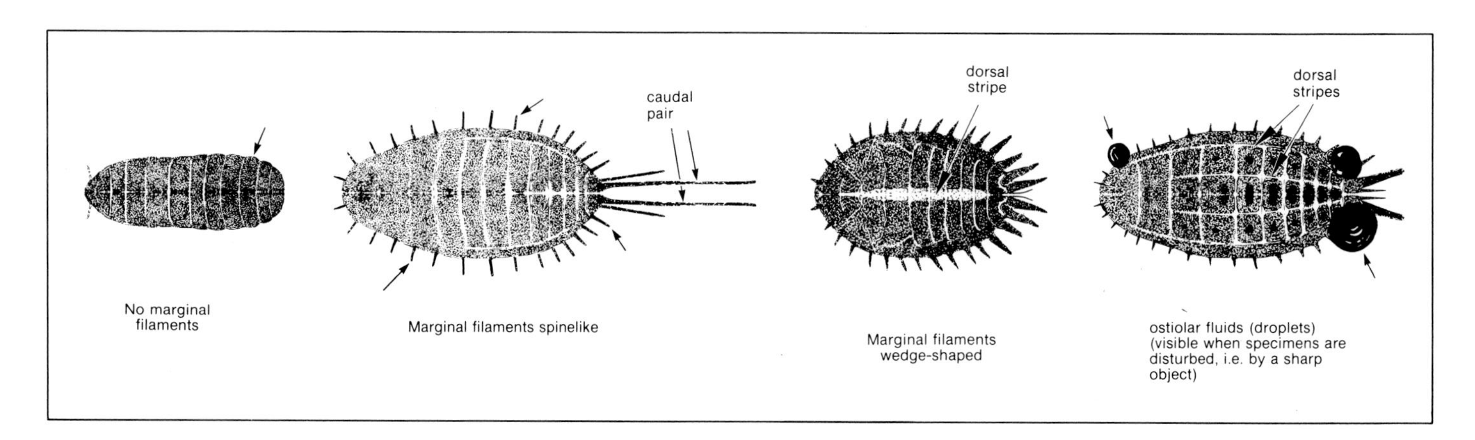

Figure 105. Mealybug shapes. (Courtesy R. J. Gill, in "Color-Photo and Host Keys to Mealybugs in California," California Dept. of Food and Agriculture, Sacramento.)

A. A mulberry tree with many woolly egg sacs of the Comstock mealybug, *Pseudococcus comstocki.* Note the knots and swollen areas at the base of the small branches.
B. A mulberry tree heavily infested with Comstock mealybugs.
C–E. Mealybugs on the bark of flowering dogwood. The insect is probably an aberrant form of the taxus mealybug (see Plate 37).

A
B
C
D
E

Subtropical Mealybugs (Plate 156)

Known only in California, the manzanita mealybug, *Puto arctostaphyli* Ferris, appears primarily on the leaves of its only recorded host plant, manzanita (*Arctostaphylos* spp.) The adult female is large (more than 5 mm long) and attractive. She is covered with long white filaments that radiate outward from her body. Like most mealybugs, this species is covered also with cottony white waxy material. The male of *P. arctostaphyli* is a tiny winged individual with slender rodlike appendages that are three to four times as long as the body. These appendages protrude from his posterior end. Virtually nothing is known about the biology or seasonal history of the manzanita mealybug.

P. arctostaphyli may be confused with *P. albicans* McKenzie, another California mealybug that infests manzanita and that is commonly called the white mealybug. The white mealybug lacks the long body filaments possessed by the manzanita mealybug. Instead, it has short platelike projections that radiate outward from the body of the female. Both species, however, may so cover the foliage of manzanita that the leaves appear to be lightly powdered with snow.

The manzanita mealybug reduces the vigor of the host plant by withdrawing plant fluids. The honeydew that is deposited by the feeding insects, and the sooty mold fungi that forms on the honeydew, detract from the appearance of the plant.

Other members of the genus *Puto* are frequently found on twigs and shoots of western conifers in Pacific Coast states and British Columbia. They too produce quantities of white filamentous wax, a sign of their presence. Twig dieback from large infestations may occur. (See Plate 32A for signs resembling those of *Puto* infestation.)

The citrus mealybug, *Planococcus citri* (Risso), is an omnivorous pest with a host list that seems endless. It occurs most frequently on herbaceous ornamentals and thus may become a serious greenhouse pest in the North. It is capable of overwintering outdoors only in the most southern states. Here they attack a number of woody ornamental plants, including species of *Bougainvillea, Ficus, Citrus, Gardenia, Codiaeum, Malus, Prunus,* and *Rosa.*

This insect feeds on above-ground parts, often clustering at shoot crotches and on foliage (E). Like most other mealybugs, it is covered with a fine white mealy wax. It moves slowly by crawling, until a favorable feeding site is found. Sexual reproduction is the common reproductive type for the citrus mealybug; however, parthenogenesis does occur. Females are sensitive to high temperature. About four times as many eggs are produced at 18°C as at 30°C. In practical terms outdoor populations produce the most progeny during the spring and fall. Yellow eggs are laid in an ovisac made of a loose cottony mass of wax filaments. Spring and fall ovisacs may contain 400–600 eggs.

When feeding, nymphs and adults withdraw great amounts of sap from the plant, most of which is excreted as honeydew, a sweet sticky waste product. The honeydew becomes a medium for the growth of a black fungus called sooty mold. The plant responds to mealybug injury by growth reduction, dwarf leaves, premature leaf drop, and dieback. A heavily infested plant looks anemic and dirty from dust adhering to the honeydew and sooty mold.

References: 239, 518, 760

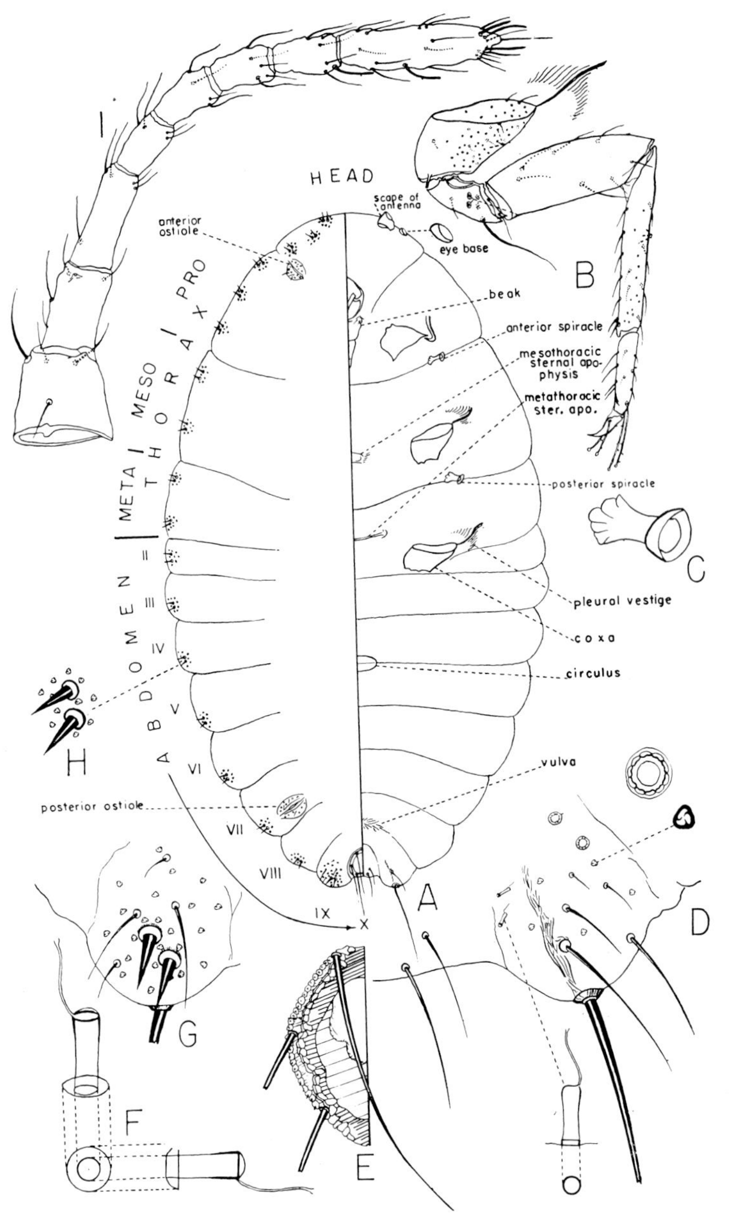

Figure 106. Diagram of a mealybug. The dorsal side is shown to the left of the center line; the ventral side to the right. Enlarged at *I* is an antenna; *B,* a leg; *C,* the posterior spiracle; *D* and *G,* the last abdominal segment. All of the major identifying characters are illustrated. Many species of mealybugs must be mounted on a microscope slide and viewed through a high-power microscope to obtain a positive identification. (Courtesy Univ. of Maryland Agricultural Experiment Station, College Park.)

A. The spots of white on the upper leaf surfaces of this manzanita bush are manzanita mealybugs, *Puto arctostaphyli.*
B. Close-up of manzanita mealybug adults showing their long waxy filaments.
C, D. Adult female manzanita mealybugs and young nymphs.
E. A cluster of citrus mealybugs feeding on a house plant. (Courtesy W. A. Sinclair, Cornell Univ.)

A
B
C
D
E

Noxious Bamboo Mealybug (Plate 157)

The noxious bamboo mealybug, *Antonina pretiosa* Ferris, is a subtropical insect commonly found in California and Florida. It is likely to be found in all of the Gulf states. Bamboo plants shipped north from these states may be infested with mealybugs. The mealybugs, however, are not likely to overwinter except in greenhouses.

The adult female mealybug and her eggs are entirely covered by a cottony white sac. She excretes honeydew that is colonized by sooty mold fungi, resulting in blackened plant parts and occasionally discolored egg sacs. When the eggs hatch, the young crawlers select nodes and the bases of leaves as feeding sites (B–D). Female crawlers lose their legs after the first molt and thus remain stationary for the balance of their lives. Bamboo is the only known host of the noxious bamboo mealybug.

References: 240, 518

A. Stems of bamboo showing dead leaf sheaths and blackened nodes. White woolly patches are present but difficult to see.

B–D. Noxious bamboo mealybugs collect under sheaths and at the bases of leaves. They feed on the stalks.

E, F. With the cottony wax removed from their bodies, noxious bamboo mealybugs are nearly black. [Pseudococcidae]

A
B
C
D
E
F

Beech Scale and Cypress Bark Mealybug (Plate 158)

From 1850 until the early 1900s a number of very important pests and diseases of trees were introduced into eastern America. These included the gypsy and satin moths, the European pine shoot moth, numerous sawflies, the balsam woolly adelgid, chestnut blight, and Dutch elm disease. *Cryptococcus fagisuga* Lindinger (=*fagi* Baer), commonly called the beech scale, is another pest that came to America from abroad.

The beech scale population, like that of the balsam woolly adelgid (Plate 30), consists only of females that produce other females. The pale yellow adult (F) is less than 1 mm in diameter and wingless, and has a spherical body covered with white woollike threads (D). The scales live only on the bark of native or introduced beeches (*Fagus* spp.).

Yellow eggs are laid in early July in the woolly filaments at the posterior of the female's body and hatch about 4 weeks later. The newly hatched nymphs, or crawlers, wander over the bark surface in search of a place to insert their long, tubelike mouthparts. The crawlers are spread from tree to tree by the wind. One generation of the beech scale is produced each year. Huge populations may develop on a tree in 2–3 years.

Heavily infested trees are covered with a white mass of "wool" (A). The beech scale itself seems to be of little consequence to the health of the tree, but in combination with the fungus *Nectria coccinea faginata,* it has killed many forest and ornamental beeches. The fungus cannot be controlled once it has colonized on the tree. Both fungus and insect are found north of Toronto and in eastern Canada and the northeastern United States south to West Virginia and western Virginia. The spread of this insect and concomitant disease seriously jeopardize shade tree and forest beech stands. The beech scale has spread much more slowly than the balsam woolly adelgid, a species that has similar habits.

The cypress bark mealybug, *Ehrhornia cupressi* Ehrhorn, sometimes referred to as the cypress bark scale or the cottony cypress scale, is found in California and Oregon on Monterey cypress, *Cupressus macrocarpa*; smooth Arizona cypress, *C. arizonica* var. *glabra*; and other *Cupressus* species. It also occurs on incense cedar, *Calocedrus decurrens*; and California juniper, *Juniperus californica.*

This mealybug is sessile and about 1.5 mm long when fully grown. Those shown in panel E are immature and therefore are just starting to cover themselves with a white waxy secretion. The waxy secretion completely hides the adult insect, which is red. Adults are found beneath bark flakes and in cracks on the trunk and larger branches of the host tree.

Infestations are often first noticed when white cottony wax protrudes from cracks in the bark and from beneath bark flakes. Dead twigs and small branches at the top of infested trees are the first visible symptoms of damage. These symptoms spread downward from limb to limb until the tree is dead. Yellow and red areas may appear on infested cypress hedges. These areas later die.

The cypress bark mealybug overwinters as a mature female. She begins laying eggs in the spring and continues through the summer. There is only one generation each year.

References: 87, 193, 275, 352, 518, 691, 898

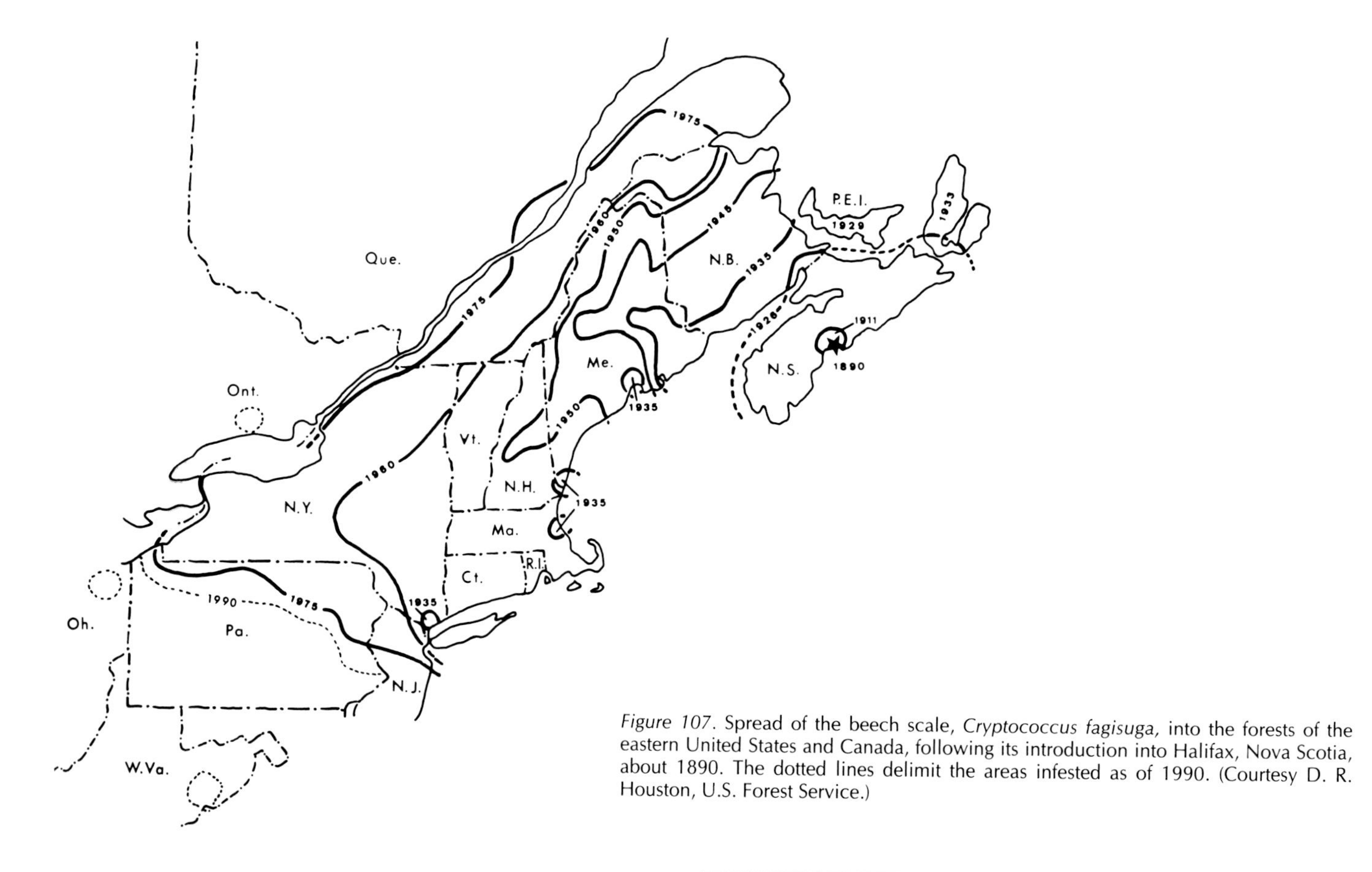

Figure 107. Spread of the beech scale, *Cryptococcus fagisuga,* into the forests of the eastern United States and Canada, following its introduction into Halifax, Nova Scotia, about 1890. The dotted lines delimit the areas infested as of 1990. (Courtesy D. R. Houston, U.S. Forest Service.)

A–C. Beech stems covered with the white woollike material produced by the beech scale.

D. Close-up of individual patches of wool. Each patch covers a single female. [Eriococcidae]

E. Young cypress bark mealybugs, *Ehrhornia cupressi,* on the trunk of Monterey cypress. Loose bark was removed to reveal the insects.

F. A fully exposed beech scale.

G. Yellowish eggs and crawlers of the beech scale are intertwined in the white woolly egg sac.

A
B
C
D
E
F
G

Sycamore Scale (Plate 159)

The sycamore scale, *Stomacoccus platani* Ferris, is the most important pest of both native and introduced sycamore (*Platanus* species) in California. The insect also occurs in Arizona. Infested leaves become disfigured by yellow spots that develop at each feeding site. These spots gradually turn brown as the affected tissue dies (G). Young infested leaves are often smaller than uninfested ones and often become distorted. Twig dieback may also result from the feeding of sycamore scales on the bark of branches and twigs. The sycamore scale can cause continual and premature leaf drop on sycamore. Leaf drop, however, may also be caused by other factors.

The scales overwinter on the tree beneath plates of bark on the trunk and limbs (C). In late winter the scales mature, and the females begin to lay eggs in masses of cottony wax. In the early spring the eggs hatch, and the tiny crawlers leave the woody parts of the tree and migrate to the new foliage. There, they insert their mouthparts into the leaves and begin feeding. At maturity, the majority of the scales migrate again to woody parts of the tree and lay eggs to begin the cycle again. There may be three to five generations each year.

The sycamore scale is unusual among scale insects .because it is mobile while in the adult stage. Female scales of most other species lose the use of their legs after feeding begins and do not move again during the remainder of their lives. Male scales fly about during their brief 2- or 3-day adult life.

References: 86, 721, 723, 724

A. Twig of sycamore showing egg-laying sites of sycamore scales, *Stomacoccus platani*. The cottony wax in the cracks of the bark indicates the presence of eggs. [Margarodidae]

B. Close-up of twig shown in A. Note orange-yellow eggs barely visible within masses of cottony wax. See arrows in B and D.

C. Overwintering sites of sycamore scales beneath bark plates. Females lay the eggs in cottony wax.

D. Newly hatched crawlers of the sycamore scale leaving egg masses. Both eggs (*arrow*) and crawlers appear as yellow spots within the white cottony material.

E. Developing sycamore scales on a leaf of sycamore. Note the leaf discoloration beneath sites where scales are feeding.

F. A female sycamore scale laying eggs. Sycamore scales are about 1.6 mm long. Note the cluster of eggs at arrow.

G. A sycamore leaf with yellow spots, which indicate scale feeding sites.

H. Distorted leaves (*upper right*) are characteristic of heavy sycamore scale attack.

A
B
C
D
E
F
G
H

Azalea Bark Scale (Plate 160)

The azalea bark scale, *Eriococcus azaleae* Comstock, was described in 1881 from specimens found on azalea in a greenhouse in Washington, DC. It was discovered in Connecticut in 1917, and since then has been reported from Florida, Alabama, Indiana, Ohio, Michigan, and New York. Eventually this species is likely to occur throughout the eastern United States. It feeds on the bark of twigs and stems and tends to settle in branch crotches. In addition to azalea, it has been found on rhododendron, andromeda, hawthorn, poplar, willow, and *Lyonia fruticosa.*

The azalea bark scale is closely related to, and in some ways resembles, mealybugs. It does not, however, belong to the mealybug group. The female scale is dark purple and is covered with a white waxy sac (E) but lacks the felted strands of wax characteristic of a mealybug. A mature female measures about 3 mm long, including her egg sac. Males are similar but are less than half the size of females. This scale might be confused with *Eriococcus borealis* Cockerell (some taxonomists say they are the same), which occurs in the western United States on species of *Salix, Populus, Ribes, Algama, Grossularia, Celtis,* and *Liquidambar.*

In Connecticut one generation of *E. azaleae* occurs annually. Overwintering scales mature in the spring, at which time they lay eggs. The eggs hatch in the latter part of June through mid-July. Crawlers tend to settle in bark crevices, in branch crotches, or on the axils of leaves. Two generations have been reported in the southern states, where first-generation eggs are laid in March and the second brood of crawlers occurs in September.

The azalea bark scale is a potentially serious pest, but to date its infestations have been minor and infrequent. About one-third of most populations of the azalea bark scale dies before winter for, as yet, unexplained reasons. Comstock, in his original description of the scale, reported that the majority of the specimens he collected had been parasitized by the chalcid wasp *Coccophagus immaculatus* Howard.

References: 511, 679

A. Typical appearance of azalea bark scales, *Eriococcus azaleae,* on azalea.
B. Azalea bark scale females in bark crevices. [Eriococcidae]
C. Azalea bark scales on a rhododendron twig.
D. The appearance of the feltlike sac of the azalea bark scale.
E. Side view of the azalea bark scale with its wax removed. Note the eggs, the white waxy covering, and the color of the female scale, which is about 2 mm long.

A
B
C
D
E

Cottony Cushion Scale (Plate 161)

The cottony cushion scale, *Icerya purchasi* Maskell, is one of the largest and most conspicuous scales that attack woody ornamental plants (A). It was accidentally introduced into California from Australia in 1868 and now also occurs in North Carolina, Arizona, Texas, Florida, and all the Gulf states. The most striking feature of this scale is the large, elongated, cottony white and fluted egg sac that protrudes from the end of the adult female's body. Inside the sac are hundreds of bright red oblong eggs.

Upon hatching, the scale crawlers (E) move to the leaves and twigs of the host plant and begin to feed. Later, the insects withdraw their mouthparts and migrate to larger twigs and branches. Unlike most scales, the cottony cushion scale retains its legs and its mobility throughout life. Damage to the host plant results from the extraction of sap by the insect. The subsequent reduction in vigor of the plant sometimes leads to defoliation or twig dieback. Also, plant parts become blackened by the sooty mold fungi that colonize the honeydew produced by the scales as they feed. The cottony cushion scale attacks a wide variety of woody ornamental plants: *Abies*—fir; *Acacia*—acacia, wattle; *Acer*—maple; *Aesculus californica*—California buckeye; *Aloysia* —verbena; *Buxus*—boxwood; *Carya*—pecan; *Casuarina*; *Cedrus libani*—cedar of Lebanon; *Celtis occidentalis*—hackberry; *Chaenomeles*—quince; *Citrus*; *Cupressus*—cypress; *Brahea edulis*—Guadaloupe palm; *Erythrina fusca*—coral tree; *Garrya*; *Juglans*—walnut; laurel; *Liquidambar*; locust; *Lyonothamnus*—ironwood; *Magnolia*; *Malus*—apple; *Nandina; the palm Chaedoria erumpens*; *Parthenocissus tricuspidata*—Boston ivy; *Pinus*–pine; *Pittosporum*; *Prunus*—almond, apricot, peach; *Punica*—pomegranate; *Pyrus*—pear; *Quercus*—oak; *Rosa*—rose; *Salix*—willow; *Salvia*—sage; *Schinus*—pepper tree.

The cottony cushion scale is a classic example of an insect that has been brought under biological control by the introduction of a natural enemy from the native home of the pest. Soon after it was discovered that the scale caused extremely serious losses to the California citrus industry, a small red and black lady beetle known as the vedalia, *Rodolia cardinalis* (Mulsant), was purposefully introduced into the California citrus groves from Australia. Within 2 years the cottony cushion scale ceased to exist as an economic pest. This was the first recorded instance of the successful introduction of a beneficial insect into any nation for the purpose of destroying an insect pest. The vedalia was introduced into other areas where the cottony cushion scale was causing problems on citrus or other plants, and the results were similarly effective. Where there have been noticeable reinfestations of this pest, the cause has often been the destruction of large numbers of vedalia by the use of pesticides on host plants.

References: 202, 217

A. Adult female cottony cushion scales, *Icerya purchasi,* on a branch of maple. [Margarodidae]
B. Orange crawlers leaving the egg sac of female scale in search of a suitable feeding site.
C. An adult female cottony cushion scale (*circle*) and immature scales on a leaf of ivy.
D. Immature cottony cushion scales on the underside of a leaf.
E. Close-up of an immature scale.
F–H. Partly grown cottony cushion scales. Note that in H legs persist.

A
B
C
D
E
F
G
H

Cottony Maple Scale (Plate 162)

The *Pulvinaria* scales are classified as soft scales and are characterized by their white cottony ovisacs. At least eight species are found in the United States. Four of these—the cottony maple scale, the cottony maple leaf scale, the cottony azalea scale, and the cottony camellia scale—are shown in Plates 162–164.

The cottony maple scale, *Pulvinaria innumerabilis* (Rathvon), is one of the largest and most conspicuous of the many scale insects that attack ornamental trees in the United States. It has been reported from virtually every state and from several Canadian provinces. As the common name implies, the favored host of this insect is maple—particularly soft maple—but it also occurs on a wide variety of other woody ornamental plants: *Acer negundo*—boxelder; *Alnus*—alder; *Celtis*—hackberry; *Cornus*—dogwood; *Crataegus*—hawthorn; *Euonymus*; *Fagus*—beech; *Maclura pomifera*—osage orange; *Malus*—apple; *Morus*—mulberry; *Parthenocissus quinquefolia*—Virginia creeper; *Platanus*—sycamore; *Populus*—poplar; *Prunus*—peach, plum; *Pyrus*—pear; *Quercus*—oak; *Robinia pseudoacacia*—black locust; *Rosa*—rose; *Rhus*—sumac; *Salix*—willow; *Syringa*—lilac; *Tilia*—linden; *Ulmus*—elm.

The cottony maple scale overwinters as an immature flat, inconspicuous female on the twigs of its host. With the onset of warm spring temperatures, it grows rapidly. By late spring the characteristic white egg sac of the female is evident (C). Each egg sac contains up to 1000 eggs.

In late June and July tiny motile crawlers begin to appear. These young scales migrate to the undersides of the leaves of the host plant, where they insert their slender mouthparts in or near veins and feed by withdrawing sap from the plant (B).

The cottony maple scale spends the summer months on either surface of the leaves. Male scales reach maturity in late summer and emerge as tiny winged individuals, which mate with the immature females. The males, which have nonfunctional mouthparts and cannot feed, live for only a day or two. Unfertilized females produce eggs, but all progeny are males.

Before the leaves begin dropping in the fall the females migrate back to the twigs, where they attach themselves for overwintering. A single generation of the cottony maple scale occurs each year.

Damage to the host tree is caused in several ways. Withdrawal of plant sap by heavy populations causes the dieback of twigs and branches and, under extreme conditions, may kill the entire tree. During the time the scales are feeding on leaves and on twigs, a large quantity of honeydew is produced by the insects. This material is colonized by sooty mold fungi, which imparts a blackened appearance to leaves, twigs, and branches. Some premature loss of foliage results from scale attack.

A number of natural enemies of the cottony maple scale are quite important in regulating populations of this species. Many wasp and fly parasites and various lady beetles attack the immature scale stages while they are feeding on the leaves (see Figure 112, accompanying Plate 167). The English sparrow is believed to be an important factor in reducing the populations of fully grown females.

The cottony maple scale may be confused with the cottony maple leaf scale, *P. acericola* (Walsh & Riley) (Plate 163), which it resembles. The cottony maple leaf scale produces its cottony egg sacs on the leaves—not the twigs and small branches—of its host.

References: 69, 193, 325, 412, 615, 670

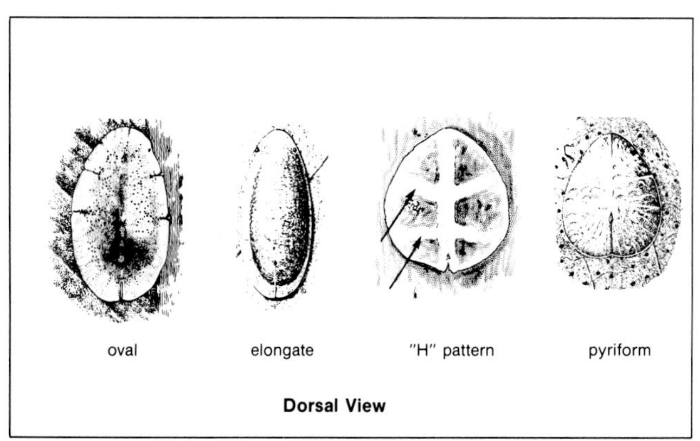

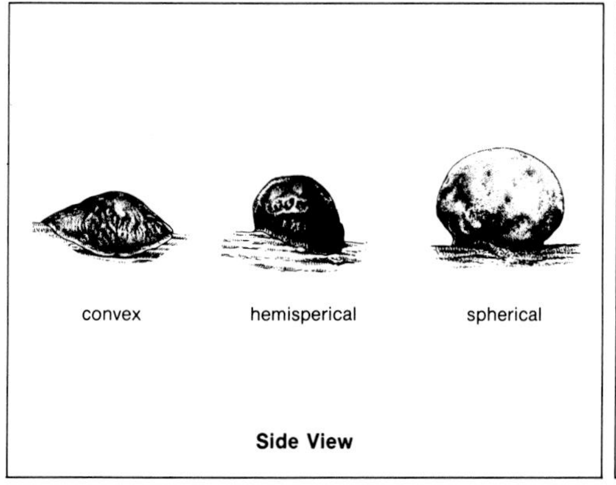

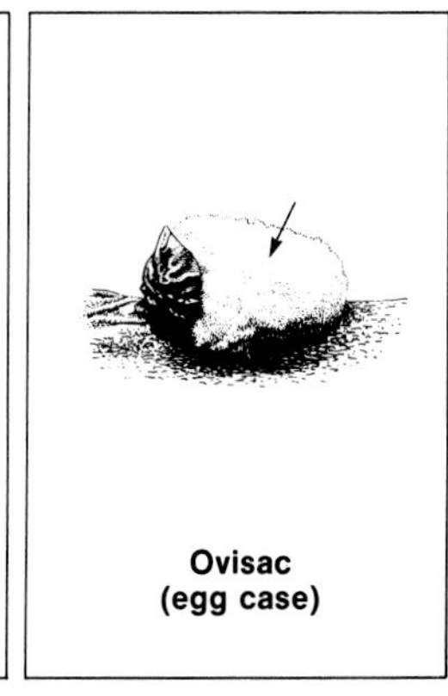

Figure 108. Soft scale shapes. (Courtesy R. J. Gill, in "Color-Photo and Host Keys to Soft Scales in California," California Dept. of Food and Agriculture, Sacramento.)

A. Adult cottony maple scale females on the underside of a branch of silver maple, *Acer saccharinum*. Mature females are about 5 mm long.

B, D. Close-ups of immature cottony maple scales on the underside of a maple leaf.

C. Dead females of cottony maple scale with their white egg sacs. [Coccidae]

E. The cottony maple scale egg sac has been pulled apart. The brownish spots inside the cottony mass are eggs. The brown structure at the top of the photograph is the dead female skeleton.

F. Mature cottony maple scale females look like popcorn strung out along twigs of maple.

A
B
C
D
E
F

Cottony Scales and Mealybugs (Plate 163)

Some species of scales, mealybugs, whiteflies, and aphids produce white waxy secretions that cover their bodies, making field identification confusing. On eastern maples, for example, it is common to find "cottony" mealybugs, scales, and aphids. The cottony maple scale, *Pulvinaria innumerabilis,* is illustrated in Plate 162.

The cottony maple leaf scale, *Pulvinaria acericola* (Walsh & Riley), occurs primarily on maples throughout most of the eastern United States, as well as in Iowa. It has also been reported in some western states and in southern Canada. Hosts other than the maple include dogwood; the hollies, *Ilex cornuta, I. crenata,* and *I. opaca*; andromeda; redbay; honeysuckle; and sourgum, *Nyssa sylvatica*. The cottony maple leaf scale may become numerous enough to cause premature leaf drop and sometimes even the death of twigs and branches.

Before egg sacs begin to form, the adult female is maroon, with a middorsal keel that is yellow-brown. Partially grown nymphs feed on twigs and branches of the host tree and later overwinter at their feeding sites. In April the males mature and mate with immature females. In May the mature females migrate to the host's leaves, where they lay their eggs in a white ovisac. On the average, their egg masses (ovisacs) contain more than 2500 eggs. After hatching, usually by the end of June, the nymphs scatter over several leaves and settle along midribs and larger veins. The young scales are pale green. Since they are about the same color as the leaf and lie snugly against the vein, they are not easily seen by the unaided eye. They remain on the leaves until autumn, when they migrate to the bark. One brood occurs each year.

Several parasites and predators of *P. acericola* have been recorded, but their effectiveness in controlling pest populations has not been studied.

The maple mealybug, or maple phenacoccus, *Phenacoccus acericola* King, is found from the northeastern United States west to Minnesota and south to Tennessee. Throughout its life this insect is covered with a flaky wax secretion (B, C). The maple mealybug may become abundant in an area, but infestations are usually more injurious to the appearance than to the health of the host.

White nymphs overwinter in crevices in the bark of the host's trunk and branches. In early spring the nymphs move to the leaves, where they feed until they become mature. In May immature males crawl to the bark, where they pupate. In June the females migrate to the bark and, after mating, return to the leaves to lay eggs. During late June and early July, large white masses of 500 or more eggs are the most conspicuous sign of the infestation. Eggs hatch in July, and the young maple mealybugs crawl to the bark, where they spend the winter. A report from Ohio indicates that two or three generations occur there annually.

The cottony azalea scale, *Pulvinaria ericicola* McConnell, was described in 1949, after being observed in Maryland on wild *Rhododendron nudiflora* that was found infested at its base, below the ground litter. Other rhododendrons and azaleas in Maryland and New York have been found to be infested on their upper stems (D). This scale has also been reported in Florida on huckleberry and in Alabama, Virginia, and New Hampshire.

The cottony azalea scale overwinters as a fertilized immature female. In Maryland oviposition occurs annually in June.

References: 27, 325, 359, 511, 854

A. Egg sacs (ovisacs) of the cottony maple leaf scale, *Pulvinaria acericola,* on sycamore maple. The egg sacs were collected on Long Island in July.

B. Maple mealybugs, *Phenacoccus acericola,* on sugar maple.

C. Immature female maple mealybugs.

D, E. Egg sacs of the cottony azalea scale, *Pulvinaria ericicola,* collected in New York in July. [Coccidae]

A
B
C
D
E

Cottony Camellia Scale (Plate 164)

Pulvinaria floccifera (Westwood) has two common names, the cottony camellia scale and the cottony taxus scale. One generation is produced each year throughout its range. Overwintering occurs in the second nymphal stage, usually on twigs. When the female reaches maturity, she is mottled tan to yellowish in color and shaped like a slightly convexed long oval, about 3 mm long (Figure 109). (For comparison with adult forms of *Pulvinaria* species, see also Figures 108, 110, and 111.) She may migrate to the underside of a leaf where she lays her eggs in a white fluted egg sac. After all her eggs are laid, she dies. In a short time the dried-up body falls away, leaving only the white egg sac attached to the leaf. The egg sacs are 5–10 mm long and may contain more than 1000 eggs. The eggs begin to hatch in June in Connecticut. The crawlers settle on the leaves and extract sap through their piercing-sucking mouthparts. The settled crawlers shown in panel B were photographed about mid-July on Long Island. Immature scales were abundant on the undersides of *Ilex verticillata* leaves as late as September in Hershey, Pennsylvania. Damage to the host plant occurs primarily in early spring and late summer, and it takes the form of off-colored, light green foliage. The presence of the cottony camellia scale becomes obvious only after the ovisacs appear (panel A; Plate 165A, B).

Host plants include camellia, holly, yew, *Jasminum* species, rhododendron, the maple *Acer palmatum*, hydrangea, *Abutilon* species, English ivy, mulberry, pittosporum, euonymus, and *Callicarpa americana*. The cottony camellia scale occurs along the Atlantic Coast from Massachusetts south to Florida and in the Pacific coastal states, Texas, and Indiana.

One parasite has been reported. It is a tiny wasp, *Coccophagus lycimnia* (Walker).

References: 679, 854

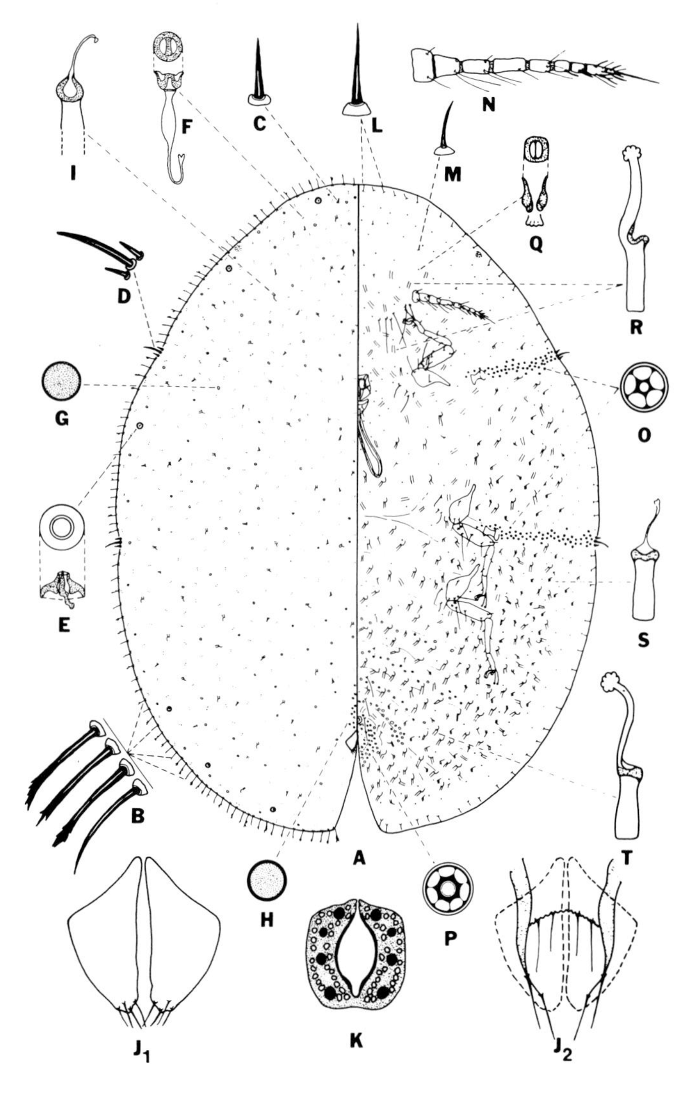

Figure 109. Diagram of a mature cottony camellia scale, *Pulvinaria floccifera,* of the type used by an insect taxonomist to compare with slide-mounted scale specimens. The upper (dorsal) side of the scale is shown to the left of the center line; at right is the lower or ventral side. Note the tiny but well-developed legs. The structures identified by letters around the border of the illustration are important characteristics for species identification: bristles, glands, pores, etc. (From Williams and Kosztarab, ref. 854, courtesy M. L. Williams.)

A. The underside of a yew branch showing numerous ovisacs of the cottony camellia scale, *Pulvinaria floccifera.* [Coccidae]

B. Close-up of two ovisacs and settled crawlers (*arrows*) of the cottony camellia scale.

C. A holly leaf showing several cottony camellia scale egg sacs. Note the dead female scale at arrow. Lined up along the main vein are many tiny yellowish crawlers.

D. Eggs of the cottony camellia scale (*left*) and settled crawlers along leaf vein of camellia.

E. Egg sacs of the cottony camellia scale on the undersides of holly leaves, *Ilex aquifolium.*

F. Cottony camellia scales in various stages of development on a camellia leaf. Arrow points to a female whose egg sac is beginning to form (note the white fringe at the posterior end; this is the egg sac).

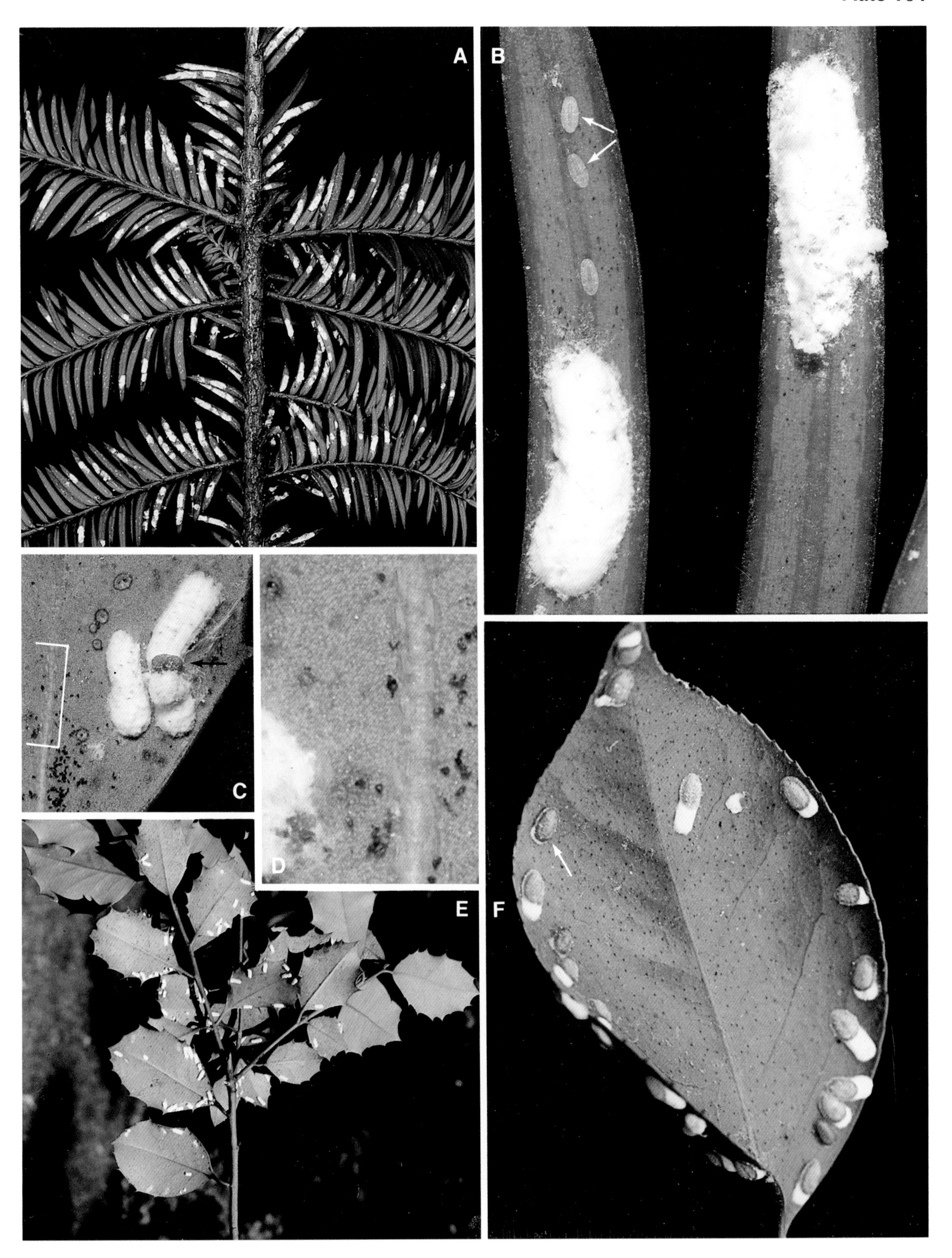
A
B
C
D
E
F

Cottony Scales (Plate 165)

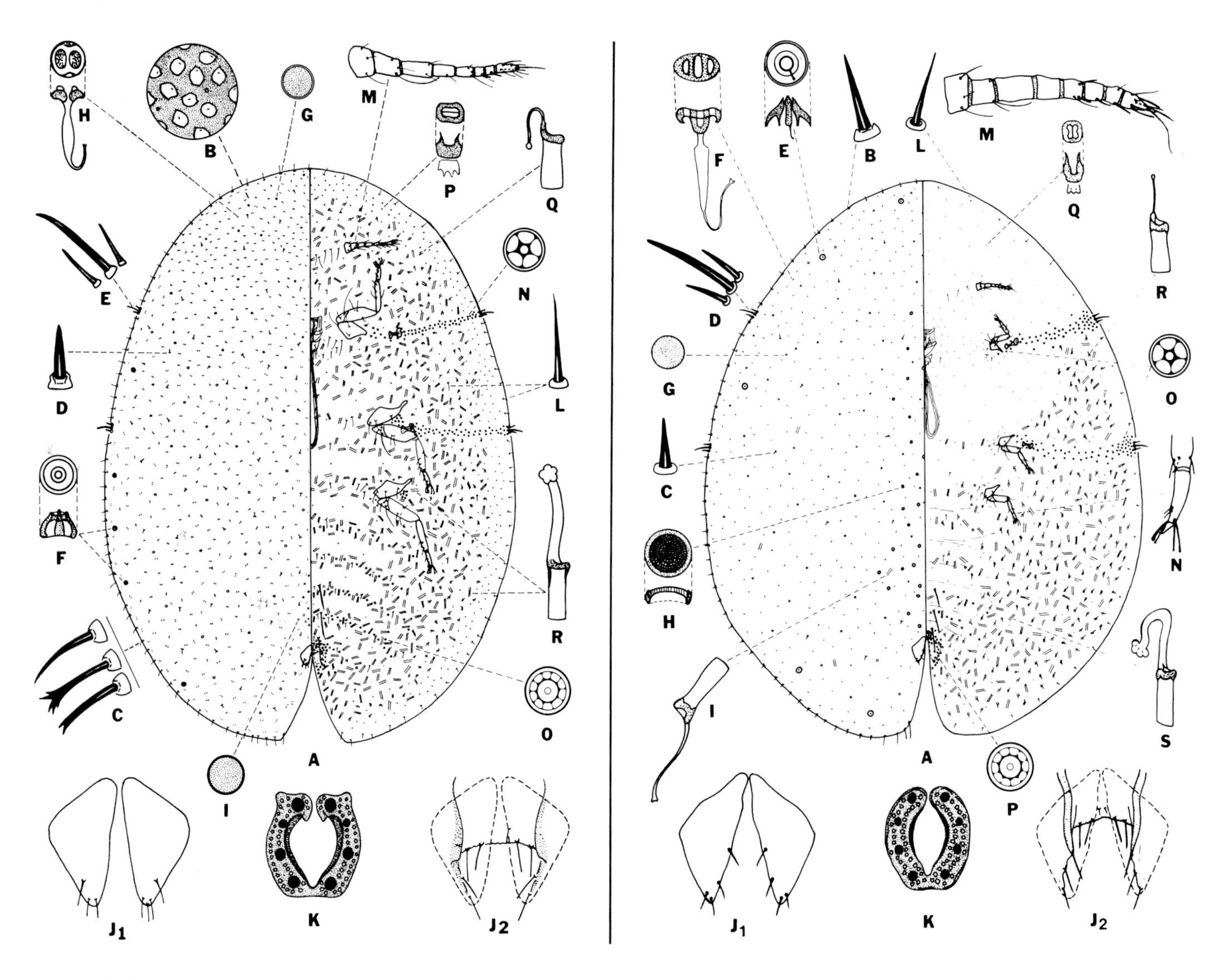

Figure 110 (left). Diagram of a mature cottony maple leaf scale, *Pulvinaria acericola,* as used by an insect taxonomist to compare with slide-mounted scale specimens. (From Williams and Kosztarab, ref. 854, courtesy M. L. Williams.)

Figure 111 (right). Diagram of a mature cottony azalea scale, *Pulvinaria ericicola,* like that used by an insect taxonomist to compare with slide-mounted specimens. (From Williams and Kosztarab, ref. 854, courtesy M. L. Williams.) To be positively identified, many scale insects must first be prepared and mounted on slides and then examined with the aid of a microscope. The prepared specimen is compared with drawings such as those shown here in order to identify the species. Compare the structures at *F* and *K* in Figure 111 with those at *H* and *K* in Figure 110. These characters differentiate the two species.

A. Viburnum leaves showing a heavy infestation of *Pulvinaria* scale egg masses. Eggs have begun to hatch; the rustlike spots near the arrow point are crawlers. These specimens were collected in mid-July in southeastern New York.
B. A *Pulvinaria* scale with protruding white egg sac.
C. Close-up of one of the scales shown in A. Part of an egg sac has been opened to show the pale brownish orange eggs. A number of crawlers are present on the leaf surface.
D. Young mealybugs, possibly *Phenacoccus acericola,* on viburnum. [Pseudococcidae]
E. Flowering dogwood leaves infested with cottony maple scales, *Pulvinaria innumerabilis.* The brown part at one end of the egg sac is the dead female. The white mass is the egg sac.
F. A female *Pulvinaria* scale and her protruding egg sac. The dead, shriveled female is at the bottom of the photograph.
G. An adult female *Pulvinaria* scale just beginning to produce her egg sac (note the white upper edge of the scale).
H. A female *Pulvinaria* scale prior to the development of an ovisac. [Coccidae]
I. Egg sacs of a *Pulvinaria* scale on flowering dogwood. Many tiny crawlers have settled along the leaf veins.

A
B
C
D
E
F
G
H
I

Pit-making Pittosporum Scale (Plate 166)

The pit-making pittosporum scale, *Asterolecanium arabidis* (Signoret), is known in the eastern, north central, and Pacific Coast states. The list of hosts subject to attack by this insect is quite extensive, including many woody and herbaceous plants. Its favored hosts, however, are mock orange, *Pittosporum tobira;* privet; and green ash.

In the adult stage the scales are oval when viewed from above and are about 3–4 mm long by 2 mm wide. The color of the insect's shell (D) ranges from white to brown, and this covering is strongly convex on its dorsal surface. Very little is known about the biology and seasonal history of *A. arabidis*.

The scales infest the stems, twigs, and leaf petioles of the host. Apparently they inject a toxic substance into the plant during feeding, which causes a retardation in normal plant growth as well as enlarged and distorted shoot development. The tips of branches may die as a result of attack. Depressions or pits are commonly formed beneath the feeding insects, which account for their common name.

The following are additional woody ornamental hosts of the pit-making pittosporum scale: *Arctostaphylos*—manzanita; *Aralia*; *Berberis*—barberry; *Carpenteria californica*—bush anemone; *Ceanothus*; *Cestrum*; *Chaenomeles japonica*—Japanese flowering quince; *Cistus*; *Correa harrisii*; *Cotoneaster horizontalis*—rock cotoneaster; *Daubentonia tripetii*—glory pea; *Deutzia scabra*—fuzzy deutzia; *Eriophyllum confertiflorum*; *Fraxinus americana*—white ash; *Fremontodendron*; *Hedera helix*—English ivy; *Helianthemum nummularium*; *Jasminum*—jasmine; *Lotus scoparius*—deerweed; *Penstemon*; *Philadelphus mexicanus*—evergreen mock orange; *Pyracantha*; *Rosa*—rose; *Salvia*—sage; *Spartium*; *Syringa*—lilac; *Veronica*; *Vitex*; *Weigela florida*.

References: 24, 216, 667

A, B. Swellings in twigs of *Pittosporum tobira* caused by the pit-making pittosporum scale, *Asterolecanium arabidis*.
C. Close-up of B, showing the scale insects.
D. Pit-making pittosporum scales. × 14. [Asterolecaniidae]

A
B
C
D

Bamboo Scale (Plate 167)

The bamboo scale, *Asterolecanium bambusae* (Boisduval), is found only on bamboo. It is widely distributed throughout the tropical areas of the world, but in the United States it is known outdoors only in Florida and California. It occurs on greenhouse plants in many states.

This insect has a soft body, typical of the scale family Asterolecaniidae to which it belongs. Its colors range from pale yellow to amber (B) to grayish black, and its form is gently oval, flattened, with a slightly convex shell surface. The posterior end of the body is slightly drawn out into a blunt point. The adult averages about 2 mm long by 1.5 mm wide. Males are unknown.

The bamboo scale infests both surfaces of the leaves as well as the stem of its host plants. Because it does not produce noticeable amounts of honeydew, the plants it attacks are not blackened by sooty molds. There does not appear to be any dramatic loss of color or reduction in growth of infested plants, so this insect must be considered a pest of only minor importance. Lady beetles are its main predators in most of the tropical countries in which it occurs (Figure 112).

References: 212, 213, 327, 530, 667

Figure 112. The adult of a scale-feeding lady beetle, *Chilocorus orbus* Casey. (Courtesy F. E. Skinner, Univ. of California, Berkeley.)

A. A bamboo stem heavily infested with bamboo scale, *Asterolecanium bambusae*. Scales usually collect at a node under a sheath. [Asterolecaniidae]

B–D. Increasing magnification to show adult female bamboo scales. Note the fringe of "hairs" outlining the body.

A
B
C
D

Pit-making Scales of Oak and Holly (Plate 168)

Asterolecanium puteanum Russell, sometimes called the holly pit scale, is a pest of American, Chinese, and Japanese hollies. It occurs along the eastern seaboard from Delaware to Florida. English hollies seem to be free of this pest. The injury caused by the holly pit scale, like that of many scales, develops slowly and subtly on host plants. A large scale population may exist before the injury symptoms are noticed. Twigs become roughened (A), pitted (C), and somewhat distorted. Growth is reduced and foliage becomes sparse. Adult scales (B) may be more yellowish green than those shown on this plant specimen. The adult female is about 1.5 mm in diameter. Scale crawlers appear in late June. Little is known about the biology of this insect.

Three species of *Asterolecanium* pit-making scales attack the twigs of oak both in the eastern and western United States. They are *A. variolosum* (Ratzenburg), known as the golden oak scale (F), *A. minus* Lindinger, and *A. quercicola* (Bouché). Their appearance, life history, and host plant preferences are so similar that it is not necessary to distinguish among them for arboricultural purposes. A wide variety of both white and black oaks are affected.

Males of these four *Asterolecanium* species are not known, or occur only rarely, in North American populations. The female, which varies from 1 to 2 mm in diameter depending on the species, is green, golden, or brown. She usually is found in a depression, the result of inhibited growth at the site where the scale is attached; hence the common name pit-making scale. After overwintering in a mature state on the twigs, females produce living young (crawlers) during the spring and summer months. An individual female may produce young for as long as 5 months. The highest crawler production in the West occurs in early summer, with a second, though smaller period of substantial crawler production in late summer. Crawlers do not move far from the parent and, for the most part, tend to colonize current-season and 1-year-old wood. After settling, the young scales suck fluids from the tree and do not move from that place for the duration of their lives. There is one generation each year.

Damage to the tree is caused by the removal of plant fluids and possibly by the injection of toxins into twigs by the feeding scales. Poor growth and dieback of twigs is a common result of the infestation. Dieback first becomes noticeable during the summer and early fall. Affected twigs of deciduous oaks retain dead leaves throughout the winter. A severe infestation can delay the leafing out of deciduous oaks in the spring by as much as 3 weeks. Young trees may be killed by *Asterolecanium* scales when heavy attacks occur year after year. On the East Coast the golden oak scale is often associated with anthracnose, a fungal disease of oak (see Sinclair et al., *Diseases of Trees and Shrubs*). White oaks are particularly susceptible. Together, the scales and the disease can quickly bring about the demise of well-established trees.

In Oregon the parasite *Habrolepis dalmanni* (Westwood) reportedly destroys as much as 20% of mixed populations of *A. minus* and *A. quercicola*. This parasite occurs also in the eastern United States.

References: 85, 216, 331, 583, 593, 636, 667

A. A stem of a Japanese holly, *Ilex crenata,* with the holly pit scale, *Asterolecanium puteanum*.
B. Scales in typical depressions or pits on the bark of a holly.
C. The holly pit scale in an unusually deep pit.
D. A valley oak, *Quercus lobata,* with dead leaves and twigs resulting from attack by a West Coast oak pit scale.
E. A heavy infestation of a pit scale on an oak twig.
F. Golden oak scales, *A. variolosum*. × 6.

A
B
C
D
E
F

Magnolia Scale, Calico Scale, and Frosted Scale (Plate 169)

The common names *soft scales* and *coccids* properly refer only to those scale insects in the family Coccidae. Before taxonomists erected the superfamily Coccoidea, the Coccidae included a number of subfamilies, all referred to as soft scales, that now have their own proper common family name. Some of the more important are giant scales, or margarodids, Margarodidae; gall-like scales, or kermesids, Kermesidae; bark crevice scales, or cryptococcids, Cryptococcidae; felt scales, or eriococcids, Eriococcidae; false pit scales, or lecanodiaspidids, Lecanodiaspididae; ornate pit scales, or cerococcids, Cerococcidae; and pit scales, or asterolecaniids, Asterolecaniidae.

The soft scales, or coccids, include the lecanium scales in the genera *Eulecanium, Sphaerolecanium, Parthenolecanium,* and *Mesolecanium.* Other soft scales are in the genera *Ceroplastes, Coccus, Eucalymnatus, Parasaissetia, Protopulvinaria, Pseudophilippia, Pulvinaria, Saissetia, Toumeyella,* and *Vinsonia.*

Most of the soft scales have a transparent coating of wax covering the body, much like the waxy layer on a polished car. Some of these, notably *Pulvinaria* species, produce a conspicuous white ovisac of cottony wax to protect the eggs (Plates 162–165). In this plate (B, I, J) the powdery or mealy wax of the frosted scale and magnolia scale is shown. Wax scales, *Ceroplastes* species, have a very thick white test that completely covers the body (Plate 170). Most of the soft scales excrete copious amounts of syrupy honeydew that becomes blackened when colonized by mats of sooty mold fungi (Plates 142E, 171B).

The magnolia scale, *Neolecanium cornuparvum* (Thro), is the largest scale insect found in the United States. The adult female sometimes measures up to 13 mm or more long (D). It is a native insect distributed throughout the eastern United States. It is known only from magnolia, and its principal hosts are *Magnolia stellata, M. acuminata, M. quinquepeta,* and *M. soulangeana.* Other species may be attacked but usually with less frequency and degree; it has been reported on *M. grandiflora* in Florida. Great amounts of honeydew are produced, giving the affected plant an untidy, unthrifty appearance because of the sooty mold growing on the honeydew.

The overwintering first-instar nymphs are elliptical, dark slate gray with a red-brown median ridge. On each lateral margin there are two spots of white wax that mark the ends of the spiracular furrows. They are usually massed on the undersides of the 1- and 2-year old twigs; as many as 278 overwintering nymphs were found on a 5-cm section of a twig 8 mm in diameter. In New York the first molt occurs in late April or May and the second in early June, by which time the insects have turned a deep purple. Stems of the host plant that are normally light green appear enlarged and purple from a massive encrustation. The nymphs secrete a white powdery layer of wax over their bodies (I). By late August most of the females have produced nymphs that wander about for a short time before settling on the new growth of a twig, where they start the cycle again. In the northeastern United States a single generation occurs each year; females produce living young.

The calico scale, *Eulecanium cerasorum* (Cockerell), is a colorful white and dark brown calico, from which it gets its name (A). The coloration is brightest when the scale reaches maturity, after which time it quickly darkens with age (E). It is globular and about 6–8 mm in diameter. At maturity, its color, size, and shape are distinctive enough for positive field identification. In winter immature females are oval, flattened, and light to dark brown.

Like other lecanium scales, only one generation of the calico scale occurs each year. Calico scales overwinter on twigs as partially grown nymphs. In California and the coastal areas of Virginia the nymphs have matured and have begun to produce eggs by early May. Upon hatching, the crawlers move to the leaves, where they settle and remain during the summer months. Before leaves drop in the fall they move back to the twigs to spend the winter. They produce a large amount of honeydew in the spring. Injury at that time is not due to sap loss but to reduced photosynthesis caused by heavy sooty mold growth on the honeydew.

Calico scale is a pest of all stone fruits and their ornamental cultivars, as well as Persian walnut, elm, zelkova, maple, pyracantha, pear, *Liquidambar* species, Boston ivy, Virginia creeper, dogwood, buckeye, wisteria, bayberry, and flowering crabapple. Entomologists in California first reported the calico scale. It is now known to occur in the Pacific Coast states and several eastern states, including Pennsylvania, Maryland, Virginia, Delaware, and on Long Island.

On the Pacific Coast birds are important predators of the calico scale. Among them, the yellow-rumped warbler, *Dendroica coronata* Linnaeus (the audubon form), is the most notable. The wasp parasite, *Blastothrix longipennis* Howard, also a West Coast species, is known to aid in control of the calico scale population.

The frosted scale, *Parthenolecanium pruinosum* Coquillett, is presumably limited to California and Arizona. The mature female is a large, convex brown scale 6–8 mm in diameter, and it is covered with a white frostlike wax (B). After the insect reaches maturity, its wax covering tends to weather away, leaving a color and form much like that of the European fruit lecanium (Plate 174D).

The frosted scale has one generation each year. Eggs are produced during April and May and hatch as brownish crawlers in June and sometimes July. The crawlers migrate to leaf blades and petioles as well as to the current season's shoots. Individuals that settle on leaves move to twigs during the early autumn. Maturation of the frosted scale continues through winter and, by early spring, the young females become covered with a powdery white wax. When dense populations develop, the foliage of the host tree becomes covered with hundreds of small clear droplets of syrupy honeydew that soon becomes overgrown with sooty mold fungi. Frosted scales may be found on crabapple; hawthorn; mountain ash; apricot, prune, and other species of *Prunus*; birch; elm; laurel; locust; rose; sycamore; and walnut. It has been reported on grape in New York.

In California the wasp *Metaphycus californicus* (Howard) is the most important and abundant parasite of the frosted scale. With proper management, this parasite has the capability of reducing the scale to nondestructive levels.

References: 325, 358, 533, 534, 561, 854

A. Mature females of the calico scale on walnut.
B. The frosted scale on Persian walnut. The whitish areas indicate where individuals had been attached.
C. Adult females of the magnolia scale. [Coccidae]
D. Close-up of the magnolia scale.
E. Dead calico scale females on pyracantha.
F. Live crawlers of the calico scale (*arrows*).
G. Calico scale eggs scattered on bark when a female "shell" was disturbed and removed.
H. Tuliptree scales on magnolia. [Coccidae]
I. Immature magnolia scales with typical waxy covering.
J. Nearly mature magnolia scale females with wax beginning to wear away.

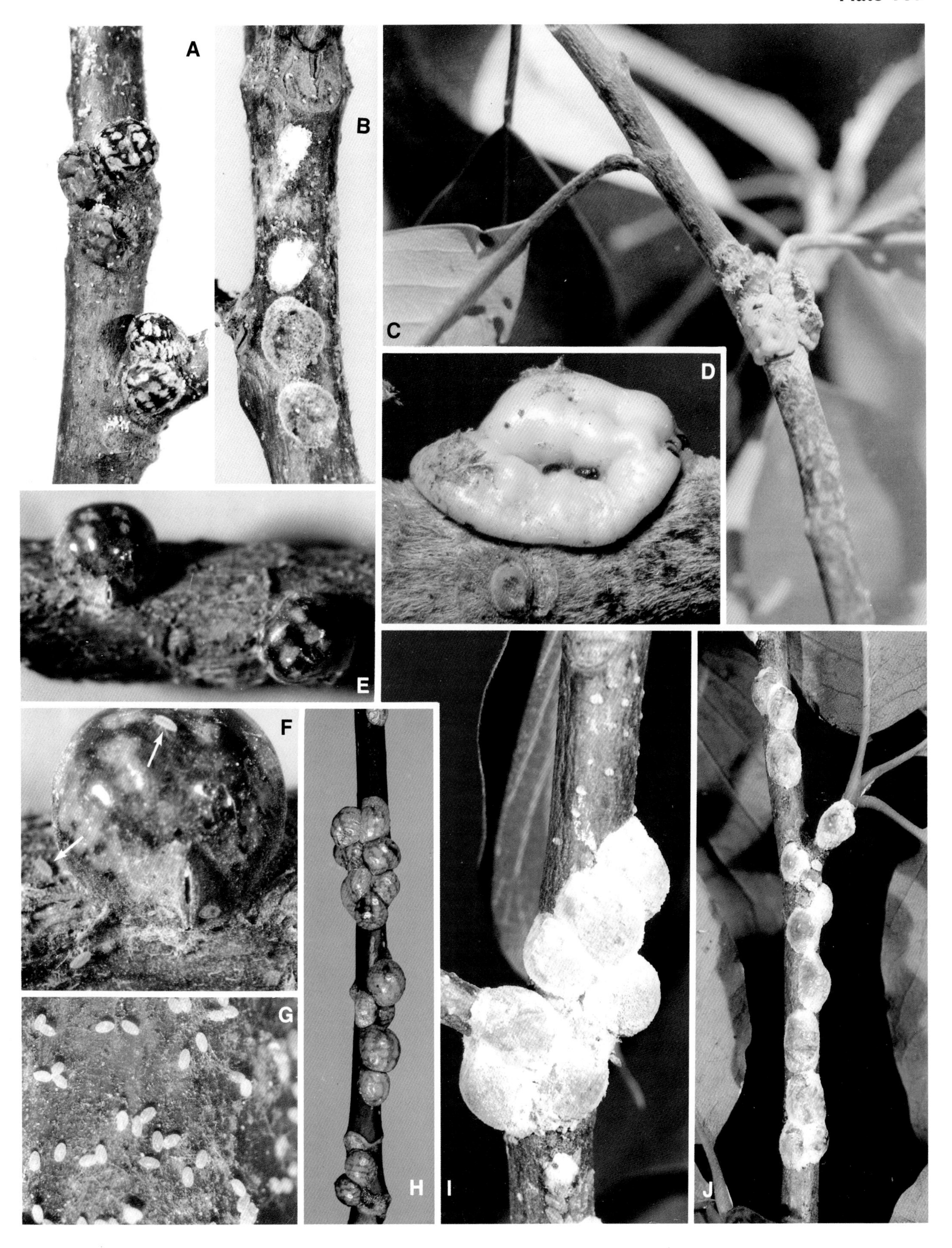
A
B
C
D
E
F
G
H
I
J

Wax Scales (Plate 170)

Wax scales belong to the genus *Ceroplastes* in the family Coccidae, known as soft scales. They are found in warmer climates around the world and are unique in having a very thick, heavy covering of wax (often called a test) over their bodies. Well over 100 species have been named, including more than 20 in North America, and 13 of them from the United States. Most of the species in the United States are common in the deep South, but a few have been reported from the mid-Atlantic and central states and the far West. *C. ceriferus* (Fabricius) became widespread and abundant in eastern Virginia in the 1960s and has spread northward through the District of Columbia and Maryland. Its present distribution extends from New York and south to the Gulf states.

Several common names have been used and suggested for *C. ceriferus*: Indian wax scale, East Indian wax scale, Japanese wax scale, and ceriferus wax scale; we use the first one. This species has a forward-projecting wax horn (H) that distinguishes it from other wax scales, although it is less discernible in older females late in the season (B). In more temperate climates there is one generation per year, with adult females overwintering. In Virginia there is an occasional report of settled crawlers and young nymphs in the early fall. In the deep South there are two or more generations. Oviposition has been reported in April for Virginia and in May for Maryland. Eggs hatch about 3 weeks later over a period of 2–3 weeks. Females each lay as many as 3000–5000 eggs. Crawlers settle only on twigs and stems, not on foliage. Soon after molting the nymphs begin to produce wax that creates a cameo appearance (G). As more wax is secreted, the body is soon covered and the waxy horn resembles a dunce cap (H). Females have four instars and males, five; however, males are rare. Reproduction is primarily parthenogenic.

The host range of the Indian wax scale is very broad. Well over 50 species of plants are known to support the pest, but it is not generally found in large numbers on shade trees. It occasionally is found on lower branches of large trees and some perennial weeds growing in severely infested areas. Most commonly infested are the Japanese and Chinese hollies, particularly Burford holly, as well as euonymus, pyracantha, boxwood, hemlock, camellia, spirea, flowering quince, podocarpus, barberry, and many others.

The size of mature females depends on the host plant and its condition or vigor. The scales produce copious amounts of honeydew resulting in dense mats of sooty mold fungi on foliage, stems, and fruits, especially holly berries, and the wax of the insects themselves.

Parasites of the Indian wax scale are uncommon in its northern range. More commonly the predator *Laetilia coccidivora* (Comstock) has been found feeding on *C. ceriferus* and also on the magnolia scale, *Neolecanium cornuparvum* (Thro), but seldom in numbers large enough to reduce a severe infestation. The caterpillar of this pyralid moth spins a web along a twig (F), enveloping scales and consuming them as it tunnels through the wax. In 1956 Heinrich (347) published a detailed description of this predator and listed more than 15 species of unarmored scales that it attacks.

The Florida wax scale, *C. floridensis* Comstock, was first reported in Florida in 1828. It has an extensive host list, including citrus and many ornamental plants. Preferred hosts are species of *Ilex, Citrus, Laurus, Ficus, Psidium,* and *Nerium*. The insects settle and remain on leaves, branches, and stems and are somewhat smaller than other common species of wax scales. Younger females have pinkish wax that darkens to dirty white. Males have not been recorded in the United States, although the stage was described by Kuwana in 1923. There are three generations each year in Florida. The scale does not overwinter in Washington, DC, Maryland, or Virginia, where it has been reported to occur. Its distribution also includes the Gulf states, the Carolinas, New York, Texas, Missouri, and New Mexico. Parasites of Florida wax scale include *Microterys clauseni* Compere, *Aneristus youngi* Girault, *Aphycus ceroplastis* Howard, and *Scutellista cyanea* Motschulsky.

The barnacle scale, *C. cirripediformis* Comstock, is widespread through the South from California to Florida, and north to Missouri, Ohio, and Pennsylvania. It overwinters as an adult female (Figure 113 and panel E) in its more southern range but can survive only indoors farther north. Males are known but very rare. First-instar nymphs settle on the upper surfaces of leaves but move to the branches and stems after molting to the third instar. Adult females are rectangular to oval and beige to grayish white. There are several generations per year in the South. The host list is extensive. It is a minor pest of citrus and ornamentals in Florida but becomes quite unsightly wherever sooty mold develops. Gardenia is a favored host.

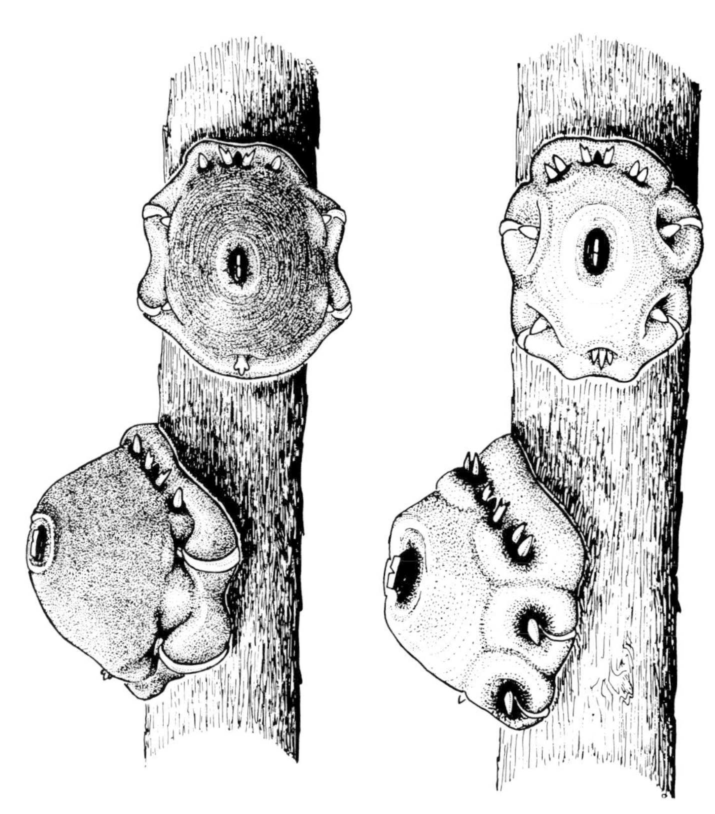

Figure 113 (left). Drawing of adult barnacle scales. (From Gimpel et al., ref. 294.)

Figure 114 (right). *Ceroplastes sinensis,* the Chinese wax scale, illustrated on *Ilex crenata*. (From Gimpel et al., ref. 294.)

The external egg parasite, *Scutellista cyanea* Motschulsky, is an effective natural control, but only where small numbers of eggs, 500 or fewer, are laid by females. In Hawaii the barnacle scale is preyed upon by the coccinellid predators *Orcus chalybeus* (Boisduval) and *Azya luteipes* Mulsant.

The Chinese wax scale, *C. sinensis* Del Guercio (Figure 114), has been reported from California, North Carolina, and Virginia in the United States. It has extensive distribution and a broad host list worldwide, so it may be a potential pest of ornamental plants. The waxy covering or test is rectangular in dorsal view, but smaller than that of the Indian wax scale. It is pinkish white darkening to reddish brown with age. Crawlers settle on the leaves in late June to early July and move to the stems and branches in the third instar. Males have not been found in the United States but have been reported elsewhere. Mature females overwinter in Virginia and begin laying eggs in late May. In the East this scale occurs on Japanese hollies and in California, on citrus.

Several parasites have been reported worldwide, and *Anicetus toumeyellae* Milliron was reared from the scale in Virginia.

A. Adult females of the Indian wax scale on pyracantha.
B. The typical appearance of the Indian wax scale. The adult is about 5 mm in diameter.
C. Adult females of the barnacle scale on euonymus.
D. An overturned female of the Indian wax scale (*top*) beginning to lay eggs. (*Bottom*), the appearance of a dead female under wax.
E. Close-up of a mature female barnacle scale. [Coccidae]
F. A stem of Burford holly with webbing occupied by the predacious caterpillar *Laetilia coccidivora*.
G. Immature Indian wax scales showing development of waxy projections and typical cameo appearance (*arrows*).
H. Mature Indian wax scale nymph showing the "dunce cap" appearance.

A
B
C
D
E
F
G
H

The red wax scale, *C. rubens* Maskell, occurs in southern Florida and Hawaii in the United States but has worldwide distribution and a broad host list. In Florida the most common host in nurseries is *Aglaonema* species. It also feeds on mango, sapodilla, species of *Dizygotheca, Aralia,* and *Viburnum,* and other plants. More varied hosts are reported in Hawaii, including citrus, mango, *Anthurium* species, and *Alyxia olivaeformis.* The wax covering or test is distinctly red with conspicuous spiracular wax bands. It is the only species with distorted and reduced legs, which are visible when a specimen is mounted on a microscope slide. Crawlers settle on leaves, branches, and stems and remain in place without moving to another site. Adult females overwinter.

The Duges wax scale, *C. dugesii* Lichtenstein, occurs in southern Florida on a limited number of host plants, including *Bursera simaruba* (gum elemi) and persimmon. It is polyphagous in Mexico, Cuba, Panama, the Virgin Islands, and the West Indies, and thus may be a potential pest in parts of North America. The scales infest the stems, but little information exists on the biology of this species.

The brachyuran wax scale, *C. brachyurus* Cockerell, occurs in Alabama and Arizona and resembles *C. sinensis.* It infests the stems of *Bouvardia* and *Ilex* species, but its biology is poorly known. The wax test is oval and grayish white.

The Nakahara wax scale, *C. nakaharai* Gimpel, occurs entirely on branches and stems. The test is rectangular and yellowish brown without a marginal flange. It was first reported from Florida in 1895 and has been found on *Coccoloba floridana, Phoradendron flavescens, Eugenia myrtifolia,* and *Ixora acuminata.* Elsewhere it is known from Cuba. Gimpel et al. (294) suggest that it may occur in other Gulf states and the Greater Antilles.

The candle wax scale, *C. utilis* Cockerell, is not considered to be economically important but has a varied host list of ornamental plants. It settles on branches and stems and remains in place. Young females are translucent green or yellow; the wax gradually darkens to a slate gray. It resembles the Nakahara wax scale. Little is published on its biology and habits.

Two other species of wax scale include the tortoise wax scale, *C. cistudiformis* Cockerell, and the desert wax scale, *C. irregularis* Cockerell. The tortoise wax scale was first reported from southern California in 1897 and is similar to the barnacle scale. It is found on citrus, passion flower, and pepper tree (*Schinus* spp.) in San Diego. It is not considered economically important. The desert wax scale is considered to be native to Mexico and the arid western United States. Saltbush (*Atriplex* spp.) is its preferred host.

References: 294, 325, 347, 621, 854

Black Scale (Plate 171)

The black scale, *Saissetia oleae* (Olivier), is one of the most common and most destructive of the soft scales in California. This species is known in the Southwest, Florida, and the Gulf states. The black scale is a common pest of certain greenhouse-grown plants in many states.

The mature female scale insect is brown to black and nearly hemispherical in profile; it ranges 3–5 mm in diameter. For many years the presence of raised ridges that form the letter H on the back of the insect (D) was considered one of the most helpful features to distinguish *S. oleae* from similar species. However, Hamon and Williams (325) report the occurrence of two related species, *S. miranda* (Cockerell & Parrott) and *S. neglecta* Delotto, which also have the raised letter H on the back. Very little is known about the biology, host plants, or potential for damage of these two species. *S. miranda* seems to have about the same geographic distribution as *S. oleae. S. neglecta,* known as the Caribbean black scale, is "possibly widely distributed in the Gulf states . . . , but because of past confusion with *S. oleae,* the distribution is unknown" (325, p. 114). Only detailed research will elucidate the important differences and similarities between these three species of *Saissetia.*

In coastal California the black scale goes through two complete generations each year. Inland there is only one annual generation. In areas where more than one generation develops, overlapping of the generations occurs, and all developmental stages can be found during the growing season.

The black scale can overwinter in any life stage, although it usually overwinters as a partially grown female. In the spring the females complete their development and deposit as many as 2000 eggs beneath their shells. The young crawlers issue from the dead body of their parent and migrate to twigs, leaves, and branches of the host, where they insert their mouthparts and settle to feed. In inland California the adult stage is attained in 8–10 months. Although male black scales are known, they are far outnumbered on the plant by the females. Most reproduction, therefore, takes place parthenogenically, that is, without mating.

Injury to the host plant is caused by the loss of sap through feeding by the scales. Severe infestations on leaves, twigs, or branches result in premature loss of foliage or dieback of the woody parts of the plant. Also injurious to the host plant is the fouling of foliage by the vast quantities of honeydew excreted by the scales. This material collects on plant parts and soon turns black as a result of the sooty mold fungi that colonize the honeydew.

The black scale is attacked by many parasites and predators. The most important parasites are tiny wasps. They insert their eggs into the bodies of the developing scales (Figure 37, accompanying Plate 46). The developing larva feeds internally on the scale. After killing the host insect and completing their development inside the scale, the adult wasps leave by way of tiny holes (H) chewed in the dorsum of the host scale's cover. An important predator of the black scale is a lady beetle, which in the larval and adult stages feeds on young scales. Under certain conditions these natural enemies are able to prevent the development of severe infestations of black scale.

The host list of *Saissetia oleae* is extremely long and includes the following plants: *Acer*—maple; *Aralia*; *Arbutus unedo*—strawberry tree; *Artemisia*; *Asparagus sataceus*—asparagus fern; *Camellia*; *Cedrus deodara*—deodar cedar; *Cestrum*; *Chaenomeles japonica*—Japanese flowering quince; *Choisya ternata*—Mexican orange; *Citrus*; *Codiaeum variegatum*—croton; *Duranta*; *Eucalyptus*; *Fagus*—beech; *Ficus*—fig; *Feijoa sellowiana*—pineapple guava; *Fraxinus dipetala*—two-petal ash; *Fuchsia*; *Gleditsia triacanthos*—honeylocust; *Grevillea*; *Grewia*; guava; *Hibiscus syriacus*—rose of sharon; *Ilex*—holly; jasmine; *Juglans regia*—Persian walnut; *Juniperus communis* cv. Stricta—Irish juniper; laurel; *Ligustrum*—privet; *Magnolia*; *Malus*—apple; *Myrtus*—myrtle; *Nerium*—oleander; *Olea*—olive; palms; *Pittosporum*; *Platanus*—sycamore; *Populus*—poplar; *Prunus*—almond, apricot, plum, prune; *Punica*—pomegranate; *Pyrus*—pear; *Rhamnus*—buckthorn; *Rhus*—sumac; *Robinia pseudoacacia*—black locust; *Rosa*—rose; *Schinus*—pepper tree; *Solanum jasminoides*—nightshade; *Torreya californica*—California nutmeg tree; *Umbellularia*—California bay; *Vitis*—grape.

References: 202, 217, 325

A. Note damage to oleander leaves caused by a black scale infestation.
B. Close-up of A, showing scales on shoot and sooty mold on leaf.
C. Immature black scales on undersides of oleander leaves.
D. A maturing black scale female, showing raised ridges outlining the letter H on dorsum. [Coccidae]
E. Mature black scales on Mexican orange.
F. Black scales on a twig of holly.
G. Close-up of scales shown in F.
H. Encrustation of black scale on citrus. Some of these scales have been parasitized. Close observation will reveal holes where the parasite has emerged.
I. Mature black scales on a citrus twig. The scales vary from 3 to 5 mm in diameter.
J. Young black scales on a California bay leaf.

A
B
C
D
E
F
G
H
I
J

Nigra Scale and Hemispherical Scale (Plate 172)

The nigra scale, *Parasaissetia nigra* (Nietner), was a serious pest of ornamental plants during the early decades of this century but subsequently has become relatively unimportant. First detected in Washington, DC, in the early 1900s, this insect now occurs outdoors in Florida and California, and indoors in greenhouses in many states. It is believed that the nigra scale also survives outdoors in all the Gulf states.

The dramatic decline in populations of the nigra scale in California in recent decades is attributed to its natural enemies. Some 24 or 25 species of parasitic wasps, many of which have been deliberately introduced into the United States for biological control purposes, destroy large numbers of *P. nigra*. Other important natural enemies include the larvae of lacewings and the larvae and adults of lady beetles.

In California one generation of *P. nigra* occurs each year. Adults are found in April, May, and June. In the adult stage the insect is brown to deep black, usually elongated, and 3–5 mm long (C). The shape of the adult scale is quite variable, depending on the substrate on which it develops. On leaves, for example, the insect is broad, whereas on slender twigs it tends to be narrow and elongated. Eggs are laid beneath the body of the females from May until the February of the following year. The majority of eggs are deposited from May to September. After an incubation period of 1–3 weeks the young crawlers migrate away from the body of the parent and settle on the leaves or woody parts of the host. Scales molt twice before the adult stage is reached. Unlike many scale insects, *P. nigra* is capable of movement at any time during the 8- or 9-month period of development from crawler to adult. In California no males of the nigra scale have ever been recorded. Reproduction, therefore, takes place entirely without mating.

Damage to the host plant is caused by withdrawal of plant juices by the developing scales. This may result in premature loss of foliage or in twig or branch dieback. Blackening of foliage that results from the development of sooty mold on honeydew, which is excreted by the scales, is another form of damage to plants caused by *P. nigra*.

The nigra scale attacks 161 species of plants in California, many of which are important woody ornamentals. Its favored hosts are said to include *Fatsia japonica*—Japanese aralia; *Hedera helix*—English ivy; *Ilex aquifolium*—English holly; *Nerium oleander*—oleander; *Pittosporum undulatum*—orange-berry pittosporum; and *Hibiscus* species.

The hemispherical scale, *Saissetia coffeae* (Walker), is a brown scale about 3–5 mm in diameter when mature (H). In profile, it has a hemispherical shape; from above it appears to be gently oval. This species is far more important as a greenhouse pest than as a threat to outdoor-grown ornamental plants. Because it is a tropical species, it is found outdoors only in the more southern portions of the United States, such as Florida and California. In these two states there may be two generations per year. In California eggs are laid over a relatively long period (May–July), and in the warmest regions where two generations occur, generations are likely to overlap. If males occur, they apparently are not important for survival of the species because females reproduce without mating. They lay 600–700 eggs beneath their bodies, then die. These eggs hatch to produce crawlers that disperse to twigs, leaves, or fruit. Once settled, the scale is permanently attached to that site, where it feeds through three nymphal stages, matures, and lays eggs for a life span of about 6 months.

This scale removes large amounts of sap from its hosts. High populations cause foliage to wilt and will eventually kill twigs and small branches. Large amounts of honeydew are excreted and drop on lower leaves, making them shiny and sticky. Sooty mold fungus rapidly grows on the honeydew and turns the leaves to sooty black, giving an infested plant a dirty, unthrifty appearance and reducing the ability of leaves to produce starch through photosynthesis.

Woody ornamentals attacked by *S. coffeae* include aralia; gardenia; lantana; bignonia; *Camellia*; *Citrus*; *Codiaeum variegatum*—croton; ferns; guava; *Myrtus*—myrtle; *Nerium*—oleander; palms; *Rhus*—sumac; *Schinus*—pepper tree; and *Zamia*.

References: 76, 202, 217, 325, 722, 759

A. Nigra scales on leaves and a shoot of *Daphne* sp.
B. Nigra scales on a stem of *Maytenus boaria*.
C. Close-up of a scale in B, showing crawlers and adult females.
D. Nigra scales on leaves of *Daphne* sp. [Coccidae]
E. Hemispherical scale encrustation on twig.
F. Hemispherical scales on fern, a very susceptible host.
G. Adult female *Parthenolecanium* scale insects.
H. Close-up of hemispherical scale adults and crawlers.

A
B
C
D
E
G
H

Tuliptree Scale (Plate 173)

The tuliptree scale, *Toumeyella liriodendri* (Gmelin), is a serious pest of the tuliptree (or yellow poplar), *Liriodendron tulipifera*, and the magnolias *Magnolia soulangeana* and *M. stellata*. The insect is so prolific that twigs and branches are often completely covered with scales (B), resulting in death of infested parts and rapid decline of the tree or shrub. It also attacks *M. grandiflora* and linden, *Tilia* species. In Florida the tuliptree scale has been observed on *Michelia figo*—banana shrub; *Gardenia augusta*—cape jasmine; *Gordonia lasianthus*—loblolly bay; *Cephalanthus* species—buttonbush; *Persea borbonia*—redbay; and *Juglans* species—walnut. Recently it has also been collected there from *Carya cordiformis, Ascyra tetrapetalum, A. hypericoides, A. edisonianum, Hypericum cistifolium, Magnolia virginiana,* and *Cassia fasciculata.* In the United States tuliptree scale is found east of the Mississippi River from Michigan south to Alabama and east to New York and Connecticut south along the Atlantic seaboard to Florida. It also occurs in California on ornamental tuliptrees and magnolias. It could be found wherever these trees are grown.

The tuliptree scale is one of the largest of all the scale insects, and on magnolia it is often mistaken for the magnolia scale, which is about the same size and has a similar life cycle but which attacks only magnolias (Plate 169). A mature female is 6–12 mm in diameter and hemispherical (C). It may vary from grayish green to pink-orange mottled with black. It is oval but irregular in outline when crowded on twigs. Adult males emerge in June and mate with the ovoviviparous females. Eggs develop within the body of the female, and the young crawlers are born alive in late August and September. One generation usually occurs each year, although there may be several in the deep South. A single female can produce as many as 3000 or more crawlers over several weeks. Reproduction occurs from late August through September. Crawlers are the size of a pinhead and settle on twigs in the fall. Second-instar nymphs overwinter, and the insects actively feed the next season from spring until late summer. They deplete the inner bark tissue of sap with their piercing-sucking mouthparts. Severe dieback of branches and even death of the entire plant results from repeated heavy populations. Infestations are often most conspicuous when adult females are present, after much feeding and damage has occurred. During the summer severe infestations may be detected earlier when considerable honeydew excretions encourage the growth of sooty mold fungi and attract ants, wasps, and hornets.

Natural enemies are common in populations of tuliptree scale, but these enemies are seldom able to control infestations until the host plant is so weakened that the scale is on the verge of starvation. The lady beetles *Hyperaspis signata, H. bipunctata* (Say), and *Chilocorus stigma* (Say) (Plate 188) readily feed on nymphs. Caterpillars of the phycitid moth *Laetilia coccidivora* (Comstock) spin silk webs 2.5–7.5 mm long on infested twigs. This silk covers the scales, and the caterpillar consumes these insects under the protection of the webs (Plate 170F). The caterpillar is especially common in Virginia. A number of parasitic wasps attack tuliptree scale. For example, Hamon and Williams (325) cited *Anicetus toumeyellae* Milliron as a common natural enemy in Virginia (this parasite also has been reared from the Chinese wax scale—see Plate 170); *Coccophagus flavifrons* Howard from Pennsylvania to Florida and Texas; *C. lycimnia* (Walker), cosmopolitan; and *Metaphycus flavus* (Howard) from New York to Florida to Texas.

Another potentially important pest in this genus is the buttonbush scale, *Toumeyella cerifera* Ferris, first reported in the United States from Virginia. It first was known only in Mexico on *Albizia* species but now is reported also from Alabama, Arizona, Florida, Louisiana, and North Carolina on buttonbush, *Cephalanthus occidentalis* Linnaeus. The life cycle is similar to that of the tuliptree scale. Females give birth to living young, and crawlers are present in late August and in September. The scales prefer cracks and crevices on the bark of the host, and often *Crematogaster* ants construct protective coverings over the developing coccids.

Other important species of *Toumeyella,* described in the text accompanying Plates 40–41, are found on conifers, including the pine tortoise scale, striped pine scale, irregular pine scale, and the Virginia pine scale.

References: 183, 325, 854

A. Yellowing of tuliptree foliage and dieback of lower branches resulting from severe tuliptree scale infestation. Note the dead twigs and small branches in the circle.

B. A large twig encrusted with nearly mature tuliptree scales.

C, D. The shapes and coloration of mature female tuliptree scales. [Coccidae]

A
B
C
D

Lecanium Scales and Kermes Scales (Plate 174)

The lecanium scales, which are common pests in the United States and Canada, include approximately a dozen species of soft scales, formerly classified in the genus *Lecanium,* family Coccidae. In 1981 Nakahara (561) revised the genus, which resulted in the current use of several new generic names for this group. The lecanium scales are difficult to distinguish from one another, and positive identification requires microscopic study by a scale specialist. Even coccidologists find it difficult to detect differences consistently between some species when using slide-mounted preparations. Certain species exhibit host-induced variation that may be as great as interspecific differences between them. Some species are highly polyphagous, whereas others occur on only a few hosts. Adequate field identification for purposes of pest management can often be made for some of the more common pests, based on a knowledge of size, hosts, habits, and life cycle.

Lecanium scales have no discernible waxy covering or test. After the female lays eggs, her hemispherical body dries, becomes brittle, and turns brown. The length varies from 3 to 12 mm, depending on the species. Many people mistakenly think that the dried, brown female (C) is a scale cover, because it covers and protects the eggs (D) during oviposition and remains in place until the eggs hatch.

Most species of lecanium scales have similar life histories. The period of greatest growth, most active feeding, and severe injury to the host occurs from spring to early summer. Eggs are deposited beneath the females beginning in late spring or early summer. Crawlers begin to emerge from beneath the female "shell" and migrate to the leaves, usually during June and July in more temperate areas. Except in the case of the terrapin scale, second-instar nymphs (see arrows in C) return to the twigs in late summer, where they overwinter. Generally, one generation of lecanium scales occurs each year.

The large hickory lecanium, *Eulecanium caryae* (Fitch), is huge in comparison to other species, often measuring more than 12 mm long. On larger limbs where it is commonly found, it is rather flat; on smaller limbs it is smaller and more convex. The rear two-thirds of the body is broader than the front third. The body of nymphs and females is partially and thinly coated with purplish white mealy wax. Mature nymphs and adult females are uniformly brown, at times with an orange tint. Second-instar nymphs overwinter on the branches of the host and mature in the spring. A hundred or more eggs are laid in May and June and hatch in late June or early July. Crawlers emerge from beneath the female and migrate to leaves, where they settle and feed. In August they move back to the bark of the branches to spend the winter in hibernation. There is one generation per year. In New York the hickory lecanium is consistently found on elm branches 4–12 cm in diameter, whereas the European fruit lecanium is predominantly found on the smaller twigs. Other host plants of *E. caryae* include *Carya* species, *Castanea dentata, Platanus occidentalis, Pyrus* species, *Quercus rubra,* and certain species of *Betula, Fagus, Salix, Celtis, Juglans, Gleditsia, Malus, Morus,* and *Prunus.* The large hickory lecanium scale occurs throughout the eastern United States.

The European fruit lecanium, *Parthenolecanium corni* (Bouché), (B, C, D, I) is the most common and abundant species of lecanium scale. Its geographic range is wide and its host list virtually unlimited. It occurs on shade trees, fruit trees, shrubs, and other woody ornamental plants throughout the United States. It varies widely in shape and form, which makes positive identification difficult even for scale specialists. Its life cycle is generally like that described above. Williams and Kosztarab (854) could not consistently differentiate *P. corni* from the Fletcher scale or the oak lecanium on the basis of structural characters. These three species must be referred to as the *corni* complex until further taxonomic research can clarify species determination. Whereas the European fruit lecanium is polyphagous, the host plants of other species in the complex apparently are more restricted. Hamon and Williams (325) suggest that host and seasonal development patterns make the species more easy to distinguish in the field.

The Fletcher scale, *Parthenolecanium fletcheri* (Cockerell), occurs on arborvitae, yew, and juniper. It is described and illustrated in Plate 42.

The oak lecanium, *Parthenolecanium quercifex* (Fitch), (E) has been associated with species of *Quercus, Platanus, Diospyros, Carya, Castanea, Betula,* and *Zanthoxylum.* In New York eggs hatch and crawlers are active in mid-July. Females are 4–7 mm long and 3–5 mm wide, light to dark brown, and nearly always with a pair of lateral humps much like those in panel H. It overwinters in the second instar and has one generation per year.

The native terrapin scale, *Mesolecanium nigrofasciatum* (Pergande), is chiefly found on *Prunus* (especially peach) and *Acer* species as well as linden, species of *Mimosa, Platanus,* and *Sassafras,* hawthorn, apple, birch, redbud, mulberry, poplar, live oak, and species of *Vaccinium* and 12 other genera. Also called the blackbanded scale, this species occurs in every state east of the Mississippi River (except Maine, New Hampshire, and Vermont) and in Ontario. It also occurs in the Gulf states, New Mexico, and parts of the Midwest. The terrapin scale gives birth to living young in early spring. The crawlers migrate to the leaves and settle along the larger veins on the leaf undersurfaces. In late summer and early autumn the scale migrates back to twigs and small branches. Unlike other lecanium scales, the native terrapin scale spends winter as adult females. Partially grown females are easily identified in the field by their red-orange color, which is intersected by black radiating lines. Its name is derived from its resemblance to the common land terrapin in body shape and markings. The globular females, when fully grown, measure 2–4 mm in diameter. There is one generation each year. Numerous parasites are listed by Peck (605) and Krombein et al. (447). Hamon and Williams (325) include 11 parasites important in natural control of the terrapin scale. A chalcid wasp, *Coccophagus lecanii* Fitch, is the most abundant parasite recorded in Maryland.

The European peach scale, *Parthenolecanium persicae* (Fabricius), attacks many hosts, including species of *Albizia, Berberis, Elaeagnus, Euonymus, Lonicera, Ulmus,* and *Vitis.* It occurs throughout the United States, chiefly on ornamentals. When severe populations develop, it can kill branches or entire plants. Numerous parasites have been recorded, including *Coccophagus lycimnia* (Walker). Nymphs and adults may feed on the trunk, branches, twigs, and leaves, and often on the callus tissue around wounds. It spends winter in the second instar, and females mature and lay eggs in early May. Crawlers migrate to the undersides of the leaves to feed during the summer. Nymphs move back to the twigs to overwinter.

Other lecanium scales are described in other plates, including the magnolia scale, calico scale, and frosted scale (Plate 169) and the globose scale (Plate 175).

Lecanium scale populations are subject to rapid increase and decline. No doubt, the changes in the physical environment play an important role in these fluctuations, but a great many parasites and predators frequently account for the collapse of infestations (C, G, and H show parasitized specimens). However, the manipulation of parasites to achieve control of scale insects on ornamental plants has not been adequately developed. Because a number of the natural enemies are efficient in natural control, they should be exploited in pest management. Predators are particularly effective in destroying the nymphs, which are exposed on leaves all summer. Peck (605) and Krombein et al. (447) have published lists of parasite records for scale insects.

The genera *Allokermes* and *Nanokermes,* in the family Kermesidae,

---

A. Adult females of *Nanokermes pubescens,* a kermes scale, on the bark of red oak.
B. Nearly mature females of the European fruit lecanium.
C. Dried, brittle bodies of dead European fruit lecanium females on redbud in the fall, long after eggs were laid and crawlers emerged. Note the overwintering second-instar nymphs at arrows.
D. European fruit lecanium females during or just after oviposition. Arrow indicates where a female was removed to reveal the eggs.
E. A twig encrusted with females of the oak lecanium, *Parthenolecanium quercifex.*
F. Females of a lecanium scale on the underside of a leaf of California holly, *Heteromeles arbutifolia.* Whitish areas show where the females had been attached.
G. A parasitized female of the frosted scale on walnut. Note parasite exit holes in the "shell."
H. A lecanium scale (*Parthenolecanium* sp.) on elm showing the coloration and shape of a female approaching maturity. This specimen is probably parasitized.
I. European fruit lecanium females on southern wax myrtle.

A
B
C
D
E
F
G
H
I

represent a small group of scale insects that are known only on oak in the United States, except for one record on chinquapin in California. Except for *Nanokermes pubescens* and *Allokermes kingi*, kermesids, or the gall-like scales, have not been well studied. Old females of kermesids are globular, have hard-shelled bodies, resemble oak galls, and are from light beige to almost black. They range from 3 to 6 mm in diameter. In Maryland females (A) of the most common oak kermes, *N. pubescens* (Bogue), lay eggs over a month-long period beginning in late June (512). Eggs begin to hatch early in July. Crawlers migrate to cracks in the bark of the trunk to spend the winter. In the spring after the first molt the nymphs migrate to new twigs. They frequently settle near leaf axils and insert their mouthparts to feed. By early June the globular females mature, resembling small tan galls or buds. Adult males mature on the trunk without migrating to new growth. Distribution records of *N. pubescens* include Kansas, Maryland, and Virginia, but it is likely to occur throughout the eastern United States.

Another species, *A. kingi* (Cockerell), has been studied by Hamon et al. (326) on *Quercus rubra* and *Q. velutina* in Virginia. It has one generation per year. Adult females are located at the base of leaf petioles and resemble small galls. In July and August they lay eggs that hatch in September. The first instars overwinter in bark crevices and molt in April of the following year. Female nymphs migrate to the new shoot and mature in June. Several individuals feeding side by side may kill the shoot. Male nymphs crawl down the tree and mature while on the trunk or in debris under the tree.

Parasitic wasps in the family Encyrtidae, for example, *Homalotylus* species, have been reared from male scales. A caterpillar, *Euclemensia bassettella* Clemens, in the family Heliodinidae has been reported as a predator of adult females in New York. Lady beetles (Coccinellidae) frequently feed on the first instars. *Chilocorus stigma* (Say) has been observed in Virginia. The reported distribution of *A. kingi* includes Massachusetts, New York, Maryland, Pennsylvania, and Virginia, but it likely occurs throughout the eastern United States. Old females are globular, about 5 mm long, and bright light to dark yellow, marbled with a reddish tint.

References: 178, 235, 325, 326, 447, 512, 513, 561, 605, 698, 774, 854

Rose Scale and Globose Scale (Plate 175)

Some of the differences between soft and armored scale insects can be seen in the facing plate, in which both the globose scale and the rose scale are shown. The males of these soft and armored scales are superficially similar (B, F). They are both small and white. From a distance groups of them look like a whitish flaky crust on the branches of host plants. Close examination with a magnifying lens shows that the immature male armored scale insect (rose scale) is covered with a definite waxy scale cover. The male breaks through the cover when it emerges. However, the cover remains attached to the plant for most of the season. When the winged males of soft scales (globose scale) are about ready to emerge, they likewise turn white. Following emergence, the empty cast "skin" (skeleton) remains attached to the twig. The winged males of both soft and armored scales tend to be numerous but inconspicuous, for only a few days of the year.

The females of armored and soft scales are seldom confused. Armored scale females such as those of the rose scale (C) have a characteristic scale cover, composed of waxy secretions, that incorporates the cast "skins" of earlier instars. The so-called scale in the soft-scale group is the body of the female itself. No cover is produced. When the insect molts, the very delicate cast "skin" is broken and sloughed off as the body expands.

The rose scale, *Aulacaspis rosae* (Bouche), is a pest of brambles and roses but has also been found on the fringe tree, *Chionanthus virginicus*, on peach, and on *Geranium* species in Florida. It also occurs throughout California. The rose scale is the only species of *Aulacaspis* in North America. Infestation by this pest is not regarded as a serious problem. Canes, twigs, or stems are the sites of attack on their host plants. Two or more generations occur in a single season, depending on the climate. In New York state two generations occur. The first-generation eggs hatch in late May or early June; the second-generation eggs hatch in August. Details of the biology of the rose scale have not been studied.

The globose scale, *Sphaerolecanium prunastri* (Fonscolombe), has been recognized since 1738 in Europe, where it is known as the plum lecanium scale (E). It now occurs in the north Atlantic and north central states. The globose scale can injure peach and plum trees—it is most commonly found on the purple leaf plum, *Prunus cerasifera* cv. *Atropurpurea*—and has also been reported on apricot, quince, sweet cherry, and pyracantha. It appears to be a pest of minor importance. There is one generation per year. Overwintering takes place in the second-instar stage. The adult stage (males and females) is reached by early May. Mating takes place and the female produces up to 3000 eggs; eggs hatch a few hours after oviposition. By the end of June the crawlers have settled on twigs and branches, where they stay for the remainder of their lives. Heavy populations result in wilted foliage and sometimes premature leaf drop.

References: 167, 443, 517, 854

A. A light infestation of rose scales, *Aulacaspis rosae*, on rose canes. [Diaspididae]
B. Male scale covers of the rose scale, an armored scale.
C. A single female scale cover of the rose scale.
D. Predominantly empty "pupal skins" of male globose scales.
E. A colony of mature female globose scales on a purple plum tree.
F. Brown male globose scale "pupae" and empty "pupal skins." *At arrows:* two males ready to emerge.
G. Female globose scales with crawlers emerging. [Coccidae]

A
B
C
D
E
F
G

European Elm Scale and Other Bark-feeding Scales (Plate 176)

Elms are susceptible to infestation by a variety of soft and armored scales. As many as five or more species may attack a single elm at the same time. Species of *Gossyparia, Pulvinaria, "Lecanium"* (now reclassified into four genera), *Clavaspis, Diaspidiotus, Quadraspidiotus, Chionaspis,* and *Lepidosaphes* have been observed on elm.

The European elm scale, *Gossyparia spuria* (Modeer), arrived on American soil in New York in 1884. In less than 10 years it reached Nevada. Now it is likely to be found wherever elms grow. European elms are the primary host, but all of the several species of native elms are susceptible hosts. We have not seen it on Asiatic elms.

Injury occurs when excessive amounts of internal fluids are removed from phloem tissue. These fluids supply the scale with nutrients and water, but the insect removes much more fluid than it can use. The excess is excreted as honeydew, and accumulations of this material become one of the symptoms of scale presence, particularly in the summer. Scales also cause premature yellowing of leaves, especially on lower branches. These may also be covered with sooty mold that develops on the honeydew. Leaves may wilt, and in extremely heavy infestations twigs may die by midsummer. Severely injured leaves do not drop in autumn but remain on the tree all winter.

The European elm scale produces one generation a year. It overwinters as a second-instar nymph tightly adhering to the base of buds, often surrounded by dead females, in cracks of elm bark (B), or at the base of twigs. These hibernating individuals resemble small mealybugs because of their oval outline and the short white waxy filaments covering their bodies. Male and female forms are distinctly different. In late winter the immature males attached to bark make white felty cocoons from material secreted from wax glands. They soon resemble grains of rice. Some of the cocoons have a pair of long filaments that protrude from the posterior end. Upon eclosing from the pupal "skin," some males have wings and a pair of white posterior filaments longer than their bodies.

In the spring females produce tough felted white fibers that ring their oval bodies, contrasting with their deep brown body color (A). By the end of June the female is laying eggs that hatch in a few hours into bright yellow crawlers. The crawlers occupy positions in grooves along the midrib and other prominent veins on the underside of leaves. Pubescence along the leaf veins partially conceals the crawlers (C), which soon molt into the second instar. Once a feeding site is established, they remain in place throughout the remainder of the summer. At this stage they, too, look much like mealybugs. Before the normal drop of foliage in autumn, many nymphs drop to the ground with leaves and are rarely able to survive.

The elm scurfy scale, *Chionaspis americana* Johnson, was widely distributed and commonly injurious to elm (H) in the East before the advent of Dutch elm disease and elm yellows. It has also been reported on *Celtis* and *Platanus* species. It overwinters as red eggs under the armor of females. Crawlers, second-instar nymphs, and adults feed on fluids extracted from bark phloem. In the Midwest there are two generations per year. In addition to *C. americana,* elm is attacked by another scurfy scale, *C. furfura* (Fitch), although this species occurs primarily on mountain ash, *Crataegus* species, butternut, *Prunus* species, cottonwood, and pyracantha. It, too, overwinters as an egg and produces two generations per year in the midwestern states. Parasite populations develop naturally in populations of both *Chionaspis* species. The most commonly found parasites are wasps belonging to the genus *Aphytis*.

The dogwood scale, *Chionaspis corni* Cooley, looks much like *C. furfura* and *C. americana. C. corni* feeds only on *Cornus* species. Heavy populations result in dead twigs and general unthriftiness of the entire plant. Overwintering takes place in the egg stage. The biology of this species has not been studied.

References: 81, 167, 186, 193, 442, 517

A. European elm scale females; about 1.3 mm long. [Eriococcidae] Arrow points to a developing *Parthenolecanium* female.
B. Overwintering second-instar European elm scale nymphs in a crack of bark.
C. Settled European elm scale crawlers beside a vein on an elm leaf.
D. A twig showing all stages of the scurfy scale. Immature scale cover at upper arrow; mature female at lower arrow. [Diaspididae]
E. Typical elm scurfy scale covers of mature females; about 2 mm long.
F. A female cover turned up to show reddish oval eggs (*arrows*).
G. Dogwood scales on a twig of *Cornus racemosa*. Scale cover about 2.2 mm long.
H. A dead elm twig encrusted with elm scurfy scale. [Diaspididae]

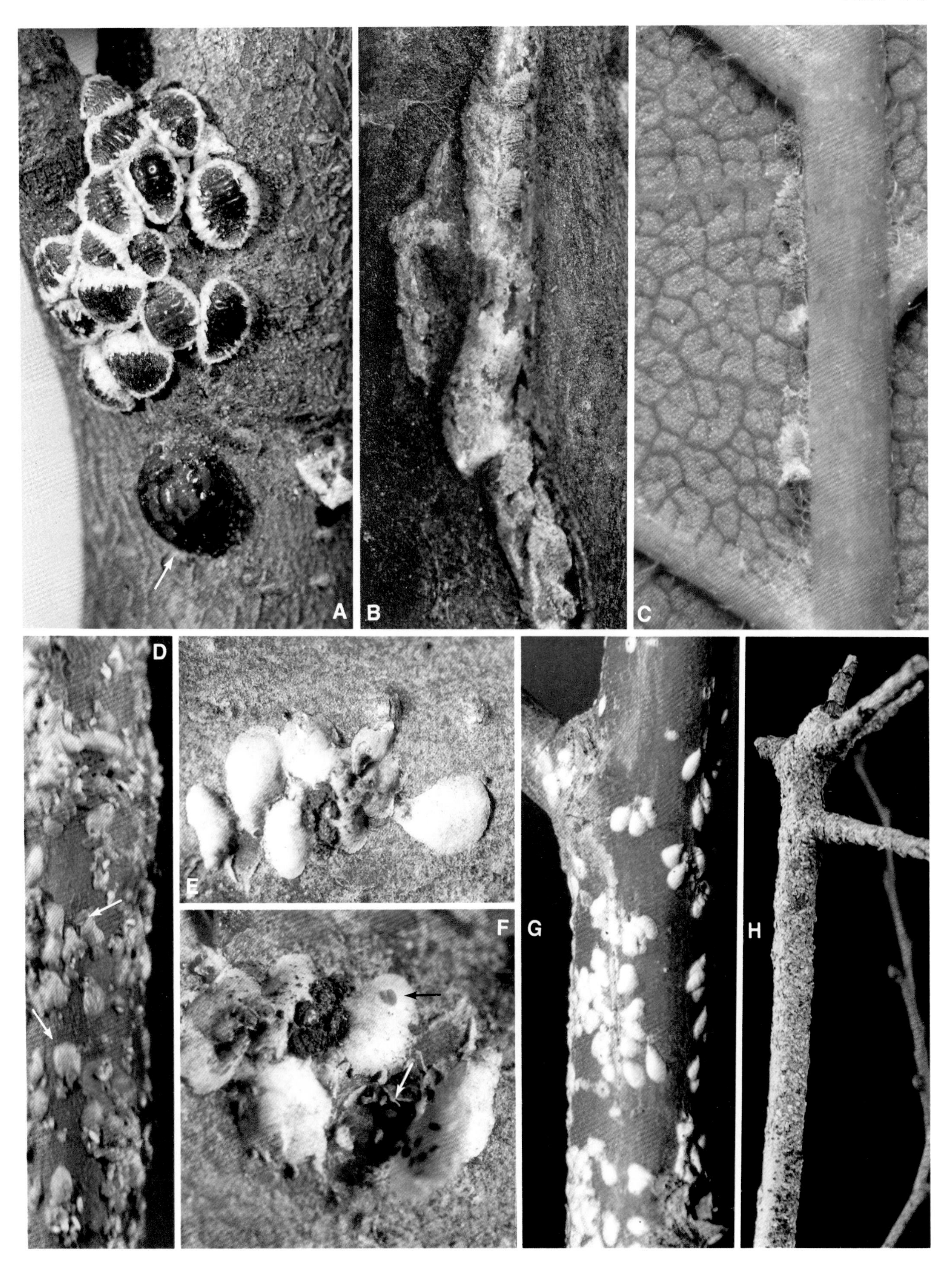
A
B
C
D
E
F
G
H

Oystershell Scale and Camellia Scale (Plate 177)

The living armored scale insect is hidden under a shell made of its own cast "skins" and hardened waxy secretions. It can be exposed by gently lifting this armor. Underneath is an animal that seems to have no resemblance to an insect. It has no legs, no eyes, and no readily apparent antennae. It is attached to its food plant by hairlike mouthparts that are embedded into the host tissue for a length of three or more times that of its body. The oystershell scale and those insects illustrated in the following 12 plates are examples of such armored scale insects.

The oystershell scale, *Lepidosaphes ulmi* (Linnaeus), is a common and important pest of trees and shrubs and is probably more widely known than any other scale insect. In 1738 Reaumur noted the resemblance of this scale to a sea shell. In Europe it is still known commonly as the mussel scale. Linnaeus first described this species in 1758. A letter dated December 1794, held among the official documents of the still active Massachusetts Society for Promoting Agriculture, gives the first known description of oystershell scale in the United States. In it, Enoch Perley included observations of its seasonal development that hold true nearly 200 years later. Oystershell scale is distributed throughout the world except in arctic and tropical areas. In the United States it occurs in every state, but more commonly in the North than in the South. In northern New England and New Brunswick during the mid-1980s acute infestations occurred in hardwood forests where beech, birch, maple, and ash were the dominant species. Twig mortality was the most obvious injury. The scale was much more common before the advent of synthetic organic insecticides. In 1916, 128 species were recorded as hosts. In 1925 Griswold (308) recognized two forms of the scale in Ithaca, New York, and found them locally on plants representing 19 genera in 12 families. These forms differ in appearance and in seasonal development. The "lilac," or banded, form is slightly larger and somewhat later in developing than the "apple," or brown, form (C, D).

Although well over 128 host plants are known, oystershell scale is a serious pest on relatively few of them. The hosts most commonly infested by the lilac form include lilac, ash, willow, poplar, and maple. The apple form infests apple and dogwood as well as poplar. Other frequently infested hosts include boxwood, birch, beech, cotoneaster, elm, horsechestnut, linden, mountain ash, pachysandra, pear, plum, sycamore, tuliptree, viburnum, Virginia creeper, and walnut. Typically the oystershell scale first involves only certain branches or parts of a host. Entire branches may become encrusted with scales before other sections of the plant become infested. If this pest is not controlled early, large portions of trees and shrubs are killed.

The oystershell scale overwinters in the egg stage under the female scale's cover. From this point in development the life history varies according to host and geographic location. In Maryland populations on lilac and maple have two generations per year, with crawlers found in May and June and again in mid-July; on poplar and willow only one generation per year occurs, and on boxwood there may be either one or two generations. In New England and Canada there is one generation per year regardless of host. Under some conditions in Canada eggs are produced asexually.

Numerous parasites and predators have been recorded from populations of the oystershell scale. They play an important role in naturally reducing heavy populations, but usually they are too late to prevent serious damage. Predatory mites feed on eggs and are particularly helpful in the spring. The twicestabbed lady beetle, *Chilocorus stigma* (Say), is an important predator, and when it is abundant in a local area, it is influential in restricting scale build-up. Generally, however, natural enemies are attracted to and develop populations in areas where large numbers of scales have already become established.

In 1989 Cooper and Oetting (888) reported much new information about the biology of the camellia scale, *Lepidosaphes camelliae* Hoke. In Georgia four to five generations are produced each year, and these generations overlap to the degree that the crawler stage (first instar) may be found almost anytime during the growing season. The crawlers are cream-colored and sometimes stippled with purple. This mobile stage of the first instar may last no more than 24 hours. Camellia scales usually feed on the undersides of leaves. Orange and yellow splotches on the upper surfaces are evidence of scale colonies underneath. It is common to find the tea scale (Plate 189) competing with the camellia scale for space. Hosts other than camellia include holly, particularly Burford holly, privet, *Cleyera, Ternstroemia,* and *Rhaphiolepsis*. The camellia scale is found from Texas to Missouri east, and northeast to New York and Massachusetts.

References: 193, 217, 283, 359, 660, 888

A. The *Viburnum* species (*left*) is near death because of the activities of the oystershell scale. The plant on right shows effects of a lesser infestation.
B. A crawler (*circle*) and a settled nymph (*arrow*) initiating its waxy cover.
C. The banded, or lilac, form on a *Syringa* species; adult female scales.
D. The brown, or apple, form of the oystershell scale; females and male (*arrow*). Note scale covers with exit hole of a parasite. [Diaspididae]
E. Four old female scale covers and numerous partially developed females. Adult females are up to 3 mm long.

A
B
C
D
E

Greedy Scale and Peony Scale (Plate 178)

The greedy scale, *Hemiberlesia rapax* (Comstock), is the most common and most widely distributed species of armored scale in California, according to E. O. Essig (217), one of California's best known entomologists. J. H. Comstock (123), who provided one of the first descriptions of this insect, said "I have named this the greedy scale on account of the great number of plants upon which the species subsists." It has been found on 45 ornamental species in Florida alone, including abelia, avocado, azalea, bay, cactus, camphor, *Castanopsis* species, chinaberry, cedar, species of *Elaeagnus* and *Euonymus*, fig, geranium, guava, gumbolimbo, holly, honeysuckle, huckleberry, juniper, loquat, magnolia, morning glory, mulberry, myrtle, oak, palm, pear, pecan, persimmon, plumeria, poisonwood, privet, sapodilla, seagrape, and tecoma. Additional hosts (found in California) include *Acacia* species, almond, apple, bladderpod, boxwood, broom, *Ceanothus* species, camellia, *Cestrum* species, chapparal broom, cherry, cissus, *Cotoneaster* species, cottonwood, English holly, English ivy, English laurel, species of *Eucalyptus, Fuchsia,* and *Genista,* giant sequoia, grape, hakea, heath, mountain holly, Japanese quince, California laurel, *Lavatera* species, locust, manzanita, mistletoe, nightshade, olive, orange, Oregon grape, palms, passion vine, pepper tree, *Pittosporum* species, pomegranate, pyracantha, quince, redbud, sage, sedum, strawberry tree, *Strelitzia* species, tanbark oak, umbrella tree, Persian walnut, and willow.

The distribution of the greedy scale includes most of the United States, except the prairie states and New England. It is a rather common insect in greenhouses in the North.

In the warmer climates the greedy scale probably reproduces continually, several generations occurring each year. The greedy scale is closely related to the latania scale (Plate 180) and may also be confused with the Putnam scale, *Diaspidiotus ancylus* (Putnam). The scale armor, or test, is usually light gray; it is circular and highly convex, measuring about 1–1.5 mm in diameter (E, H–J) when fully grown. As in many of the armored, or diaspid, scales, the cover of the greedy scale shows concentric color lines when viewed from above. These rings are formed by the discarded skeletons and secretions of the juvenile form that, as the insect grows, are pushed upward and incorporated into the scale insect's armor.

H. rapax feeds by sucking sap and cell contents from woody stems, twigs, leaves, or fruit. Yellow foliage, poor growth, and dead twigs or branches are all symptomatic of persistent attack by this insect (A, F, G).

Pseudaonidia paeoniae (Cockerell) is sometimes called the peony scale, although it is primarily a pest of azalea, camellia, and species of *Myrica, Ilex,* and, occasionally, *Ligustrum*. It is presumed to be a southern species and has been found in Florida, Georgia, Mississippi, South Carolina, and Virginia, as well as on Long Island west to Missouri and Texas. Little is known about its biology. Eggs are lavender (C) and are probably laid in April. In parts of Virginia the purple crawlers appear in late May. They feed on bark and in severe infestations can kill twigs and branches. The female armor, or test, is 3–4 mm in diameter and is usually embedded in the epidermis and partially covered by flakes of bark, giving it a grayish cast. The high point of the armor, called the first exuviae, is orange-yellow (B). This feature can usually be relied on in making field identification. The peony scale has one generation each year.

The camphor scale, *P. duplex* (Cockerell), is sometimes confused with the peony scale and may occur on both the leaves and twigs of camellia, as well as on other mutual hosts. The camphor scale was accidentally introduced into the United States from Japan and has become a pest of camphor in Texas, Mississippi, and Alabama. It was recently reported in Virginia from *Ilex cornuta* 'Burfordii' and represents a potential threat to many subtropical trees and shrubs in the southern United States. In Florida 43 host plants are listed for it. The camphor scale has three generations each year south of Virginia.

References: 19, 123, 167, 217, 517, 530, 786

A. Symptoms of injury caused by the greedy scale on California laurel.
B. Peony scales (*arrows*) on bark of *Camellia*. The first exuviae (skeleton) is orange-yellow. Bark flakes partially cover part of the armor, giving the scale an effective camouflage.
C. Peony scales at a crotch of *Camellia*. A cluster of eggs is located in the circle. [Diaspididae]
D. Aucuba injured by greedy scales.
E, H–J. Close-ups of greedy scales on various hosts. Inside the circle in E are young scales, eggs, and crawlers. The arrow in E points to the test of a mature female.
F, G. An English holly severely injured by greedy scales. Premature leaf drop is the major symptom of injury.

A
B
C
D
E
F
G
H
I
J

Oleander Scales (Plate 179)

The oleander scale, *Aspidiotus nerii* Bouché (=*A. hederae* Vallot) resembles the greedy scale (Plate 178). It is a subtropical species commonly found on oleander and occurs outdoors in the warmer parts of the United States. Its white to yellowish cover makes it highly visible, especially on the upper surface of a leaf. This scale is also found on the undersides of leaves. If leaves are hairy (trichomes), the scale may be nearly invisible. High populations are common, which suggests that there is more than one generation each year. In the greenhouse expect feeding stages throughout the year. It attacks the bark, leaves, and fruit of its host plants, which include *Acacia* species, aloe, *Aucuba* species, avocado, azalea, blueberry, Boston ivy, boxwood, buddleia, buckthorn, broom, cactus, California laurel, camellia, carob, *Ceanothus* species, cherry, coffeeberry, currant, daphne, dogwood, English ivy, *Elaeagnus* species, *Eucalyptus* species, ferns, *Genista* species, grape, grapefruit, Grecian laurel, guava, hakea, heath, holly, holly oak, hypericum, lemon, magnolia, manzanita, maple, mistletoe, Monterey pine, mulberry, myrsine, myrtle, nightshade, olive, *Osmanthus* species, orange, palms, pepper tree, pistache, poinsettia, pomegranate, privet, redbud, redwood, rose, rubber, sago palm, sumac, umbrella plant, umbrella tree, viburnum, *Vinca* species, vitex, yew, and yucca. Little is known about the biology of this insect.

Pseudaulacaspis cockerelli (Cooley) is sometimes called the false oleander scale—not a good name because it has such a wide host range. It was first described in California from palms taken in quarantine from China. Now it is widespread in Florida, Georgia, and Alabama. It probably occurs in all of the Gulf states. This scale confines its feeding to foliage; rarely does it attack tender shoots and fruits. It causes chlorotic spots, visible from the upper surface, that are several times larger than the scale cover (test). Heavy infestations cause the entire leaf to turn yellow and drop prematurely.

In south Georgia all stages of *P. cockerelli* are present throughout the year. Greenhouse studies indicate that a generation can develop in five weeks. Females (B) are relatively long lived and have a pear-shaped armor. The elongate male armor is about 1 mm long. Males tend to segregate themselves as crawlers, resulting in clusters of the elongate exuviae. Dekle (167) listed more than 100 plant species as hosts, most of which were ornamentals and include *Magnolia grandiflora, M. virginiana, Aucuba japonica, Hedera helix,* flowering dogwood, *Taxus* species, oleander, banana shrub (*Michelia figo*), *Elaeagnus* species, and palmetto.

The size of the armor of females varies with the host: on palmetto, 2–3 mm long; on aucuba, 4 mm.

References: 217, 322, 517, 562, 785

A. Oleander scales, about 1–2 mm in diameter, look like grayish spots on the leaves of *Magnolia grandiflora*.

B. A cluster of *Pseudaulacaspis cockerelli* on magnolia.

C. Clusters of oleander scales on a variegated Virginia creeper (*Parthenocissus* sp.) leaf.

D, G. Oleander scales on *Viburnum rhytidophyllum* leaves. On this host the scale prefers the underside of the leaf and causes small yellowish spots that are visible from the upper surface (*arrows*). The scales shown in G are covered by the leaf hairs, making them almost invisible.

E. Oleander scales on a species of *Pittosporum*. [Diaspididae]

F. At the arrow points are two adult female oleander scales with their shells removed.

H. Close-up of the oleander scale on magnolia.

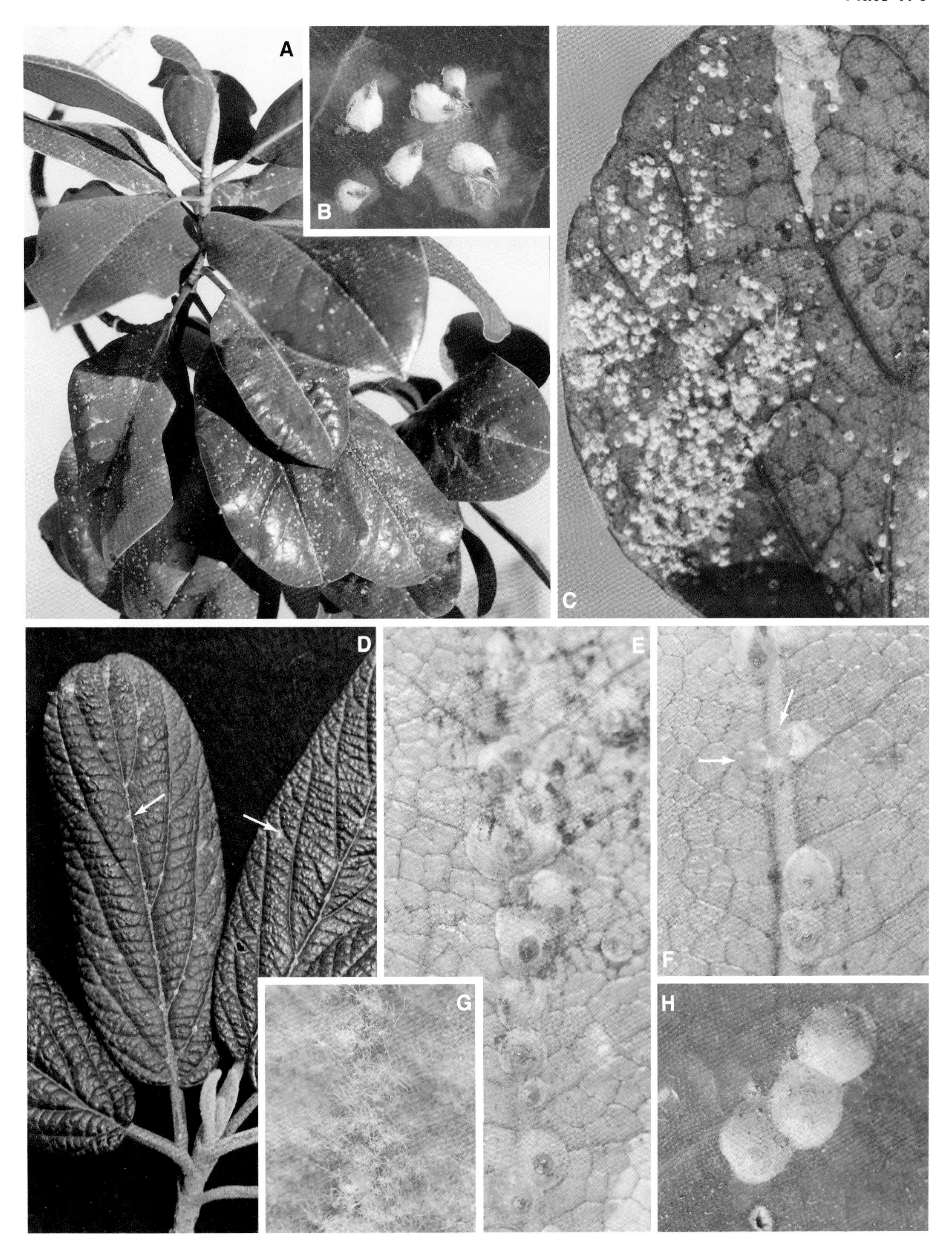
A
B
C
D
E
F
G
H

Subtropical Armored Scales (Plates 180–182)

The latania scale, *Hemiberlesia lataniae* (Signoret), is known primarily in the coastal states but only as far north as Maryland in the East. It feeds on a great many plants, including a wide range of ornamentals. In California some of the latania scale's hosts are species of *Acacia, Cedrus, Euonymus, Fatsia, Fuchsia,* and *Grevillea,* as well as *Hedera helix,* species of *Howea, Olea europaea,* and species of *Persea, Rosa, Rubus, Salix,* and *Yucca.* In Florida its most common hosts are *Casuarina* species, loquat, palms, and rose; in Maryland this pest often feeds on holly.

H. lataniae is a circular armored scale (D) about 1.5–2 mm in diameter. The scale cover varies from brown to gray and is almost indistinguishable from the greedy scale, *H. rapax* (Plate 178). Only a microscopic examination of specially prepared stained specimens will reveal the difference between these two species.

Female latania scales lay eggs beneath their covers. Upon hatching from the eggs, the young crawlers migrate a short distance from the parent scale. Here they settle, insert their mouthparts into the host plant, and begin to feed. The legs of the crawler become functionless after it settles. In southern California scales undergo two molts within 2 months, become fully mature, and begin to lay eggs. Most reproduction takes place without mating. In Maryland there are two generations per year. This species overwinters as second-instar males and females on leaves or stems (Figure 115).

The majority of a scale population feeds on the twigs and branches of its host, although individuals may be found on any aerial part of the plant. If the population is great, dieback of twigs and smaller branches occurs. No honeydew deposits or blackening of plant parts results from an infestation as occurs with soft scales. The latania scale has many natural enemies, among them lady beetles, lacewings, predaceous mites, and tiny parasitic wasps. These provide a measure of biological control. Wasp parasites are most important. They emerge in mid-May, in mid-June, during the second week of August, and in early September in Maryland. Using Maryland as a reference point, one can estimate the time of parasite emergence for other areas. Do not apply pesticides for scale control at the time of parasite emergence.

The cocos scale, *Diaspis cocois* Lichtenstein, is a small armored scale (A) that infests the leaves of species of *Chamaerops, Cocos, Kentia, Latania, Livistona, Phoenix,* and *Roystonea* in California. This scale was found in Florida in 1961 but was reported to have been eradicated. The adult female is white and circular, whereas the male scale, also white, is quite elongate. The armor of *D. boisduvalii* Signoret looks like that of *D. cocois.* The two cannot be distinguished in the field. The boisduval scale is common on palm and yucca in subtropical areas of the United States and is frequently found in greenhouses elsewhere, especially on orchids.

Another armored scale, *Lindingaspis rossi* (Maskell), attacks oleander, olive, *Hedera helix,* several palms, and species of *Araucaria, Euonymus, Ficus, Gardenia, Camellia, Citrus,* and *Eucalyptus.* It is a tropical Australian species, and its distribution in the United States is limited to California. A related species, *Lindingaspis floridana* Ferris, is found in Florida, where it infests mango, avocado, *Ficus* and *Ligustrum* species, palms, and other plants. The scale covers of both species look very much alike. Both live only on leaves.

Pseudaonidia trilobitiformis (Green), sometimes called the trilobe scale, is often found on *Citrus, Ixora,* and *Nerium.*

References: 167, 212, 517, 767

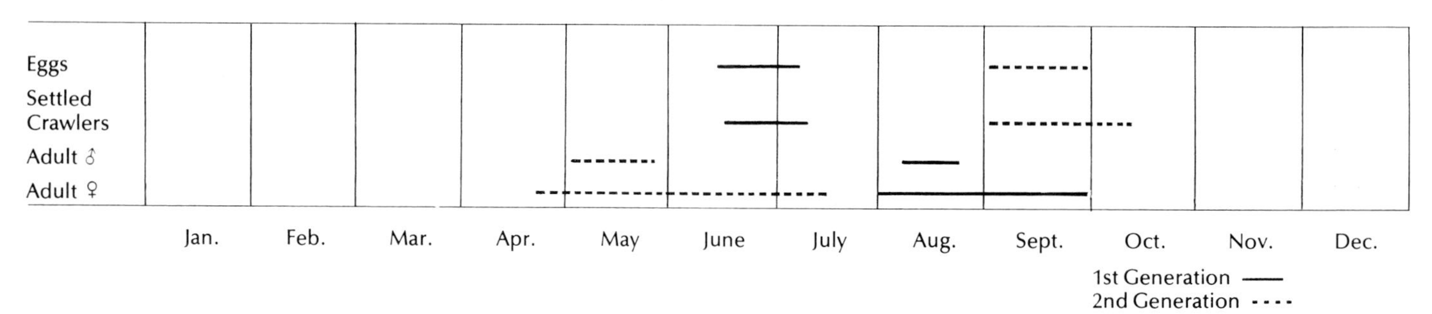

Figure 115. Seasonal history of *Hemiberlesia lataniae* on holly in Maryland. (After Stoetzel and Davidson, ref. 767.)

A. A palm frond heavily infested with cocos scales, *Diaspis cocois.*
B. Close-up of male (*arrow*) and female *Diaspidiotus* sp. scale covers on the palm *Livistona* sp.
C. Latania scales, *Hemiberlesia lataniae,* on a small branch of the handflower tree, *Chiranthodendron pentadactylon.*
D. Latania scales on a handflower tree leaf. Armor about 1.5 mm in diameter.
E. Latania scale females reared on a potato tuber. Note how dark the peak (exuviae) of the cover appears after the thin white wax layer is rubbed off.
F. Male (*arrow*) and female *Lindingaspis rossi* scales on *Araucaria* sp. foliage.

A
B
C
D
E
F

The California red scale, *Aonidiella aurantii* (Maskell), is considered to be the most important pest of citrus in the world. It also occurs on a large number of other woody plants. This scale, a native of China, occurs in the Gulf states and in California.

The California red scale attacks all parts of the host plant. Symptoms of infestation include yellowing of foliage, defoliation, and the death of twigs and branches. In extreme cases death of the entire plant may result. The scales feed by sucking the sap of their host plant. It is believed that the scale may introduce a toxic substance into plant tissues, causing injury to the plant beyond that resulting from feeding alone (D–F).

In southern California there are from two to four generations of the California red scale each year. More generations develop in inland areas of California than on coastal sites. The adult female does not lay eggs but gives birth to living young. The crawlers settle on any part of the host plant and begin to feed within a day of emerging from beneath the scale cover of the female parent. There is a great deal of overlapping of generations, so all stages may be found at any time of the year.

The following are common woody ornamental hosts of the California red scale: *Acacia*—wattle; *Acer negundo*—boxelder; *Aspidistra*—iron plant; *Buxus*—boxwood; *Chaenomeles*—quince; *Citrus*; *Coprosma*; *Eucalyptus*; *Euonymus*; *Ficus*—fig; *Fuchsia*; *Ilex*—holly; *Juglans*—walnut; *Leptospermum*—tea tree; *Ligustrum*—privet; *Magnolia virginiana*—sweet bay; *Malus*—apple; *Mangifer indica*—mango; *Morus*—mulberry; *Olea*—olive; palms; *Passiflora*—passion vine; *Pistacia*—pistachio; *Podocarpus*; *Pyrus*—pear; *Quercus*—oak; *Rosa*—rose; *Salix*—willow.

For many years the yellow scale, *Aonidiella citrina* (Coquillett), was considered to be a variety of the California red scale. However, differences in parasite preference and several subtle anatomical differences between these insects have prompted taxonomists to regard them as separate species. More recently, the discovery that female yellow and California red scales produce chemically different sex pheromones adds evidence that they are in fact two species.

The two scales have very similar life cycles but have different habits. When the yellow scale infests *Citrus* species, it tends to feed on the fruit and leaves instead of the woody plant parts. The California red scale, on the other hand, infests all parts of its host plant.

The yellow scale occurs in California, Texas, and Florida, and it attacks a number of plants in addition to species of *Citrus*. Many of these are important ornamentals. Its hosts are fewer than those of *A. aurantii* and include species of *Aucuba, Daphne,* and *Euonymus, Ficus elastica*—Indian rubber tree, *Hedera helix*—English ivy, palms, and *Pandanus utilis*—screw pine.

In California the most important natural enemies of the California red scale on *Citrus* plants are the parasitic wasps *Aphytis melinus* Debach, *A. lignanensis* Compere, *Comperiella bifasciata* Howard, and *Prospaltella perniciosi* Tower. Important parasites of the yellow scale are *A. melinus* and *C. bifasciata*. It is believed that these natural enemies provide substantial biological control of these scales on ornamental plants.

References: 202, 217, 656

A. Yellow spotted leaves of privet that have been damaged by the yellow scale, *Aonidiella citrina*.

B, C. Yellow scales on the lower surface (B) and upper surface (C) of privet leaves, *Ligustrum lucidum*.

D–F. Close-ups of the California red scale, *A. aurantii*, on lemon fruit. Males at arrows. The red or yellow colors are not satisfactory field identification characters.

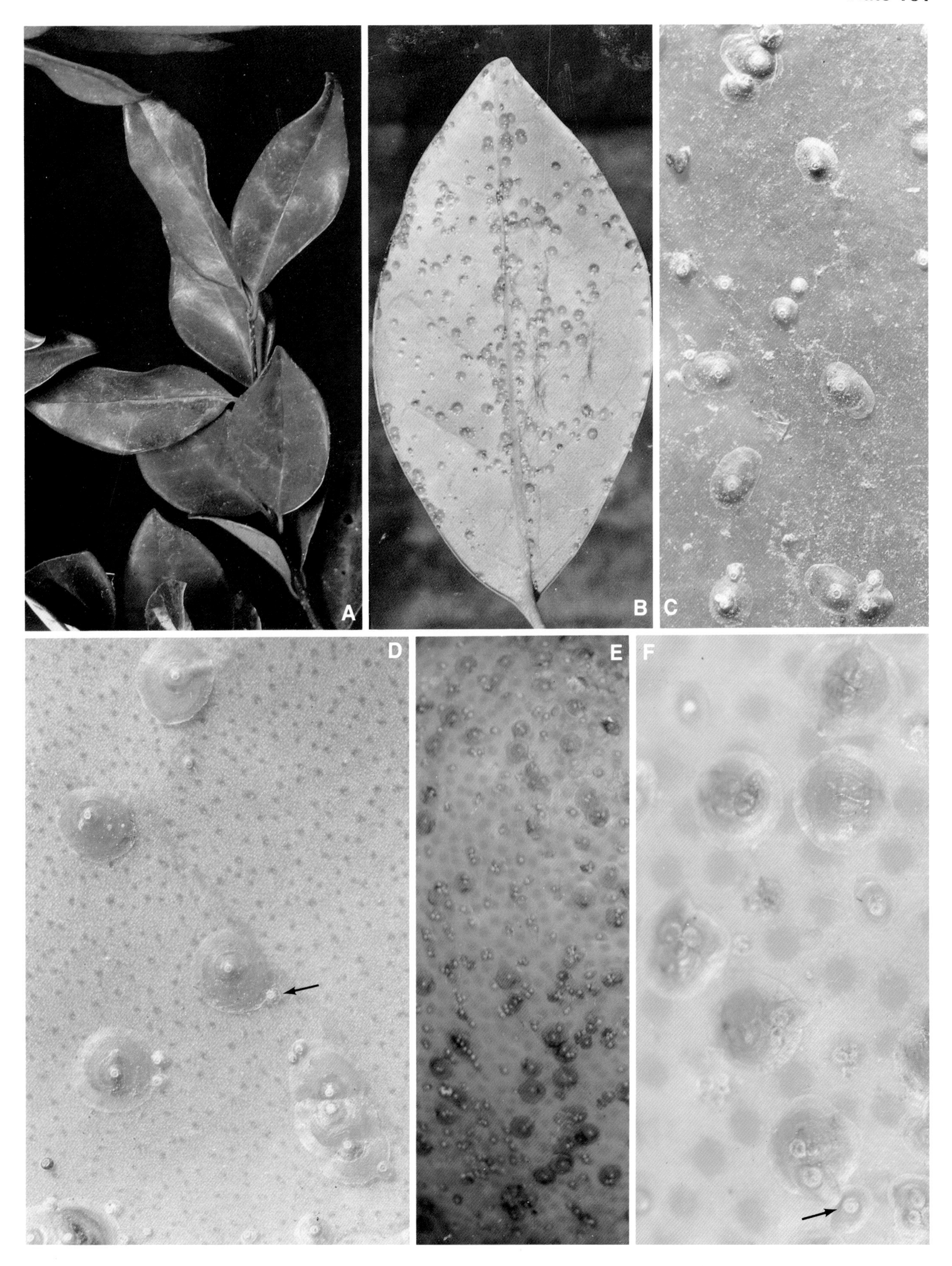
A
B
C
D
E
F

(Subtropical Armored Scales, continued, Plate 182)

The Florida red scale, *Chrysomphalus aonidum* (Linnaeus), infests only the leaves of its hosts. It occurs outdoors in all of the citrus-growing areas of the world, and in greenhouses and plant conservatories everywhere. The list of hosts upon which it feeds seems endless. Dekle (167) lists about 260 genera of known host plants. The Florida red scale (B) is one of the most serious pests on citrus and ornamental plants in Florida. In California it is of little importance as a pest of citrus but is present on 27 genera of ornamental plants. Four to five generations occur each year. Most generations overlap. Females produce eggs over an extended period of time. Eggs hatch several hours after they are laid. All stages are present at any time during the season.

The Glover scale (E), *Lepidosaphes gloveri* (Packard), is a relative of the oystershell scale. Its range covers all citrus-growing regions. It occurs on citrus, magnolia, euonymus, arborvitae, boxwood, ivy, cherry laurel, cypress, mulberry, myrtle, orchid, cabbage palmetto, *Podocarpus* species, and privet. In California, it is reported on willow and umbrella pine as well as on euonymus, magnolia, and palms. It is often found in association with the purple scale and has a similar life cycle. It occurs mainly on twigs and branches but is also found on leaves.

Another cosmopolitan species found in citrus-growing areas all over the world is the purple scale (G), *Lepidosaphes beckii* (Newman). Dekle (167) has listed 43 genera of hosts in Florida, including citrus. McKenzie (517) noted that in addition to citrus in California, boxwood, eleagnus, and holly are attacked. The purple scale feeds on the bark, leaves, and fruit of its host plant. It strongly resembles the oystershell scale. Its seasonal development is continuous and occurs in three annual, overlapping generations. The females lay 60–80 eggs that hatch over a period of 2 or more weeks. About 6–8 weeks is required for maturation of an individual scale insect.

First specimens of *Duplaspidiotus claviger* (Cockerell), sometimes called the camellia mining scale, were collected from *Camellia sasanqua* in Florida in 1962. By that time the scale was well established as a foreign introduction. The common name, which implies a mining capability, is misleading. It is a typical armored scale that feeds from phloem tissue while permanently affixed to the bark of twigs and small branches. As it develops on camellia, its armor, or test, is almost covered by the outer layer of bark that grows around it (D). This makes recognition in the field very difficult. Damage by the scale is expressed by very limited growth and twig dieback.

The insect's armor is approximately circular, gray, and about 2 mm in diameter. The life cycle has not been studied, but it is presumed to produce several generations per year. The eggs are located under the armor and have the additional protection of the bark. Upon hatching, the crawlers may be trapped, but one or two individuals may survive by simply inserting their mouthparts into the bark and continuing to develop, their own armor soon giving additional protection. Dead scales often remain attached to the host for long periods.

Plants most frequently infested are *Camellia japonica, C. sasanqua,* and *Ligustrum japonicum*. Other hosts include *Callicarpa americana, Callistemon* species, pecan, *Cestrum* species, *Cornus florida* and *C. stricta, Eugenia species, Ficus* species, azalea, *Gardenia* species, *Ilex opaca,* crape myrtle, wax myrtle, *Osmanthus* species, cherry, laurel, and pyrancantha. This scale is primarily spread through the movement of infested nursery stock. We have records of the scale in Florida and Georgia. It can be expected to occur wherever its host plants grow.

Other armored scales found on camellia and often mistaken for the camellia mining scale are *Pseudaonidia paeoniae* (Cockerell) (Plate 178), *D. tesseratus* (Grandpre & Charmoy), and *Howardia biclavis* (Comstock), species that are also given the common name mining scale. The natural enemies of these scales include parasitic wasps, predaceous mites, and lady beetles.

References: 167, 168, 517

A. Citrus foliage infested with Florida red scales.
B. Florida red scales in various stages of development. Arrow at bottom points to a yellow-bodied scale whose cover was removed. Arrow at top points to an elongate male.
C. Nearly mature and mature females of the Florida red scale. Arrow points to a crawler.
D. Several camellia mining scales in a bark fissure of camellia. [Diaspididae]
E. Glover scale females on a citrus leaf. [Diaspididae]
F. Various stages of Glover scale; elongate forms are adult females (*arrow*).
G. Immature females of the purple scale.

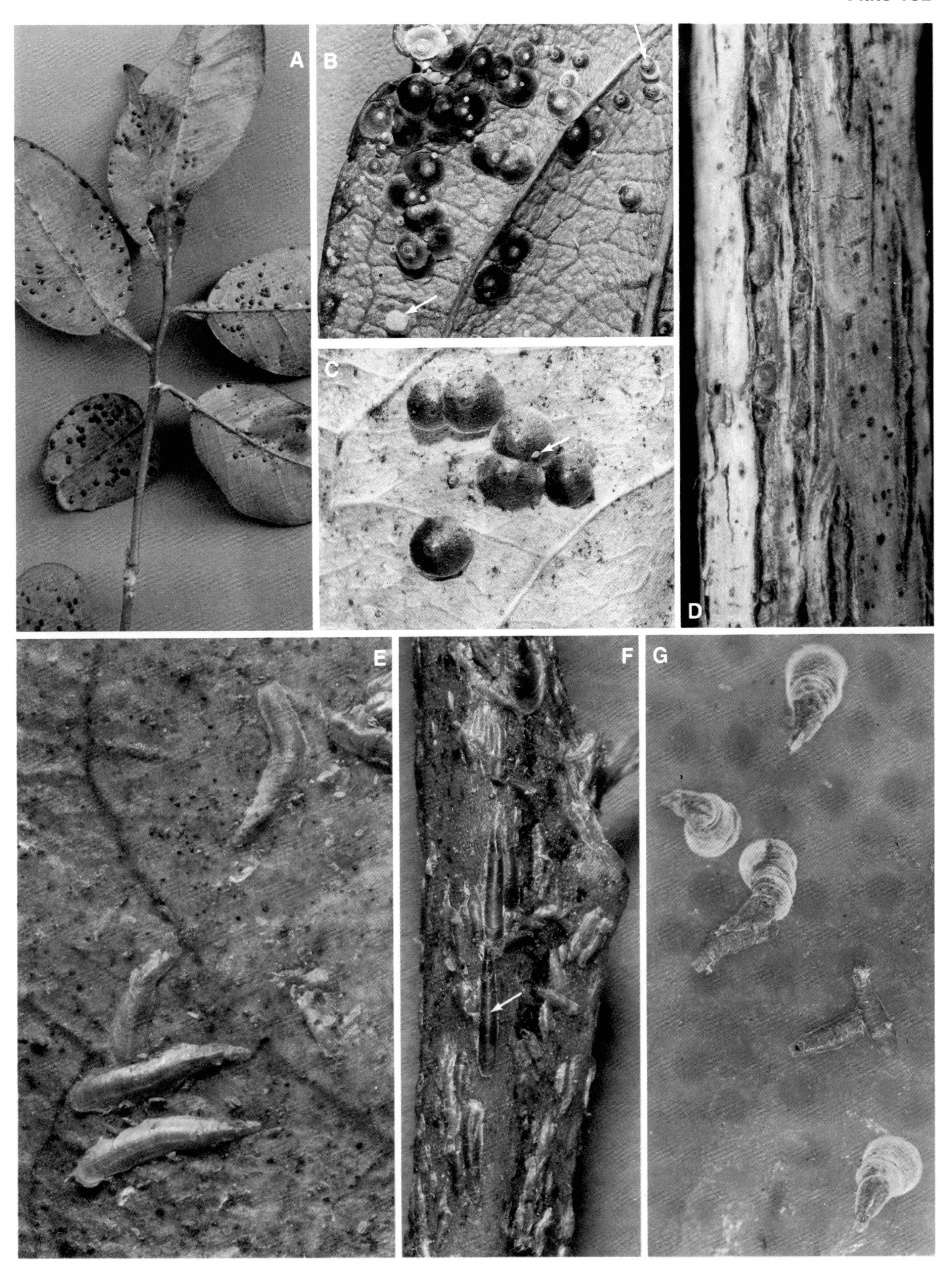
A
B
C
D
E
F
G

Dictyospermum Scale and Gall-forming Armored Scales (Plate 183)

The dictyospermum scale, *Chrysomphalus dictyospermi* (Morgan), is found in subtropical areas of the United States. It feeds chiefly on leaves (A) but may also be found on bark or fruit. When abundant, it brings about the slow decline and eventual death of its host. Because of their small size (less than 2 mm in diameter), the scales may be unnoticed until the infested host is beyond recovery. Some common hosts of the dictyospermum scale are species of the following plants: *Acacia, Aralia, Araucaria,* arborvitae, *Aucuba,* avocado, *Bauhinia,* boxwood, *Callistemon, Camellia, Cinnamomum, Citrus, Clematis, Cotoneaster, Cupressus, Elaeagnus, Eucalyptus, Eugenia, Euonymus, Feijoa, Ficus, Gardenia, Ilex, Kalmia, Laurus, Ligustrum,* mulberry, myrtle, oleander, olive, *Osmanthus,* palm, *Pyracantha, Raphiolepis,* rose, *Taxus, Vinca,* and willow, as well as cherry laurel.

In subtropical climates generations of this scale overlap. Thus, their seasonal history is somewhat confusing. In southern California there may be five or six generations each year. In the spring a generation may be completed in 7 weeks.

Female dictyospermum scales lay eggs that remain under their shells or scale covers. Upon hatching, the young crawlers leave the protection of the scale cover and rove about, seeking a suitable place to settle and begin feeding. The crawlers are very tiny (less than 0.5 mm long) and are yellow (Plate 186C depicts a typical armored scale crawler). Several days after a crawler settles, it assumes a circular shape and produces a fine mass of white cottony threads through pores. These white threads eventually cover the entire insect. When the insect molts, the cast "skin" forms a nipple-shaped prominence at the top of the female cover. With the first molt, the crawler's legs and antennae are shed. Hence, the insect becomes a saclike creature with mouthparts as its only appendages. Upon reaching maturity, it produces eggs to bring forth another generation.

Sweetgum, *Liquidambar styraciflua,* is a primary host of *Diaspidiotus liquidambaris* (Kotinsky), sometimes called the sweetgum scale. The summer population of this scale insect is found in small pits on the underside of the host's leaves. These pits are often located at the junction of the leaf veins. Individuals are not easily noticed (C). An early symptom of their presence, however, are yellow spots on the leaf's upper surface (E). The affected leaf eventually forms a bluntly conical gall containing a single scale. A second generation feeds on buds and twigs, where the scales overwinter as fertilized adult females (Figure 116). In Maryland and Ohio eggs hatch in early June, and at this time the crawlers migrate to leaves. The main damage done by this insect is to foliage. Both males and females feed on the underside of leaves, where their feeding stimulates the formation of a small gall. The armor of the adult female is circular, white to gray, and about 1.2 mm in diameter. *D. liquidambaris* has been found in the southeastern quadrant of the United States, in New York state, and, less frequently, in California. The pest also occurs on *Acer rubrum.*

Abgrallaspis liriodendri Miller & Howard causes similar gall-like deformities on the upper side of tuliptree (*Liriodendron tulipifera*) leaves. The female resides in a pit on the underside, associated with the gall-like structure on the upper side of the leaf. This scale was first described in 1981. As no biological studies have been made to date, its overwintering stage and location are not known. Its distribution is presumed to be limited to Louisiana and surrounding states. Only these two armored scales in North America are known to cause galls on foliage.

References: 167, 442, 517, 538, 767

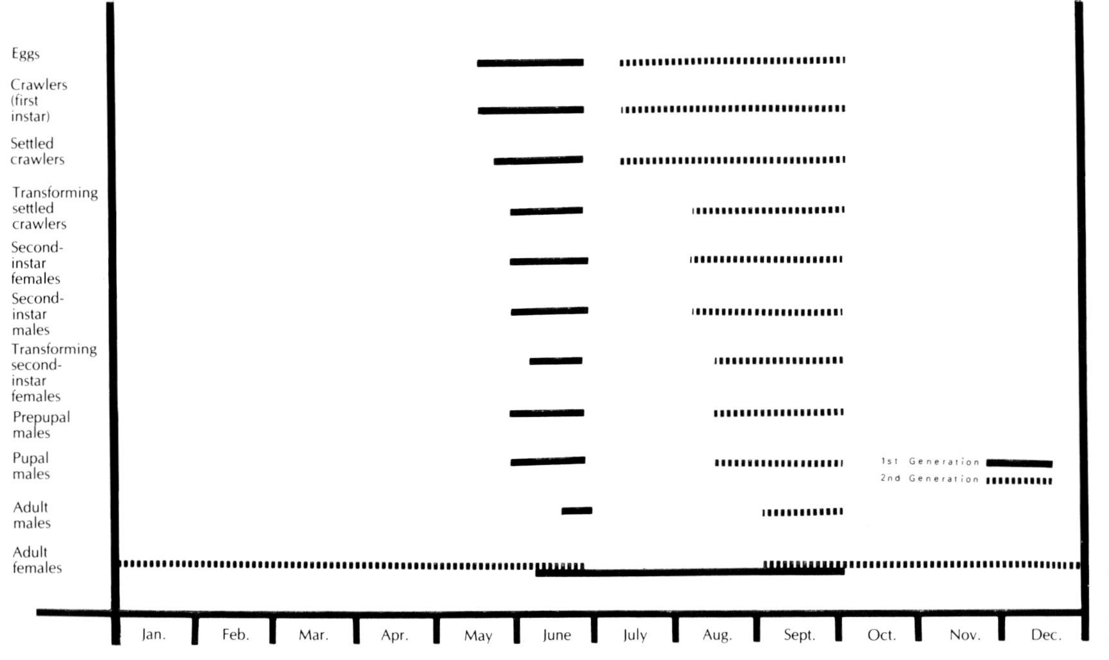

Figure 116. Seasonal history of *Diaspidiotus liquidambaris* on *Liquidambar styraciflua* in Maryland. (From Stoetzel and Davidson, ref. 767, courtesy *Annals of the Entomological Society of America.*)

A. *Ficus pumila* leaves, heavily infested with dictyospermum scales, *Chrysomphalus dictyospermi.*
B. Mature and immature female dictyospermum scales.
C. Male sweetgum scales, *Diaspidiotus liquidambaris,* at the junction of veins on a sweetgum leaf.
D. The roughened upper surface of sweetgum leaves marks the beginning of gall-like enlargements caused by the sweetgum scale.
E. Early symptoms of the presence of sweetgum scales on the upper surface of a sweetgum leaf.
F. The undersurface of a sweetgum leaf showing young sweetgum scales in pits.

A
B
C
D
E
F

Obscure Scale and Gloomy Scale (Plate 184)

Comstock described both the obscure scale and the gloomy scale in 1911 as *Aspidiotus obscura* and *A. tenebricosa*, respectively. In 1941 Ferris (240) placed both species in the genus *Melanaspis*. Thus, these two pests of shade trees are referred to under the genus *Aspidiotus* in much of the old literature. Both species attack the trunk and larger limbs of young and old trees. Dieback of branches, limbs, and sometimes of entire trees occurs where these infestations are not controlled. The scale insects suck sap from the phloem cells, depriving the tree of food manufactured in the leaves.

The obscure scale, *Melanaspis obscura* (Comstock), is widely distributed in the United States. Although it is most prevalent in the South, it is also common in certain areas of the middle Atlantic states and has been reported from one site in California. The obscure scale's principal hosts are species of *Quercus, Castanea,* and *Carya,* and it occurs on beech, Persian walnut, willow, maple, hickory, grape, dogwood, wild myrtle, chinquapin, hackberry, and the hog plum, *Spondias mombi,* although some of these records are questionable. In California, the pest also attacks apricot, peach, and plum. In recent years, the obscure scale has been regarded as the most destructive pest to ornamental plants in at least two eastern states. It is not a pest in forests.

Detailed studies in Maryland by Stoetzel and Davidson (766) have provided life history data on the obscure scale. On white oaks it completes growth about mid-August, 1 month later than on red oaks. On pin oak males and females overwinter until about the first of May, when they begin to mature. Eggs are laid starting in July. By late July egg production is greatly reduced but continues into early September. The population of crawlers peaks during July, then declines, with another slight surge occurring in August. One generation develops during the summer.

Three important factors make it difficult to effectively control the obscure scale. First, the insects tend to settle close together. This results in layers of scales as they grow and enlarge (C). Second, egg laying occurs over a relatively long period, which results in an extended period of crawler activity. Third, crawlers often settle beneath old clusters of scales where eggs were laid and consequently are never exposed to insecticides that may be applied for their control. The waxy scale cover of the live insect also provides protection. There are, however, numerous parasites and predators of this scale that may be effective as biological control agents.

The biology of the gloomy scale (E), *Melanaspis tenebricosa* (Comstock), is similar to that of the obscure scale. It is chiefly a pest of silver maple, *Acer saccharinum,* and red maple, *A. rubrum.* It has also been found on sugar maple, elm, hackberry, boxelder, buckthorn, sweetgum, gallberry, *Ilex glabra,* mulberry, and soapberry (*Sapindus* spp.). It is highly destructive and may kill trees if not controlled. Currently, the few recommended control measures are only partially effective. The gloomy scale occurs from Florida north to New York, west to Illinois and Missouri, and south to Texas. Trees in nurseries as well as those in landscape plantings are often severely injured by this pest. Like the egg-laying, egg-hatching, and crawler activities of the obscure scale, those of the gloomy scale are long-lasting. Sometimes, populations of this pest increase to the extent that food supplies in the host are depleted. This condition, along with build-up of parasites and predators, eventually leads to population collapse.

References: 167, 240, 764, 766

A. Scale covers of the gloomy scale on a limb of red maple, *Acer rubrum.*

B. Clusters of the obscure scale on a pin oak limb. Note bark depressions. [Diaspididae]

C. A typical cluster of female covers of the obscure scale.

D. Individual scale covers of immature female obscure scales. The black nipple is the first-instar exuviae.

E. Nearly mature gloomy scale females on red maple. The cover of one has been lifted to expose the insect's body.

F. A "çovey" of second-instar obscure scales on red maple with covers lifted to expose bodies. Arrow points to the living scales.

A
B
C
D
E
F

San Jose Scale and Walnut Scale (Plate 185)

The two species of scales shown here are widely distributed geographically and have many hosts. The San Jose scale (G), *Quadraspidiotus perniciosus* (Comstock), is highly injurious and destructive to its host plant. Until the advent of the newer organic insecticides after World War II, it was one of the most common, most destructive, and most widely distributed scale insect pests of fruit, shade, and ornamental trees. It is still common throughout the United States, Canada, and in many other parts of the world. Although first described from North America, the San Jose scale originated in Asia.

The literature pertaining to the San Jose scale is voluminous. The pest appeared in the United States in 1870, on a shipment of ornamental plants from the Orient. It is still a major pest and, if not controlled, is capable of killing trees. Well over 60 kinds of fruit and ornamental trees are infested by the San Jose scale. Pyracantha and cotoneaster are especially susceptible to severe damage. Numerous other hosts may support small populations and provide for perpetuation of this species. These include boxwood, dogwood, hawthorn, privet, walnut, linden, rose, mountain ash, and Japanese photinia, *Photinia glabra*, sometimes called red top.

Partially grown males and females of the San Jose scale overwinter and mature in the spring; females produce living young. Feeding and development proceed rapidly, resulting in as many as five overlapping generations in a single season. In northern states generations overlap from mid-May through September. In southern Ontario there may be two to three generations per year.

Damage is caused when nymphs and adult females pierce plant tissues with their long threadlike microscopic mouthparts and suck out plant fluids. The greenish host tissue of shoots and leaves around the scale often turns red. On twigs and small branches the red color extends deeply into the inner bark (F) to the xylem, but the color is not visible at the surface of thick bark.

A high population will cause twig and branch dieback, and if some degree of control by natural or artificial means is not exercised, the entire plant will die. All parts of the host plant except roots may be attacked. However, the greatest populations occur on twigs and branches where the build-up of scale covers forms a gray crust.

New biological and behavioral studies have resulted in technologies that improve monitoring, pinpointing the proper time for chemical application. Females produce a sex pheromone (650). A synthetic pheromone is now commercially available and is used to trap males. Though it is not the purpose of this book to provide details about monitoring and other technologies, we encourage readers to learn about the value and use of pheromone traps and the growing degree-day system of timing pest management. Such information is available through professional consultants, cooperative extension programs, and state land grant universities.

The walnut scale (D), *Quadraspidiotus juglansregiae* (Comstock), occurs throughout the United States. This insect is found primarily on deciduous trees and shrubs, although it is also known to occur on coniferous trees such as hemlock, Virginia pine, or Scots pine, and on broad-leaved evergreens. A partial host list includes ash, birch, black locust, boxelder, buckeye, dogwood, elm, hackberry, hawthorn, hickory, holly, honeylocust, horsechestnut, Kentucky coffee tree, linden, maple, mountain ash, poplar, privet, sour cherry, sweetgum, tuliptree, and witch-hazel. It has not been reported on any of the *Juglans* species except Persian walnut.

The walnut scale attacks the bark of the host tree, causing serious injury to the phloem, or living bark tissue. There are usually two generations each year. In the East the scales overwinter as adults, in California as second-instar females and males. After maturing and mating, the female lays eggs (June in Ohio and again in August; in mid-May and again in mid-August in California). Eggs hatch in 2–3 days. The crawlers move to a feeding site on the bark, insert their mouthparts, and remain sessile for the remainder of their lives. The second-stage nymph produces a white cover over its body that turns to light gray or grayish brown in about 1 week (B). First-generation individuals complete their development by the end of July. These scales are found in round clusters (D) in small populations and completely encrust twigs and small branches during outbreaks. Naked females can be identified by their yellow body and two deep constrictions that separate the thoracic segments.

Several small wasps parasitize armored scales. The most important parasites belong to the genera *Aphytis* and *Prospaltella*. The twice-stabbed lady beetle (black with two red dots on its wing covers), *Chilocorus stigma* (Say), is an important predator.

References: 250, 359, 416, 491, 517, 650, 764

A. Newly settled first-instar walnut scale nymphs. Note white waxy covers.
B. Second-instar nymphs initially produce whitish to light brown wax covers. Later they are grayish.
C. Waxy covers on a mature female.
D. Walnut scales in all stages on a twig.
E. A rose stem showing typical symptoms of infestation by the San Jose scale. Dark discoloration also appears on the green bark and fruit of other hosts.
F. Phloem and xylem discoloration on a crabapple branch heavily infested with San Jose scales.
G. The effects of a serious San Jose scale infestation on flowering quince twigs.
H. Second-instar San Jose scale covers. Arrows point to one scale insect from which the cover has been removed. The fully grown scale produces a cover about 1–2 mm in diameter.

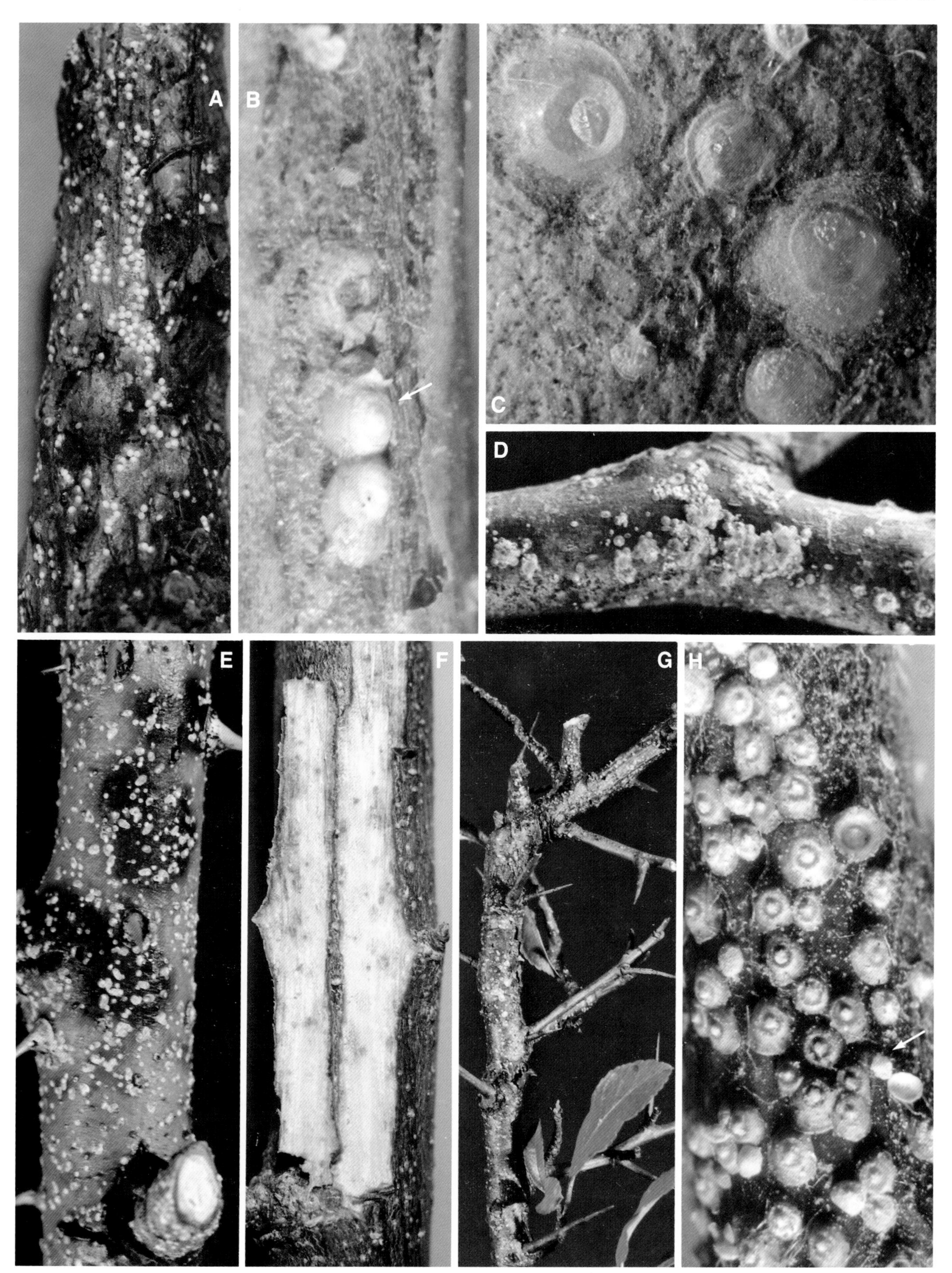
A
B
C
D
E
F
G
H

Euonymus Scales (Plate 186)

The euonymus scale, *Unaspis euonymi* (Comstock), is an important pest in all temperate regions of the world except Australia. It attacks a number of hosts including species of *Euonymus, Camellia, Buxus, Celastrus, Daphne, Eugenia, Hedera, Hibiscus, Ilex, Jasminum, Ligustrum, Lonicera, Olea, Paxistima, Pachysandra, Solanum,* and *Prunus*. Apparently, this scale was introduced from Japan or China. In the United States it often attacks evergreen euonymus such as *Euonymus japonica* and can cause complete defoliation or death of the plant. The tree forms such as *E. europaea* are equally susceptible. *E. kiautschovica* (=*sieboldiana*) appears to resist heavy attacks of euonymus scale even when grown among heavily infested *E. japonica*. This scale is likely to be found throughout the United States and in many of the Canadian provinces.

The euonymus scale is often overlooked until it has caused serious damage. One symptom of a light attack is the occurrence of yellowish or whitish spots on the leaves (A, C). The female scales are usually found along the stems and leaf veins of the host plant (D, E). At times, however, the whole plant is whitened by the covers of the smaller male scales. When this occurs, the plant's leaves may drop and sometimes a normally green plant becomes bare by midsummer. Plants growing close to buildings seem to be damaged more than those growing where there is free air circulation. This pest frequents the stems at ground level, where they are well protected.

The scales overwinter as fully grown fertilized, grayish to brown females, which are easily distinguished from the smaller whitish males (E, G). Eggs are deposited in early spring, beneath the dark-colored female scale covering. The eggs hatch over a period of 2 or 3 weeks in early June in the Northeast, and in late May in Virginia. The nymphs crawl to other parts of the host plant or are blown to other susceptible hosts. A second generation of crawlers develops by mid-July in Virginia. There may be up to three generations per year in southern locations.

Frequent treatments with certain pesticides are usually required for control. Because of apparent resistance to this pest, *E. kiautschovica* should be used as a substitute for *E. japonica* where low-maintenance plants are desired. *E. alatus, E. sachalinensis,* and *E. sanguinea* also appear to be resistant to the euonymus scale. However, their landscape use may be different from the evergreen forms.

Recently, two predators from Korea—a lady beetle and a nitidulid—have been released at various sites in the eastern United States. The lady beetle, *Chilocorus kuwanae* (Silvestri), is black with two red elytral spots and closely resembles the native twicestabbed lady beetle, *C. stigma* (Say) (see Plate 188G). Once released, *C. kuwanae* noticeably reduced euonymus scale populations. By the third year following release of the original beetles, the beetle populations had increased to the point that they are now considered well established. The nitidulid, *Cybocephalus* sp. (probably *nipponicus* Endrody-Youngs), is also an important predator of diaspine scales. It is now established and has reduced the number of euonymus scales in the metropolitan Washington, DC, area.

Lepidosaphes yanagicola Kuwana, an armored scale pest of *E. alatus*, has also been recorded on basswood, elm, and willow. It does not occur on any evergreen euonymus. It is confined to twigs and is most abundant between the bark ridges that are so characteristic of its host. Heavy infestations cause premature leaf drop, twig dieback, and decreased resistance to winter injury.

L. yanagicola overwinters as an adult female. Oviposition begins (in eastern Pennsylvania) in mid-June and continues for about a month. All progeny mature and mate before frost. There is one generation each year. The adult females, which resemble miniature oystershells, are about 2 mm long. When the test is removed, their white bodies are exposed. This insect is found mostly in the Northeast, with scattered populations as far south and west as Oklahoma.

References: 102, 288, 761, 894, 895

A, C. Yellowish or whitish spots on *Euonymus* species caused by the euonymus scale.

B. The second-stage cover of *Lepidosaphes yanagicola* on *Euonymus alatus* twigs.

D. Orangish euonymus scale crawlers and newly settled nymphs on a stem.

E. Narrow whitish euonymus scale male covers with wider and darker female covers, as well as some newly settled nymphs. Females are about 2 mm long.

F, G. Leaves of *E. japonica* heavily infested with euonymus scale. F shows yellowish spot symptoms as seen from the upper surface of the leaf; G shows both male and female scale covers on the lower surface of the leaf.

H. Characteristic clustering of scales along twigs and leaf midribs.

A
B
C
D
E
F
G
H

Armored Snow Scales (Plate 187)

Color is not often used by taxonomists as a basis for classifying animals. However, it is convenient to group several scale species together on the basis of their color. Some armored scales have white covers, and a few of them have been collectively called snow scales. Those scale insects placed in the genera *Unaspis* (Plate 187), *Aulacaspis* (Plate 175), *Chionaspis* (Plate 187), *Quernaspis* (Plate 187), and *Pinnaspis* Plate 187) may fit the description of snow scales or scurfy scales. They have white or sometimes grayish white covers (tests). The test venter, although also white, is very thin. This plate illustrates *Pinnaspis strachani* (Cooley) and *Quernaspis quercus* (Comstock). Table 17 lists most of the *Chionaspis* species common to trees and shrubs. See Plate 176 for illustrations of other *Chionaspis* (scurfy scale) species.

P. strachani, the lesser snow scale (A), is a general pest of ornamentals. It is also found on a number of subtropical plants. It has been recorded as an economic species in both Florida and California, and occurs on more than 200 plant species or cultivars. Some of its more common woody ornamental hosts include species of *Hibiscus, Acacia, Cassia, Citrus, Ficus, Pittosporum,* rose, *Schinus,* and palm. Although it is common and omnivorous, its biology has not been studied in detail. In subtropical climates the lesser snow scale engages in continuous reproduction and produces several generations each year. The male scale is a different shape than the female. Panel D is a photograph of empty male scale covers. The female, about 2 mm long (F), has the form of an oyster shell. It feeds on the bark, fruit, and leaves of the host plant. In Florida many individuals of this species are killed by parasitic wasps.

The citrus snow scale, *U. citri,* occurs only on *Citrus* species. Like the lesser snow scale, the male of this species is white; the female is brown and sometimes nearly black. It, too, is found only in Florida and California.

The fern scale, *P. aspidistrae* (Signoret), is a common greenhouse pest on fern and other perennials in the eastern half of the United States and adjacent Canada. However, in Gulf Coast states where subtropical conditions exist, it occurs outdoors on such woody plants as croton, *Magnolia virginiana, Camellia japonica,* and species of *Acacia, Citrus, Cocos, Ficus, Hibiscus,* and *Liriope.* The scale may occur on leaves, bark, and fruit. Adult male scales resemble the greenhouse white fly. They are short lived and mostly active at night. The pear-shaped adult female cover is pale brown and about 1.8–2.6 mm long. Eggs are found under the scale cover. Newly hatched crawlers are pale yellow to light brown. The crawlers are usually active for less than 24 hours. If they settle on leaves, they usually select the undersurface. Under outdoor conditions they produce one generation each year.

Quernaspis (=*Chionaspis*) *quercus* (Comstock) occurs in New Mexico and California (E). It is found on the bark of eight species of Western oaks such as *Quercus lobata, Q. agrifolia,* and tanbark oak, *Lithocarpus densiflorus.* This species causes little damage to its host, though the bark of some trees may appear white because of the abundance of the old male scale covers. The biology of this scale is unknown.

References: 163, 167, 517, 833, 900

Table 17. **Bark-feeding *Chionaspis* species common to trees and shrubs**

Scientific name	Common name	Hosts
Chionaspis acericola	Maple scurfy scale	Red and silver maple, ash, river birch
C. caryae	Hickory scurfy scale	Hickory, pecan
C. floridensis	Florida scurfy scale	Ash, birch
C. gleditsia	Honeylocust scurfy scale	Ash, *Albizia* spp., *Carpinus* spp., *Gleditsia* spp., *Ostrya* spp.: bark and leaves
C. hamoni	Florida willow scale	Willow
C. kosztarabi	Ash scurfy scale	Ash, hophornbeam
C. lintneri	Lintneri scurfy scale	Birch, alder, dogwood, hazel, ash, lilac, serviceberry, walnut, willow, viburnum
C. longiloba	Longiloba scurfy scale	Willow, poplar
C. nyssae	Sourgum scurfy scale	*Nyssa* spp.: bark and leaves
C. ortholobis	Cottonwood scurfy scale	Willow, *Arbutus* spp., *Arctostaphylos* spp., *Ceanothus* spp., *Cytisus* spp., Rhamnus spp.
C. platani	Sycamore scurfy scale	*Platanus* spp.: bark and leaves
C. salicisnigrae	Willow scurfy scale	Willow, *Populus* spp., ash
C. sassceri	Western scurfy scale	Citrus, *Ceanothus* spp., *Fremontodendron* spp.
C. styracis	Styrax scurfy scale	*Styrax* spp.
C. wisteriae	Wisteria scurfy scale	*Wisteria* spp.: bark and leaves

Source: Data from Bullington et al., ref. 887.

A. Stems of the shrub *Leucophyllum frutescens* encrusted with the lesser snow scale, *Pinnaspis strachani.*
B. A stem killed by the lesser snow scale.
C. Lesser snow scale on the trunk of *Citrus sinensis* in Winterhaven, Florida. (Courtesy W. A. Sinclair, Cornell Univ.)
D. Close-up of a cluster of lesser snow scale males. Female scales blend so well with the bark that they are not visible in this photograph.
E. Male scales of *Quernaspis quercus* on *Quercus agrifolia* in Belmont, California.
F. Hidden female scale. The arrow points to a lesser snow scale. The circle outlines an unidentified female scale. Male scales have emerged from the white scale covers.
G. Two adult female scales with their covers removed. A female lesser snow scale is shown at the arrow point.

A
B
C
D
E
F
G

White Peach Scale and White Prunicola Scale (Plate 188)

The white peach scale, *Pseudaulacaspis pentagona* (Targioni-Tozzetti), was first described in 1886 in Italy. It has been a serious pest in the United States since the early 1900s. The white peach scale is also known as the West Indian peach scale. It has been one of the most serious pests of fruits, especially peach and cherry. It is also very destructive to ornamental trees and shrubs, including flowering cherry, plum, peach, privet, aucuba, lilac, walnut, catalpa, chinaberry, persimmon, *Hibiscus* species, chinese elm, golden raintree, honeysuckle, redbud, spirea, dogwood, oleander, and many others. Dekle (167) lists 97 hosts in Florida. Mulberry is a favorite host, and it is commonly collected on *Prunus persica, Callicarpa* species, and *Melia azedarch*. The white peach scale is common on the East Coast from Maryland southward. In 1930 peach production in Florida came to a standstill because of damage done by the white peach scale. In the Carolinas and Virginia it is an important stonefruit pest.

In northern Florida the white peach scale has four generations each year. From the Carolinas north through Maryland there are three generations.

The white peach scale overwinters as an adult female. In North Carolina the female begins to deposit her eggs in early April. Each female produces more than 100 eggs. The peak of the egg-hatching season for first-generation eggs is during the first week of May in North Carolina and Virginia. The peak egg-hatching season for second-generation eggs is during the first week of August; for third-generation eggs, during the first week of September. Since the eggs of any one generation are laid during a period that extends over 1 month, the crawlers also emerge over a similar length of time. Female embryos are coral; male embryos are whitish pink. The color difference remains distinct through the crawler stage until the end of the first instar. In the adult stage the cover of the male scale is bright white (D) and about 2.0–2.5 mm long. The female scale is larger but gray and thus less conspicuous. Infestations appear more intense during periods when males mature. Large masses of cottony secretions are quite apparent.

The white peach scale feeds primarily on the bark of the trunk and larger limbs. It is not uncommon for the bark of entire trees to be thoroughly encrusted with scales. Populations build up rapidly, resulting in the death of large branches and, frequently, entire trees.

Smith (707) observed that considerable natural mortality occurs in the first two generations in North Carolina and that the scale spreads mostly during the third generation in the fall.

Studies in 1983 by Davidson et al. (145) demonstrated the existence of a second species, *P. prunicola* (Maskell), so closely related to *P. pentagona* that for years no one was aware that the white peach scale was not one but two species. The common name white prunicola scale has been given to the second species. The two species cannot be separated by casual field identification. In addition to the several morphological characters that allow for taxonomic distinction, there are biological and behavioral differences that aid in field separation. The prunicola scale's seasonal activities occur 2 weeks or more later than those of *P. pentagona*, and the delay persists for the entire season. In Maryland eggs are found in late April–early May, again in early July, and again in late-August. There are three generations per year, except in northeast Pennsylvania and farther north, where there are two.

Of the two species, the white prunicola scale is more common in temperate climatic zones; for example, scales found in upstate New York and New England will most likely be *P. prunicola*. Overlap of the two species occurs in Maryland and Virginia. A few collections have been made in Florida, Alabama, Mississippi, and Louisiana, and single collections have been made in California and New Mexico. Its host range is not nearly as wide as that of the white peach scale. The white prunicola scale is most frequently encountered on Japanese flowering cherry (*Prunus serrulata*), *Ligustrum* species, and *Syringa* species.

Parasitic and predaceous insects are numerous. Coccinellids include the twicestabbed lady beetle, *Chilocorus stigma* (Say), *Lindorus lophanthae* (Blaisdell), and *Exochomus childreni* Mulsant in Florida. The wasp parasites *Prospaltella berlesei* (Howard) and *Aspidiotiphagus citrinus* (Craw), along with a predaceous thrips and a mite species in the family Belbidae, have also been reported there. A sex pheromone has been identified for the white peach scale.

References: 58, 68, 145, 167, 450, 707

A. Males of the white prunicola scale, *Pseudaulacaspis prunicola*, heavily encrusted on the bark of flowering cherry.
B. Female white peach scales in profile. [Diaspididae]
C. Heavily encrusted white peach scales on lilac.
D. Male and female (*arrow*) white peach scales on flowering cherry.
E. A privet hedge damaged by white peach scales.
F. Catalpa trees severely injured by white peach scales.
G. A lady beetle emerging from its pupa. The larva had been feeding on young prunicola scales.
H. An adult twicestabbed lady beetle feeding on scale eggs and crawlers.

A
B
C
D
E
F
G
H

Fiorinia Scales (Plate 189)

Five species of *Fiorinia* are known to occur in the United States. They are the tea scale, *Fiorinia theae* Green; the palm fiorinia scale, *F. fioriniae* (Targioni-Tozzetti); the juniper fiorinia scale, *F. juniperi* (Bouche), which also occurs on hemlock; the Japanese fiorinia scale, *F. japonica* Kuwana; and the elongate hemlock scale, *F. externa* Ferris (Plate 45). The palm fiorinia scale and the tea scale, the two species that commonly occur in the South, are discussed here.

Dekle (167) considers the tea scale one of the 10 pests most harmful to ornamentals in Florida nurseries. It is widely distributed throughout the deep South and is very difficult to control. Twenty-five host species are known in Florida. Of these, camellia, Chinese holly, and Burford holly are frequently infested. Species in the following genera are also hosts in Florida: *Aucuba, Callistemon, Camellia, Citrus, Cornus, Euonymus, Eurya, Gardenia, Gordonia, Ilex, Malpighia, Mangifera, Melaleuca, Osmanthus, Poncirus, Rubescens, Senecio, Symplocos,* and *Thea*.

The tea scale was first described from specimens found on the tea bush in India. It is now widely distributed in that country on tea bushes as well as on *Citrus* species. In Japan it occurs on *Eurya japonica*. In the United States it is frequently intercepted on nursery stock shipped northward from the South. However, it is not established north of the Carolinas and southeastern Virginia, except in greenhouses, where it is occasionally found on camellias. It has not been a serious pest in California.

The tea scale occurs on the leaves of its host (B). It is a small scale; the adult female is about 1.3 mm long and the male is two-thirds that length. Females are elongate oval, and their coloring ranges from dark brown or dark gray to almost black (D). After molting, the second-instar scale cover completely covers the insect. Male scales are narrow and snow white (C). When populations of this pest become dense, long white waxy filaments are produced so profusely that the undersides of infested leaves take on a conspicuous cottony appearance (A).

The female lays 10–16 yellow eggs that remain beneath her body until they hatch, in 1–3 weeks. The crawlers are also bright yellow and settle permanently in a feeding position on the leaf, 2 or 3 days after they hatch. This species occurs on the undersides of the leaves and requires 40–65 days to complete its life cycle. A number of overlapping generations occur in a year, resulting in the presence of all stages of the scale throughout the growing season.

In Florida the palm fiorinia scale, *F. fioriniae,* has been reported on 29 species of plants in 23 genera. These hosts are chiefly various palms and bays (*Persea, Gordonia, Camellia,* and *Podocarpus* species, and *Magnolia virginiana*). The palm fiorinia scale does not cause any severe economic losses in Florida. The female is elliptical, about 1.3 mm long, and translucent yellow to orange-brown (F). Males are absent or rare; no filamentous wax is associated with this species.

Little is known of the life history of the palm fiorinia scale, but all stages may be present at the same time. This suggests the existence of a number of overlapping generations. The insects infest both the upper and lower surfaces of the leaves.

It has been found in California, Oklahoma, Texas, Georgia, and Florida.

References: 144, 167, 240, 813, 836

A. Cottony appearance and thin foliage of a camellia that is infested with *Fiorinia theae,* the tea scale.
B. Typical appearance of a tea scale infestation on a Burford holly leaf.
C. Abundance of filamentous wax from male tea scales.
D. Females and a male (*arrow*) of the tea scale. [Diaspididae]
E. Typical appearance of camellia infested with *Fiorinia fioriniae,* the palm fiorinia scale.
F. Settled crawlers, immatures, and adult females of the palm fiorinia scale on *Persea* sp.

A
B
C
D
E
F

Fourlined Plant Bug and *Lygocoris* Plant Bugs (Plate 190)

Fourlined plant bug is an appropriate common name for *Poecilocapsus lineatus* (Fabricius), an insect of some beauty in both adult and nymph stages. Nymphal coloring varies from bright red to yellow (D). Forewings of the adult are yellow (C) but may turn bright green. However, the four black stripes that give the insect its name remain distinct.

The fourlined plant bug has been recorded to feed on 250 plant species, most of which are herbaceous plants. The woody ornamental plants that serve as food include azalea, deutzia, dogwood, forsythia, viburnum, weigela, rose, amur maple, and sumac. Feeding injury to leaves by both the nymph and adult generally takes the form of discrete spotted discoloration. Injured areas may turn black (A) or become translucent (B) because of the extraction of chlorophyll, and after several weeks the remaining necrotic tissue may drop out, leaving small holes.

This plant bug overwinters as an egg in lineal clusters that are inserted into a slit near the top of tender shoots or water sprouts. The slit is cut by the female with her ovipositor as part of the egg-laying process. Egg clusters are easily seen after leaves drop in autumn. In southeastern Pennsylvania eggs begin to hatch in mid to late April. This species requires about 30 days to complete nymphal development. The late hatching of eggs accounts for the presence of adults in the middle of July. Adults are voracious feeders, attacking the upper surface of the leaf. They are timid insects and hide under leaves or fly when disturbed. Because a few fourlined plant bugs can cause great damage to a host plant, it is best to eradicate these insects at their first appearance on ornamentals. On shrubby plants the topmost leaves are the first to be injured. The fourlined plant bug is widely distributed throughout the northern quadrant of the United States and parts of eastern Canada.

Lygocoris viburni (Knight) is a small, yellowish brown plant bug, 5 mm in length, that feeds on *Viburnum lentago.* In New York state this bug occurs in early June and often in such numbers that host foliage is badly injured. The leaves become ragged with holes throughout. Its distribution ranges from Minnesota, Michigan, and Illinois east through Pennsylvania, Connecticut, Ontario, Quebec, and Prince Edward Island. There is one generation each year.

L. communis (Knight) is sometimes called the pear plant bug, but its primary hosts are *Cornus* species, where it lays its eggs. *Ilex verticillata* is also a host in the insect's southern range. Its color is greenish to gray with a reddish lateral stripe on its body; legs are greenish to yellowish and its body is about 5 mm long. There is one generation per year. The eggs overwinter under the bark of young twigs. The green nymphs mature in about 24 days, usually by the middle of June in Illinois. *L. communis* is found from California north to Alberta and Yukon, east to Ontario and Nova Scotia, Maine, and south to North Carolina.

References: 79, 80, 702, 703, 836, 843

A, B. Injury to amur maple (A) and forsythia caused by fourlined plant bugs.

C. An adult fourlined plant bug; about 7 mm long. The forewing color may vary from yellow to green. [Miridae]

D. A fourlined plant bug nymph.

A
B
C
D

Tarnished Plant Bug and Boxelder Plant Bugs (Plate 191)

The plant bugs shown in this plate cause two types of injury, as illustrated in panels A–D. *Lygus lineolaris* (Palisot de Beauvois) is commonly known as the tarnished plant bug (E). It is perhaps the most numerous and harmful of the true bugs (order Hemiptera).

Historically, the tarnished plant bug has been an economically important pest in fruit tree nurseries, causing injury to apple, pear, peach, plum, cherry, and quince. Today, we find them injurious to a number of species grown in ornamental and forest tree nurseries. Pine seedlings can be severely damaged or even killed, and white spruce, Douglas-fir, and larch all can be damaged. Other host plants include weeds, vegetables, and a wide variety of ornamental flowers. At least 385 host plants have been recorded. Problems on woody plants occur mainly in the early part of the growing season. Adults that have overwintered attack swollen and opening buds. Bud feeding may distort foliage or cause disbudding that forces lateral buds to develop and ultimately causes the plant to be bushy. Woody nursery and landscape plants with indeterminant growth or multiple flushes of growth are most vulnerable. If the attack takes place after shoot elongation begins, the tip may die (they often turn black), or the meristem is so damaged that shoot stunting or distortion occurs. On poplar seedlings or cuttings a symptom called split stem lesion develops on young stems and occasionally on short branches, growing tips, and leaf petioles. The lesion is an irregular elongated split with a swollen flair of necrotic bark and exposed xylem. Frequently, stems break at or above the injured area. Injury is brought about by stylet (mouthpart) irritation, cell destruction through removal of cell contents, and the toxic effect of insect saliva. In some cases apical dominance is lost.

Eggs are laid mainly in the stems and flowers of herbaceous plants. After hatching, the young nymphs readily move around but usually remain to feed on the plant selected by the parent until they mature. Adults are capable flyers and readily move from place to place. There may be two to five generations each year. In midsummer a life cycle may be completed in about 25 days. By late summer this pest usually becomes very abundant. It was originally described from eastern United States but now is considered to be the most widely distributed mirid in North America, from Florida to Alaska and from California to the Canadian Maritimes. Traps, used in the spring, are good monitors; if properly used, they aid in reducing the population.

The boxelder bug, *Boisea* (=*Leptocoris*) *trivittatus* Say, is often an enormous nuisance around homes where boxelder, *Acer negundo,* is used as a shade tree. They are especially troublesome in the Mississippi River valley. This species is known from eastern Canada throughout the eastern United States and west to Nevada.

In the fall boxelder bug adults (G) become gregarious and assemble on the south sides of trees, buildings, and rocks exposed to the sun. After large masses of boxelder bugs congregate, they fly to nearby buildings to hibernate for the winter. They are a nuisance to occupants and are considered a common household pest. In the spring they emerge from hibernation when the buds of the boxelder trees open and fly back to their host plants, where they lay their eggs, usually in late April to early May. They place egg clusters on stones, leaves, grasses, shrubs, and trees, especially crevices in the bark of boxelder trees. When freshly laid, eggs are straw yellow. As the embryo develops, the egg appears red. The nymphs (F), upon hatching, are also red. The first-generation nymphs and adults feed on fallen boxelder seeds. They later migrate to female trees when seeds begin to form and feed on newly developing leaves, which grow in distorted shapes (B, C), and on seeds. The primary host plant is boxelder, but the bug also occurs on seed-bearing silver maples, and occasionally they feed on the fruit of apple and plum. There are two generations per year in the warmer part of its range.

The western boxelder bug, *Boisea* (=*Leptocoris*) *rubrolineatus* Barber, (I) is similar to the boxelder bug but is restricted to Nevada, Arizona, and Texas and to states bordering the Pacific Ocean and British Columbia. It feeds on boxelder and other maples. Leaf distortion and discoloration (C) result from the feeding activities of *B. rubrolineatus.*

Future studies may reveal greater variation in the habits of this species. A boxelder plant bug, presumably *B. rubrolineatus,* reportedly feeds on almond and other fruit trees in California. Adults lay their eggs on leaves and nuts. When feeding, the nymphs and adults pierce the husk of the nut with their sharp mouthparts, thereby causing a black spot to form on the kernel. Injury cannot be seen until the kernel is removed from the shell.

References: 79, 99, 217, 341, 359, 428, 432, 480, 671, 702, 842, 891

A. Injury to *Ilex crenata* believed to have been caused by the fourlined plant bug.
B, C. Leaves of boxelder, *Acer negundo,* injured by the western boxelder bug, *Boisea rubrolineatus.*
D. A *Raphiolepsis* sp. leaf injured by a plant bug. An injured leaf becomes distorted and covered with spots.
E. An adult tarnished plant bug, *Lygus lineolaris*; about 5 mm long.
F. A nymph of the boxelder bug, *Boisea trivittatus.*
G. An adult boxelder bug.
H. The small milkweed bug *Lygaeus kalmii* Stål. Notice the resemblance between this species and the boxelder bug. [Lygaeidae]
I. An adult western boxelder bug. [Rhopalidae]
J. Empty egg shells of the boxelder bug.

A
B
C
D
E
F
G
H
I
J

Plant Bugs and a Beetle Affecting Sycamore (Plate 192)

Plagiognathus albatus (Van Duzee), the sycamore plant bug, feeds on foliage of London plane, *Platanus acerifolia,* and American plane, *P. occidentalis,* in all nymphal stages and as an adult. Where the insect's mouthparts pierce the leaf tissue, chlorotic areas develop. Over time, irregular brown spots develop where the chlorosis was most intense, later to fall out, leaving a ragged and tattered leaf (A, B). If a very young leaf is attacked, it will also be distorted.

This insect overwinters as an egg embedded in woody tissue at the base of leaf buds. Eggs begin to hatch after the first-flush leaves reach a length of about 4 cm. Feeding takes place on both sides of the leaf. Nymphs mature in southeastern Pennsylvania from late May to early June. Adults (C, D) are present till mid-July. There is one generation per year. Its known distribution includes the northeast quadrant of the United States, as far south as North Carolina and west to California. It is also known in Quebec and Ontario.

The injury caused by the sycamore plant bug could be mistaken for that caused by *Neochlamisus platani* Brown, a peculiarly shaped leaf beetle that looks like the frass of large caterpillars and that exhibits strange behavior. The holes made by *N. platani* (Figure 117) are generally larger than those caused by the mirid, but, of course, they are created in an entirely different way. The biology of this insect has been investigated by Neal (565). Larvae surround themselves with cases made of their own frass and stay on the leaf blade except to molt. Molting usually takes place at a leaf axil. At this time, the case looks like a leaf bud. This insect has four instars. Adults cover their eggs with a bell-shaped case also made from frass. Their feeding is limited to sycamore and London plane leaves. They are found from Massachusetts to Illinois and west to Colorado and south from Florida to Texas. There is one generation per year (Figure 118). See page 224 for a related species, *Neochlamisus cribripennis.*

Kunzeana kunzii (Gillette) is a very small leafhopper, approximately 2 mm long. It is predominantly dull green and occurs in all of the southwestern states and Colorado. It is believed to feed primarily on legumes. Panels E and F show the injury that this pest causes to *Acacia decurrens. K. kunzii* feeds on the undersides of leaves. It defoliates huisache in Texas and also feeds on citrus, causing "oil" spots on fruit. Little is known about the biology and seasonal history of this species. It is often abundant in northern California during October.

References: 432, 502, 565, 702, 839

Figure 117 (left). A sycamore leaf showing the pattern of holes chewed by sycamore leaf beetle larvae and adults. (Drawing by J. R. Baker, North Carolina State Univ. Cooperative Extension.)

Figure 118 (right). Adult, larva, and pupal case of a sycamore leaf beetle, *Neochlamisus platani.* Adult about 4.5 mm long. [Chrysomelidae] (Drawings by J. R. Baker, North Carolina State Univ. Cooperative Extension.)

A. Leaves of *Platanus acerifolia* injured by the plant bug *Plagiognathus albatus.*
B. A London plane tree leaf showing injury typical of that caused by the sycamore plant bug.
C, D. *P. albatus* adults on the upper and lower surfaces of a London plane tree leaf; about 5 mm long. [Miridae]
E, F. *Acacia decurrens* showing yellow stippled foliage caused by a leafhopper, *Kunzeana kunzii,* in California. [Cicadellidae]

A
B
C
D
E
F

Ash Plant Bugs (Plate 193)

Fraxinus species and the shade-tree cultivars of green, white, blue, moraine, and Modesto ash are hosts to several species of plant bug. The best-known species of these plant bugs once occupied discrete geographic areas. Now, largely through horticultural commerce, the eastern species, *Tropidosteptes amoenus* Reuter, is widely distributed and has been recorded from British Columbia. A western species, *T. pacificus* (Van Duzee), was imported into Pennsylvania with ash nursery stock. Injury to ash takes the form of yellowish white leaf stippling visible from the upper surface that eventually coalesces to form broad chlorotic areas. In the course of a growing season chlorotic spots turn brown, and the entire leaf dries and drops prematurely. Juvenile leaves that have been fed upon remain dwarfed or are severely deformed (A, B). The undersurface of injured leaves is marked with black specks of excrement from the bugs.

T. amoenus is believed to produce two generations per year throughout its range. Eggs are laid under the bark of host trees in autumn and overwinter there (Figure 119). In south central Pennsylvania eggs begin to hatch in early May (mid-May in Ontario). The nymphs crawl to the undersides of leaves and commence feeding. About 5 weeks are required to complete the nymphal stage. First-generation adults (D) appear in late July and deposit their eggs on the midrib of the leaf. Second-generation nymphs likewise feed on the undersides of leaves, and second-generation adults remain on the host tree until they are killed by frost.

Another eastern species that feeds on ash is *T. brooksi* Kelton. It closely resembles *T. amoenus* but is mostly greenish; the scutellar triangle is pale yellowish green, and the eyes appear to bulge. The biology of *T. brooksi* is poorly known but is presumed to closely parallel that of *T. amoenus*.

Two plant bugs, *T. illitus* (Van Duzee) and *T. pacificus* (Van Duzee), damage ash foliage in the western United States. *T. illitus* nymphs are a uniform light brown, and the adults (F) are black and yellow or brown and yellow. One generation is produced each year. *T. pacificus* adults are light brown (G); the nymphs are green with black spots. Two generations are produced each year. Both species lay their eggs in ash shoots or water sprouts; *T. pacificus* lays eggs for the second generation in the midrib of leaves as well. The egg is the overwintering stage for both species. In early spring, soon after bud break, the eggs hatch and nymphs feed on new leaves, causing a yellow stippling. Varnish-like excrement of the insect spots the lower surface of infested leaves. Feeding by these pests on stems and petioles leads to the wilting or curling and eventual drying of the leaf blade.

The western species of plant bugs occasionally defoliate ash in the Sacramento Valley of California. However, defoliation caused by the anthracnose fungus *Gloeosporium aridum* is far more severe and important than that resulting from ash bug infestations (see Sinclair et al., *Diseases of Trees and Shrubs*). Whatever the cause of defoliation, affected trees put out new foliage after the causal agent has disappeared. By June or July both *T. illitus* and *T. pacificus* have completed their activities for the year and are not seen until the following spring.

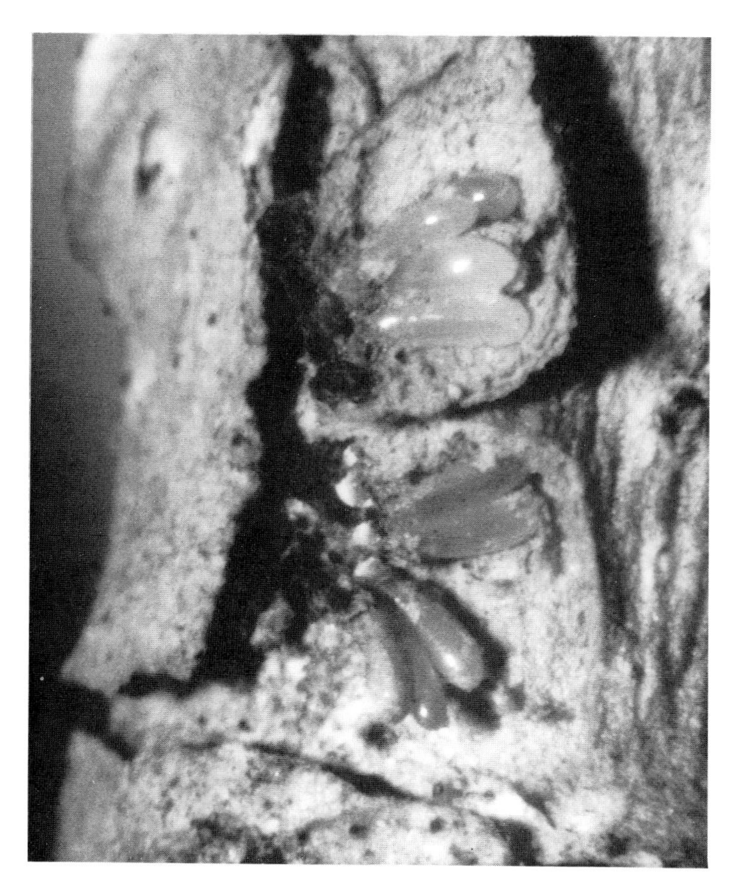

Figure 119. A cluster of oblong ash plant bug eggs under a loose bark flap. (Courtesy A. G. Wheeler, Jr., Pennsylvania Dept. of Agriculture.)

References: 176, 349, 429, 432, 795, 841

A. *Fraxinus velutina* leaves showing symptoms of injury caused by *Tropidosteptes illitus*. Note dead leaves at the growing tip. This injury occurs early in the season.

B. A nursery cultivar of *Fraxinus americana*. The mottled yellow color of the leaves resulted from feeding injury caused by *T. amoenus*.

C. *Fraxinus velutina* leaflets. The yellow discoloration, dead leaf margins, and spots of excrement on both upper and lower surfaces of the leaf were caused by *T. illitus*.

D. A *T. amoenus* adult. The eastern species is about 5 mm long.

E. A western species of plant bug in its final molt. Arrow points to black bands on the legs, a useful feature to confirm identification.

F. A *T. illitus* adult, a western species. [Miridae]

G. A *T. pacificus* adult, a western species.

H. An immature *T. amoenus*.

A
B
C
D
E
F
G
H

Plant Bugs, Leafhoppers, and a Treehopper of Honeylocust (Plate 194)

Prior to the 1950s the honeylocust, *Gleditsia triacanthos,* was considered a tree generally free of serious insect pests. The greatly expanded use of the honeylocust as an urban shade tree and the development of popular thornless cultivars, however, has brought much attention to their considerable number of arthropod pests. There are no less than seven species of plant bugs (Miridae), an equal number of leafhoppers (Cicadellidae), at least two species of treehoppers (Membracidae), and a number of eriophyid and spider mites (Plate 228) that feed on this popular shade tree.

Diaphnocoris chlorionis (Say) has received enough attention by entomologists to be assigned the common name honeylocust plant bug (D). One generation develops each year. The young as well as adults feed on foliage, causing severe leaf distortion, dwarfed leaflets, chlorosis, and yellow to brownish spots. Tiny irregular holes result when necrotic tissue drops out of affected leaves (A, C). When there is a large population of bugs, defoliation may occur. Light and moderately damaged foliage persists throughout the growing season.

Honeylocust plant bugs overwinter as eggs just beneath the bark surface of 2- and 3-year-old twigs, with only the operculum exposed. The eggs hatch after the vegetative buds of honeylocust start to open. The young nymphs crawl into the unfolding leaflets and commence feeding. The most serious damage occurs at this time, when the bug is hidden from view. The immature leaves are very sensitive to irritation, and cells are killed by the penetration of the insects' stylets. Nymphal development requires about 30 days. In south central Pennsylvania adults appear near the end of May and disappear by mid-July. Shortly after females become adults, they lay eggs in lineal clusters under the bark of twigs.

Many of the distribution records of the honeylocust plant bug include states with large commercial tree nurseries. With the movement of plant material in normal commerce, it is likely that this insect is now or soon will be found wherever thornless honeylocusts grow.

Macropsis fumipennis (Gillette & Baker) (E) is a leafhopper that feeds on leaflet stalks, petioles, and rachises (Figure 120). Its green color, like that of the honeylocust plant bug, makes it difficult to see this insect when it is feeding or resting on honeylocust foliage. It overwinters as an egg inserted into shoots near a bud or in 1- or 2-year-old twigs. Eggs hatch when leaves begin to open in the spring, and adults are present in host trees, in Pennsylvania, until early July. This leafhopper causes little damage to honeylocust, its major host. There is one generation per year. It occurs from Alberta east to Quebec and Vermont and south to Arizona and Georgia. These widespread collection records suggest that its actual distribution probably coincides with that of honeylocust. *M. ocellata* Provancher is a leafhopper that frequently injures *Salix alba* in the eastern quadrant of the United States and the adjacent Canadian provinces. *M. graminea* (Fabricius) feeds on various species of *Populus* and is found in the Pacific Northwest, the states and provinces bordering the Great Lakes, and the mid-Atlantic states.

Figure 120. Nymphs of the leafhopper *Macropsis fumipennis* feeding on the petiolules of honeylocust. (Courtesy A. G. Wheeler, Jr.)

Another sucking insect, *Micrutalis calva* (Say), a treehopper, feeds on honeylocust and black locust, as well as other plant species. It is one of the smallest treehoppers in the United States, appearing in honeylocust trees about the same time as the honeylocust plant bug and the leafhopper *M. fumipennis.* All of the nymphal stages of *M. calva* feed on leaflets and petioles. The biology of this insect has not been much studied. There are probably two generations per year. It is not clear what plants host the second generation. Overwintering eggs are laid in twigs of honeylocust in late October (61) in Ohio. Though the adult is primarily black, the abdomen is yellowish.

Panel B shows injury to *Juglans nigra* caused by the plant bug *Plagiognathus punctatipes* Knight. It has a developmental cycle similar to the honeylocust plant bug and is found throughout the natural range of black walnut.

References: 61, 79, 321, 702, 799

A. Injury on *Gleditsia triacanthos* caused by *Diaphnocoris chlorionis.*
B. Injury to *Juglans nigra* caused by *Plagiognathus punctatipes,* a plant bug.
C. A normal leaflet and a severely injured leaf caused by the honeylocust plant bug, *Diaphnocoris chlorionis.*
D. The honeylocust plant bug; 3.7 mm long. [Miridae]
E. The leafhopper *Macropsis fumipennis*; 4.5 mm long. [Cicadellidae]

A
B
C
D
E

Treehoppers (Plate 195)

An interesting group of insects called membracids or treehoppers injure many woody plants. Treehoppers are strange-looking creatures with bizarre forms; some have been described as resembling miniature dinosaurs (see Figure 121, accompanying Plate 196). They feed by inserting their sharp needlelike mouthparts into plant tissues and sucking out cell fluids. Injury to the plant is primarily caused by females that cut slits in the young bark in preparation for egg laying. They then deposit their eggs through these slits in the bark.

The buffalo treehopper, *Stictocephala bisonia* Kopp & Yonke, may be found throughout much of the United States and Canada. It lays eggs in the early fall under the bark (C) of many kinds of trees and shrubs including maple, crabapple, elm, hawthorn, cherry, locust, poplar, ash, and quince. These eggs (D) remain under the bark throughout the winter and hatch late in the spring. The young nymphs are green and their bodies are very spiny. After hatching, they drop to the ground and feed on various weeds and grasses until they reach maturity. It takes about 6 weeks to complete all of the nymphal stages, after which adults fly to deciduous trees. However, egg laying does not begin until early August. Oviposition is facilitated by a sharp, knifelike ovipositor that the female uses to cut slits in the bark. Eggs are forced to the right and left under the bark through the slit (C, D). Oviposition continues until frost. Scar tissue later forms in the shape of a double crescent, a characteristic of past injury by this pest (B). The bark of such injured twigs is roughened and thereafter never grows very vigorously. Canker-causing fungi and other pathogenic fungi gain entrance through these slits. Small twigs that have been thus injured may die from desiccation. It is through this type of injury that the buffalo treehopper has become the best-known member of the treehopper family. There is one generation per year. Other hosts include sycamore, butternut, willow, hickory, peach, hazel, and black locust.

Glossonotus acuminatus (Fabricius) is common, but not abundant, in late summer (E). It is found on black, white, and red oak; locust; chestnut; pear; and probably many other trees and shrubs. Nymphs frequently conceal themselves in the crotches of small branches. Little is known of its biology and life history. It is distributed throughout the eastern quadrant of the United States and as far south as Oklahoma and South Carolina.

Platycotis vittata (Fabricius) might well be called the oak treehopper, for it is primarily associated with at least six species of oak, including live oak. It also occurs on *Betula alba, B. nigra,* and chestnut. It has no great notoriety as a pest but when abundant it becomes a concern to those responsible for landscape trees. The damage it causes is limited to small slits in the bark of shoots; the slits callus over, leaving scars.

P. vittata produces two generations annually throughout its geographic range. The adult female overwinters, and in the spring (mid-April in Ohio) she lays eggs in parallel longitudinal slits about 2 mm long in the bark of twigs. After laying a clutch of eggs, she, much like a bird, sits on them or moves occasionally but remains close by. Before the eggs hatch, the female makes a series of slits below the egg mass. When the eggs hatch, the nymphs aggregate around these bark slits, and the female remains nearby. The female remains with the nymphs until they mature, in about mid-June. Without this parental care the brood of 50–100 nymphs will not survive. A second generation begins in August and early September. The brooding adults are brown, a protective coloration against the background of a twig. Newly matured adults are illustrated in panel H. The color and horn shape of this species is extremely variable. This treehopper is closely related to those illustrated in Plate 197.

P. vittata is found from New York, eastern Pennsylvania, and New Jersey south to Florida, west to Texas, Arizona, and California and north to Vancouver Island, BC. Interior states in which it has been recorded include West Virginia, Ohio, Illinois, Tennessee, Missouri, and Arkansas.

References: 85, 217, 441, 489, 524, 589, 872

A. The bark of a red maple, *Acer rubrum,* twig with numerous oviposition scars of *Stictocephala bisonia,* the buffalo treehopper. Arrow points to old eggshells.
B. An elm twig damaged by the buffalo treehopper. Arrow points to a double crescent scar.
C. The same twig as in B. Bark has been peeled away from the wood to reveal egg clusters.
D. Close-up of egg clusters of the buffalo treehoppers in elm. The bark has been peeled back to show the eggs.
E. An adult treehopper, *Glossonotus acuminatus.* [Membracidae]
F, G. Nymphs of *Platycotis vittata* in various stages of development: (*horizontal arrow*) a newly molted nymph, (*vertical arrow*) brooding adult.
H. An adult treehopper, *P. vittata.* Length, including horn, 9–12 mm.

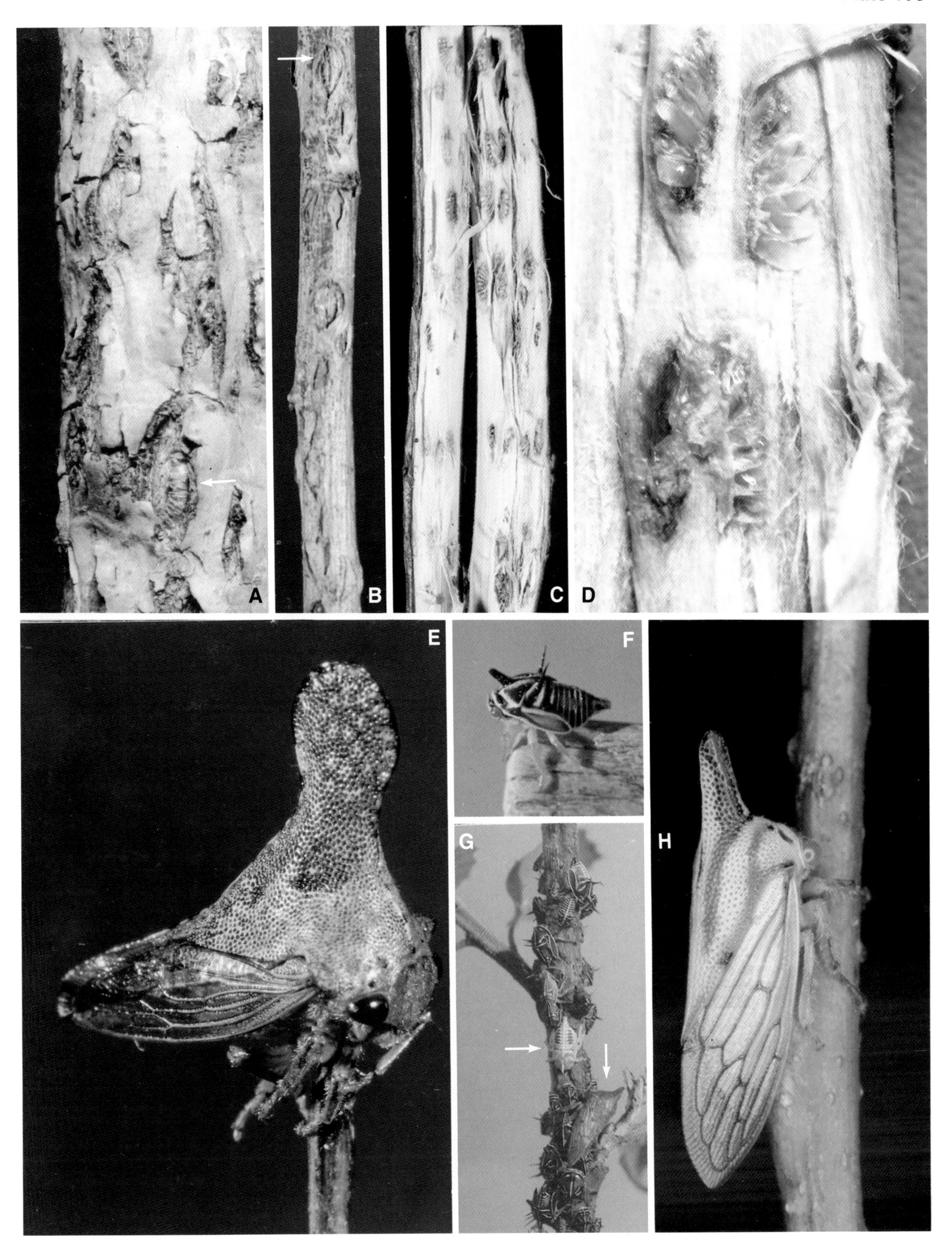
A
B
C
D
E
F
G
H

Thornbug and Relatives (Plate 196)

The thornbug, *Umbonia crassicornis* (Amyot & Serville), is a subtropical treehopper that is abundant in Florida. The thornbug is aptly named, for as it clings to the stem of a plant it closely resembles a thorn, thereby gaining some protection from its enemies. The naturalist takes special interest in the thornbug because of its striking color and bizarre form. The male and female have different forms (C). This species has developed a chemical communication between parent and offspring that aids in the defense of the young. Other chemicals that flow onto the exoskeleton from gland openings cause potential predators to reject these insects because of the taste.

The female lays her eggs in a mass under the tender bark of twigs. The eggs hatch in about 20 days, and she tends to her brood, actively maintaining the colony. Nymph clusters, or colonies, range from 15 to 50 individuals. The young (nymphs) have three horns rather than one (B). Nymphs, as well as adults, injure plants by sucking sap from the bark of twigs. Females also cut slits in the bark as a feeding aid for newly hatched nymphs and as a place to deposit their eggs. Four generations of thornbugs occur per year. Most females lay a single clutch of eggs. Thornbugs may injure humans too. There are reports of barefoot children stepping on the spines of thornbugs that have dropped from trees. Wounds are slow to heal and sometimes become infected.

Some trees (*Cassia* spp., for example) prematurely drop their leaves when injured by the thornbug. Dead terminal twigs are often a result of its attack on *Pithecellobium* species. The thornbug produces large quantities of honeydew, which becomes a nuisance when it drops on cars or outdoor living areas such as patios. Dense clusters of nymphs and adults are frequently seen on *Hibiscus* species, *Albizia lebbeck, Acacia* species, *Jacaranda acutifolia, Delonix regia, Callistemon* species, and *Calliandra* and *Mimosa* species. The range of some of the host plants exceeds the present known range of the thornbug.

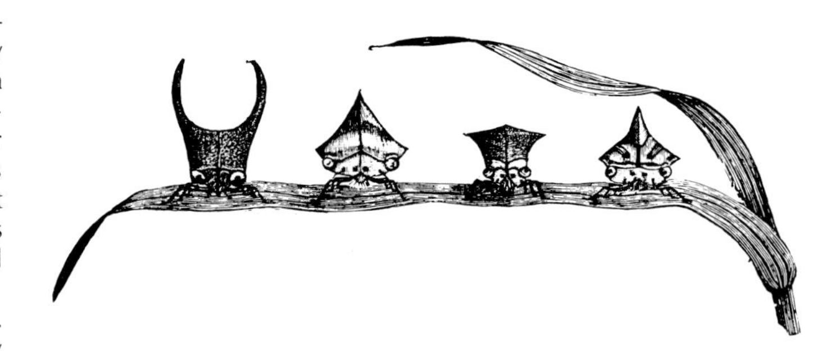

Figure 121. Not a cartoon, but true likenesses of treehoppers. (Drawing by Anna Botsford Comstock, Cornell University Press.)

Figure 121 illustrates other forms of treehoppers.

References: 521, 873

A. A cluster of thornbugs, *Umbonia crassicornis,* on *Albizia lebbeck,* commonly called woman's tongue tree. Adults are 8–10 mm long. [Membracidae]

B. Preserved specimens. *From left:* a male, female, and two nymphal stages.

C, D. Mostly female thornbugs. A male is indicated by the arrow in C.

E. A treehopper believed to be the brooding *Platycotis vittata.* Note oviposition slit at arrow.

F. An adult leafhopper, *Smilia camelus.*

G. An adult buffalo treehopper (see Plate 195).

A
B
C
D
E
F
G

Twomarked Treehopper (Plate 197)

The twomarked treehopper, *Enchenopa binotata* (Say), is more a curio of natural history than a pest, although a potential for serious injury to its hosts does exist. It overwinters as an egg, embedded under the bark of young twigs of deciduous trees or shrubs. In New York eggs hatch during late May and early June. The nymphs congregate on shoots, where they feed by sucking sap. They do not resemble their parents. Their bodies are dark gray to brown and have several spine-like structures that extend from their abdomens. After about 5 weeks, as nymphs, they go through their final molt and become adults (E). The gregarious behavior is carried over to the adult stage. It is common to see adults arrange themselves one behind the other in rows on twigs.

Adult twomarked treehoppers have a relatively long life. They begin to lay eggs in early August and continue laying eggs until killed by frost. To deposit eggs, the female uses the ovipositor to cut a slit parallel with the long axis of a twig; she forces 4–6 eggs one at a time through the slit and under the bark. After laying a clutch of eggs, she covers the slit with a sticky white, frothy lipid that hardens in about 5 days. It has been postulated that the froth acts as insulation from cold and helps to hold the eggs at a high humidity. Egg parasites are believed to be attracted to the volatile chemicals in this material.

The food plants of *E. binotata* include hoptree, *Ptelea trifoliata*; black walnut; butternut; black locust; viburnum; redbud, *Cercis canadensis*; and bittersweet, *Celastrus scandens*. This insect has also been reported on *Wisteria* species. It occurs east of a line that extends from Texas north to South Dakota, and it has also been found in Arizona, Colorado, and southern Ontario.

References: 440, 871, 874

A. *Viburnum lentago* (hybrid). The numerous oviposition scars caused by the twomarked treehopper, *Enchenopa binotata*, will remain visible for years.
B. An oviposition slit in the bark of *Juglans nigra*.
C. The same slit shown in B cut open to show eggs of an unidentified treehopper.
D. Hoptree, *Ptelea trifoliata*, twigs covered with froth that has been deposited over the slits where eggs were laid.
E. Two female twomarked treehoppers alongside froth masses; about 5 mm long.

A
B
C
D
E

Rose Leafhopper (Plate 198)

The rose leafhopper, *Edwardsiana rosae* (Linnaeus), is widely distributed throughout the United States and southern Canada. Its chief food plants belong to the rose family, including the cultivated rose. The rose, however, serves as the food plant for but one of the several annual generations of this leafhopper. Nymphs of summer generations feed on the foliage of species of *Cornus* (Plate 200), oak, *Prunus, Crataegus, Malus, Populus, Ulmus, Acer,* and other plants.

The fall brood of eggs is deposited within the bark of canes of wild or cultivated rose. Bush fruits such as blackberry and raspberry are also hosts of the rose leafhopper. Egg deposition often occurs before the first frost or onset of cool weather. The presence of eggs is indicated by dark, pimple-shaped spots on the cane. As many as 577 eggs have been observed per lineal inch of rose cane (115). Eggs hatch in the spring after the threat of frost is past. The young nymphs are white and have red eyes. They wander about the plant until they find a suitable leaf and then establish themselves on the underside. If food is plentiful, they may complete nymphal development on a single leaf. By the fourth molt the red coloring in the eyes disappears. The first-brood adults may vacate rose leaves for other woody plants (listed above). They insert eggs into the plant tissue on the undersides of the leaves. The eggs soon hatch and the first summer generation develops. Rose leafhopper nymphs always move forward, never sideways, as is the case with many leafhoppers.

Plant injury occurs three ways: destruction of the chlorophyll, removal of plant fluids, and winter egg deposition. An insect that feeds continually on an apple leaf during its nymphal development, for example, will remove or destroy from one-third to one-half of the green chlorophyll. Injured leaves often drop prematurely. Fungal diseases (often stem cankers) enter rose canes when the tissues are cut by the leafhopper's egg-laying apparatus. Such diseases often kill the less hardy cultivated rose.

Roses are the primary ornamental plant affected by the rose leafhopper (B). Feeding and egg laying by the rose leafhopper on rose may be sufficient to kill the plants. This insect is an immigrant from Europe. It is not appreciably affected by predaceous and parasitic enemies, although the wasp egg parasite, *Anagrus armatus* (Ashmead), can reduce populations to the level where no other control measures need be taken. Lacewings such as *Chrysopa californica* Coquillett and *Hemerobius pacificus* Banks are predators.

The white apple leafhopper, *Typhlocyba pomaria* McAtee, also feeds on rose foliage (Plate 200).

References: 52, 115, 682, 827, 836

A. An adult rose leafhopper. [Cicadellidae]
B. Typical leaf injury caused primarily by the feeding of nymphs.
C. A fifth-instar nymph.
D. A molted "skin" of a leafhopper nymph found on the underside of a leaf.

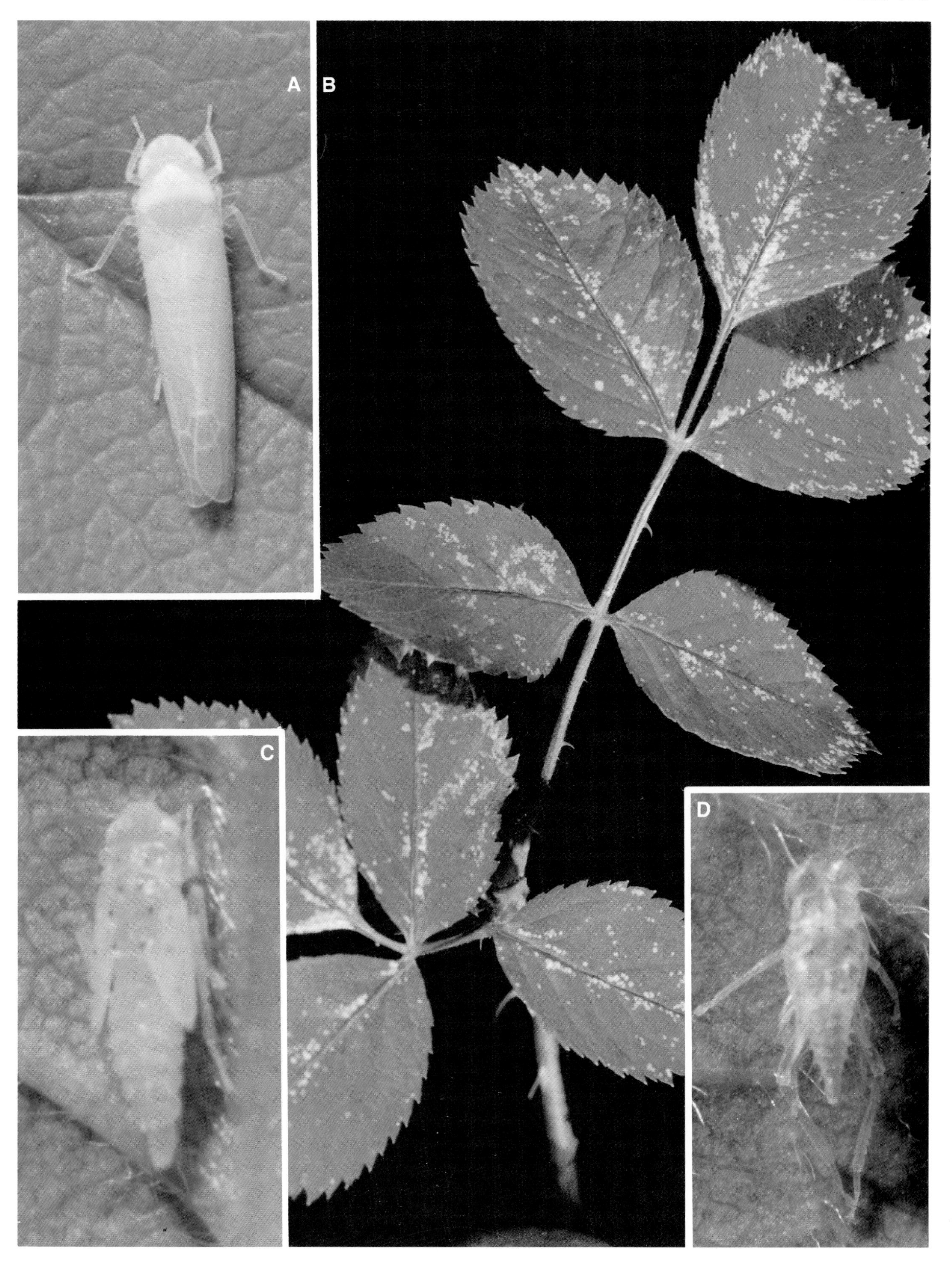
A
B
C
D

Potato Leafhopper and Whitebanded Elm Leafhopper (Plate 199)

All leafhoppers feed on foliage and/or shoots, and all embed their eggs under the epidermis of leaves, in the cortex of shoots, or in the cork of twigs or stem. This plate illustrates shoot and foliage damage caused by the piercing feeding activities of the potato leafhopper, *Empoasca fabae* (Harris), (E, F). It is easy to be misled by the name "leafhopper" and seek other causes for damage done to shoots (A–C).

The potato leafhopper is an example of the many species that feed in vascular tissue, primarily the phloem. It is also one of the many homopterans that produce a proteinaceous fluid that lines the holes produced by the probing stylet mouthparts. The fluid solidifies in the hole, forming what is called a sheath. The sheath remains after the insect removes its stylets. Because the feeding site and sheath are invisible to the unaided eye, there is no primary injury symptom.

As the stylet probes the plant tissue, internal cells are lacerated and cell fragments, or parts of the stylet sheath, clog the sieve tube. Salivary toxins injected into tissues along with the physical disruption by stylets may cause cells to collapse or contents to coagulate. This causes marginal burn called hopperburn, a disease symptom characterized by the browning of leaf edges. If wilting should occur, xylem vessels were punctured and plugged by sheath material or cell debris.

When vascular tissue is damaged in shoots, the adjacent tissue swells (B), causing abnormal physiological reactions that reduce the shoots' ability to winterize (C). Many shoots thus affected do not survive the winter; those that do survive are weak and never regain their vigor.

The potato leafhopper cannot overwinter in the North because of the sensitivity of the eggs to cold temperatures. It may be found year round near the Gulf of Mexico. In the North it appears rather suddenly in late May and early June, blown north by prevailing winds. Upon arriving in northern regions, it seems to be attracted to leguminous hosts such as alfalfa. The potato leafhopper occurs predominantly east of the 100th meridian (a line from Corpus Christi, Texas, to Winnipeg, Canada).

Eggs are deposited in slits cut in veins (E) on the underside of the leaf blade or in petioles. Nymphs undergo five instars. Six generations occur in Virginia, and four or five in Ohio. Vigorously growing Norway and sugar maples, birch, apple, chestnut, Persian walnut, and probably other tree species are attractive feeding sites for the potato leafhopper. Studies by Morette (545) and Carlson (103), working on Norway and sugar maple, showed a marked reduction in the length of developing internodes resulting in severe stunting, extensive foliage damage (D), and the development of multiple tops. Feeding on the above-mentioned tree species is essentially confined to the most succulent tissue.

An exotic fungal pathogen, *Zoophthora radicans*, has been successfully used on an experimental basis to control the potato leafhopper.

Scaphoideus luteolus Van Duzee, the whitebanded elm leafhopper, is another example of a phloem-feeding leafhopper; it appears to feed only on leaves and shoots of elm. Eggs are laid in the cork of twigs in late summer and hatch the following spring after leaves develop. The nymphal stage is known only from elm. It is never abundant but, when present, it feeds principally on the midrib and larger leaf veins, always from the underside. Excessive feeding may cause apical portions of leaves to die, an injury also called hopperburn. As the nymphs grow, they disperse. Nymphs are dark brown with a transverse white band behind the thorax. Adults are capable fliers and readily move from tree to tree, spending most time in the inner crown of the tree. This species would be of no economic consequence if it were not a vector of the elm yellows disease (phloem necrosis). This disease, caused by a mycoplasmalike organism, and Dutch elm disease have killed most of the mature American elms in the Northeast. Elm yellows occurs from Oklahoma, Kansas, and Nebraska east to Georgia, New York, and southern Ontario. The leafhopper occurs over the natural range of American elm (see Sinclair et al., *Diseases of Trees and Shrubs*).

References: 32, 38, 52, 103, 117, 160, 409, 501, 502, 545, 587, 614, 622, 638, 772, 867, 880

A. A nursery-grown birch tree showing scant foliage and dead twigs and shoots caused by the feeding of the potato leafhopper.
B. A swollen birch shoot and distorted foliage caused by the potato leafhopper. Symptoms are most evident in early September in New York.
C. The winter condition of a birch shoot and twig damaged by the potato leafhopper.
D. Potato leafhopper injury to second-flush foliage of Norway maple.
E. A potato leafhopper nymph. Notice the white eyes. The round swollen bumps on the leaf vein are the sites of eggs that have been inserted into the vascular tissue.
F. An adult potato leafhopper.

A
B
C
D
E
F

Foliage Injury by Leafhoppers (Plate 200)

Hundreds of species of leafhoppers feed on the leaves of woody plants. They are difficult to identify from external features; most females cannot be reliably identified to species. Because of the large number of species and identification difficulties, the ostensibly most important species selected here include those that damage a wide range of host plants, have an extensive geographic range, represent a severe localized problem, or are known vectors of woody plant diseases. Pictures on the facing page show typical injury symptoms caused by leafhoppers in the genera *Erythroneura, Empoasca, Edwardsiana, Typhlocyba,* and *Alebra*. Most ornamental plants are attacked by at least one species of *Erythroneura* or *Empoasca* (160).

Leafhoppers overwinter either as eggs or adults. The overwintering egg is embedded in the cortex of shoots or in the cork of twigs or stems. Adults hibernate under leaf litter or in bark crevices. Where second and subsequent generations occur, eggs are laid in underside leaf tissue either in a petiole or in veins (Plate 199E), often in proximal rows. In this position the incubation period is passed in saturated humidity. Nymphs, as well as adults, feed on the undersides of leaves, although the typical symptom, a whitish stipple, is visible from the upper surface. This habitat places these insects in a highly humidified atmosphere. (Plant transpiration occurs through stomata, and the undersides of leaves contain most of the stomata.) High humidity is necessary for the existence of many leafhopper species.

Leafhoppers feed on mesophyll or in vascular tissues (phloem and xylem), which helps to explain injury symptoms. With the exception of species in the genus *Empoasca,* all other genera listed here are mesophyll feeders. Mouthparts are inserted from the lower leaf surface, pass through the spongy mesophyll, and tap the palisade cells. The cells are torn open and emptied of their contents. The results of such feeding cause yellowish white stipple spots (A, B) that often coalesce into large whitish blotches in mature leaves. The total plant responds by being stunted or exhibiting reduced vigor.

Leafhoppers that feed in vascular tissue cause leaves to become curled or otherwise distorted (Plate 199D). The digestive saliva of some leafhoppers is toxic to plant tissues and may diffuse through cell walls, affecting adjacent cells. In some cases the toxin may become systemic and move to foliage far removed from the feeding site. If leaf margins become necrotic and turn brown, the symptom is called hopperburn.

Leafhopper females are distinctly selective in their choice of plants for the deposition of eggs. The nymphs are almost always found on plants where egg laying occurred (i.e., on preferred hosts). In the late growing season some adults tend to be polyphagous.

Alebra albostriella (Fallen), sometimes called the maple leafhopper, is bright yellow, about 4 mm long, and ubiquitous east of the Rocky Mountains and in California. It feeds on elm (*Ulmus americana*) (E, F), beech, oak, cherry, hawthorn, basswood, sumac, *Carpinus* species, hickory, *Asimina* species, and probably others. *A. albostriella* is an early season leafhopper in the East but in central California it is most abundant in midsummer. There is one generation per year. The females lay eggs singly in longitudinal bark slits in twigs. Sometimes these slits completely encircle the twig, which may severely injure or kill it. Eggs hatch when leaves are about half grown, and nymphs mature in about 1 month. In the Washington, DC, area, adults have disappeared by the middle of July.

The California buckeye, *Aesculus californica,* is subject to severe injury by certain leafhoppers. The injury shown in panel D is caused by a species of *Empoasca*. This buckeye, a favorite host, is a common ornamental tree in central and northern California. Because of the dry climate, the leaves of this tree normally drop by early August. When the *Empoasca* leafhopper is present, hopperburn is likely to be noticed and leaf drop may occur a month earlier than normal. The symptoms shown in panel D are manifestations of injury caused by the leafhopper and by drought. The species attacking this tree sometimes bite humans. The bite is sharp, but the pain is brief.

Typhlocyba pomaria McAtee is about 3.5 mm long and has an approved common name, white apple leafhopper. This species feeds in the mesophyll, producing the same symptoms described for *A. albostriella*. Two generations occur each year. Rosaceous species, such as apple and rose, are preferred hosts. Other food plants include crabapple, cherry, hawthorn, birch, alder, and elm. Geographic distribution of the leafhopper coincides with its host plants.

On apple, overwintering eggs may be found in internodes up to 4 years old. There are five nymphal instars, with the first adults appearing in early June in New York. They choose to feed on fully expanded leaves. *Anagrus epos* Girault is an important egg parasite.

The genus *Erythroneura* contains many species that feed on trees and cause the stipple symptom. *E. lawsoniana* Baker, *E. arta* Beamer, and *E. usitata* Beamer commonly occur on sycamore and presumably on London plane. These leafhoppers may be found from New York west to Texas, Kansas, and Iowa. *E. lawsoniana* often occurs in swarms about host trees. The opaque white adults are about 33 mm long and are attracted to lights at night. They produce two broods each year and overwinter under leaves and bark fissures. Additional hosts include *Carpinus* species, apple, willow, and grape. Adults occasionally stab human skin, producing a sharp but short-lived pain. *E. gledistia* Beamer causes stippled leaves on *Robinia* species, *Aesculus* species, *Cercis canadensis,* and hawthorn. *E. ziczac* Walsh causes severe injury to Boston ivy and Virginia creeper. This leafhopper is common in older residential areas; it may become a nuisance in dwellings during evening hours.

Edwardsiana commisuralis (Stål) is a light yellow to light orange leafhopper with a dark margin on the forewing. Although it is distributed across the United States, damage has been limited to the Rocky Mountains or farther west. Dogwood (*Cornus florida*) in California (A) and alder in British Columbia may be quite noticeably injured. The same type of injury occurs in the East by other species of *Edwardsiana* and *Erythroneura*.

In the Northwest *Keonolla confluens* Uhler, a large (5–8 mm) dark-colored leafhopper, devitalizes willows, cherry, peach, apple, balsam poplar, and *Ceanothus velutinus*.

References: 38, 52, 117, 241, 788

A. The light stippled spots on the upper side of this dogwood leaf were probably caused by the feeding of *Edwardsiana* leafhoppers.

B. The stippled spots on the blades of red maple, *Acer rubrum,* leaves were probably caused by *Alebra albostriella*.

C. Stipple spots magnified to show the snowflakelike symptom characteristic of leafhoppers that feed on mesophyll. This symptom always appears on the upper leaf surface.

D. Buckeye leaves severely injured by an *Empoasca* phloem-feeding leafhopper.

E. Cast "skins" of the leafhopper *A. albostriella* on American elm. "Skins" characteristically remain attached to the leaf by the partially embedded stylets.

F. Severe injury to elm leaves by *A. albostriella*.

A
B
C
D
E
F

Redbanded Leafhopper and Other Sharpshooter Leafhoppers (Plate 201)

Rhododendrons, including azalea, are one of the most valued flowering shrubs. Though several species of leafhoppers may occasionally "taste" its foliage and buds, two species are of major concern. *Graphocephala coccinea* Forster (C) and *G. fennahi* Young, sometimes called the redbanded leafhopper and rhododendron leafhopper, respectively, are likely to be found wherever rhododendron grows.

G. coccinea feeds on rosebay rhododendron (*Rhododendron maximum*), catawba rhododendron (*R. catawbiense*) and their cultivars, mountain laurel (*Kalmia latifolia*), evergreen and deciduous azaleas, rose, forsythia, privet, and approximately 50 other plant species. This brilliantly colored insect (C) may be found throughout the growing season. Eggs are laid in the spongy mesophyll. Visible through what appear to be swollen internal tissues and the plant's lower epidermis, the egg appears oval and flattened (D). Nymphs feed primarily, if not exclusively, on the undersides of leaves, where molted "skins" (E) may be attached and serve as clues to the cause of leaf injury. Fully grown nymphs are about 8 mm long and creamy white; they are good jumpers. Two generations are produced each year. The first generation in southern Pennsylvania appears in late April, the second about mid-July. They are either long-lived or egg hatching occurs over several weeks, because adults may be found from mid-June through October or the first killing frost. They are among the largest of the eastern North American leafhoppers.

Injury may occur from both feeding and egg-laying activities, and the symptoms remain grossly visible throughout the life of the leaf. Oviposition injury appears in the leaf as shown in panel D and, after the egg has hatched, as a scar or hole about 4 mm in diameter. Feeding injury on catawba rhododendron is illustrated in panels A and E, and all of it was caused by first-generation nymphs. Gross injury is caused by nymphs when they feed on immature foliage and elongating shoots. The pictures were taken in late July to illustrate the long-term nature of the injury.

G. fennahi was not described until 1977, but it no doubt has a long history in North America. It looks nearly identical to *G. coccinea,* but close observation will reveal more green and less red or orange in the forewings of *G. fennahi.* The two species have distinct morphological differences, apparent with the aid of a microscope. Their life history is similar in many respects, too, except that the eggs of *G. fennahi* are deposited in bud sepals. It probably has a more restricted host range (azalea, rhododendron), but Wheeler and Valley (845) imply that its geographic distribution may be greater than that of *G. coccinea,* based on reports from Oregon and British Columbia.

Injury to azalea is typically that of a leafhopper (B), where stylets penetrate cells on the upper side of the leaf and withdraw cell contents, leaving yellowish stipple marks that often coalesce.

Several related species commonly called sharpshooters are found in the southwestern states and into northern California (G, I). *Graphocephala atropunctata* (Signoret), also called the blue-green sharpshooter (recent synonyms—*Hordnia circellata* and *Graphocephala circellata*) has many host plants but prefers vines, shrubs, and trees. Hosts commonly used for oviposition include rose, Boston ivy (*Parthenocissus tricuspidata*), periwinkle (*Vinca major*), Algerian ivy (*Hedera canariensis*), Mexican bush sage (*Salvia leucantha*), Sydney golden wattle (*Acacia longifolia*), *Myoporum laetum,* Australian bush cherry (*Syzygium paniculatum*), *Fuchsia hybrida,* sea lavender (*Limonium perezii*), tarata (*Pittosporum eugenioides*), mirrow plant (*Coprosma repens*), *Cestrum elegans,* and sycamore (*Platanus* spp.). It will also feed on Persian walnut (*Juglan regia*), birch, and various species of willow.

Eggs are laid in a slit cut in the petiole, rarely in the midrib, early in the growing season. Nymphs are white with a yellow tinge along the sides of the abdomen. In the northern part of their range, adults are basically green or bluish green; in the southern range they are often bright blue. There is one generation each year, with the adult overwintering. *G. atropunctata* sucks sap and cell contents from leaves, which causes leaves to curl or cup and to become spotted in a mosaic fashion. *Homalodisca* species, also called sharpshooters, are found in the South and feed on a variety of trees such as *Prunus* and *Celtis* species.

These insects may be of greater importance than injury symptoms indicate because they are efficient vectors of several microorganisms that cause plant diseases. The saliva they inject while feeding may also be toxic to some plants. Saliva toxicity expresses itself by the clearing of veins and veinlets, yellow vein branching, and chlorosis. *Graphocephala* species have been associated with bud blast disease in rhododendron.

The aster leafhopper (H), *Macrosteles quadrilineatus* Forbes, is a serious virus vector. It spreads certain viruses to various vegetable and annual flowers. Periwinkle (*Vinca major*) and *Thunbergia* species are woody plants that are susceptible to the aster yellows virus; the aster leafhopper feeds on both. Viral disease symptoms on periwinkle include chlorotic foliage and malformed flowers. *Thunbergia* responds to the viral infection by showing dwarfed leaves and chlorotic veins, and the plant loses its normal ability to climb.

The aster leafhopper is widespread throughout North America. It migrates from south to north each year, overwintering in the egg stage in the North. Four to five generations occur during the growing season in southern Ontario. The greenish yellow adult is about 4 mm long.

References: 86, 132, 575, 687, 845, 846, 880

A. A rhododendron shoot and leaves showing gross foliage distortion caused from feeding by *Graphocephala coccinea.* [Cicadellidae]

B. Feeding injury on azalea cultivar Peachy Keen caused by *G. fennahi.*

C. The adult leafhopper *G. coccinea*; 8.5 mm long.

D. *G. coccinea* eggs embedded in a mountain laurel leaf. (Courtesy J. F. Stimmel, Bureau of Plant Industry, Pennsylvania Dept. of Agriculture.)

E. The underside of a rhododendron leaf showing leaf distortion and cast leafhopper "skins."

F. A rose leaf with crinkling thought to be caused by the feeding of sharpshooter leafhoppers.

G. A sharpshooter leafhopper, *G. atropunctata*; 7 mm long.

H. The aster leafhopper, sometimes called the six-spotted leafhopper; about 4 mm long. (Courtesy New York State Agricultural Experiment Station, Geneva.)

I. *G. atropunctata* leafhoppers seen from the underside.

A
B
C
D
E
F
G
H
I

Spittlebugs on Angiosperms (Plate 202)

The term *spittlebug* applies to a group of sapsucking insects whose immature stages cover themselves with frothy salivalike masses (B) composed of air bubbles trapped in fluids discharged from the alimentary canal and special epidermal glands. The spittle mass protects the nymphs. There may be from one to several nymphs per spittle mass. Only nymphs produce spittle. Adults are usually inconspicuous and usually brownish; they readily jump or fly when disturbed. It is not uncommon for adults to feed on one plant group, for example, trees, and nymphs to feed on another, say, grasses. Most species have one generation per year and winter as eggs under or in the bark of shoots or twigs. They are related to the cicadas and leafhoppers.

Plant damage results from excessive extraction of plant juices by nymphs in order to maintain the protective spittle masses, from excessive extraction of sap by adults to meet their food requirement, from injection of toxic substances while feeding, from transmission of viruses or mycoplasmalike organisms, and from feeding punctures that may provide points of entry for other pathogens. Plants respond with loss of thriftiness and with wilted, stunted, distorted, and discolored foliage; necrotic spots on leaves may also appear.

The dogwood spittlebug, *Clastoptera proteus* Fitch (A, B, F), occurs east of the prairies wherever dogwoods grow. It also feeds on species of *Vaccinum* and *Aesculus*. Adults and nymphs feed on the same host. Eggs are laid in punctures made beneath the bark of outer twigs. The females place 1–3 eggs in each puncture. This is the overwintering stage. Nymphs are present in June and early July. Very little is known about the biology of this species. Although its spittle is very conspicuous, we have never seen populations high enough to require control measures. The spittle, of course, is obnoxious on ornamental plants.

Distribution of the alder spittlebug, *C. obtusa* (Say), is transcontinental across the northern states and adjacent Canadian provinces. There are two generations per year. Eggs of the spring generation are laid in September in the northern parts of its range but as late as October farther south. Clusters of 3 or fewer of these eggs are laid under the bark of outer twigs, often at the base of buds, in slits the female makes with her ovipositor. Second-generation eggs are laid in slits cut in shoots that elongated in the spring. Alder, birch, witch-hazel, and hickory are the only known hosts and food source for nymphs (C). Collection records indicate that adults probably feed on a wide range of trees and shrubs such as oak, *Aesculus* species, walnut, willow, and blueberry.

Prosapia bicincta (Say) has no common name. It is included here because of the damage caused to holly foliage by the adult. When this bug feeds on old leaves, blotches appear on the underside. If young or immature leaves are attacked, the result may be distortion, stunting, discoloration, or necrosis at the site of stylet penetration. Injury to holly is rarely encountered in its northern range. Adults have also been reported to feed on redbud and cherry.

The adult *P. bicincta* spittlebug has two distinct bands (D, E), although they are often partially or completely absent. There is probably one generation per year at least in its northern range. Eggs are laid on grasses, where nymphs develop. Their feeding activities may cause serious problems on turf grasses in Florida and the Gulf states. Spittle masses are found on the lower stems and roots. Little is known about its biology.

This spittlebug is recorded from Massachusetts and New York west to northern Michigan, south to Kansas and Texas and Florida. There is at least one northern subspecies, *P. bicincta ignipecta,* that lives beyond the range of native or exotic holly. We find no confirming records but presume that it feeds on wild cherry and perhaps redbud.

The most abundant, widespread, and economically important spittlebug is the meadow spittlebug, *Philaenus spumarius* (Linnaeus). The adult is practically omnivorous. Nymphs (H) feed mostly on herbaceous plants. We have no records of studies about this insect on woody ornamental plants; however, it is a potential threat to young nursery stock that is being forced with excess nitrogen, especially if the stock is next to weedy fields and pastures. Adults are variably brown and about 6 mm long.

References: 184, 328, 520

A, B. Spittle masses on dogwood shoots containing nymphs of the dogwood spittlebug.

C. Spittle masses on an alder shoot (*Alnus tenuifolia*) made by nymphs of the alder spittlebug. (Courtesy D. A. Leatherman, Colorado State Forest Service.)

D, E. The adult of the spittlebug *Prosapia bicincta,* a pest of holly; up to 9 mm long.

F. An adult dogwood spittlebug; about 4 mm long.

G. Spittle masses on twigs of *Acacia cardiophylla* caused by the nymphs of *Clastoptera arizonana*. Adults (not illustrated), about 3.5 mm long, are brown with a distinct broad white band on the wings.

H. Nymphs of the meadow spittlebug. (Courtesy Ray Kriner, Hendersonville, NC.)

A
B
C
D
E
F
G
H

Planthoppers and Plant Bugs (Plate 203)

The planthoppers shown in this plate are related to leafhoppers and spittlebugs. *Metcalfa pruinosa* (Say), illustrated in panel B, belongs to a group called flatids. It is abundant throughout the United States and southern Canada but is probably more common in the southern states. *M. pruinosa* feeds on many different plants, such as hickory, pecan, oak, elm, birch, linden, ash, privet, redbud, camellia, maple, azalea, viburnum, magnolia, holly, seagrape, dogwood, cherry, laurel, and all the flowering fruit trees, such as flowering peach. The injury caused to ornamental plants is rather insignificant, although oviposition punctures may kill small twigs. Also, the white filaments of flocculent material (A) made by nymphs affects the ornamental value of a plant.

The insect overwinters as an egg inside the twigs of its host plant. No particular oviposition pattern is evident, as in treehoppers. Eggs are scattered singly, each inserted through a slit in the bark cut by the ovipositor. In the Weslaco area of Texas eggs hatch in late March; in southern Illinois they begin to hatch in late May. Nymphs develop on twigs, feeding on the undersides of succulent leaves or on shoots. They also feed on the fruit of trees such as *Citrus* species. The nymphs, which could easily be mistaken for mealybugs, are white and usually surrounded or completely covered by long filaments of white waxy material that may resemble cotton (A). If alarmed, they jump like leafhopper nymphs.

In the South about 9 weeks are necessary for *M. pruinosa* nymphs to develop into adults; adult activity occurs over a period of 2 months. The adult shown in panel B was collected in New York state on August 20. Adults vary from brown to gray, depending on the presence or absence of a bluish white waxy powder on the cuticle. Adults are attracted to ultraviolet light at night. There is one generation each year.

Several related species such as *Anormenis chloris* and *Ormenoides venusta* are often confused in the field. Adults of both species are snow white to pale green. Nymphs are either white or greenish and, like those of *M. pruinosa,* are covered with long white waxy filaments. Eggs are laid in longitudinal slits cut into the bark of shoots (Figure 122). Their biologies, life history, and host plants closely resemble those of *M. pruinosa. A. chloris* ranges from Connecticut to New York, south to Florida, and west to Iowa and Arizona. *O. venusta* ranges from Pennsylvania south to Florida and west to Texas and Missouri. Some woody landscape plants that serve as hosts of these two species are *Acer* species, bittersweet (*Celastrus scandens*), hawthorn, *Prunus* species, *Quercus rubra,* black walnut, and blueberry.

Acanalonia conica (Say) is another flatid that feeds on the twigs and shoots of black walnut, *Alnus glutinosa,* mulberry, sassafras, sweetgum, redbud, black locust, boxelder, dogwood, and ash. Adults are pale green and about 10 mm long. Eggs are laid in live twigs in about the same manner as those of *M. pruinosa*. In southern Illinois eggs hatch from mid-May to mid-June. Nymphs are also coated with a white flocculence. This species ranges from New Jersey south to Florida and west to Texas and Nebraska.

Adults and nymphs of the rhopalid plant bug, *Niesthrea louisianica* Sailer, feed on flower buds and seeds of *Abutilon* species and rose of sharon (*Hibiscus syriacus*). The adult is bright orange. Eggs are laid in masses, deposited on the undersides of leaves. This bug is mainly a southern species that produces three or four generations a year in the Gulf states; it occasionally is found as far north as Long Island.

Euthochtha galeator (Fabricius) (E) belongs to the family Coreidae. This bug feeds on a variety of wild and cultivated plants in both rural and urban landscapes, and it is common in the eastern half of the United States from central New York to Florida and west to the Great Plains. In our experience it is never found in large numbers. However, because the female can lay more than 250 eggs, the potential for high populations exists (879). *E. galeator* is relatively large, and its mouthparts penetrate deeply into shoot and bud tissues, causing shoot terminals to wilt and often to die; flower and vegetative buds also may be grossly distorted. This bug usually makes few punctures but tends to feed from a single site for several hours unless disturbed. The longer it feeds in one site, the more pronounced the injury. Injury results not only from the mechanical action of the stylets but also from saliva, which probably contains enzymes toxic to plant tissues. Damage is reported most frequently on rose when *E. galeator* feeds on both the closed bud and opening flower. Other hosts include shoots of hickory, oak, sumac, *Citrus* species, and lychee as well as a wide range of ornamental shrubs.

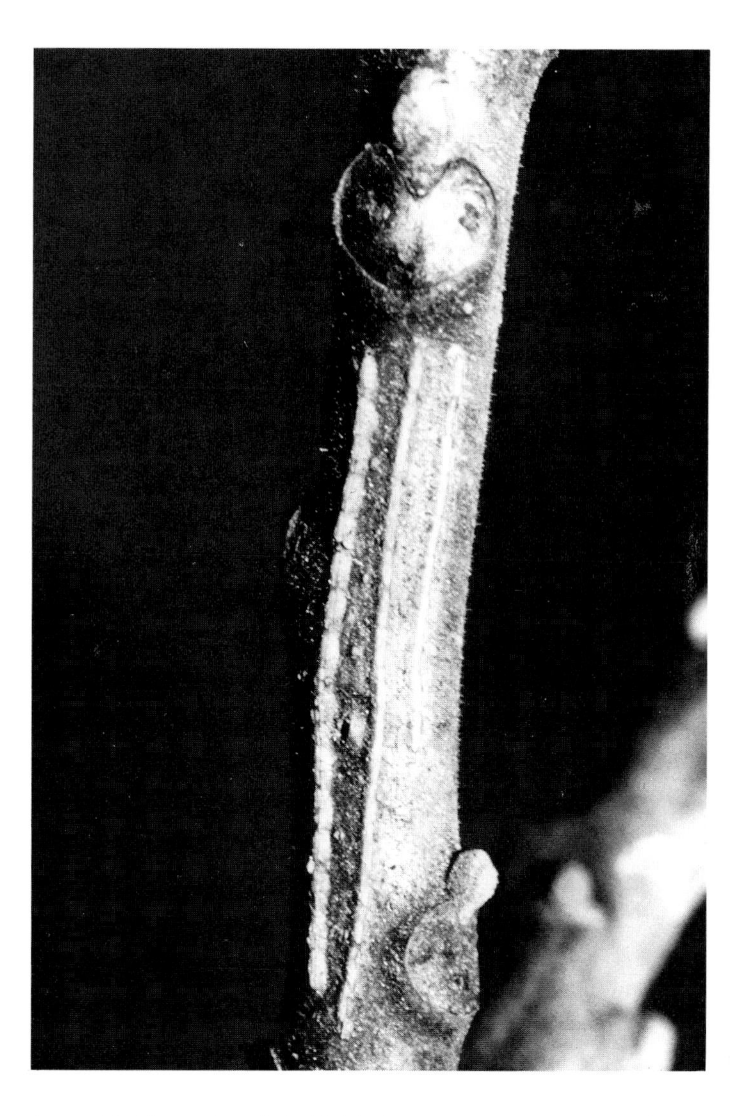

Figure 122. The vertical slit in this black walnut twig contains eggs of *Anormenis cloris*. (From Wilson and McPherson, ref. 864, courtesy S. W. Wilson and *Annals of the Entomological Society of America*.)

E. galeator hibernates as an adult and in the North becomes active as buds begin to open. Because of its broad range of food plants, activity begins long before the spring frost-free date. It lays eggs in small clusters on leaves of almost any green plant, including grasses. The eggs are usually bright gold and turn red just before hatching. Nymphs, which frequently aggregate, are characterized by an abundance of spines (D). The adult population generally peaks in the spring after hibernation and again in late August.

Leaf-footed bugs are so named because of the peculiar shape of their hind tibiae. The specimens shown in panels C and F are *Leptoglossus phyllopus* (Linnaeus). Leaf-footed bugs are found primarily in the southern half of the United States, although they have been reported

---

A. White cottony puffs produced by nymphs of the planthopper *Metcalfa pruinosa* on a water sprout.
B. A lateral view of *M. pruinosa*. The insect is 5.5–8.0 mm long.
C. An adult leaf-footed bug on lychee foliage and flower. (Courtesy F. W. Mead, Florida Dept. of Agriculture.)
D. A nymph of *Euthochtha galeator*. [Coreidae] (Courtesy F. W. Mead, Florida Dept. of Agriculture.)
E. An adult *E. galeator* feeding on a rose petal; about 15 mm long. (Courtesy of F. W. Mead, Florida Dept. of Agriculture.)
F. An adult leaf-footed bug, *Leptoglossus phyllopus,* on *Schinus terebinthifolius*; about 20 mm long. [Coreidae]
G. The plant bug *Acanthocephala femorata*; about 30 mm long.

A
B
C
D
E
F
G

on Long Island and in Kansas, Iowa, Texas, Colorado, Utah, and California. In the South adults may be collected throughout the year, although peak populations occur in warmer months.

Because the mouthparts of leaf-footed bugs are particularly long, they can penetrate the husk of nut fruits when the fruit is in an early stage of development. Leaf-footed bugs are among the insects that cause black pit in pecan kernels. Nymphs resemble adults but do not acquire leaf-shaped legs until they are nearly grown. Eggs, which resemble a series of longitudinal segments, are golden brown and are laid in a single row along a stem or leaf midrib.

These bugs are likely to be a concern to the homeowner with yard trees, primarily fruit and nut trees. Both adults and nymphs feed on *Citrus* species, peach, plum, pecan, and walnut, causing blemishes, undersized fruit, and premature fruit drop. They also feed on the fruit and flowers of blueberry, *Hibiscus* species, loquat, persimmon, crape myrtle, *Ligustrum* species, and rose.

The related species *L. zonatus* (Dallas), *L. occidentalis* Heidmann, and *L. clypealis* Linnaeus, found in California, also have leaf-shaped tibiae. Their biologies and feeding behavior are essentially the same as those of the eastern species. *L. clypealis* feeds on buds, seeds, and fruits of ornamental nut trees, for example, pistachios. These three species look much like those illustrated in panels C and F.

Acanthocephala femorata (Fabricius) is a larger, more robust insect than leaf-footed bugs (G). It, too, is primarily a southern species and one of the relatively few insects that is said to prey on other insects as well as plants. This bug, a predator of the cottonworm, damages the fruits of cherry, *Citrus* species, peach, plum, and pecan; it also feeds on flowers and flower buds. It is a rather fearsome-looking creature but less important economically than any of the other bugs mentioned. Both males and females have femoral spines, which are larger in the males. Both sexes are presumed to use them defensively against predators; in addition, males use them to battle with other males to establish territorial rights.

A. terminalis (Dallas) is a showy dark brown coreid up to 22 mm long and found most frequently late in the growing season. It, too, has a leaflike expansion of the hind leg (tibia). Distribution and host records show it to be primarily a foliage feeder on many trees and shrubs in most of the eastern states, including the Great Lakes states.

References: 528, 837, 838, 864, 865, 879

Lace Bugs of Broad-leaved Evergreens and Sages (Plate 204)

In terms of the injury they do to ornamental trees and shrubs, the Tingidae, or lace bugs, are the most important family in the order Hemiptera. In 1965 Drake and Ruhoff (190) published the monograph *Lacebugs of the World,* cataloguing 1820 species in 236 genera. Of these, there are three leading economic species in the genus *Stephanitis* that feed on ornamental broad-leaved evergreens in the United States.

Lace bug damage on broad-leaved evergreens is most common and severe on andromeda, azalea, laurel, pyracantha, and rhododendron. High populations of lace bugs, and consequently higher degrees of damage, occur on azalea and rhododendron when they grow in sunny rather than shady locations; andromeda lace bug is damaging in both sun and shade.

The feeding of lace bugs is distinctively characteristic. It can be readily distinguished from that of other insects and mites. Lace bugs feed on the undersides of leaves, but the damage is very apparent on the upper surface. Damage symptoms bear a strong similarity to leafhopper damage, but lace bugs produce varnishlike spots on the undersides of the leaves. An occasional shed "skin" of a leafhopper nymph is evidence of the cause of that damage. Lace bug damage may resemble mite injury from a distance. Feeding by mites causes chlorotic flecks in the leaves that are much finer than those caused by lace bugs. Close examination reveals that large numbers of contiguous cells are chlorotic where lace bugs have fed. Positive identification of lace bug damage is confirmed by the presence of brown patches or black droplets of excrement on the undersides of damaged leaves. Frequently, the cast "skins" of nymphs remain attached to the undersides of leaves. A nymph is shown in Figure 123.

Species that occur on broad-leaved evergreens overwinter as eggs. The eggs are either inserted into veins or cemented to the leaves with a brown crusty material. Eggs hatch in May in Virginia and usually before the end of May further north. At least two generations occur each year in the North and three or more in the South. It is especially important to prevent damage on broad-leaved evergreens early in the season because foliage will retain the unsightly injury and be less functional for more than a year.

Stephanitis pyrioides (Scott), the azalea lace bug, a native of Japan, is a primary pest of azalea (D). It is found in New York and Massachusetts south to Florida and Alabama and west to Missouri. In Maryland four generations occur each year. *S. rhododendri* Horvath is a pest of rhododendron (C) and *Kalmia latifolia* and is found from southern Canada to Georgia, West Virginia, Ohio, and Michigan; it is known also in Washington and Oregon.

S. takeyai is primarily a pest of andromeda and *Leucothoe,* although there are records of its being taken on *Styrax* species and willow. In eastern states it is most commonly encountered on andromeda. Its life history is similar to that of *S. pyrioides*. Eggs are deposited into the midveins on the undersides of leaves and covered with a brownish substance that hardens into a scablike protective covering. There are five nymphal stages. In Connecticut there are four generations per year. This lace bug was introduced into Connecticut from Japan about 1945. Since then it has been found from western North Carolina to Maine.

Figure 123. An immature azalea lace bug, *Stephanitis pyrioides*. Note its dark color and spines. Wings have not yet developed. (See also Plate 205.)

A. Lace bug injury on andromeda, caused by *Stephanitis takeyai*.
B. Varnishlike deposits from lace bugs on the undersides of andromeda leaves.
C. Leaves of rhododendron, injured by *S. rhododendri*. Range of damage is from none to severe. [Tingidae]
D. "Varnish" spots on azalea, indicating a severe infestation. They were caused by *S. pyrioides*.
E. Typical appearance of the underside of a previously infested rhododendron leaf.
F. Purple sage (*Salvia* sp.) injury caused by the lantana lace bug, *Teleonemia scrupulosa*. Note that the black fecal matter is on the upper surface of leaves.

A
B
C
D
E
F

(Plate 204, continued)

The lantana lace bug, *Teleonemia scrupulosa* Stål, is a subtropical and tropical species native to Central and South America. Its principle hosts are the sages *Lantana* species and *Leucophyllum* species. *Lantana* species have become serious weed pests of rangelands and regeneration forests, and for this reason the bug is of interest as a candidate for their biological control.

In our most southern states numerous cultivars of *Lantana* and *Leucophyllum* species (purple sage, Texas sage) are prized ornamentals for landscaping. On *Lantana,* the lace bug feeds on the undersurface of leaves and greedily attacks newly opened buds and flowers. On purple sage it feeds on both the upper and lower surface of foliage. When on the upper surface, the black excrement spots (F) are readily visible, often throughout the winter months.

The adult is a small (3–4 mm long) brown bug, and its rather narrow wings have an intricate, raised lacework pattern. Adults are active fliers during the summer months, dispersing readily if their food plant becomes less palatable. Overwintering may occur in either the adult or egg stage. On *Lantana* species eggs are laid on the undersides of leaves embedded in the midrib or major veins. Upon hatching, the nymphs, which have many dark spines, aggregate into colonies and usually feed on the undersides of leaves.

In the United States the lantana lace bug occurs in Florida and Texas and probably exists in all of the states along the Gulf Coast.

References: 190, 196, 311, 334, 431, 645, 678, 848

Lace Bugs of Deciduous Plants (Plates 205–206)

There are at least 27 species of lace bugs in the genus *Corythucha* that feed on deciduous trees and shrubs. Most have very specific host preferences, which aids field identification. The adult's (B) wings and thorax are beautifully sculptured and resemble an intricate, lacy network. They range from 3 to 8 mm long.

All the lace bugs of this genus overwinter as adults on or near their host in bark crevices, branch crotches, tufts of grass, or similarly protected areas. All generally have the same sequence of events in life cycle development.

Corythucha pallipes Parshly, sometimes called the birch lace bug, feeds on *Betula* and *Acer* species, *Fagus grandifolia,* mountain ash, and to a lesser extent on *Ostrya virginiana* and *Salix* species. Distribution of *C. pallipes* is transcontinental throughout the northern states and southern Canada.

Adults overwinter among fallen leaves on the ground and other protected places close to their host plants. Spring-generation eggs are laid in small groups on the lower surfaces of leaves in the axils of veins. Part of the egg is inserted into leaf tissue. Pubescence along the veins of yellow birch help to conceal the brown eggs.

There are five nymphal instars, which require a minimum of 26 days to reach the adult stage. Summer adults are found in New England from mid-July until mid-September. There are two generations per year. Injury caused by the feeding of nymphs and adults on mountain ash is illustrated in panels A, F, and G.

The hawthorn lace bug, *Corythucha cydoniae* (Fitch), occurs throughout most of the United States, southern Canada, and northern Mexico. Most of its host plants belong to the family Rosaceae, but occasionally plants outside that family are attacked. Hawthorn, pyracantha, cotoneaster, and quince are favored host plants. Occasionally, *Amelanchier* foliage may be severely injured. Schultz (684) tested the relative preference of this lace bug for five species or cultivars of container-grown *Cotoneaster* and *Pyracantha,* with the following results. *Cotoneaster horizontalis* and *Pyracantha atalantioides* cv. Aurea were the least preferred, and *C. dammeri* cv. Royal Beauty and *P. koidzumii* cv. Ingleside Crimson were most preferred. Intermediate in preference were *C. dammeri* cv. Lofast, *C. buxifolius, C. salicifolius, P. coccinea* cv. Orange Glow and cv. Loboy, *P. augustifolia* cv. Variegata. In laboratory tests he found that the hawthorn lace bug preferred white beam, *Sorbus aria,* over *Crataegus phaenopyrum, Pyracantha crenato-serrata,* or *Cotoneaster dammeri.* Schultz (685) evaluated four additional *Cotoneaster* species, *C. lacteus, C. nitens, C. acutifolius,* and *C. watereri* cv. Brandekeir, for resistance to the hawthorn lace bug. Significantly fewer nymphs completed development on *C. lacteus.* The presence of dense pubescence on the underside of *C. lacteus* foliage is suggested as a factor in lace bug resistance.

One or more generations of the hawthorn lace bug occur each year, depending on climatic conditions. In New England there may be one generation each year, whereas in Maryland four generations occur. Development from egg to adult requires 5–7 weeks under outdoor conditions in New England.

References: 31, 189, 190, 527, 606, 684, 685, 782, 807, 840

Table 18. **Lace bugs and their host plants**

Name	Preferred host
Primarily Eastern Species	
Alder lace bug,* *Corythucha pergandei*	Alder, hazel, elm, birch, crab-apple
Avocado lace bug, *Pseudocysta persea*	Avocado
Basswood lace bug, *Gargaphia tiliae*	Linden, basswood
Birch lace bug, *C. pallipes*	Yellow birch, white birch, beech, eastern hophornbeam, willow, mountain ash, maple
Buckeye lace bug, *C. aesculi*	Buckeye
Cherry lace bug (Plate 206), *C. pruni*	Wild cherry
C. associata	Wild cherry
C. bellula	Hawthorn
C. marmorata	Basswood
Elm lace bug, *C. ulmi*	American elm (only)
Hackberry lace bug, *C. celtidis*	Hackberry
Hawthorn lace bug,* *C. cydoniae*	*Crataegus* spp. (hawthorn), pyracantha, *Cydonia* spp. (quince)
Oak lace bug, *C. arcuata*	Oak (subspecies *C. a. mali* found also on maples)
Sycamore lace bug,* *C. ciliata*	Sycamore
Walnut lace bug,* *C. juglandis*	Butternut, black walnut, linden (Plate 205)
Willow and poplar lace bug, *C. elegans*	Balsam poplar, quaking aspen, bigtooth aspen, willow
Willow lace bug (no spines on margins), *C. mollicula* (=*canadensis*)	Willow
Primarily Western Species	
Angulate tingid, *C. angulata*	*Ceanothus* spp.*
Apple lace bug, *C. coelata*	Apple
Arizona ash tingid, *Leptoypha minor*	Ash, poplar
California Christmas berry tingids, *C. bullata, C. heteromelecola, C. incurvata*	California Christmas berry (toyon) (Plate 206)
Ceanothus tingid, *C. obliqua*	*Ceanothus* spp.* (Plate 206)
Chokecherry tingid, *C. padi*	Western choke cherry
Eastern willow tingid,* *C. salicis*	Willow
L. costata	Ash, hazel
Western sycamore lace bug, *C. confraterna*	Sycamore
Western willow tingid, *C. salicata*	Willow, apple

*Found throughout the United States and parts of Canada.

A, F, G. A range of injury symptoms, from a white stipple to brown blotches, on the upper and lower surfaces of mountain ash leaves. The damage was caused by *Corythucha pallipes.*

B. An adult walnut lace bug, *C. juglandis,* on the underside of a basswood leaf. [Tingidae]

C. The undersurface of a basswood leaf. Leaf injury and excrement spots of the walnut lace bug are apparent.

D. The upper surface of a hawthorn leaf showing severe discoloration. All discolored tissue is dead.

E. The lower surface of the leaf in D showing the excrement spots deposited by the hawthorn lace bug, *C. cydoniae.*

A
B
C
D
E
F
G

Lace bugs, or tingids, are common on certain western trees and shrubs, but they have not attracted as much attention as eastern species. Western lace bugs are found predominantly on wild or native vegetation, and usually populations are low.

The adult *Corythuca incurvata* Uhler, the so-called California Christmas berry lace bug, is a flat insect with a distinctive reticulated dorsal surface that resembles lace. It measures about 3.3 mm long and has a yellowish brown back. Adults overwinter beneath loose bark of toyon (*Heteromeles arbutifolia*) or under fallen leaves or other plant refuse on the ground. In the spring it returns to the foliage, where females insert eggs into the underside of the leaf. In about 1 month eggs hatch, and the tiny nymphs insert their sucking mouthparts into the undersides of leaves and begin to remove the contents of cells. The nymphs are oval, motile insects and are heavily armed with long spines on their bodies (similar nymphs are shown in B). After five molts over 6 weeks, the adults appear. There are three annual broods in coastal central California. Two other *Corythuca* lace bugs also occasionally attack toyon (see Table 18, accompanying Plate 205).

The ceanothus tingid, *C. obliqua* Osborn & Drake, is slightly larger than *C. incurvata* and has a pale upper surface with many small brown dots. It occurs on *Ceanothus* species in California, Oregon, and Idaho.

There are two species of cherry tingids (see Table 18). *C. pruni* Osborn & Drake is illustrated in panels C and D and in Figure 124. It is primarily a northeastern species but has been recorded as far west as Michigan and south to the mountains of North Carolina. *Prunus serotina* is probably its only host.

Adults (D) overwinter among leaf litter and coarse grasses. Emergence from hibernation is linked to the spring phenology of its host; adults become active about the time cherry leaves are half-grown and flower racemes are in early bud. Females begin to lay eggs in early May (Massachusetts). Brown eggs are deposited in circular patches on the undersides of leaf blades, partially inserted into leaf tissue. Hatching begins in early June. Nymphal development requires about 4 weeks—from egg to adult takes about 7 weeks. There is one generation per year in its northern range, four generations in Maryland and the mid-Atlantic states.

The typical scorched appearance of cherry foliage (E) results from feeding by numerous nymphs and adults. Most, if not all, feeding occurs on the undersides of leaves. Injury is most conspicuous when foliage is viewed from the upper surface. The entire contents of both spongy and palisade parenchyma cells are withdrawn through the lace bug's sucking mouthparts. This accounts for the loss of foliage color and accompanying drying and shrinking of cell walls.

The sycamore lace bug, *C. ciliata* (Say), is a common pest of sycamore, especially east of the Rocky Mountains. It overwinters as an adult under the flakes of exfoliated sycamore bark. Adults become active about the time leaves begin to develop in the spring. Eggs are attached to the undersides of leaves with a brown sticky substance and hidden in the leaf pubescence. Within a few days the eggs hatch, and nymphs begin feeding on the underside of the leaf. With sucking mouthparts, they pierce the epidermis and withdraw fluids and cell contents, causing the characteristic chlorotic flecks that are visible on the upper side of the leaf. The spiny black nymphs (G) only remotely resemble the adults (F). A complete cycle, from egg to adult, requires approximately 30 days. Two or more generations occur each year, the exact number depending on the length of the growing season. In Maryland there are five generations each year. In late summer all stages may be found feeding together. By this time dark varnishlike spots of excrement may cover most of the undersurface of leaves. Although *Platanus* species are their primary host, the sycamore lace bug has been collected from ash and hickory in Missouri.

Figure 124. An adult cherry lace bug, *Corythucha pruni*. The wing markings are characteristic of this species. (See Plate 206D for color pattern.)

Natural enemies of the sycamore lace bug include lacewings, assassin bugs, minute pirate bugs, spiders, and predaceous mites. In North Carolina as much as one-fourth of the sycamore lace bug eggs may be inviable. This suggests that there is an unusual and, at this point, unexplained mortality factor at work in this region.

References: 29, 217, 384, 385, 606, 807

A. The brown spots on these *Ceanothus* leaves are symptoms of severe injury by the ceanothus lace bug, *Corythucha obliqua*.
B. Ceanothus lace bug nymphs on the underside of a *Ceanothus* leaf.
C. A *Corythucha pruni* nymph on *Prunus* sp.
D. A *C. pruni* adult on *Prunus* sp.; 4 mm long.
E. Injury to a leaf by *C. pruni*; lower and upper leaf surface.
F. An adult sycamore lace bug, *C. ciliata*; about 3.75 mm long. [Tingidae]
G. An immature sycamore lace bug.
H. Injury to sycamore leaves caused by the sycamore lace bug.

A
B
C
D
E
F
G
H

Greenhouse Thrips (Plate 207)

Long a pest of certain greenhouse-grown plants, the greenhouse thrips, *Heliothrips haemorrhoidalis* (Bouché), also damages outdoor-grown woody ornamentals in California, Florida, and several other southern states. The list of host plants for this insect is very extensive and includes *Acer*—maple; *Azalea*; *Citrus*; *Codiaeum variegatum*—croton; *Cornus*—dogwood; *Ficus*; *Fuchsia*; *Heteromeles arbutifolia*—toyon; *Hypericum*—St. Johnswort; *Laurus nobilis*—Grecian laurel; *Magnolia*; *Mangifera indica*—mango; *Persea americana*—alligator pear; *Rhododendron*; and *Viburnum*.

Unlike many other thrips, the greenhouse thrips openly feeds on leaves, beginning on the lower surface first, rather than on blossoms, buds, or growing shoots. It prefers shaded conditions. Hot, sunny, dry weather is not suitable for a build-up of this pest. Both immature and adult greenhouse thrips (both of which live in dense colonies) feed by puncturing plant cells with their specialized mouthparts and withdrawing cell sap. Such activity causes a flecking, bleaching, or silvering of the affected leaves; young foliage may become distorted. Damaged foliage becomes papery and wilts, then drops prematurely. Fully as important as direct feeding injury by *Heliothrips* is the production of vast quantities of varnishlike excrement, which collects on the foliage and creates an unsightly appearance (C).

The adult greenhouse thrips measures only about 1 mm and is dark black with a silver sheen. (A photograph of an adult rose thrips, which resembles the adult greenhouse thrips, appears in Plate 208E.) Eggs are deposited within leaf tissues and hatch in 2–3 weeks. Thrips larvae are translucent and white except for a drop of dark acrid feces that accumulates on the dorsum as a deterrent to predators. Larvae feed for only about 2–3 weeks. There are two resting stages (prepupa and pupa) that occur on foliage. Adults appear about a week after pupation. They are active feeders and persist for some time. In California five to seven generations occur each year. The thrips overwinters in the egg stage.

Although it has wings, the adult greenhouse thrips is not a strong flier. Movement from place to place therefore proceeds slowly. We have seen an area of damage in a planting of the ground cover *Hypericum* grow daily, as the peripheral plants came under attack by the advancing infestation of thrips.

References: 30, 666

A. Field view of greenhouse thrips injury to Grecian laurel. Commonly, only one section of a large plant is injured by this pest.
B. Close-up of greenhouse thrips damage to Grecian laurel.
C. Injury to viburnum by greenhouse thrips. Note bleached leaves and varnishlike spots of excrement.
D. A leaf of a rhododendron damaged by greenhouse thrips.

A
B
C
D

Thrips (Plates 208–209)

Thrips are relatively small insects, 1–2 mm long, and unique among insects because of their narrow, nearly veinless wings fringed with relatively long hairs (D, E). They are weak fliers but aerially dispersed by drifting in wind currents, often for many miles. The few that are troublesome in the temperate United States arrive in the summer from a tropical or subtropical site; a few species, however, hibernate in the North. Species associated with North American ornamentals injure flowers, stipple or bleach foliage (Plate 207), or cause structural abnormalities in the form of callus tissue (Plate 209) galls, leaf malformation (Plate 209), leaf fold, leaf roll, or leaf blisters.

On deciduous trees the pear thrips, *Taeniothrips inconsequens* (Uzel), is probably the most notable. Adults are slender, brownish black, and up to 1.5 mm long. They appear in spring (February in California; April and May in the Northeast) and feed on the tender tissues of opening flower and vegetative buds. Shortly thereafter, they lay their eggs in flower stems and leaf petioles. Small brown scars develop soon after eggs are laid. Both larvae and adults feed on foliage by piercing and rasping the epidermis, allowing plant juices to escape to the surface where the fluid is sucked into the mouth. Injury symptoms are expressed as tattered, distorted, dwarfed, and mottled foliage; and sometimes defoliation. Blisterlike scars develop along the veins and petioles. Larvae are pale white and translucent. Feeding is finished by early June. Pupation occurs in the ground, where the insect remains over winter. There is one generation per year. Control of the pear thrips is particularly challenging because adults appear before foliation begins and because they enter the developing buds to feed. The North American population consists of females only; they reproduce parthenogenically. Shade tree hosts of pear thrips include maple, birch, beech, ash, and cherry. They are of major economic concern to apple, pear, cherry, and plum growers. Pear thrips occur in most of the northeast quadrant of the United States from West Virginia to Maine, New Brunswick, and Quebec, plus all of the Pacific Coast states and British Columbia.

Thrips calcaratus Uzel, the introduced basswood thrips, is a European species that was first found and identified in the United States (i.e., New York) in 1925. It took nearly 60 years to recognize it as a serious pest of basswood, *Tilia americana*. The damage was found in Wisconsin. In 1986, 260,000 acres of basswood was defoliated there.

Little is known about the biology of this insect in either Europe or the United States, and it cannot yet be explained why the American *Tilia* is so sensitive to its feeding. While basswood is not used as an ornamental, the potential for this thrips to injure the several ornamental European linden cultivars concerns nurseryworkers and those responsible for landscape maintenance.

Like most other northern species of thrips, *T. calcaratus* overwinters in the soil. In Wisconsin adults emerge in late April to early May, a period that coincides with basswood bud break. If the host plant tissue is in the proper condition, the adults begin to feed immediately. Even a small amount of feeding on buds may kill them and cause them to drop. The female lays eggs along the main rib on the underside of the leaf. Eggs hatch and larvae appear in early June. Feeding by the larvae and adults causes the newly opened leaves to drop. The combination of bud damage and leaf damage may lead to total defoliation. Because defoliation occurs only in the spring and refoliated trees are unaffected, researchers have inferred that only one generation of thrips appears each year.

The assumed distribution of *T. calcaratus* is from New England west to all of the Great Lake states. It has been collected from *Acer, Carya, Fagus, Fraxinus, Malus, Ostrea, Quercus,* and *Sambucus* species. At present, serious damage is limited to basswood. The larvae and adults possess no features that would allow for positive field identification.

California privet, *Ligustrum ovalifolium*, and regel privet, *L. obtusifolium regelianum*, are not among the more valuable ornamental plants. They are, however, commonly used for hedges. They are attacked by two pests that are frequently overlooked: the privet thrips, *Dendrothrips ornatus* (Jablonowski), and the privet rust mite, *Aculus ligustri* (Keifer) (Plate 231E).

The larvae and adults of the privet thrips cause conspicuous chlorotic flecks and a dusty grayish appearance to their hosts' leaves (A). The privet thrips number from 20 to 30 per leaf when abundant, but are larger than the mites and cause more damage per individual. The adult privet thrips (D) is winged and is an active flier. Little is known of its life history and seasonal development. Adults lay eggs on the leaves in late spring, and several generations develop during the summer until plants are rendered unfit for feeding by the thrips. The immature forms are spindle-shaped, light yellow, and wingless (C). They occur primarily on the undersides of the leaves. Adults are about 1 mm long. Larvae are smaller but two or three times larger than the eriophyid mites. Thrips are spindle-shaped, whereas eriophyid mites tend to be carrot-shaped or wormlike (see Figure 135, accompanying Plate 232).

The redbanded thrips, *Selenothrips rubrocinctus* (Giard), is a tropical species found throughout Florida. In south Florida it breeds continuously and generations overlap. It can overwinter as an adult and therefore may be able to survive in south Georgia and along the Gulf Coast.

Eggs of the redbanded thrips are inserted singly under the epidermis of the undersurface of leaves. Each egg site is covered with a drop of excreta that hardens and provides protection. Larvae and adults feed on foliage by rupturing the epidermis with their mouthparts, killing the cells at the feeding site. Where these thrips are abundant, premature defoliation occurs and much honeydew is produced.

Only immature thrips have the red bands (F). The adult female has a dark brown to black body that is about 1.2 mm long. Its host plants include both fruit and shade trees and numerous woody ornamental shrubs and vines, such as avocado, lychee nut, tung, acacia, persimmon, sweetgum, Brazil pepper tree, and pyracantha.

The adult thrips shown in panel E is *Frankliniella tritici* (Fitch), the flower thrips. This species is the most abundant and widely distributed thrips in the United States. It has many overlapping generations. Under favorable conditions a complete life cycle may occur in 2 weeks. Thrips feed by rasping the soft plant tissue and sucking up the plant fluid. This causes necrotic spots or blotches on flower petals (G) or rough callous tissue on leaves (Plate 209H). Immature thrips are yellow and are usually hidden under flower petals. The western flower thrips, *Frankliniella occidentalis* (Pergande), has been a pest in California for about 100 years and is now well established in eastern United States and discontinuously across the country. It is primarily a pest of herbaceous ornamentals but has become a problem on certain field crops such as cotton. It is potentially a threat to late or continuously flowering shrubs such as rose.

Thrips breed primarily on various grasses, weeds, clover, and alfalfa. Garden flowers and flowering woody ornamentals are generally infested by thrips that have come from these plants. *F. tritici* attracts attention when it damages rose buds. White roses seem to be more attractive to migrating thrips than those of any other color. Hawthorns are often attacked by flower thrips, which injure both flowers and flower buds. Sometimes the feeding injury is so severe that the bud will not open.

Control of thrips is difficult, especially during the migration period. In the Washington, DC, area migration occurs in May and June. Thrips are subject to predation by aphid lions, by the larvae of lacewing flies, and by other predators. Late in the growing season these predators often prevent populations of thrips from becoming larger.

References: 4, 30, 174, 466, 836, 903

A. Privet leaves showing injury caused by the privet thrips.
B. Thrips injury to *Ligustrum ovalifolium*. Note immature thrips at arrows.
C. Privet thrips larvae.
D. An adult privet thrips, *Dendrothrips ornatus*.
E. An adult flower thrips. [Thripidae]
F. A cluster of immature redbanded thrips on viburnum. (Courtesy H. A. Denmark, Florida State Dept. of Agriculture and Consumer Services, Div. of Plant Industry.)
G. A rose flower injured by thrips of the genus *Frankliniella*. White flowers are injured most severely.

A
B
C
D
E
F
G

Laurel fig (Indian laurel), *Ficus microcarpa,* particularly the popular variety *nitida,* is subject to severe injury by the Cuban laurel thrips, *Gynaikothrips ficorum* (Marchal). It has been found in the United States only in California, Florida, Hawaii, and Texas.

Severe infestations of this thrips greatly reduce the ornamental value of the host plants but apparently do not cause serious or lasting injury. Adults and nymphs feed on the tender expanding leaves, creating purplish red spots on the undersurfaces and causing foliage to severely curl or roll (A). The curled leaves are ineffective photosynthetically and soon turn yellow and drop prematurely.

In California five generations of the Cuban laurel thrips occur each year. On hot days the rapid-flying adults are active. They are black and about 1 mm long. Breeding is almost continuous. The female lays her eggs (C) on the upper surface of curled leaves. Here eggs hatch and develop with some protection from adverse weather conditions. All stages may be found on a single leaf. Under optimum conditions the complete development of one generation may occur in as few as 30 days. Populations are most abundant from October through December.

The Cuban laurel thrips is of tropical origin. The potential role of leading predators from the Carribean region are now being studied in California for control of this pest. Common native predators such as lady beetles and lacewings are not effective. Other species of *Ficus* such as *F. aurea* and *F. benjamina* appear to be secondary hosts and are attacked if the insects are extremely abundant. *G. ficorum* also has been recorded on species of *Viburnum* and *Citrus* in Florida.

The toyon thrips, *Rhynchothrips ilex* (Moulton), is of importance in California where toyon, or Christmas berry, an evergreen shrub, is used widely as an ornamental. Although the insect has been collected from a number of other plants, it is believed that it reproduces only on toyon. Damage to the host is caused by both adults and nymphs. By rasping the newly expanding foliage with their mouthparts, they effectively prevent the leaves from developing normally. The result is misshapen and badly distorted terminal growth (D, E). Damage may be so severe that parts of the leaf may die or turn black, or the entire leaf may fall. Injury by the toyon thrips is more serious in coastal locations than in inland California.

The adult thrips (Plate 208) is black, narrow, and 1 mm long with silvery white wings folded lengthwise over the abdomen. In late winter and early spring adults leave the deformed foliage they damaged the year before and begin laying eggs on new unfolding leaves. The pale yellow larvae feed on the new growth. When mature, the larvae drop to the ground, tunnel into the soil, and pupate there. Upon reaching the adult stage, these individuals return to the toyon plants, where a partial second generation develops. By late summer breeding and reproduction cease. The adults remain in hiding on the foliage until the next year.

Thrips of other species on other plants, and in other geographic areas, may cause similar injury. Several species of thrips, particularly those of the genus *Frankliniella,* cause widespread injury to the blossoms, growing shoots, and other vegetative parts of many woody ornamental plants throughout the United States. All are very small insects—less than 1.5 mm long—and look like tiny black or straw-colored slivers of wood as they move about on plants. Thrips are most numerous during late spring and midsummer. In the western states their destructiveness to ornamentals increases dramatically when uncultivated vegetation begins to dry up, forcing the thrips to migrate by flying to areas where artificial watering maintains plants in an attractive state during the remainder of the growing season. The plants susceptible to thrips injury are too numerous to list here, but rose, peony, privet, and herbaceous plants such as daisy, gladiolus, and chrysanthemum are among the favored hosts.

Generally, thrips prefer to feed in sites such as flower buds, open blossoms including pollen, vegetative growing points, and behind leaf sheaths, where a measure of protection is provided by the tightness of plant parts. When flower feeding occurs, the blossoms become streaked with discolored areas and bloom life is shortened. If growing points are attacked, the foliage, as it opens out, may be twisted (F) or otherwise distorted and may become flecked with yellow or scarred (G, H). Foliage may also become bleached or silvered owing to the feeding activities of thrips (Plate 208). Thrips cause damage by rupturing plant cells with their conelike rasping mouthparts and imbibing the plant sap that escapes.

Adult flower thrips lay eggs, inserting them into plant tissue by means of the ovipositor. The tiny immature thrips feed in the same manner as the adults. On completion of the juvenile stage, a resting stage occurs on the plant or in or on the soil below. The entire life cycle, from egg to egg, may require as few as 3 weeks for some species. As many as eight generations occur annually.

In southern regions thrips may be so abundant that they become a nuisance in and around homes. When thrips crawl on people, they may cause a minor skin irritation.

References: 4, 30, 169, 632

A. The growing tip of *Ficus microcarpa* var. *nitida* showing how the Cuban laurel thrips, *Gynaikothrips ficorum,* injures its host.
B. Close-up of the curled undersides of leaves injured by thrips.
C. An injured laurel fig leaf opened to show eggs (*circle*), and young thrips. A, B, and C all show damage done by the Cuban laurel thrips.
D. The young and growing twig of toyon injured by *Rhynchothrips ilex.*
E. Several toyon leaves showing degrees of thrips injury.
F. A *Viburnum odoratissimum* shoot with leaves injured by thrips feeding on the underside.
G. The undersides of *V. odoratissimum* leaves showing brown callus scar tissue. The callus is a response to the injury caused by thrips.
H. Close-up of the scar tissue on a single leaf. F, G, and H show injury by thrips believed to have been caused by members of the genus *Frankliniella.*

A
B
C
D
E
F
G
H

Dogwood Club Gall (Plate 210)

The dogwood club gall is caused by the larva of *Resseliella clavula* (Beutenmueller), which is a small fly, or midge. These galls may be found throughout the normal range of the flowering dogwood, *Cornus florida,* from Connecticut to Florida.

R. clavula is primarily an eastern species. Its larvae, or maggots, overwinter on the ground. They are protected by sod, decayed grass, and duff. Pupation occurs in the spring, and in Connecticut adults emerge from the pupae in June. They emerge much earlier in the Carolinas and Georgia. The midges are about 2 mm long and are delicate and extremely wary. They deposit their eggs in the tiny terminal leaves. Upon hatching, the maggots work their way to newly developing shoots at the base of the petiole. Here the shoot tissue grows around the larva. The first symptom of their presence is a wilted, deformed leaf. As the maggots grow, a cavity develops in the shoot and the surrounding tissue swells, eventually producing the club-shaped gall. The galls taper at both ends and vary from less than 1 cm to about 5 cm long and may contain up to 60 maggots. As the maggots mature, they become yellow or orange (D). More than one gall may occur on a single shoot. During the late summer or early fall the maggots inside the gall make one or more exit holes and drop to the ground to overwinter.

A heavy infestation of clubgall midges may seriously stunt a tree. A light infestation in a large tree will hardly be noticed. Galled twigs with their terminal leaves usually die prematurely. The leaves on the galled twig may cling to the tree for a year or more. Most of the flower and leaf buds that develop beyond the apex of the gall die, thus greatly reducing the number of flowers the following spring.

A small chalcid parasite of the maggot has been identified as a species of *Platygaster.* The degree of control caused by this parasite is unknown.

Reference: 677

A. Twigs of *Cornus florida*. Galls were caused by the gall midge *Resseliella clavula*.

B. A dead twig with galls that have remained on the tree for several years.

C. A fresh gall cut open to show the developing maggots. When fully grown, maggots are about 2 mm long.

D. Close-up of the maggots. [Cecidomyiidae]

A
B
C
D

Bud, Shoot, and Stem Galls (Plate 211)

There are 30 species of cynipid gall wasps in the United States and Canada that belong to the genus *Diplolepis*. All of them induce galls on *Rosa* species, and all the organs are attacked by one species or another, causing galls on leaf, stem, bud, or root. All species of *Diplolepis* appear to be parthenogenic. Larvae and their secretions stimulate tissues of the host plant to form a gall; each gall has characteristic size, shape, and structure that allow for identification of the gall maker.

The mossyrose gall wasp, *Diplolepis rosae* (Linnaeus), causes a large spherical hairy mass (D), which is 25 mm or more in diameter, to form on year-old rose twigs in the spring. The galls are light green and mosslike at first. Later they turn brown. Inside each gall is a cluster of hard kernellike cells from which the mosslike filaments originate. Each cell contains the larva of a tiny gall wasp. Adults emerge the spring following gall development. One generation occurs each year.

The long rose gall (C), formed by *D. dichlorcera* Harris, is spindle-shaped and is covered with many spines. The gall ranges from 12 to 50 mm long; its diameter may reach 20 mm.

Mossyrose galls and long rose galls are commonly found along the East Coast, but they also occur in western Ontario. In Utah, Idaho, and Oregon the gall wasp *D. bassetti* (Beutenmueller) causes a mossy, filamentous gall that is much smaller than the mossyrose gall.

The spiny rose gall wasp, *D. bicolor* (Harris), causes the gall shown in panel E. Galls range from 5 to 10 mm in diameter, grow in clusters, and originate on leaf primordia. The gall has a rather thick wall with a spherical cavity inside where one larva develops.

The insect overwinters as a mature larva in the gall, which falls to the ground before the first frost of the season. In early spring, about the time rose leaf buds break, wasps chew their way out of the old gall and search for expanding leaf buds, where they lay one to several eggs. The gall is initiated after the egg hatches. When the larva begins to feed, gall tissue grows rapidly around it, and in 2 months the larva is fully grown. This insect occurs coast to coast in southern Canada and the northern United States.

Oriental chestnut trees have become substitutes for the American chestnut. In the mid-1970s *Dryocosmus kuriphilus* Yasumatsu, sometimes called the oriental chestnut gall wasp, was found in Georgia and since then has become a threat to the chestnut industry in the southeastern United States. In time, it may be a threat to every oriental chestnut tree planted in urban or suburban yards. The adult is about 3 mm long and its body is dark brown. Males have never been found. Female wasps emerge from old galls late in May or early June (Georgia) and lay clusters of eggs inside the new but dormant buds. Eggs hatch in late July, but the larvae grow very little until the following spring. When the infested bud breaks (March), larvae induce the formation of a somewhat spherical gall that, upon reaching full size, may be 15 mm in diameter; it is smooth and strawberry colored. Fully grown larvae inside the gall are whitish, legless, and about 2.5 mm long. Leaves that develop at the tip of the gall are dwarfed or grossly distorted. Trees with severe infestations lose their vigor and often die. Old galls are brown and remain on the tree for up to 2 years.

Spindle-shaped swellings along the stems of *Ceanothus* species are caused by *Periploca ceanothiella* (Cosens), the ceanothus stem gall moth. This pest has a stunting effect on the new growth, since galls may number more than 20 per linear foot of growth. More serious damage, however, occurs on the bloom, which may be reduced to about one-fourth its normal size. Some twigs may be killed as a result of the injury. This pest is of particular concern in California. Its distribution in the United States probably corresponds to the distribution of *Ceanothus* species. Specimens of this insect have been reported, however, in Kansas, New York, Ontario, and Texas as well as in California and Oregon. Research data from Munro (557) in southern California have shown striking differences in susceptibility to stem gall moth, based on the relative abundance of galls observed on species and cultivars of *Ceanothus*:

Severely infested	Not infested
Ceanothus griseus	*C. americanus*
C. griseus horizontalis	*C. cuneatus*
Moderately infested	*C. foliosus*
C. cyaneus	*C. gloriosus*
C. thyrsiflorus	*C. gloriosus exaltatus*
C. cv. Marie Simon	*C. greggi perplexus*
C. cv. Ray Hartman	*C. jepsonii*
Lightly infested	*C. impressus*
C. arboreus	*C. insularis*
C. diversifolius	*C. masonii*
C. integerrimus	*C. megacarpus*
C. lemmonii	*C. papillosus*
C. leucodermis	*C. parryi*
C. lobbianus	*C. prostratus*
C. oliganthus	*C. purpureus*
C. cv. Concha	*C. ramulosus fascicularis*
C. cv. Mary Lake	*C. rigidus albus*
C. cv. Mountain Haze	*C. spinosus*
C. cv. Royal Blue	*C. verrucosus*
C. cv. Sierra Blue	C. cv. Blue Cloud
C. cv. Treasure Island	C. cv. Lester Rowntree

The ceanothus stem gall moth overwinters in the larval stage in its gall on the plant. Pupation and moth emergence from galls take place during the spring and early summer. Eggs are laid on the plants, and newly hatched larvae tunnel directly into the buds and flower inflorescences, causing the formation of galls. There is never more than one larva in each gall, and only a single generation of the ceanothus stem gall moth occurs each year.

References: 233, 557, 600, 692–94

A. A *Ceanothus thyrsiflorus* twig showing the galls and the resultant dwarfed and dead foliage.

B. A gall cut open to show the larva and pupa of *Periploca ceanothiella*.

C. The long rose gall, formed by *Diplolepis dichlorcera*. [Cynipidae]

D. The winter coloration of the mossyrose gall, which is caused by *D. rosae*.

E. A cluster of mature spiny rose galls caused by *D. bicolor*. (Courtesy Reis Collection, Entomological Society of America.)

A
B
C
D
E

Oak Leaf and Twig Galls (Plates 212–216)

Wherever oaks occur, they are attacked by a group of small insects called gall makers. These insects cause deformities, known as galls, of various shapes, sizes, and colors on leaves, twigs, bark, flowers, buds, acorns, and even the roots of the tree. The galls are better known than the insects that produce them. The majority of gall makers that attack oak are wasps, but in some cases flies are responsible.

Galls are caused by powerful plant growth-regulating chemicals or other stimuli produced by the insect that react with plant hormones. The inner walls of the gall are rich in protein and thus provide the larvae residing inside the gall with an abundance of concentrated food. The larvae are somewhat, but by no means completely, protected from parasites and predators by the abnormal plant tissue that surrounds them.

Each gall maker incites its own distinctive gall, which is unlike those caused by other species. It is from the shape, location, and appointments of these galls that the insect that produced it can most easily be identified. Some galls of oak are globular or dish-shaped, whereas others look like thorns or spiny balls. The size may vary from 1 mm to over 50 mm. Galls are so commonly associated with oaks that many people regard them as typical structures of the plant. Some of the earlier botanical drawings of oaks actually show galls as representative of the normal plant. Galls are specific as to the kind of oak on which they occur. For example, those that are found on a member of the black oak group do not occur on white oaks.

A characteristic of many gall wasps is a phenomenon known as heterogamy, or alternation of generations. The offspring of a gall wasp may produce galls quite unlike those produced by the parent but identical with those made by their grandparents. Also, the alternating generations of gall wasps often make their galls on different parts of the oak—for example, leaves and buds, or flowers and leaves. Individual wasps of the two generations often differ in body structure to the extent that they appear to be two distinct insect species. In fact, the 717 species of gall wasps reported to occur on oaks in North America undoubtedly number far fewer than that, but biologists have yet to unravel the identity of many pairs of alternating generations of gall wasps. As these complicated pairs are identified, a single species name must be assigned.

An example of alternating generations occurs in the gall wasp *Dryocosmus dubiosus* (Fullaway), which attacks California live oak, *Quercus agrifolia* (Plate 214). The insect overwinters on the ground in fallen galls. In the early spring the wasps emerge, fly to the trees, and lay eggs in the male flowers. Instead of a normal flower, a tiny saclike gall forms in its place. Development proceeds quickly within this gall, and the adult soon emerges and lays her eggs in the veins of the leaves. Some 3 months later a tiny green gall with a blunt spine on each end—known as the two-horned oak gall—appears on the lower leaf surface. The gall becomes brown when mature and drops to the ground, where the insect overwinters. Leaves appear scorched during years when *D. dubiosus* is present in large numbers, for the process of egg deposition or gall formation apparently disrupts normal leaf functions. The gall wasp life cycle is repeated every year (see Figure 101, accompanying Plate 214). It would be very difficult for the casual observer to conclude that the flower gall had any connection with the leaf gall or leaf scorch, yet the same species of insect caused all of this damage.

The oak apple is familiar to most children who have found and popped open the dry galls, some of which are as large as apples. Actually, several species of gall wasps cause oak apples. A common one in the East is *Amphibolips confluenta* (Harris), which has two distinct generations, one consisting of only females and the other of both sexes. The large galls are attached to the midrib or petiole of a leaf and are filled with a spongy mass that is easily sliced with a knife. However, the single larva in each "apple" is inside a small and very hard center seedlike cell (Plate 215D). When the gall dries, the spongy interior becomes a mass of fibers radiating from the central cell to the now thin and papery shell of the gall. This gall does no measurable harm to its oak host.

The jumping oak gall, caused by the gall wasp *Neuroterus saltatorius* Hy. Edwards, is a tiny globular seedlike gall on the underside of California white oak leaves (D). The galls, each of which contains a single larva, drop to the ground when they have matured. The activity of the insect inside actually makes the gall jump repeatedly a few centimeters off the ground. The reason for this movement is not entirely clear, but it has been suggested that those individuals that succeed in locating themselves in a crevice in the soil are more likely to overwinter successfully than those that do not. These galls found on oaks in the West may at times become so numerous that they cause serious discoloration (C) or premature loss of leaves.

After overwintering, a population of female *N. saltatorius* wasps emerges from the galls. Eggs are then laid in the newly opening buds. A few weeks later blisterlike swellings appear on the young white oak leaves. These are galls of wasps formerly known as *N. decipiens,* but which are now recognized as the alternating generation of *N. saltatorius* (Plate 214A). Wasps that emerge from these blisterlike galls are both males and females. After mating, females lay their eggs in the leaves and galls appear about 40 days later.

The question sometimes arises about why certain oaks bear large numbers of galls while adjacent trees of the same species have very few. In studies on *N. saltatorius* and a few other gall wasps it was found that trees whose buds opened earlier than those on nearby trees were much more heavily galled. Trees without opening buds afford the wasps no opportunity to deposit their eggs. Gall makers must attack the plant at a very precise phenological moment if normal plant tissue is to be successfully stimulated to form a gall.

The majority of oak leaf galls do little or no harm to their host. However, twig galls may cause severe injury and bring about the demise of the tree. The horned oak gall, caused by *Callirhytis cornigera* (Osten Sacken), and the gouty oak gall, formed by *C. quercuspunctata* (Bassett), may be particularly injurious or even fatal to shade trees. *C. cornigera* occurs from southern Canada to Georgia and attacks the twigs of pin, scrub, black, blackjack, and water oak. *C. quercuspunctata* is found in the same general geographic area but infests twigs of scarlet, red, pin, and black oak. The galls of these two species may be as large as 50 mm in diameter. They often grow side by side to form a mass half a meter long that extends along the length of a small branch. They are solid and woody, with many larval cells near the center.

Horned and gouty oak galls are similar except that horns protrude through the surface of the former. These horns develop the second or third year after the gall maker's eggs are laid, when the larvae inside are nearing their full size (Plate 213C). One adult wasp emerges from each horn. The biology of the insects that form these galls is complicated. The following is a biological account of the horned oak gall in western New York.

In May and June tiny female wasps emerge from the fully developed twig galls. These females lay eggs in the larger veins on the undersides of leaves. Eggs hatch, and the larvae cause tiny galls that develop in the veins. The galls appear as oblong blisters and are noticeable from late May through June. In early July this wasp generation matures, and males and females emerge from these galls. After mating, the females lay eggs in young oak twigs. Galls of this generation do not usually appear until the following spring. Two years, or more, are required for the wasps living in the woody galls to mature. The branches of a heavily infested tree may droop from the weight of these knotty excrescences (see Figure 125, accompanying Plate 213; Plate 213D).

Besides the gall maker, a surprisingly large number of organisms are found in insect-caused galls. In one study of the live oak woolly leaf gall maker, *Andricus laniger* Ashmead (Plate 215C), 16 species of insects were associated with its galls. They included five species of wasps that were parasites of the gall maker larvae, two that were inquilines inside galls, and a small caterpillar that fed on the gall tissue.

One of the common fly-caused galls on oak, known as the vein pocket gall, is made by *Macrodiplosis quercusoruca* (Felt). This midge

---

A. Galls of the cynipid wasp *Andricus kingi* Bassett on leaves of valley oak. Galls occur also on blue oak and Oregon oak.

B. Close-up of galls shown in A. White galls redden as they mature. Galls here are about natural size.

C. Upper leaf surface of valley oak infested with galls of the cynipid wasp *Neuroterus saltatorius*. Galls are on undersides of leaves.

D. Close-up of the underside of a leaf infested with galls of *N. saltatorius*. Galls magnified about 5 times. (Also see plate 214A.)

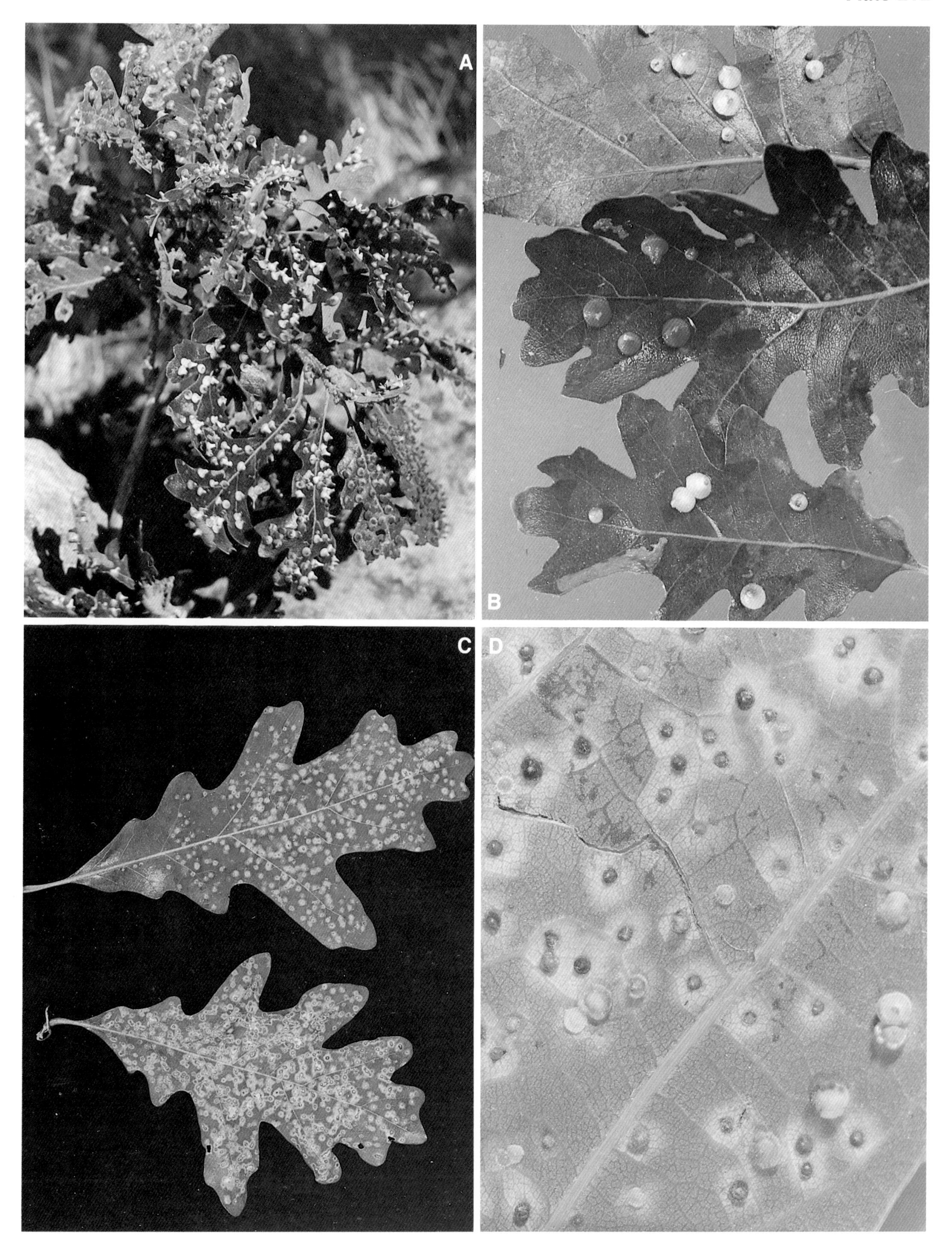
A
B
C
D

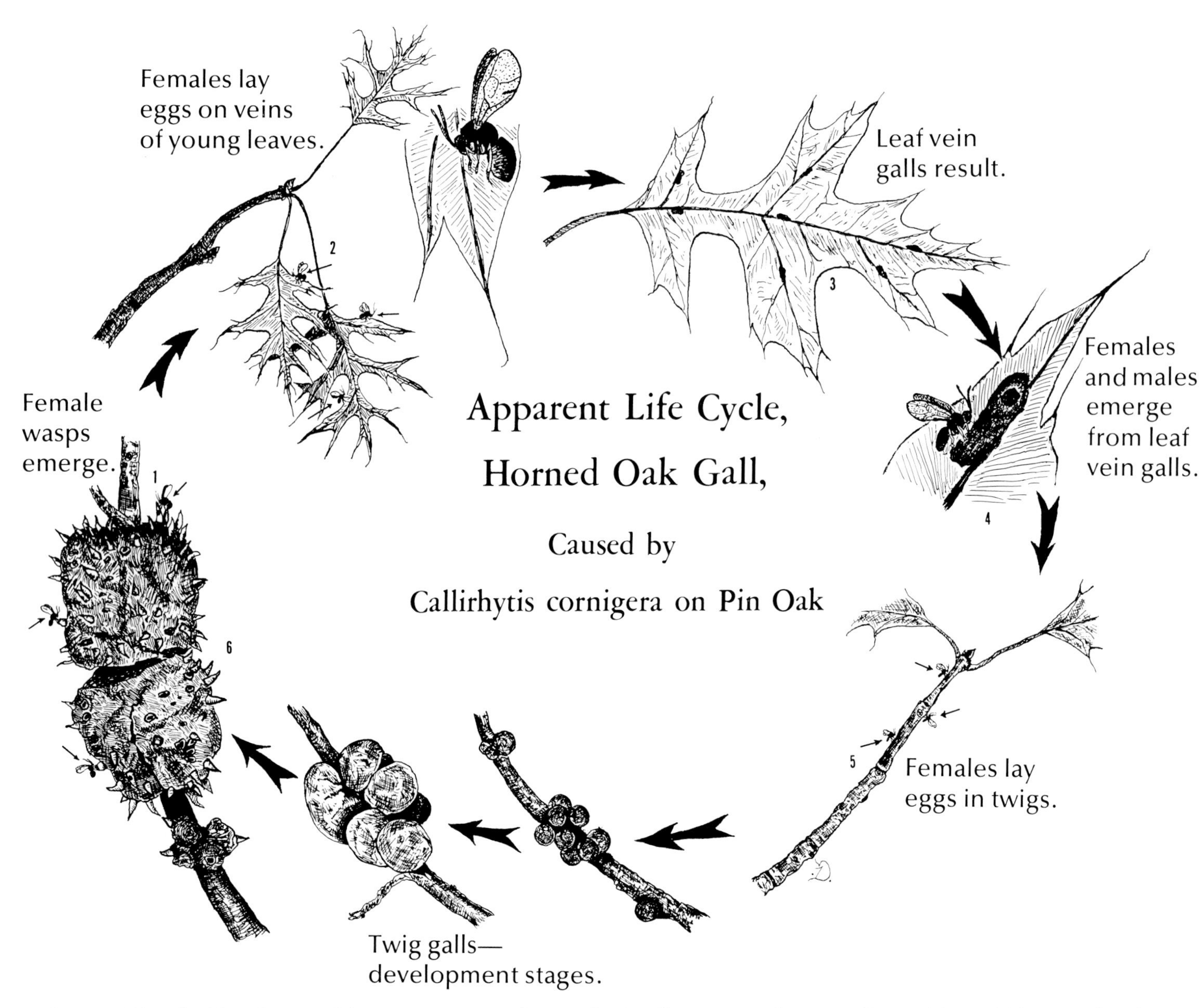

Figure 125. Life cycle of the gall wasp *Callirhytis cornigera* on pin oak. In May female gall wasps emerge from twig galls (*1*) and fly to oak foliage (*2*), where they deposit their eggs in the leaf veins. The larvae cause vein galls (*3*) to form. During midsummer the insects in the vein galls mature and emerge as male or female wasps (*4*). After mating, the females lay eggs in oak twigs (*5*). Twig galls form slowly at the place where eggs were laid, reaching maximum size (*6*) in about 2 years. (Drawing by R. Dirig.)

occurs in the eastern half of the United States, where it causes elongate, pocketlike swellings on the lateral veins and midribs of scrub and pin oak leaves (Plate 215A, B).

The insect attack begins when the newly unfolding leaf begins to flatten out. After the gall midge eggs hatch, the larvae migrate to mid and lateral veins where, in a few days, gall tissue grows around each larva. When fully grown, the larva is whitish and about 2 mm long. Development is complete by midspring, when larvae emerge from the gall, drop to the ground, and diapause there until the next spring.

M. erubescens (Osten Sacken) causes a marginal fold or leaf pocket on pin and red oak. Several larvae feed inside the tubular roll and swollen area. The galls are narrow, about 3 mm, and vary from 5 to 25 mm long.

References: 187, 219, 233, 317, 399, 457, 495, 662, 663, 669, 830, 831

A. Young horned oak galls on pin oak twigs. Photograph was taken in the spring.
B. Gouty oak galls on pin oak twigs. One gall has been cut open to show the developing larvae.
C. Mature horned oak galls with horns. Where horns are missing, the adult wasps have emerged.
D. A pin oak tree heavily infested by galls. Photograph was taken in the winter.
E, F. Twigs of black oak, *Quercus velutina*, showing cells of the gall wasp *Callirhytis crypta* Ashmead (*right*). Bark has been removed to expose cells.

A
B
C
D
E
F

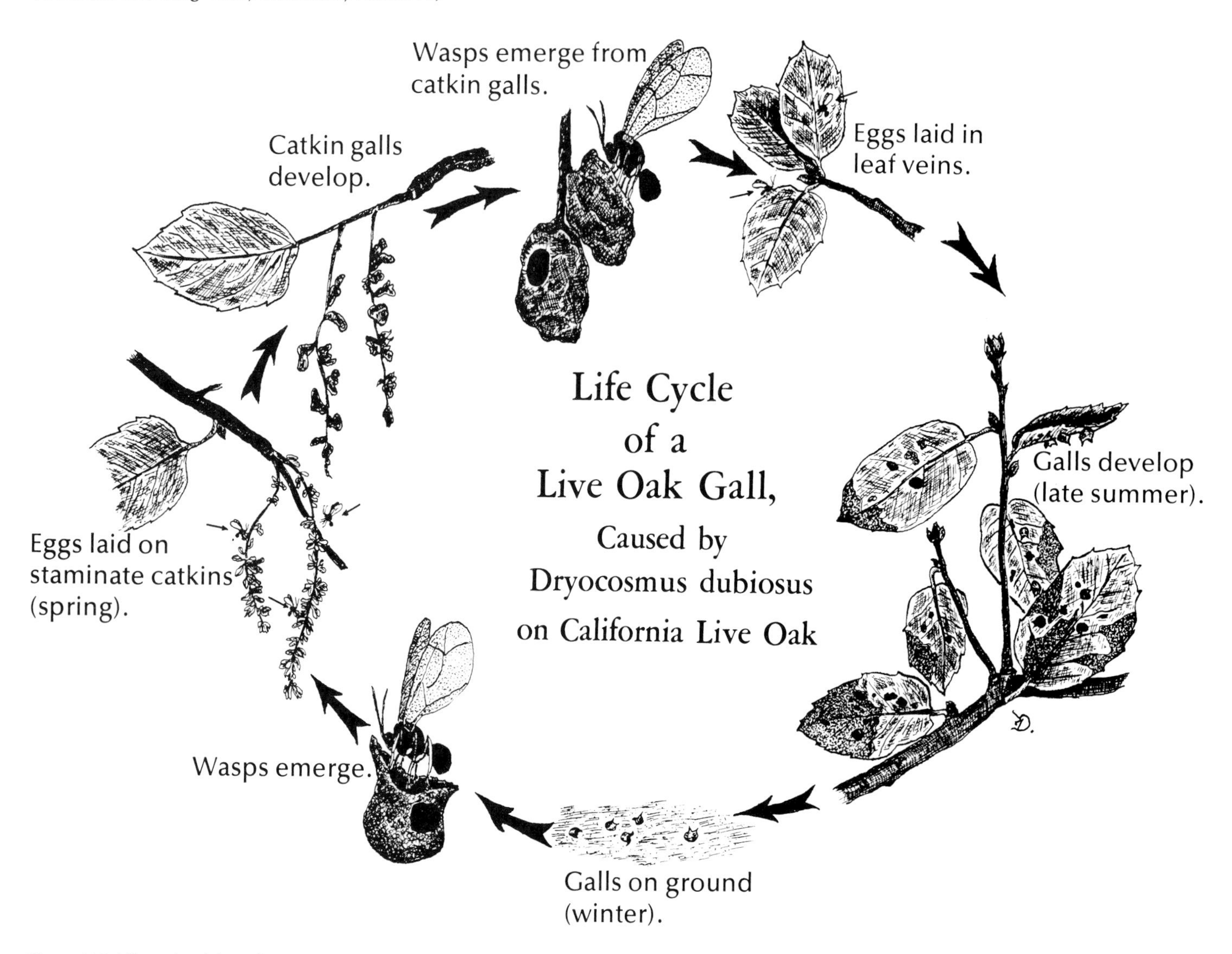

Figure 126. Life cycle of the gall wasp *Dryocosmus dubiosus* on California live oak. This species goes through one complete cycle, involving two generations, each year. (Drawing by R. Dirig.)

Figure 127. Coast live oak twig galled by *Bassettia herberti* (Weld), a western gall maker. Cross section (*right*) shows wasp cells in the wood beneath the bark. Birds have shredded the bark (*left*) to feed on the wasps in the cells. (Courtesy C. S. Koehler.)

A. Galled and deformed valley oak leaves. Galls were caused by the cynipid wasp *Neuroterus saltatorius* (Edwards) (*decipiens* form). (Also see Plate 212C, D.)
B. The wool sower gall, caused by *Callirhytis seminator* Harris, on white oak. An eastern gall. *Left:* wool pulled apart to show seedlike structures, each of which contains a gall-making insect.
C. The hedgehog gall, formed by *Acraspis erinacei* Beutenmueller, on upper and lower surface of a white oak leaf. Three stages of growth are shown. An eastern gall.
D. Coast live oak leaves with symptoms of injury by the gall wasp *Dryocosmus dubiosus*. A western gall.
E. Undersides of leaves from the same tree shown in D.

A
B
C
D
E

Figure 128. Midge galls on oak foliage. *1.* Veinfold galls on *Quercus alba*: (*above*) lower surface; (*below*) upper surface with gall opened to reveal larvae of *Macrodiplosis niveipila*. *2.* Marginal leaf rolls on *Q. vaccinifolia* (northwestern United States) caused by *Contarinia* sp. *3.* Leaf galls on *Q. rubra*, lower surface: (*a*) hairy vein pocket caused by *M. ?niveipila*; (*b*) smooth, thin-walled pocket with lateral cross section showing larva of *M. majalis*; (*c*) short, striate vein pocket caused by *Contarinia* sp.; (*d*) marginal leaf roll caused by *M. erubescens*; (*e*) smooth vein pocket caused by *M. qoruca*. (From R. J. Gagné, *The Plant-Feeding Gall Midges of North America* [Ithaca, N.Y.: Cornell University Press, 1989]; drawing by Deborah Leather Roney. Courtesy R. J. Gagné.)

A. Gall on pin oak, sometimes called the vein pocket gall. It is caused by the midge *Macrodiplosis quercusoruca*.

B. Gall tissue removed to show the maggots (larvae) of *M. quercusoruca*. [Cecidomyiidae]

C. Live oak woolly leaf galls on leaves of *Quercus virginiana*, caused by the cynipid wasp *Andricus laniger*.

D. Oak apples on red oak. Left gall has been cut open to show the larva of *Amphibolips confluenta* in central cell. The gall turns brown as it ages.

A
B
C
D

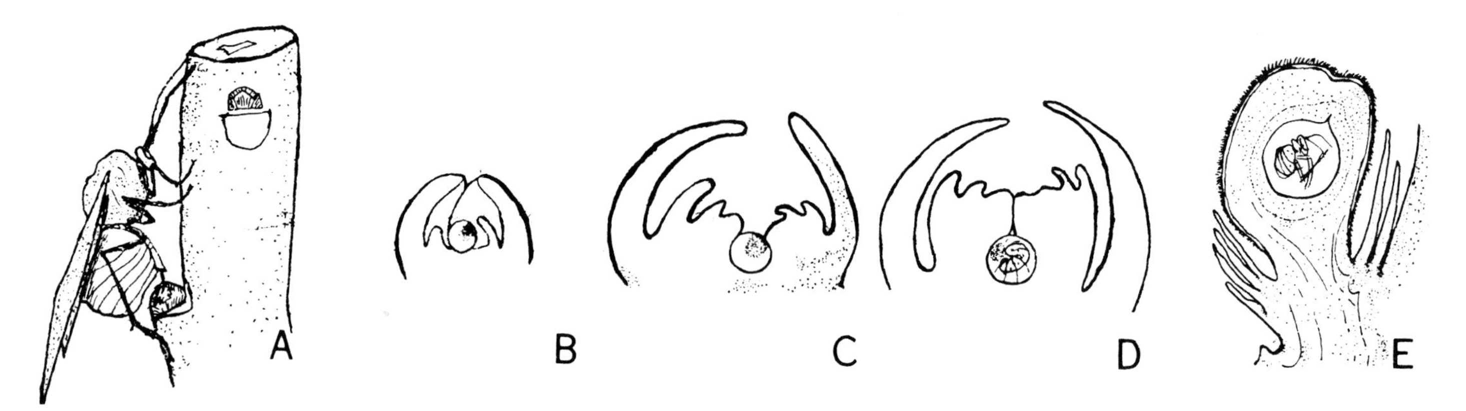

Figure 129. The sequence of bud gall development: *A,* an adult gall wasp lays an egg in an immature bud; *B,* the position of the egg inside the bud scales; *C,* as the bud grows the plant cells enclose the egg; *D,* the egg hatches and the larva develops, stimulating the growth of abnormal plant tissue; *E,* gall tissue has now replaced the bud and the insect will soon cut a channel to the outside. (Drawings modified from G. S. Winterringer, Illinois State Museum Story of Illinois Series, no. 12, with permission.)

A. Valley oak with spined turban galls on undersides of leaves, caused by the cynipid wasp *Antron douglasii* (Ashmead), a western species. Galls illustrated are about natural size. Galls also occur on blue oak and California scrub oak.

B. A woody twig gall on California black oak, caused by the cynipid wasp *Callirhytis perdens* (Kinsey). A bird has shredded parts of the gall on the right to get at the insects inside. A western gall. [Cynipidae]

C. A spined turban gall, caused by *A. douglasii,* enlarged about 3 times.

D. A red oak tree with most of its leaves replaced by "roly-poly" galls.

E. A woolly leaf gall on *Quercus engelmannii,* caused by the cynipid wasp *Andricus fullawayi* Beutenmueller.

F. Close-up of galls on the tree in D.

A
B
C
D
E
F

Aphid and Psyllid Galls (Plate 217)

The witch-hazel leaf gall aphid, *Hormaphis hamamelidis* (Fitch), and the spring witch-hazel bud gall aphid, *Hamamelistes spinosus* Shimer (Plate 140), are common on witch-hazel, *Hamamelis* species, and their alternate host, birch (*Betula* spp.). Both overwinter in the egg stage on the twigs of witch-hazel.

Hormaphis hamamelidis could also be called the witch-hazel cone gall aphid. It causes conical galls on the upper sides of leaves early in spring, when young nymphs begin feeding. A single aphid initiates gall formation. The aphid matures rapidly and is contained within the developing gall. The adult female produces great numbers of young that pack the hollow interior of the gall. When the young mature, some develop wings and then emerge through the opening or orifice (C) on the undersides of leaves. These aphids fly to the foliage of birch in late spring and early summer. Others remain wingless and stay on the witch-hazel plant, presumably overwintering on twigs. These aphids are black and oval in outline, looking more like a scale crawler than a typical aphid. Large numbers of this morph may be seen in August on the undersides of galled leaves, where they continue to feed. Several wingless generations occur during the summer. With the approach of autumn a winged generation develops and migrates back to witch-hazel. Eggs are laid on the twigs, where they spend the winter and the life cycle is completed.

The cone galls are green (B), sometimes tipped with red. They are about 1 cm high and are fully developed by June (somewhat earlier in the latitude of Washington, DC). The galls persist on leaves for the remainder of the season but are empty. The galls are unsightly but are not notably injurious or deleterious to the trees. Feeding by these aphids does not stimulate abnormal growth reactions or gall formations on birch.

So far as is known, all of the horticultural cultivars of witch-hazel are equally susceptible to attack by these aphids. *Hormaphis hamamelidis* occurs in the northeastern quadrant of the United States from Illinois to the East Coast, south to North Carolina, and north to Canada. Recent studies (unpublished) have shown that the life cycle of this species varies according to geographic elevation. In the higher elevations of Maryland and Virginia birch is not a required secondary host.

The galls shown in panels D–G were found in Louisiana on sugarberry, *Celtis laevigata*. They are caused by a psyllid in the genus *Pachypsylla*. Little is known of the biology, habits, and occurrence of species in the genus in the Gulf states.

For more discussion on psyllids, see the text for Plate 218.

Reference: 607

A. A witch-hazel branch with leaf galls caused by the aphid *Hormaphis hamamelidis*.
B. Galls on the upper leaf surface of witch-hazel.
C. The gall opening on the undersurface of a witch-hazel leaf.
D. Sugarberry leaves with crater-shaped *Pachypsylla* galls on the upper surface.
E. Crater-shaped galls. View is of undersurface of the leaf.
F, G. Apple- or quince-shaped galls on sugarberry. View is of underside of leaf. In F the top of a gall has been removed to expose a young *Pachypsylla* nymph.

A
B
C
D
E
F
G

Hackberry Galls (Plate 218)

Hackberry trees and shrubs are not planted as frequently as most other woody plants. They are used extensively as windbreaks and shade trees in the Midwest and as an occasional specimen tree in landscape plantings in the East. Wherever they occur, hackberry psyllids are also likely to be found. Adults look much like miniature cicadas. The adult psyllid is 4–5 mm long and has hind legs adapted for springing into flight from a resting position. For this reason insects in the family Psyllidae are also called jumping plant lice.

Ten species of *Pachypsylla* have been reported to attack hackberry. The three most common species are the hackberry blister gall maker, *P. celtidisvesicula* Riley; the hackberry nipple gall maker, *P. celtidismamma* (Riley); and the hackberry bud gall psyllid, *P. celtidisgemma* Riley. The life cycle and habits of the hackberry blister gall maker and the hackberry nipple gall maker are similar. The bud gall psyllid, however, has a different life cycle.

The hackberry blister gall (F) measures 3–4 mm in diameter and is only slightly raised from the leaf surface. The hackberry nipple gall (J) is about 4 mm in diameter and nearly 6 mm high. Both species overwinter as adults in the crevices of rough bark, or—all too often—inside houses. One generation of these insects occurs each year. Mating and egg laying occur over a period of weeks, beginning when new leaves unfold from the buds. Eggs hatch in a week to 10 days, and nymphs begin to feed on the leaves. At the feeding site rapid and abnormal plant growth takes place on the leaf, and a pouch or gall forms around the nymph(s). They live inside the galls throughout the summer and emerge as adults in September. The emerging adults can be extremely annoying to people in late summer. They sometimes alight by the hundreds on cars, buildings, and even on objects in homes near large hackberry trees. The host trees apparently do not suffer seriously from the galls, although severe infestations over a period of years may weaken them. Hackberry blister gall is most common on tree forms of hackberry, whereas the nipple gall tends to be more prevalent on shrubby types. The nipple gall has been reported to cause premature leaf drop on tree forms in Illinois.

The same trees infested with leaf galls may also support a large population of hackberry bud gall psyllids. *P. celtidisgemma* adults occur for 2–3 weeks during the latter part of June. Eggs are laid on the leaves and nymphs crawl to the newly formed buds where gall formation occurs. By fall, nymphs have molted to the fifth instar, which remains in the gall throughout the winter and early spring. Both the leaf and bud-inhabiting psyllids feed on plant cells with their piercing-sucking mouthparts. Even when bud galls are abundant, damage to host plants is not severe. A sufficient number of buds generally remains uninfested for the tree to maintain its vigor.

Parasites are common and important in the natural control of hackberry psyllids. The chalcid wasp, *Psyllaephagus pachypsyllae* (Howard), has been reported to destroy 30% of the bud gall nymphs. Studies of the nipple gall have shown 47–51% of the nymphs may be destroyed by the internal parasites *Torymus pachypsyllae* (Ashmead), *P. pachypsyllae, Eurytoma semivenae* Bugbee, and three other less abundant species. The weevil *Conotrachelus buchanani* Schoof was found to be predaceous on psyllids. It also feeds in the galls.

Hackberry is also afflicted by 10 species of gall midges (Cecidomyiidae). The thorn gall caused by *Celticecis spiniformis* (Patton) is the most widespread gall on hackberry. The gall almost always occurs on the lower surface of the leaf. The shape of the gall is somewhere between a cone and a thorn and reaches a height of about 3 mm when fully developed. The base is always round. Galls are present on leaves for about 3 months but drop out of the leaf about a month before normal leaf fall. When the larvae are full grown in summer, they spin a cocoon within the gall to overwinter. Adults emerge the following spring and begin the cycle again.

References: 134, 815

A. Psyllid eggs on the undersurface of a newly developed hackberry leaf.
B. Hackberry psyllid eggs magnified about 10×. Note eyespots of embryo (*arrows*).
C. Newly emerged nymphs before initiation of gall formation.
D. An adult female psyllid.
E. An unidentified twig gall on hackberry.
F. Hackberry blister galls on upper leaf surfaces.
G. Lower surfaces of same leaves shown in F.
H. The beginning of gall formation around the hackberry nipple gall maker, *Pachypsylla celtidismamma*.
I. A petiole gall on hackberry caused by a psyllid.
J. Galls caused by the hackberry nipple gall maker on the underside of a hackberry leaf.

A
B
C
D
E
F
G
H
I
J

Yaupon Psyllid Gall (Plate 219)

Feeding by many psyllid species causes galls to form on their host plant. *Gyropsylla ilicis* (Ashmead), for example, is responsible for a leaf gall on yaupon, *Ilex vomitoria,* a holly that the southeastern Indians at one time used ceremonially. Because it is the only known gall-making psyllid on yaupon, this pest is easily identified.

There is one generation of *G. ilicis* each year. In northern Florida adults begin to emerge from the galls in November, and peak numbers occur in February. Eggs are laid in small clusters when the buds of yaupon have opened enough to show green leaf tissue. Oviposition usually ceases sometime in April. When eggs hatch, the nymphs begin to feed immediately and rarely move from the point where they initially insert their stylets. Galls form rather rapidly and soon cover the clusters of nymphs. The nymphs require about 10 months to complete their development. Gall size varies greatly; the largest are about 14 mm long. Likewise, the number of nymphs per gall is variable. The inner surface of the gall is covered with white waxy filaments that appear to be woven together. One or more ball-like structures, slightly larger than a nymph, are found inside the gall. They are covered with white waxlike material. If a ball is punctured, clear fluid will run out and the ball will collapse. This material is probably honeydew excreted by the psyllids but kept contained in an ingenious way.

This insect occurs along the coastal plain from North Carolina to Florida and along the Gulf Coast to Louisiana.

Reference: 529

A. Yaupon twigs with galls on the youngest leaves caused by *Gyropsylla ilicis.*

B, C. Close-up of several galls.

D, E. Galls cut open to show the young psyllids. The arrow in D points to a ball of honeydew.

F. An adult *G. ilicis*; about 5 mm long. [Psyllidae]

A
B
C
D
E
F

Psyllid Gall of *Persea* (Plate 220)

Trioza magnoliae Ashmead is a common inhabitant of Florida and occurs in Georgia, Alabama, and Mexico. It has also been reported from New Jersey. Mead (522) summarized the known information on this species in 1963. This psyllid was described by Ashmead in 1881 and discussed in 1914 by Crawford (134) in a monograph on New World species of Psyllidae.

The principal host of *T. magnoliae* is redbay, *Persea borbonia*. It has also been found on other species of *Persea*: silkbay, *P. humilis*; swampbay, *P. palustris*; and shorebay, *P. littoralis*. Large unsightly galls form on the leaves but apparently cause little damage to the tree, because a large proportion of leaves remain unaffected. The galls are about 25 mm long and are green. They are, however, covered with a whitish to bluish bloom (A). Galls first form on the leaf margins and as they grow cause the leaf blade to roll and the midvein to curl (B). The psyllids are enclosed in the pouchlike gall. The gall's hollow interior also contains shed "skins," white waxy material, and drops of honeydew. When the galls mature, a slit opens along the side of the gall, permitting the adult psyllids to emerge. In Florida nymphs have been found from April to June, and in November and December. The majority of adults emerge in mid-May.

Adults look like miniature cicadas. They are about 4 mm long and are greenish to light brown. Sometimes brown stripes appear on the thorax. Little is known about the biology of this insect.

References: 134, 522

A. Young bay leaves with galls caused by the psyllid *Trioza magnoliae*. Such symptoms are often found by late March in central Florida. The galls shown in this photograph range from 4 to 11 mm long.
B. Leaf malformations on redbay caused by *T. magnoliae*.
C. Newly developing galls as they appear in late March.
D. Appearance of redbay leaves infested with *T. magnoliae*, a psyllid gall maker.
E. A gall cut open to reveal nymphs. [Psyllidae]

A
B
C
D
E

Laurel, Eugenia, Ceanothus, and Pepper Tree Psyllids (Plate 221)

The laurel psyllid, *Trioza alacris* Flor, is an Old World insect. It was first reported in the United States in Oakland, California, in 1911. As nearly as can be determined, this species was introduced into America on nursery stock that originated in Belgium. In Europe it feeds on Grecian laurel, *Laurus nobilis,* and English laurel, *Prunus laurocerasus,* yet in the United States only the former host is attacked. This is surprising in view of the fact that English laurel is far more common in California than is Grecian laurel. The laurel psyllid occurs in the New World only in California and New Jersey.

Damage to *L. nobilis* foliage is characterized by inward-rolled leaf margins that eventually form a gall within which the immature psyllids, called nymphs, develop (C). Some thickening of the affected leaf portions also occurs. When newly formed, the galls are the same color as the unaffected leaves. Later, however, they turn reddish and eventually become brown or black as the leaf tissue dies (B). Defoliation of part or all of the plant has been reported in severe infestations. Some foliage may turn black because sooty mold fungi colonize the honeydew produced by the insects. In addition, the cast "skins" of the nymphs persist on the foliage and contribute to the unsightliness of infested laurel plants.

T. alacris overwinters as an adult on the plant. In early spring egg laying begins on newly formed terminal leaves. Eggs are pale yellow when newly deposited and are covered with a grayish white powder. Like most psyllid eggs, those of *T. alacris* are attached to the leaf surface by a slender tube called a stipe. This appendage is necessary to maintain a proper moisture balance in the egg; incubation will not proceed if the eggs are removed from the leaf surface and transferred to another substrate.

On hatching, the tiny flattened nymphs insert their mouthparts into leaves to withdraw plant sap. This feeding causes foliage to curl and thicken; eventually a discrete gall is formed into which the nymphs move to continue feeding. Nymphs are yellow to orange, and their bodies are covered by white cottony wax. At maturity, they leave the galls, attach themselves to the surface of an open leaf, and transform into winged adults. The laurel psyllid has several generations each year, the last one maturing in the fall. During most of the growing season, all stages of the psyllid are commonly found at the same time on the same tree.

Another *Trioza* psyllid, *T. eugeniae,* attacks eugenia, *Eugenia myrtifolia,* in California. A native of Australia, the insect was first discovered in Los Angeles County in 1988. Eugenia is a waxy-leaved ornamental shrub commonly grown as a hedge or small tree in zones that are normally frost-free.

Female eugenia psyllids deposit elongate yellow eggs on the new leaves and shoots of eugenia. Eggs hatch into nymphs that crawl to and settle on the undersides of leaves, where they feed by sucking sap. Feeding causes a pit to develop around each yellowish, scalelike nymph, and so many of these pits may occur on a single leaf that it becomes severely blistered and distorted. The eugenia psyllid reproduces year-round in coastal California. Populations are higher on the coast than inland, where high temperatures kill the immature insects.

On several species of *Ceanothus* small white flocculent spheres are often associated with the leaves. These structures indicate the presence of another psyllid, *Euphalerus vermiculosus* Crawford. The cottony white material is formed by the nymphs. With rare exceptions, they occur on the undersides of the leaves. Waxen threads, produced by the psyllid, curl around the nymph and cover its body. A cell or cavity is formed inside this mass by the movements of the nymph. As the insect molts, its cast "skin" becomes entangled in the waxen threads and becomes a part of the flocculent covering. The waxy mass is not attached to the leaf except in a small area that surrounds the feeding site. When the nymph has completed feeding, it leaves the cottony mass and develops into a free-flying adult. Noticeable injury to *Ceanothus* species does not result from an infestation by *Euphalerus,* although the flocculence on leaves is unsightly.

In 1984 a psyllid native to South America and new to North America was discovered in Long Beach, California, attacking the California pepper tree, *Schinus molle.* Over the next several years the insect, regionally known as the pepper tree psyllid, *Calophya rubra* (Blanchard), became widely distributed coastally both in southern and northern California. The Brazilian pepper tree, *S. terebinthifolia,* is not affected.

The pepper tree psyllid reproduces throughout the year. Feeding by the nymphs causes formation of pit galls similar to those shown in Plate 217D. These craters usually occur on the lower leaf surface, but the upper leaf surface, petioles, and small green shoots also are affected. Nymphs develop within the pits. Severe infestations cause discoloration and distortion of affected plant parts, cessation of growth, and loss of foliage.

References: 133, 214, 217, 287, 892, 893

A. Field view of a Grecian laurel plant showing an infestation of the laurel psyllid.
B. Galled leaf margins and discoloration of leaves caused by the laurel psyllid.
C. Severely galled terminal of *Laurus nobilis.* Note nymphs (*arrow*) of *Trioza alacris* inside the partly opened gall.
D. Close-up of rolled leaf margins and cast "skins" of last-instar psyllid nymphs.
E. A twig of *Ceanothus gloriosus.* The conspicuous white cottony material on the leaves is produced by the psyllid *Euphalerus vermiculosus.*
F. Cast "skins" of *T. alacris* nymphs on a leaf of Grecian laurel.

A
B
C
D
E
F

Petiole and Leaf Stipule Galls (Plate 222)

The phylloxerans are closely related to the aphids. Twenty-nine species in the genus *Phylloxera* cause galls on *Carya* species. One of the common galls (A) and perhaps the largest (up to 25 mm in diameter) is caused by *Phylloxera caryaecaulis* (Fitch), sometimes called the hickory leafstem gall phylloxera. The globose galls form primarily from the cambial tissues in shoot bark and petioles of hickory, with an occasional gall forming on the midvein of a leaflet. The gall maker can be identified by the size and shape of the gall and the location of the exit orifice (A). The gall, which becomes hard and woody, often remains on the tree for several years.

The life cycle is completed in 1 year on a single tree. The overwintering egg hatches into a stage called the fundatrix, or stem mother. In central Indiana eggs begin to hatch in mid to late April, and the young fundatrices begin to feed at or near the swelling bud. Small pits form around each insect, and by early May the tiny insect is completely enclosed by green gall tissue. The gall is fully formed by mid-May. While the gall is growing, the cavity inside is being filled with a thousand or more eggs. These hatch, and by the end of May the insects are mature, at which time the galls split open and the winged adults emerge. Adults move to the underside of leaflets and lay yellow eggs. These eggs hatch to become males and females; they mate, and the female deposits a single amber-brown egg inside old galls or in crevices on the surface of the gall. All of the cycle to this point has occurred in about 2 months. No further development occurs until the following year.

P. caryaecaulis is found from West Virginia, west to Nebraska and Ontario, and south to Florida. *P. devastatrix* Pergande, *P. russellae* Stoetzel, and *P. notabilis* Pergande are economically important gall formers on pecan. *P. devastatrix* attacks woody shoots and spurs; *P. russellae* and *P. notabilis* cause galls on foliage. The leaf galls produced by *P. russellae* are circular and flattened, with a diameter of about 5.6 mm; they cause a reticulated pattern on the leaflet surface between secondary veins. The gall opening on the undersurface of the leaflet is marked by dense, short, white pubescence. When the gall is mature, the opening breaks into two to five bracks. Defoliation occurs when infestations are heavy. The seasonal history of *P. russellae* is much like that of other phylloxerans. Figure 130 illustrates the sexupara form that develops inside the gall. When mature, it is golden yellow with gray appendages. All feeding and reproductive activities are completed by the first of June. There is one generation of the gall maker per year. This phylloxeran occurs from Georgia west to Oklahoma and Texas.

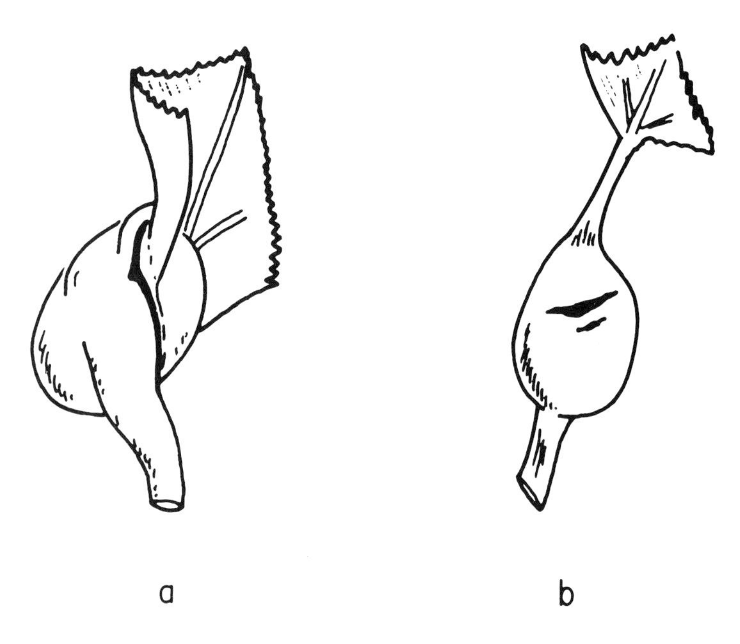

Figure 131. A typical *Pemphigus populicaulis* gall (a) involves both petiole and blade and has an oblique hiatus; *P. populitransversus* galls (*b*) form entirely from the petiole and have a transverse hiatus. (From Faith, ref. 225, courtesy *Journal of the New York Entomological Society* and D. P. Faith.)

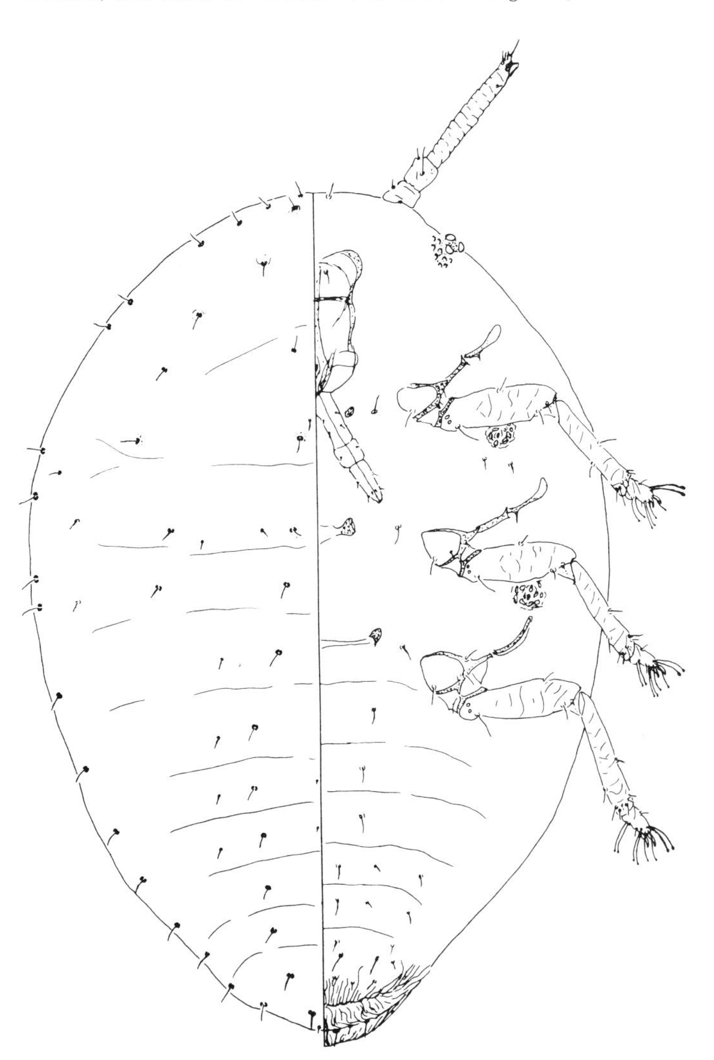

Figure 130. A generalized drawing of a mature sexupara of *Phylloxera russellae*. The artist's midline separates the dorsal half (*left*) and the ventral half (*right*). Length, about 800 μm. (From Stoetzel, ref. 763, courtesy M. B. Stoetzel.)

Pemphigus populicaulis Fitch and *P. populitransversus* Riley are gall-forming aphids common to eastern cottonwood (*Populus deltoides*) and plains cottonwood (*P. sargentii*). These aphids induce globose galls on the leaf petiole and differ only in the location of the slit (Figure 131) that opens to allow asexual forms to fly to their summer hosts. No leaf blade distortion results from galls formed by *P. populitransversus*.

The life cycle of both gall-making species is essentially the same and takes place in one calendar year. Eggs, the product of sexual union, are laid in autumn in crevices of exfoliating bark. They hatch in the spring, and nymphs feed on newly developing leaf petioles. In a few days the upfolding of gall tissue completely encloses the stationary fundatrix, or

---

A. Galls on hickory leaves and shoots caused by the leafstem gall adelgid *Phylloxera caryaecaulis*. Galls may be found on petioles and midveins of leaflets. Arrow points to exit orifice.
B. Dried, nutlike galls at the base of a petiole of *Caryae ovata* caused by a *Phylloxera* species.
C. First-instar nymphs (stem mother stage) of *Pemphigus betae* on a leaf of *Populus angustifolia*. Photo taken at the time of bud break. (Courtesy Thomas Whitham.)
D. A pecan leaf gall cut open to show the huge population of phylloxerans inside.
E. A *Populus fremontii* branch with leaf galls caused by the aphid *Pemphigus populicaulis*.
F. A *P. populicaulis* gall cut open to show adelgids inside and a syrphid fly maggot feeding on the adelgids. (Courtesy Thomas Whitham.)

A
B
C
D
E
F

stem mother. The fundatrix gives birth to young, parthenogenically, while confined inside the gall in a hollow chamber. These young, when mature, have wings. During midsummer a slitlike orifice develops, allowing the winged aphids to escape and fly to the roots of plants in the families Cruciferae or Compositae, where they produce several generations. In late summer winged aphids develop on the secondary host, whereupon they fly back to a species of *Populus*, the primary host. Mating occurs and the female lays the eggs that overwinter.

P. populitransversus occurs in 30 states, Alberta, and Manitoba but not west of the Continental Divide. *P. populicaulis* occurs in 20 states, British Columbia, Alberta, and Quebec.

The poplar vagabond aphid, *Mordvilkoja vagabunda* (Walsh), produces large galls on *Populus* species that transform the meristem of leaf stipules into a dense irregular mass up to 12 cm in diameter (Figure 132). They are most noticeable in late summer after leaves drop. Like most other leaf galls, they arouse much biological curiosity. They cause little besides aesthetical injury to the tree.

Galls develop in early spring (in mid-May in Minnesota), after a newly hatched fundatrix begins to feed on a leaf stipule. Gall growth continues as the aphid feeds, and soon the insect is completely covered with gall tissue. Galls at first are bright red, become yellow-green, and finally turn dark brown at maturity. The interior of the gall contains a large central main chamber and many small peripheral chambers, all filled with asexually produced young. *M. vagabunda*'s development on its primary host is completed by early July. The winged adults then migrate to a secondary host plant, *Lysimachia* species, members of the primrose family, where they feed on leaves and roots. In North Carolina migration from the galls begins about May 1 and ends about July 1. The return migration to poplar occurs from September to November. These returning individuals give birth to males and females. The sexual forms mate, and each female lays one egg coated with waxy filaments. Eggs from several females are often grouped together in old galls. Old galls may remain attached to the tree for 2 or 3 years.

Figure 132. Sequence of gall development from the stipules of a cottonwood leaf; caused by the vagabond gall aphid. (From Ignoffo and Granovsky, ref. 402, courtesy Carlo Ignoffo.)

The known distribution includes Maine, Connecticut, Ohio, Illinois, Minnesota, Iowa, Missouri, Nebraska, Kansas, Colorado, and Wyoming.

References: 100, 225, 335, 401, 402, 709, 762, 763

Eyespot Galls (Plate 223)

The most colorful and conspicuous leaf gall on red maple, *Acer rubrum*, is caused by a gall midge *Acericecis* (=*Cecidomyia*) *ocellaris* (Osten Sacken). The original description likened the gall to an eyespot. Only on rare occasions is the gall abundant enough to cause injury, and it is illustrated and discussed here only because it attracts attention and arouses curiosity about its natural history.

The gall midge emerges from the soil early in May (Long Island) and lays its eggs in leaf tissue on the undersides of leaves. Here the larva develops quickly, causing a slight swelling or elevation on the upper leaf surface. In response to the gall maker and the growth-regulating hormone it injects into the leaf, the plant develops bright red and yellow rings around the gall (A, B). The color is most intense in early June and over time turns brown. The larva (C) grows rapidly and completes its development in 8–10 days. When ready to pupate, it vacates the gall, drops to the ground, and burrows into the soil. There is one generation each year. According to Gagné (277) fresh galls occasionally occur in midsummer and may also appear on sugar, silver, and striped maple. This insect probably occurs throughout the range of red maple. The damage at the feeding site may superficially look like that of a leaf mine. The larva lives in a pit and does not change its position. It is attached to the plant by its mouthparts and by the surface tension of plant secretions. The total leaf area affected is 8–10 mm in diameter.

Other tree species are also subject to eyespot galls produced by related cecidomyid midges. For example, such galls occur on dogwood (*Cornus* spp.) (E) and are made by *Parallelodiplosis subtruncata* (Felt). There is one larva (maggot) per gall, which like *A. ocellaris* develops rapidly. The circle of abnormal color in the leaf covers a diameter of about 6–8 mm. This insect also behaves like a leaf miner; its presence can be detected by holding the leaf in front of a light source. Several generations probably occur every year. Collection records available to us show that its distribution is limited to New York, Maryland, and Minnesota, but we suspect that it is more widely distributed. *Resseliella liriodendri* (Osten Sacken) makes an eyespot on the leaves of tuliptree (F). In the center of the spot is a dark, rather spherical gall of much greater density than surrounding tissue. The gall tissue has fooled many diagnosticians into thinking it a fungus fruiting body. As the leaf and spot age, the necrotic tissue may drop out, leaving a round hole. The spot measures from 4–7 mm in diameter. *R. liriodendri* has several generations each year. Full-grown larvae force a slit in the epidermis of the leaf and drop to the duff or soil to pupate. Very little is known about the biology and life history of these eyespot gall makers.

References: 231, 277, 495

A, B. Eyespot galls on red maple, *Acer rubrum*. Galls were caused by *Acericecis ocellaris*, a midge. Three color phases are shown. Each spot is about 8 mm in diameter.

C. The lower epidermis of a leaf removed to expose the larva of *A. ocellaris*. Larva about 2 mm long. [Cecidomyiidae]

D. The undersurface of a red maple leaf showing the opening of the maple eyespot gall.

E. Eyespot gall on a *Cornus florida* leaf caused by the midge *Parallelodiplosis subtruncata*.

F. Eyespot gall on a tuliptree leaf caused by the midge *Resseliella liriodendri*.

A
B
C
D
E
F

Leaf Galls of Ash and Elm (Plate 224)

Galls are pathologically developed plant tissues induced by a parasitic organism. The stimulus is believed to be a growth-regulating chemical produced by the parasite that causes the enlargement and proliferation of cells. If the parasite leaves the host or dies, normal cells are again produced, but, once formed, gall tissue does not change. Each gall-making species causes an excrescence structurally different from all others. Thus one can identify the gall-making organism without actually seeing it, by observing the structure of the gall and by noting the species of the host plant.

About 95% of the known types of galls of the world are caused by nematodes, mites, and insects. The remaining 5% are caused by bacteria, fungi, and viruses. Galls may occur on any of the actively growing parts of plants, from root tips to the growing points of shoots. The facing plate shows leaf galls induced by an aphid, a fly, and a mite. Plates 210–227 show examples of some of the most common galls of woody plants.

The galls in panel E are elm cockscomb galls caused by the elm cockscombgall aphid, *Colopha ulmicola* (Fitch). In many respects these galls do resemble a cockscomb. The galls are not all the same size or shape, but in general they are about 25 mm long, 6 mm high, and are irregularly gashed and toothed. Newly matured galls are reddish along the sides and top, especially if they have exposure to the sun. As the galls age, they turn dull brown and may become dry and hard. On the underside of the leaf there is a long, slitlike opening. If this orifice is opened when the gall is still fresh, one female aphid and a number of her young will be found. The young aphids are able to leave the gall through the orifice and are sometimes seen crawling about on the outside of the leaf. These aphid nymphs are pale white, with blackish legs and antennae. *C. ulmicola* is one of the woolly aphids closely related to *Eriosoma* species. It has two hosts: elm, where it causes the leaf gall in the spring, and grasses, where it feeds underground during the summer on roots. Its life history is complicated.

Elm cockscomb galls occur on American elm, *Ulmus americana,* and *U. fulva* wherever these trees grow. A similar cockscomb-type gall occurs on elm foliage in the West and is produced in the spring by the stem mothers of the pear root aphid, *Eriosoma languinosa* (Hartig). The summer host is pear. These aphids do little harm to the plant but attract attention because of the leaf galls they produce. In the spring *Eriosoma pyricola* Baker & Davidson (Plate 145), another woolly aphid, causes a spiral gall on the foliage of American elm. The aphid migrates to pear roots during the summer. It is found in the Pacific Coast states and British Columbia.

Eriophyes ulmi Gar, a mite, produces a gall (D) on leaves of American elm. The presence of the galls causes some curling and deformation of the foliage. The gall always occurs on the upper leaf surface but has an opening on the underside of the leaf. The mites feed and reproduce inside the gall, migrating to new growth as new leaves develop. The mite illustrated in panel D is the early summer form. Many eriophyid mites have a summer form that is annulate and wormlike, and a winter form much coarser in structure. In either form they cannot be seen readily without magnification because of their very small size (0.2 mm). This gall is likely to be found wherever elms grow. If numerous, it can disfigure the leaves somewhat but is seldom seriously injurious. The gall midge *Dasineura ulmi* (Beutenmueller) makes a leaf gall similar to the one illustrated in panel D.

The ash midrib gall midge, *Contarinia canadensis* Felt, causes a disfiguring leaf gall (A–C) on American ash, *Fraxinus americana.* The galls vary in size from 10 to 30 mm. They are succulent and thick walled, and each has a small cavity that is occupied by one or more white maggots.

The adult midge of this species lays its eggs in the vein of very young ash leaflets in the spring. The gall forms soon after the egg is laid. Little is known about the biology and habits of this insect or how it overwinters. Galls produced by this midge occur on forest as well as shade trees. They are, however, not very common. As with most galls, their unique appearance is often cause for concern of tree owners and others curious about unusual growths on plants. *C. canadensis* is known only from the eastern half of the United States and Canada.

References: 140, 231, 233, 275, 399, 495, 710

A–C. Ash midrib gall caused by the midge *Contarinia canadensis.* [Cecidomyiidae]

D. Elm leaves with bladder-type galls caused by the mite *Eriophyes ulmi.* [Eriophyidae]

E. An elm cockscomb gall, caused by the aphid *Colopha ulmicola,* on American elm. The newly mature gall has a reddish "comb." [Aphididae]

A
B
C
D
E

Honeylocust Pod Gall (Plate 225)

Honeylocust trees, especially the thornless trade-name cultivars such as Imperial and Shademaster, are being widely planted in various parts of the United States. At one time, they were thought to be relatively free of insect pests. Now, however, it is known that several insect and mite species damage these shade trees.

The honeylocust pod gall midge, *Dasineura gleditchiae* (Osten Sacken), causes leaf deformation. Leaflets become globular or podlike (C), and within each podlike gall are one to several small, whitish (or slightly yellowish) larvae that are about 6 mm long (E). The larvae mature and pupate within the pods and then emerge as small, delicate flies or midges that are about 3 mm long. The male is generally black. The female is black with a red abdomen. Flies first appear when new growth begins in the spring. The female lays microscopic kidney-shaped, lemon-colored eggs in the young leaflets. The larvae hatch in a day or two and immediately start to feed, at the same time causing the leaf to develop into a pod. In Connecticut from five to seven generations may occur in a single year. The insect apparently over-winters as an adult somewhere outside the pod.

Galled leaflets may dry up and drop prematurely. Repeated attacks may cause death of small branches, but new shoots usually form at the base of the dead twigs. The trees are not killed, but they lose their ornamental qualities in yard or street tree plantings.

The pod gall midge is now believed to be transcontinental and is likely to occur wherever honeylocust shade tree cultivars are grown. The cultivar Sunburst is particularly susceptible to injury. See Plates 223, 226, and 227 for other cecidomyiid gall midges.

Numerous woody ornamental plants act as hosts for various species of midge larvae. Some of their larvae (maggots) act as intermediate gall makers on leaves (Figure 133), intermediate in that the plant tissues are not grossly modified, but maggot feeding may cause the leaf margins to fold, curl, or blacken and sometimes become distorted.

Reference: 676

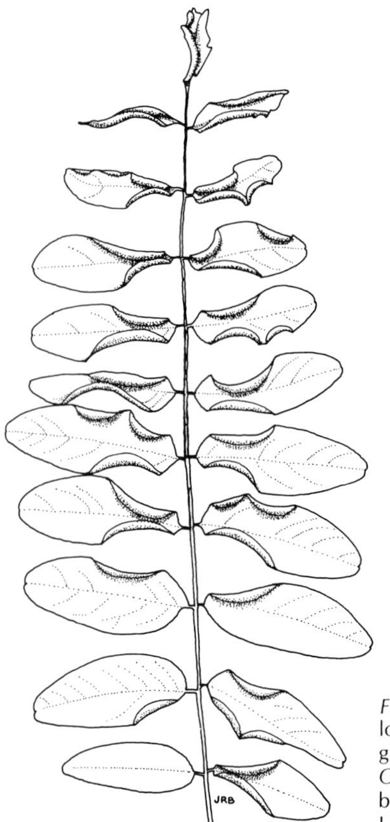

Figure 133. Compound leaf of black locust with leaflets showing marginal rolls caused by a gall midge, *Obolodiplosis robiniae.* (Drawing by J. R. Baker, North Carolina State Univ. Cooperative Extension.)

A, B, D. General damage caused by the honeylocust pod gall midge, *Dasineura gleditchiae.* [Cecidomyiidae]

C. Well-developed podlike galls replace the leaflets.

E. A gall opened to reveal the maggotlike larvae.

A
B
C
D
E

Midge and Sawfly Galls of Poplar and Willow (Plate 226)

Mayetiola rigidae (Osten Sacken) is a gall-forming fly whose common name, willow beaked-gall midge, gives a fair description of the abnormal plant structure that it causes (A). Its feeding and hormonelike chemicals in its saliva contribute to the conversion of a bud into gall tissue. The amount of injury caused is probably insignificant unless there are several succeeding years of high gall midge populations. Extensive injury can change the shape of the tree in much the same way excessive pruning does.

The willow beaked-gall midge produces one generation each year. In Michigan adults (B) emerge in April, mate, and lay eggs on swollen willow buds. A female lays a single egg on a bud but may visit and lay eggs on many buds. Eggs hatch and the larvae penetrate the soft bud tissues. The gall begins to develop in mid-May. At that time the gall is barely discernible as a swollen leafy bud. By early July the gall has attained over 85% of its growth but is still shrouded with green leaves. During July the gall hardens, turns red, and drops most of the attending leaves. The gall is fleshy with a hollow cylinder or chamber in the middle where the larva develops. The larva (in its third instar) attains its full growth by September, at which time it is about 4.5 mm long. It overwinters in this stage and pupates inside the gall in mid-March.

This insect occurs on several willow species but is most troublesome on pussy willow. It is most likely to be found in the northern half of the United States and the bordering provinces of Canada.

Several sawflies of the genus *Pontania* cause fleshy, subspherical galls on the leaves of willow. The sawflies may occur wherever willows grow, but the species will vary depending on the geographic region. *Pontania proxima* (Lepeletier) is thought to be an eastern species. It causes bright red excrescences (C), which often occur in two parallel rows, one row on either side of the midvein. The gall is 7–10 mm long and projects through both surfaces of the leaf.

Pontania californica Marlatt and *Phyllocolpa parva* (Cresson) are western species. The galls produced by these species are of about the same size and color as those of *Pontania proxima*. Little is known about the biology of these sawflies. The amount of damage caused by them is negligible.

Insects in the sawfly genus *Euura* make galls in leaf petioles or in shoots and twigs of willow. They are very host specific and may even show preference for certain individual willow clones. Except for the adult, all life stages are found in the gall. The galls are technically different from most other galls described in this book because they are largely engendered by the parent and not by the immature. During oviposition colleterial fluid is injected along with eggs through the ovipositor into meristematic tissue; this initiates gall formation. The gall continues to grow so long as the larva remains active.

Panel D illustrates two types of shoot galls caused by two species of *Euura* from Colorado. The galls to the left, tapered at both ends, are caused by *E. exiguae* Smith. The mature galls are uniformly brown to russet and 20–70 mm long. Adults emerge from old galls, and in California oviposition begins in early March and continues through April. These galls are common from the Rocky Mountains west to California. The galls on the right were caused by *E. scoulerianae* Smith. The gall varies from 20 to 40 mm long. Each *Euura* gall contains one insect.

Another fly, *Rhabdophaga strobiloides* Walsh, produces the willow cone gall (E). The adult fly looks like a gnat. It appears in late April or early May and deposits a single egg in the swelling terminal bud. When the egg hatches, the bud ceases to develop into a normal shoot with leaves. Instead, the bud continues to get larger and larger. Because the gall asserts apical dominance over the lateral buds, most nutrients go to support gall tissue. It reaches a maximum length of 25 mm by early July. The larva remains in a cavity inside the gall through the winter, and in early spring it transforms to a pupa. A fully developed fly emerges shortly thereafter. Other species of cecidomyiid flies may occupy the old gall, feeding as saprophytes on aging gall tissue. The gall occurs on several species of willow throughout the northeastern and north central states, and as far south as Maryland and Virginia. Distribution records are not complete. Injury to the host plant is minimal. Control may be effected by removing and destroying the galls before the end of summer.

Figure 134. A, Shoot galls, at an early stage of development, caused by the poplar twig gall fly, *Hexomyza schineri. B,* Galls on poplar twigs; galls are most apparent when the tree is dormant. *C,* Full-size galls. Note the pupa nearly out of the gall. (Photos courtesy Whitney Cranshaw, Colorado State Univ.)

Although the poplar twig gall fly, *Hexomyza schineri* (Giraud), cannot properly be called a midge, it is a gall maker. It belongs to the family Agromyzidae, members of which are primarily known as leaf miners. Apparently limited to the Rocky Mountain region, it attacks poplars but is most prevalent and damaging on aspen. It overwinters within living gall tissue as full-grown yellow-green maggots. Pupation occurs in early spring in the gall, but the pupae work their way out of the gall and drop to the ground. When new shoots begin to elongate, adults emerge and mate; the female lays her eggs in the new shoot tissue. After the eggs hatch, gall formation begins (Figure 134). One to three larvae may be found in a single gall. In Colorado there is generally one generation each year.

Active galls are green but change color after the insects vacate. Galls permanently disfigure the tree but do not seem to threaten the life of any part of the tree. The poplar twig gall fly is subject to high rates of parasitism by the chalcid wasp *Eurytoma contractura* Bugbee (Whitney Cranshaw, Colorado State Univ., pers. comm.).

References: 217, 231, 495, 682, 715, 828, 858, 859

A. Willow-beaked galls, caused by the willow beaked-gall midge, *Mayetiola rigidae*. Average length of the galls is 25 mm.

B. A willow beaked-gall midge resting at the tip of a gall. Midge length about 6.5 mm. (Courtesy Malcolm M. Furniss.)

C. A fleshy willow leaf gall on *Salix fragilis* or *S. blanda*. The gall was caused by *Pontania proxima*, a sawfly. The discolored leaves at the bottom of the photograph are not related to the gall but are a symptom of a fungal disease.

D. Shoot galls caused by sawflies in the genus *Euura*. [Tenthredinidae] (Courtesy D. A. Leatherman, Colorado State Forest Service.)

E. A willow cone gall, caused by *Rhabdophaga strobiloides*.

A
B
C
D
E

Rhododendron Gall Midge, Rose Midge, and Other Gall Midges (Plate 227)

Rhododendron catawbiense and its many hybrids, as well as *R. maximum,* are subject to feeding injury by *Clinodiplosis rhododendri* (Felt), the rhododendron gall midge. Distorted foliage is the typical symptom of gall formation. Several kinds of damage occur. Leaves that are attacked in the bud stage (A) may be so badly injured that they die. Young leaves attacked when newly freed of their bud scales may be dwarfed and deformed (B). All infested leaves develop an inrolling of one or both margins. Damaged areas become necrotic after fully grown larvae drop to the soil. Certain stages of damage may resemble leafhopper injury (Plate 201).

The insect normally overwinters in the soil as a prepupa; pupation occurs in the spring. Midge emergence is closely synchronized with the spring phenological development of the host plant (see Plate 226B for a picture of a gall midge). Peak emergence probably occurs during full flower bloom, or as vegetative buds begin to swell. Because rhododendrons often produce second and occasionally third flushes of growth, two additional generations of midges may occur.

Eggs are most often deposited on the undersurfaces or rolled edges of leaves as soon as they are free from the bud scales but before the leaves have fully separated from each other. Eggs are often attached to each other in clutches (D). They hatch within 3 days, and the larvae seek the undersurface to begin feeding. Feeding induces a strong inrolling of the leaf margin. The inner surface of the leaf roll provides a moist, protected environment necessary for successful larval development. The larva matures in about 7 days. It then drops to the ground and burrows into loose soil where it makes a flimsy silk cocoon.

The rhododendron gall midge occurs in New York, New Jersey, eastern Pennsylvania, Delaware, Maryland, Virginia, and North Carolina. It is a native insect and is probably present throughout the natural range of *R. maximum. Asphondylia azaleae* Felt causes a bud gall in native pinxter azalea, *R. periclymenoides* (=*nudiflorum*). Infested flower buds become swollen and never produce flowers.

The rose midge, *Dasineura rhodophaga* (Coquillett), does not cause a gall but is covered here for the convenience of dealing with the family Cecidomyiidae, whose members are collectively called gall midges.

Developing larvae (maggots) of the rose midge destroy vegetative shoots, kill flower buds, and/or cause abnormal flower development. This insect overwinters as a pupa in a white silky cocoon in loose soil under plants infested during the previous season. Adult emergence the following spring depends on soil temperature, which may remain cold enough to permit roses to produce their first crop of flowers without injury. Female midges lay their eggs under the sepals of flower buds, or in opening leafbuds and elongating shoots. In warm weather the eggs hatch in 2 days, and the hungry larvae make numerous slashes in the tissue with their sickle-shaped mouthparts. From these cuts they extract sap, causing the injured tissue to die and turn brown (F) then black. Larvae may be very abundant, 15 or 20 on a single flower bud.

The creamy white larvae (G) mature in 5–7 days, reaching a length of 1.8 mm. At this time they normally drop to the ground to pupate, but in summer months pupation occasionally occurs in the injured rose tip. The midge is about 1 mm long and is rarely recognized or even seen by the average gardener. The complete life cycle in warm weather takes 12–16 days. Adults, which do not feed, live only 1–2 days. Many generations occur each year. Serious injury is noticed in late June in the Washington, DC, area, then continues throughout the growing season.

No cultivated or wild roses are immune, although some may escape infestation because of a lack of synchronous phenological development between the plant and insect.

The rose midge is a native pest but has been reported only in widely scattered areas of the northeastern quadrant of the United States, Wisconsin, and Ontario.

Many other midge species occasionally cause economic damage or mar the beauty of plants. *Prodiplosis morrisi* Gagné feeds on cottonwood and hybrid poplars from Ontario to Mississippi. Feeding prevents leaves from unfolding and eventually causes damaged leaves to turn black and drop. Terminal (shoot) growth may be severely stunted. Maggots of the hollyberry gall midge, *A. ilicicola* Foote, feed within the carpellary cavities of American holly fruit and cause the berries to remain green during autumn and winter. The larva overwinters in the injured fruit. An unidentified fungus associated with the maggot forms a part of its diet and also contributes to the lack of berry color. Distribution of *A. ilicicola* appears to be limited to the natural range of American holly. There appears to be a distinct difference in cultivar susceptibility. Early-blooming trees are more subject to attack. Midges begin to emerge from the green fruit in mid-May in Delaware and lay their eggs in the ovules of open blossoms. The toyon berry midge, *A. photiniae* (Pritchard), feeds on the berries of *Heteromeles arbutifolia* and prevents the fruit from developing. The locust midge *Obolodiplosis robiniae* (Haldeman) causes black locust leaflets to roll under from the sides. Here the maggot feeds in the protection of the roll (see Figure 133, accompanying Plate 225).

Two recently described gall midges that feed on honeylocust are being observed with increasing frequency in the East. *Neolasioptera brevis* Gagné causes shoot swellings on native and all ornamental cultivars of honeylocust. The swollen gall tissue is about twice the normal shoot diameter and results in tip dieback and bushiness from the development of lateral buds. Larvae overwinter in the gall tissue and emerge as flies from late April to late May (Michigan). During this period pupal molt "skins" may be found protruding from emergence holes. Eggs are laid on the elongating shoots. Symptoms soon develop at the site of egg deposition as necrotic spots about 2 mm in diameter. Fully grown larvae are about 3 mm long. There is one generation per year.

Meunieriella aquilonia Gagné makes a spot gall on leaflets. There is one larva per gall. Larvae may be found from late June through mid-September, which suggests that there may be several generations each year (see also Plate 225).

For gall midges on conifers, see Plate 51.

References: 274, 277a, 348, 631, 672, 717, 744, 833a

A. Very young leaves of *Rhododendron catawbiense* damaged by larvae of the rhododendron gall midge, *Clinodiplosis rhododendri.* Larvae are present in the rolled leaf margins. Note swollen and necrotic areas.

B. Severe leaf damage caused by the rhododendron gall midge (contrast with normal foliage).

C. Rhododendron gall midge damage that has become necrotic.

D. Eggs of the rhododendron gall midge on the undersurface of an expanding leaf.

E. Mature larvae of the rhododendron gall midge on the outer scale of a flower bud. (All photos of rhododendron gall midge and damage courtesy D. R. Specker, Cornell Univ.)

F. Rose bush showing rose gall midge injury to leaves and shoots. (Courtesy F. Smith and R. Webb, USDA.)

G. Rose gall midge injury to rosebud. Note larva (*arrow*). (Courtesy F. Smith and R. Webb, USDA.)

A
B
C
D
E
F
G

Spider Mites, False Spider Mites, and Tarsonemid Mites (Plates 228–229)

The common name *spider mites* properly applies only to mite species in the family Tetranychidae, which, along with more than 100 other families, belongs to the order Acarina. Mites are not insects but are related to spiders, scorpions, harvestmen (daddy longlegs), and other orders, all of which make up the class Arachnida. Spider mites, like all arachnids, have four pairs of legs, except in the first post-embryonic stage, when they have three pairs. All spider mites develop through five stages: egg, six-legged larva, protonymph, deutonymph, and adult. The common name is derived from their ability to produce silk. A pair of silk glands associated with the mouthparts enables the mites to produce mats of webbing on foliage or silk strands on which they can spin down from infested leaves. Although wingless, spider mites often are wafted by convection currents or blown by the wind when they are suspended on threads of silk. The silk-producing habit varies, and some spider mites make very little webbing.

Each spider mite has a pair of needlelike mouthparts, called chelicerae, which can be thrust in and out to rupture cells of the host leaf tissue. They are not hollow, like insect stylets, but more like flexible needles that stab foliage cells. After the cells are pierced, the oral opening is placed over the wound and the fluids are sucked out. The feeding action causes fine stippling or flecking of the foliage (I). Where large populations attack, injured and chlorotic areas coalesce, and large portions of the leaf, often the entire leaf, become yellow or bronzed (A, F, H, I). It is not uncommon to find hundreds of egg shells, cast "skins," eggs, and mites on a single leaf (D).

Many generations may occur in a single season. Adult females may live for several weeks, laying many eggs daily and several hundred in a lifetime. An individual mite can develop from egg to adult in just a few days, which accounts for the very rapid build-up of extremely large and damaging populations. A single female and her progeny may occupy only 6.5 sq cm of leaf surface in a single generation. It has been observed that the accumulation of dust and dirt on the foliage tends to favor an increase in many mite populations. The decimation of mite predators by certain chemical treatments also contributes to mite population build-up. Nutritional condition of foliage, particularly high nitrogen levels, also enhances populations and suggests a potential method for population management. Severe infestations have been observed to develop on plants treated with certain insecticides, notably DDT and carbaryl, even where elimination of predators was not the cause. Reasons for this phenomenon are not yet well understood.

The very small size (0.25–1 mm long) makes spider mites difficult to see. A hand magnifier or dissecting microscope is necessary to determine if mites are present, what stages exist, and if they are alive or dead. In the field and armed with good magnification, one can with reasonable accuracy identify the more common pests, basing identification on mite size, color, location on the plant, time of year found, and the host. Though some spider mites are polyphagous, many have restricted host preferences, which aids in field identification. It should be remembered that because there are hundreds of species and species complexes of spider mites that have been described from ornamental plants, positive identification is best made by a trained acarologist, who depends on the use of a sophisticated microscope.

The major genera of spider mites found on deciduous plants include *Bryobia, Panonychus, Eurytetranychus, Eotetranychus, Schizotetranychus, Oligonychus,* and *Tetranychus*. Important or common species are discussed below.

The clover mite, *Bryobia praetiosa* Koch, is regarded by most acarologists as a species group that includes many variants. Most entomologists and many references are likely to treat them all as "clover mites" under this name. Some specialists have named a number of species within the group, based on significant differences in life cycles and host preferences. The clover mite that is familiar to most people is a nuisance in homes and other buildings, where they overwinter. This mite, designated as *B. borealis* Oudemans and called the dwelling bryobia, primarily feeds and lives on grasses and weeds during the growing season. It has five to six generations each year, overwinters as eggs and adults, and lays its eggs on tree bark or buildings. The apple bryobia, *B. arborea* Morgan & Anderson, is destructive to *Malus* species and has three generations a year. Eggs hatch early in the spring, and population density peaks in late spring and early summer. It spends winter as an egg on twigs and branches. The elm bryobia, *B. ulmophila* Peck, can cause severe damage on the lower branches of elm trees and has two to three generations per year during late spring and early summer. Eggs of the honeysuckle bryobia, *B. lonicerae,* overwinter on twigs of shrub honeysuckle and hatch as soon as new growth starts in the spring. After two generations, by the middle of June, most overwintering eggs for the next generation have been deposited on the twigs. The gooseberry bryobia, *B. ribis* Thomas, with one generation per year, and the ivy bryobia, *B. kissophila* Van Eyndhoven, with four to six generations, have also been recognized as separate species. In adddition to these species in the "clover mite" complex, there are numerous other species of *Bryobia*. In Arizona, for one example, Tuttle and Baker (792) described a new species, *B. ephedrae,* from joint fir, *Ephedra trifurca*.

Clover mites, somewhat flattened dorsally, are about 0.75 mm long; their first pair of legs is longer than the other three. The longer front legs are constantly, almost nervously, used to probe, much like the antennae of insects. Eggs and newly hatched larvae are bright red, but after they feed, the body becomes greenish to greenish black with an orange tint and orange legs. When crushed, the mites leave a reddish or orange-red stain.

Two major fruit pests, the citrus red mite and European red mite, are also important pests of ornamental trees and shrubs. Both have long dorsal setae that are borne on prominent, light-colored tubercles, unlike other common spider mites. Eggs are characteristically flattened on top and have a bristlelike stalk that arises from the center (B, C).

The citrus red mite, *Panonychus citri* (McGregor), is a very serious pest that occurs widely in citrus-growing areas from Florida to California; however, it was not found in Arizona until 1967. Elsewhere it is common in greenhouses and interior plantscapes on plants imported from citrus areas. It has been reported on *Eleagnus* species in Florida. Eggs and young mites are brick red and darken to a purplish or almost black color in the adult stage. In Florida it is often called the purple mite.

The European red mite, *Panonychus ulmi* (Koch), occurs widely in the Northeast and Northwest, primarily on plants in the Rosaceae family, especially species of *Crataegus, Malus,* and *Prunus*. It has also been reported as a pest of almond, black locust, buckthorn, elm, mountain ash, rose, and walnut. It may have as many as seven generations a year in New York fruit orchards. Eggs overwinter on the bark of branches and larger limbs, sometimes by the tens of thousands on apple trees. Eggs begin to hatch in late April in New York. Subsequent generations lay eggs on the foliage, where the mites feed until late summer. Eggs, young mites, and adults are brick red.

There are many species of *Eotetranychus* throughout the United States, with variable distribution from region to region. In 1963 Reeves (648) discussed 13 species from New York, only 1 of which was included among the 16 reported from Arizona by Tuttle and Baker (792) in 1968. Although some species have a wide host range, most have a limited number of hosts or are host specific, which facilitates identification. Mites in this genus are small to medium (about 0.3–0.5 mm long); they are yellow or yellowish green and feed on the undersides of the leaves. They produce very little webbing, and the leaves they infest may appear cupped, distorted, chlorotic or may drop prematurely. The yellow-orange adult females overwinter on and are protected by the bark of the host or debris on the ground.

A. Red oak leaves severely injured by the oak spider mite, *Oligonychus bicolor*.

B, C. Highly magnified European red mite eggs showing a characteristic center stalk. (Courtesy New York State Agricultural Experiment Station, Geneva.)

D, E. Close-up of a portion of a red oak leaf showing empty white eggshells of mites (D) and viable eggs ready to hatch (E). Note the leaf discoloration.

F. Bleached pyracantha leaves, the result of spider mite feeding.

G. An adult oak spider mite and four newly laid eggs. Arrow points to an eriophyid mite to show relative sizes. [Tetranychidae]

H. Honeylocust leaflets injured by the honeylocust spider mite, *Eotetranychus multidigituli*.

I. An *Amelanchier arborea* leaf injured by the mite *Tetranychus mcdanieli*.

A
B
C
D
E
F
G
H
I

The honeylocust spider mite, *Eotetranychus multidigituli* (Ewing), is a serious pest of *Gleditsia triacanthos,* its only known host. It often causes all of the foliage on a tree to turn brownish by July (H). It has been reported in New England, New York, and Ohio, as far west as Illinois and as far south as Louisiana and North Carolina. Adult females overwinter in bark crevices and under bud scales. Panel G in Plate 229 shows unusually large clusters of adult mites on the branches. Egg laying begins as early as mid-April in Illinois, but oviposition usually occurs in early to mid-June in the Northeast. An individual can develop from egg to adult in 4 days in the summer, and in 11 days in cooler weather. Populations rapidly build up after several generations, but by midsummer, where injury is severe, they collapse, and new growth on the trees remains conspicuously green. When infestations are light, mites may continue to reproduce until autumn, when females move to the bark to hibernate.

E. tiliarium (Hermann) has caused severe discoloration on linden on Long Island and is established throughout the Northeast and mid-Atlantic states. Other hosts are sycamore, hazel, buckeye, and hawthorn. The mites feed on the undersides of the leaves between the veins.

Aponychus spinosus (Banks), a species formerly in the genus *Eutetranychus,* may also be a pest of *Tilia* species. In the Northeast *Tilia tomentosa* may be functionally defoliated by late August. The mites feed on the upper leaf surfaces and produce several generations per year. Other recorded hosts include *Morus alba* and *Ulmus rubra*. This mite has wide distribution in its range.

Eotetranychus matthyssei Reeves has caused severe damage to elms in New York state and also has been reported from redbud, hackberry, and black locust. Not described until 1963, it likely was one of the several species of spider mites and rust mites (Eriophyidae) that caused complete browning of elms that were intensively sprayed with DDT for elm bark beetle and elm leaf beetle in the Northeast in the 1940s and 1950s. Population outbreaks of spider mites have long been observed to occur after intensive use of DDT and carbaryl. The yellow spider mite, *E. carpini borealis* (Ewing), is a subspecies of *E. carpini* (Oudemans) and is an important pest of apple, pear, and sometimes cherry, raspberry, blueberry, spirea, alder, and willow in the western United States. Thewke and Enns (780) reported it on silver and sugar maple in Missouri. *E. carpini carpini* (Oudemans) is a European subspecies. A third form has been recognized in Connecticut and New York on beech as another, but as yet unnamed, subspecies of *E. carpini*. The Garman spider mite, *E. uncatus* Garman, was described as an apple pest in Connecticut but has been reported on white birch in Utah, on pin oak in Missouri, and on alder, birch, ironwood (hornbeam), linden, sour cherry, and apple in New York. *E. querci* Reeves, a species closely related to *E. uncatus,* was found in New York on birch and oak with severe damage to pin oak. *E. populi* (Koch), sometimes called the poplar spider mite, is found in the eastern United States on willow and poplar, as well as on poplar in Missouri. The pecan leaf scorch mite, *E. hicoriae* (McGregor), is an important pest of pecan and chestnut groves in the Southeast and has been reported also from hickory, horsechestnut, beech, and oak. The sixspotted mite, *E. sexmaculatus* (Riley), is primarily a citrus pest in California and Florida but is also found on citrus in Arizona and may occur in other citrus areas. *E. lewisi* McGregor is a citrus pest on the West Coast, and the Yuma spider mite, *E. yumensis* (McGregor), generally occurs in citrus-growing areas. Many other species of *Eotetranychus* have been reported and described on various hosts in the United States.

Mites in the genus *Schizotetranychus* resemble *Eotetranychus* species because of their yellow to yellowish green color, size, and habits. They feed on the undersides of leaves but generally overwinter as eggs on the twigs. *Schizotetranychus garmani* Pritchard & Baker, the so-called willow spider mite, known only from willow, occurs in Connecticut, Massachusetts, New York, and Missouri and probably elsewhere. Another species, *S. schizopus* Zacher, was commonly found on willow in New York by Reeves (648), causing severe yellowing and browning of the foliage. *S. spireafolia* Garman has been reported on spirea from Connecticut, Pennsylvania, and New York. The bamboo spider mite, *S. celarius* Banks, occurs only on bamboo and has been found in Florida, Georgia, and California. Thewkes and Enns (780) reported only one species in this genus from Missouri. Tuttle and Baker (792) reported six known and four newly described species from Arizona, most of which were found on grasses. Further intensive studies will likely add new species and expand the host list and distribution of this genus.

References for Plate 228: 12, 210, 568, 633, 648, 780, 792, 825

The boxwood mite, *Eurytetranychus buxi* (Garman), which is common and widespread, frequently damages highly valuable plants, especially European, common, and English boxwood (C, H). Japanese boxwood appears to be much less susceptible. The mites overwinter as yellowish green eggs on the undersides of leaves. Eggs hatch in late April or early May in Michigan, and slightly earlier in Virginia. These mites have legs somewhat longer than those of other common tetranychids, and they more closely resemble spiders. They feed on the upper and lower surfaces of the leaves, and depending on the severity of the infestation, produce varying degrees of stippling (C, H), yellow or bronze streaking, or premature leaf drop. As many as eight generations a year occur in Michigan, but generally mite activity is more prevalent in the spring and early summer.

The genus *Oligonychus* includes a large number of species. One of the most important, the spruce mite, *Oligonychus ununguis* (Jacobi), and several other species that also feed on the needles of conifers are described with Plates 52 and 53. On broad-leaved evergreens, *Oligonychus* mites feed on both the upper and lower leaf surfaces, but they lay eggs mostly on the undersurfaces. On deciduous hosts most feed on the upper leaf surfaces, except the maple spider mite, *O. aceris* (Shimer), and *O. propetes* Pritchard & Baker, which feed only on the undersides of the foliage. *Oligonychus* mites are usually reddish or reddish brown but occasionally green with beige to orange legs. Adults are medium sized (0.4–0.75 mm long) and oval. This group overwinters as eggs, which may be red, amber, light green, or translucent pearl.

The southern red mite, *O. ilicis* (McGregor), is the most important, widespread, and destructive spider mite on broad-leaved evergreens, especially Japanese hollies, camellia, and azalea. It is also a pest of other hollies, laurel, rhododendron, and, generally, plants in the families Ericaceae and Aquifoliaceae. In New York it has been reported on sweet pepperbush, *Clethra alnifolia,* by Reeves (648). Baker (31) included elaeagnus, eucalyptus, eugenia, hibiscus, photinia, pyracantha, and viburnum as hosts in North Carolina. Injury primarily occurs as stippling of foliage, but leaves may be distorted when they are young and expanding (D). Red eggs overwinter on the undersides of leaves. Summer eggs are darker red (F) and often extremely abundant (E) on preferred hosts when the mites are not controlled. Numerous generations occur each year, but the mites are most prolific in cooler weather. Population densities are greatest in the spring and fall. Although some individuals may be active during the summer, the majority of the population is generally dormant (in aestivation) in the egg stage during the heat of midsummer.

The oak spider mite, *O. bicolor* (Banks), is dark reddish brown, not unlike the southern red mite (Plate 228G). It infests only the upper leaf surfaces of oak, birch, chestnut, beech, elm, and hickory. Severe leaf bronzing (Plate 228A) is most apparent in midsummer, predominantly on lower branches of larger trees. Beech foliage may be severely bronzed in June. In the fall eggs are laid around twig axes, where they overwinter. They are sufficiently exposed to make them vulnerable to dormant control treatments. The oak spider mite is widespread, occurring from New England south to North Carolina and west to Kansas.

The platanus spider mite, *O. platani* (McGregor), is common on sycamore in the hot, dry valleys of California and the Southwest. It has also been reported from loquat, oak, pyracantha, cotoneaster, and several other hosts.

O. newcomeri (McGregor) was first described from Yakima, Washington, on pear, but is believed to be indigenous to the Northeast, where it is common on *Amelanchier, Crataegus,* and *Malus* species. Reeves (648) also found it on ironwood (hornbeam), ash, wild cherry, and mountain ash. It overwinters in the egg stage and in New York infests its hosts from May to September.

The maple spider mite, *O. aceris* (Shimer), occurs on maple and feeds on the undersides of the leaves. It is found in the eastern United States and has also been reported from Kansas, Missouri, and Washington. The adult mite may be red or green. This species can cause severe damage to foliage when infestations are severe, usually during midsummer.

O. viridis (Banks) causes a bronze discoloration on the upper surface of *Carya* foliage similar to oak spider mite damage; it feeds on the upper leaf surfaces. Reeves (648) collected this species on shagbark hickory in New York, but other records are only from pecan in Florida, Georgia, and Louisiana.

O. propetes Pritchard & Baker is seldom a serious pest but is widespread. The mites cause characteristic chlorotic flecking along the main veins on the upper surface of white oak leaves, although they live and feed only on the undersurface. This species occurs in the Northeast and has been observed in Virginia on white oak. It is reported from Missouri on hawthorn, hackberry, and oak, and it was collected from *Quercus arizonica* in Arizona.

Species of *Tetranychus* are less common and less diverse on trees and shrubs than on herbaceous and crop plants. Several are important fruit pests, but most are found on low-growing plants. When they inhabit trees, they occur on lower branches or saplings. The two-spotted spider mite, named *Tetranychus bimaculatus* (Harvey) and *T. telarius* (Linnaeus) in older literature, has long been controversial among taxonomists and is considered a species complex. At present, the twospotted spider mite has been split into separate species, two of which are predominantly accepted: the common twospotted spider mite, *T. urticae* (Koch), of more northern climates, and its counterpart red form in the South, the carmine spider mite, *T. cinnarbarinus* (Boisduval). These two species represent the most widespread, destructive, ubiquitous spider mites in the entire family. They attack weeds, vegetables, flowers, field and forage crops, brambles and other small fruits, tree fruits, foliar and greenhouse crops, and certain trees and shrubs. Adult females in this genus are medium to large mites, 0.7–1.0 mm long.

The twospotted spider mite, *T. urticae* Koch, is green or greenish yellow with two lateral dark spots that are visible when the mite is viewed from above. Overwintering females are orange to orange-red and hibernate in ground litter or under bark of shrubs and trees. Their spots may vary in intensity depending on how recently the mite has molted. The spots result from internal accumulation of waste food; hence, recently molted mites may lack or have very faint spots. In addition to infesting a great many nonwoody hosts, this species can be a serious pest of roses, flowering fruits, azalea, and several other shrubs. It may cause damage to maple, elm, redbud, and has been reported from ash, black locust, and poplar. It is occasionally found on other trees. The twospotted mite feeds and reproduces whenever conditions are favorable for plant growth, from earliest spring to late fall. Warm and hot weather favors rapid development and increased feeding and reproduction. At 75°F, for example, development from egg hatch to adults takes only 5 days. A female lives 2–4 weeks and produces about 100–300 eggs.

The habits, hosts, life history, and destructiveness of the carmine spider mite, *T. cinnarbarinus,* are much the same as for *T. urticae.* In the North it may be found in greenhouses and interior plantscapes but is predominantly associated with more southerly climates.

The magnolia spider mite, *T. magnoliae* Boudreaux (A, B), occurs on *Magnolia grandiflora* and tuliptree in Louisiana, and may be found elsewhere in the deep South.

The fourspotted spider mite, *T. canadensis* (McGregor), is primarily a fruit tree pest in the eastern and southwestern United States and southeastern Canada. It has also been reported on elm, linden, horsechestnut, osage orange, poplar, rose, and umbrella tree, but it does not cause serious damage to these hosts.

The Schoene spider mite, *T. schoenei* McGregor; the Pacific spider mite, *T. pacificus* McGregor; and the McDaniel spider mite, *T. mcdanieli* McGregor, are other species that attack fruit trees and may also be found on woody ornamental plants.

T. yuccae Tuttle & Baker was described from Yucca in Arizona. *T. homorus* Pritchard & Baker occurs on ash and hickory in North Carolina, and it has been reported from ash in Missouri. In California *T. hydrangeae* Pritchard & Baker is an important pest of greenhouse hydrangea, and it can also survive on bean, violet, and strawberry.

The false spider mites, often called flat mites, in the family Tenuipalpidae, are close relatives of spider mites, but they do not produce silk and are much less convex. Before the superfamily Tetranychoidea was created, the tenuipalpids were included as a tribe or subfamily of the spider mites. About a dozen genera include over 100 species worldwide that feed on an extensive variety of hosts, including conifers, orchids, cacti, flowers, and herbaceous as well as woody plants. Although many species occur in North America, relatively few are of economic importance on trees and shrubs.

The false spider mites are oval, flattened, and range from light lemon yellow to dark red. They are slow moving, even when disturbed, and tend to remain in the same restricted area for extended periods. Although the more common species on trees and shrubs feed on the undersurfaces of leaves, some are found primarily on shoots, twigs, and stems, where they feed on bark tissue. Their development is similar to that of spider mites; stages include an egg, larva, protonymph, deutonymph, and adult female and male. They usually hibernate in the adult female stage but may overwinter as eggs.

The privet mite, *Brevipalpus obovatus* Donnadieu, has been reported on more than 50 diverse hosts from Maine west to Oregon and California and south to Florida. This mite is orange to dark red and feeds on the undersides of leaves. It may cause severe bronzing and reddening of the foliage, but privet tends to turn yellowish bronze in severe infestations. In Missouri the privet mite was collected from June to December, and in greenhouses, where it is a serious pest, it is present all year. (See Plate 231.)

The phaelenopsis mite, *Tenuipalpus pacificus* Baker, is a pest of orchids. The citrus flat mite, *B. lewisi* (McGregor), is a serious fruit pest and has also been a pest of ornamental plants on the East and West coasts.

B. californicus (Banks) has been reported from California, Texas, Kansas, Louisiana, Florida, and Maryland. It feeds on more than 60 genera of fruit and ornamental plants, causing necrotic areas on the undersides of infested leaves. Immature mites are reddish and become rufous or amber as adults. Adults are less than 0.25 mm long.

Because many similar species of false spider mites occur on a wide variety of host plants, positive identification should be made by an acarologist. Much remains to be learned about the occurrence, hosts, habits, life histories, and distribution of these mites.

The most notorious member of the family Tarsonemidae is the common cyclamen mite, *Phytonemus* (=*Steneotarsonemus*) *pallidus* (Banks), a serious greenhouse pest of African violet, cyclamen, strawberry, and a number of other flowering plants. All tarsonemid mites are extremely small, about 0.15 mm long, and cannot be seen without a magnifier. They distort buds, new growth, foliage, and stems. The broad mite, *Polyphagotarsonemus latus* (Banks), is a frequent associate of the cyclamen mite, and both have wide distributions and extensive host ranges. Feeding by the broad mite malforms terminal leaves and flower buds; blooms abort and plant growth is distorted. The broad mite and the false spider mite, *B. californicus,* are serious pests of *Pittosporum* species in Florida.

There are no known insect parasites of mites, but mites are susceptible to fungal and viral diseases. Insect and mite predators are the most important group of natural enemies. A number of mites prey on spider mites and eriophyid mites, most notably species in the family Phytoseiidae. Some investigators consider the phytoseiids to be the most effective and widespread predators of injurious plant-feeding mites. They prevent increases of potentially destructive mites because they are most effective at low population densities. They have been used as effective biological control agents for mite pests in greenhouses and on some crops. *Amblyseius fallacis* (Garman), a phytoseiid mite, has been shown to be a particularly effective predator in orchards and nurseries, in part because of its ability to develop resistance to the commonly used organophosphate insecticides.

Insect predators include lady beetles and certain thrips, true bugs, and lacewing larvae. *Stethorus* species, for example, round black lady beetles about 1 mm long, are capable of virtually eliminating spider mite populations but usually only after mite populations have peaked and mite damage is severe. Applying pesticides against mites after damage is severe (and, presumably, when populations of natural enemies are high) is especially deleterious to lady beetles and other predators. Some of the more recent pesticides are not as destructive to predaceous mites as they are to predaceous insects.

References for Plate 229: 31, 172, 173, 407, 633, 635, 648, 780, 792, 825

A, B. *Magnolia grandiflora* leaves showing injury on the upper surface caused by the feeding of the spider mite *Tetranychus magnoliae.* B. Injury may be concentrated at the leaf tip along the midvein.

C. Boxwood mite injury on common boxwood.

D. Foliage distortion on azalea caused by early-season injury by the southern red mite.

E, F. Southern red mite eggs on the undersides of Japanese holly leaves.

G. Clusters of overwintering adult honeylocust spider mites. They are most abundant at twig junctions.

H. English boxwood leaves injured by the boxwood mite.

A
B
C
D
E
F
G
H

Bud and Rust Mites (Plates 230–231)

Some mites in the family Eriophyidae are called leaf vagrants (Plate 231G) because they feed freely on the foliage of plants. Leaf vagrants are also called rust mites if their damage causes russeting of the leaves. Other mites in the same family incite the proliferation of plant tissue and produce galls (Plates 232–233). Still others cause internal bud injury. A number of species induce their host plants to produce blisters on the foliage (Plates 233–234). Thus, eriophyid mites may also be called gall mites, bud mites, and blister mites. Fortunately, the appearance of the plant reaction to these mites provides, along with knowledge of the host type, a reliable means of identification of many species. All eriophyid mites measure less than 0.5 mm long. They can be easily overlooked, even with a 10× magnifier, and even if one is trying to observe them.

Certain gall mites and bud mites are described and shown here and on Plates 231–235.

Aceria (=*Eriophyes*) *parapopuli* (Keifer) has been given the common name poplar bud gall mite. In the spring some of the mites that have overwintered migrate to swelling new buds. There they initiate new galls, which stops further elongation of new shoots. The galls are succulent and dark green, and some writers have described them as cauliflowerlike swellings (Figure 135). By late summer the galls are hard and brick red; they may stay active for up to 4 years and grow to 4 cm in diameter. Abandoned galls are grayish and may remain on the tree for 5 years or more. Beyond the aesthetic damage, the galls cause branches to be stunted and crooked, often resulting in death of branches. The tree is seldom killed.

The mites that live inside the gall feed on plant fluids and produce a new generation every 2–3 weeks. An active gall may contain more than 3000 mites. Control is difficult because only those mites that migrate are exposed to predators and pesticides. All northern poplars are subject to attack. Damage is lightest on such native species as trembling aspen, plains cottonwood, and balsam poplar. The most severe damage occurs on the numerous hybrid poplars developed for use as ornamentals and in shelterbelts. Another species, *A. populi* Nalepa, causes massive bud proliferation. Neither of these poplar mites has been studied to determine its life cycle.

Cecidophyes betulae (Nalepa) causes proliferation of bud tissue on birch like that shown in panel B. A growth similar to that shown in A may also result from its activities.

The taxus bud mite (E), *Cecidophyopsis psilaspis* (Nalepa), is occasionally a serious pest of yew, *Taxus baccata*. It causes enlargement, blasting, and death of buds. It has damaged yew in several of the northeastern states. A northern European species, *C. psilaspis* is capable of becoming more prevalent in the United States if infested plants or propagated materials are distributed in the nursery trade.

Adult taxus bud mites crawl between the bud scales, where they overwinter. Feeding and reproduction occur during the late summer and fall. Secondary microorganisms cause the decay of the buds. As many as 1000 mites may live in a single bud. Heavily infested buds do not grow in the spring. Light infestations result in distorted needles and shoots. When new buds are formed in the summer, mites migrate to them and carry on the life cycle.

Taxomyia taxi (Inchbald) is a midge that causes a bud gall on yew; the gall is called an artichoke gall. The taxus bud mite causes galls with some similarities. *T. taxi* is not known to occur in the United States, but because its galls resemble those of the taxus bud mite, it may have been overlooked.

On hackberry, *Celtis occidentalis,* a witches'-broom gall is occasionally associated with the presence of an eriophyid mite, *Aceria celtis* (Kendall), and a powdery mildew fungus, *Sphaerotheca phytoptophila* (see Sinclair et al., *Diseases of Trees and Shrubs*.)

Huge numbers of mites are usually found in the buds. The first eggs of the growing season are laid in May, and colonies continue to develop until fall. Only fragments of the life history of the mites are known. Witches'-brooms have been found from New York to Illinois, south to Alabama and west to Mississippi. They are likely to be found wherever hackberry grows and especially on specimen or shade trees. Trees are most unsightly after foliage has dropped in the fall (G).

A common and equally unsightly gall is produced by *Aceria fraxiniflora* (Felt), the ash flower gall. *A. fraxiniflora* causes tremendous proliferation of flower buds on male ash trees.

Figure 135. Various stages of gall formation on poplar caused by the poplar bud gall mite, *Aceria parapopuli*. Early stages are dark green and succulent. (Courtesy Northern Forestry Centre, Edmonton, Alberta.)

Highbush blueberry and its cultivars are occasionally used in landscape plantings. Blueberry has a wide array of arthropod pests, including the blueberry bud mite, *Acalitus vaccinii* (Keifer). This eriophyid mite is spindle-shaped and about 200 μm long. It attacks the fruit bud and feeds on the inner surfaces of bud scales and on the developing leaf and floral parts enclosed by the bud scales. Persistent feeding by large numbers of mites prevents the bud from expanding properly and kills the flowers or, in the extreme, may completely kill the bud. Distinctive red blisters on bud scales are reliable symptoms of the presence of bud mites. In North Carolina these symptoms are most apparent from February through April.

In the southern part of the blueberry bud mite's range it is active all year. Mites present when the buds open in the spring gradually migrate along the stem to the basal scales of newly forming buds. These mites work their way under the outer scales, where they feed, mate, and lay eggs. As the colony increases, the mites move to the center of the bud. By early September in North Carolina some mites can be found deep within infested buds. Maximum populations are reached by the end of December and January.

The blueberry bud mite occurs from the Canadian Maritimes south to Florida and Mississippi. Injury is greatest in the southern half of its range. It may be found on native as well as cultivated *Vaccinium* species. Selective pruning of old canes helps to reduce bud mite populations.

References: 233, 407, 423, 424, 426, 537, 647, 726

A. *Populus tremuloides* with massive bud proliferation galls caused by an as yet unnamed eriophyid mite.
B. A *Cecidophyes betulae* bud gall on birch.
C. Male flower buds on yew; uninfested natural growth.
D. A blasted bud severely infested with taxus bud mites. × 5.
E. Immature (white) and adult (orange) taxus bud mites.
F. The typical appearance of blasted buds infested with taxus bud mites.
G. A witches'-broom found on hackberry and caused by an eriophyid mite, winter condition.
H. Close-up of a witches'-broom with foliage caused by eriophyid mites on hackberry.

A
B
C
D
E
F
G
H

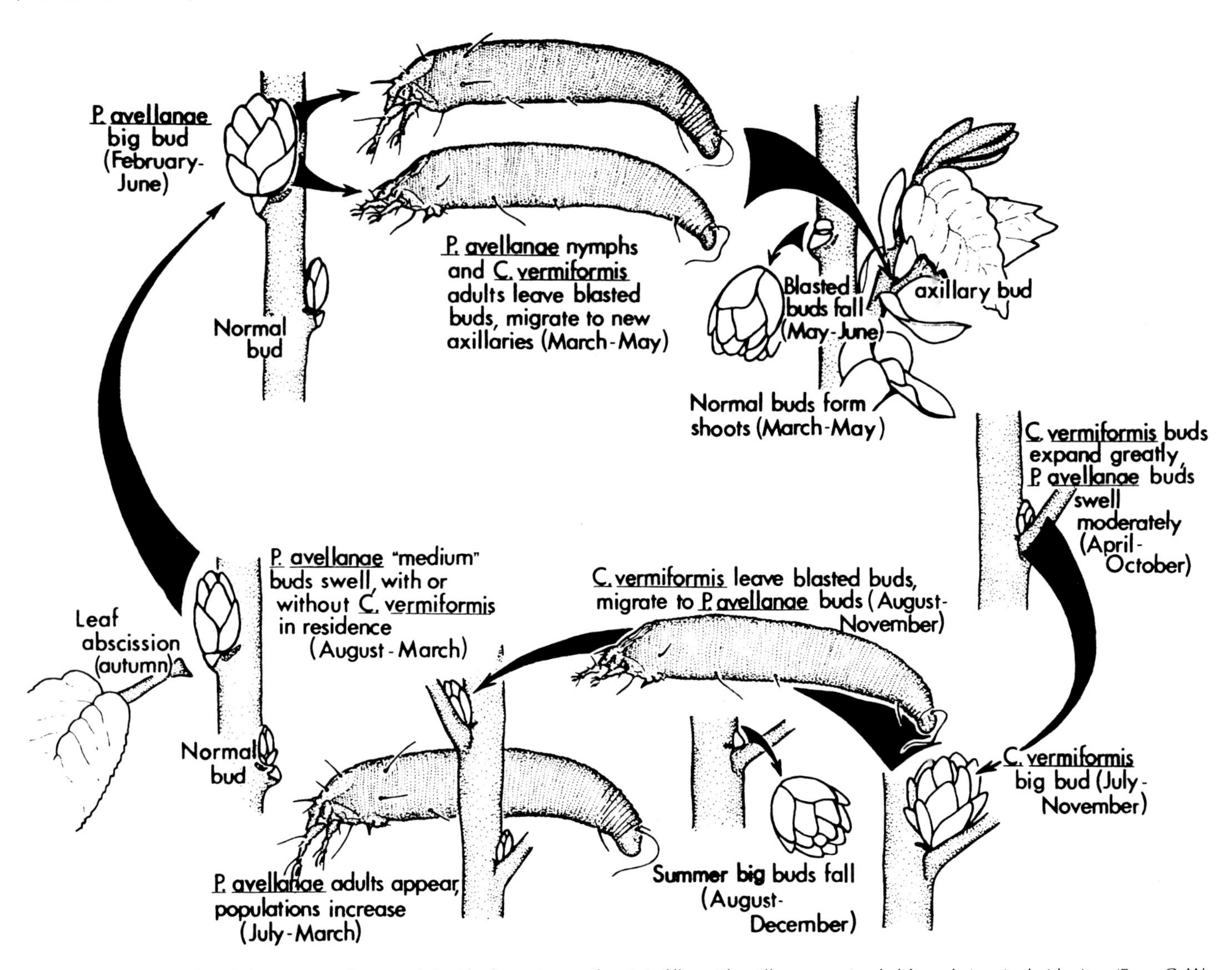

Figure 136. Activity cycles of *Phytoptus avellanae* and *Cecidophyopsis vermiformis* in filbert. They illustrate a simple life cycle in eriophyid mites. (From G. W. Krantz, *A Manual of Acarology,* 2nd ed., Oregon State Univ., Corvallis, 1978, courtesy G. W. Krantz and Oregon State Univ. Bookstores.)

Two leaf vagrant mites, *Acaphylla steinwedeni* Keifer and *Calacarus carinatus* (Green), injure camellia leaves. Heavily infested leaves develop a bronze cast that makes the leaves look rusty. These mites occur in Gulf Coast states and in California. In California and Florida they may be found the year round. *Cosetacus camelliae* (Keifer) inhabits flower buds of camellia. Symptoms of its presence are not detectable until early February (in northern California), when heavily infested buds show brown edges on the bud scales (A). If reproduction of the mites is not retarded, the flower buds turn brown and drop before blooming. The mite feeds on the inner side of the bud scale and sometimes on the developing flower petals. It cannot be readily seen without the aid of a microscope or a 20× hand lens. The mite is colorless and is sometimes described as wormlike. Life stages of this species are shown (magnified about 50 times) in panel C. Little is known about its biology.

The eriophyid bud mites *Phytoptus* (=*Phytocoptella*) *avellanae* (Nalepa)—the filbert bud mite—and *Cecidophyopsis vermiformis* (Nalepa) injure both flower and vegetative buds of filbert (Figure 136). These species can occur wherever European filberts or their hybrids grow. The mites make their way into newly forming axillary buds during late spring and feed internally on the undeveloped bud tissue, which causes the bud to enlarge. Reproduction takes place inside the infested bud. If the mite population is large enough, the bud will be killed, and the number of nuts produced during the following year will be reduced. Normal and swollen buds are shown in panel F.

Aculus (=*Vasates*) *ligustri* (Keifer), sometimes called the privet rust mite, was described from *Ligustrum* species in Pasadena, California. It has adapted to cool temperatures and initiates activity as soon as new leaves develop. Overwintering females migrate from under bud scales to the leaves and begin laying eggs. Numerous overlapping generations occur from early spring into early summer. This eriophyid may number more than 2000 individuals per leaf on more than 70% of the foliage. During a cool summer some mites may be active during the middle of the growing season, but generally activity is nil until fall. In southern California this species may be found all winter on privet leaves. The life cycle varies with climatic conditions. The new leaves of infested *Ligustrum ovalifolium* curl and turn brown. Leaves of infested *L. armurense* become severely cupped and drop from the bush while still green (G). Privet rust mite is common and widespread, from California east to New York and Massachusetts and south at least through Virginia and Tennessee.

All eriophyid mites have but two pairs of legs. Only one other group of arthropods, the hair follicle mites, is similar. All other mites and spiders have four pairs of legs. All other arthropods have three or more

---

A, B, D. Progressive browning of bud scales on camellia buds caused by the eriophyid mite *Cosetacus* (=*Aceria*) *camelliae*.
C. The inside of a camellia bud scale showing a high population of the mite *C. camelliae*. All stages are present. Highly magnified.
E. A highly magnified photograph of *Aculus ligustri,* a mite.
F. Normal buds and "big buds" of filbert. The large buds have been damaged by the feeding of eriophyid mites inside the bud.
G. *Ligustrum armurense* showing symptoms of injury caused by the eriophyid mite *A. ligustri* (or a closely related species).

A
B
C
D
E
F
G

pairs of legs. Since the eriophyid mites are so small, their needlelike mouthparts are able to penetrate only the plant cells at the surface. Consequently, the damage appears as russeting rather than as coarse chlorotic flecks. The complete symptom includes pale green and cupped foliage (G). Positive identification of the species is difficult and is based on body setae, the configuration of the genital cover flap, and body microtubules (see Figure 135).

Predaceous mites in the family Phytoseiidae are common among infestations of the privet rust mite but do not occur in sufficient numbers to keep pace with the reproductive rate of the rust mite. Predaceous mites are more effective when the density of their prey is low.

Another mite, *Brevipalpus obovatus* Donnadieu (Tenuipalpidae), has been given the common name privet mite. It is a bit larger (0.25 mm) than eriophyid mites and has four pairs of legs. It feeds on the foliage not only of privet but of citrus, fuchsia, aucuba, azalea, Boston ivy, and about 50 other genera of ornamental plants. It is literally cosmopolitan because it is known on every continent. Eggs are elliptical and bright orange-red when first laid, becoming darker as hatching time approaches. The color of adult females varies, ranging from light orange to dark red. The females congregate in sheltered places on the host plant to overwinter. Males occur, but parthenogenic reproduction is the rule.

References: 426, 446

Eriophyid Gall Makers (Plates 232–235)

Mites that cause plants to produce bladder or spindle galls, or dense masses of hairy or beadlike growths called erinea, are extremely small (about 0.15 mm) and belong to the family Eriophyidae. Several species of eriophyid mites induce maples to produce three types of galls: bladder, spindle, and erineum galls. The brilliant red color often associated with the bladder galls makes them spectacular and conspicuous (A). Consequently, they cause much concern among tree owners, who are not aware that the colorful growths on the foliage are rarely detrimental to the health of the tree. Early spring foliage is affected, particularly those leaves next to the trunk and larger branches. When mites inside the galls mature, they move from the galls to newly developing leaves and initiate more galls. Mite activity decreases as the growing season progresses. Sufficient foliage unaffected by galls is produced during a growing season to sustain the tree without serious harm.

The maple spindle gall occurs most frequently on sugar maple (B) and is caused by *Vasates aceriscrumena* (Riley). The maple bladder-gall mite, *V. quadripedes* (Shimer), is most common on silver maple (A) and red maple, *Acer rubrum*. Sometimes leaf distortion occurs in heavy infestations. Both spindle and bladder galls occur on the upper leaf surfaces. Several species of erineum-producing mites cause red or green patches on the lower and upper sides of maple leaves (C, D). These include *Aceria* (=*Eriophyes*) *elongatus* Hodgkiss, a red erineum, *A. modestus* (Hodgkiss), a green erineum, and *Aculops maculatus* (Hodgkiss), among others. *Aculops maculatus* occurs as an inquiline in the red (crimson) erineum or as a leaf vagrant. Erineum galls may look like felt patches or, when magnified, like the beaded surface on a movie projection screen. *Aceria calaceris* Keifer causes a red erineum on Rocky Mountain maple and *A. negundi* (Hodgkiss) causes a small round depression filled with white felty hairs on the undersides of boxelder leaves. Plate 234H shows the upper side of a boxelder leaf with white erineum patches. Although they are in an abnormal location, these too are believed to be caused by *A. negundi*.

Generally, maple gall mites overwinter as adults under bark scales, where they are able to withstand severe weather conditions. Early in the spring they move to unfolding leaves and begin feeding. Mites that cause bladder galls and spindle galls feed on the undersides of the leaves. Initially, a slight depression results. The leaf then rapidly produces a pouchlike gall at that spot, enclosing the mite. An opening remains on the underside of the leaf. The mite continues to feed and lays numerous eggs within the gall. Reproduction is prolific, and as mites mature, they leave the gall and continue infesting new foliage. By July, mite activity on the foliage ceases.

Pearlbush, *Exochorda racemosa,* is subject to foliar disfiguration by the eriophyid mite *Phyllocoptes exchordae* Keifer. The species is believed to be new to science, and little is known about its biology and distribution. Because the eriophyids are so host specific, we expect that this mite will only be found on pearlbush and that the mite and its gall may be found wherever pearlbush grows. Currently the mite is known only in New York state.

In central New York galls on pearlbush are formed in early May before all of the spring-flush leaves are fully formed. *P. exchordae* makes a lineal gall along the margins of lateral veins (E–G). The gall may be found on either side of the leaf. When galls are present, the leaf may be deformed (G) or dwarfed. In late August some galls begin to open. At this stage a gall contains 50 or more tiny reddish orange mites (F), similar in form to that depicted in Figures 137–138. Related gall-forming mites overwinter under bud scales. A similar gall may occasionally be found on crabapple foliage.

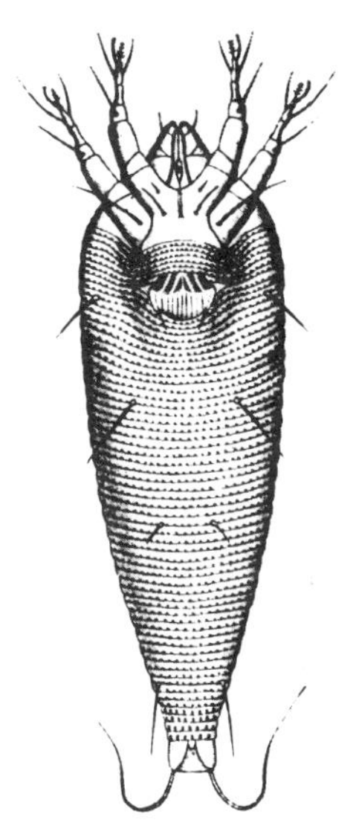

Figure 137. A drawing of *Aculops maculatus* to show an enlarged perspective of an eriophyid mite form. The ventral view shows the genital flap of the female. This species is an inquiline leaf vagrant that occurs on sugar maple leaves; less than 200 μm long.

The life cycle described above for *P. exchordae* is that of a species having a simplified, abbreviated development. Most of the eriophyids go through a complicated alternation of generations in which summer, winter, asexual, and sexual forms are produced in a programed sequence, each with a unique behavior.

Eriophyes (=*Phytoptus*) *emarginatae* Keifer causes a green pouch gall on the upper surface of *Prunus* species. The gall form is similar to the maple bladder gall. *Aceria parulmi* Keifer produces a spindle gall on several *Ulmus* species that is similar to the maple spindle gall. Similarly shaped galls are found on cherry and linden. Erineum galls also commonly occur on beech and poplar.

References: 150, 372, 424, 426

A. *Vasates quadripedes* galls, commonly called maple bladder galls, on silver maple.
B. A maple spindle gall, caused by *Vasates aceriscrumena,* on sugar maple.
C. An *Aceria elongatus* erineum gall on the upper surface of sugar maple leaves.
D. An *Aceria aceris* erineum on the lower surface of a silver maple leaf.
E–G. Leaf galls on pearlbush caused by *Phyllocoptes exchordae.* F shows top of gall cut away to expose the cavity and the tiny reddish mites inside. Mites are about 200 μm long.

A
B
C
D
E
F
G

Representative types of erineum, blister, and marginal leaf roll galls associated with several eriophyid mites are illustrated in this plate.

Wartlike leaf galls on hickory, *Carya* species (A), are caused by an as yet unidentified eriophyid species collected in Ottawa. A similar gall is produced on mountain ash, *Sorbus* species. Leaf blisters on mountain ash (B, E) are caused by *Tetraspinus* (=*Plataculus*) *pyramidicus* (Keifer), which may be the species shown in panel B. American beech, *Fagus grandifolia,* commonly produces light green to yellowish erineum galls on the upper and lower leaf surfaces (C). These are caused by *Acalitus fagerinea* (Keifer). In late summer, when erinea start to dry, the mites migrate to buds, where they overwinter (Figure 136). Other trees—such as the European white elm, *Ulmus laevis*—exhibit the same type of erineum gall. Under high magnification (F) all erineum-type galls appear as a continuous mass of beadlike growths. Normally the mites migrate to buds, where they overwinter (Figure 138). Other galls turn brown and start to dry, mites move out over the leaves (see arrows).

Hairy or feltlike erineum galls are also common on alder (D). An immature eriophyid mite is shown in panel H. The extremely small size of these mites makes them barely discernible with a 10× hand lens. Another type of gall induced by eriophyid mites is the marginal fold gall. Mites that feed along the edges of leaves are common on plants such as wild cherry, sourgum, and hawthorn, *Crataegus* species (G).

Eriophyid mites are host specific. Hundreds of species have been described and many are yet unknown. Very careful studies by well-qualified specialists are necessary to identify the species of eriophyid mites encountered on trees and shrubs. In many cases their biologies and habits are complex and all too often unknown, leaving a tremendous amount of biological information yet to be discovered.

References: 407, 424

Figure 138. Scanning electron micrograph of *Acalitus fagerinea,* an eriophyid mite that causes an erineum gall on American beech; about 230 μm long. (Courtesy U.S. Forest Service.)

A. An unidentified mite gall on hickory, *Carya* sp.
B, E. Eriophyid mite galls on mountain ash, *Sorbus* sp.
C. The yellow, feltlike pads are erineum mite galls on *Fagus grandifolia* caused by *Acalitus fagerinea.*
D. Feltlike patches caused by eriophyid mites on *Alnus* sp.
F. A magnified view of the feltlike gall of the beech leaf shown in C.
G. A marginal fold gall on hawthorn, *Crataegus* sp.
H. An immature eriophyid mite.

A
B
C
D
E
F
G
H

(Eriophyid Gall Makers, continued, Plate 234)

One unique type of injury produced by certain eriophyids is the leaf blister (A). A number of these species are associated with pomaceous hosts. They cause serious injury and frequently brown and curl nearly all of the foliage on infested trees. These species also attack *Amelanchier, Cotoneaster, Crataegus,* and *Sorbus* species. The pearleaf blister mite, *Eriophyes* (=*Phytoptus*) *pyri* (Pagenstecher) (A) is a common pest of pear and apple; first reported in the United States in 1872, it was widespread and abundant in major pear-growing areas of the United States and Canada in the late 1800s and early 1900s. Now it occurs predominantly on fruit trees around the home, where few pest control measures are applied.

Pear and apple blister mites overwinter under bud scales and migrate to the leaves in the spring. Mites feed on undersurfaces of the leaves. The epidermis ruptures where feeding occurs and mites enter the opening to deposit eggs. As eggs hatch, more and more mites destroy increasing numbers of cells. A pimple-shaped swelling appears on the upper surface and develops into a blister, which enlarges to 3 mm or so in diameter. Many blisters coalesce, discoloring large continuous patches on the leaves. As internal leaf tissue deteriorates, infested areas become brown or black, then shrink and dry out.

E. tiliae (Pagenstecher) incites an erineum gall (B, D) on little leaf linden, *Tilia cordata.* It is not known to attack the native linden, *T. americana. E. tiliae* has been reported from New York and California, but little is known of its life history and habits. Its biology is probably similar to that of species found on beech, elm, maple, alder, and other plants.

Panels E and G illustrate strange but interesting galls on shoots and leaf petioles of black walnut. The newly formed gall is brightly colored, changing from greenish to pink to crimson and eventually to reddish brown. The gall is a solid mass with erineumlike hairs on its surface. The mite causing this injury, *Aceria* (=*Eriophyes*) *caulis* Cook, is about 200 μm long and overwinters in buds as an adult called a deutogyne. The gall is initiated by the deutogyne, and her progeny (called protogynes) keep it growing. The life history is complicated and involves alternation of generations. The mite is probably distributed throughout the range of its host.

White erineum patches (H) on the foliage of boxelder, *Acer negundo,* occur in the presence of *Aceria negundi* (Hodgkiss). This mite is widespread throughout much of the United States. Several other species of eriophyid mites also have been reported from boxelder.

Quercus agrifolia, California live oak, is subject to an erineum maker, *A. mackiei* Keifer (not illustrated). Colonies of this mite attack spring foliage, causing large blisterlike swellings similar in size and shape to the injury caused by the oak leaf blister fungus, *Taphrina coerulescens* (see Sinclair et al., *Diseases of Trees and Shrubs*). The blisters are visible on the upper surface; they are oblong and up to 12 mm in their longest dimension. At first they are glossy green and later become brown. On the underside a dense greenish to yellowish erineum pocket or concavity forms, which later turns brown.

Mites may be collected from these pockets in June. Leaves may also roll or curl under or become grossly distorted under the pressure of large populations of mites. In the East leaves of white oak, *Q. alba,* may have similar blisters and erineum pockets, but they are caused by *A. triplacis* Keifer. There are several other species that cause similar symptoms on such plants as litchi and mountain ash.

Q. virginiana, the live oak of the Southeast, hosts the vagrant eriophyid mite *Johnella virginiana* Keifer. Its feeding causes the leaf to curl. No erineum is formed.

Aculops fuchsiae Keifer, known locally as the fuchsia gall mite, attacks *Fuchsia* species in coastal California, causing growing points, young leaves, and blossoms to become twisted and stunted, grotesquely swollen and blistered, and often reddened. The mites live and reproduce within the folds of galled tissue and among plant hairs. Like some other eriophyids that have been studied in detail, the fuchsia gall mite probably is spread locally by wind, as well as by bees and hummingbirds that visit flowers. Evaluation of 72 species and cultivars of fuchsia for their relative susceptibility to the fuchsia gall mite revealed the following:

Immune or highly resistant to infestation
'Baby Chang', 'Chance Encounter', 'Cinnabarina', *Fuchsia boliviana, F. microphylla* subsp. *hidalgensis, F. minutiflora, F. radicans, F. regia* var. *alpestris, F. tincta, F. thymifolia, F. venusta,* 'Isis', 'Mendocino Mini', 'Miniature Jewels', 'Ocean Mist', and 'Space Shuttle'.

Moderately resistant to infestation
'Couquet', 'Dollar Princess', 'Englander', *F. arborescens, F. denticulata, F. gehrigeri, F. macrophylla, F. procumbens, F. triphylla,* 'Golden West', 'Gordon's China Rose', 'Indian Maid', 'Joni', 'Lena', 'Little Ronnie', 'Machu Picchu', 'Pink Marshmallow', 'Postiljon', 'Psychedelic', and 'Snowy Summit'.

Highly susceptible to infestation
'Angel's Dream', 'Angel's Flight', 'Bicentennial', 'Capri', 'China Doll', 'Christy', 'Dark Eyes', 'Display', 'Firebird', 'First Love', *F. magellanica,* 'Golden Anne', 'Golden Anniversary', 'Hula Girl', 'Jack Shahan', 'Jingle Bells', 'Jubie-Lin', 'Kaleidoscope', 'Kathy Louise', 'Lisa', 'Louise Emershaw', 'Marinka', 'Novella', 'Papoose', 'Pepper', 'Pink Jade', 'Raspberry', 'Red Spider', 'South Gate', 'Stardust', 'Swingtime', 'Tinker Bell', 'Troubador', 'Vienna Waltz', 'Voodoo', and 'Westergeist'.

Numerous species of predatory mites have been associated with eriophyids, most commonly, *Seius* species. They not only feed on leaf vagrants and erineum mites but frequently enter galls to prey upon mites as they develop inside.

References: 407, 424, 426, 457, 532, 889, 899

A. Pearleaf blister mite injury on pear.
B, D. Erineum galls on little leaf linden.
C, F. Black walnut pouch gall.
E, G. Black walnut leaf petiole galls caused by the mite *Aceria caulis.*
H. Erineum galls caused by *A. negundi* on boxelder.

A
B
C
D
E
F
G
H

Several eriophyid mite species attack walnut. In the West the walnut blister mite, *Aceria* (=*Eriophyes*) *erinea* (Nalepa), has been recognized as a pest of Persian (English) walnut for many years. It causes yellowish, densely hairy, and feltlike pads to form on the undersurfaces of leaflets. On the upper surface galls may be as long as 18 mm; they are light green and resemble blisters. Mites overwinter under bud scales and become active when the buds begin to swell in the spring. In the East *A. cinereae* Keifer makes a yellowish feltlike erineum on both surfaces of the leaf. On the top the area containing the erinea becomes swollen and the tissue is more dense (A). The erinea on the undersides of leaves are in depressions that fit between the lateral veins (B). As the leaf and erinea age, the erinea darken and by late August may be nearly black. Heavily infested leaves are twisted or otherwise distorted. When leaflets are covered with these galls, their effectiveness as food producers is greatly curtailed. The mites also feed on the nut husk, causing a russeting of its surface. The range of this mite is probably the same as that of its host.

A. brachytarsus (Keifer) causes pouch galls (C) on *Juglans californica* leaves and, if abundant, will distort the foliage. Like most eriophyids, they overwinter under bud scales. An eriophyid mite, *A. nyssae* (Trotter), produces yellow, pimple-shaped blisters on the upper leaf surfaces of black tupelo, *Nyssa sylvatica* (E). Little is known about its biology and habits. It occurs throughout much of the southeastern United States.

The cause of a disease of oak branches (D) has not been definitely established, but eriophyid mites as well as bacteria and viruses are all suspected. Bud proliferation is often associated with eriophyid mite infestations on many woody plants. Panel D, for example, shows hundreds of dead buds on oak around rough, calloused areas. Bud formation was induced, but buds died before they were able to grow. This condition may be controlled by removing affected parts well below the evidence of symptoms.

Hundreds of different eriophyid mites feed on all kinds of trees and shrubs. Each causes characteristic symptoms on its particular host. Because of their minute size, relatively few have yet been collected or studied. Much remains to be learned about eriophyid mites, their habits, and the cause-and-effect relationship between the mites and their hosts.

References: 424, 426, 533

A. Butternut leaves and nuts affected by *Aceria cinereae*. An erineum gall is produced on the upper side of the leaf.
B. The same butternut leaves in A showing the erineum gall from the lower surface.
C. Galls of the walnut purse gall mite, *A. brachytarsus*. These galls are about 5 mm in diameter.
D. Bud proliferation on the branch of an oak, possibly induced by eriophyid mites.
E. Leaf galls caused by *A. nyssae* on black tupelo, *Nyssa sylvatica*.

A
B
C
D
E

Katydids, Grasshoppers, and Periodical Cicadas (Plate 236)

Some of our most common and universally known insects are katydids and grasshoppers. The broadwinged katydid, *Microcentrum rhombifolium* (Saussure), is a minor pest that feeds on both leaves and flowers of trees and shrubs. It is slow moving and rather heavy bodied, capable of only short flights. When fully grown, it may measure up to 47 mm from the head to the end of the wing tip. Katydids may be found throughout much of the United States but are rare in the northernmost states. They are usually of little economic importance, though occasionally they do defoliate young citrus trees and ornamental plants. The injury sustained by the host usually takes the form of chewed-out, semicircular holes along the leaf margins. In warmer areas of the United States broadwinged katydids may be found from March through November.

The eggs are laid in a characteristic manner (A, B) either on twigs or leaves. When the young hatch, they usually feed on the nearest foliage. Their green bodies provide excellent camouflage. The wings of an adult (A) look much like a leaf. Katydids are arboreal and are only rarely found on the ground.

Grasshoppers may feed on all types of green terrestrial plants. The eastern lubber grasshopper, *Romalea guttata* (Latreille) is common all along the south Atlantic and Gulf coasts. During the summer literally thousands of these grasshoppers may be seen sunning themselves on the roads of Florida and Georgia. The eastern lubber is one of the largest grasshoppers in the United States. With a heavy body and short, stubby wings, it is incapable of flight (E). It is sluggish and clumsy, has weak hind legs, and can leap only short distances.

Lubber grasshoppers overwinter in the egg stage in the ground. In Florida eggs begin to hatch in February. The nymphs are almost solid black, with yellow and occasionally red markings (F); young nymphs are gregarious—50 to 100 may be found on a single plant. They mature in June and oviposition begins in early July. These eggs do not hatch until the following spring. Adults are present through August. Nymphs and adults occasionally damage nursery stock. One generation occurs each year.

According to Simons (700) there are at least 150 species of cicadas in the United States; somewhat less than half of them occur in the West. For example, in New York there are 9 species, and Kansas and Colorado have about 25 species. The periodical cicadas (*Magicicada* spp.) of the eastern states are the most spectacular because of their long life cycle, number of broods, and the folklore associated with them. Before Europeans colonized America, some Indian tribes thought that the appearance of these insects had an evil significance, and in many rural areas people still view their appearance superstitiously. Many tribes collected them as food, and some of the Onondaga Indians of central New York continue to use them for this purpose. They fry them in butter and eat them like popcorn.

Periodical cicadas have two distinct races, based on the time required to complete their life cycle: a 17-year northern race and a 13-year southern race. In each of the races there are three distinct species. Each of the six species may be distinguished by size, color, song, and oviposition site. *Magicicada septendecim* (Linnaeus) is the most common cicada, and along with its sibling species, it is the longest-lived insect in North America.

The various populations are described as broods. Most of the broods are separated geographically, but some overlap. Thirty broods have been located, and the years when each will appear can be accurately predicted (Table 19). Some are quite small; others have apparently been eradicated by humans. Emergence of some of the larger broods is a striking event; thousands of cicadas seem to appear overnight and begin singing at dawn. Only the male has a sound-producing apparatus.

Brood X, the largest (numerically and geographically) of the 17-year cicadas, occurs over much of the northeastern quarter of the United States. Brood XIX is the largest of the 13-year cicadas. It occurs over much of the southern part of the United States but extends into southern Illinois and northern Missouri.

In some southern areas cicadas begin to emerge from the ground in late April. In northern regions emergence begins in late May or early June. The fact that cicadas have either a 13- or 17-year life cycle should not be interpreted to mean that cicadas are seen only at 13- or 17-year intervals. Broods of these insects emerge somewhere almost every year.

Adult cicadas (G) do not feed on leaves. If they feed at all it is by sucking sap from young twigs. Severe injury to host plants is caused by the cicada's egg-laying habits. The female has a sawlike egg-laying apparatus that she uses to cut the bark of twigs and to splinter the sapwood. In the splintered sapwood, called an egg nest, she lays 24–48 eggs. A single female may make up to 20 of these splinter "pockets" (G) and lay up to 600 eggs. The adult lives for about 20–23 days. After hatching, the nymphs drop to the ground, burrow into the soil, and search out a suitable root from which to suck sap. This is the beginning of a 13- or 17-year underground existence.

A severely injured tree damaged by egg-laying females will have a large number of dead twigs, as illustrated in panel C. These dead twigs (flags) represent only a small portion of the twigs that contain egg nests. Twigs that survive develop scars that remain as symptoms for many years. Trees that show the most conspicuous symptoms of oviposition injury are oak, hickory, ash, maple, hawthorn, apple, black locust, birch, and dogwood. Periodical cicadas may occasionally oviposit on herbaceous perennials such as delphinium.

No one has attempted to measure the damage caused by nymphs, but it is considered to be minimal because of the insects' slow development. Their food comes from xylem fluids in tree roots. Nymphs occur at soil depths ranging from 7 to 37 cm. Their forelegs are developed into large clawlike appendages that are useful for tunneling.

Cicadas have many natural enemies. Insect-feeding birds appreciably reduce their numbers. Predatory insects and mites attack the eggs, and a fungal disease caused by *Massospora cicadina* kills many of them. The fungal infection is confined to the terminal portion of the abdomen.

Cicadas belonging to the genus *Tibicen* (all eastern species) are often called dog-day cicadas. They are also often erroneously called annual cicadas. The life cycle of dog-day cicadas is unknown but probably requires 2–5 years. The broods overlap so that some adults appear in the same area every year. There are at least seven species. One of the most common eastern species is *Tibicen linnei* (Smith & Grossbeck) (D). *T. cinctifer* (Uhler) is one of the destructive species found in the Southwest. So far as is known, all species deposit their eggs in the same manner as the periodical cicadas, and in deciduous, broad-leaved evergreens or conifers. A few species belonging to the genus *Okanagana* injure conifers and may be a problem for West Coast Christmas tree growers.

Egg punctures can be extremely damaging to nursery stock as well as to specimen plants. Twigs and small branches with many bark slits wilt and die, or are broken off by winds. The symmetry of ornamental plants may be destroyed. Native deciduous forest trees may be damaged and ornamentals, such as azaleas, are sometimes killed by the cicada's activities.

References: 127, 151, 304, 320, 462, 492, 604, 700, 743, 847

A, B. A broadwinged katydid, *Microcentrum rhombifolium,* on a *Bougainvillea* twig. Arrows point to eggs lined up in typical fashion on the twig and an azalea leaf.

C. Periodical cicada injury to a maple tree. (Plates 124 and 125 illustrate similar symptoms caused by other kinds of insects.)

D. *Tibicen linnei,* a dog-day cicada; 40 mm long. [Cicadidae]

E. The eastern lubber grasshopper, *Romalea guttata.* [Acrididae]

F. A nymphal stage of the eastern lubber grasshopper.

G. Periodical cicadas, *Magicicada septendecim*; about 35 mm long. Note cut bark where eggs have been laid (*arrow*).

A
B
C
D
E
F
G

(Plate 236, continued)

***Table 19.* The coincidence of periodical cicada broods for the years 1906–2048**

Brood	I	II	III	IV	V	VI	VII	VIII	IX	X	XI	XII	XIII	XIV	XV	XVI	XVII
XVIII		1945				1932				1919				1906			
XIX			1946				1933				1920				1907		
XX				1947				1934				1921				1908	
XXI	1961				1948				1935				1922				1909
XXII		1962				1949				1936				1923			
XXIII			1963				1950				1937				1924		
XXIV				1964				1951				1938				1925	
XXV	1978				1965				1952				1939				1926
XXVI		1979				1966				1953				1940			
XXVII			1980				1967				1954				1941		
XXVIII				1981				1968				1955				1942	
XXIX	1995				1982				1969				1956				1943
XXX		1996				1983				1970				1957			
XVIII			1997				1984				1971				1958		
XIX				1998				1985				1972				1959	
XX	2012				1999				1986				1973				1960
XXI		2013				2000				1987				1974			
XXII			2014				2001				1988				1975		
XXIII				2015				2002				1989				1976	
XXIV	2029				2016				2003				1990				1977
XXV		2030				2017				2004				1991			
XXVI			2031				2018				2005				1992		
XXVII				2032				2019				2006				1993	
XXVIII	2046				2033				2020				2007				1994
XXIX		2047				2034				2021				2008			
XXX			2048				2035				2022				2009		

Source: USDA, Plant Pest Control Div., *Cooperative Economic Insect Report,* 18(1968): 321–322.

Crickets and Walkingstick (Plate 237)

Crickets are probably best known for their songs (chirps), most likely noticed in the early evening after a hot August day. Some people think the tree cricket's song is the most musical of all insect sounds. Though tree crickets may have some aesthetic value, they have proven to be serious economic pests, especially under nursery conditions. The injury they cause comes not from their feeding but from their egg-laying habits.

Tree crickets inhabit trees, bushes, and high weeds. Of the 16 species found north of Mexico, the snowy tree cricket, *Oecanthus fultoni* T. J. Walker (A), the blackhorned tree cricket, *O. nigricornis* Walker (D), and *O. angustipennis* Fitch, sometimes called the narrow-winged cricket, are the most important.

Tree crickets feed on a great variety of plant and animal material, including aphid and scale insects, foliage, flower parts, ripe and decaying fruits, and fungus fruiting bodies. However, their feeding is never economically important. Under some circumstances they are predators of pest insects on annual crops. Normally they lay eggs in shoots and in 2- to 4-year-old twigs or stems up to 5 cm in diameter of a wide range of woody plants, from brambles to trees. This may have one of three results: a heavily calloused scar that, in time, completely covers the wound; a permanently disabled plant part; or the death of the affected parts if the initial wounds become infected with one of several canker-forming fungi (see Sinclair et al., *Diseases of Trees and Shrubs*).

Oviposition injury and subsequent disease can be disastrous in nurseries. Such injury always begins in autumn, when there is little to be feared from most pest species. The initial opening in the bark, a 1–4 mm hole, is made by the tree cricket's mandibles. The female inserts her ovipositor into the fresh wound, deepens the hole, and embeds her eggs individually in the pith in a row. Following oviposition, she caps the embedded eggs with chewed bark mixed with external debris and often including fragments of fungus fruiting bodies, or with her own feces, depending on the tree cricket species.

The eggs overwinter, well protected from bad weather. They begin to hatch about mid-June in western New York, and earlier farther south. There are five nymphal instars.

All the tree crickets named in Table 20 are nocturnal. They pair, oviposit, and feed mostly at dusk or at night. Certain sphecid (thread-waisted) wasps are important predators of the tree cricket. These hunters have the uncanny ability to find their prey, which they anesthetize by stinging and then carry back to their nests as food for their brood.

Tree cricket ovipositor slits have been implicated in shoot dieback of *Taxus*. The female makes only one or two slits per shoot and deposits a single egg in each slit. There are wasp parasites of these cricket eggs.

The mormon cricket, *Anabrus simplex* Haldeman (G), is a wingless longhorn grasshopper (Tettigoniidae) that occurs in dry areas of the western states, Canadian prairie provinces, and the interior of British Columbia (for related eastern species, see Plate 236). It is large, heavy bodied, and omnivorous. When high populations occur, *A. simplex* is often a pest of food production agriculture. When migrating, it feeds on woody plant foliage and shoots, including ponderosa pine. Feeding occurs mainly in the evening. Eggs are laid singly in the soil, with one generation per year.

Stick insects are strange-looking wingless creatures whose bodies blend well with their hardwood host plants. They truly look like sticks and feed, as late-instar nymphs and adults, throughout the tree crown. *Diapheromera femorata* (Say), commonly called walkingstick (F), varies in color from brown to green, sometimes mottled with shades of gray and red. Walkingsticks are slow moving and may remain motionless for long periods. They are primarily forest defoliators and, except for small towns near large forests, they are not likely to be of concern in the urban forest.

Eggs are laid in autumn until the arrival of cold weather. The oval

A. An adult snowy tree cricket; 14 mm long. [Gryllidae] (Courtesy Gray Collection, Oregon State Univ.)
B. A 3-year-old tuliptree twig showing holes made by the blackhorned tree cricket and 1-year-old callus tissue. (Courtesy W. H. Hoffard, U.S. Forest Service.)
C. A longitudinal section of a viburnum twig showing yellow curved eggs (*arrow*) and hole plugs produced by the blackhorned tree cricket.
D. An adult blackhorned tree cricket; 14 mm long. (Courtesy W. H. Hoffard, U.S. Forest Service.)
E. Egg-laying damage on apple caused by the snowy tree cricket. (Courtesy New York State Agricultural Experiment Station, Geneva.)
F. A walkingstick adult; 5 cm long. [Phasmatidae] (Courtesy Reis Collection, Entomological Society of America.)
G. A mormon cricket adult; 4 cm long. [Tettigoniidae] (Courtesy Gray Collection, Oregon State Univ.)

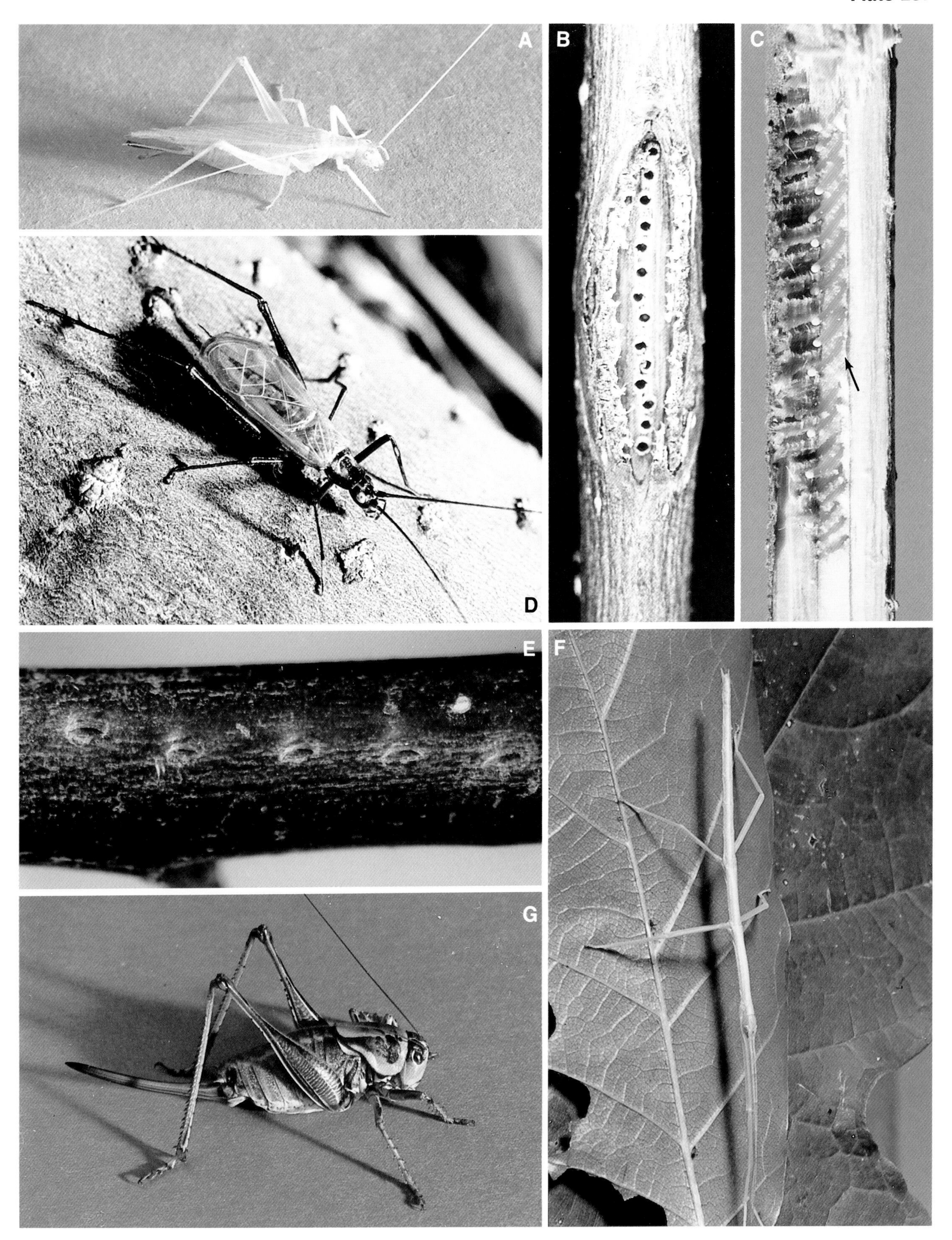
A
B
C
D
E
F
G

***Table 20.* Tree crickets: Hosts, injury, and distribution**

Common name	Oviposition: Hosts	Oviposition: Pattern	Distribution
Snowy tree cricket	Wide range: apple, ash, elm, hawthorn, lilac, walnut, rose, willow	Single puncture, or widely dispersed rows (E) in pith	Transcontinental; north of Mexico
Blackhorned cricket	Cherry, dogwood, poplar, viburnum, others	Vertical row of punctures (B) in bark phloem	Transcontinental; north of Mexico
Narrowwinged cricket	Alder, apple, maple, oak, lilac, dogwood, others	Single puncture in bark phloem associated with a lenticel	Texas to Ontario, and east

eggs are about 1.5 mm in diameter; they are black or brown with an olive-colored band on one side. The female ejects her eggs indiscriminately from high in the tree (producing an average of three per day), and they drop randomly to the litter below, where they overwinter. Eggs hatch early in June, and the nymphs feed on understory shrubs until midsummer, and thereafter in the tree crowns.

One generation occurs every 2 years in the North. In the southern part of its range one generation per year is the rule. Outbreak populations are very cyclic. The preferred hosts of older nymphs and adults are oak, basswood, and wild cherry, but they also feed on aspen, birch, hackberry, hickory, locust, dogwood, and apple. *D. femorata* ranges over the eastern half of the United States and adjacent Canada.

In the far West the stick insect *Timema californica* Scudder attacks conifers. It looks somewhat like an earwig and varies from pink to green. Though extensive defoliation is rare, in 1980 a small acreage of Douglas-fir was damaged in northern California.

References: 53, 55, 271, 272, 286, 378, 686, 777, 832, 856

Wasps and Bees as Plant Pests (Plate 238)

The European hornet, *Vespa crabro germana* Christ, is the largest hornet found in the United States (E). It resembles the well-known cicada killer wasp but is stouter, hairier, and more reddish brown. It injures plants not by feeding but by chewing bark from twigs and branches for use in building its nest. This results in severe girdling (A), which kills the distal portion of branches. The injury occurs most extensively in August and September, when large colonies of hornets have developed. Squirrels are often falsely accused of causing this damage (Plate 241).

The trees and shrubs most frequently attacked by this giant hornet are lilac, birch, willow, boxwood, mountain ash, poplar, and, occasionally, rhododendron. European hornets may be found all along the Atlantic Coast from Massachusetts to Georgia. Their presence has also been reported in Pennsylvania and Ohio. Nests may be located up to 90 meters away from injured plants. The nests usually are located in partitions of buildings, tree holes, or holes in the ground.

The work of leafcutter bees, *Megachile* species, is usually little more than a curiosity. The bee is black and looks somewhat like a small bumble bee with rows of white hairs that fringe the abdomen. The female cuts circular pieces of foliage (D), usually from rose, azalea, redbud, or ash, and uses the discs as a lining and plug for her egg cells. The bees are solitary but they make numerous cells in hollowed-out twigs and other protected natural cavities. These bees are known throughout the United States.

A group of wasplike insects commonly called horntails (F) are more of a naturalist's curiosity than a pest. They derive their common name from the hornlike tail at the end of the abdomen.

One of the common horntails is the pigeon tremex, *Tremex columba* Linnaeus. The larva of this species infests maple, elm, oak, hickory, sycamore, beech, apple, and pear. It is whitish and is deeply segmented with fleshy, poorly formed thoracic legs. It grows to a length of about 50 mm. Larvae bore round tunnels, about the size of a lead pencil, through the heartwood of host trees. They maintain no opening to the outside.

The adult female horntail has the unusual ability to thrust her sharp ovipositor into solid wood to a depth of 13 mm, where she deposits a single egg. The ovipositor frequently becomes wedged in the wood. In such cases the trapped female dies. The female prefers to lay her eggs in unhealthy and dying trees. She is the vector of a vigorous tree-rotting fungus, *Cerrina* (=*Daedalea*) *unicolor* (see Sinclair et al., *Diseases of Trees and Shrubs*). Several races of horntails occur. They are found primarily in the northern part of the United States from coast to coast.

The extremely large parasite *Megarhyssa macrurus macrurus* (Linnaeus) lays its eggs in horntail larval galleries. The female has an ovipositor nearly 8 cm long that she drills deep into wood infested with horntail larvae. Often the ovipositor cannot be withdrawn from the tree and the female dies.

References: 689, 756

A. Fresh injury to a birch twig caused by the European hornet, *Vespa crabro germana*.
B. Year-old injury to birch caused by the European hornet.
C. Close-up to show the texture of the wood after the European hornet has removed birch bark.
D. Rose leaves. Note typical circular cuts made by leafcutter bees.
E. An adult European hornet.
F. An adult horntail, *Tremex columba*. [Siricidae]
G. Side view of a living horntail.

A
B
C
D
E
F
G

Ants (Plate 239)

Ants belonging to the genus *Camponotus* are called carpenter ants because they make galleries in the softer part of wood to provide cradles and nesting space for their broods. Nests may be found in the interior of living trees (A) such as sugar maple, sweetgum, *Prunus* species, northern white cedar, and balsam fir, in standing dead trees, in stumps, or in the wooden parts of buildings. There are some soil-nesting species.

Of the many species in this group, the black carpenter ant, *Camponotus pennsylvanicus* (De Geer), has the widest range and is the most common. It occurs from Quebec and Ontario west to North Dakota and south to Texas and Florida. It is a wood-nesting species; the nest is usually a few feet from ground level, but sometimes colonies reside high in a tree. The ants do not eat wood but continually cut chips of wood, usually from between annual rings, to extend their galleries. The wood chips are usually carried from the nest and dropped at the base of the host tree. Over time, the galleries may weaken the xylem core to such an extent that trees along city streets or in any urban setting become a hazard to buildings and people. A colony may betray its presence by its foraging trails, which are well established and may be readily visible in lawns. Trails also run underground, coming to the surface a distance away from the nest. Colonies can also be detected by sawdustlike wood fragments that accumulate at or near nest openings.

The food of carpenter ants consists largely of dead and live insects, honeydew obtained from aphids or other sapsucking insects, the juice of ripe fruit, and sap from certain plants.

A colony is made up of a reproductive female, males, workers, eggs, larvae, and pupae. The workers range from 6 to 13 mm long. The adults and workers are typically black, but some individuals are reddish black. They have no sting, but large workers can inflict a painful bite with their strong mandibles (C), and the wound can be aggravated by the injection of, or contact with, the formic acid that they produce.

Fire ants have gained fame because of their damage to farm crops and livestock. They get their name because, when they bite, they inject a venom that causes a fiery itching. There are several species of fire ants, but the one of greatest concern is the red imported fire ant, *Solenopsis invicta* Buren. All fire ants are predaceous on insects and other small invertebrates. They also are attracted to sugar and honeydew. When foraging, they may gnaw holes in roots and buds, and in the spring they seek sweet sap from trees. They cannot be tolerated in plant nurseries. Fire ants are mound builders. On home lawns mounds are commonly built around trees or shrubs (D).

Fire ant colonies consist of three castes: queens, drones, and workers. Workers are most abundant; an average colony contains more than 100,000 of them. Workers vary from 1.5 to 6 mm long. They are currently distributed from the Carolinas to Florida and west to Texas.

The Allegheny mound ant, *Formica exsectoides* Forel, is probably most troublesome in Christmas tree plantations and in conifer reforestation areas. Conifers are jeopardized because of this ant's unique behavior. It will not tolerate shade over the mound, although it does allow vines and grass to grow around its base. Most mounds are 35–50 cm high (E). These ants are most intolerant of young trees that grow adjacent to the nest and attack this vegetation at the base of the stem. Formic acid, produced by special glands, is injected into or rubbed on exposed live phloem tissue (H). This kills the tree by preventing the downward movement of food through the inner bark.

Workers are about 3–6 mm long and reddish brown. Their food consists of living and dead insects, honeydew from various sapsucking insects, and overripe fruits. *F. exsectoides* occurs in the Alleghenies of Virginia, West Virginia, Maryland, and Pennsylvania and north into New England.

Louisiana and Texas is the home territory of the Texas leafcutting ant, *Atta texana* (Buckley). This species is notable not only because it injures plants but also because of its natural history. These ants live in large colonies that may exceed 2 million individuals. They do not eat plants but cultivate their own food, a fungus, that grows on the foliage brought to the nest. The ants are selective in their choice of foliage: hackberry, beech, redbud, oak, hickory, and sweetgum in summer, and pine needles during the winter. They may completely defoliate trees. Ant workers always start at the top of a tree and work their way down. Foraging for leaves usually occurs at night in summer and during daylight in winter.

Leafcutting ants nest in the soil. A colony is marked by the presence of several crater-shaped mounds, with the walls 13–35 cm high and 30–38 cm in diameter. At the bottom of each crater is an entrance hole that goes into a labyrinth of tunnels and chambers. Leaf pieces are placed in fungus garden chambers, where they serve as the medium for the growth of a fungus. Clearly defined foraging trails that lead to the desired plant are established on the surface of the ground. These trails are marked by chemicals that the ant produces, called trail pheromones. Hundreds of ants may be seen carrying round leaf fragments back to the nest as they follow their chemical road signs. Worker ants are 1.5–12 mm long and rusty red.

References: 258, 259, 549, 551, 720, 862, 866

A. A black willow tree that has been wind thrown. Note the extent of years of carpenter ant damage.
B. A black carpenter ant that has crawled from one of its galleries; about 12 mm long.
C. The head of a black carpenter ant showing the sharp cusps of its mandibles.
D. A fire ant mound near the base of a pittosporum shrub.
E. The mound of an Allegheny mound ant colony.
F. A carpenter ant among wood chips that it and other colony members have removed from a nest tree. [Formicidae]
G. An Allegheny ant mound in cross section, showing a maze of tunnels.
H. A white pine seedling killed by formic acid injected into the soft, smooth bark by Allegheny mound ants. Note the bark necrosis at the root crown.

A
B
C
D
E
F
G
H

Slugs and Snails (Plate 240)

Slugs and snails often devour the foliage of plants in a manner similar to that of insects. They leave a trail of mucus that glistens in light for extended periods, even after it dries. They produce large holes in fruit, flowers, roots, and tubers, as well as in foliage. On woody plants they feed only on the foliage. Woody ornamentals that lie on the ground—like cotoneaster and vines—are most susceptible. Certain snails are reported to be particularly injurious to citrus trees and privet shrubs.

Slugs and snails are terrestrial relatives of oysters and clams. A slug is a snail with only an internal shell remnant. Both slugs and snails have tentacles or feelers on their heads. Each has a filelike mouthpart that rasps at the soft plant tissue and draws particles into its mouth opening. Feeding occurs primarily at night; during the day both slugs and snails remain concealed, usually on the ground where it is dark and damp. A large mucous or slime gland is located behind the mouth of slugs. There are, however, many mucous glands interspaced over much of the body that serve to keep the entire body surface moist.

Slugs lay eggs in damp places in masses of 25 or more (B). The newly hatched slug is dull white. Under favorable conditions it may grow to a length of 25 mm in a month. When it is fully grown it is brown. It takes more than a year for slugs to mature, at which time they are 80–120 mm long.

The spotted garden slug, *Limax maximus* Linnaeus, was introduced from Europe (C). Urban and suburban gardens are its most common haunts. It is found from Massachusetts to Virginia and west to Oregon and California. The spotted garden slug feeds on foliage and flowers close to the ground, but it is also known to climb trees.

The snail shown in panel A is probably *Cepaea nemoralis* (Linnaeus), known as the banded wood snail (=brown-lipped snail). The breadth of the mature shell is about 22 mm. As a rule it takes a year for this species to reach maturity. It also is an introduced species and occurs in Massachusetts, New York, New Jersey, Virginia, Tennessee, Pennsylvania, and Wisconsin. It has been recorded in Ontario and is believed to occur in California.

Helix aspersa Müller, sometimes called the brown garden snail, is one of the edible European snails. It is seldom, if ever, used for that purpose in the United States. This species can be a noxious pest in many areas where the humidity remains high for at least part of the year. It feeds on such plants as boxwood, rose, hibiscus, magnolia, and peach. When the snail is fully grown, its brown shell is flecked with yellow, wrinkled, and up to 38 mm in diameter. The round, white eggs are laid in masses about 25 mm beneath the soil surface. From 10 to 20 eggs may be laid at a time. Young snails grow slowly. They often require 2 or 3 years to reach maturity. The brown garden snail is believed to be found throughout most of the deep South as well as in Arizona and California. It was successfully eradicated in Florida during the 1960s.

Many snails feed on the foliage of trees and shrubs in tropical and subtropical regions. *Bulimulus alternatus mariae* (Albers) is found in southeast Texas, where it feeds on legume foliage, especially the acacias (E). They remain on trees except for the time required for egg deposition, which takes place in leaf mold, under or near rocks. Its shell is uniformly white. Most of the tree-climbing snails are not abundant enough to become pests. In fact, a few are listed as endangered species because of habitat destruction. Only two families of tree-climbing snails, Bulimulidae and Pupillidae, are found in the United States mainland. Of continuous regulatory concern is the white garden snail, *Theba pisana* (Müller) (Helicidae), native to England, Ireland, France, and Mediterranean countries. Considered by some biologists to be the worst potential agricultural pest among the helicid snails introduced into North America, it is currently known only in San Diego County, California. It is a tree-climbing defoliator with an astounding biotic potential. Many species of ornamental shrubs and trees are subject to its attack. The shell is a somewhat depressed spiral, with an ivory background color marked by a variable number of narrow dark brown spiral bands that may be solid or broken. The adult shell size ranges from 12 to 15 mm in diameter. These snails commonly aestivate, during hot dry weather, on practically any vertical surface.

Natural enemies of snails and slugs include toads, several parasitic flies, and carabid and lampyrid beetles.

References: 13, 90, 159, 166, 296, 417, 477

A. A fully grown snail believed to be *Cepaea nemoralis*.

B. Eggs of the gray garden slug, *Agriolimax reticulatus* (Müller). The species is common throughout much of the United States. (Courtesy New York State Agricultural Experiment Station, Geneva.)

C. The spotted garden slug, *Limax maximus*.

D. Clematis leaves supporting two snails believed to be *Helix aspersa*. The foliage shows typical slug and snail injury.

E. *Bulimulus alternatus mariae*, one of the few tree-climbing snails in the United States. Mature snails are about 3 cm long.

A
B
C
D
E

Birds and Small Mammals That Injure Trees (Plate 241)

Birds and small mammals on diverse occasions damage trees and shrubs in city residential areas or, more commonly, in suburbs. The most common offenders are woodpeckers, squirrels, mice, or voles, and rabbits. It is common for the uninformed to attribute their damage symptoms to arthropods.

There are at least seven species of woodpeckers that can make wounds in bark deep enough to reach vascular tissue and allow sap to flow. The sapsuckers are of most concern. These birds rarely, if ever, dig through bark to capture wood-boring insects; rather, they feed on sap and inner bark. Sapsucker feeding results in a distinct pattern of uniformly drilled holes through the bark (A–C). These holes are characteristically arranged in horizontal rings, or bands, and are sometimes aligned in vertical rows. The holes may be round or rectangular and the injury may be slight or severe. These birds usually select a favorite tree and feed on it year after year.

Table 21. Types of food tap holes drilled by the yellow-bellied sapsucker

	Primary bands			Columns	
	Sap bands	Bast bands	Spiral bands	Sap	Bast
Season					
Summer	Rare	Rare	None	Abundant	As drilled
Fall	Common	Abundant	None	Rare	Rare
Winter	Rare to common	Rare to common	None	Rare to common	Common
Spring	Common	Common	Common	As available	Abundant
Preferred plants	Pines Hemlocks Maples Many others	Maples Beeches Elms Few others	Aspens Others?	Birches Aspens Beeches Many others; nonsummer	
Location on plant	Most of bole Angiosperms: major limbs	Bole	Uppermost branches	Bole Angiosperms: limbs	
Group configuration	Line	Line	Spiral	Band below columns	
Shape of hole	Round	Ragged	Rectangular	Round	
Cross section	V-shaped	Inverted V	Squared	Inverted V	

Source: Tate, ref. 776a, courtesy *The Auk*.

Sapsuckers are handsome birds. The red-breasted sapsucker, *Sphyrapicus varius ruber* (Gmelin), is common west of the Sierra Nevada mountain range and north to Alaska. The yellow-bellied sapsucker (A), *S. varius varius* (Linnaeus), is common throughout much of the United States and Canada and breeds in conifer and deciduous forests. It is a migratory bird that usually winters in the mid-latitudes of the United States or farther south to the Gulf of Mexico. During warm winter days sapsuckers drill holes in many species of trees and tend these holes, taking as food the small amount of sap that accumulates. In the spring during northward migration, they make holes in conifers, where they obtain most of their food. When the buds of angiosperms begin to swell, sapsuckers pluck them off and eat the soft inner portion. During summer months their food consists of the soft inner bark (including phloem) and sap, and occasional insects such as spruce budworm, forest tent caterpillar moths, and carpenter ants. Birch trees are by far the most commonly used summer food source. It has been estimated that one pair of nesting sapsuckers may kill one or two trees a year.

Birds in migration may visit the same tree twice annually, and with an average longevity of about 6 years, a single bird may severely injure many trees at intermediate stops. Apple orchards have been destroyed by these birds. Fungi that cause tree diseases often gain access to a host tree by way of the bird's beak. Fungus spores are sometimes blown or washed into the tree wounds. The shake defect phenomenon (the separation of annual rings in lumber), of concern to foresters, has been attributed in part to initial damage by sapsuckers.

The tree species most commonly attacked by sapsuckers are hemlock, pine, red spruce, fir, Douglas-fir, red maple (*Acer rubrum*), sugar maple, aspen, beech, birch, larch, willow, magnolia, apple, certain palms, and species of *Acacia, Grevillea,* and *Casuarina*. There are records of at least 275 species of trees and shrubs that are food sources for sapsuckers.

Squirrels—particularly the red squirrel, *Tamiasciurus hudsonicus* Erxleben—are fond of maple sap in the spring. With their chisellike teeth, they make V-shaped wounds (E) on bark and then lap up the sap. Dust and fungus spores become trapped in this sticky sap. If canker disease fungi colonize the wound, a large scar develops and the tree may die or never again grow vigorously (see Sinclair et al., *Diseases of Trees and Shrubs*).

In conifer-growing regions of the Northwest, and probably elsewhere, red squirrels eat cambium and phloem tissue, causing severe bark injury. Stems may be completely girdled. Such injury occurs largely on young trees from 6 to 20 cm DBH (diameter at breast height) and during the early part of the growing season. Squirrels frequently attack the bases of trees, where vascular nutrients are at their highest volume. Some wildlife specialists believe that phloem tissue is not a preferred food and that where damage occurs, it is because alternate foods (seeds, nuts) are not available.

Conifers are also injured by other foraging activities of the squirrel. In the spring, in suburban communities where squirrels abound, it is common to see hundreds of spruce twigs littering the ground. Studies of the feeding habits of the tassel-eared squirrel, *Sciurus aberti* Woodhouse, on ponderosa pine indicate that these squirrels feed predominantly on cortical tissues of small twigs. To obtain cortical tissues, squirrels remove terminal twigs in the upper portions of the tree by biting through a branch stem several centimeters from the terminal needle cluster. The needle cluster is clipped from the end of the twig and drops to the ground. The squirrel then removes the rough outer bark with the teeth to expose the phloem, cambium, and some of the current year's xylem. Feeding studies suggest that trees that produce small amounts of monoterpenes, a class of organic compounds that are toxic to certain mammals and insects, are most severely attacked. Large amounts of monoterpenes, particularly α-pinene, act as a feeding deterrent. α-Pinene is the same chemical that attracts certain bark beetles.

In the spring and again in late autumn the gray squirrel, *Sciurus carolinensis* Gmelin, occasionally strips bark from a variety of conifer

---

A. The yellow-bellied sapsucker, a member of the woodpecker family, injures many kinds of coniferous and deciduous trees. Sapsuckers range throughout North America. (Courtesy U.S. Forest Service.)
B. Sapsucker damage to birch.
C. Sapsucker damage to *Acacia decurrens*. Pitch flow is a defense mechanism of the tree.
D. Girdling damage on *Ligustrum* sp. caused by rabbits during the winter.
E, G. The V-shaped notches on maple bark are caused by squirrels that bite the bark to get sap.
F, H. Pellet gun injury to alder. Note (*arrow*) pellet embedded in the bark.
I. Girdling injury to a young maple caused by mice.

A
B
C
D
E
F
G
H
I

and hardwood trees. Some bark is probably eaten, but most seems to be used to construct nests. Severe tree damage may result.

Squirrels also eat the buds of maple, elm, beech, ironwood, birch, willow, poplar, and spruce in the spring. Deuber of Yale University observed two pairs of squirrels, in little more than a day, cut 2866 twigs from elm trees to obtain the seeds (49). These seeds made up their entire diet during the latter part of May and early June. One way to avoid such tree vandalism is to trap the offending squirrels and relocate them in rural areas; however, this in no way ensures the survival of the animals or that other squirrels will not take their places.

Rabbits (*Sylvilagus* species) and mice, also known as voles (*Microtus pennsylvanicus* Ord), may girdle trees and shrubs by eating the bark during the winter. Mice construct mazes of tunnels under snow or matted grass to reach sources of food. They girdle plants under the snow (I). Bark is not ordinarily mouse food, but if normal food is scarce and the mouse population is high, trees and shrubs may be seriously damaged. Rabbits girdle trees at the snow line, which at times may be several feet above the base of the tree or shrub (D). Both rabbits and mice feed on thin-barked trees and shrubs such as holly, ligustrum, pine, birch, apple, and young maples.

People injure trees, sometimes thoughtlessly. A young alder (H) served as a support for a pellet gun target. The damage, shown as bark scars, was caused by the BBs. Other common sources of injury by humans are improperly used lawnmowers and weed trimmers.

References: 49, 82, 143, 226, 540, 776a

Sources of Information on Pests and Pest Control

Most of the states and provinces in the United States and Canada have governmental agencies or educational services that provide pest management recommendations and consultation about pest problems. Pest control recommendations frequently change as new pesticides and regulations are developed and our understanding of the biology of pests expands. Nonchemical means of pest management also become more practical as biological information increases. Research laboratories and institutions that develop pest control recommendations include agricultural and forest experiment stations, colleges of agriculture and forestry (often located at state or provincial universities), and state, provincial, or federal departments of agriculture.

Information, technical assistance, and recommendations are available directly from county, state, or federal extension services. The Cooperative Extension Service and various state extension divisions are represented by local county extension agents or agricultural advisers. The local county or city extension offices should be contacted for assistance related to plant culture, maintenance, and protection, and for recommendations for pest control. They serve not only commercial agricultural interests, but consumers, home horticulturists, and other residents within their governmental units. Leaflets, bulletins, and other publications on the subject of pest control are generally available from the extension agents' offices.

It is best to direct inquiries to local representatives. An extension or agricultural agent is usually listed in the telephone directory under the *city* or *county government*, but sometimes under the *United States Department of Agriculture*. In Canada call or write your nearest federal or provincial Forestry Office or write to Information Services, Agriculture Canada, Ottawa K1A 0C7.

Other sources of information include botanical gardens, arboretums, horticultural consultants, arborists, nurserymen, garden store managers, and garden editors of newspapers and magazines.

Land-Grant Institutions and Agricultural Experiment Stations in the United States

For help in finding who is the local extension representative write to the extension entomologist or plant pathologist at the college of agriculture of your state university or to your state experiment station.

Alabama: Auburn University, Auburn 36849
Alaska: University of Alaska, Fairbanks 99701; Experiment Station, Palmer 99645; Cooperative Extension, 2651 Providence Avenue, Anchorage
Arizona: University of Arizona, Tucson 85721
Arkansas: University of Arkansas, Fayetteville 72701; Cooperative Extension Service, P.O. Box 391, Little Rock 72203
California: University of California, Berkeley 94720; Riverside 92521; Davis 95616
Colorado: Colorado State University, Fort Collins 80523
Connecticut: University of Connecticut, Storrs 06268; Connecticut Agricultural Experiment Station, New Haven 06504
Delaware: University of Delaware, Newark 19717
District of Columbia: Extension Service 20002
Florida: University of Florida, Gainesville 32611
Georgia: University of Georgia, Athens 30601; Agricultural Experiment Station, Experiment 30212; Coastal Plain Station, Tifton 31794
Hawaii: University of Hawaii, Honolulu 96822; Hilo 96720
Idaho: University of Idaho, Extension Service, Boise 83702; Agricultural Experiment Station, Moscow 83843
Illinois: University of Illinois, Urbana 61820
Indiana: Purdue University, Lafayette 47907
Iowa: Iowa State University, Ames 50010
Kansas: Kansas State University, Manhattan 66506
Kentucky: University of Kentucky, Lexington 40546
Louisiana: Louisiana State University, University Station, Baton Rouge 70803
Maine: University of Maine, Orono 04473
Maryland: University of Maryland, College Park 20742
Massachusetts: University of Massachusetts, Amherst 01002; Suburban Experiment Station, 240 Beaver Street, Waltham 02154
Michigan: Michigan State University, East Lansing 48823
Minnesota: University of Minnesota, St. Paul 55101
Mississippi: Mississippi State University of Applied Arts & Sciences, Mississippi State 39762
Missouri: University of Missouri, Columbia 65211
Montana: Montana State University, Bozeman 59715
Nebraska: University of Nebraska, Lincoln 68583
Nevada: University of Nevada, Reno 89557
New Hampshire: University of New Hampshire, Durham 03824
New Jersey: Rutgers, The State University, New Brunswick 08903
New Mexico: New Mexico State University, Las Cruces 88003
New York: Cornell University, Ithaca 14853; Agricultural Experiment Station, Geneva 14456
North Carolina: North Carolina State University, Entomology Extension, Raleigh 27695
North Dakota: North Dakota State University, Fargo 58102
Ohio: Ohio State University, Columbus 43210; Ohio Agricultural Research & Development Center, Wooster 44691
Oklahoma: Oklahoma State University, Stillwater 74074
Oregon: Oregon State University, Corvallis 97331
Pennsylvania: Pennsylvania State University, University Park 16802
Puerto Rico: University of Puerto Rico, Mayaguez 00708
Rhode Island: University of Rhode Island, Kingston 02881
South Carolina: Clemson University, Clemson 29631
South Dakota: South Dakota State University, Brookings 57006
Tennessee: University of Tennessee, Knoxville 37901
Texas: Texas A & M University System, College Station 77841; Agricultural Experiment Station, Lubbock 79414; Cooperative Extension, 17360 Coit Road, Dallas 75252
Utah: Utah State University, Logan 84321
Vermont: University of Vermont, Burlington 05401
Virginia: Virginia Polytechnic Institute and State University, Blacksburg 24061
Virgin Islands: College of the Virgin Islands, St. Croix Campus 00850
Washington: Washington State University, Pullman 99163; Western Washington Research and Extension Center, Puyallup 98371
West Virginia: West Virginia University, Morgantown 26506
Wisconsin: University of Wisconsin, Madison 53706
Wyoming: University of Wyoming, Laramie 82070

U.S. Department of Agriculture, Forest Pest Control Offices

USDA Forest Service
Federal Office Building
P.O. Box 1628
Juneau, AK 99802

USDA Forest Service
11177 West 8th Avenue
P.O. Box 25127
Lakewood, CO 80225

USDA Forest Service
630 Sansome Street
San Francisco, CA 94111

USDA Forest Service
1720 Peachtree Road NW
Suite 800
Atlanta, GA 30367

USDA Forest Service
2500 Shreveport Highway
Pineville, LA 71360

USDA Forest Service
1992 Folwell Avenue
St. Paul, MN 55108

USDA Forest Service
Federal Building
P.O. Box 7669
Missoula, MT 59807

USDA Forest Service
P.O. Box 5895
Asheville, NC 28803

USDA Forest Service
Louis Wyman Forestry Sciences Lab
P.O. Box 640
Durham, NH 03824

USDA Forest Service
Federal Building
517 Gold Avenue SW
Albuquerque, NM 87102

USDA Forest Service
P.O. Box 3623
Portland, OR 97208

USDA Forest Service
370 Reed Road
Broomall, PA 19008

USDA Forest Service
Federal Office Building
324 25th Street
Ogden, UT 84401

USDA Forest Service
180 Canfield Street
Morgantown, WV 26505

Regional Pest Control Information in Canada

British Columbia: Department of Biological Sciences, Pestology Centre, Simon Fraser University, Burnaby V5A 1S6
Forest Protection Branch, Ministry of Forests, 1450 Government Street, Victoria V8W 3E7
Faculty of Forestry, University of British Columbia, 191-2357 Main Mall, Vancouver V6T 2H1
Pacific Forest Research Centre, Department of Forestry, 506 West Burnside Road, Vancouver V8Z 1M5
Maritime Provinces: Department of Agriculture and Forestry, P.O. Box 2000, Charlottetown, P.E.I. C1A 7N8
Dept. of Lands and Forests, P.O. Box 68, Truro, Nova Scotia B2N 5B8
Dept. of Natural Resources, Forest Management Branch, Box 6000, Fredericton, N.B. E3B 5H1
Maritimes Forest Research, Department of Forestry, P.O. Box 4000, Fredericton, N.B. E3B 5P7
Faculty of Forestry, The University of New Brunswick, Bag Service No. 44555, Fredericton, N.B. E3B 6C2
Newfoundland: Department of Forest Resources and Lands, Building 810, Pleasantville, St. John's A1A 1P9
Newfoundland Forest Research Centre, Department of Forestry, Building 305, P.O. Box 6028, Pleasantville, St. John's A1C 5X8
Ontario: Ministry of Natural Resources, Maple L0J 1E0
Environment Canada, Ottawa K1A 0E7
Fisheries and Oceans Canada, 240 Sparks Street, Ottawa K1A 0G6
Pesticides Division, Plant Products and Quarantine Directorate, Agriculture Canada, Room 1127A, K.W. Neatby Building, Ottawa K1A 0C6
Forest Pest Management Institute, Department of Agriculture, Sault Ste. Marie P6A 5M7
Great Lakes Forest Research Centre, Department of Forestry, P.O. Box 490, 1219 Queen Street East, Sault Ste. Marie P6A 4M7
Petawawa National Forestry Institute, Department of Agriculture, Chalk River K0J 1J0
The University of Toronto, Faculty of Forestry, 203 College Street, Toronto M5S 1A1
Faculty of Forestry, Lakehead University, Thunder Bay P7B 5E1
Prairie Provinces: Alberta Forest Service, Department of Energy and Natural Resources, 8th Floor, 9915-108 Street, Edmonton, Alberta T5K 6C9
Department of Parks and Renewable Resources, Box 3003, Prince Albert, Saskatchewan S6V 6G1
Department of Natural Resources, 300-530 Kenaston Boulevard, Winnepeg, Manitoba R3N 1Z4
Northern Forest Research Centre, Department of Forestry, 5320-122nd Street, Edmonton, Alberta T6H 3S5
The University of Alberta, Forestry Faculty, Faculty of Agriculture and Forestry, Edmonton T6G 2H1
Quebec: Minister de l'Energie et des Ressources, Service d'Entomologie et de Pathologie, 175 rue St.-Jean, Quebec, P.Q. G1R 1N4
Laurentian Forest Research Centre, Department of Forestry, P.O. Box 3800, Ste.-Foy, Quebec G1V 4C7
L'Université Laval, Faculté de foresterie et de Géodésie, Ste.-Foy, Quebec G1K 7P4

Glossary

Abiotic—Without life or ever being alive.
Adventitious buds—Buds that arise or occur sporadically in other than the usual location. Their formation and continued growth are usually stimulated by pruning or by insects.
Aestivation—To rest, sleep, or be inactive (dormant) during the hot summer months. An aestivating insect may be inactive for 1 or 2 months.
Anal shield—A hard plate on the terminal segment of a caterpillar and certain other immature insects.
Asexual—Literally meaning without sex. Reproducing without the union of sperm and egg.
Biotic—Relating to living things.
Brood—The individuals that hatch from the eggs of one mother, or individuals that hatch at about the same time and normally mature about the same time.
Bud blast—A developed bud that may appear to have died suddenly. Dead buds may be partially opened.
Chlorosis—A common symptom of disease marked by yellow or blanched leaves.
Chrysalis—The pupa of a butterfly.
Clypeus—A face plate (sclerite) on the head of an insect located above all of the mouth parts.
cm—Centimeter, a unit of length. 2.5 cm = 1 inch.
Commensalism—A relation between two kinds of organisms in which they live together but neither one at the expense of the other.
Cornicles—Dorsal tubular appendages on the posterior part of the abdomen of certain aphids.
Crawler—An immature stage of an adelgid or scale insect. The active crawling stage.
Crochets—Sharp hooks on the "foot" of prolegs of lepidopterous larvae; pronounced "croshays."
Cultivar (cv.)—A cultural variety of an economic plant, maintained by asexual propagation or controlled breeding.
Declivity—The usually steep descending posterior face of the elytra (wing covers) of bark beetles.
Deutogyne—A reproductive stage of certain eriophyid mites. A distinct female that characterizes one of the alternate generations. (*See* Protogyne.)
Deutonymph—The third instar of false spider mites, spider mites, and tarsonemid mites.
Diapause—A period of rest in the life of an insect when growth is nil and metabolic activity is low.
Dieback—The dying backward from bud tip to branch.
Distal—Farthest from the body or main stem.
Dorsal—Relating to the upper surface; the back.
Elytra—The hard or leathery front wings of beetles that may act as covers for the folded, functional wings.
Epidermis—The outer layer of plant tissue; the layer of an insect that produces the cuticula of an insect's exoskeleton.
Erineum—A gall made up of tiny plant hairs with round heads. The gall develops in the form of a velvety and sometimes glossy pad. These galls are caused by eriophyid mites.
Eriophyid—Any of a group of tiny mites characterized by having a slender body and two pairs of legs, rather than four.
Excrescence—A gall or abnormal growth that disfigures a plant. A tumor.
Exfoliate—To cast off or flake off, as bark.
Exocarp—The outer layer of a fruit; the fleshy part of a peach or husk of a walnut or hickory nut.
Exoskeleton—The external skeleton of insects and other arthropods. The exoskeleton is discarded (molted) at various times. The molted skeleton is sometimes called a "skin."
Exuviae—The cast exoskeleton of an insect or other arthropod.
Flocculence—A woolly, sometimes fuzzy, covering of a soft waxy substance often resembling wool.
Frass—Wood fragments mixed with excrement produced by an insect.
Free-living—Living on and free to move about as opposed to living within a gall, under bark, or in a leaf mine.
Fundatrix—A parthenogenic, often wingless female that develops from an egg. One of several adult forms of an adelgid or aphid.
Gall—An abnormal growth or excrescence on a plant. May be caused by insects, mites, bacteria, or other living entities. A plant tumor.
Generation—A measurement of time relative to birth and reproductive stage of an individual. The period between the birth of one generation and that of the next.
Growing degree-days—Arithmetic accumulation of daily mean temperatures above a designated threshold temperature. Used to relate plant and arthropod growth and maturation to environmental air temperature.
Hibernaculum—Silken shelter in which larvae of some Lepidoptera hibernate.
Honeydew—A sweet, sticky fluid excreted by aphids, certain scales, mealybugs, whiteflies, and some leafhoppers.
Hopperburn—Foliar discoloration, usually brown and usually around the periphery of a leaf, caused by the feeding of certain leafhoppers.
Inquiline—A gall inhabitant that lives in and feeds on gall tissue but is incapable of initiating galls on its own. Some may directly or indirectly kill the gall maker.
Instar—The stage of an insect between molts.
Internode—The part of a stem between two successive nodes. (*See* Node.)
Kernel—The edible portion of a nut or seed.
Larva—An immature stage between egg and pupa of an insect with complete metamorphosis. A caterpillar, maggot, or grub are examples.
Life cycle—The series of changes in the form of a life beginning with the egg, ending in the adult reproductive stage.
μm—Micrometer, a unit of length. 1000 μm = 1 mm. (*See* mm.)
Mesophyll—The middle or internal components of a leaf. (*See* Parenchyma.)
Midrib—The central vein of a leaf.
mm—Millimeter, a unit of length. 25 mm = 1 inch.
Molt—A process of shedding the exoskeleton (sometimes improperly called a "skin").
Morphologist—A scientist who studies the form and structure of living organisms.
Mouthparts—The structures of an insect that aid in obtaining food.
Necrosis—The death of tissue.
Node—A slightly enlarged portion of a stem (twig) where leaves and buds arise, and where branches and twigs originate.
Nymph—A young insect with simple metamorphosis. A young leafhopper or true bug (Hemiptera) is called a nymph.
Operculum—A lid or cover at the anterior end of an egg that opens to allow the immature insect to emerge.
Orifice—An opening in a plant part, often a gall, for the exit of insects developing inside.
Ovipositor—The external egg-laying apparatus of an insect.
Ovisac—An extension of the body wall, or specialized structure of a female, where certain kinds of insects deposit their eggs.
Ovoviviparous—Producing eggs that hatch within the maternal body.
Parasite—An animal or plant that lives in or on another living organism (host) for at least a part of its life cycle. The host is injured or killed by the relationship.
Parenchyma—Soft tissue between the epidermal layers of a leaf.
Parthenogenesis—Reproduction without fertilization.
Pathogenic—Disease producing.
Petiole—The leaf stem or stalk that attaches to a twig or shoot.
Phenology—A branch of science that deals with the relationship between climate and periodic biological phenomena, e.g., flowering, egg laying.
Pheromone—A complicated biochemical substance produced by insects to communicate with the same species through the sense of smell.
Polyphagous—An animal that successfully feeds (and reproduces) on many plant species.
Prolegs—Fleshy abdominal legs of certain insect larvae.
Protogyne—A reproductive stage of certain eriophyid mites; a malelike female. One of the two forms of the same mite species. (*See* Deutogyne.)
Protonymph—The second instar of false spider mites, spider mites, and tarsonemid mites.
Pseudogall—A false gall; appearing like a gall but not necessarily an abnormal growth.
Pubescent—Covered with short, fine hairs.
Pupa—A resting stage of an insect; the stage between the larva and the adult in insects with complete metamorphosis.
Rasping mouthparts—Insect mouthparts for filing or rasping plant tissue; typical of thrips.

Russeting—A brownish superficial roughening of the "skin" of fruits, leaves, and other plant parts.
Scale insect—A group of insects characterized as having a hard convex covering over their body with no visible appendages or segmentation. Usually small, ranging from 1 to 5 mm in length or diameter.
Scutellum—A plate (sclerite) on the dorsum of the thorax that appears more or less triangular.
Semiochemicals—Complex volatile biochemicals made by arthropods and used to convey biological messages, e.g., sex pheromones.
Sessile—Term used to describe insects, in certain stages, that are normally immobile.
Seta(e)—A bristlelike hair.
Setaceous—Bristlelike in form or texture.
Sexuparae—Winged parthenogenic female adelgids or aphids that fly from the summer host to the overwintering host.
Sign—The actual organism, skeleton, or product of a pest, used to associate or identify the cause of a symptom.
Skeletonize—Term used to describe the feeding pattern of certain leaf-feeding insects. Only the veins or basic matrix remain.
Snout beetle—A weevil. An insect with a long proboscis or snout that bears chewing mouthparts.
Sooty mold—A dark, often black, fungus growing in insect honeydew.
Spicule—A small, pointed, and rigid part of an external organ or structure.
Spiracle—The external opening of the respiratory organ of "air-breathing" insects.
Stomata—Tiny openings in leaves where respiration or exchange of oxygen and carbon dioxide takes place and water vapor is lost.
Stylet—A needlelike part of the piercing-sucking mouthparts of certain insects.
Symptom—The injury by, or plant response to, an offender.
Tarsus—The foot of an insect, consisting of one to five segments.
Taxonomist—A scientist trained in the classification of plants or animals.
Test—The external shell or covering of many invertebrate animals. Applies mainly to armored scale insects.
Thorax—The middle region of an insect's body where the legs and wings are attached.
Tubercle—A small knoblike or rounded protuberance of the derm.
Urticating hair—Bristles or hairs on the bodies of certain insects that when touched release rash-causing chemicals.
Ventral—The lower surface or undersurface; opposite the back.
Vespid wasp—A social wasp. One that builds nests made of bark, transformed into "paper" elaborated through the process of mastication.
Witches'-broom—An abnormal cluster of twigs about a common focus. A brushlike growth of small branches on trees and shrubs.

References

1. Abrahamson, L. P., Chu, H., and Norris, D. M. 1967. Symbiotic interrelationships between microbes and ambrosia beetles. II. The organs of microbial transport and perpetuation in *Trypodendron betulae* and *T. retusum* (Coleoptera: Scolytidae). Ann. Entomol. Soc. Am. 60:1107–1110.

1a. Akers, R. C., Herms, D. A., and Nielsen, D. G. 1986. Emergence and adult biology of *Agrilus difficilis* (Coleoptera: Buprestidae), a pest of honeylocust, *Gleditsia triacanthos*. Great Lakes Entomol. 19:27–30.

2. Aliniazee, M. T. 1974. Evaluation of *Bacillus thuringiensis* against *Archips rosanus* (Lepidoptera: Tortricidae). Can. Entomol. 106:393–398.

3. Allen, D. C. 1979. Observations on biology and natural control of the orangehumped mapleworm, *Symmerista leucitys* (Lepidoptera: Notodontidae), in New York. Can. Entomol. 111:703–708.

4. Ananthakrishnan, T. N. 1984. Bioecology of thrips. Indira Publishing House, Oak Park, MI. 205 pp.

5. Ananthakrishnan, T. N. 1984. Biology of gall insects. Edward Arnold, Baltimore. 362 pp.

6. Anderson, J. F., Ford, R. P., Kegg, J. D., and Risley, J. H. 1976. The red pine scale in North America. Conn. Agric. Exp. Stn. Bull. 765:1–6.

7. Anderson, J. F., and Kaya, H. K. 1978. Field and laboratory biology of *Symmerista canicosta*. Ann. Entomol. Soc. Am. 71:137–142.

8. Anderson, J. F. 1960. Forest and shade tree entomology. John Wiley & Sons, New York. 428 pp.

9. Anderson, J. F., and Kaya, H. K. 1974. Parasitism of the elm spanworm by *Telenomus alsophilae* and *Actia ontario* in Connecticut. Pages 267–276 in: Twenty-fifth anniversary memoirs. Conn. Entomol. Soc., New Haven. 322 pp.

10. Anderson, L. D., and Papp, C. S. 1961. The larger elm leaf beetle, *Monocesta corli* (Say) (Coleoptera: Chrysomelidae). Proc. Entomol. Soc. Wash. 63:203–207.

11. Annand, P. N. 1928. A contribution toward a monograph of the Adelginae (Phylloxeridae) of North America. Stanford Univ. Publ. Biol. Sci. 6. 146 pp.

12. Anonymous. 1953. Insects of importance in New Jersey nurseries. N.J. Dep. of Agric., Div. Plant Industry, Circ. 390. 175 pp.

13. Anonymous. 1959. Land slugs and snails and their control. USDA Farmers' Bull. 1895.

14. Anonymous. 1960. Pests and diseases of trees and shrubs. Wisc. State Dep. of Agric. Bull. 351. 87 pp.

15. Anonymous. 1960. The elm leaf beetle. USDA Leaflet 184. 4 pp.

16. Anonymous. 1961. Mimosa webworm and its control. Purdue Univ. Dep. Entomol. Agric. Ext. Serv. Mimeo E-11. 2 pp.

17. Anonymous. 1965–1974. Cooperative economic insect report. USDA weekly publication of APHIS.

17a. Anonymous. 1979. Spruce budworms in the Pacific Northwest. USDA For. Serv. CANUSA Newsletter 6.

18. Anonymous. 1982. Tussockosis: the tussock moth itch. Pestopics 5. British Columbia Ministry of Forests. 2 pp.

19. Anonymous. 1980. Camphor scale (*Pseudaonidia duplex*) Virginia. New state record. USDA Coop. Plant Pest Rep. 5:367.

20. Antonelli, A. L., Byther, R. S., Maleike, R. R., Collman, J. S., and Davison, A. D. 1984. How to identify rhododendron and azalea problems. Wash. State Univ. EB 1229. 28 pp.

21. Appleby, J. E., and Neiswander, R. B. 1966. Life history and control of the juniper tip midge, *Oligotrophus apicis* Appleby and Neiswander. Ohio Agric. Res. Dev. Cent. Res. Bull. 980. 26 pp.

22. Appleby, J. E. 1975. Control of the spring generation of Nantucket pine tip moth with insecticides. J. Arboric. 1:91–92.

23. Archibold, K. D. 1956. Forest Aphidae of Nova Scotia. Proc. Nova Scotia Inst. Sci. Vol. 24, Part 2. 269 pp.

24. Armitage, H. M. 1944. Twenty-fifth annual report. Bur. of Entomol. and Plant Quarantine. Calif. State Dep. Agric. Bull. 33:228–275.

25. Arnott, D. A. 1957. Occurrence of *Trirhabda pilosa* Blakee (Coleoptera: Chrysomelidae) on sagebrush in British Columbia, with note on life history. Proc. British Columbia Entomol. Soc. 53:14–15.

26. Atkins, M. D. 1968. Scolytid pheromones—ready or not. Can. Entomol. 100:1115–1117.

27. Baerg, W. J. 1947. The biology of the maple leaf scale. Univ. of Ark. Agric. Exp. Stn. Bull. 470. 14 pp.

28. Bailey, L. H., and Bailey, E. Z., 1976. Hortus third (rev. by the staff of the Bailey Hortatorium, Cornell Univ.). Macmillan Co., New York. 1290 pp.

29. Bailey, N. S. 1951. The Tingoidea of New England and their biology. Entomol. Am. 31:1–140.

30. Bailey, S. F. 1938. Thrips of economic importance in California. Calif. Agric. Exp. Stn. Circ. 346. 77 pp.

31. Baker, J. R. 1980. Insects and related pests of shrubs. N.C. State Agric. Ext. Serv. AG-189. 199 pp.

32. Baker, W. L. 1949. Studies on the transmission of the virus causing Phloem necrosis of American elm with notes on the biology of its insect vector. J. Econ. Entomol. 42:729–732.

33. Baker, W. L. 1972. Eastern forest insects. USDA For. Serv. Misc. Publ. 1175. 642 pp.

34. Balch, R. E. 1952. Studies of the balsam woolly aphid, *Adelges piceae* Ratz., and its effects on balsam fir. Can. Dep. Agric. Publ. 867. 76 pp.

35. Balch, R. E., and Underwood, G. R. 1950. The life history of *Pineus pinifoliae* (Fitch) (Homoptera: Phylloxeridae) and its effects on white pine. Can. Entomol. 82:117–123.

36. Balduf, W. V. 1929. The life history of the goldenrod beetle, *Trirhabda canadensis* Kirby (Coleoptera: Chrysomelidae). Entomol. News 40:35–39.

37. Balduf, W. V. 1965. Observations on *Archips cerasivoranus* (Fitch) (Tortricidae: Lepidoptera) and certain parasites (Diptera: Hymenoptera). Ohio J. Sci. 65:60–67.

38. Ball, E. D. 1926. The life histories of two leafhoppers, a study in adaptation. J. Econ. Entomol. 19:95–99.

39. Ball, J. J., and Simmons, G. A. 1984. The Shigometer as a predictor of bronze birch borer risk. J. Arboric. 10:327–329.

40. Balsbaugh, E. U. Jr., and Hays, K. L. 1972. The leaf beetles of Alabama (Coleoptera: Chrysomelidae). Agric. Exp. Stn., Auburn Univ. Bur. 443. 223 pp.

40a. Baranowski, R. M., and Slater, J. A. 1986. Coreidae of Florida (Hemiptera: Heteroptera). Vol. 12 of Arthropods of Florida and neighboring land areas. Fla. Dep. of Agric. and Consumer Serv., Gainesville. 82 pp.

41. Barrett, R. E. 1932. An annotated list of the insects and arachnids affecting the various species of walnuts or members of the genus *Juglans*. Univ. Calif. Publ. Entomol. 5:275–309.

42. Barter, G. W. 1957. Studies of the bronze birch borer, *Agrilus anxius* Gory in New Brunswick. Can. Entomol. 89:12–36.

43. Barter, G. W. 1965. Survival and development of the bronze poplar borer, *Agrilus liragus* Barter and Brown (Coleoptera: Buprestidae). Can. Entomol. 97:1063–1068.

44. Basinger, A. J. 1938. The orange tortrix, *Argyrotaenia citrana*. Hilgardia 11:635–669.

45. Beal, J. A., and Massey, C. L., 1945. Bark beetles and ambrosia beetles (Coleoptera: Scolytidae) with special reference to species occurring in North Carolina. Duke Univ. Sch. of Forestry Bull. 10. 178 pp.

46. Beal, J. A., Haliburton, Wm., and Knight, F. B. 1952. Forest insects of the southeast: with special reference to species occurring in the piedmont plateau of North Carolina. Duke Univ. Sch. of Forestry Bull. 14. 168 pp.

47. Bean, J. L., and Godwin, P. A. 1955. Description and bionomics of a new red pine scale, *Matsucoccus resinosae*. For. Sci. 1:164–176.

48. Becker, E. C. 1979. *Pyrrhalta viburni* (Coleoptera: Chrysomelidae), a Eurasian pest of viburnum recently established in Canada. Can. Entomol. 111:417–419.

49. Becker, W. B. 1938. Leaf-feeding insects of shade trees. Mass. Agric. Exp. Stn. Bull. 353. 82 pp.

50. Beckwith, R. C. 1963. An oak leaf tier, *Croesia semipurpurana* (Lepidoptera: Tortricidae), in Connecticut. Ann. Entomol. Soc. Am. 56:741–744.

51. Beckwith, R. C. 1978. Larval instars of the Douglas-fir tussock moth. USDA Agric. Handb. 536. 14 pp.

52. Beirne, B. P. 1956. Leafhoppers (Homoptera: Cicadellidae) of Canada and Alaska. Can. Entomol. 88 (suppl. 2). 180 pp.

53. Beirne, B. P. 1972. Pest insects of annual crop plants in Canada. IV. Hemiptera-Homoptera. V. Orthoptera. VI. Other groups. Mem. Entomol. Soc. Can. 85. 73 pp.

54. Bell, H. T., and Clarke, R. G. 1978. Resistance among rhododendron species to obscure root weevil feeding. J. Econ. Entomol. 71:869–870.

55. Bell, P. D. 1979. Rearing the blackhorned tree cricket, *Oecanthus nigricornis* (Orthoptera: Gryllidae). Can. Entomol. 111:709–712.

56. Belyea, R. M., and Sullivan, C. R. 1956. The white pine weevil: a review of current knowledge. For. Chronicle 32:58–67.

57. Benham, G. S. Jr., and Farrar, R. J. 1976. Notes on the biology of *Prionus laticollis* (Coleoptera: Cerambycidae). Can. Entomol. 108:569–576.
58. Bennett, F. D., and Brown, S. W. 1958. Life history and sex determination in the diaspine scale, *Pseudaulcaspis pentagona* (Targ.) (Coccidea). Can. Entomol. 110:317–352.
59. Bennett, W. H. 1955. The pupal morphology of the locust twig borer (Lepidoptera: Olethreutidae). Proc. Entomol. Soc. Wash. 57:243–244.
59a. Benyus, J. M. 1983. Christmas tree pest manual. USDA For. Serv., North Central For. Exp. Stn. 108 pp.
60. Berisford, C. W. 1982. Pheromones of *Rhyacionia* spp.: identification, function, and utility. J. Ga. Entomol. Soc. 17(2nd suppl.):23–30.
61. Betsch, W. D. Jr. 1978. A biological study of three hemipterous insects on honeylocust in Ohio. Master's thesis, Ohio State Univ., Columbus.
62. Billings, R. F., and Pase, H. A. III. 1979. A field guide for ground checking southern pine beetle spots. USDA For. Serv. Agric. Handb. 558. 19 pp.
63. Blackman, M. W. 1915. Observations on the life history and habits of *Pityogenes hopkinsi* Swaine. N.Y. State Coll. Forestry, Syracuse Univ. Tech. Publ. 16:11–66.
64. Blair, L. M., Enns, W. R., and Kearby, W. H. 1980. The morphology of the larva of *Conotrachelus retentus*, the black walnut curculio (Coleoptera: Curculionidae). J. Kans. Entomol. Soc. 53:325–332.
65. Blake, D. H. 1931. Revision of the species of beetles of the genus *Trirhabda* north of Mexico. Proc. U.S. Natl. Mus. 79:1–36.
65a. Blake, E. A., and Wagner, M. R. 1987. Collection and consumption of Pandora Moth, *Coloradia pandora lindseyi* (Lepidoptera: Saturniidae), larvae by Owens Valley and Mono Lake Paiutes. Bull. Entomol. Soc. Am. 33:23–27.
66. Blakeslee, E. B. 1915. American plum borer. USDA Bull. 262. 13 pp.
67. Bobb, M. L. 1972. Influence of pheromones on mating behavior and populations of Virginia pine sawfly. Environ. Entomol. 1:78–80.
68. Bobb, M. L., Weidhaas, J. A. Jr., and Ponton, L. F. 1973. White peach scale: life history and control. J. Econ. Entomol. 66:1290–1292.
69. Boerg, W. J. 1947. The biology of the maple leaf scale. Ark. Coll. Agric. Bull. 470. 14 pp.
70. Boisver, J., Cloutier, C., and McNeil, J. 1981. *Hyadaphis tataricae* (Homoptera: Aphididae). A pest of honeysuckle new to North America. Can. Entomol. 113:415–418.
71. Bolsinger, M., and Flückinger, W. 1984. Effect of air pollution at a motorway on the infestation of *Viburnum opulus* by Aphis fabae. Eur. J. For. Pathol. 14:256–260.
72. Borden, J. H., and Dean, W. F. 1971. Observations on *Eriocampa ovata* L. (Hymenoptera: Tenthredinidae) infesting red alder in southwestern British Columbia. J. Econ. Entomol. Soc. 68:26–28.
73. Borror, D. J., and White, R. E. 1970. A field guide to the insects. Houghton Mifflin, Boston. 404 pp.
74. Bratley, H. E. 1932. The oleander caterpillar, *Syntomedia epilais* Walker. Fla. Entomol. 15:57–67.
75. Bray, D. F. 1957. Matsucoccus. Sci. Tree Top. 2:10–13.
76. Brewer, B. S., and Howell, J. O. 1981. Description of the immature stages and adult female of *Saissetia coffeae*. Ann. Entomol. Soc. Am. 74:548–555.
77. Brewer, J. W. 1971. Biology of the pinyon stunt needle midge. Ann. Entomol. Soc. Am. 64:1099–1102.
78. Bright, D. E. Jr. 1976. The insects and arachnids of Canada. Part 2. The bark beetles of Canada and Alaska, Coleoptera: Scolytidae. Can. Dep. Agric. Biosystematics Res. Inst. Publ. 1576. 241 pp.
79. Britton, W. E. 1923. Guide to the insects of Connecticut. Part IV. The Hemiptera or sucking insects of Connecticut. Conn. State Geol. and Natural Hist. Survey Bull. 34:335–345.
80. Britton, W. E., and Zappe, M. P. 1927. Some insect pests of nursery stock in Connecticut. Conn. Agric. Exp. Stn. Bull. 292:119–173.
81. Britton, W. E., and Friend, R. B. 1935. Insect pests of elms in Connecticut. Conn. Agric. Exp. Stn. Bull. 369:265–307.
82. Brockley, R. P., and Elmes, E. 1985. Barking damage by red squirrels in juvenile-spaced lodgepole pine stands in the Kamloops forest region. B.C. Ministry of Forests, Research Branch, Rep. RR 85003-HQ. 17 pp.
83. Brooks, H. L., and Warren, L. O. 1964. Biology of a pine bark aphid, *Cinara watsoni*, and its response to temperature. J. Kans. Entomol. Soc. 37:310–316.
84. Brown, A. F. 1972. Tischeriidae of America north of Mexico (Micro-Lepidoptera). Mem. Am. Entomol. Soc. 28. Acad. Natural Sci., Philadelphia. 148 pp.
85. Brown, L. R., and Eads, C. O. 1965. A technical study of insects affecting the oak tree in southern California. Calif. Agric. Exp. Stn. Bull. 810. 105 pp.
86. Brown, L. R., and Eads, C. O. 1965. A technical study of insects affecting the sycamore tree in southern California. Calif. Agric. Exp. Stn. Bull. 818. 38 pp.
87. Brown, L. R., and Eads, C. O. 1967. Insects affecting ornamental conifers in southern California. Calif. Agric. Exp. Stn. Bull. 834. 72 pp.
88. Brown, L. R. 1983. Recent impact of two moths on woody landscape plants. Proc. Turfgrass and Landscape Inst. Univ. Calif. Coop. Ext. and Southern Calif. Turfgrass Council, Anaheim, Calif. 158 pp.
89. Bryant, D. G. 1976. Distribution, abundance, and survival of the balsam woolly aphid, *Adelges piceae* (Homoptera: Phylloxeridae), on branches of balsam fir, *Abies balsamea*. Can. Entomol. 108:1097–1111.
90. Burch, J. B. 1962. How to know the eastern land snails. Wm. C. Brown Co., Dubuque, IA. 214 pp.
91. Burke, H. E. 1936. Important insect enemies of cypress trees. Western Shade Tree Conf. Proc. 3:70–79.
92. Burke, H. E., and Boving, A. G. 1929. The Pacific flathead borer. USDA Tech. Bull. 83. 35 pp.
93. Burke, H. E., and Herbert, F. B. 1920. California oak worm. USDA Farmers' Bull. 1076. 14 pp.
94. Burke, J., and Percy, J. 1983. Viruses infecting five species of lepidoptera. Can. For. Serv. Bi-mon. Res. Notes. 3:1–2.
95. Burkot, T. R., and Benjamin, D. M. 1979. The biology and ecology of the cottonwood leaf beetle, *Chrysomela scripta* (Coleoptera: Chrysomelidae), on tissue cultured hybrid *Aigeiros* (*Populus* × *Euramericana*) subclones in Wisconsin. Can. Entomol. 111:551–556.
96. Burns, D. P. 1971. Yellow poplar weevil. USDA For. Serv. Forest Pest Leaflet 125. 5 pp.
97. Butcher, J. W., and Carlson, R. B. 1962. Zimmerman pine moth biology and control. J. Econ. Entomol. 55:668–671.
98. Butler, L. 1985. Food plant studies for the half-wing geometer, *Phigalia titea*. Can. Entomol. 117:547–551.
99. Caldwell, D. L. 1981. The control and impact of *Lygus lineolaris* (P. & B.) on first-year peach scion stock in Missouri. Ph.D. diss., Univ. of Missouri, Columbia. 141 pp.
100. Caldwell, D. L., and Schuder, D. L. 1979. The life history and description of *Phylloxera caryaecaulis* on shagbark hickory. Ann. Entomol. Soc. Am. 72:384–390.
101. Cannon, W. N. Jr. 1971. Distribution records of the locust leafminer, *Odontota* [*Chalepus*] *dorsalis* (Thun.), in the United States. Entomol. News 81:115–120.
102. Cantelo, W. W. 1953. Life history and control of euonymus scale in Massachusetts. Mass. Agric. Exp. Stn. Bull. 471. 31 pp.
103. Carlson, K. D. 1980. Early testing for resistance to verticillium wilt and potato leafhopper feeding in Norway and sugar maples. Master's thesis, State Univ. of N.Y. College of Environmental Science and Forestry, Syracuse.
104. Carolin, V. M., and Knopf, J. A. E. 1968. The pandora moth. USDA For. Serv. Forest Pest Leaflet 114. 7 pp.
105. Carothers, W. A., and Ghent, J. 1980. Biological evaluation of the cypress looper outbreak on the Big Cypress national preserve in Florida. Southeastern Area, State and Private Forestry, USDA, Atlanta, Ga. Rep. 81-1-14.
106. Carter, C. I. 1971. Conifer woolly aphids (Adelgidae) in Britain. For. Comm. Bull. 42. 51 pp.
107. Cary, L. R. 1923. Plant-house aleyrodes, *Aleyrodes vaporariorum* Westw. Maine Agric. Exp. Stn. Rep. 19, pp. 125–144.
108. Chamberlin, W. J. 1915–1920. Flat-headed borers which attack orchard trees and cane fruits in Oregon. Pages 103–108 in: Third crop pest and horticultural report. Agric. Exp. Stn.
109. Champlain, A. B. 1924. *Adirus trimaculatus* Say: a rose pest. J. Econ. Entomol. 17:648–650.
110. Chapman, P. J., and Lienk, S. E. 1971. Tortricid fauna of apple in New York. Special Publication N.Y. State Agric. Exp. Stn. Geneva, NY. 122 pp.
111. Chapman, P. J., and Lienk, S. E. 1974. Green fruitworms. N.Y. Food and Life Sci. Bull. 50. Plant Sci. 6. N.Y. State Agric. Exp. Stn. Geneva, NY. 15 pp.
112. Chapman, R. N. 1915. Observations on the life history of *Agrilus bilineatus*. J. Agric. Res. 3:283–294.
113. Chellman, C. W. 1971. Insects, diseases, and other problems of Florida's trees. Fla. Dep. of Agric. Bull. 196. 156 pp.
114. Chen, C., and Appleby, J. E. 1984. Chemical control of the cypress twig gall midge, *Taxodiomyia cupressiananassa*, in central Illinois. J. Environ. Hortic. 2:38–40.
115. Childs, L. 1918. The life history and control of the rose leafhopper. Oregon Agric. Exp. Stn. Bull. 148. 32 pp.
116. Chittenden, F. H. 1910. The oak pruner. USDA. Bur. Entomol. Circ. 130. 7 pp.
117. Christian, P. J. 1953. Revision of the North American species of *Typhlocyba*. Kans. Univ. Sci. Bull. 35:1103–1277.
118. Clafin, S. H., and Allen, D. C. 1981. Biology of *Ancylis discigerana* (Lepidoptera: Tortricidae). Can. Entomol. 113:265–270.
119. Clark, R. C., and Raske, A. G. 1974. The birch casebearer. Newfoundland Forestry Notes 7. Newfoundland For. Res. Centre, St. John's. 5 pp.
120. Clemens, W. A. 1916. The pine bark beetle. Cornell Univ. Agric. Exp. Stn. Bull. 383:287–298.
121. Clepper, H. E. 1944. Hemlock: the state tree of Pennsylvania. Pa. Dep. of Forests and Waters Bull. 52 (rev.). 27 pp.
122. Collman, S. J. 1971. Insect pests of ornamental trees and shrubs of the University of Washington Arboretum. Master's thesis. Univ. of Washington, Seattle.
123. Comstock, J. H. 1916. Reports on scale insects. Cornell Univ. Agric. Exp. Stn. Bull. 372:276–603.

124. Condrashoff, S. F. 1962. Bionomics of three closely related species of *Contarinia* Rond. (Diptera: Cedidomyiidae) from Douglas-fir needles. Can. Entomol. 94:376.
125. Condrashoff, S. F. 1964. Bionomics of the aspen leafminer, *Phyllocnistis populiella* Cham. (Lepidoptera: Gracillariidae). Can Entomol. 96:857–874.
126. Cook, J. R., and Solomon, J. D. 1976. Damage, biology, and natural control of insect borers in cottonwood, *Populus deltoides*. Pages 272–279 in: Proceedings of a symposium on eastern cottonwood and related species. Louisiana State Univ., Baton Rouge. 485 pp.
127. Cory, E. N., and Knight, P. 1937. Observations on brood X of the periodical cicada in Maryland. J. Econ. Entomol. 30:287–294.
128. Cote, W. A., and Allen, D. C. 1973. Biology of the maple trumpet skeletonizer *Epinotia aceriella* (Lepidoptera: Olethreutidae) in New York. Can. Entomol. 105:463–470.
129. Cote, W. A. III, and Allen, D. C. 1980. Biology of two-lined chestnut borer, *Agrilus bilineatus*, in Pennsylvania and New York. Ann. Entomol. Soc. Am. 73:409–413.
130. Coulson, R. N., and Witter, J. A. 1984. Forest entomology, ecology, and management. John Wiley & Sons, New York. 669 pp.
131. Craig, W. S. 1954. Biology and control of the woolly apple aphid in Iowa nurseries. Iowa State Coll. J. Sci. 28:296–297.
131a. Craighead, F. C. 1923. North American cerambycid larvae. A classification and the biology of North American cerambycid larvae. Can. Dep. Agric. Entomol. Branch Bull. 27. 239 pp.
132. Crane, P. S. 1970. The feeding behavior of the blue-green sharpshooter, *Hordina circellata* (Baker) (Homoptera: Cicadellidae). Ph.D. diss., Univ. of California, Davis.
133. Crawford, D. L. 1912. A new insect pest (*Trioza alacris* Flor.). Monthly Bull. Calif. State Commission Hortic. 1:86–87.
134. Crawford, D. L. 1914. A monograph of the jumping plant-lice or Psyllidae of the new world. Smithsonian Inst., U.S. Natl. Museum Bull. 85. 186 pp.
135. Creighton, J. T. 1929. The biology and life history of the palm-leaf skeletonizer (palm leaf miner), *Homaledra sabelella* Chambers. Unpublished monograph, Univ. of Florida, Gainesville.
136. Cumming, M. E. P. 1953. Notes on the life history and seasonal development of the pine needle scale, *Phenacaspis pinifolia* (Fitch) (Homoptera: Diaspididae). Can. Entomol. 85:347–352.
137. Cumming, M. E. P. 1959. The biology of *Adelges cooleyi* (Gill) (Homoptera: Phylloxeridae). Can. Entomol. 91:601–617.
138. Cumming, M. E. P. 1962. The biology of *Pineus similis* (Gill) (Homoptera: Phylloxeridae) on spruce. Can. Entomol. 94:395–408.
139. Cumming, M. E. P. 1968. The life history and morphology of *Adelges lariciatus* (Homoptera: Phylloxeridae). Can Entomol. 100:113–126.
140. Cutwright, C. R. 1925. Subterranean aphids of Ohio. Ohio Agric. Exp. Stn. Bull. 387:175–238.
141. Darling, C. D., and Smith, D. R. 1985. Description and life history of a new species of *Nematus* (Hymenoptera: Tenthredinidae) on *Robinia hispida* (Fabaceae) in New York. Proc. Entomol. Soc. Wash. 87:225–230.
142. Daviault, L., and Ducharme, R. 1966. Life history and habits of the green spruce leaf miner. *Epinotia nanana* (Treitschke) (Lepidoptera: Tortricidae). Can. Entomol. 98:693–699.
143. Davidson, A. M., and Adams, W. 1973. The grey squirrel and tree damage. Q. J. For. 67:237–247.
144. Davidson, J. A., and McComb, C. W. 1958. Notes on the biology and control of *Fiorinia externa* Ferris. J. Econ. Entomol. 51:405–406.
145. Davidson, J. A., Miller, D. R., and Nakahara, S. 1983. The white peach scale, *Pseudaulacapsis pentagona* (Targioni-Tozzetti): evidence that current concepts include two species. Proc. Entomol. Soc. Wash. 85:753–761.
146. Davidson, J. A. 1964. The genus *Abgrallaspis* in North America (Homoptera: Diapididae). Ann. Entomol. Soc. Am. 57:638–643.
147. Davidson, R. H., and Peairs, L. M. 1966. Insect pests of farm, garden, and orchard. John Wiley & Sons, New York. 675 pp.
148. Davis, C. S., Black, J. H., and Carlson, C. V. 1968. Controlling Pacific flatheaded borer. Calif. Agric. 22:6–7.
149. Davis, D. R. 1964. Bagworm moths of the Western hemisphere. Smithsonian Inst., U.S. Natl. Museum Bull. 244. 233 pp.
150. Davis, R., Flechtmann, C. H. W., Boczek, J. H., and Barké, H. E. 1982. Catalogue of eriophyid mites (Acari: Eriophyoidea). Warsaw Agric. Univ. Press, Warsaw. 224 pp.
151. Davis, W. T. 1926. The cicadas or harvest flies of New Jersey. N.J. Dep. of Agric. Circ. 97. 27 pp.
152. Dean, H. A., and Bailey, J. C. 1961. A flatid planthopper, *Metcalfa pruinosa*. J. Econ. Entomol. 54:1104–1106.
153. DeBach, P., ed. 1964. Biological control of insect pests and weeds. Reinhold, New York. 844 pp.
154. DeBoo, R. F. 1966. Investigations of the importance, biology, and control of *Eucosma gloriola* Heinrich (Lepidoptera: Olethreutidae) and other shoot and tip moths of conifers in New York. Ph.D. diss., Cornell Univ., Ithaca, NY. 185 pp.
154a. DeBoo, R. F., Campbell, L. M., LaPlante, J. P., and Daviault, L. P. 1973. Plantation research. VIII. The pine needle midge, *Contarina baeri* (Diptera: Ceciodomyiidae), a new insect pest of Scots pine. Environ. Can. For. Serv., Chemical Control Res. Inst. Inf. Rep. CC-X-41. 31 pp.
155. DeBoo, R. F., and LaPlante, J. P. 1975. *Contarinia baeri,* a pest of Scots pine. Bi-monthly Res. Notes, Dep. of Environ. Can. For. Serv. 31:1.
156. DeBoo, R. F., and Weidhaas, J. A. 1967. Phenological notes on a non-migrating population of *Pineus floccus* (Homoptera: Chermidae). Can. Entomol. 99:765–766.
157. DeBoo, R. F., Sippell, W. L., and Wong, H. R. 1971. The eastern pine-shoot borer, *Eucosma gloriola* (Lepidoptera: Tortricidae), in North America. Can. Entomol. 103:1473–1486.
158. DeClerck, R. A., and Shorthouse, J. D. 1985. Tissue preference and damage by *Fenusa pusilla* and *Messa nana* (Hymenoptera: Tenthredinidae), leaf-mining sawflies on white birch (*Betula papyrifera*). Can. Entomol. 117: 351–362.
159. Deisler, J. E., and Strange, L. A. 1985. The white garden snail, *Theba pisana* (Muller) (Gastropoda: Helicidae). Fla. Dep. of Agric. and Consumer Serv. Entomol. Circ. 2. 4 pp.
160. DeLong, D. M. 1965. Ecological aspects of North American leafhoppers and their role in agriculture. Bull. Entomol. Soc. Am. 11:9–26.
161. Dekle, G. W. 1962. Azalea caterpillar (*Datana major* G. & R.) (Lepidoptera: Notodontidae). Fla. Dep. of Agric. Div. of Plant Industry Entomol. Circ. 6. 1 p.
162. Dekle, G. W. 1963. Yellow oleander caterpillar (*Palpita flegia* Cramer) (Lepidoptera: Pyralidae). Fla. Dep. of Agric. Div. of Plant Industry Entomol. Circ. 12. 1 p.
163. Dekle, G. W. 1965. Snow scales on Florida citrus. Fla. Dep. of Agric. Div. Plant Industry Entomol. Circ. 39. 2 pp.
164. Dekle, G. W. 1966. Azalea leafminer (*Gracillaria azaleella* Brants) (Lepidoptera: Gracillariidae). Fla. Dep. of Agric. Div. of Plant Industry Entomol. Circ. 55. 2 pp.
165. Dekle, G. W. 1968. Cabbage palm caterpillar (*Litoprosopus futilis* [G & R]) (Noctuidae: Lepidoptera). Fla. Dep. of Agric. Div. of Plant Industry Entomol. Circ. 75. 2 pp.
166. Dekle, G. W. 1969. The brown garden snail (*Helix aspersa* Muller). Fla. Dep. of Agric. Div. of Plant Industry Entomol. Circ. 83. 2 pp.
167. Dekle, G. W. 1976. Florida armored scale insects. Revised. Vol. 3 of Arthropods of Florida and neighboring land areas. Fla. Dep. of Agric. and Consumer Serv., Gainesville. 345 pp.
168. Dekle, G. W., and Kuitert, L. C. 1975. Camellia mining scale, *Duplaspidiotus claviger* (Cockerell) (Diaspididae: Homoptera). Fla. Dep. of Agric. and Consumer Serv. Entomol. Circ. 152. 2 pp.
169. Denmark, H. A. 1967. Cuban laurel thrips, *Gynaikothrips ficorum,* in Florida. Fla. Dep. of Agric. Div. of Plant Industry Entomol. Circ. 59. 2 pp.
170. Denmark, H. A. 1969. A podocarpus aphid, *Neophyllaphis podocarpi* Tak. Fla. Dep. of Agric. Div. of Plant Industry Entomol. Circ. 84. 1 p.
171. Denmark, H. A. 1970. A juniper gall mite, *Trisetacus quadrisetus juniperinus* (Nal.). Fla. Dep. of Agric. and Consumer Serv. Div. of Plant Industry Entomol. Circ. 94. 1 p.
172. Denmark, H. A. 1980. Broad mite, *Polyphagotarsonemus latus* (Banks), (Acarina: Tarsonemidae) on pittosporum. Fla. Dep. of Agric. and Consumer Serv. Entomol. Circ. 213. 2 pp.
173. Denmark, H. A. 1982. *Brevipalpus californicus* (Banks), a pest of woody ornamentals (Acarina: Tenuipalpidae). Fla. Dep. of Agric. and Consumer Serv. Entomol. Circ. 240. 2 pp.
174. Denmark, H. A., and Wolfenbarger, D. O. 1971. The red-banded thrips, *Selenothrip rubro-cinctus* (Giard) in Florida (Thysanoptera: Thripidae). Fla. Dep. of Agric. and Consumer Serv. Entomol. Circ. 108. 2 pp.
175. Dewey, J. E. 1975. Pine looper. USDA For. Serv. Forest Pest Leaflet 151. 5 pp.
176. Dickerson, E. L., and Weiss, H. B. 1916. The ash leaf bug, *Neoborus amoenus* Reut. (Hem.). J. N.Y. Entomol. Soc. 24:302–306.
177. Dickson, R. C. 1950. The Fuller rose beetle. Calif. Agric. Exp. Stn. Bull. 719. 8 pp.
178. Dietz, H. F., and Morrison, H. 1916. The Coccidae or scale insects of Indiana. Eighth Annu. Rep. Indiana State Entomol., pp. 195–321.
179. Dixon, W. N. 1982. *Anacamptodes pergracilis* (Hulst). A cypress looper (Lepidoptera: Geometridae). Fla. Dep. of Agric. and Consumer Serv. Div. of Plant Industry Entomol. Circ. 244. 2 pp.
180. Dixon, W. N., and Woodruff, R. E. 1982. The black twig borer *Xylosandrus compactus* (Eichhoff) (Coleoptera: Scolytidae). Fla. Dep. of Agric. Div. of Plant Industry Entomol. Circ. 250. 2 pp.
181. Doane, C. C., and Manus, M. L., eds. 1981. The gypsy moth: Research toward integrated pest management. Forest Service, APHIS, Tech. Bull. 1584. 757 pp.
182. Donley, D. E., and Acciavatti, R. E. 1980. Red oak borer. USDA For. Serv. Forest Insect and Disease Leaflet 163.
183. Donley, D. E., and Burns, D. P. 1965. The tulip tree scale. USDA For. Serv. Forest Pest Leaflet 92. 5 pp.
184. Doring, K. C. 1942. Host plant records of Cercopidae in North America, North of Mexico (Homoptera). J. Kans. Entomol. Soc. 15:65–72; 15:73–92.
185. Doss, R. P. 1980. Investigation of the basis of resistance of selected

Rhododendron species to foliar feeding by the obscure root weevil (*Sciopithes obscurus*). Environ. Entomol. 9:549–552.
186. Doten, S. B. 1906. The European elm scale. Univ. of Nevada Agric. Exp. Stn. Bull. 85. 34 pp.
187. Doutt, R. L. 1959. Heterogony in *Dyrocosmus* (Hymenoptera: Cynipidae). Ann. Entomol. Soc. Am. 52:69–74.
188. Dozier, H. L. 1928. The Fulgoridae or planthoppers of Mississippi, including those of possible occurrence: a taxonomic, biological, ecological, and economic study. Tech. Bull. Miss. Agric. Exp. Stn. 14:112–114.
189. Drake, C. J. 1922. The life history of the birch tingid, *Corythucha pallipes* Parshley. N.Y. State Coll. of Forestry, Syracuse Univ. Tech. Publ. 16. 5 pp.
190. Drake, C. J., and Ruhoff, F. A. 1965. Lacebugs of the world: a catalogue (Hemiptera: Tingidae). Smithsonian Inst., U.S. Natl. Museum Bull. 243. 634 pp.
191. Drooz, A. T. 1960. The larch sawfly: its biology and control. USDA For. Serv. Tech. Bull. 1212. 52 pp.
192. Drooz, A. T. 1960. White pine shoot borer (*Eucosma gloriola* Heinrich). J. Econ. Entomol. 53:248–251.
193. Drooz, A. T. 1985. Insects of eastern forests. USDA For. Serv. Misc. Publ. 1426. 608 pp.
194. Drooz, A. T., Doggett, C. A., and Coppel, H. C. 1979. The introduced pine sawfly, a defoliator of white pine new to North Carolina. USDA, Southeast For. Exp. Stn. Asheville, N.C., For. Serv. Res. Note SE-273. 3 pp.
194a. Duckworth, W. D., and Eichlin, T. D. 1978. The clearwing moths of California. Occasional Papers Entomol. 27. 80 pp.
195. Duffield, J. W. 1985. Inheritance of shoot coatings and their relation to resin midge attack on ponderosa pine. For. Sci. 31:427–429.
196. Dunbar, D. M. 1974. Bionomics of the andromeda lace bug, *Stephanitis takeyai*. Pages 277–289 in: Twenty-fifth anniversary memoirs. Conn. Entomol. Soc., New Haven. 322 pp.
197. Dunbar, D. M., and Stephens, G. R. 1975. Association of two-lined chestnut borer and shoestring fungus with mortality of defoliated oak in Connecticut. For. Sci. 21:169–174.
198. Duncan, R. W. 1982. Silver spotted tiger moth. Pacific For. Res. Centre, Can. Forestry Serv. Pest Leaflet 5. 4 pp.
199. Dyar, H. G. 1897. On the larvae of certain sawflies (Tenthredinidae). J. N.Y. Entomol. Soc. 5:18–30.
200. Eaton, C. B. 1955. The Saratoga spittle bug. USDA For. Serv. Forest Pest Leaflet 3. 4 pp.
201. Eaton, C. B., and Yuill, J. S. 1971. Gouty pitch midge. USDA For. Serv. Forest Pest Leaflet 46. 8 pp.
202. Ebeling, W. 1959. Subtropical fruit pests. Univ. of Calif. Div. of Agric. Sci., Berkeley. 436 pp.
203. Edmunds, G. F. Jr. 1973. Ecology of black pineleaf scale (Homoptera: Diaspididae). Environ. Entomol. 2:765–777.
204. Eichlin, T. D. 1980. *Stenolechia bathrodyas* Meykrick, a recently introduced pest of ornamental conifers in southern coastal California (Lepidoptera: Gelechiidae). Pan-Pac. Entomol. 56:213–219.
205. Eide, P. M. 1966. The life history and control of *Nemocestes incomptus* (Horn), a native root weevil attacking strawberries in western Washington. J. Econ. Entomol. 59:1004–1005.
206. Eidt, D. C. 1969. The life histories, distribution and immature forms of the North American sawflies of the genus *Cephalcia* (Hymenoptera: Pamphiliidae). Mem. Entomol. Soc. Can. 59. 56 pp.
207. Eidt, D. C., and Embree, D. G. 1968. Distinguishing larvae and pupae of the winter moth *Operoptera brumata,* and the Bruce spanworm, *O. bruceata* (Lepidoptera: Geometridae). Can. Entomol. 100:536–539.
208. Eidt, D. C., and Nichols, J. O. 1970. An outbreak of a sawfly, *Pamphilius phyllisae* (Hymenoptera: Pamphiliidae), on northern red oak, with notes on larval morphology. Can. Entomol. 102:53–63.
209. Engelhardt, G. P. 1946. The North American clearwing moths of the family Aegeriidae. Smithsonian Inst., U.S. Natl. Museum Bull. 190. 222 pp.
210. English, L. L. 1962. Illinois trees and shrubs: their insect enemies. Ill. Natural History Circ. 47. 92 pp.
211. Essig, E. O. 1912. Plant lice affecting citrus trees. Calif. State Comm. Hortic. Monthly Bull. 1:115–133.
212. Essig, E. O. 1913. Injurious and beneficial insects of California. Calif. State Comm. Hortic. Monthly Bull. 2:1–351.
213. Essig, E. O. 1916. The soft bamboo scale. Calif. State Comm. Hortic. Monthly Bull. 5:72–73.
214. Essig, E. O. 1917. The tomato and laurel Psyllids. J. Econ. Entomol. 10:433–444.
215. Essig, E. O. 1932. A genus and species of the family Aphididae new to North America. Univ. Calif. Publ. Entomol. 6:1–8.
216. Essig, E. O. 1945. The pit-making Pittosporum scale. Calif. State Dep. of Agric. Bull. 34:134–136.
217. Essig. E. O. 1958. Insects and mites of western North America. Macmillan Co., New York. 1050 pp.
218. Essig, E. O., and Hoskins, W. H. 1944. Insects and other pests attacking agricultural crops. Revised. Calif. Agric. Ext. Serv. Circ. 87. 197 pp.
219. Evans, D. 1972. Alternate generations of gall cynipids (Hymenoptera: Cynipidae) on Garry oak. Can. Entomol. 104:1805–1818.
220. Evans, David. 1982. Pine shoot insects common in British Columbia. Can. For. Serv. Pacific For. Res. Centre. BC-X-233. 56 pp.
221. Evans, W. G. 1972. The pear slug and its control. Univ. of Alberta. Dep. of Entomol. Leaflet 833. 2 pp.
222. Ewan, H. G. 1957. Jack-pine sawfly. USDA For. Serv. Forest Pest Leaflet 17. 4 pp.
223. Ewan, H. G. 1961. The Saratoga spittlebug: a destructive pest in red pine plantations. USDA For. Serv. Tech. Bull. 1250. 52 pp.
224. Eyer, J. R. 1942. Life history and control of the giant apple root borer. N.M. Agric. Exp. Stn. Bull. 295. 14 pp.
225. Faith, D. P. 1979. Strategies of gall formation in *Pemphigus* aphids. N.Y. Entomol. Soc. 87:21–37.
226. Farentinos, R. C., Capretta, P. J., Kepner, R. E., and Littlefield, V. M. 1981. Selective herbivory in tassel-eared squirrels: role of monoterpenes in ponderosa pines chosen as feeding trees. Science 213:1273–1275.
227. Farrar, R. J., and Kerr, T. W. 1968. A preliminary study of the life history of the broad-necked root borer in Rhode Island. J. Econ. Entomol. 61:563–564.
228. Farris, M. E., Appleby, J. E., and Weber, B. C. 1982. Walnut caterpillar. USDA For. Serv. Forest Insect and Disease Leaflet 41. 6 pp.
229. Fasoranti, J. O. 1984. The life history and habits of a *Ceanothus* leaf miner, *Tischera immaculata* (Lepidoptera: Tischeriidae). Can. Entomol. 116: 1441–1448.
230. Fedde, G. F. 1973. Impact of the balsam woolly aphid (Homoptera: Phylloxeridae) on cone and seed produced by Fraser fir. Can. Entomol. 105:673–680.
230a. Felt, E. P. 1898. Elm-leaf beetle in New York State. Bull. N.Y. State Museum 5:1–43.
231. Felt, E. P. 1906. Insects affecting park and woodland trees. N.Y. State Museum Memoir 8. 2 vols. 877 pp.
232. Felt, E. P. 1924. Manual of tree and shrub insects. L. H. Bailey, ed. Macmillan Co., New York. 382 pp.
233. Felt, E. P. 1940. Plant galls and gall-makers. Comstock Publ. Co., Ithaca, NY. 364 pp.
234. Felt, E. P., and Joutel, L. H. 1904. Monograph of the genus *Saperda*. N.Y. State Mus. Bull. 74. 86 pp.
235. Fenton, F. A. 1917. Observations on *Lecanium corni* Bouché and *Physokermes piceae* Schr. Can. Entomol. 49:309–320.
236. Fenton, F. A. 1939. Control of shade tree borers. Okla. Agric. Exp. Stn. Circ. 84. 28 pp.
237. Ferguson, D. C. 1972. Bombycoidea: Saturniidae. In: The moths of America north of Mexico. Fascicle 20.2A and 20.2B. R. B. Dominick et al., eds. E. W. Classey Ltd., London. 275 pp.
238. Ferguson, D. C. 1978. Noctuoidea: Lymantriidae. In: The moths of America north of Mexico. Fascicle 22.2. R. B. Dominick et al., eds. E. W. Classey Ltd., London. 110 pp.
239. Ferris, G. F. 1950. Atlas of the scale insects of North America. Ser. V. The Pseudococcidae (Part 1). Stanford Univ. Press, Stanford, Calif. 278 pp.
240. Ferris, G. F. 1953. The Pseudococcidae. Part II. Pages 279–506 in: Atlas of the scale insects of North America. Vol. VI. Stanford Univ. Press, Stanford, CA.
241. Filer, T. H., Solomon, J. D., McCracken, F. I., Oliveria, F. L., Lewis, R. Jr., Weiss, M. J., and Rogers, T. J. 1977. Sycamore pests, a guide to major insects, diseases and air pollution. USDA For. Serv. Southern Forest Exp. Stn. 36 pp.
242. Finnegan, R. J. 1958. The pine weevil *Pissodes approximatus* Hopkins in southern Ontario. Can. Entomol. 90:285–286.
243. Finnegan, R. J. 1962. The pine root-collar weevil, *Hylobius radicus* Buch., in southern Ontario. Can. Entomol. 94:11–17.
244. Finnegan, R. J. 1965. The pine needle miner, *Exoteleia pinifoliella* (Chamb.) (Lepidoptera: Gelechiidae), in Quebec. Can. Entomol. 97:744–750.
245. Finnegan, R. J. 1967. Notes on the biology of the pitted ambrosia beetle, *Corthylus punctatissimus* (Coleoptera: Scolytidae), in Ontario and Quebec. Can. Entomol. 99:49–54.
246. Fiori, B. J., and Doland, D. D. 1984. Resistance of *Betula davurica* to the birch leafminer, *Fenusa pusilla* (Hymenoptera: Tenthredinidae). Can. Entomol. 116:1275–1276.
247. Fiori, B. J., and Craig, D. W. 1987. Relationship between color intensity of leaf supernatants from resistant and susceptible birch trees and rate of oviposition by the birch leafminer *Fenus pusilla* (Lepeletier) (Hymenoptera: Tenthredinidae). J. Econ. Entomol. (in press)
248. Fisher, W. W. 1942. A revision of the North American species of Buprestid beetles belonging to the tribe Chrysobothrini. USDA Misc. Publ. 470. 274 pp.
249. Fitzgerald, T. D. 1973. Coexistence of three species of bark-mining *Marmara* (Lepidoptera: Gracillariidae) on green ash and descriptions of new species. Ann. Entomol. Soc. Am. 66:457–464.
249a. Fleming, W. E. 1962. The Japanese beetle. USDA Agric. Handb. 236. 30 pp.

250. Flint, M. L. 1982. Integrated pest management for walnuts. State Integrated Pest Management Project. Dir. of Agric. Sci. Publ. 3270. 96 pp.
251. Flint, W. P., and Farrar, M. D. 1940. Protecting shade trees from insect damage. Ill. Agric. Exp. Stn. and Ext. Serv. Circ. 509. 60 pp.
252. Fontaine, M. S., and Foltz, J. L. 1985. Adult survivorship, fecundity, and factors affecting laboratory oviposition of *Pissoides nemorensis* (Coleoptera: Curculionidae). Can. Entomol. 117:1575–1578.
253. Forbes, A. R., and Chan, C. 1978. The aphids (Homoptera: Aphidae) of British Columbia. 7. A revised host plant catalogue. J. Entomol. Soc. B. C. 75:53–67.
254. Forbes, R. S., and Daviault, L. 1964. The biology of the mountain ash sawfly, *Pristophora geniculata* (Htg.) (Hymenoptera: Tenthredinidae), in eastern Canada. Can. Entomol. 96:1117–1133.
255. Forbes, W. T. M. 1954. Lepidoptera of New York and neighboring states. Part 3. Noctuidae. Cornell Agric. Exp. Stn. Mem. 329. 433 pp.
256. Forbush, E. H., and Fernald, C. H. 1896. The gypsy moth. Mass. State Board of Agric., Boston. 495 pp.
257. Fowells, H. A. 1965. Silvics of forest trees of the United States. USDA For. Serv. Agric. Handb. 271. 762 pp.
258. Fowler, H. G., and Roberts, R. B. 1980. Foraging behavior of the carpenter ant *Camponotus pennsylvanicus* (Hymenoptera: Formicidae) in New Jersey. J. Kans. Entomol. Soc. 53:295–304.
259. Fowler, H. G., and Roberts, R. B. 1982. Carpenter ant (Hymenoptera: Formicidae) induced wind breakage in New Jersey shade trees. Can. Entomol. 114:114–117.
260. Frankie, G. W. 1969. Investigations on the ecology of the cypress bark moth, *Laspeyresia cupressana* (Kearfott) (Lepidoptera: Olethreutidae). Ph.D. diss., Univ. of California, Berkeley. 126 pp.
261. Frankie, G. W., and Koehler, C. S. 1967. Cypress bark moth on Monterey cypress. Calif. Agric. 21:6–7.
262. Frankie, G. W., and Koehler, C. S. 1971. Studies on the biology and seasonal history of the cypress bark moth, *Laspeyresia cupressana* (Lepidoptera: Olethreutidae). Can. Entomol. 103:947–961.
263. Freeman, T. N. 1958. The Archipinae of North America (Lepidoptera: Tortricidae). Can. Entomol. Suppl. 7. 89 pp.
264. Freeman, T. N. 1960. Needle-mining Lepidoptera of pine in North America. Can. Entomol. 92, suppl. 16. 51 pp.
265. Freeman, T. N. 1967. Annotated keys to some nearctic leaf-mining Lepidoptera on conifers. Can. Entomol. 99:419–435.
266. Frick, K. E. 1959. Synopsis of the species of agromyzid leafminers described from North America (Diptera). Proc. U.S. Natl. Mus. 108:347–465.
267. Friend, R. B. 1927. The biology of the birch leaf skeletonizer, *Bucculatrix canadensisella* Chamb. Conn. Agric. Exp. Stn. Bull. 288:395–486.
268. Frost, S. W. 1923. A study of the leaf-mining Diptera of North America. N.Y. (Cornell) Agric. Exp. Stn. Mem. 78. 228 pp.
269. Frost, S. W. 1924. The leaf-mining habit in the Coleoptera. Ann. Entomol. Soc. Am. 17:457–467.
270. Frost, S. W. 1959. Insect life and insect natural history. 2nd rev. ed. Dover, New York. 522 pp.
271. Fulton, B. B. 1915. The tree crickets of New York: life history and bionomics. N.Y. Agric. Exp. Stn. Tech. Bull. 42. 47 pp.
272. Fulton, B. B. 1925. Physiological variations in the snowy tree cricket, Oecanthus niveus De Geer. Ann. Entomol. Soc. Am. 18:363–383.
273. Furniss, M. M., and Barr, W. F. 1975. Insects affecting important native shrubs of northwestern United States. USDA For. Serv. Gen. Tech. Rep. INT-19, 64 pp. Intermt. For. and Range Exp. Stn., Ogden, Utah.
274. Furniss, R. L. 1942. Biology of *Cylindrocopturus furnissi* Buchanan on Douglas-fir. J. Econ. Entomol. 35:853–859.
275. Furniss, R. L., and Carolin, V. M. 1977. Western forest insects. USDA For. Serv. Misc. Publ. 1339. 654 pp.
276. Gagné, R. J. 1978. A systematic analysis of the pitch pine midges, *Cecidomyia* spp. (Diptera: Cecidomyiidae). USDA Tech. Bull. 1575. 18 pp.
277. Gagné, R. J. 1983. *Acericecis* Gagné, a new genus for *Cecidomyia ocellaris* Osten Sacken (Diptera: Cecidomyiidae), the maple leaf ocellate gall maker in North America. Proc. Entomol. Wash. 85:704–709.
277a. Gagné, R. J., and Valley, K. 1984. Two new species of Cecidomyiidae (Diptera) from honeylocust, *Gleditsia triacanthos* L. (Fabaceae), in eastern United States. Proc. Entomol. Soc. Wash. 86:543–549.
278. Galford, J. R. 1977. Evidence for a pheromone in the locust borer. USDA For. Serv. Res. Note NE-240. 3 pp.
279. Galford, J. R. 1980. Bait bucket trapping for red oak borers (Coleoptera: Cerambycidae). USDA For. Serv. Res. Note NE-293. 2 pp.
279a. Galford, J. R. 1985. Role of predators on an artificially planted red oak borer population. USDA For. Serv. Res. Note NE-331. 2 pp.
280. Gargiullo, P. M., Berisford, C. W., and Godbee, J. F. Jr. 1985. Predication of optimal timing for chemical control of the Nantucket pine tip moth, *Rhyacionia frustrana* (Comstock) (Lepidoptera: Tortricidae), in the southeastern coastal plain. J. Econ. Entomol. 78:148–154.
281. Garman, H. 1916. The locust borer and other insect enemies of the black locust. Ky. Agric. Exp. Stn. Bull. 200:99–135.
282. Garman, P. 1923. Notes on the life history of the spruce mite. Conn. Agric. Exp. Stn. Bull. 247:240–242.
283. Garrett, W. T. 1972. Biosystematics of the oystershell scale, *Lepidosaphes ulmi* (L.) Homoptera: Diaspididae in Maryland. Ph.D. diss., Univ. of Maryland, College Park. 108 pp.
284. Gibbons, C. F., and Butcher, J. W. 1961. The oak skeletonizer, *Bucculatrix ainsliella*, in a Michigan woodlot. J. Econ. Entomol. 54:681–684.
285. Gibson, A. 1924. The occurrence of the tortricid *Cacoecia rosana* L. in Canada. J. Econ. Entomol. 17:51–59.
286. Giese, R. L., and Knauer, K. H. 1977. Ecology of walkingsticks. For. Sci. 23:45–63.
287. Gill, R. 1984. Peppertree psyllid, *Calophya schini*. Calif. Plant Pest and Disease Rep. 3:119–121.
288. Gill, S. A., Miller, D. R., and Davidson, J. A. 1982. Biology and taxonomy of the euonymus scale, *Unaspis euonymi* (Comstock), and detailed biological information on the scale in Maryland (Homoptera: Diaspididae). Publ. MP 969. Md. Agric. Exp. Stn. College Park. 36 pp.
289. Gillespie, A. M. 1932. The birch casebearer in Maine. Maine For. Serv. Bull. 7. 23 pp.
290. Gillespie, D. R., and Beirne, B. P. 1982. Leafrollers (Lepidoptera) on berry crops in the lower Fraser Valley, British Columbia. J. Entomol. Soc. B. C. 79:31–36.
291. Gillette, C. P. 1908. New species of Colorado Aphididae, with notes upon their life habits. Can. Entomol. 40:61–68.
292. Gillette, C. P. 1909. American snowball louse, *Aphis viburnicola*. Entomol. News 20:280–285.
293. Gillette, C. P. 1909. *Phyllaphis coweni* Ckll. Can. Entomol. 41:41–45.
294. Gimpel, W. F. Jr., Miller, D. R., and Davidson, J. A. 1974. A systematic revision of the wax scales, genus *Ceroplastes*, in the U.S. (Homoptera; Coccoidea; Coccidae). Univ. of Md. Agric. Exp. Stn. Misc. Bull. 841. 85 pp.
295. Glass, E. H. 1944. Feeding habits of two mealybugs, *Pseudococcus comstocki* (Kuw.) and *Phenacoccus colemani* (Ehrh.). Va. Polytechnic Inst., Blacksburg, Va. Agric. Exp. Stn. Tech. Bull. 95. 16 pp.
296. Godan, D. 1983. Pest slugs and snails: biology and control. Translated by S. Gruber. Springer-Verlag, New York. 445 pp.
297. Godbee, J. F. Jr., and Franklin, R. T. 1976. Attraction, attack patterns, and seasonal activity of the black turpentine beetle. Ann. Entomol. Soc. Am. 69:653–655.
298. Goeden, R. D., and Norris, D. M. 1965. Some biological and ecological aspects of ovipositional attack in *Carya* spp. by *Scolytus quadrispinosus*. (Coleoptera: Scolytidae). Ann. Entomol. Soc. Am. 58:771–777.
299. Gouger, R. J. 1971. Control of *Adelges tsugae* on hemlock in Pennsylvania. Bartlett Tree Research Laboratories. Sci. Tree Top. 3:1.
300. Grant, J. 1966. The hosts and distribution of the root weevils *Hylobius pinicola* (Couper) and *H. warreni* Wood in British Columbia. J. Econ. Entomol. Soc. B. C. 63:3–5.
301. Grant, G. G., and Frech, D. 1980. Disruption of pheromone communication of the rusty tussock moth, *Orgyia antiqua* (Lepidoptera: Lymantriidae), with (Z)-6-heneicosen-11-one. Can. Entomol. 112:221–222.
302. Grant, G. G., MacDonald, L., Frech, D., Hall, K., and Slessor, K. N. 1985. Sex attractants for some eastern species of *Rhyacionia*, including a new species, and *Eucosma gloriola* (Lepidoptera: Tortricidae). Can. Entomol. 117:1489–1496.
303. Green, C. T. 1914. The cambium miner in river birch. J. Agric. Res. 1:471–474.
304. Griffiths, J. T., and Thompson, W. L. 1957. Insects and mites found on Florida citrus. Fla. Agric. Exp. Stn. Bull. 591. 96 pp.
305. Grigorov, S. 1965. *Hyadaphis tataricae* (Aizenberg) (Homoptera: Aphididae)—biology and means of control. Hortic. Vitic. Sci. (Sofia) 2:493–501.
306. Grimble, D. G., and Newell, R. G. 1972. Saddle prominent oviposition studies in New York and adjacent states. State Univ. Coll. of Environmental Sci. and Forestry, Syracuse, Applied Forestry Res. Inst. Rep. 12. 38 pp.
307. Grimble, D. G., Nord, J. C., and Knight, F. B. 1969. Oviposition characteristics and early larval mortality of *Saperda inornate* and *Oberea schaumii* in Michigan. Ann. Entomol. Soc. Am. 62:308–315.
308. Griswold, G. H. 1937. Common insects of the flower garden. Cornell Ext. Bull. 371. 59 pp.
309. Grobler, J. H. 1962. The life history and ecology of the woolly pine needle aphid, *Schizolachnus pini-radiatae* (Davidson) (Homoptera: Aphididae). Can. Entomol. 94:35–45.
310. Haack, R. A., and Benjamin, D. M. 1982. The biology and ecology of the twolined chestnut borer, *Agrilus bilineatus* (Coleoptera: Buprestidae) on oaks, *Quercus* spp., in Wisconsin. Can. Entomol. 114:385–389.
311. Habeck, D. H., and Mead, F. W. 1975. Lantana lace bug, *Teleonemia scrupulosa* Stal. (Hemiptera: Tingidae). Fla. Dep. of Agric. and Consumer Serv. Entomol. Circ. 156. 2 pp.
312. Haegele, R. W. 1936. The elm leaf beetle. Univ. Idaho Ext. Circ. 52. 8 pp.
313. Hajek, A. E. 1986. Aphid host preference used to detect a previously unrecognized birch in California. Environ. Entomol. 15:771–774.
314. Hajek, A. E., and Dahlsten, D. L. 1986. Coexistence of three species of

leaf feeding aphids (Homoptera) on *Betula pendula*. Oecologia (Ber.) 68: 380–386.
315. Hall, R. W. 1986. Preference for and suitability of elms for adult elm leaf beetle (*Xanthogaleruca luteola*) (Coleoptera: Chrysomelidae). Environ. Entomol. 15:143–146.
316. Hall, R. W., and Ehler, L. E. 1980. Population ecology of *Aphis nerii* on oleander. Environ. Entomol. 9:338–344.
317. Hamel, D. R. 1973. The biology and ecology of the live oak woolly leaf gall, *Andricus laniger* Ashmead (Homoptera: Cynipidae). Master's thesis, Texas A&M Univ., College Station. 146 pp.
318. Hamilton, C. C. 1925. The boxwood leaf miner. Univ. of Md. Agric. Exp. Stn. Bull. 272:143–170.
319. Hamilton, C. C. 1942. The taxus mealybug, *Pseudococcus cuspidatae* Rau. J. Econ. Entomol. 35:173–175.
320. Hamilton, D. W. 1966. Periodical cicadas. USDA Leaflet 540. 8 pp.
321. Hamilton, K. G. A. 1983. Revision of the Macropsini and Neopsini of the new world (Rhynchota: Homoptera: Cicadellidae), with notes on intersex morphology. Mem. Entomol. Soc. Can. 123:1–223.
322. Hamon, A. B. 1980. False oleander scale, *Pseudaulacaspis cockerelli* (Cooley) (Homoptera: Coccoidea: Diaspididae). Fla. Dep. of Agric. and Consumer Serv. Entomol. Circ. 95. 2 pp.
323. Hamon, A. B. 1981. Woolly whitefly, *Aleurothrixus floccosus* (Maskell) (Homoptera: Aleyrodidae: Aleyrodinae). Fla. Dep. of Agric. and Consumer Serv. Entomol. Circ. 232. 2 pp.
324. Hamon, A. B. 1985. Personal communications, collection records from Fla. Dep. of Agric. Div. of Plant Industry.
325. Hamon, A. B., and Williams, M. L. 1984. The soft scales of Florida (Homoptera: Coccoidea: Coccidae). In: Arthropods of Florida and neighboring land areas. Vol. 11. Fla. Dep. Agric. and Consumer Serv. Div. of Plant Industry. Contrib. 600. 194 pp.
326. Hamon, A. B., Lambdin, P. L., and Kosztarab, M. 1976. Life history and morphology of *Kermes kingi* in Virginia (Homoptera: Coccoidea: Kermesidae). In: Morphology and systematics of scale insects. No. 8. Va. Polytechnic Inst. and State Univ. Res. Div. Bull. 111:1–31.
327. Hamon, A. B. 1980. Bamboo pit scale, *Asterolecanium bambusae* (Boisduval) (Homoptera: Coccoidea: Asterolecaniidae). Fla. Dep. of Agric. Div. of Plant Industry. Entomol. Circ. 220. 2 pp.
327a. Hanisch, M. A., Brown, H. D., and Brown, E. A., eds. 1983. Dutch elm disease management guide. USDA For. Serv. and Ext. Serv. Bull. 1. 23 pp.
328. Hanna, M. 1970. An annotated list of spittlebugs of Michigan (Homoptera: Cercopidae). Mich. Entomol. 3:2–16.
329. Hansell, D. E., ed. 1970. Handbook of hollies. Am. Hortic. Mag. 49:234–255.
330. Hanson, J. B., and Benjamin, D. M. 1967. Biology of *Phytobia setosa*, a cambium miner of sugar maple. J. Econ. Entomol. 60:1351–1355.
331. Hanson, P. E., and Miller, J. C. 1984. Scale insects on ornamental plants: a biological control perspective. J. Arboric. 10:259–264.
332. Hantsbarger, W. M., and Brewer, J. W. 1970. Insect pests of landscape plants. Colo. State Univ. Coop. Ext. Bull. 472A. 65 pp.
333. Hard, J. S., Torgersen, T. R., and Schmiege, D. C. 1976. Hemlock sawfly. USDA For. Serv. Forest Insect and Disease Leaflet 31. 7 pp.
334. Harley, K. L. S., and Kassulke, R. C. 1971. Tingidae for control of *Lantana camara* (Verbenaceae). Entomophaga 16:389–410.
335. Harper, A. M. 1959. Gall aphids on poplar in Alberta. I. Description of galls and distribution of aphids. Can. Entomol. 91:489–496.
336. Harman, D. M., and Berisford, C. W. 1979. Host relationships and determination of larval instars of the locust twig borer, *Ectytolopha insiticiana*. Environ. Entomol. 8:19–23.
337. Harman, S. W. 1940. The cranberry rootworm as an apple pest. N.Y. State Agric. Exp. Stn. Geneva, Bull. 692. 11 pp.
338. Harris, J. W., and Coppel, H. C. 1967. The poplar-and-willow borer, *Sternochetus* (=*Cryptorhynchus*) *lapathi* (Coleoptera: Curculionidae) in British Columbia. Can. Entomol. 99:411–418.
339. Hartzell, A. 1943. Biology of the holly leaf miner. Contrib. Boyce Thompson Inst. 13:17–28.
339a. Hartzell, A. 1957. Red pine scale, with special reference to its host plants and cold-hardiness. Contrib. Boyce Thompson Inst. 18:421–428.
340. Harville, J. P. 1955. Ecology and population dynamics of the California oak moth. Microentomology 20:83–166.
341. Haseman, L. 1918. The tarnished plant-bug and its injury to nursery stock. Univ. of Mo. Agric. Exp. Stn. Res. Bull. 29. 26 pp.
342. Haseman, L., and McLane, S. R. 1940. The history and biology of the juniper midge (*Contarinia juniperina* Felt). Ann. Entomol. Soc. Am. 33:612–615.
343. Haviland, E. E. 1943. Hibernation and survival of the locust leaf-miner. J. Econ. Entomol. 36:639–640.
344. Hay, C. J. 1972. Red oak borer (Coleoptera: Cerambycidae) emergence from oak in Ohio. Ann. Entomol. Soc. Am. 65:1243–1244.
345. Hay, C. J., and Morris, R. C. 1961. Carpenterworm. USDA For. Serv. Forest Pest Leaflet 64. 8 pp.
346. Head, R. B., Neal, W. W., and Morris, R. C. 1977. Seasonal occurrence of the cottonwood leaf beetle *Chrysomela scripta* (Fab.) and its principal insect predators in Mississippi and notes on parasites. J. Ga. Entomol. Soc. 12:157–163.
347. Heinrich, C. 1956. American moths of the subfamily Phycitinae. Smithsonian Inst. U.S. Natl. Museum Bull. 207. 58 pp.
348. Heinrichs, E. A., Burgess, E. E., and Matheny, E. L. Jr. 1973. Control of leaf-feeding insects on yellow poplar. J. of Econ. Entomol. 66:1240–1241.
349. Henry, T. J., and Smith, C. L. 1979. An annotated list of the Miridae of Georgia (Hemiptera—Heteroptera). J. Ga. Entomol. Soc. 14:212–220.
350. Henry, T. J., and Signarovits, L. L. 1979. A bark beetle as a potential pest of ornamental white pine (Coleoptera: Scolytidae). Pa. Dep. of Agric. Bur. Plant Industry Regulatory Hortic. Entomol. Circ. 42:15–16.
351. Henry, T. J., and Bennett, L. W. 1980. Pine root collar weevil, *Hylobius radicis* Buchanan. Pa. Dep. of Agric. Entomol. Circ. 47. 2 pp.
352. Herbert, F. B. 1920. Cypress bark scale. USDA Dep. Bull. 838. 22 pp.
353. Hering, E. M. 1951. Biology of the leaf miners. Dr. W. Junk Publishers, The Hague. 420 pp.
354. Heriot, A. D. 1938. Biological and morphological differences between *Eriosoma crataegi* (Oestlund) and *Eriosoma lanigera* (Haus.) Proc. Entomol. Soc. B. C. 34:22–32.
355. Herrick, G. W. 1910. The snow-white linden moth. Cornell Univ. Agric. Exp. Stn. Bull. 286:51–64.
356. Herrick, G. W. 1912. The larch casebearer. Cornell Univ. Agric. Exp. Stn. Bull. 322:39–54.
357. Herrick, G. W. 1923. The maple casebearer. Cornell Univ. Agric. Exp. Stn. Bull. 417. 15 pp.
358. Herrick, G. W. 1931. The magnolia scale (*Neolecanium cornuparvum* Thro). Ann. Entomol. Soc. Am. 24:302–305.
359. Herrick, G. W. 1935. Insect enemies of shade trees. Comstock Publishing Co., Ithaca, NY. 417 pp.
360. Herrick, G. W., and Tanaka, T. 1926. The spruce gall aphid. Cornell Univ. Agric. Exp. Stn. Bull. 454. 17 pp.
361. Hess, A. D. 1940. Biology and control of the round-headed apple-tree borer, *Saperda candida* Fabricius. N.Y. State Agric. Exp. Stn. Bull. 688:5–93.
362. Hidaka, T. 1977. Adaptation and speciation in the fall webworm. Kodansha Ltd., Tokyo. 179 pp.
363. Highland, H. A. 1964. Life history of *Asphondylia ilicicola* (Diptera: Cecidomyiidae), a pest of American holly. J. Econ. Entomol. 57:81–83.
364. Hildahl, V., and Peterson, L.O.T. 1974. Fall and spring cankerworms in the prairie. Northern For. Res. Cent. Edmonton, Alberta. Information Rep. NOR-X-100.
365. Hille Ris Lambers, D. 1966. Notes on California aphids, with description of new genera and new species (Homoptera: Aphididae). Hilgardia 37:569–623.
366. Hille Ris Lambers, D. 1968. The mystery of the woolly apple aphid. Aphidol. Newsletter 7:4–5.
367. Hillman, G. H. 1890. A serious rose pest. Nev. Agric. Exp. Stn. Bull. 9. 4 pp.
368. Hinckley, A. D. 1972. Comparative ecology of two leafminers on white oak. Environ. Entomol. 1:358–361.
369. Hixon, E. 1941. The walnut Datana. Oklahoma A&M Coll. Exp. Stn. Bull. B–246. 29 pp.
370. Hodges, R. W. 1971. Sphingoidea. In: The moths of America north of Mexico. Fascicle 21. R. B. Dominick et al., eds. E. W. Classey Ltd., London. 158 pp.
371. Hodges, R. W. 1983. Check list of the lepidoptera north of Mexico. E. W. Classey Ltd., London. 274 pp.
372. Hodgkiss, H. E. 1930. The Eriphyidae of N.Y. Vol. II. The maple mites. N.Y. Agric. Exp. Stn. Tech. Bull. 163. 45 pp.
373. Hodson, A. C. 1942. Biological notes on the basswood leaf-miner, *Baliosus ruber* (Weber). J. Econ. Entomol. 35:570–573.
374. Hoebeke, E. R., and Johnson, W. T. 1984. A European privet sawfly, *Macrophyla punctumalbum* (L.): North American distribution, host plants, seasonal history, and description of the immature stages (Hymenoptera: Tenthredinidae). Proc. Entomol. Soc. Wash. 87:25–33.
375. Hoebeke, E. R. 1987. *Yponomeuta cagnagella* (Lepidoptera: Yponomeutidae): A palearctic ermine moth in the United States, with notes on its recognition, seasonal history, and habits. Ann. Entomol. Soc. Am. 80:462–467.
376. Hoerner, J. L. 1936. Western rose curculio. Colo. Agric. Exp. Stn. Bull. 432. 19 pp.
377. Hofer, G. 1920. The aspen borer and how to control it. USDA Farmers' Bull. 1154. 11 pp.
378. Hoffard, W. H., and Anderson, R. L. 1979. How to identify and control tree cricket damage. Forestry Bull. SA-FB/P3. 4 pp.
379. Hoffman, C. H., Hepting, G. H., and Roth, E. R. 1947. Droop of white pine caused by *Pineus*. J. Econ. Entomol. 40:229–231.
380. Holland, W. J. 1903. The moth book. Doubleday, Page, New York. 479 pp.

381. Holms, J., and Ruth, D. S. 1968. Spruce aphid in British Columbia. Dep. Fisheries and Forestry, For. Res. Lab., Victoria, Forest Pest Leaflet 16. 5 pp.
382. Hood, C. E. 1940. Life history and control of the imported willow leaf beetle. USDA Circ. 572. 9 pp.
383. Hopping, G. R. 1937. Sawfly biologies. No. 2. *Hemichroa crocea* Geoffroy. Can. Entomol. 69:243–249.
384. Horn, K. F., Wright, C. G., and Farrier, M. H. 1979. The lacebugs (Hemiptera: Tingidae) of North Carolina and their hosts. N.C. Agric. Exp. Stn. Tech. Bull. 257. 22 pp.
385. Horn, K. F., Farrier, M. H., and Wright, C. G. 1983. Some mortality factors affecting the eggs of the sycamore lacebug, *Corythucha ciliata* (Say) (Hemiptera: Tingidae). Ann. Entomol. Soc. Am. 76:262–265.
386. Hosley, N. W. 1938. Damage to forest trees by wild mammals. Mass. For. and Park Assoc. Tree Pest Leaflet 30. 4 pp.
387. Hosten, E. H., and Hard, J. 1985. Efficacy of *Bacillus thuringiensis* Berliner for supressing populations of large aspen tortrix in Alaska. Can. Entomol. 117:587–591.
388. Hottes, F. C., and Frison, T. H. 1931. The plant lice or Aphididae of Illinois. Ill. Natural History Survey Bull. Vol. XIX, Article III, pp. 121–447.
389. Hough, W. S. 1925. Biology and control of Comstock's mealybug on the umbrella catalpa. Va. Agric. Exp. Stn. Tech. Bull. 29. 27 pp.
390. Houser, J. S. 1918. Destructive insects affecting Ohio shade and forest trees. Ohio Agric. Exp. Stn. Bull. 332:165–487.
391. Houseweart, M. W., and Brewer, J. W. 1972. Biology of a piñon spindle gall midge (Diptera: Cecidomyiidae) Ann. Entomol. Soc. Am. 65:331–336.
392. Howard, L. D., and Chittenden, F. H. 1916. The leopard moth: a dangerous imported insect enemy of shade trees. USDA Farmers' Bull. 708. 11 pp.
393. Howard, L. D., and Chittenden, F. H. 1916. The bagworm, an injurious shade tree insect. USDA Farmers' Bull. 701. 11 pp.
394. Hoyt, S. C., and Madsen, H. F. 1960. Dispersal behavior of the first-instar nymphs of the woolly apple aphid. Hilgardia. 30:267–299.
395. Hunt, R. S., and Funk, A. 1983. Common pests of arbutus. Revised. Pac. For. Res. Centre, Pest Leaflet 63. 6 pp.
396. Hussey, N. H. 1952. A contribution to the bionomics of the green spruce aphid (*Neomyzaphis abietina* Walker). Scott. Forestry 6:121–130.
397. Hussey, N. H., and Scopes, N. 1985. Biological pest control. Cornell Univ. Press, Ithaca, NY. 240 pp.
398. Hutchings, C. B. 1924. The lesser oak carpenter worm and its control. Can. Dep. Agric. Entomol. Branch Circ. 23.
399. Hutchins, Ross E. 1969. Galls and gall insects. Dodd, Mead, & Co., New York. 128 pp.
400. Hyche, L. L. 1982. Observations on the biology of the sawfly *Eriocampa juglandis* in Alabama. J. Ga. Entomol. Soc. 17:417–421.
401. Ignoffo, C. M., and Granovsky, A. A. 1961. Life history and gall development of *Mordvilkojo vagabunda* (Homoptera: Aphidae) on *Populus deltoides*. Ann. Entomol. Soc. Am. 54:486–499.
402. Ignoffo, C. M., and Granovsky, A. A. 1961. Life history and gall development of *Mordvilkojo vagabunda* (Homoptera: Aphidae) on *Populus deltoides*. Part 11. Gall development. Ann. Entomol. Soc. Am. 54:634–641.
403. Jaynes, Richard A. 1979. Nut tree culture in North America. Northern Nut Growers Assoc., Hamden, CT. 466 pp.
404. Jensen, D. D. 1951. The North American species of *Psylla* from willow, with descriptions of new species and notes on biology (Homoptera: Psyllidae). Hilgardia 20:299–324.
405. Jensen, D. D. 1957. Parasites of the Psyllidae. Hilgardia 27:71–99.
406. Jensen, G. L., and Koehler, C. S. 1969. Biological studies of *Scythropus californicus* on Monterey pine in northern California. Ann. Entomol. Soc. Am. 62:117–120.
407. Jeppson, L. T., Keifer, H. H., and Baker, E. W. 1975. Mites injurious to economic plants. Univ. of California Press, Berkeley. 614 pp.
408. Jessen, E. 1964. Life history of the Monterey pine needle miner, *Argyresthia pilatella* (Lepidoptera: Yponemeutidae). Ann. Entomol. Soc. Am. 57:332–341.
409. Johnson, D. M. 1935. Leafhoppers of Ohio. Ohio Biological Survey Bull. 39:39–122.
410. Johnson, E. L. 1978. Biology of *Caliroa quercuscoccinea* (Dyer) (Hymenoptera: Tenthredinidae) in Kentucky. Master's thesis, Univ. of Kentucky, Lexington.
411. Johnson, N. E. 1965. Reduced growth associated with infestations of Douglas-fir seedlings by *Cinara* sp. (Homoptera: Aphidae). Can. Entomol. 97:113–119.
412. Johnson, S. A. 1906. The cottony maple scale. Coll. Agric. Exp. Stn. Press Bull. 27. 4 pp.
413. Johnson, W. T. The Asiatic oak weevil and other insects causing damage to chestnut foliage in Maryland. J. Econ. Entomol. 49:717–718.
414. Johnson, W. T., and Russell, M. L. 1962. The black citrus aphid *Toxoptera aurantii* (Fonscolombe) in Maryland. Proc. Entomol. Soc. Wash. 64:90.
415. Jones, T. H., and Schaffner, J. V. 1959. Cankerworms. USDA For. Serv. Leaflet 183. 8 pp.
416. Jorgensen, C. D., Rice, R. E., Hoyt, S. C., and Westigard, P. H. 1981. Phenology of the San Jose scale (Homoptera: Diaspididae). Can. Entomol. 113:149–159.
417. Judge, F. D. 1972. Aspects of the biology of the gray garden slug (*Deroceras reticulatum* Muller). Search Agric. 2:1–18.
418. Kattoulas, M. E., and Koehler, C. S. 1965. Studies on the biology of the irregular pine scale. J. Econ. Entomol. 58:727–730.
419. Kaya, H. K., and Lindegren, J. E. 1983. Parasitic nematode controls western poplar clearwing moth. Calif. Agric. 37:31–32.
420. Kearby, W. H., and Benjamin, D. M. 1964. The biology and ecology of the red-pine needle midge and its role in fall browning of red pine foliage. Can. Entomol. 96:1313.
421. Keen, F. P. 1952. Insect enemies of western forests. USDA Misc. Publ. 273. 271 pp.
422. Keifer, H. H. 1936. California micro-lepidoptera VII. Bull. So. Calif. Acad. Sci. 35:9–29.
423. Keifer, H. H. 1938–1975. Eriophyid studies. Vols. I–XXVI (1938–1959) Bull. Calif. Dep. of Agric. Vols. XXVII–XXVIII (1959) Bur. Entomol. Calif. Dep. of Agric., Occasional Papers 1 & 2. B Series 1–21 (1969–1974) Bur. Entomol. Calif. Dep. of Agric. Special Publ.
424. Keifer, H. H. 1946. A review of North American economic eriophyid mites. J. Econ. Entomol. 39:563–570.
425. Keifer, H. H., and Saunders, J. L. 1972. *Trisetacus campnodus*, n. sp. (Acarina: Eriophidae), attacking *Pinus sylvestris*. Ann. Entomol. Soc. Am. 65:46–49.
426. Keifer, H. H., Baker, E. W., Kono, T., Delfinado, M., and Styer, W. E. 1982. An illustrated guide to plant abnormalities caused by eriophyid mites in North America. USDA Agric. Handb. 573. 178 pp.
427. Keifer, H. H. 1930. California microlepidoptera IV. Pan-Pac. Entomol. 7:27–34.
428. Kelton, L. A. 1975. The lygus bugs (genus *Lygus* Hahn) of North America. Heteroptera: Miridae. Mem. Entomol. Soc. Can. 95. 101 pp.
429. Kelton, L. A. 1978. A new *Tropidosteptes* Uhler (1908), synonym of *Xenoborus* Reuter (1878), with description of a new species (Heteroptera: Miridae). Can. Entomol. 110:471–473.
430. Kerr, T. W. Jr. 1949. The arborvitae weevil *Phyllobius intrusus* Kono. R.I. Agric. Exp. Stn. Bull. 305. 30 pp.
431. Khan, A. H. 1944. On the lantana bug (*Teleonemia scruprilosa* (Stal.). Indian J. Entomol. 6:149–161.
432. Knight, H. H. 1941. The plant bugs or Miridae of Illinois. Ill. Natural Hist. Survey. Vol. 22 (art. 1). 234 pp.
433. Knull, J. N. 1946. The long-horned beetles of Ohio (Coleoptera: Cerambycidae). Ohio Biol. Surv. Bull. 39. 238 pp.
434. Koehler, C. S., and Tauber, M. 1964. *Periploca nigra*, a major cause of dieback of ornamental juniper in California. J. Econ. Entomol. 57:563–566.
435. Koehler, C. S., Kattoulas, M. E., and Frankie, G. W. 1966. Biology of *Psylla uncatoides*. J. Econ. Entomol. 59:1097–1100.
436. Koehler, C. S., Frankie, G. W., Moore, W. S., and Landwehr, V. R. 1983. Relationship of infestation by the sequoia pitch moth (Lepidoptera: Sesiidae) to Monterey pine injury. Environ. Entomol. 12:979–981.
437. Koehler, C. S., and Moore, W. S. 1983. Resistance of several members of the Cupressaceae to the cypress tipminer, *Argyresthia cupressella*. J. Environ. Hortic. 1:87–88.
438. Koehler, C. S., Moore, W. S., and Coate, B. 1983. Resistance of *Acacia* to the acacia psyllid, *Psylla uncatoides*. J. Environ. Hortic. 1:65–67.
439. Koerber, T. W., and Struble, G. R. 1971. Lodgepole needleminer. USDA For. Serv. Forest Pest Leaflet 22. 8 pp.
440. Kopp, D. D., and Yonke, T. R. 1973. The treehoppers of Missouri. Part 1. Subfamilies Centrotinae, Hoplophorioninae, and Membracinae (Homoptera: Membracidae). J. Kans. Entomol. Soc. 46:42–64.
441. Kopp, D. D., and Yonke, T. R. 1974. The treehoppers of Missouri. Part 4. Subfamily Smiliinae; Tribe Telamonini (Homoptera: Membracidae). J. Kans. Entomol. Soc. 47:80–130.
442. Kosztarab, M. 1963. The armored scale insects of Ohio (Homoptera: Coccoidea: Diaspididae). Ohio Biological Survey. Vol. II, no. 2. 120 pp.
443. Kosztarab, M., and Kozar, F. 1987. Scale insects of central Europe. Kluwer Academic Publishers, Boston. (in press)
444. Kotinsky, J. 1921. Insects injurious to deciduous shade trees and their control. USDA Farmers' Bull. 1169. 100 pp.
445. Kraft, S. K., and Denno, R. F. 1982. Feeding responses of adapted and nonadapted insects to the defensive properties of *Baccharis halimifolia* L. (Compositae). Oecologia (Berl.) 52:156–163.
446. Krantz, G. W. 1979. The role of *Phytocoptella avellanae* (Nalepa) and *Cecidophyopsis vermiformis* (Nal.) (Eriophyoidea) in big bud of filbert. Pages 201–208 in: Proceedings of the IV International Congress of Acarology. Akadémiai Kiadó, Budapest. 751 pp.
447. Krombein, K. V., Hurd, P. D. Jr., Smith, D. R., and Burks, B. D. 1979. Catalogue of Hymenoptera in America north of Mexico. Vol. I and indices. Smithsonian Inst. Press, Washington, DC. 1198 pp.

448. Kucera, D. R., and Orr, P. W. 1981. Spruce budworm in the eastern United States. USDA Forest Serv. Forest Insect and Disease Leaflet 160. 7 pp.
449. Kuitert, L. C. 1958. Insect pests of ornamental plants. Fla. Agric. Exp. Stn. Bull. 595. 51 pp.
450. Kuitert, L. C. 1967. Observations on the biology, bionomics, and control of white peach scale, *Pseudaulacaspis pentagona* (Targ.). Proc. Fla. State Hortic. Soc. 80:376–381.
451. Kulp, L. A. 1968. The taxonomic status of dipterous holly leaf miners (Diptera: Agromyzidae). Univ. of Md. Agric. Exp. Stn. Bull. A-55:4.
452. Kusch, D. S. 1977. Large aspen tortrix. Pest Leaflet Northern For. Res. Cent. Edmonton, Alberta.
453. Langford, G. S. 1937. Biology and control of the juniper webworm in Maryland. J. Econ. Entomol. 30:320–323.
454. Langlois, T. H., and Langlois, M. H. 1964. Notes on the life-history of the hackberry butterfly, *Asterocampa celtis,* on South Bass Island, Lake Erie (Lepidoptera: Nymphalidae). Ohio J. Sci. 64:1–11.
455. Lanier, G. N. 1982. Behavior modifying chemicals in Dutch elm disease vector control. Pages 371–394 in: Proceedings of the Dutch elm disease symposium and workshop. E. S. Kondo, Y. Hiratsuka, and B. W. G. Denyer, eds. Manitoba Dep. Natural Resources, Winnipeg.
456. Lanier, G. N. 1986. *Pissodes nemorensis* (=*Pissodes approximatus*). Personal communication.
457. Larew, H., and Capizzi, J. 1983. Common insect and mite galls of the Pacific northwest. Oregon State Univ. Press, Corvallis. 80 pp.
458. Latta, R. 1937. The rhododendron white fly and its control. USDA Circ. 429. 8 pp.
459. LaBonte, G. A., and Lipovsky, L. J. 1967. Oak-leaf shot hole caused by *Japanagromyza viridula.* J. Econ. Entomol. 60:1266–1270.
460. Lehman, R. D. 1983. Strawberry root weevil, *Otiorhynchus ovatus* (L.), as a pest of conifer seedlings. Pa. Dep. of Agric., Bur. of Entomol. Circ. 79. 2 pp.
461. Lehman, R. D. 1984. Azalea stem borer, *Oberea myops.* Pa. Dep. of Agric. Regulatory Hortic. Entomol. Circ. 86. 2 pp.
461a. Lehman, R. D. 1986. *Agrilus difficilis,* a potential pest of honeylocust. Pa. Dep. of Agric. Entomol. Circ. 106. 3 pp.
462. Leonard, D. E. 1964. Biology and ecology of *Magicicada septendecim* (L.) (Hemiptera: Cicadidae). J. N.Y. Entomol. Soc. 72:19–23.
463. Leonard, M. D. 1963. A list of aphids of New York. Proc. Rochester Acad. Sci. 10:289–428.
464. Leonard, M. D., and Bissell, T. L. 1970. A list of the aphids of District of Columbia, Maryland, and Virginia. Univ. of Md. Agric. Exp. Stn. MP770. 129 pp.
465. Lewis, H. C. 1939. Fruit spotting by the green leafhopper. Calif. Citrogr. 24:108–109.
466. Lewis, T. 1973. Thrips: their biology, ecology and economic importance. Academic Press, New York. 349 pp.
467. Lindegren, J. E., and Barnett, W. W. 1982. Applying parasitic nematodes to control carpenterworms in fig orchards. Calif. Agric. 36:7–8.
468. Lindquist, O. H. 1982. Keys to lepidopterous larvae associated with the spruce budworm in northeastern North America. CANUSA Spruce Budworm Program. Great Lakes For. Res. Centre, Canadian Forestry Service. 18 pp.
469. Lindquist, O. H. 1971. The adelgids (Homoptera) on forest trees in Ontario with key to galls on spruce. Proc. Entomol. Soc. Ont. 102:23–27.
470. Lindquist, O. H., and Jackson, G. G. 1965. A leaf-mining sawfly on oak. Can. Dep. Forestry Bi-mon. Prog. Rep. 21:1–4.
471. Lindquist, O. H., and MacLeod, L. S. 1967. A biological study of *Epinotia solandriana* (Lepidoptera: Olethreutidae), a leaf roller on birch in Ontario. Can. Entomol. 99:1110–1114.
472. Lindquist, O. H., and Syme, P. D. 1981. Insects and mites associated with Ontario forests. Can. Forestry Serv., Sault Ste. Marie, Ont. Information Report O-X-333. 116 pp.
473. Linsley, E. G. 1940. Notes on *Oncideres* twig girdlers. J. Econ. Entomol. 33:561–563.
474. Linsley, E. G. 1959. Ecology of Cerambycidae. Annu. Rev. Entomol. 4:99–138.
474a. Linsley, E. G. 1961. The Cerambycidae of North America. Part I. Introduction. Univ. Calif. Publ. Entomol. 18:1–135.
474b. Linsley, E. G. 1962. The Cerambycidae of North America. Part II. Univ. Calif. Publ. Entomol. 19:1–102.
474c. Linsley, E. G. 1963. The Cerambycidae of North America. Part IV. Univ. Calif. Publ. Entomol. 21:1–165.
475. Little, E. L. Jr. 1979. Checklist of United States trees (native and naturalized). USDA For. Serv. Agric. Handb. 541. 375 pp.
476. Loerch, C. R., and Cameron, C. A. 1984. Within-tree distributions and seasonality of immature stages of the bronze birch borer, *Agrilus anxius* (Coleoptera: Buprestidae). Can. Entomol. 116:147–152.
477. Lovett, A. L., and Black, A. B. 1920. The gray garden slug (with notes on allied forms). Bull. Ore. Agric. Exp. Stn. 170. 43 pp.
478. Lowe, J. H. 1965. Biology and dispersal of *Pineus pinifoliae* (F.). Master's thesis, Yale Univ., New Haven. 104 pp.
479. Luck, R. F., and Scriven, G. T. 1979. The elm leaf beetle, *Pyrrhalta luteola,* in southern California: its host preference and host impact. Environ. Entomol. 8:307–313.
480. Lugger, O. 1900. Bugs injurious to our cultivated plants. Univ. of Minn. Agric. Exp. Stn. Bull. 69: 257 pp.
481. Mackay, M. R. 1959. Larvae of the North American Olethreutidae. Can. Entomol. Suppl. 10. 330 pp.
482. Mackay, M. R. 1962. Larvae of the North American Tortricinae (Lepidoptera: Tortricidae). Can. Entomol. Suppl. 28. 182 pp.
482a. MacKay, M. R. 1968. The North American Aegeriidae (Lepidoptera): a revision based on late-instar larvae. Entomol. Soc. Can. Mem. 58. 112 pp.
483. Macleod, D. M., and Tyrrell, D. 1979. *Entomophthora crustoosa* n. sp. as a pathogen of the forest tent caterpillar *Malacosoma disstria* (Lepidoptera: Lasiocampidae). Can. Entomol. 111:1137–1144.
484. MacAloney, H. J. 1961. Pine tortoise scale. USDA For. Serv. Forest Pest Leaflet 57. 7 pp.
485. MacAloney, H. J., and Schmiege, D. C. 1962. Identification of conifer insects by type of tree injury. USDA, Lake States Forest Exp. Stn. Paper 100. 41 pp.
486. MacAloney, H. J., and Ewan, H. G. 1964. Identification of conifer insects by type of tree injury, north central region. U.S. For. Serv. Res. Paper LS-11. 70 pp.
487. MacAloney, H. J., and Wilson, L. F. 1964. The red-headed pine sawfly. USDA For. Serv. Forest Pest Leaflet 14. 5 pp.
488. MacGown, M. W., and Osgood, E. A. Jr. 1971. Two new species of *Platygaster* (Hymenoptera: Proctotrupoidea, Platygastridae) parasitic on balsam gall midge, *Dasineura balsamicola* (Diptera: Cecidomyiidae). Can. Entomol. 103:1143–1146.
489. Madsen, H. F., and Barnes, M. M. 1959. Pests of pear in California. Calif. Agric. Exp. Stn. Circ. 478. 40 pp.
490. Madsen, H. F., and Hoyt, S. C. 1957. *Schizura ipomaeae* Dbldy. attacking plums in California. J. Econ. Entomol. 50:284–287.
491. Mague, D. L., and Reissig, W. H. 1983. Phenology of the San Jose scale (Homoptera: Diaspididae) in New York State apple orchards. Can. Entomol. 115:717–722.
492. Maier, C. T. 1980. A mole's eyeview of seventeen-year periodical cicada nymphs, *Magicicada septendecim* (Hemiptera: Homoptera: Cicadidae) Ann. Entomol. Soc. Am. 73:147–152.
493. Maier, C. T. 1984. Seasonal development and flight activity of the spotted tentiform leafminer *Phyllonorycter blancardelle* (Lepidoptera: Gracillariidae), in Connecticut. Can. Entomol. 116:435–441.
494. Mains, E. B. 1958. North American entomogenous species of *Cordyceps.* Mycologia 50:169–222.
495. Mani, M. S. 1964. Ecology of plant galls. Junk Publishers, The Hague. 434 pp.
496. Marcovitch, S. 1916. The red rose beetle. Office of the Minn. State Entomol. Circ. 36. 4 pp.
497. Marrone, P. G., and Zepp, D. B. 1979. Descriptions of the larva and pupa of *Callirhopalus* (subg. *Pseudocneorhinus*) *bifasciatus,* the twobanded Japanese weevil, with new host plant records. Ann. Entomol. Soc. Am. 72:833–836.
498. Martineau, R. 1984. Insects harmful to forest trees. Agric. Canada. Gov. Publishing Centre, Supply and Services, Ottawa. 261 pp.
499. Massey, C. L. 1940. The pandora moth (*Coloradia pandora*), a defoliator of lodgepole pine in Colorado. Master's thesis, Duke Univ., Durham, NC.
500. Matheson, R. 1917. The poplar and willow borer. Cornell Agric. Exp. Stn. Bull. 388:457–483.
501. May, C., and Baker, W. L. 1952. Insects and spread of forest-tree diseases. Pages 677–682 in: Insects. The Year Book of Agriculture. USDA, Washington, D.C.
502. McAtee, W. L. 1924. Dikraneura in the United States (Homoptera: Eupterygidae). Wash. Entomol. Soc. Proc. 26:75–76.
503. McCambridge, W. F. 1974. Pinyon needle scale. USDA For. Serv. Forest Pest Leaflet 148. 4 pp.
504. McClintock, E., and Leiser, A. T. 1979. An annotated checklist of woody ornamental plants of California, Oregon, and Washington. Univ. of Calif. Div. of Agric. Sci. Publ. 4091. 134 pp.
505. McClure, M. S. 1978. Seasonal development of *Fiorinia externa, Tsugaspidiotus tsugae* (Homoptera: Diaspididae), and their parasite, *Aspidiotiphagus citrinus* (Hymenoptera: Aphelinidae): importance of parasite-host synchronism to the population dynamics of two scale pests of hemlock. Environ. Entomol. 7:863–870.
506. McClure, M. S. 1979. Spatial and seasonal distribution of disseminating stages of *Fiorinia externa* (Homoptera: Diaspididae) and natural enemies in a hemlock forest. Environ. Entomol. 8:869–873.
507. McClure, M. S. 1982. Distribution and damage of two *Pineus* species (Homoptera: Adelgidae) on red pine in New England. Ann. Entomol. Soc. Am. 75:150–157.
507a. McClure, M. S. 1987. Controlling hemlock scales with least environmental impact. Conn. Agric. Exp. Stn. Bull. 844. 8 pp.
508. McClure, M. S., and Fergione, M. B. 1977. *Fiorinia externa* and *Tsugaspidiotus tsugae* (Homoptera: Diaspididae). Distribution, abundance, and

new hosts of two destructive scale insects of eastern hemlock in Connecticut. Environ. Entomol. 6:807–811.
509. McClure, M. S., Dahlsten, D. L., DeBarr, G. L., and Hedden, R. L. 1983. Control of pine bark scale in China. J. For. 81:440, 475–478.
510. Mc Comb, C. W. 1986. A field guide to insect pests of holly. Holly Soc. of Am. Inc., Baltimore, MD. 122 pp.
511. McConnell, H. S. 1949. A new North American species of *Pulvinaria*. Proc. Entomol. Soc. Wash. 51:29–34.
512. McConnell, H. S., and Davidson, J. A. 1959. Observations on the life history and morphology of *Kermes pubescens* Bogue (Homoptera: Coccoidea: Dactylopiidae). Ann. Entomol. Soc. Am. 52:463.
513. McDaniel, E. I. 1930. Soft scales injurious to deciduous ornamentals. Mich. State Agric. Exp. Stn. Circ. Bull. 133. 17 pp.
514. McDaniel, E. I. 1933. Important leaf-feeding and gall-making insects infesting Michigan's deciduous trees and shrubs. Mich. State Agric. Exp. Stn. Spec. Bull. 243. 70 pp.
515. McDaniel, E. I. 1933. Some wood borers attacking the trunks and limbs of deciduous trees and shrubs. Mich. State. Agric. Exp. Stn. Spec. Bull. 238. 38 pp.
516. McKenzie, H. L. 1933. Observations on the genista caterpillar, *Tholeria reversalis* Guenée (Lepidoptera: Pyralidae). Calif. Dep. Agric. Monthly Bull. 22:410–412.
517. McKenzie, H. L. 1956. The armored scale insects of California. Bull. Calif. Insect Survey. Vol. V. Univ. of California Press, Berkeley. 209 pp.
518. McKenzie, H. L. 1967. Mealybugs of California. Univ. of California Press, Berkeley. 526 pp.
519. McLeod, J. M. 1963. Life history and habits of a spruce needle-miner, *Eucordylea ducharmei* Free. (Lepidoptera: Gelechiidae). Can. Entomol. 95: 443–447.
520. Mead, F. W. 1962. A spittlebug (*Prosapia bicincta*) (Say) (Homoptera: Cercopidae). Fla. Dep. of Agric. Div. of Plant Industry Entomol. Circ. 7. 1 p.
521. Mead, F. W. 1962. The thorn bug, *Umbonia crassicornis*. Fla. Dep. of Agric. Div. of Plant Industry Entomol. Circ. 8. 2 pp.
522. Mead, F. W. 1963. A psyllid, *Trioza magnoliae* (Ashmead) (Homoptera: Psyllidae). Fla. Dep. of Agric. Div. of Plant Industry Entomol. Circ. 15. 2 pp.
523. Mead, F. W. 1963. The spittlebug genus *Aphrophora* in Florida. Fla. Dep. of Agric. Div. of Plant Industry Entomol. Circ. 20. 2 pp.
524. Mead, F. W. 1967. The oak treehopper, *Platycotis vittata* (Fabricius) (Homoptera: Membracidae). Fla. Dep. of Agric. Div. of Plant Industry Entomol. Circ. 57. 2 pp.
525. Mead, F. W. 1969. Citrus flatid planthopper, *Metcalfa pruinosa* (Say) (Homoptera: Flatidae). Fla. Dep. of Agric. Div. of Plant Industry Entomol. Circ. 85. 2 pp.
526. Mead, F. W. 1970. *Ctenodactylomia watsoni* Felt, a gall midge pest of seagrape, *Coccoloba uvifera* L., in Florida (Diptera: Cecidomyiidae). Fla. Dep. of Agric. and Consumer Serv. Entomol. Circ. 97. 2 pp.
527. Mead, F. W. 1972. The hawthorn lace bug, *Corythucha cydoniae* (Fitch), in Florida (Hemiptera: Tingidae). Fla. Dep. of Agric. and Consumer Serv. Entomol. Circ. 127. 2 pp.
528. Mead, F. W. 1981. The coreid bug, *Euthochtha galeator* (Fabricius) in Florida (Hemiptera: Coreidae). Fla. Dep. of Agric. and Consumer Serv. Entomol. Circ. 222. 4 pp.
529. Mead, F. W. 1983. Yaupon psyllid, *Gyropsylla ilicis* (Ashmead) (Homoptera: Psyllidae) Fla. Dep. of Agric. and Consumer Serv. Entomol. Circ. 247. 2 pp.
530. Merrill, G. B. 1953. A revision of the scale insects of Florida. State Plant Board of Fla. Bull. 1. 143 pp.
531. Merrill, G. B., and Chaffin, J. 1923. Scale-insects of Florida. Fla. State Plant Board Quarterly Bull. 7:177–298.
532. Metcalf, C. L., Flint, W. P., and Metcalf, R. L. 1951. Destructive and useful insects. McGraw-Hill, New York. 1071 pp.
533. Michelbacher, A. E., and Ortega, J. C. 1958. A technical study of insects and related pests attacking walnuts. Calif. Agric. Exp. Stn. Bull. 764. 86 pp.
534. Michelbacher, A. E., and Hitchcock, S. 1956. Frosted scale on walnuts. Calif. Agric. 10:11–14.
535. Middlekauff, W. W. 1958. North American sawflies of the genera *Acantholyda, Cephalcia,* and *Neurotoma* (Hymenoptera, Pamphiliidae). Univ. Calif. Publ. Entomol. 14:51–174.
536. Middleton, W. 1922. Sawflies injurious to rose foliage. USDA Farmers' Bull. 1252. 14 pp.
537. Milholland, R. D., and Meyer, J. R. 1984. Diseases and arthropod pests of blueberries. N.C. Agric. Res. Serv., N.C. State Univ. Bull. 468. 33 pp.
538. Miller, D. R., and Howard, F. W. 1981. A new species of *Abgrallaspis* (Homoptera: Coccoidea: Diaspididae) from Louisiana. Ann. Entomol. Soc. Am. 74:164–166.
538a. Miller, W. E. 1987. Guide to the olethreutine moths of midland North America (Tortricidae). USDA For. Serv. Agric. Handb. 660. 104 pp.
539. Millers, I., and Wallner, W. E. 1975. The redhumped oakworm. USDA For. Serv. Forest Pest Leaflet 153. 6 pp.
540. Mills, E. M. 1938. Tree injury by squirrels. Mass. Agric. Exp. Stn. Bull. 353. 2 pp.
541. Mitchell, R. G., Amman, G. D., and Waters, W. E. 1970. Balsam woolly aphid. USDA For. Serv. Forest Pest Leaflet 118. 10 pp.
542. Mitton, J. B., and Sturgeon, K. B eds. 1982. Bark beetles in North American conifers: a system for the study of evolutionary biology. Univ. of Texas Press, Austin. 527 pp.
543. Monroe, H. S. 1970. Biosystematics of the genus *Trirhabda* Leconte of America north of Mexico (Chrysomelidae: Coleoptera). Ph.D. diss., Univ. of Idaho, Moscow.
544. Mordvilko, A. 1924. *Eriosoma lanigera*: biology and distribution. Trans. Dep. of Applied Entomol., Leningrad, Vol. XII, No. 3. 110 pp.
545. Morette, M. F. D. 1959. Studies on the potato leafhopper, *Empoasca fabae,* as a nursery pest. Ph.D. diss., Rutgers University, New Brunswick, NJ. 223 pp.
546. Morris, R. C., and Oliveria, F. L. 1976. Insects of periodic importance in cottonwood. Pages 280–285 in: Proceedings of a symposium on eastern cottonwood and related species. Louisiana State Univ., Baton Rouge. 485 pp.
547. Morris, R. C., Filer, T. H., Solomon, J. D., McCracken, F. I., Overgaard, N. A., and Weiss, M. J. 1975. Insects and diseases of cottonwood. USDA For. Serv. Gen. Tech. Rep. SO-8. 37 pp. New Orleans.
548. Morris, R. F. 1967. Factors inducing diapause in *Hyphantria cunea*. Can. Entomol. 99:522–529.
549. Moser, J. C. 1963. Content and structure of *Atta texana* nest in summer. Ann. Entomol. Soc. Am. 56:286–291.
550. Moser, J. C. 1965. The interrelationships of three gall makers and their natural enemies on hackberry (*Celtis occidentalis* L.) N.Y. State Museum and Sci. Serv., Albany, N.Y. Bull. 402. 95 pp.
551. Moser, J. C. 1967. Trails of the leaf cutters. Nat. Hist. 76:33–35.
552. Mote, D. C. 1936. Strawberry root weevil control in Oregon. Oregon State Agric. Exp. Stn. Circ. 115. 11 pp.
553. Moulton, D. 1907. The Monterey pine scale, *Physokermes insignicola* (Craw.). Proc. Davenport Acad. Sci. 12:1–25.
554. Moznette, G. G. 1922. The avocado: its insect enemies. USDA Farmers' Bull. 1261. 31 pp.
555. Muesebeck, C. F. W., Krombein, K. V., and Townes, H. K. 1951. Hymenoptera of America north of Mexico. USDA Agric. Monograph 2. 1420 pp.
556. Mumma, R. O., and Zettle, A. S. 1977. Observations on immature oak leaf roller, *Archips semiferanus*. Ann. Entomol. Soc. Am. 70:641–646.
557. Munro, J. A. 1963. Biology of the Ceanothus stem-gall moth, *Periploca ceanothiella* (Cosens). J. Res. Lepid. 1:183–190.
558. Munro, J. A. 1965. Occurrence of *Psylla uncatoides* on *Acacia* and *Albizia* with notes on control. J. Econ. Entomol. 58:1171–1172.
559. Munro, J. A., and Fox, A. C. 1934. Carpenter worm biology and control. North Dakota Agric. Exp. Stn. Bull. 278. 23 pp.
560. Murdoch, R. 1970. Lilac leaf miner. Dep. of Entomol., Univ. of Alberta, Edmonton. Leaflet 833a. 2 pp.
561. Nakahara, S. 1981. The proper placements of the nearctic soft scale species assigned to the genus *Lecanium* Burmeister (Homoptera: Coccoidea). Proc. Entomol. Soc. Wash. 83:283–286.
562. Nakahara, S. 1982. Checklist of the armored scales (Homoptera: Diaspididae) of the conterminous United States. USDA, APHIS, Plant Protection and Quarantine. 110 pp.
563. Neal, J. W. Jr. 1981. Timing insecticide control of rhododendron borers with pheromone trap catches of males. Environ. Entomol. 10:264–266.
564. Neal, J. W. Jr. 1982. Rhododendron borer: a worthy competitor. J. Am. Rhododendron Soc. 36:57–60.
565. Neal, J. W. Jr. 1985. *Neochlamisus platani,* Chrysomelidae. Personal communication.
566. Needham, J. G., Frost, S. W., and Tothill, B. H. 1928. Leaf-mining insects. Williams & Wilkins, Baltimore. 351 pp.
567. Neiswander, C. R. 1941. *Coryphista meadii*: a new pest of Japanese barberry. J. Econ. Entomol. 34:386–389.
568. Neiswander, R. B. 1966. Insect and mite pests of trees and shrubs. Ohio Agric. Res. and Development Cent. Res. Bull. 983. 54 pp.
569. Nettleton, W. A., and Hain, F. P. 1982. The life history, foliage damage, and control of the balsam twig aphid, *Mindarus abietinus* (Homoptera: Aphididae), in Fraser fir Christmas tree plantations of western North Carolina. Can. Entomol. 114:155–165.
570. Neunzig, H. H. 1972. Taxonomy of *Acrobasis* larvae and pupae in eastern North America. USDA Tech. Bull. 1457. 158 pp.
571. Newton, W. G., and Allen, D. C. 1982. Characteristics of trees damaged by sugar maple borer, *Glycobius speciosus* (Say). Can. J. For. 12:738–744.
572. Ngoan, N. D., Wilkinson, R. C., Short, D. E., Moses, C. S., and Mangold, J. R. 1976. Biology of an introduced ambrosia beetle *Xylosandrus compactus* in Florida. Ann. Entomol. Soc. Am. 69:872–879.
573. Nichols, J. O. 1961. The gypsy moth in Pennsylvania. Pa. Dep. of Agric. Misc. Bull. 4404. 82 pp.
574. Nichols, J. O. 1968. Oak mortality in Pennsylvania. J. For. 66:681–684.
575. Nielson, M. W. 1968. The leafhopper vectors of phytopathogenic viruses. USDA, ARS Tech. Bull. 1382. 386 pp.
576. Nord, J. C. 1970. *Saperda inornata* Say 1824 (Insecta, Coleoptera):

proposed use of plenary powers to designate a neotype to stabilize the nomenclature. Z.N.(S) 1921. Bull. Zool. Nomencl. 27, Part 2:123–128.
577. Nord, J. C., Grimble, D. G., and Knight, F. B. 1972. Biology of *Saperda inornata* [*S. concolor*] (Coleoptera: Cerambycidae) in trembling aspen, *Populus tremuloides*. Ann. Entomol. Soc. Am. 65:127–135.
578. Nord, J. C., Grimble, D. G., and Knight, F. B. 1972. Biology of *Oberea schaumii* (Coleoptera: Cerambycidae) in trembling aspen, *Populus tremuloides*. Ann. Entomol. Soc. Am. 65:114–119.
579. Nord, J. C., Ragenovich, I., and Doggett, C. A. 1984. Pales weevil. USDA Forest Insect and Disease Leaflet 104. 11 pp.
580. Nordin, G. L., and Johnson, E. L. 1983. Biology of *Caliro quercuscoccineae* (Dyer) (Hymenoptera: Tenthredinidae) in central Kentucky. II. Development and behavior. J. Kans. Entomol. Soc. 57:569–579.
581. Nordin, G. L., and Appleby, J. E. 1969. Bionomics of the juniper webworm. Ann. Entomol. Soc. Am. 62:287–292.
582. Ohmart, C. P., and Dahlsten, D. L. 1977. Biological studies of budmining sawflies, *Pleroneura* spp. (Hymenoptera: Xyelidae), on white fir in the central Sierra Nevada of California. Can. Entomol. 109:1001–1007.
583. Okiwelu, S. N. 1977. Studies on a pit-making scale, *Asterolecanium minus*, on *Quercus lobata*. Ann. Entomol. Soc. Am. 70:615–621.
584. Oliver, A. D. 1964. A behavioral study of two races of the fall webworm, *Hyphantria cunea* (Lepidoptera: Arctiidae) in Louisiana. Ann. Entomol. Soc. Am. 57:192–194.
585. Oliver, A. D., and Chapin, J. B. 1980. The cranberry rootworm: adult seasonal history and factors affecting its status as a pest of woody ornamentals in Louisiana. J. Econ. Entomol. 73:96–100.
586. Ollieu, M. M. 1971. Damage to southern pines in Texas by *Pissoides nemorensis*. J. Econ. Entomol. 64:1456–1459.
587. Osborn, H. 1915. Leafhoppers of Maine. Maine Agric. Exp. Stn. Bull. 238. Entomol. Paper 78. 160 pp.
588. Osborn, H. 1928. The leafhoppers of Ohio. Ohio Biol. Surv. Bull. 14, Vol. 3, No. 4, pp. 199–374.
589. Osborn, H. 1940. The Membracidae of Ohio. Ohio Biol. Surv. Bull. 37, Vol. 7, No. 2, pp. 51–101.
590. Osgood, E. A., and Gagné, R. J. 1978. Biology and taxonomy of two gall midges (Diptera: Cecidomyiidae) found in galls on balsam fir needles with description of a new species of *Paradiplosis*. Ann. Entomol. Soc. Am. 71:85–91.
591. Packard, A. S. 1890. Insects injurious to forest and shade trees. USDA Fifth Rep. Entomol. Commission Bull. 7. 955 pp.
592. Parker, D. A., and Moyer, M. A. 1972. Biology of a leaf roller *Archips negundanus* in Utah (Lepidoptera: Tortricidae). Ann. Entomol. Soc. Am. 65:1415–1418.
593. Parr, T. 1940. *Asterolecanium variolosum* Ratzeburg, a gall-forming coccid, and its effect upon the host tree. Yale Univ. School of Forestry Bull. 46. 49 pp.
594. Parrott, P. J., and Fulton, B. B. 1915. The cherry and hawthorn sawfly leafminer. N.Y. State Agric. Exp. Stn. Bull. 411:551–580.
595. Parrott, P. J., and Hodgkiss, H. E. 1916. Miscellaneous notes on injurious insects. N.Y. State Agric. Exp. Stn. Bull. 423:376–380.
596. Patch, E. M. 1908. The saddle prominent. Maine Agric. Exp. Stn. Bull. 161:311–350.
597. Patch, E. M. 1913. Woolly aphid of the elm. Maine Agric. Exp. Stn. Bull. 220:259–298.
598. Patch, E. M. 1915. Woolly aphid of elm and juneberry. Maine Agric. Exp. Stn. Bull. 241:197–204.
599. Patch, E. M. 1916. Elm leaf rosette and woolly aphid of the apple. Maine Agric. Exp. Stn. Bull. 256:330–344.
600. Payne, J. A. 1978. Oriental chestnut gall wasp: new nut pest in North America. Pages 86–88 in: Proceedings of the American chestnut symposium, West Virginia Univ., Morgantown. 122 pp.
601. Payne, J. A., Polles, S. G., Sparks, D., and Wehunt, E. J. 1976. The distribution, economic importance, and chemical control of the tilehorned prionus (Coleoptera: Cerambycidae) in Georgia. J. Ga. Entomol. Soc. 11:9–16.
602. Payne, J. A., Tedders, W. L., Cosgrove, G. E., and Foard, D. 1972. Larval mine characteristics of four species of leafmining Lepidoptera in pecan. Ann. Entomol. Soc. Am. 65:74–84.
603. Pechuman, L. L. 1938. A preliminary study of the biology of *Scolytus sulcatus* LeC. J. Econ. Entomol. 31:537–543.
604. Pechuman, L. L. 1968. The periodical cicada, brood VII (Homoptera: Cicadidae: Magicicada). Trans. Am. Entomol. Soc. 94:137–153.
605. Peck, O. 1963. A catalogue of the Nearctic Chalcidoidea (Insecta: Hymenoptera). Can. Entomol. Suppl. 30. 1092 pp.
606. Pemberton, C. 1911. The California Christmas-berry tingid. J. Econ. Entomol. 4:339–343.
607. Pergande, T. 1901. The life history of two species of plant-lice. USDA Div. of Entomol. Tech. Series 9.
608. Pergande, T. 1912. The life history of the alder blight aphid. USDA Bur. of Entomol. Tech. Series 24. 28 pp.
609. Peterson, A. 1962. Larvae of insects. Part I. Lepidoptera and plant infesting hymenoptera. Privately published, Ann Arbor, MI. 315 pp.
610. Peterson, L. O. T. 1958. The boxelder twig borer, *Proeoteras willingana* (Kearfott), (Lepidoptera: Olethreutidae). Can. Entomol. 90:639–646.
611. Peterson, L. O. 1939. The poplar borer, *Saperda calcarata* Say, in the park-land regions of the prairie provinces of Canada. Northcentral States Entomol. Conf. Proc., pp. 76–79.
612. Peterson, L. O., and DeBoo, R. F. 1969. Pine needle scale in the prairie provinces. Can. Dep. of Fisheries and For. Note MS-L-5. 9 pp.
613. Peterson, L. O. T., and Hildahl, V. 1969. The spruce spider mite in the prairie provinces. Northern Forest Res. Cent., Edmonton, Alberta. 9 pp.
614. Phillips, J. H. H. 1951. An annotated list of Hemiptera inhabiting sour cherry orchards in the Niagara peninsula, Ontario. Can. Entomol. 83:194–205.
615. Phillips, J. H. H. 1962. Description of the immature states of *Pulvinaria vitis* (L.) and *P. innumerabilis* (Rathron) (Homoptera: Coccoidea) with notes on the habits of these species in Ontario, Canada. Can. Entomol. 94:497–502.
615a. Phillips, T. W., Teale, S. A., and Lanier, G. N. 1987. Biosystematics of *Pissodes* Germar (Coleoptera: Curculionidae): seasonality, morphology, and synonymy of *P. approximatus* Hopkins and *P. nemorensis* Germar. Can. Entomol. 119:465–480.
616. Pierce, H. F., and Appleby, J. E. 1973. Biology of the leaf crumpler, *Acrobasis indigenella*, in central Illinois. Ann. Entomol. Soc. Am. 66:501–504.
617. Pilon, J. G. 1965. Bionomics of the spruce bud moth, *Zeiraphera ratzeburgiana* (Ratz.) (Lepidoptera: Olethreutidae). Phytoprotection 46:5–13.
618. Pinnock, D. E., Hagen, K. S., Cassidy, D. V., Brand, R. J., Milstead, J. E., and Tassan, R. L. 1978. Integrated pest management in highway landscapes. Calif. Agric. 32:33–34.
619. Pless, C. D., and Stanley, W. W. 1967. Life history and habits of the dogwood borer *Thamnosphecia scitula* (Lepidoptera: Aegeriidae) in Tennessee. J. Tenn. Acad. Sci. 42:117–123.
620. Plumb, G. H. 1953. The formation and development of the Norway spruce gall caused by *Adelges abietis* L. Conn. Agric. Exp. Stn. Bull. 566. 77 pp.
621. Pollet, D. K. 1972. The morphology, biology, and control of *Ceroplastes ceriferus* (Fab.) and *C. floridensis* Comstock (Homoptera: Coccoidae: Coccidae). Ph.D. diss., Virginia Polytechnic Institute and State Univ., Blacksburg. 208 pp.
622. Poos, F. W., and Wheeler, N. H. 1943. Studies on host plants of the leafhoppers of the genus *Empoasca*. USDA Tech. Bull. 850. 51 pp.
623. Potter, D. A. 1985. Population regulation of the native holly leafminer, *Phytomyza ilicicola* Loew (Diptera: Agromyzidae) on American holly. Oecologia (Berl.) 66:499–505.
624. Potter, D. A., and Timmons, G. M. 1981. Factors affecting predisposition of flowering dogwood trees to attack by the dogwood borer. HortScience 16:677–679.
625. Potter, D. A., and Timmons, G. M. 1983. Flight phenology of the dogwood borer (Lepidoptera: Sesiidae) and implications for control in *Cornus florida*. J. Econ. Entomol. 76:1069–1074.
626. Potter, D. A., and Timmons, G. M. 1983. Forecasting emergence and flight of the lilac borer (Lepidoptera: Sesiidae) based on pheromone trapping and degree-day accumulations. Environ. Entomol. 12:400–403.
627. Powell, J. A. 1964. Biological and taxonomic studies on tortricine moths, with reference to the species in California. Univ. Calif. Publ. Entomol. 32:1–317.
628. Powell, J. A. 1985. Occurrence of the cotoneaster webworm, *Athrips rancidella*, in California (Lepidoptera: Gelechiidae). Pan-Pac. Entomol. 61: 40–41.
629. Powell, J. A., and Miller, W. E. 1978. Nearctic pine tip moths of the genus *Rhyacionia*: biosystematic review. USDA For. Serv. Agric. Handb. 514. 51 pp.
630. Prentice, R. M. 1965. Forest lepidoptera of Canada. Forest insect survey. Vol. 4. Microlepidoptera. Can. Dep. For. Publ. 1142:545–834.
631. Pritchard, A. E. 1953. The gall midges of California. Bull. Calif. Insect Survey 2:225–250.
632. Pritchard, A. E. 1949. California greenhouse pests and their control. Calif. Agric. Exp. Stn. Bull. 713. 71 pp.
633. Pritchard, A. E., and Baker, E. W. 1955. A revision of the spider mite family Tetranychidae. Pacific Coast Entomol. Soc. Mem. Ser., Vol. 2. 472 pp.
634. Pritchard, A. E., and Powell, J. A. 1959. *Pyramidobela angelarum* Keifer on ornamental *Buddleia* in the San Francisco Bay area. Pan-Pac. Entomol. 35:82.
635. Pritchard, A. E., and Baker, E. W. 1959. The false spider mites (Acarina: Tenuipalpidae). Univ. Calif. Publ. Entomol. 14:175–274.
636. Pritchard, A. E., and Beer, R. E. 1950. Biology and control of *Asterolecanium* scales on oak in California. J. Econ. Entomol. 43:494–497.
637. Purrington, F. F., and Nielsen, D. G. 1977. Biology of *Podosesia* (Lepi-

doptera: Sesiidae) with description of a new species from North America. Ann. Entomol. Soc. Am. 70:906–910.

638. Putman, W. L. 1941. The feeding habits of certain leafhoppers. Can. Entomol. 73:39–53.

639. Quaintance, A. L. 1899. New or little known Aleyrodidae. I. Can. Entomol. 31:1–4.

640. Quaintance, A. L., and Baker, A. C. 1914. Classification of the Aleyrodidae. Part II. Bur. Entomol. USDA. Tech. Ser. 27:95–109.

641. Quednau, F. W. 1967. Ecological observations on *Chrysocharis laricinellae* (Hymenoptera: Eulophidae), a parasite of the larch casebearer (*Coleophora laricella*). Can. Entomol. 99:631–641.

642. Rabkin, F. B., and Lejune, R. R. 1954. Some aspects of the biology and dispersal of the pine tortoise scale. Can. Entomol. 86:570–575.

643. Raske, A. G., and Hodson, A. C. 1964. The development of *Pineus strobi* (Hartig) (Adelginae, Phylloxeridae) on white pine and black spruce. Can. Entomol. 96:599–616.

644. Rau, G. J. 1942. The Canadian apple mealybug, *Phenacoccus aceris* Signoret, and its allies in northeastern America. Can. Entomol. 74:118–125.

645. Raupp, M. J. 1984. Effects of exposure to sun on the frequency of attack by the azalea lace bug. J. Am. Rhododendron. Soc. 38:189–190.

646. Raupp, M. J., and Denno, R. F. 1984. The suitability of damaged willow leaves as food for the leaf beetle *Plagiodera versicolora*. Ecol. Entomol. 9:443–448.

647. Redfern, M. 1975. The life history and morphology of the early stages of the yew gall midge, *Taxomyia taxi* (Inchbald) (Diptera: Cecidomyiidae). J. Nat. Hist. 9:513–533.

648. Reeves, R. M. 1963. Tetranychidae infesting woody plants in New York State and a life history study of the elm spider mite, *Eotetranychus matthyssei* n. sp. Cornell Univ. Agric. Exp. Stn. Mem. 380. 99 pp.

649. Retnakaran, A., and Tomkins, W. 1982. Effectiveness of a moult-inhibiting insect growth regulator in controlling the oak leaf shredder. Can. For. Serv. Res. Notes 2:5–6.

650. Rice, R. E. 1974. San Jose scale: field studies with a sex pheromone. J. Econ. Entomol. 76:561–562.

651. Richards, W. R. 1967. The *Pterocomma* of Canada and Greenland with notes on the phyletic position of the Pterocommatini (Homoptera: Aphididae). Can. Entomol. 99:1015–1040.

652. Riedl, H., Weires, R. W., Seaman, A., and Hoying, S. A. 1978. Seasonal biology and control of the dogwood borer, *Synanthedon scitula* (Lepidoptera: Sesiidae) on clonal apple rootstocks in New York. Can. Entomol. 117:1367–1377.

653. Rings, R. W. 1970. Contributions to the bionomics of green fruitworms: the life history of *Orthosia hibisci*. J. Econ. Entomol. 63:1562–1568.

654. Rings, R. W. 1973. Contributions to the bionomics of green fruitworms: the life history of *Lithophane antennata*. J. Econ. Entomol. 66:364–368.

655. Roden, D. B. 1981. Potential for selection for freezing-tolerance in an Ontario population of *Scolytus multistriatus* (Coleoptera: Scolytidae). Can. For. Serv. Res. Notes 1:17–18.

656. Roelofs, W. L., Gieselmann, M. J., Mori, K., and Moreno, D. S. 1982. Sex pheromone chirality comparison between sibling species—California red scale and yellow scale. Naturwissenschaften 69:384.

657. Rose, A. H., and Lindquist, O. H. 1973. Insects of eastern pine. Can. For. Serv. Publ. 1313. 127 pp.

658. Rose, A. H., and Lindquist, O. H. 1977. Insects of eastern spruces, fir, and hemlock. Can. For. Serv. Forestry Tech. Rep. 23. 159 pp.

659. Rose, A. H., and Lindquist, O. H. 1980. Insects of eastern larch, cedar, and juniper. Can. For. Serv. Forestry Tech. Rep. 28. 100 pp.

660. Rose, A. H., and Lindquist, D. H. 1982. Insects of eastern hardwood trees. Can. For. Serv. Forestry Tech. Rep. 29. 309 pp.

661. Rosenthal, S. S., Frankie, G. W., and Koehler, C. S. 1969. Biological studies of *Argyresthia franciscella* and *A. cupressella* on ornamental Cupressaceae. Ann. Entomol. Soc. Am. 62:109–112.

662. Rosenthal, S. S., and Koehler, C. S. 1971. Heterogony in some gall-forming Cynipidae (Hymenoptera) with notes on the biology of *Neuroterus saltatorious*. Ann. Entomol. Soc. Am. 64:565–570.

663. Rosenthal, S. S., and Koehler, C. S. 1971. Intertree distributions of some cynipid (Hymenoptera) galls on *Quercus lobata*. Ann. Entomol. Soc. Am. 64:571–574.

664. Ross, D. A. 1962. Bionomics of the maple leaf cutter, *Paraclemensia acerifoliella* (Fitch), (Lepidoptera: Incurvariidae). Can. Entomol. 94:1053–1063.

665. Ruggles, A. G. 1915. Life history of *Oberea tripunctata* Swed. J. Econ. Entomol. 8:79–85.

666. Russell, H. M. 1912. The greenhouse thrips. USDA Bur. Entomol. Circ. 151. 9 pp.

667. Russell, L. M. 1941. A classification of the scale insect genus *Asterolecanium*. USDA Misc. Publ. 424. 322 pp.

668. Russell, L. M. 1982. The genus *Neophyllaphis* and its species (Hemiptera: Homoptera: Aphididae). Fla. Entomol. 65:538–573.

669. Russo, R. A. 1979. Plant galls of the California region. Boxwood Press, Pacific Grove, CA. 208 pp.

670. Sanders, J. G. 1905. The cottony maple scale. USDA Bur. Entomol. Circ. 64. 6 pp.

671. Sapio, F. J., Wilson, L. F., and Ostry, M. E. 1982. A split-stem lesion on young hybrid *Populus* trees caused by the tarnished plant bug, *Lygus lineolaris* (Hemiptera [Heteroptera]: Miridae). Great Lakes Entomol. 15:237–246.

672. Sasscer, E. R., and Borden, A. D. 1919. The rose midge. USDA Bull. 778. 8 pp.

673. Saunders, J. L. 1969. Occurrence and control of the balsam twig aphid on *Abies grandis and A. concolor*. J. Econ. Entomol. 62:1106–1109.

674. Saunders, J. L., and Harrigan, W. R. 1976. Chemical control of eriophyoid mites on Scots pine. J. Econ. Entomol. 69:333–335.

675. Schneski, W. 1966. Biology and control of the balsam twig aphid, *Mindarus abietinus* Koch. Unpublished research, Cornell Univ., Ithaca, NY.

676. Schread, J. C. 1959. Pod gall of honeylocust. Conn. Agric. Exp. Stn. Circ. 206. 4 pp.

677. Schread, J. C. 1964. Dogwood club gall. Conn. Agric. Exp. Stn. Circ. 225. 6 pp.

678. Schread, J. C. 1968. Control of lacebugs on broadleaved evergreens. Conn. Agric. Exp. Stn. Bull. 684. 7 pp.

679. Schread, J. C. 1970. Control of scale insects and mealybugs on ornamentals. Conn. Agric. Exp. Stn. Bull. 710. 27 pp.

680. Schread, J. C. 1971. Control of borers in trees and woody ornamentals. Conn. Agric. Exp. Stn. Circ. 241. 11 pp.

681. Schread, J. C. 1971. Leafminers and their control. Conn. Agric. Exp. Stn. Bull. 693. 19 pp.

682. Schuh, J., and Mote, D. C. 1948. Insect pests of nursery and ornamental trees and shrubs in Oregon. Oreg. Agric. Exp. Stn. Bull. 449. 164 pp.

683. Schultz, D. E., and Allen, D. C. 1975. Biology and description of the cherry scallop shell moth, *Hydria prunivorata* (Lepidoptera: Geometridae) in New York. Can. Entomol. 107:99–106.

684. Schultz, P. B. 1983. Evaluation of hawthorn lace bug (Hemiptera: Tingidae) feeding preference on *Cotoneaster* and *Pyracantha*. Environ. Entomol. 12:1808–1810.

685. Schultz, P. B. 1985. Evaluation of selected *Cotoneaster* spp. for resistance to hawthorn lace bug. J. Environ. Hortic. 3:156–157.

686. Severin, H. H. P., and Severin, H. C. 1911. The life history of the walkingstick, *Diapheromera femorata*. J. Econ. Entomol. 4:307–320.

687. Severin, H. H. P. 1949. Life history of the blue-green sharpshooter, *Neokolla circellata*. Hilgardia 19:171–206.

688. Sharplin, J. 1964. The mourning cloak butterfly. Univ. of Alberta, Dep. of Entomol. Leaflet 475. 2 pp.

689. Shaw, F. R., and Weidhaas, J. A. Jr. 1956. Distribution and habits of the giant hornet in North America. J. Econ. Entomol. 49:275.

690. Shenefeld, R. D., and Benjamin, D. M. 1955. Insects of Wisconsin forests. Univ. of Wisc., Coll. Agric. Circ. 500. 110 pp.

691. Shigo, A. L. 1970. Beech bark disease. USDA For. Serv. Forest Pest Leaflet 75. 7 pp.

692. Shorthouse, J. D. 1974. Inducing oviposition by a cynipid rose gall wasp (*Diplolepis* [Harr.]) in the laboratory. Marcellia 38:95–98.

693. Shorthouse, J. D. 1982. Resource exploitation by gall wasps of the genus *Diplolepis*. Pages 193–198 in: Proceedings of the fifth international symposium on insect-plant relationships. Wageningen, Netherlands.

694. Shorthouse, J. D., and Richie, A. J. 1984. Description and biology of a new species of *Diplolepis* Foureroy (Hymenoptera: Cynipidae) inducing galls on the stems of *Rosa acicularis*. Can. Entomol. 116:1623–1636.

695. Shorthouse, J. D., and West, R. J. 1987. Role of the inquiline *Dasineura balsamicola* (Diptera: Cecidomyiidae) in the balsam fir needle gall. Proc. Entomol. Soc. Ont. (in press)

696. Silver, G. T. 1957. Studies on the arborvitae leaf miners in New Brunswick (Lepidoptera: Yponomeutidae and Gelechiidae). Can. Entomol. 89:171–182.

697. Silver, G. T. 1958. Studies on the silver-spotted tiger moth, *Halisidota argentata* Pack. (Lepidoptera: Archtiidae), in British Columbia. Can Entomol. 90:65–80.

698. Simanton, F. L. 1916. The terrapin scale, an important insect enemy of peach orchards. USDA Bull. 351. 96 pp.

699. Simmons, G. A. 1983. The obliquebanded leafroller and *Cenopis pettitana* infesting maple buds in Michigan. Ann. Entomol. Soc. Am. 66:1166–1167.

700. Simons, J. N. 1954. The cicadas of California, Homoptera: Cicadidae. Calif. Insect Surv. Bull. 2:153–192.

701. Skuhravy, V. 1973. "Needle blight" and "needle droop" on *Pinus silvestris* L. in North America (Diptera, Cecidomyiidae). Inst. Entomol., Czechoslovak Acad. Sci., Prague. Z. Angew. Entomol. 72:421–428.

702. Slater, J. A., and Baranowski, R. M. 1978. How to know the true bugs. Wm. C. Brown Co., Dubuque, IA. 256 pp.

703. Slingerland, M. V. 1893. The four-lined leaf-bug. Cornell Univ. Agric. Exp. Stn. Bull. 58:207–239.
704. Slingerland, M. V. 1906. The bronze birch borer. Cornell Agric. Exp. Stn. Bull. 234:63–78.
705. Smereka, E. P. 1965. The life history and habits of *Chrysomela crotchi* Brown (Coleoptera: Chrysomelidae) in northwestern Ontario. Can. Entomol. 97:541–549.
706. Smith, C. C. 1952. The life history and galls of a spruce gall midge, *Phytophaga piceae* Felt (Diptera: Cecidomyiidae). Can. Entomol. 84:272–275.
707. Smith, C. F. 1969. Controlling peach scale. Res. and Farming (N.C. Agric. Res. Serv.) 28:12.
708. Smith, C. F. 1969. Pemphiginae associated with roots of conifers in North America (Homoptera: Aphididae). Ann. Entomol. Soc. Am. 62:1128–1152.
709. Smith, C. F. 1971. The life cycle and redescription of *Mordvilkoja vagabunda* (Homoptera: Aphididae). Proc. Entomol. Soc. Am. 73:359–367.
710. Smith, C. F., and Parron, C. S. 1978. An annotated list of Aphididae (Homoptera) of North America. N.C. Agric. Exp. Stn. Tech. Bull. 255. 428 pp.
711. Smith, C. F., and Denmark, H. A. 1984. Life history and synonomy of *Grylloprociphilus imbricator* (Fitch) (Homoptera: Aphididae). Fla. Entomol. 67:430–434.
712. Smith, D. R. 1971. Nearctic sawflies. III. Herterarthinae: adults and larvae (Hymenoptera: Tenthredinidae). USDA Agric. Res. Serv. Tech. Bull. 1420. 84 pp.
713. Smith, D. R. 1974. Azalea sawflies and a new species of *Nematus* Panzer (Hymenoptera: Symphyta). Proc. Entomol. Soc. Wash. 76:204–207.
714. Smith, D. R., Ohmart, C. P., and Dahlsten, D. L. 1977. The fir shoot-boring sawflies of the genus *Pleroneura* in North America (Hymenoptera: Xyelidae). Ann. Entomol. Soc. Am. 70:761–767.
715. Smith, E. L. 1968. Biosystematics and morphology of Symphyta. I. Stem-galling *Euura* of the California region and a new female genitalic nomenclature. Ann. Entomol. Soc. Am. 61:1389–1407.
716. Smith, F. F. 1955. Notes on the biology and control of *Pseudocneorhinus bifasciatus*. J. Econ. Entomol. 48:628.
717. Smith, F. F., and Webb, R. E. 1976. The rose midge. Am. Rose Annu. 61:57–73.
718. Smith, F. F., and Fisher, J. H. 1930. The boxwood leaf miner. Pa. Dep. Agric. Bull. 13:1–14.
719. Smith, L. M. 1932. The shothole borer. Calif. Agric. Ext. Serv. Circ. 64. 13 pp.
720. Smith, M. R. 1965. House-infesting ants of the eastern United States. USDA Tech. Bull. 1326. 105 pp.
721. Smith, R. H. 1941. Spraying for the sycamore scale. Western Shade Tree Conf. Proc. 8:30–39.
722. Smith, R. H. 1944. Bionomics and control of the nigra scale, *Saissetia nigra*. Hilgardia 16:255–288.
723. Smith, R. H. 1944. Insects and mites injurious to sycamore trees in western North America. Arborist's News 9:9–15.
724. Smith, R. H. 1945. Scale important pest of sycamore trees. Pac. Coast Nurseryman 3:7, 8, 13.
725. Smith, S. G., and Sugden, B. A. 1969. Host trees and breeding sites of native North American *Pissodes* bark weevils with a note on synonymy. Ann. Entomol. Soc. Am. 62:146–148.
726. Snetsinger, R., and Himelick, E. B. 1957. Observations on witches' broom of hackberry. Plant Dis. Rep. 41:541–544.
727. Solomon, J. D. 1972. Biology and habits of the living beech borer in red oaks. J. Econ. Entomol. 65:1307–1310.
728. Solomon, J. D. 1975. Biology of an ash borer, *Podosesia syringa* in green ash in Mississippi. Ann. Entomol. Soc. Am. 68:325–328.
729. Solomon, J. D. 1977. Biology and habits of the oak branch borer (*Goes debilis*). Ann. Entomol. Soc. Am. 70:57–59.
730. Solomon, J. D. 1977. Frass characteristics for identifying insect borers (Lepidoptera: Cossidae and Sesiidae; Coleoptera: Cerambycidae) in living hardwoods. Can. Entomol. 109:295–303.
731. Solomon, J. D. 1983. Lilac borer (*Podosesia syringae*) discovered causing terminal mortality and resulting forks in young green ash trees. J. Ga. Entomol. Soc. 18:320–323.
732. Solomon, J. D., and Donley, D. E. 1983. Bionomics and control of the white oak borer. USDA For. Serv. Southern For. Exp. Stn. Res. Paper SO-198. 5 pp.
733. Solomon, J. D., and Dix, M. E. 1979. Selected bibliography of the clearwing borers (Sesiidae) of the United States and Canada. USDA For. Serv. Gen. Tech. Rep. 50-22. 18 pp.
734. Solomon, J. D., and Morris, R. C. 1966. Clearwing borer in red oaks. USDA For. Serv. Southern For. Exp. Stn., New Orleans, LA. Res. Note SO-39. 3 pp.
735. Solomon, J. D., and Randall, W. K. 1978. Biology and damage of the willow shoot sawfly in willow and cottonwood. Ann. Entomol. Soc. Am. 71:654–657.
736. Solomon, J. D. 1980. Cottonwood borer (*Plectrodera scalator*): a guide to its biology, damage, and control. USDA For. Serv. Res. Paper SO-157. 10 pp.
737. Solomon, J. D., and Neel, W. W. 1972. Emergence behavior and rhythms in the carpenterworm moth, *Prionoxystus robiniae*. Ann. Entomol. Soc. Am. 65:1296–1299.
738. Solomon, J. D., Oliveria, F. L., Tumlinson, J. H., and Doolittle, R. E. 1982. Occurrence of clearwing borers (Sesiidae) in west central Mississippi. J. Ga. Entomol. Soc. 17:4–12.
739. Solomon, J. D., Newsome, L., and Darwin, W. N. 1972. Carpenterworm moths and cerambycid hardwood borers caught in light traps. J. Ga. Entomol. Soc. 7:76–79.
740. Solomon, J. D., and Payne, J. A. 1986. A guide to the insect borers, pruners, and girdlers of hickory and pecan. USDA For. Serv. Southern For. Exp. Stn. Gen. Tech. Rep. SO-64. 31 pp.
741. Solomon, J. D., Vowell, T. E. Jr., and Horton, R. C. 1975. Hackberry butterfly, *Asterocampa celtis*, defoliates sugarberry in Mississippi. J. Ga. Entomol. Soc. 10:17–18.
742. Solomon, J. D. 1974. Biology and damage of the hickory borer, *Goes pulcher*, in hickory and pecan. Ann. Entomol. Soc. Am. 67:257–260.
743. Soper, R. S. 1974. The genus *Massospora*, entomopathogenic for cicadas. Part I. Taxonomy of the genus. Mycotaxon 1:13–40.
744. Specker, D. R. 1983. Biology of the rhododendron gall midge, *Clinodiplosis rhododendri* (Felt) (Diptera: Cecidomyiidae) on Long Island. Master's thesis, Cornell Univ., Ithaca, NY.
745. Speers, C. F. 1941. The pine spittlebug (*Aphrophora parallela* Say). N.Y. State Coll. For. Tech. Pub. 54. 65 pp.
746. Spencer, K. A. 1969. The Agromyzidae of Canada and Alaska. Mem. Entomol. Soc. Can. 64. 311 pp.
747. Spencer, K. A. 1981. A revisionary study of the leaf-mining flies (Agromyzidae) of California. Div. Agric. Sci., Calif. Special Publ. 3273. 489 pp.
748. Sreenivasam, D. D., Benjamin, D. M., and Walgenbach, D. D. 1972. The bionomics of the pine tussock moth. Univ. of Wisc. Res. Bull. 282. 36 pp.
749. Stein, J. D., and Kennedy, P. C. 1972. Key to shelterbelt insects in the northern great plains. USDA For. Serv. Res. Paper RM-85. Rocky Mt. For. and Range Exp. Stn. 153 pp.
750. Stein, J. D. 1974. Elm sawfly. USDA Forest Pest Leaflet 142. 6 pp.
751. Stehr, F. W., and Cook, E. F. 1968. A revision of the genus *Malacosma* Hubner in North America (Lepidoptera: Lasiocampidae): systematics, biology, immatures and parasites. Smithsonian Inst., U.S. Natl. Museum Bull. 276. 321 pp.
752. Stelzer, M. J. 1971. Western tent caterpillar. USDA For. Serv. Forest Pest Leaflet 119. 5 pp.
753. Stevens, R. E. 1959. Biology and control of the pine needlesheath miner, *Zelleria haimbachi* Busck (Lepidoptera: Hyponomeutidae). USDA For. Serv. Pac. S.W. For. and Range Exp. Stn., Berkeley, CA. Tech. Paper 30. 20 pp.
754. Stevens, R. E., Sartwell, C., Koerber, T. W., Daterman, G. E., Sower, L. L., and Powell, J. A. 1980. Western *Rhyacionia* (Lepidoptera: Tortricidae, Olethreutinae) pine tip moths trapped using synthetic sex attractants. Can. Entomol. 112:591–603.
755. Stewart, P. A., and Lam, J. E. Jr. 1970. Capture of forest insects in traps equipped with blacklight lamps. J. Econ. Entomol. 63:871–873.
756. Stillwell, M. A. 1967. The pigeon tremex, *Tremex columba* (Hymenoptera: Siricidae), in New Brunswick. Can. Entomol. 99:685–689.
757. Stimmann, M. W., Kaya, H. K., Burlando, T. M., and Studdert, J. P. 1985. Black vine weevil management in nursery plants. Calif. Agric. 39:25–26.
758. Stimmel, J. F. 1975. Longtailed mealybug, *Pseudococcus longispinis* (Targ.-Tozz.) Homoptera: Pseudococcidae. Pa. Dep. Agric. Entomol. Circ. 7. 2 pp.
759. Stimmel, J. F. 1978. Hemispherical scale, *Saissetia coffeae* (Walker) Homoptera: Coccidae. Pa. Dep. Agric. Entomol. Circ. 34. 2 pp.
760. Stimmel, J. F. 1979. Citrus mealybug. *Planococcus citri* Risso. Homoptera: Pseudococcidae. Pa. Dep. Agric. Entomol. Circ. 45. 2 pp.
761. Stimmel, J. F. 1983. "Euonymus alatus scale," *Lepidosaphes yanagicola* Kuwana (Homoptera: Diaspididae). Pa. Dep. Agric. Entomol. Circ. 78. 2 pp.
761a. Stimmel, J. F. 1986. *Aspidiotus cryptomeriae* Kuwana, an armored scale pest of conifers. Pa. Dep. Agric. Regulatory Hortic. Entomol. Circ. 108. 2 pp.
762. Stoetzel, M. B., and Tedders, W. L. 1981. Investigation of two species of *Phylloxera* on pecan in Georgia. J. Ga. Entomol. Soc. 16:144–150.
763. Stoetzel, M. B. 1981. Two new species of *Phylloxera* (Phylloxeridae: Homoptera) on pecan. J. Ga. Entomol. Soc. 16:127–144.
764. Stoetzel, M. B. 1975. Seasonal history of seven species of armored scale insects of the *Aspidiotini* (Homoptera: Diaspididae). Ann. Entomol. Soc. Am. 68:489–492.
765. Stoetzel, M. B. 1976. Scale cover formation in the Diaspididae (Homoptera: Coccidae). Proc. Entomol. Soc. Wash. 78:323–332.
766. Stoetzel, M. B., and Davidson, J. A. 1971. Biology of the obscure scale, *Melanapsis obscura* (Homoptera: Diaspidae), on pin oak in Maryland. Ann. Entomol. Soc. Am. 64:45–50.
767. Stoetzel, M. B., and Davidson, J. A. 1974. Biology, morphology, and

taxonomy of immature stages of 9 species in the *Aspidiotini* (Homoptera: Diaspididae). Ann. Entomol. Soc. Am. 67:475–509.
768. Streu, H. T., and Vasvary, L. M. 1970. Pests of holly in the eastern United States. Nat. Hortic. Mag. 39:234–243.
769. Struble, G. R., and Johnson, P. C. 1964. Black pine leaf scale. USDA Forest Pest Leaflet 91. 6 pp.
770. Styer, W. E., Nielsen, D. G., and Balderston, C. P. 1972. A new species of *Trisetacus* (Acarina: Eriophyoidea: Nalepellidae) from Scotch pine. Ann. Entomol. Soc. Am. 65:1089–1091.
771. Swenk, M. H. 1927. The pine tip moth in the Nebraska National Forest. Univ. of Neb. Agric. Exp. Stn. Res. Bull. 40. 50 pp.
772. Swingle, R. U., Witten, R. R., and Young, H. C. 1949. The identification and control of elm phloem necrosis and Dutch elm disease. Ohio Agric. Exp. Stn. Spec. Circ. 80. 11 pp.
773. Syme, P. D. 1981. Occurrence of the introduced sawfly *Acantholyda erythrocephala* (L.) in Ontario. Can. For. Serv. Res. Notes 1:4–5.
774. Symons, T. B., and Cory, E. H. 1910. The terrapin scale. Md. Agric. Exp. Stn. Bull. 149:83–92.
775. Tahvanainen, J., Julkunen-Tiitto, R., and Kettunen, J. 1985. Phenolic glycosides govern the food selection pattern of willow-feeding leaf beetles. Oecologia (Berl.) 67:52–56.
776. Tashiro, H. 1973. Evaluation of soil applied insecticides on insects of white birch in nurseries. Search Agric. 3:1–10.
776a. Tate, J. Jr. 1973. Methods and annual sequence of foraging by the sapsucker. Auk 90:840–856.
777. Taylor, G. S. 1983. Cryptosporiopsis canker of *Acer rubrum*: Some relationships among host, pathogen, and vector. Plant Dis. 69:984–986.
778. Thatcher, R. C. 1967. Pine sawfly, *Neodiprion excitans*. USDA For. Serv. Forest Pest Leaflet 105. 4 pp.
779. Thatcher, R. C., Searcy, J. L., Coster, J. E., and Hertel, G. D., eds. 1980. The southern pine beetle. USDA For. Serv. Tech. Bull. 1631. 266 pp.
780. Thewke, S. E., and Enns, W. R. 1970. The spider-mite complex (Acarina: Tetranychoidea) in Missouri. Univ. of Mo. Museum Contrib., Monograph 1. 106 pp.
781. Thielges, B. A., and Campbell, R. L. 1972. Selection and breeding to avoid the eastern spruce gall aphid. Am. Christmas Tree J. 16:3–6.
781a. Thomas, M. C., and Dixon, W. N. 1992. Pine shoot beetle, *Tomicus piniperda* (Linnaeus): a potential threat to Florida pines (Coleoptera: Scotylidae). Fla. Dep. of Agric. and Consumer Serv. Entomol. Circ. 354. 2 pp.
782. Thompson, B. G., and Wong, K. L. 1933. Western willow tingid, *Corythuca salicata* Gibson, in Oregon. J. Econ. Entomol. 26:1090–1095.
783. Thompson, H. E. 1962. Controlling the elm calligrapha beetles. Kans. Agric. Exp. Stn. Circ. 385. 7 pp.
784. Tilden, J. W. 1951. The insect associates of *Baccharis pilularis* DeCandolle. Microentomology 16:149–185.
785. Tippins, H. H. 1968. Observations on *Phenacaspis cockerelli* (Cooley) (Homoptera: Diaspididae), a pest of ornamental plants in Georgia. J. Ga. Entomol. Soc. 3:13–15.
786. Tippins, H. H., Clay, H., and Barry, R. M. 1977. Peony scale: a new host and biological information. J. Ga. Entomol. Soc. 12:68–71.
787. Tonks, N. V. 1974. Occurrence of a midge *Oligotrophus betheli* Felt, on juniper on Vancouver Island, British Columbia. (Diptera: Cecidomyiidae). J. Entomol. Soc. B. C. 71:33.
788. Trammel, K. 1974. The white apple leafhopper in New York: insecticide resistance and current control status. Search Agric. 4. 10 pp.
789. Tunnock, S., and Ryan, R. B. 1985. Larch casebearer in western larch. USDA For. Serv. Forest Insect and Disease Leaflet 96. 8 pp.
790. Turgeon, J. 1985. Life cycle and behavior of the spruce budmoth, *Zeiraphera canadensis* (Lepidoptera: Olethreutidae), in New Brunswick. Can. Entomol. 117:1239–1247.
791. Turnock, W. J. 1953. Some aspects of the life history and ecology of the pitch nodule maker, *Petrova albicapitana* (Busck) (Lepidoptera: Olethreutidae). Can. Entomol. 85:233–243.
792. Tuttle, D. M., and Baker, E. W. 1968. Spider mites of the southwestern United States and a revision of the family Tetranychidae. Univ. of Arizona Press, Tucson. 143 pp.
793. Underhill, G. W. 1935. The pecan tree borer in dogwood. J. Econ. Entomol. 28:393–396.
794. U.S.D.A. 1985. Insects of eastern forests. U.S. For. Serv. Misc. Publ. 1426. 606 pp.
795. Usinger, R. L. 1945. Biology and control of ash plant bugs in California. J. Econ. Entomol. 38:585–591.
796. Valley, K. 1979. Obliquebanded leafroller as a pest of rhododendrons in Pennsylvania. Pa. Dep. Agric. Bur. of Plant Industry. Entomol. Circ. 38. 2 pp.
797. Valley, K. 1983. Native holly leafminer, *Phytomyza ilicicola* Loew. Pa. Dep. of Agric. Bur. of Plant Industry. Entomol. Circ. 80. 2 pp.
798. Valley, K. 1985. *Altica ignita* Illiger, a potential pest of cultivated rhododendron. Pa. Dep. Agric. Bur. of Plant Industry. Entomol. Circ. 92. 2 pp.
799. Valley, K., and Wheeler, A. G. Jr. 1985. Leafhoppers (Hemiptera: Cicadellidae) associated with ornamental honeylocust: seasonal history, habits, and description of eggs and fifth instar. Ann. Entomol. Soc. Am. 78:709–716.
800. Varty, I. W. 1967. Erythroneura leafhoppers from birches in New Brunswick. Can. Entomol. 99:570–573.
801. Varty, I. W. 1967. Leafhoppers of the subfamily Typhlocybinae from birches. Can. Entomol. 99:170–180.
802. Voegtlin, D. J. 1976. A biosystematic study of *Cinara* spp. (Homoptera: Aphididae) of the conifers of the westside Sierra forests. Ph.D. diss., Univ. of California, Berkeley. 208 pp.
803. Voegtlin, D. J. 1981. Notes on a European aphid (Homoptera: Aphididae) new to North America. Proc. Entomol. Soc. Wash. 83:361–362.
804. Volck, W. H. 1956. Strawberry root weevils—control them! Utah State Coop. Ext. Bull. Leaflet 7.
805. Volck, W. H. 1907. The California tussock moth. Calif. Agric. Exp. Stn. Bull. 183:191–216.
806. Wade, M. J., and Breden, F. 1986. Life history of natural populations of the imported willow leaf beetle, *Plagiodera versicolora* (Coleoptera: Chrysomelidae). Ann. Entomol. Soc. Am. 79:73–79.
807. Wade, O. 1917. The sycamore lace-bug (*Corythucha ciliata* Say). Okla. Agric. Exp. Stn. Bull. 116. 16 pp.
808. Wagener, N. W. 1936. The cypress bark canker and other western cypress diseases. Western Shade Tree Conf. Proc. 3:79–85.
809. Walgenbach, D. D., and Benjamin, D. M. 1964. Biology of the pine tussock moth. North Central Branch Entomol. Soc. Proc. 19:21–22.
810. Walgenbach, D. D., and Benjamin, D. M. 1966. Biology of the black pine leaf scale. Univ. of Wisc. Res. Bull. 265. 15 pp.
811. Wallace, P. P. 1945. Biology and control of the dogwood borer *Thamnosphecia (Synanthedon) Scitula* Harris. Conn. Agric. Exp. Stn. Bull. 488: 373–395.
812. Wallesz, D. P., and Benjamin, D. M. 1960. The biology of the pine webworm, *Tetralopha robustella* in Wisconsin. J. Econ. Entomol. 53:587–589.
813. Wallner, W. E. 1965. The biology and control of fiorinia hemlock scale, *Fiorinia externa* Ferris. Ph.D. diss., Cornell University, Ithaca, NY. 166 pp.
814. Wallner, W. E. 1969. Insects affecting woody ornamental shrubs and trees. Mich. State. Univ. Ext. Bull. 530. 45 pp.
815. Walton, B. C. J. 1960. The life cycle of the hackberry gall-former, *Pachypsylla celtidisgemma* (Homoptera: Psyllidae). Ann. Entomol. Soc. Am. 53: 265–277.
816. Warren, L. O., and Tadic, M. 1970. The fall webworm, *Hyphantria cunea* (Dryry). Univ. of Ark. Agric. Exp. Stn. Bull. 795. 106 pp.
817. Washburn, R. I., and McGregor, M. D. 1974. White fir needle miner. USDA For. Serv. For. Pest Leaflet 144. 5 pp.
818. Watson, J. R. 1915. The woolly whitefly. Fla. Agric. Exp. Stn. Bull. 126:81–102.
819. Weaver, C. R., and King, D. R. 1954. Meadow spittlebug. Ohio Agric. Exp. Stn. Res. Bull. 741. 99 pp.
819a. Weaver, J. E., and Dorsey, C. K. 1967. Larval mine characteristics of five species of leaf-mining insects in black locust, *Robinia pseudoacacia*. Ann. Entomol. Soc. Am. 60:172–186.
820. Weaver, J. E., and Dorsey, C. K. 1965. Parasites and predators associated with five species of leaf-mining insects in black locust. Ann. Entomol. Soc. Am. 58:933–934.
821. Webb, F. E., and Forbes, R. S. 1951. Notes on the biology of *Pleroneura borealis* Felt (Xyelidae: Hymenoptera). Can. Entomol. 83:181–183.
822. Webster, H. V., and St. George, R. A. 1947. Life history and control of the webworm, *Homadaula albizziae*. J. Econ. Entomol. 40:546–553.
823. Weidhaas, J. A. Jr. 1959. Investigations of the effects of some modern pesticides on certain coccinellid predators of aphids and mites in Massachusetts. Ph.D. diss., Univ. of Massachusetts, Amherst. 259 pp.
824. Weidhaas, J. A. Jr., and Shaw, F. R. 1956. Control of the taxus mealybug with notes on its biology. J. Econ. Entomol. 49:273–274.
825. Weidhaas, J. A. Jr. 1979. Spider mites and other *Acarina* on trees and shrubs. J. Arboric. 5:9–15.
826. Weidman, R. H., and Robbins, G. T. 1947. Attacks of pitch moth and turpentine beetle on pines in the Eddy Arboretum. J. For. 45:428–433.
827. Weigel, C. A., and Baumhofer, L. G. 1948. Handbook on insect enemies of flowers and shrubs. USDA Misc. Publ. 626. 115 pp.
828. Weis, A. E. 1984. Apical dominance asserted over lateral buds by the gall of *Rhabdophaga strobiloides* (Diptera: Cecidomyiidae). Can. Entomol. 116: 1277–1279.
829. Weiss, H. B., and Lott, R. B. 1922. The juniper webworm, *Dichomeris marginellus* Fabr. Entomol. News 33:80–82.
830. Weld, L. H. 1957. Cynipid galls of the Pacific slope. Privately published, Ann Arbor, MI. 64 pp.
831. Weld, L. H. 1959. Cynipid galls of the eastern United States. Privately published, Ann Arbor, MI. 124 pp.
832. Wellhouse, W. H. 1922. The insect fauna of the genus *Crataegus*. Cornell Univ. Agric. Exp. Stn. Mem. 56:1041–1136.
833. Werner, W. H. R. 1930. Observations on the life history and control of the

fern scale, *Hemichionaspis aspidistrae* Sign. Pap. Mich. Acad. Sci. 13:517–541.

833a. Wertheim, C. G., and Morton, H. L. 1986. Honeylocust twig-gall midge (Diptera: Cecidomyiidae) in Michigan. Great Lakes Entomol. 19:169–173.

834. West, R. J., and Shorthouse, J. D. 1982. Morphology of the balsam fir needle gall induced by the midge *Paradiplosis tumifex* (Diptera: Cecidomyiidae). Can. J. Bot. 60:131–140.

835. Westcott, C., ed. 1960. Handbook on biological control of plant pests. Vol. 16, No. 3. Plants and gardens. Brooklyn Botanic Garden, New York. 97 pp.

836. Westcott, C. 1964. The gardener's bug book. Doubleday & Co., New York. 625 pp.

837. Wheeler, A. G. Jr. 1977. Life history of *Niesthrea louisianica* (Hemiptera: Rhopalidae) on rose of sharon in North Carolina. Ann. Entomol. Soc. Am. 70:631–634.

838. Wheeler, A. G. Jr. 1978. Planthoppers on ornamentals in Pennsylvania (Homoptera: Fulgoroidea). Reg. Hortic., Pa. Dep. Agric. Entomol. Circ. 35. 2 pp.

839. Wheeler, A. G. Jr. 1980. Life history of *Plagiognathus albatus* (Hemiptera: Miridae) with a description of the fifth instar. Ann. Entomol. Soc. Am. 73:354–356.

840. Wheeler, A. G. Jr. 1981. Hawthorn lace bug, first record of injury to roses, with a review of host plants. Great Lakes Entomol. 14:37–43.

841. Wheeler, A. G. Jr. 1982. Ash plant bug, *Tropidosteptes amoenus* Reuter (Hemiptera: Miridae). Reg. Hortic., Pa. Dep. Agric. Entomol. Circ. 68. 2 pp.

842. Wheeler, A. G. Jr. 1982. Bed bugs and other bugs. In: Handbook of pest control. 6th ed. Franzak & Foster Co., Cleveland. 1101 pp.

843. Wheeler, A. G. Jr., and Miller, G. L. 1981. Fourlined plant bug (Hemiptera: Miridae). A reappraisal: life history, host plants, and plant response feeding. Great Lakes Entomol. 14:23–35.

844. Wheeler, A. G. Jr., and Snook W. A. II. 1986. Biology of *Sumitrosis rosea* (Coleoptera: Chrysomelidae), a leafminer of black locust, *Robinia pseudoacacia* (Leguminosae). Proc. Entomol. Soc. Wash. 88:521–530.

845. Wheeler, A. G. Jr., and Valley, K. R. 1980. *Graphocephala coccinea*: seasonal history and habits on ericaceous shrubs, with notes on *G. fennahi* (Homoptera: Cicadellidae). Melsheimer Entomol. Ser. 29:23–27.

846. Wheeler, A. G. Jr., and Valley, K. R. 1980. A rhododendron leafhopper: field recognition and habits. Bull. Am. Rhododendron Soc. 34:202–205.

847. White, J. 1981. Flagging: host defenses versus oviposition strategies in periodical cicadas (*Magicicada* spp., Cicadidae, Homoptera). Can. Entomol. 113:727–738.

848. White, R. P., and Hamilton, C. C. 1935. Diseases and insect pests of rhododendron and azalea. N.J. Agric. Exp. Stn. Circ. 350. 23 pp.

849. Whitham, T. G. 1979. Territorial behavior of *Pemphigus* gall aphids. Nature 279:324–325.

850. Wickman, B. E., Mason, R. R., and Trostle, G. C. 1981. Douglas-fir tussock moth. USDA For. Serv. Forest Insect and Disease Leaflet 86. 10 pp.

851. Wilford, H. B. 1937. The spruce gall aphid in southern Michigan. Univ. of Mich. School of Forestry. Conserv. Circ. 2. 35 pp.

852. Wilkinson, C., and Scoble, M. J. 1979. The Nepticulidae (Lepidoptera) of Canada. Mem. Entomol. Soc. Can. 107. 129 pp.

853. Wilkinson, R. C., and Foltz, J. L. 1980. A selected bibliography (1959–1979) of three southeastern species of *Ips* engraver beetles. Bull. Entomol. Soc. Am. 26:375–380.

854. Williams, M. L., and Kosztarab, M. 1972. Morphology and systematics of the Coccidae of Virginia with notes on their biology. Homoptera: Coccoidae. Va. Polytechnic Inst. and State. Univ. Res. Div. Bull. 74. 215 pp.

855. Wilson, L. F. 1962. Yellow-headed spruce sawfly. USDA For. Serv. Forest Pest Leaflet 69. 4 pp.

856. Wilson, L. F. 1964. Walkingstick. USDA For. Serv. Forest Pest Leaflet 82. 4 pp.

857. Wilson, L. F. 1966. Introduced pine sawfly. USDA For. Serv. Forest Pest Leaflet 99. 4 pp.

858. Wilson, L. F. 1966. Life history and habits of the willow beaked gall midge, *Mayetiola rigidae* (Diptera: Cecidomyiidae), in Michigan. Can. Entomol. 100:202–206.

859. Wilson, L. F. 1968. Life history and habits of the pine cone willowgall midge, *Rhabdophaga strobiloides* (Diptera: Cecidomyiidae), in Michigan. Can. Entomol. 100:430–433.

860. Wilson, L. F. 1970. Red-headed pine sawfly. USDA For. Serv. Forest Pest Leaflet 14. 6 pp.

861. Wilson, L. F. 1972. Life history and outbreaks of an oak leaf roller, *Archips semiferanus* (Lepidoptera: Tortricidae), in Michigan. Great Lakes Entomol. 5:71–77.

862. Wilson, L. F. 1977. A guide to insect injury of conifers in the lake states. USDA For. Serv. Agric. Handb. 501. 218 pp.

863. Wilson, L. F., and Schmiege, D. C. 1970. Pine root collar weevil. USDA For. Serv. Forest Pest Leaflet 39. 7 pp.

864. Wilson, S. W., and McPherson, J. E. 1981. Life histories of *Anormenis septentrionalis, Metcalfa pruinosa,* and *Ormenoides venusta*, with descriptions of immature stages. Ann. Entomol. Soc. Am. 74:299–301.

865. Wilson, S. W., and McPherson, J. E. 1981. Life histories of *Acanalonia bivittata* and *A. conica*, with descriptions of immature stages. Ann. Entomol. Soc. Am. 74:289–298.

865a. Witter, J. A., and Fields, R. D. 1977. *Phyllobus oblongus* and *Sciaphillus asperatus* associated with sugar maple production in northern Michigan. Environ. Entomol. 6:150–154.

866. Wojcik, D. P., Buren, W. F., Grissell, E. E., and Cardysle, T. 1976. The fire ants (Solenopsis) of Florida (Hymenoptera: Formicidae). Fla. Dep. Agric. and Consumer Serv. Entomol. Circ. 173. 4 pp.

867. Wolfe, H. R. 1955. Leafhoppers of the state of Washington. Wash. Agric. Exp. Stn. Circ. 277. 37 pp.

868. Wong, H. R., Drouin, J. A., Szlabey, D. L., and Dang, P. T. 1983. Identification of three species of *Proteoteras* (Lepidoptera: Tortricidae) attacking shoots of Manitoba maple in the Canadian prairies. Can. Entomol. 115:333–339.

868a. Wong, H. R., and Szlabey, D. L. 1986. Larvae of the North American genera of Diprionidae (Hymenoptera: Symphyta). Can. Entomol. 118:577–587.

869. Wong, H. R., Drouin, J. A., and Rentz, C. L. 1985. *Petrova albicapitana* (Busk) and *P. metallica* (Busk) (Lepidoptera: Tortricidae) in *Pinus contorta* Dougl. stands of Alberta. Can. Entomol. 177:1463–1470.

870. Wong, H. R., Melvin, J. C. E., and Drouin, J. A. Damage by a willow shoot-boring sawfly in Alberta. Tree Planters' Notes 27:18–20.

870a. Wood, S. L. 1982. The bark and ambrosia beetles of North and Central America. A taxonomic monograph. Great Basin Naturalist Mem. 6. Brigham Young Univ., Provo, Utah. 1359 pp.

871. Wood, T. K. 1968. The chemical composition, host plant variation, and overwintering survival value of the egg froth of the membracid *Enchenopa binotata* Say. Ph.D. diss., Cornell Univ., Ithaca, NY.

872. Wood, T. K. 1976. Biology and presocial behavior of *Platycotis vittata* (Homoptera: Membracidae). Ann. Entomol. Soc. Am. 69:807–811.

873. Wood, T. K. 1977. Defence in *Umbonia crassicornis*: role of the pronotum and adult aggression (Homoptera: Membracidae). Ann. Entomol. Soc. Am. 70:524–528.

874. Wood, T. K., and Patton, R. L. 1971. Egg froth distribution and deposition by *Enchenopa binotata* (Homoptera: Membracidae). Ann. Entomol. Soc. Am. 64:1190–1191.

875. Woodruff, R. E. 1979. Fuller rose beetle, *Pantomorus cervinus* (Boheman), in Florida (Coleoptera: Curculionidae). Fla. Dep. Agric. and Consumer Serv. Entomol. Circ. 207. 4 pp.

876. Woollerman, E. H. 1970. The locust borer. USDA For. Serv. Forest Pest Leaflet 71. 8 pp.

877. Wray, D. L. 1961. Biology and life history of the ligustrum weevil (Curculionidae). Coleop. Bull. 15:119–120.

878. Yates, H. O., Overgaard, N. A., and Koerber, T. W. 1981. Nantucket pine tip moth. USDA Forest Insect and Disease Leaflet 70. 7 pp.

879. Yonke, T. R., and Medler, J. T. 1969. Description of immature stages of Coreidae. I. *Euthochtha galeator*. Ann. Entomol. Soc. Am. 62:469–473.

880. Young, D. A. 1977. Taxonomic study of the Cicadellinae (Homoptera: Cicadellidae). Part 2. New world Cicadellini and the genus *Cicadella*. N.C. Agric. Exp. Stn. Tech. Bull. 239. 1135 pp.

881. Young, H. C., App, B. A., Gill, J. B., and Hollingsworth, H. S. 1950. White-fringed beetles and how to combat them. USDA Circ. 850. 15 pp.

882. Zepp, D. B. 1978. Egg pod formation by *Callirhopalus* (subg. *Pseudocneorhincus*) *bifasciatus* (Roelofs) (Coleoptera: Curculionidae: Eremninae). Coleop. Bull. 32:311–313.

883. Antonelli, A. L., LaGasa, E., and Bay, E. C. 1989. Apple ermine moth. Wash. State Univ. Coop. Ext. Bull. EB 1526. 2 pp.
884. Bellows, T. S., Paine, T. D., Arakawa, K. Y., Meisenbacker, C., Leddy, P., and Kabashima, J. 1990. Biological control sought for ash whitefly. Calif. Agric. 44:4–6.
885. Berisford, C. W., Harman D. M., Freeman, B. L., Wilkinson, R. C., and McGraw, J. R. 1979. Sex pheromone cross-attraction among four species of pine tip moths, *Rhyacionia* species. J. Chem. Ecol. 5:205–210.
886. Brooks, F. E. 1922. Curculios that attack the young fruits and shoots of walnut and hickory. USDA Agric. Bull. 1066. 16 pp.
887. Bullington, S. W., Kosztarab, M., and Jiang G.-Z. 1989. Morphology and systematics of scale insects–No. 15. II. Adult males of the genus *Chionapsis* (Homoptera: Diaspididae) of North America. Va. Agric. Exp. Stn. Bull. 88-2, pp. 127–184.
888. Cooper, R. M., and Oetting, R. D. 1989. Life history and field development of the camellia scale (Homoptera: Diaspididae). Ann. Entomol. Soc. Am. 82:730–736.
889. Costello, L. R., Koehler, C. S., and Allen, W. W. 1987. Fuchsia gall mite management. Univ. Calif. Coop. Ext. Leaflet 7179. 2 pp.
890. Denton, R. E. 1979. Larch casebearer in western larch forests. USDA For. Serv. Gen. Tech. Rep. INT-55, Intermt. For. Range Exp. Stn., Ogden, Utah. 62 pp.
891. Dixon, W. N. 1989. The tarnished plant bug, *Lygus lineolaris* (Palisot de Beauvois), in conifer nurseries (Heteroptera: Miridae). Fla. Dep. of Agric. and Consumer Serv. Div. of Plant Industry. Entomol. Circ. 320. 2 pp.
892. Downer, J. A., Kabashima, J. N., Paine, T. D., and Koehler, C. S. 1990. Eugenia psyllid. Univ. Calif. Coop. Ext. Leaflet 1791. 2 pp.
893. Downer, J. A., Sivhra, P., Molinar, R. H., Fraser, J. B., and Koehler, C. S. 1988. New psyllid pest of California pepper tree. Calif. Agric. 42:30–31.
894. Drea, J. J., and Carlson, R. W. 1987. The establishment of *Chilocorus kuwanae* (Coleoptera: Coccinellidae) in the eastern United States. Proc. Entomol. Soc. Wash. 89:821–824.
895. Drea, J. J., and Carlson, R. W. 1988. Establishment of *Cybocephalus* sp. (Coleoptera: Nitidulidae) from Korea on *Unaspis euonymi* (Homoptera: Diaspididae) in the eastern United States. Proc. Entomol. Soc. Wash. 90:307–309.
896. Gill, R. J. 1988. The scale insects of California. Part 1. The soft scales (Homoptera: Coccoidea: Coccidae). Calif. Dep. of Food and Agric. Tech. Ser. in Agric. Biosyst. and Plant Pathol. no. 1. 132 pp.
897. Hay, C. J. 1958. Life history and control of a root collar borer (*Euzophera ostricolorella* Hulst) in yellow poplar. J. Econ. Entomol. 51:251–252.
898. Houston, D. R., Parker, E. J., and Lonsdale, D. 1979. Beech bark disease: patterns of spread and development of the initiating agent *Cryptococcus fagisuga*. Can. J. For. Res. 9:336–344.
899. Koehler, C. S., Allen, W. W., and Costello, L. R. 1985. Fuchsia gall mite management. Calif. Agric. 39:10–12.
900. Liu T.-X., Kosztarab, M., and Rhoades, M. 1989. Studies on the morphology and systematics of scale insects—No. 15. I. Biosystematics of the adult females of the genus *Chionaspis* (Homoptera: Coccoidea: Diaspididae) of North America, with emphasis on polymorphism. Va. Agric. Exp. Stn. Bull. 88-2. 198 pp.
901. McClure, M. S. 1987. Biology and control of hemlock woolly adelgid. Conn. Agric. Exp. Stn. Bull. 851. 9 pp.
902. McClure, M. S. 1989. Evidence of a polymorphic life cycle in the hemlock woolly adelgid, *Adelges tsugae* (Homoptera: Adelgidae). Ann. Entomol. Soc. Am. 82:50–54.
903. Raffa, K. F., and Hall, D. J. 1988. *Thrips calcaratus* Uzel (Thysanoptera: Thripidae), a new pest of basswood trees in the Great Lakes region. Can. J. For. Res. 18:1661–1662.
904. Robinson, D. 1953. Garden bagworm, *Apterona crenuletta* (=*helix*), in Nevada and Placer Counties, California. Calif. Dep. Agric. Bull. 42:25–33.
905. Scriven, G. T., Reeves, E. L., and Luck, R. F. 1986. Beetle from Australia threatens eucalyptus. Calif. Agric. 40:4–6.
906. Suomi, D., and Akre, R. D. 1988. Snailcase bagworm. Wash. State Univ. Coop. Ext. Bull. 1485. 2 pp.
907. Svihra, P., and Koehler, C. S. 1989. Flatheaded borer in white alder. Univ. Calif. Coop. Ext. Leaflet 7187. 2 pp.
908. Taft, W. H., Smitley, D., and Snow, J. W. 1990. A guide to the clearwing borers (Sesiidae) of the north central United States. USDA For. Serv.
909. Thoneny, W. T., and Nordin, G. L. 1988. Phenology and status of adult locust twig borers (Lepidoptera: Tortricidae) in southeastern Kentucky. Environ. Entomol. 17:35–39.
910. Tissot, A. N., and Pepper, J. O. 1967. Two new species of *Cinara* (Homoptera: Aphididae) associated with pine rust lesions. Fla. Entomol. 50:1–10.
911. Warner, R. E., and Negley, F. B. 1976. The genus *Otiorhynchus* in America north of Mexico (Coleoptera: Curculionidae). Proc. Entomol. Soc. Wash. 78:240–262.
912. Wheeler, A. G., Jr. 1989. *Cinara piticornis* (Hartig), a pest of nursery-grown spruce. Regul. Hortic. Entomol. Circ. 134. 15:13–15.
913. Wheeler, A. G., Jr., and Hoebeke, E. R. 1988. *Apterona helix* (Lepidoptera: Psychidae), a palearctic bagworm moth in North America: new distribution records, seasonal history, and host plants. Proc. Entomol. Soc. Wash. 90:20–27.

Index of Insects, Mites, and Other Animals

This listing primarily consists of the scientific and common names of the pest species discussed in the text. Parasites and predators of arthropods are listed under the entries parasites of arthropods and predators of arthropods. Individual parasites and predators do not appear as main entries. Plant pathogens transmitted by arthropods are included, as are some subject entries.

Abbott's sawfly, 18
Abgrallaspis ithacae, 102; *liriodendri*, 382
acacia psyllid, 288
acacia whitefly, 320, 322
Acalitus fagerinea, 484; *vaccinii*, 478
Acanalonia conica, 422
Acanthocephala femorata, 422, 424; *terminalis*, 424
Acanthocinus nodosus, 276; *obsoletus*, 276; *princeps*, 276; *spectabilis*, 276
Acantholyda spp., 18; *burkei*, 22; *erythrocephala*, 22; *verticalis*, 22; *zappei*, 22
Acaphylla steinwedeni, 480
Acarina, 118, 120, 122, 432, 464, 472–488
Aceria aceris, 482; *brachytarsus*, 488; *calaceris*, 482; *caulis*, 486; *celtis*, 478; *cinereae*, 488; *elongatus*, 482; *erineus*, 488; *fraxiniflora*, 478; *mackiei*, 486; *modestus*, 482; *negundi*, 482, 486; *nyssae*, 488; *parapopuli*, 478; *parulmi*, 482; *triplacis*, 486
Acericecis ocellaris, 462
Acizzia uncatoides, 288
Acleris variana, 28
Acraspis erinacei, 444
Acrididae, 490
Acrobasis betulella, 212; *betulivorella*, 212; *exsulella*, 212; *indigenella*, 218; *juglandis*, 212; *minimella*, 212
Acronicta americana, 158
Aculops fuschiae, 486; *maculatus*, 482
Aculus ligustri, 432, 480
Acyrthosiphon solani, 310
Adelges abietis, 76, 112, 114; *cooleyi*, 76, 112; *lariciatus*, 76, 78, 112; laricis, 76, 78, 112; *oregonensis*, 76; *piceae*, 74, 76; *tsugae*, 76, 78, 112
Adelgidae, 74, 76, 112, 114
adelgids: *Adelges laricis*, 76, 78, 112; *A. oregonensis*, 76; balsam woolly, 74, 76; Cooley spruce gall, 76, 78, 112; eastern spruce gall, 76, 112, 114; hemlock woolly, 76, 78, 80; pine bark, 76, 78; pine leaf, 76, 78, 112; *Pineus borneri*, 76; *P. coloradensis*, 76; *P. floccus*, 76, 112; *P. similis*, 76, 112; *P. sylvestris*, 76; spruce gall, 76, 78, 112, 114
Agrilus sp., 264, 272; *acutipennis*, 266; *angelicus*, 266, 272; *anxius*, 272; *bilineatus*, 270; *burkei*, 270; *difficilis*, 270; *liragus*, 272
Agriolimax reticulatus (slug), 498
Agromyza viridula, 206
Agromyzidae, 206, 468
ailanthus webworm, 174
Alabama pine scale, 92
Albunda fraxini, 260
Alcathoe caudata, 260
alder dagger moth, 160
alder flea beetle, 228
alder lace bug, 426
alder leafbeetle, 224
alder spittlebug, 420
alder woolly sawfly, 136
Alebra sp., 416; *albostriella*, 416
Aleurocanthus wogiumi, 322
Aleurochiton forbesii, 324
Aleuroplatus berbericolus, 322; *coronatus*, 320
Aleuropleurocelus nigrans, 322
Aleurothrixus floccosus, 322; *interrogatonis*, 320
Aleyrodes amnicola, 322
Aleyrodidae, 318, 320, 322, 324
Allantus cinctus, 132
Allegheny mound ant, 496
Allokermes sp., 364; *kingi*, 366
Alsophila pometaria, 142
Altica ambiens alni, 228; *ignita*, 228
Amauronematus azaleae, 134
Amblyseius fallacis, 476
ambrosia beetles, 70, 250, 262, 284, 286; pitted, 250
American dagger moth, 158, 160
American hornet moth, 258
American plum borer, 252
Amorbia humerosana, 214
Amphibolips confluenta, 440, 446
Amphipyra pyramidoides, 148
Anabrus simplex, 492
Anacamptodes pergracilis, 26
Ancylis discigerana, 218; *platanana*, 218
Andricus fullawayi, 448; *kingi*, 440; *laniger*, 440, 446
andromeda lace bug, 424
angulate tingid, 426
Anisota senatoria, 156; *stigma*, 156; *virginiensis*, 156
Anoplonyx laricivorus, 18; *occidens*, 18
Anormenis chloris, 422
anthocorid bug, 92
anthracnose (fungal disease), 352, 402
Antispila nysaefoliella, 208
Antonina pretiosa, 330
Antron douglasii, 448
ants, 28, 84, 96, 282, 304, 308, 324, 362, 496; Allegheny mound, 496; black carpenter, 496; carpenter, 282, 496, 500; *Crematogaster clara*, 96, 262; fire, 496; red imported fire, 496; Texas leafcutting, 496
Aonidiella aurantii, 378; *citrina*, 378; *taxus*, 110
Aphididae, 80–84, 292–316, 450, 464
aphids, 13, 80, 460; *Aphis ceanothi*, 298; apple, 292, 300; apple grain, 292; *Atarsaphis agrifoliae*, 296; balsam twig, 80; bamboo, 312; bean, 300; beech blight, 296; *Betulaphis brevipilosa*, 296; birch, 296; black citrus, 232, 302; black pecan, 292; black pine, 84; bowlegged fir, 82; *Calaphis betulaecolens*, 296; California laurel, 294; *Callipternella calliptera*, 296; *Cinara atlantica*, 84; *C. pseudotsugae*, 84; *C. sabinae*, 84; *C. tujafilina*, 84; cotton, 308; cowpea, 308; crapemyrtle, 292; *Drepanaphis* sp., 302; duskyveined, 292; elm cockscomb-gall, 464; *Eriosoma crataegi*, 306; *E. rileyi*, 306; *Euceraphis betulae*, 296; fourspotted hawthorn, 314, 316; foxglove, 310; gall, 450, 460, 464; giant bark, 310; giant willow, 310; green peach, 300; green spruce, 82; honeysuckle, 314; *Illinoia rhododendri*, 308; ivy, 308; linden, 302; manzanita leafgall, 298; melon, 308; *Nearctaphis crataegifoliae*, 314; *Neophyllaphis podocarpi*, 294; Norway maple, 302; oak woolly, 304; oleander, 312; oleander and milkweed, 312; pear root, 464; *Periphyllus lyropictus*, 302; poplar petiolegall, 460; poplar vagabond, 462; potato, 300; powdery pine needle, 84; *Prociphilus americanus*, 304; *P. caryae*, 304; *Pterocomma smithiae*, 310; rose, 308; rosy apple, 296; *Sarucallis kahawaluokalani*, 292; snowball, 300; spiny witch-hazel gall, 296; spirea, 298; spotted pine, 84; spruce, 82; *Stegophylla essigi*, 304; *S. quercifolia*, 304; tuliptree, 292; viburnum, 300; *Wahlgreniella nervata*, 298; walnut, 292; white pine, 84; witch-hazel bud gall, 450; witch-hazel cone gall, 450; witch-hazel leaf gall, 450; woolly, 464; woolly alder, 304, 306; woolly apple, 306, 316; woolly beech leaf, 296; woolly elm, 306; woolly pine needle, 84
Aphis ceanothi, 298; *citricola*, 298; *craccivora*, 308; *fabae*, 300; *gossypii*, 308; *hederae*, 308; *nerii*, 312; *pomi*, 292, 300; *viburniphila*, 300
Aphrophora canadensis, 86; *cribata*, 86; *parallela*, 86; *permutata*, 86; *saratogensis*, 86
Apion nigrum, 224
Apiosporina morbosa (fungus), 258
Aponychus spinosus, 474
Apotomis spp., 218
apple-and-thorn skeletonizer, 216
apple aphid, 292, 300
apple bark borer, 258
apple blister mite, 486
apple bryobia (mite), 472
apple grain aphid, 292
apple lace bug, 426
appleleaf trumpet miner, 196
apple mealybug, 324
Apterona helix, 178
arborvitae leaf miners, 42
arborvitae soft scale, 98
arborvitae weevil, 240, 244
Archips argyrospila, 172, 202, 218; *cerasivorana*, 172; *fervidana*, 172, 214; *griseus*, 218; *negundana*, 172, 218; *rosanus*, 218; *semiferana*, 172, 214, 218
Archytas metallicas, 150
Arctiidae, 34, 158, 160, 166
Arge clavicornis, 134; *pectoralis*, 128
Argidae, 134
Argyresthia arceuthobiella, 42; *aureoargentella*, 42; *cupressella*, 42; *franciscella*, 42; *freyella*, 42; *libocedrella*, 42; *pilatella*, 40; *thuiella*, 42; *trifasciae*, 42
Argyresthiidae, 40, 42
Argyrotaenia citrana, 216; *franciscana*, 218; *juglandana*, 218; *pinatubana*, 46; *quercifoliana*, 214, 218; *velutinana*, 214
Arilus cristatus, 190
Arizona ash tingid, 426
armored scales, 100–110, 232, 366, 386–394; *Abgrallaspis liriodendri*, 382; ash scurfy, 390; Asiatic red, 110; *Aspidiotus cryptomeriae*, 110; black pineleaf, 100, 108; Boisduval, 376; California red, 378; camellia mining, 380; camphor, 372; *Carulaspis minima*, 106; *Chionaspis furfura*, 368; citrus snow, 390; cocos, 376; cottonwood scurfy, 390; dictyospermum, 382; dogwood, 368; *Duplaspidiotus tesseratus*, 380; elm scurfy, 368; elongate hemlock, 104, 394; euonymus, 388; false oleander, 374; fern, 390; *Fiorinia japonica*, 104, 394; Florida red, 380; Florida scurfy, 390; Florida willow scurfy, 390; gloomy, 384; Glover, 380; greedy, 232, 372, 374, 376; hemlock, 102; hickory scurfy, 390; honeylocust scurfy, 390; *Howardia biclavis*, 380;

Japanese fiorinia, 394; juniper, 106; juniper fiorinia, 394; latania, 376; *Lepidosaphes camelliae,* 370; *L. yanagicola,* 388; lesser snow, 390; *Lindingaspis floridana,* 376; *L. rossi,* 376; Lintneri scurfy, 390; Longiloba scurfy, 390; maple scurfy, 390; mining, 380; *Nuculaspis tsugae,* 102; obscure, 384; oleander, 374; oystershell, 370; palm fiorinia, 394; *Parlatoria pittospori,* 110; peony, 372; pine, 108; pine needle, 100, 108; purple, 322, 380; Putnam, 372; *Quernaspis quercus,* 390; rose, 366; San Jose, 386; snow, 390; sourgum scurfy, 390; Styrax scurfy, 390; sweetgum, 382; sycamore scurfy, 390; tea, 394; walnut, 386; western scurfy, 390; white peach, 392; white prunicola, 392; willow scurfy, 390; wisteria scurfy, 390; yellow, 378
artichoke gall midge, 478
Aschersonia aleyrodis (fungus), 322
Asciodes gordalis, 182
Ascocalyx abietina (fungus), 56
ash borer, 260
ash clearwing, banded, 260
ash flower gall, 478
ash midrib gall midge, 464
ash plant bugs, 402
ash scurfy scale, 390
ash whitefly, 322
Asiatic oak weevil, 240, 244
Asiatic weevils, 13, 244
Asiatic red scale, 110
aspen blotch miner, 196
aspen leaf beetle, 226
aspen leafminer, 194
aspen leaftier, 194, 214
aspen tortrix, large, 214, 216
Asphondylia azaleae, 470; *ilicicola,* 470; *photiniae,* 470
Aspidiotiphagus citrinus, 102, 104, 392
Aspidiotus cryptomeriae, 110; *hederae,* 374; *nerii,* 374; *obscura,* 384; *tenebricosa,* 384
aster leafhopper, 418
Asterocampa celtis, 152; *clyton,* 152
Asterolecaniidae, 310, 348, 350, 352
Asterolecanium sp., 310; *arabidis,* 348; *bambusae,* 350; *minus,* 352; *puteanum,* 352; *quercicola,* 352; *variolosum,* 352
Asynonychus godmani, 230, 240
Atarsaphis agrifoliae, 296
Athrips rancidella, 218
Atta texana, 496
Atteva punctella, 174
audubon warbler, 354
Aulacaspis rosae, 366
Aulacorthum solani, see *Acyrthosiphon solani*
Aureobasidium pullulans (fungus), 46
Automeris io, 164
azalea bark scale, 336
azalea caterpillar, 154
azalea lace bug, 424
azalea leafminer, 202
azalea stem borer, 288
azalea whitefly, 318

baccharis leaf beetle, 238
Bacillus (bacterial pathogen): *popilliae,* 236; *thuringiensis,* 138, 142, 146, 152, 168, 176, 216, 218
bagworms, 13, 176, 178
Baliosus ruber, 190
balsam gall midge, 116
balsam shootboring sawfly, 20
balsam twig aphid, 80
balsam woolly adelgid, 74, 76
bamboo aphid, 312
bamboo mealybug, noxious, 330
bamboo scale, 350
bamboo spider mite, 474
banded ash clearwing, 260
banded wood snail, 498
banded woollybear, 160
barberry looper, 144
barberry whitefly, 322
bark beetles, 248, 250; ambrosia, 70, 250, 262, 284, 286; black turpentine, 62, 66; cypress, 70; *Dendroctonus brevicomis,* 66; *D. pseudotsugae,* 66; Douglas-fir, 66; eastern juniper, 70; eastern larch, 62; hickory, 250, 252; *Hylesinus* sp., 252; *Ips* sp., 62, 64, 66, 70; *I. avulsus,* 64; *I. calligraphus,* 64; *I. grandicollis,* 64; *I. tridens,* 66; *Leperisinus* sp., 252; mountain pine, 66; native elm, 246, 248; northern cedar, 68, 70; peach, 250; *Phloeosinus* sp., 66; *P. cupressi,* 68, 70; *P. dentatus,* 68, 70; *P. taxodii,* 68, 70; pine engraver, 62, 64; pitted ambrosia, 250; *Pityogenes* sp., 66; *P. hopkinsi,* 62; *Pityophthorus* spp., 66; *P. orarius,* 60; red turpentine, 62, 66; *Scolytus* spp., 70, 246, 248, 250; shot hole borer, 250; smaller European elm, 246, 248; small southern pine engraver, 64; southern cypress, 70; southern pine, 64, 66; western pine, 66
bark miners, 198
barnacle scale, 356
Bassettia herberti, 444
Bassus pumilus, 36
basswood lace bug, 426
basswood leafminer, 190
bast scale, pine, 92
bean aphid, 300
Beauveria sp. (fungal pathogen), 260
bees, leafcutter, 494
beech blight aphid, 296
beech leafminer, 190
beech scale, 332
beetles: Fuller rose, 230, 240; whitefringed, 280. *See also* ambrosia beetles; bark beetles; borers; leaf beetles; longhorned beetles; predators of arthropods: lady beetles; leaf beetles
Bemisia tabaci, 322
Betulaphis brevipilosa, 296
biological control, 8
birch aphids, 296
birch casebearer, 186
birch lace bug, 426
birch leafminer, 128, 184, 296
birch-leafmining sawflies, 184
birch sawfly, 128
birch skeletonizer, 220
birch tubemaker, 212
birds, 12, 18, 24, 154, 254, 340, 354, 500; black-billed cuckoos, 152; bluejays, 154; chickadees, 24; English sparrows, 340; nuthatches, 24; robins, 154; sapsuckers, 500; woodpeckers, 12, 68, 252, 254, 258, 260, 272, 282, 286, 500; yellow-billed cuckoos, 152; yellow-rumped warblers, 354
black aleyrodid, 322
black-billed cuckoo, 152
black carpenter ant, 496
black citrus aphid, 232, 302
blackheaded ash sawfly, 134
blackheaded budworm, 28
blackheaded fireworm, 208
blackheaded pine sawfly, 16
"Black Hills" pandora moth, 26
blackhorned tree cricket, 492, 494
black knot, 258
black locust weevil, 224
black pecan aphid, 292
black pine aphid, 84
black pineleaf scale, 100, 108
black scale, 358
black turpentine beetle, 62, 66
black twig borer, 262
black vine weevil, 54, 240, 242
black walnut curculio, 234
black walnut pouch gall, 486
black walnut leaf petiole gall, 486
blister mites, 478–488; walnut, 486
blueberry bud mite, 478
blueberry case beetle, 224
blue-green sharpshooter, 418
bluejay, 154
Boisduval scale, 376
Boisea rubrolineata, 398; *trivattata,* 398
borers, 12; American plum, 252; apple bark, 258; ash, 260; azalea stem, 288; black twig, 262; boxelder twig, 200; broadnecked root, 280; cottonwood crown, 258; cottonwood twig, 200; eastern pine shoot, 52; larger shothole, 246; lesser peachtree, 258; lilac, 260; locust twig, 274; maple petiole, 200; peachtree, 258; pitch mass, 72; poplar-and-willow, 268; rhododendron, 258; *Saperda* spp., 254; seagrape, 232; shothole, 250; western pine shoot, 52; western sycamore, 258. *See also* flatheaded borers; roundheaded borers
bowlegged fir aphid, 82
boxelder bug, 398
boxelder twig borer, 200
boxwood leafminer, 204
boxwood mite, 475, 476
boxwood psyllid, 290
boxwood webworm, 166
Brachys aerosus, 190; *aeruginosus,* 190; *ovatus,* 190
brachyuran wax scale, 358
Brevipalpus californicus, 476; *lewisi,* 476; *obovatus,* 476, 482
bristly roseslug, 132
broad mites, 476
broadnecked root borer, 280
broadwinged katydid, 490
bronze birch borer, 272
bronze poplar borer, 272
brown fruitworms, 148
brown garden snail, 498
brownheaded ash sawfly, 134
brown hemlock needleminer, 38
brown-lipped snail, 498
Bruce spanworm, 146
Bryobia arborea, 472; *borealis,* 472; *ephedrae,* 472; *kissophila,* 472; *lonicerae,* 472; *praetiosa,* 472; *ribis,* 472; *ulmophila,* 472
Bucculatrix ainsliella, 220; *albertiella,* 220; *canadensisella,* 220; *pomifoliella,* 220
buckeye lace bug, 426
buck moth, 156; Nevada, 156
bud blast disease, 418
bud miner, 20
bud mites, 478–480
budworms: eastern blackheaded, 28; spruce, 28, 500; western spruce, 28
buffalo treehopper, 406, 408
Bulimulidae (snails), 498
Bulimulus alternatus mariae (snail), 498
Buprestidae, 190, 266, 270, 272
butterflies: cabbage, 24; California tortoiseshell, 152; hackberry, 152; mourningcloak, 152; pine, 24; tawny emperor, 162
butternut curculio, 234
butternut woollyworm, 136
buttonbush scale, 362

cabbage butterfly, 24
cabbage palm caterpillar, 162
Cacopsylla buxi, 290; *pyricola,* 290
Calaphis betulaecolens, 296; *juglandis,* 292
Calcarus carinatus, 480
calico scale, 354
California casebearer, 186
California Christmas berry tingids (lace bugs), 426, 428
California laurel aphid, 294
California oakworm, 146
California prionus, 280
California red scale, 378
California tortoiseshell, 152
Caliroa cerasi, 130; *quercuscoccineae,* 130
Calligrapha bigsbyana, 226; *scalaris,* 226
calligrapha, elm, 226
Callipterinella calliptera, 296
Callirhopalus bifasciatus, 240, 244
Callirhytis cornigera, 440, 442; *crypta,* 442; *perdens,* 448; *quercuspunctata,* 440; *seminator,* 444
Calomycterus setarius, 240, 244
Calophya californica, 290; *flavida,* 290; *nigripennis,* 290; *rubra,* 458; *triozomima,* 290
Caloptilia azaleella, 202; *fraxinella,* 196; *syringella,* 196

Calosoma sycophanta, 138
cambium miners, see bark miners
camellia mining scale, 380
camellia scale, 370
Camereria cincinnatiella, 192; *hamadryadella*, 192, 196; *hamameliella*, 196; *ostryarella*, 192
camphor scale, 372
Camponotus sp., 496; *herculeanus pennsylvanicus*, 28; *pennsylvanicus*, 496
Canadian pine scale, 92
candle wax scale, 358
canker fungus, 276
cankerworms, 270; fall, 142, 144; spring, 142, 144. *See also* caterpillars; loopers; spanworms
Canonura princeps, 276
Caribbean black scale, 358
carmine spider mite, 476
carpenter ants, 282, 496, 500
carpenterworm, 256, 282; little, 256
Carulaspis juniperi, 106; *minima*, 106
case bearers: birch, 186; California, 186; elm, 186; larch, 36, 186; pecan cigar, 186; pecan leaf, 212
casemaker, oak ribbed, 220
caterpillars: azalea, 154; cabbage palm, 162; California oakworm, 146; cherry sphinx, 164; eastern tent, 168; euonymus, 174; forest tent, 168, 170, 270, 500; genista, 166; oleander, 162; orangestriped oakworm, 156; Pacific tent, 170; pinkstriped oakworm, 156; redhumped, 150, 156; redhumped oakworm, 156; Sonoran tent, 170; southwestern tent, 170; sphinx, 164; uglynest, 172; unicorn, 156; walnut, 150, 154, 156; western tent, 170; yellownecked, 154. *See also* cankerworms; loopers; moths; spanworms
Catoloccus aeneoviridis, 30
Caulocampus acericaulis, 200
ceanothus lace bug, 246, 248
ceanothus psyllid, 458
ceanothus stem gall moth, 438
ceanothus tingid, 426, 428
Cecidomyia piniinopis, 44; *reeksi*, 44; *resinicola*, 44; *resinicoloides*, 44
Cecidomyiidae, 44, 116, 204, 436, 446, 452, 462–470
Cecidophyes betulae, 478
Cecidophyopsis psilaspis, 478; *vermiformis*, 480
cecropia moth, 164
Celticecis spiniformis, 452
Cepaea nemoralis (snail), 498
Cephalcia californica, 22; *distincta*, 22; *fascipennis*, 22; *frontalis*, 22; *fulviceps*, 22; *hopkinsi*, 22; *marginata*, 22; *nigra*, 22; *provancheri*, 22; *semidea*, 22
Cephidae, 286
Cerambycidae, 264, 274–288
Ceratocystis ulmi (dutch elm disease), 246
Cercopidae, 86, 420
Cerococcidae, 354
Ceroplastes sp., 354; *brachyurus*, 358; *ceriferus*, 356; *cirripediformis*, 356; *cistudiformis*, 358; *dugesii*, 358; *floridensis*, 356; *irregularis*, 358; *nakaharai*, 358; *rubens*, 358; *sinensis*, 356; *utilis*, 358
Cerrina unicolor (fungus), 494
Chaetophloeus sp., 70
chafers: European, 236; rose, 236; western rose, 236
chainspotted geometer, 24, 144, 146
cherry lace bug, 426
cherry scallop shell moth, 146
cherry sphinx caterpillar, 164
cherry tingids, 426, 428
Chilocorus spp., see predators of arthropods
Chinese wax scale, 356
Chionaspis sp., 368, 390; *acericola*, 390; *americana*, 368; *caryae*, 390; *corni*, 368; *floridensis*, 390; *furfura*, 368; *gleditsia*, 390; *hamoni*, 390; *heterophyllae*, 108; *kosztarabi*, 390; *lintneri*, 390; *longiloba*, 390; *nyssae*, 390; *ortholobis*, 390; *pinifoliae*, 100, 108; *platani*, 390; *salicisnigrae*, 390; *sassceri*, 390; *styracis*, 390; *wisteriae*, 390
chokecherry tingid, 426
Choreutidae, 216
Choreutis pariana, 216
Choristoneura conflictana, 216, 218; *fumiferana*, 28; *lambertina*, 28; occidentalis, 28; *pinus*, 28; *rosaceana*, 216
Chromaphis juglandicola, 292
Chrysobothris femorata, 270; *mali*, 270
Chrysocharis laricinellae, 36
Chrysomela aenicollis, 226; *californica*, 226; *crotchi*, 226; *interrupta*, 226; *mainensis*, 224; *scripta*, 226; *tremulae*, 226
Chrysomelidae, 190, 224–228, 238, 242, 400
Chrysomphalus aonidum, 380; *dictyospermi*, 382
Chrysopa spp., see predators of arthropods
cicadas, 490, 492; dog-day, 490; periodical, 13, 490
Cicadellidae, 400, 404, 412–418
Cicadidae, see cicadas
Cimbex americana, 136
Cinara atlantica, 84; *cronarti*, 84; *curvipes*, 82, 84; *fornacula*, 82; *piticornis*, 84; *pseudotsugae*, 84; *sabinae*, 84; *strobi*, 84; *tujafilina*, 84
Cingilia catenaria, 24, 144, 146
citrus blackfly, 322
citrus flat mite, 476
citrus mealybug, 328
citrus red mite, 472
citrus root weevil, 240
citrus snow scale, 390
Cladius difformis, 132
Clastoptera arborina, 86; *arizonana*, 86; *juniperina*, 86; *obtusa*, 420; *proteus*, 420; *undulata*, 86
Clavaspis sp., 368
clearwing moths, 72, 254–262; apple bark borer, 258; ash borer, 260; banded ash, 260; dogwood borer, 262; hornet moth, 258; lesser peach tree borer, 258; lilac borer, 260; *Paranthrene simulans*, 254; *P. tabaniformis*, 254; pitch mass borer, 72; *Podosesia* sp., 260; rhododendron borer, 258; sequoia pitch moth, 72; *Sesia tibialis*, 258; *Synanthedon novaroensis*, 72; *S. resplendens*, 258; western poplar, 256
clematis borer, 260
Clinodiplosis rhododendri, 470
cloudywinged whitefly, 322
clover mites, 472
Coccidae, 88, 94, 96, 98, 340–346, 354–366
Coccinellidae, 366
Coccus sp., 354
cockscombgall aphid, 464
coconut mealybug, 90
cocos scale, 376
codling moth, 234
Colaspis pseudofavosa, 224
Coleophora laricella, 36; *laticornella*, 186; *limnosipennella*, 186; *sacramenta*, 186; *serratella*, 186; *ulmifoliella*, 186
Coleophoridae, 36, 162, 186
Coleotechnites apicitripunctella, 38; *canusella*, 40; *ducharmei*, 32; *juniperella*, 42; *macleodi*, 38; *milleri*, 40; *piceaella*, 32; *resinosae*, 40; *stanfordia*, 42; *thujaella*, 42
Colopha ulmicola, 464
Coloradia doris, 26; *pandora*, 26
Columbian timber beetle, 286
Comperiella bifasciata, 106, 378
Compsilura concinnata, 158
Comstock mealybug, 326
conifer sawflies: Abbott's, 18; balsam shootboring, 20; blackheaded pine, 16; cypress, 18; European pine, 16, 18; European spruce, 18; hemlock, 18; Hetrick's, 18; introduced pine, 16, 18; jack pine, 16; larch, 16, 18; loblolly pine, 16; lodgepole, 18; *Neodiprion pratti paradoxicus*, 16; pine-webbing, 18; pitch pine, 18; *Pontania* sp., 468; redheaded pine, 16, 18; red pine, 16; spotted loblolly pine, 16; Swaine jack pine, 18; twolined larch, 18; Virginia pine, 16, 18; webspinning, 22, 136; western larch, 18; white-pine, 18; yellowheaded spruce, 18
Conophthorus sp., 70
Conotrachelus aratus, 234; *buchanani*, 452; *juglandis*, 234; *retentus*, 234
Contarinia baeri, 44; *canadensis*, 464; *coloradensis*, 116; *juniperina*, 46; *pseudotsugae*, 44
Cooley spruce gall adelgid, 76, 112
Coptodisca arbutiella, 194, 208; *splendoriferella*, 208
Coreidae, 422, 424
Corthylus columbianus, 286; *punctatissimus*, 250
Coryneum cardinale (fungus), 70
Coryphista meadii, 144
Corythucha aesculi, 426; *arcuata*, 426; *associata*, 426; *bellula*, 426; *bullata*, 426; *celtidis*, 426; *ciliata*, 426, 428; *coelata*, 426; *confraterna*, 426; *costata*, 426; *cydoniae*, 426; *elegans*, 426; *heteromelecola*, 426; *incurvata*, 426, 428; *juglandis*, 426; *marmorata*, 426; *mollicula*, 426; *obliqua*, 426, 428; *padi*, 426; *pallipes*, 426; *pergandei*, 426; *pruni*, 426, 428; *salicata*, 426; *salicis*, 426; *ulmi*, 426
Cosetacus camelliae, 480
Cossidae, 254, 256
cotton aphid, 308
cottonwood borer, 276, 278, 280
cottonwood crown borer, 258, 260
cottonwood leaf beetle, 226
cottonwood scurfy scale, 390
cottonwood twig borer, 200
cottony azalea scale, 342, 346
cottony camellia scale, 344
cottony cushion scale, 338
cottony cypress scale, 332
cottony manzanita whitefly, 322
cottony maple leaf scale, 340, 342, 346
cottony maple scale, 340, 342, 346
cottony taxus scale, 344
cowpea aphid, 308
cranberry rootworm, 242
crapemyrtle aphid, 292
Crematogaster sp., 362; *clara*, 96, 260
Cremona cotonestri, see *Athrips rancidella*
crickets: blackhorned tree, 492, 494; mormon, 492; narrowwinged, 492, 494; snowy tree, 492, 494; tree, 13, 492
Croesia albicomana, 172; *semipurpurana*, 172, 218
Croesus latitarsus, 128
crown whitefly, 320
crumpler, leaf, 218
Cryptococcidae, 354
Cryptococcus fagisuga, 332
Cryptorhynchus lapathi, 268
Ctenodactylomyia watsoni, 232
Cuban laurel thrips, 434
curculios: black walnut, 234; butternut, 234; hickory shoot, 234; rose, 238
Curculionidae, 190, 230, 234, 240–244, 268
curled rose sawfly, 132
cyclamen mite, 476
Cylindrocopturus sp., 54; *eatoni*, 60; *furnissi*, 60
Cynipidae, 438–448
cypress bark beetles, 70
cypress bark mealybug, 332
cypress bark moth, 60
cypress bark scale, 332
cypress canker, 60
cypress leaftier, 38
cypress looper, 26
cypress sawfly, 18
cypress tip miner, 42
cypress tip moths, 42
cypress twig gall midge, 118
Cyrtepistomus castaneus, 240, 244
Cytospora sp. (fungus), 286

dagger moth, see American dagger moth
Dasineura balsamicola, 116; *gleditchiae*, 466; *rhodophaga*, 470; *ulmi*, 464
Dasychira grisefacta, 26; *manto*, 26; *pinicola*, 26; *plagiata*, 26
Datana angusii, 150, 154; *integerrima*, 150, 154; *major*, 154; *ministra*, 154
deer brush whitefly, 320
Dendroctonus brevicomis, 66; *frontalis*, 64, 66; *ponderosae*, 66; *pseudotsugae*, 66; *simplex*, 62; *terebrans*, 62, 66; *valens*, 62, 66, 72
Dendroica coronata (bird), 354
Dendrothrips ornatus, 432

deodar weevil, 54, 56
Deraeocoris aphidiphagus, 306; *nitenatus*, 306
desert wax scale, 358
diagnosis of insect injury, 11, 12, 13
Dialeurodes chittendeni, 318; *citrifolii*, 322
Diapheromera femorata, 494
Diaphnocoris chlorionis, 404
Diaspididae, 100, 104, 106, 110, 366–394
Diaspidiotus sp., 368, 376; *ancylus*, 372; *liquidambaris*, 382
Diaspis boisduvalii, 376; *cocois*, 376
Dicerca lurida, 270, 272
Dichomeris ligulella, 30; *marginella*, 30
dictyospermum scale, 382
Dioptidae, 146
Dioryctria abietivorella, 48, 50; *zimmermani*, 48, 50
Diplolepis bassetti, 438; *bicolor*, 438; *dichlorcera*, 438; *rosae*, 438
Diprion similis, 16, 18
dog-day cicada, 490, 492
dogs, 12
dogwood borer, 258, 260, 262
dogwood club gall, 436
dogwood sawfly, 126
dogwood scale, 368
dogwood spittlebug, 420
dogwood twig borer, 262, 288
Douglas-fir beetle, 66
Douglas-fir pitch moth, 72
Douglas-fir tussock moth, 34
Douglas-fir twig weevil, 60
downy woodpecker, 278, 286
Drepanaphis sp., 302
Dryocampa rubicunda, 156
Dryocosmus dubiosus, 440, 444; *kuriphilus*, 438
Duges wax scale, 358
Duplaspidiotus claviger, 380; *tesseratus*, 380
dusky birch sawfly, 128
duskyveined aphid, 292
Dutch elm disease, 246, 248
dwelling bryobia, 472
Dysaphis plantaginea, 296
Dysmicoccus wistariae, 88

Eacles imperialis, 26; *imperialis pini*, 26
eastern blackheaded budworm, 28
eastern fivespined ips, 64
eastern juniper bark beetle, 68, 70
eastern larch beetle, 62
eastern lubber grasshopper, 490
eastern pine looper, 24
eastern pine shoot borer, 52
eastern pine weevil, 54, 56
eastern spruce gall adelgid, 76, 114
eastern tent caterpillar, 168
eastern willow tingid, 426
Ecdytolopha insiticiana, 274
Ectoedemia platanella, 196
Edwardsiana sp., 416; *commisuralis*, 416; *rosae*, 412
Ehrhornia cupressi, 332
Elaphidionoides parallelus, 264; *villosus*, 264
Elatobium abietinum, 82
elm bark beetles, 248
elm bryobia (mite), 472
elm calligrapha, 226
elm casebearer, 186
elm cockscombgall aphid, 464
elm lace bug, 426
elm leaf beetle, 222
elm leafminer, 186
elm sawfly, 136
elm scurfy scale, 368
elm spanworm, 144–146
elm yellows (disease), 246, 414
elongate hemlock scale, 104, 394
Empoasca sp., 416; *fabae*, 246, 414
Empusa fresenii (fungus), 298
Enaphalodes rufulus, 282
Enchenopa binotata, 410
Encyrtidae, 366
Endelomyia aethiops, 132
Endothenia albolineana, 32
Engelmann spruce weevil, 54
English sparrow, 340
Ennomos subsignarius, 144
Entomophaga aulicae (fungus), 138
Entomophthora sp. (fungus), 168; *aphrophora*, 86; *fumosa*, 324
Eotetranychus sp., 472; *carpini borealis*, 474; *carpini carpini*, 474; *hicoriae*, 474; *lewisi*, 474; *matthyssei*, 474; *multidigituli*, 472, 474; *populi*, 474; *querci*, 474; *sexmaculatus*, 474; *tiliarium*, 474; *uncatus*, 474; *yumensis*, 474
Epargyreus clarus, 148
Epinotia aceriella, 212; *meritana*, 38; *nanana*, 32; *solandriana*, 218; *subviridis*, 38
Episimus tyrius, 212
Erannis tiliaria, 144
erineum gall mites, 482–488. *See also* gall mites
Eriocampa juglandis, 136; *ovata*, 136
Eriococcidae, 90, 332, 336, 354, 368
Eriococcus araucariae, 90; *azaleae*, 336; *borealis*, 336
Eriophyes aceris, 482; *brachytarsus*, 234, 488; *calaceris*, 482; *caulis*, 486; *celtis*, 478; *cinereae*, 488; *elongatus*, 482; *emarginata*, 482; *erinea*, 488; *fraxiniflora*, 478; *mackiei*, 486; *modestus*, 482; *negundi*, 482, 486; *nyssae*, 488; *parapopuli*, 478; *parulmi*, 482; *populi*, 478; *pyri*, 486; *tiliae*, 486; *triplacis*, 486; *ulmi*, 464
Eriophyidae, *see* eriophyid mites
eriophyid mites, 122, 212, 234, 432, 464, 478–488; blueberry bud, 478; bud, 478, 480; taxus bud, 122, 478; walnut blister, 488
Eriosoma sp., 464; *americanum*, 306; *crataegi*, 306; *languinosa*, 464; *lanigerum*, 306, 316; *pyricola*, 464; *rileyi*, 306
ermine moths, 148, 174
Erynia neoaphidis (fungus), 84
Erythroneura sp., 416; *arta*, 416; *basilaris*, 246; *gledistia*, 416; *lawsoniana*, 416; *obliqua*, 246; *usitata*, 416; *ziczac*, 416
Eucallipterus tiliae, 302
Eucalymnatus sp., 354
Euceraphis betulae, 296
Eucosma sp., 48; *gloriola*, 52; *sonomana*, 52
eugenia psyllid, 458
Eulachnus agilis, 84; *rileyi*, 84
Eulecanium sp., 354; *caryae*, 364; *cerasorum*, 354
Eulophidae, 96
euonymus caterpillar, 174
euonymus scale, 388
Euphalerus vermiculosus, 458
European alder leafminer, 188
European chafer, 236
European corn borer, 286
European elm bark beetle, 246, 248
European elm scale, 368
European fruit lecanium, 98, 354, 364
European hornet, 258, 494
European peach scale, 364
European pine sawfly, 16, 18
European pine shoot moth, 48, 50
European red mite, 472, 474
European roseslug, 132
European spruce needleminer, 32
European spruce sawfly, 18
Eurytetranychus sp., 472; *buxi*, 475
Eurytoma contractura, 468; *semivenae*, 452
Eutetrapha tridentata, 276
Euthochtha galeator, 422
Euthoracaphis umbellulariae, 294
Euura sp., 468; *atra*, 286; *exiguae*, 468; *scoulerianae*, 468
Euzophera ostricolorella, 254; *semifuneralis*, 252
excrescence, see galls: excrescence
Exoteleia burkei, 48; *pinifoliella*, 40
eyespot galls, 462
eyespotted bud moth, 214

fall cankerworm, 142, 144
fall webworm, 160, 166
false oleander scale, 374
false spider mites, 476
Fenusa dohrnii, 188; *pusilla*, 128, 184, 296; *ulmi*, 186
fern scale, 390
filbert bud mite, 480
filbert leafroller, 218
Fiorinia sp., 394; *externa*, 104, 394; *fiorinae*, 394; *japonica*, 104, 394; *juniperi*, 394; *theae*, 394
fire ants, 496
flatheaded appletree borer, 270
flatheaded borers, 264, 266, 270, 272; *Agrilus* sp., 264, 272; *A. acutipennis*, 266; *A. burkei*, 270; *A. difficilis*, 270; bronze birch, 272; bronze poplar, 272; *Dicerca lurida*, 270, 272; dogwood, 258, 262; flatheaded, 270; flatheaded appletree, 270; Pacific flatheaded, 270; Pacific oak twig girdler, 266, 272; pecan, 262; twolined chestnut, 270
Flatidae, 232, 422
flatids, 232, 422
flat mites, 476
flea beetles, 228
Fletcher scale, 98, 364
Florida red scale, 380
Florida scurfy scale, 390
Florida wax scale, 356
Florida willow scurfy scale, 390
flower thrips, 432
forest tent caterpillar, 168, 170, 270, 500
Formica exsectoides, 496
Formicidae, 496
fourlined green fruitworm, 148
fourlined plant bug, 396
fourspined engraver, see *Ips avulsus*
fourspotted hawthorn aphid, 314, 316
fourspotted spider mite, 476
foxglove aphid, 310
Frankliniella sp., 432, 434; *occidentalis*, 432; *tritici*, 432
frosted scale, 354, 364
fruittree leafroller, 172, 202, 214, 218
fruitworms, 148; brown, 148; fourlined green, 148; green, 148; humped green, 148; speckled green, 148
Fuller rose beetle, 230, 240
fungi (insect-transmitted pathogens of plants), 236, 464; *Apiosporina morbosa*, 258; *Aureobasidium pullulans*, 46; *Ceratocystis ulmi* (dutch elm disease), 246; *Cerrina unicolor*, 494; *Coryneum cardinale*, 70; *Cronatium fusiforme*, 84; *Cytospora* sp., 286; *Fusarium moniliforme*, 56; *F. solani*, 250, 262; *Gloeosporium aridum*, 402; *Hypoxylon mammatum*, 276; *Nectria coccinea faginata*, 332; *Phomopsis* sp., 46; *Seiridium cardinale*, 60; *Sphaerotheca phytoptophila*, 478; *Taphrina coerulescens*, 486; *Verticillium* sp., 12
Fusarium moniliforme (fungus), 56; *solani*, 250, 262
fuschia gall mite, 486

Galasa nigrinodis, 166
gall insects: aphids, 450, 460–464; beetles, 276, 278; caterpillars, 438; *Phylloxera* spp., 460; sawflies, 468. *See also* adelgids; gall midges; gall mites; gall psyllids; gall wasps
gall midges, 44, 116, 232, 452, 462, 470; *Acericecis ocellaris*, 462; ash midrib, 464; *Asphondylia azaleae*, 470; *A. ilicicola*, 470; *A. photiniae*, 470; balsam, 116; *Celticecis spiniformis*, 452; *Clinodiplosis rhododendri*, 470; *Contarinia canadensis*, 464; *C. pseudotsugae*, 44; *Ctenodactylomyia watsoni*, 232; cypress twig, 118; *Dasineura gleditchiae*, 466; *D. rhodophaga*, 470; *D. ulmi*, 464; hollyberry, 470; honeylocust pod, 466; *Janetiella coloradensis*, 116; juniper tip midge, 46; locust, 470; *Macrodiplosis erubescens*, 442; *M. quercusoruca*, 440, 446; *Mayetiola piceae*, 114, 116; *M. rigidae*, 468; *Meunieriella aquilonia*, 470; Monterey pine midge, 116; *Neolasioptera brevis*, 470; *Obolodiplosis robiniae*, 466, 470; *Paradiplosis tumifex*, 116; *Parallelodiplosis subtruncata*, 462; *Pinyonia edulicola*, 116; *Prodiplosis morrisi*, 470; *Rhabdophaga strobiloides*, 468; rhododendron, 470; rose midge, 470; *Rus-*

gall midges (cont.)
seliella clavula, 436; R. liriodendri, 462; Taxodiomyia cupressiananassa, 118; Taxomyia taxi, 478; Thecodiplosis piniradiatae, 116; T. piniresinosae, 44; T. resinosae, 116; thorn, 452; vein pocket, 440, 446; willow beaked, 468; willow cone, 468
gall mites, 464, 478–488; *Acalitis fagerinea*, 484; *Aculops maculatus*, 482; *A. fuschiae*, 486; ash flower, 478; *Eriophyes aceris*, 482; *E. betulae*, 478; *E. calaceris*, 482; *E. caulis*, 486; *E. celtis*, 478; *E. cinereae*, 488; *E. elongatus*, 482; *E. mackiei*, 486; *E. modestus*, 482; *E. negundi*, 482, 486; *E. nyssae*, 488; *E. parapopuli*, 478; *E. parulmi*, 482; *E. ulmi*, 464; juniper, 122; maple bladder, 482; maple spindle, 482; *Nalepella* sp., 122; *Phyllocoptes* sp., 482; *Phytoptus emarginatae*, 482; *P. pyri*, 486; *P. tiliae*, 486; *Plataculus pyramidicus*, 484; *Taxomyia taxi*, 478; *Trisetacus quadrisetus juniperinus*, 122; walnut blister, 488; walnut purse, 234, 488
gall psyllids: *Gyropsylla ilicis*, 454; hackberry blister, 452; hackberry bud, 452; hackberry nipple, 452; laurel, 458; *Pachypsylla* sp., 450; pepper tree, 458; *Trioza magnoliae*, 456
galls, 316, 382, 436–470, 478–488; artichoke, 478; ash flower, 478; ash midrib, 464; black walnut leaf petiole, 486; black walnut pouch, 486; ceanothus stem, 438; Cooley spruce, 76, 112; dogwood club, 436; eastern spruce, 76, 114; elm cockscomb, 464; erineum, 482–488; excrescence, 464; eyespot, 462; gouty oak, 440, 442; hackberry blister, 452; hackberry bud, 452; hackberry nipple, 452; hedgehog, 444; hickory leafstem, 460; honeylocust pod, 466; horned oak, 440, 442; jumping oak, 440; live oak woolly leaf, 246, 440; long rose, 438; maple bladder, 482; maple spindle, 482; marginal fold, 484; mossyrose, 438; oak apple, 440, 446; oriental chestnut, 438; pouch, 482; rhododendron, 470; roly-poly, 448; spindle, 482; spined turban, 448; spiny rose, 438; thorn, 452; vein pocket, 440, 446; willow beaked-gall, 468; willow cone, 468; witch-hazel bud, 450; witch-hazel cone, 450; witch-hazel leaf, 450; wool sower, 444
gall wasps: *Andricus fullawayi*, 448; *A. kingi*, 440; *A. laniger*, 440, 446; *Callirhytis crypta*, 442; *C. perdens*, 448; *Diplolepis bassetti*, 438; *Dryocosmus dubiosus*, 440, 444; gouty oak, 440, 442; hedgehog, 444; horned oak, 440; jumping oak, 440; long rose, 438; mossy rose, 438; *Neuroterus saltatorius*, 440, 444; oak apple, 440, 446; oriental chestnut, 438; roly-poly, 448; spined turban, 448; wool sower, 444
garden bagworm, 178
Gargaphia angulata, 426; *tiliae*, 426
Garman spider mite, 474
Gelechia panella, 194
Gelechiidae, 30, 38, 40, 194, 218
genista caterpillar, 166
geometers: chainspotted, 24, 144, 146; half-wing, 146
Geometridae, 24, 26, 144, 146
giant bark aphid, 310
giant silkworm moths, 26, 156, 164; buck moth, 156; cecropia moth, 164; greenstriped mapleworm, 156; imperial moth, 26; io moth, 164; Nevada buck moth, 156; orangestriped oakworm, 156; pandora moth, 26; spiny oakworm, 156
giant willow aphid, 310
Gilpinia frutetorum, 18; *hercyniae*, 18
globose scale, 366
Gloeosporium aridum (fungus), 402
gloomy scale, 384
Glossonotus acuminatus, 406
Glover scale, 380
Glycobius speciosus, 276
Glypta sp., 52
Goes debilis, 282; *pulcher*, 282; *pulverulentus*, 282; *tigrinus*, 282, 284
golden mealybug, 90
golden oak scale, 352
golden woodpecker, 278
gooseberry bryobia (mite), 472
Gossyparia spuria, 368
gouty oak gall, 440, 442
Gracillariidae, 194, 196
grapeleaf skeletonizer, 164
grape mealybug, 88
grape whitefly, 322
Graphocephala atropunctata, 418; *circellata*, 418 *coccinea*, 418; *fennahi*, 418
Graphognathus sp., 280
grasshoppers, 13, 490, 494; eastern lubber, 490; longhorn, 492
gray garden slug, 498
gray squirrel, 502
gray willow leaf beetle, 224
greedy scale, 232, 372, 374, 376
green fruitworm, 148
green hemlock needleminer, 38
greenhouse thrips, 430
greenhouse whitefly, 320, 322
green peach aphid, 300
green spruce aphid, 82
green spruce leafminer, 32
greenstriped mapleworm, 156
gregarious oak leafminer, 192
growing degree days, 386
Gryllidae, 492
Grylloprociphilus imbricator, 296
Gynaikothrips ficorum, 434
Gypsonoma haimbachiana, 200
gypsy moth, 138, 140, 168, 182, 270
Gyropsylla ilicis, 454

hackberry blister gall maker, 452
hackberry bud gall psyllid, 452
hackberry butterfly, 152
hackberry lace bug, 426
hackberry nipple gall maker, 452
hackberry psyllids, 290, 450, 452
half-wing geometer, 146
Halysidota harrisii, 160; *tessellaris*, 158
Hamamelistes spinosus, 296, 450
Harrisina americana, 164; *brillians*, 164
Hartigia cressoni, 132; *trimaculata*, 132
hawthorn lace bug, 426
hawthorn leafmining sawfly, 188
hedgehog gall, 444
Helicidae, 498
Heliothrips haemorrhoidalis, 430
Heliozelidae, 194, 208
Helix aspersa, 498
Hemiberlesia lataniae, 376; *rapax*, 232, 372, 376
Hemichroa crocea, 136
Hemileuca maia, 156; *nevadensis*, 156
hemispherical scale, 360
hemlock fiorinia, 104
hemlock looper, 24
hemlock rust mite, 122
hemlock sawfly, 18
hemlock scale, 102
hemlock woolly adelgid, 76, 78, 80, 112
Herptomonas sp. (protozoan), 130
Hesperiidae, 148
Heterarthrus nemoratus, 184
Heterocampa guttivitta, 154
Heterorhabditis heliothidis (nematode), 240
Hetrick's sawfly, 18
Hexeris enhydris, 232
Hexomyza schineri, 468
hickory bark beetle, 250, 252
hickory borer, 282; painted, 274, 276
hickory leafstem gall phylloxera, 460
hickory lecanium scale, 364
hickory scurfy scale, 390
hickory shoot curculio, 234
hickory tiger moth, 160
hickory tussock moth, 160
Himella intractata, 148
hollyberry gall midge, 470
holly bud moth, 206, 208
holly leafminers, 206
holly looper, 146
holly pit scale, 352
Homadaula anisocentra, 180
Homaledra sabalella, 162
Homalodisca sp., 418
honeylocust plant bug, 404
honeylocust pod gall midge, 466
honeylocust scurfy scale, 390
honeylocust spider mite, 472, 474, 476
honeysuckle aphid, 314
honeysuckle bryobia, 472
Hordnia circellata, 418
Hormaphis hamamelidis, 450
horned oak gall, 440, 442
hornet, European, 258, 494
hornet moth, 258
horntails, 132, 494
Howardia biclavis, 380
huisache girdler, 286
humped green fruitworm, 148
husk maggot, 234
Hyadaphis tataricae, 314
Hyalophora cecropia, 164
Hydria prunivorata, 146; *undulata*, 146
Hylastes sp., 66
Hylesinus sp., 252
Hylobius pales, 56; *radicis*, 56
Hylurgopinus rufipes, 246, 248
Hyphantria cunea, 166
Hypoxylon mammatum (fungus), 276

Icerya purchasi, 338
Ichthyura inclusa, 158
Illinoia liriodendri, 292; *rhododendri*, 308
imperial moth, 26
imported fire ant, 496
imported longhorned weevil, 240
imported willow leaf beetle, 228
inchworms, 142, 144, 146
Incurvariidae, 212
Indian wax scale, 356
introduced basswood thrips, 432
introduced pine sawfly, 16, 18
io moth, 164
Ips sp., 62, 64, 66, 70; *avulsus*, 64; *calligraphus*, 64; *grandicollis*, 64; *pini*, 62, 64; *tridens*, 64
irregular pine scale, 94
Itoplectis conquisitor, 30, 176
ivy aphid, 308
ivy bryobia (mite), 472

jack pine sawfly, 16
Janetiella coloradensis, 116
Janus abbreviatus, 286
Japanese bayberry whitefly, 322
Japanese beetle, 236
Japanese fiorinia scale, 394
Japanese wax scale, 356
Japanese weevil, see twobanded Japanese weevil
Johnella virginiana, 486
jumping oak gall, 440
jumping plant lice, see psyllids
juniper fiorinia scale, 394
juniper gall mite, 122
juniper midge, 46
juniper scale, 106
juniper tip midge, 46
juniper twig girdler, 58
juniper webworm, 30

katydids, 490; broadwinged, 490
Keonolla confluens, 416
Kermesidae, 354, 364, 366
Kunzeana kunzii, 400

lace bugs, 13, 424–428; alder, 426; angulate tingid, 426; apple, 426; Arizona ash tingid, 426; azalea, 424; basswood, 426; birch, 426; buckeye, 426; California Christmas berry tingid, 426, 428; ceanothus tingid, 426, 428; cherry, 426, 428; chokecherry tingid, 426; *Corythucha associata*, 426; *C. bellula*, 426; *C. marmorata*, 426; eastern willow tingid, 426; elm, 426;

hackberry, 426; hawthorn, 426; lantana, 424, 426; *Leptoypha costata*, 426; oak, 426; *Stephanitis rhododendri*, 424; *S. takeyai*, 424; sycamore, 426, 428; walnut, 426; western sycamore, 426; western willow tingid, 426; willow, 426; willow and poplar, 426
lacewings, see predators of arthropods: lacewings
lady beetles, see predators of arthropods: lady beetles
Laetilia coccidivora, 356, 362
Lambdina athasaria, 24; *fiscellaria*, 24; *fiscellaria lugubrosa*, 24; *pellucidaria*, 24
lantana lace bug, 424, 426
larch casebearer, 36, 186
larch looper, 24
larch sawfly, 16, 18
large aspen tortrix, 214, 216
larger elm leaf beetle, 222
larger shothole borer, 246
Lasiocampidae, 168, 170
Laspeyresia cupressana, 60; *pomonella*, 234
latania scale, 376
laurel psyllid, 458
leaf beetles, 222–230, 236, 238, 242, 400; alder leaf, 224; alder flea, 228; *Altica ignita*, 228; aspen leaf, 226; baccharis leaf, 238; *Chrysomela aenicollis*, 226; *C. californica*, 226; *C. crotchi*, 226; *C. scripta*, 226; cottonwood leaf, 226; cranberry rootworm, 242; elm calligrapha, 226; elm leaf, 222; gray willow leaf, 224; imported willow leaf, 228; Japanese, 236; larger elm leaf, 222; *Pyrrhalta viburni*, 224
leaf crumpler, 218
leafcutter bees, 494
leaf-footed bugs, 422–424
leafhoppers, 13, 246, 400, 404, 408, 412–418; aster, 418; *Edwardsiana commisuralis*, 416; *Empoasca* sp., 416; *Erythroneura arta*, 416; *E. basilaris*, 246; *E. gledista*, 416; *E. lawsoniana*, 416; *E. obliqua*, 246; *E. usitata*, 416; *E. ziczac*, 416; *Keonolla confluens*, 416; *Kunzeana kunzii*, 400; *Macropsis fumipennis*, 404; *M. graminea*, 404; *M. ocellata*, 404; maple, 416; potato, 414; redbanded, 418; rhododendron, 418; rose, 412; sharpshooters, 418; six-spotted, 418; *Smilia camelus*, 408; *Typhlocyba ulmi*, 246; white apple, 412, 416; whitebanded elm, 246, 414
leaf miners, 32, 36, 38, 42, 184–210; *Agromyza viridula*, 206; appleleaf trumpet miner, 196; arborvitae, 42; *Argyresthia arceuthobiella*, 42; *A. aureoargentella*, 42; *A. cupressella*, 42; *A. franciscella*, 42; *A. freyella*, 42; *A. libocedrella*, 42; *A. thuiella*, 42; *A. trifasciae*, 42; aspen, 194; aspen blotch, 196; azalea, 202; basswood, 190; beech, 190; birch, 128, 184, 296; birch casebearer, 186; birch-leafmining sawflies, 184; boxwood, 204; *Brachys aeruginosus*, 190; *B. ovatus*, 190; brown hemlock needleminer, 38; *Caloptilia fraxinella*, 196; *Cameraria* spp., 196; *C. ostryarella*, 192; *Coleotechnites apicitripunctella*, 38; *C. canusella*, 40; *C. ducharmei*, 32; *C. juniperella*, 42; *C. macleodi*, 38; *C. piceaella*, 32; *C. resinosae*, 40; *C. thujaella*, 42; *Coptodisca arbutiella*, 194, 208; cypress leaftier, 38; *Ectoedemia platanella*, 196; elm, 186; elm casebearer, 186; *Endothenia albolineana*, 32; *Epinotia nanana*, 32; European alder, 188; *Gelechia panella*, 194; green hemlock, 38; green spruce, 32; gregarious oak, 192; holly, 206, 208; juniper webworm, 30; larch casebearer, 36, 186; lilac, 196; locust, 190; lodgepole needleminer, 40; magnolia, 210; *Marmara arbutiella*, 194; *Messa nana*, 184; Monterey pine needleminer, 40; native holly, 206; orange spruce needleminer, 32; *Paraleucoptera heinrichi*, 198; *Paraphytomyza cornigera*, 194; *Parectopa robiniella*, 190; *Phyllocnistis* sp., 194; *Phytomyza ditmani*, 206; *P. glabricola*, 206, 208; *P. ilicis*, 206, 208; *P. opaca*, 206, 208; *P. verticillata*, 206, 208; *P. vomitoriae*, 206; pine needleminer, 40; pine needle sheathminer, 38; *Profenusa canadensis*, 188; *P. platanae*, 188; resplendent shield bearer, 208; solitary oak, 192; spotted tentiform, 196; *Stenolechia bathrodyas*, 42; tentiform, 196; *Tischeria* spp., 196; tuliptree, 210; tupelo, 208; white fir needleminer, 38; willow flea weevil, 190
leaf-mining sawflies: birch leafminer, 184; birch leafmining, 184; *Fenusa dohrnii*, 188; sycamore leafmining, 188
leaf rollers, 202, 208, 214, 218; *Ancylis* spp., 218; *Archips* spp., 214, 218; azalea leafminer, 202; blackheaded fireworm, 208; *Choristoneura conflictana*, 216, 218; *Croesia semipurpurana*, 172, 218; *Epinotia solandriana*, 218; filbert, 218; fruittree, 172, 202, 214, 218; oak, 214; obliquebanded, 216; orange tortrix, 216; *Pseudosciaphila duplex*, 214; redbanded, 214; *Sparganothis acerivorana*, 216; whiteline, 214
leaf skeletonizer, palm, 162
leaf skeletonizers, see skeletonizers
leaf tiers, 172, 214; *Acrobasis* spp., 212; *Archips* spp., 214, 218; aspen, 214; azalea leafminer, 202; *Croesia albicomana*, 172; cypress, 38; *Episimus tyrius*, 212; large aspen tortrix, 214, 216; oak, 172, 214; olethreutid moths, 212; *Pandemis pyrusana*, 218; *Pyramidobela angelarum*, 214; *Sparganothis* spp., 216
lecanium scales, 354, 364; European fruit, 364; hickory, 364; oak, 364
Lecanodiaspididae, 354
leopard moth, 254
Leperisinus sp., 252
Lepidosaphes sp., 368; *beckii*, 322, 380; *camelliae*, 370; *gloveri*, 380; *ulmi*, 370; *yanagicola*, 388
Leptocoris rubrolineatus, 398; *trivittatus*, 398
Leptoglossus clypealis, 424; *phyllopus*, 422; *zonatus*, 424
Leptoypha costata, 426; *minor*, 426
lesser peachtree borer, 258
lesser snow scale, 390
Leucoma salicis, 158
ligustrum weevil, 244
lilac borer, 260
lilac leafminer, 196
Limax maximus, 498
linden aphid, 302
linden borer, 276
linden looper, 144
Lindingaspis floridana, 376; *rossi*, 376
Lintneri scurfy scale, 390
Lithophane antennata, 148; *innominata*, 148
Litoprosopus futilis, 162
little carpenterworm, 256
live oak woolly leaf gall maker, 440
living beech borer, 282
loblolly pine sawfly, 16
locust borer, 274, 278
locust leafminer, 190
locust midge, 470
locust twig borer, 274
lodgepole needleminer, 40
lodgepole sawfly, 18
lodgepole terminal weevil, 56
longhorned beetles, 276, 278, 280, 282, 284, 286, 288
longhorn grasshopper, 492
Longiloba scurfy scale, 390
Longistigma caryae, 310
long rose gall wasp, 438
longtailed mealybug, 324
loopers, 24, 142; barberry, 144; cypress, 26; hemlock, 24; holley, 146; larch, 24; pine, 24; western hemlock, 24. *See also* cankerworms; caterpillars; spanworms
Lophocampa argentata, 34; *argentata sobrina*, 34; *argentata subalpina*, 34; *caryae*, 160
Lopidea robinea, 190
Lygaeidae, 398
Lygaeus kalmii, 398
Lygocoris communis, 396; *viburni*, 396
Lygus lineolaris, 398
Lymantria dispar, 138, 140, 168
Lymantriidae, 26, 34, 158
Lyonetiidae, 220

Macremphytus tarsatus, 126; *varianus*, 126
Macrodactylus subspinosus, 236; *uniformis*, 236
Macrodiplosis erubescens, 442; *quercusoruca*, 440, 446
Macrohaltica ambiens, 228
Macrophya punctumalbum, 136
Macropsis fumipennis, 404; *graminea*, 404; *ocellata*, 404
Macrosiphum euphorbiae, 300; *liriodendri*, 292; *rosae*, 308
Macrosteles quadrilineatus, 418
madrone shield bearer, 208
Magdalis sp., 54
maggot, husk, 234
Magicicada sp., 490; *septendecim*, 490
magnolia leafminer, 210
magnolia scale, 354, 356
magnolia spider mite, 476
Malacosoma americanum, 168; *californicum pluviale*, 170; *constrictum*, 170; *disstria*, 168, 170; *incurvum*, 170; *tigris*, 170
mammals, 500, 502
manzanita leafgall aphid, 298
manzanita mealybug, 328
maple bladder gall mite, 482
maple callus borer, 260
maple leafcutter, 212
maple leafhopper, 416
maple mealybug, 342
maple petiole borer, 200
maple phenacoccus, 342
maple scurfy scale, 390
maple shoot borers, 200
maple spider mite, 475
maple spindle gall (mite), 482
maple trumpet skeletonizer, 212
mapleworms: greenstriped, 156; orangehumped, 154, 156
Margarodidae, 92, 334, 338, 354
marginal fold gall, 484
Marmara sp., 194, 198; *arbutiella*, 194; *fasciella*, 198
Massospora cicadina (fungus), 490
Matsucoccus acalyptus, 92, 94; *alabamae*, 92; *bisetosus*, 92, 94; *californicus*, 92; *degeneratus*, 92; *eduli*, 92; *fasciculensis*, 92; *gallicolus*, 92; *macrocicatrices*, 92; *matsumurae*, 92; *monophyllae*, 92; *paucicatrices*, 92; *resinosae*, 92; *secretus*, 92; *vexillorum*, 90, 92, 94
Matsucoccus scales: Alabama pine, 92; Canadian pine, 92; needle fascicle, 92; pine blast, 92; pine twig gall, 92; pinyon needle, 92, 94; ponderosa pine twig, 92, 94; Prescott, 90, 92; red pine, 92; sugar pine, 92; sycamore, 334. See also *Matsucoccus*
Mayetiola piceae, 114, 116; *rigidae*, 468
McDaniel spider mite, 476
meadow spittlebug, 86, 420
mealybugs, 88, 324–332, 346; apple, 324; citrus, 328; coconut, 90; Comstock, 326; cypress bark, 332; golden, 90; grape, 88; longtailed, 324; manzanita, 328; maple, 342; noxious bamboo, 330; obscure, 88, 324; *Oracella acuta*, 88; taxus, 88; white, 328
mealybug destroyer, 90
measuring worms, 24, 142–146
Megachile sp., 494
Megacyllene caryae, 274; *robiniae*, 274
Melanaspis coccolobae, 232; *obscura*, 384; *tenebricosa*, 384
Melanocallis caryaefoliae, 292
melon aphid, 308
Membracidae, 404–410
Merhynchites bicolor, 238
Mesolecanium sp., 354; *nigrofasciatum*, 364
Messa nana, 184
Metcalfa pruinosa, 232, 422
Meunieriella aquilonia, 470
mice, 12, 58, 500, 502

microbial parasites of arthropods, *see* parasites of arthropods: microbial
Microcentrum rhombifolium, 490
Microtus pennsylvanicus, 502
Micrutalis calva, 404
midges: juniper, 46; juniper tip, 46; locust, 470; Monterey pine resin, 44; pine needle, 44; pitch, 44; *Prodiplosis morrisi*, 470; resin, 44; rose, 470; toyon berry, 470. *See also* gall midges
milkweed bug, 398
milky disease, 236
millipedes, 12
mimosa webworm, 180
Mindarus sp., 84; *abietinus*, 80
miners: bark, 198; bud, 20, 214. *See also* leaf miners
mining scales, 380
Miridae, 396–404
mites, 12, 98, 118–122, 464, 472–488; *Brevipalpus californicus*, 476; broad, 476; citrus flat, 476; cyclamen, 476; false spider, 476; flat, 476; phaelenopsis, 476; predatory, 486, 490; privet, 476, 482. *See also* eriophyid mites; gall mites; rust mites; spider mites
moles, 12
Monarthropalus buxi, 204; *flavus*, 204
Monocesta coryli, 222
Monterey pine midge, 116
Monterey pine needleminer, 40
Monterey pine resin midge, 44
Monterey pine scale, 94
Monterey pine shoot moth, 48
Monterey pine tip moth, 48
Monterey pine weevil, 54, 56
Mordvilkoja vagabunda, 462
mormon cricket, 492
mossyrose gall wasp, 438
moths: American dagger, 158; American hornet, 258; "Black Hills" pandora, 26; buck, 156; ceanothus stem gall, 438; cecropia, 164; cherry scallop shell, 146; codling, 234; cypress bark, 60; cypress leaftier, 38; cypress tip, 42; Douglas-fir pitch, 72; Douglas-fir tussock, 34; ermine, 148; European pine shoot, 48, 50; giant silkworm, 26, 164; gypsy, 138, 140, 168, 182, 270; hickory tiger, 160; hickory tussock, 160; holly bud, 206; hornet, 258; imperial, 26; io, 164; juniper webworm, 30; larch casebearer, 36; leopard, 254; Monterey pine shoot, 48; Monterey pine tip, 48, 50; Nantucket pine tip, 48, 50; Nevada buck, 156; northern pitch twig, 72; pale tussock, 158; pandora, 26; pine needleminer, 40; pine needle sheathminer, 38; pine tube, 46; pine tussock, 26; pitch, 72; pitch pine tip, 48; rusty tussock, 158; satin, 158; sequoia pitch, 72; silk, 164; silverspotted tiger, 34; southwestern pine tip, 48; spear-marked black, 146; spruce bud, 28; spruce needleminer, 38; sycamore tussock, 160; subtropical pine tip, 50; tiger, 34; tussock, 34, 158; western pine tip, 48; western tussock, 160; whitemarked tussock, 158; winter, 146; Zimmerman pine, 48, 50. *See also* clearwing moths
mountain-ash sawfly, 128, 286
mountain pine beetle, 66
mourningcloak butterfly, 152
mulberry whitefly, 324
mycoplasmalike organism (insect-transmitted cause of elm yellows), 246
Myzus persicae, 300

nabid bugs, 300
Nakahara wax scale, 358
Nalepella tsugifoliae, 122
Nanokermes sp., 364; *pubescens*, 364, 366
Nantucket pine tip moth, 48, 50
narrowwinged cricket, 492, 494
native elm bark beetle, 246, 248
native holly leafminer, 206
Nearctaphis crataegifoliae, 314
Nectria coccinea faginata (fungus), 322
needle fascicle scale, 92
needle miners, 38; brown hemlock, 38; European spruce, 32; green hemlock, 38; lodgepole, 40; Monterey pine, 40; orange spruce, 32; pine, 40; spruce, 32; white fir, 38. *See also* leaf miners
nematodes, 12, 236, 464. *See also* parasites of arthropods: nematodes
Nematus hispidae, 134; *lipovskyi*, 134; *tibialis*, 134
Nemocestes incomptus, 240
Neoaplectana sp., 280; *carpocapsae*, 240, 256
Neoceruraphis viburnicola, 300
Neochlamisus cribripennis, 224; *platani*, 400
Neoclytus acuminatus, 278
Neodiprion abbottii, 18; *burkei*, 18; *excitans*, 16, 18; *hetricki*, 18; *lecontei*, 16, 18; *nanulus nanulus*, 16; *pinetum*, 18; *pinusrigidae*, 18; *pratti banksianae*, 16; *pratti paradoxicus*, 16; *pratti pratti*, 16; *sertifer*, 16, 18; *swainei*, 18; *taedae linearis*, 16; *taedae taedae*, 16; *tsugae*, 18
Neolasioptera brevis, 470
Neolecanium cornuparvum, 354, 356
Neophasia menapia, 24
Neophyllaphis podocarpi, 294
Nepticulidae, 196
Neuroterus decipiens, 440; *saltatorius*, 440, 444
Nevada buck moth, 156
Niesthrea louisianica, 422
nigra scale, 360
Nipaecoccus aurilanatus, 90; *nipae*, 90
nitidulids, 282
Noctuidae, 24, 148, 158
northern cedar bark beetle, 68, 70
northern pine weevil, 56
northern pitch twig moth, 72
Norway maple aphid, 302
Notodontidae, 150, 154, 156
noxious bamboo mealybug, 330
Nuculaspis californica, 100, 108; *tsugae*, 102
Nymphalidae, 152
Nymphalis antiopa, 152; *californica*, 152

oak apple, 440, 446
oak branch borer, 282
oak clearwing moth, 260
oak lace bug, 426
oak leaf blister (fungus), 486
oak leafroller, 214
oak leaftier, 172
oak lecanium (scale), 364
oak ribbed casemaker, 220
oak skeletonizer, 220
oak spider mite, 472, 475
oak treehopper, 406
oak tubemaker, 212
oak twig girdler, 266
oak webworm, 172, 214
oak woolly aphid, 304
oakworms, 270; California, 146; orangestriped, 156; pinkstriped, 156; redhumped, 156
Oberea bimaculata, 262; *delongi*, 286, 288; *myops*, 288; *schaumii*, 288; *tripunctata*, 262
obliquebanded leafroller, 216
Obolodiplosis robiniae, 466, 470
obscure mealybug, 88, 324
obscure root weevil, 240, 242
obscure scale, 384
Ochyromera ligustri, 244
Odontopus calceatus, 210
Odontota dorsalis, 190
Oecanthus angustipennis, 492; *fultoni*, 492; *nigricornis*, 492
Oiketicus abbotii, 176; *townsendi*, 176
Okanagana sp., 490
oleander and milkweed aphid, 312
oleander aphid, 312
oleander caterpillar, 162
oleander scale, 374
Olethreutes cespitana, 202
Olethreutidae, 212, 214
Oligonychus sp., 472; *aceris*, 475; *bicolor*, 475; *ilicis*, 475; *milleri*, 118; *newcomeri*, 475; *platani*, 475; *propetes*, 475; *subnudus*, 118, 120; *ununguis*, 118, 475; *viridis*, 475
Oligotrophus betheli, 46
Oncideres cingulata, 264; *pustulata*, 286
Operophtera bruceata, 146; *brumata*, 146
Oracella acuta, 88
orangehumped mapleworm, 154, 156
orange spruce needleminer, 32
orangestriped oakworm, 156
orange tortrix, 216
Orgyia antiqua, 158; *leucostigma*, 158; *pseudotsugata*, 34; *vetusta*, 160
oriental chestnut gall wasp, 438
Ormenoides venusta, 422
Orthosia hibisci, 148
Ostrinia nubilalis, 286
Otiorhynchus spp., 244; *ovatus*, 240; *rugifrons*, 240; *rugosostriatus*, 240; *sulcatus*, 54, 240
oystershell scale, 370

Pachnaeus litus, 240
Pachylobius picivorus, 56
Pachypsylla sp., 450, 452; *celtidisgemma*, 452; *celtidismamma*, 452; *celtidisvesicula*, 452
Pacific flatheaded borer, 270
Pacific oak twig girdler, 266, 272
Pacific spider mite, 476
Pacific tent caterpillar, 170
painted hickory borer, 274, 278
Paleacrita vernata, 142, 144
pales weevil, 13, 56
pale tussock moth, 158, 160
palmerworm, 8
palm fiorinia scale, 394
palm leafskeletonizer, 162
Palpita flegia, 162
Pamphiliidae, 22
Pamphilius phyllisae, 136
Pandemis limitata, 218; *pyrusana*, 218
pandora moth, 26
Panonychus citri, 472; *ulmi*, 472
Pantomorus cervinus, 230, 240
paper wasp (*Polistes* sp.), 260
Parabemisia myricae, 332
Paraclemensia acerifoliella, 212
Paradiplosis tumifex, 116
Paraleucoptera heinrichi, 198
Parallelodiplosis subtruncata, 462
Paranthrene asilipennis, 260; *dollii* 260; *robiniae*, 256; *simulans*, 254, 260; *tabaniformis*, 254, 260
Paraphytomyza cornigera, 194
Paraprociphilus tesselatus, 304, 306
Parasa indetermina, 164
Parasaissetia sp., 354; *nigra*, 360
parasites of arthropods
—insects: *Agathis pumila*, 36; *Allotropa burrelli*, 326; *A. convexifrons*, 326; *A. utilis*, 324; *Anagrus armatus*, 412; *A. epos*, 416; *Aneristus youngi*, 356; *Anicetus toumeyellae*, 356, 362; *Apanteles* sp., 24, 216, 262; *A. harrisinae*, 164; *A. hyphantriae*, 166; *Aphelinus mali*, 306; *Aphycus ceroplastis*, 356; *Aphytis* sp., 368, 386; *A. lignanensis*, 378; *A. melinus*, 378; *Aptesis segnis*, 188; *Archytas metallicas*, 150; *Aspidiotiphagus citrinus*, 102, 104, 392; *Azya luteipes*, 356; *Bassus pumilus*, 36; *Bessa selecta*, 136; *Blastothrix longipennis*, 354; *Blepharipa pratensis*, 138; *Bracon gelechiae*, 30; *B. xanthonotus*, 170; *Carcelia yahensis*, 34; *Catoloccus aeneoviridis*, 30; *Chirotica thyridopteryx*, 176; *Chrysocharis laricinellae*, 36; *Clausenia purpurea*, 326; *Closterocerus tricinctus*, 190; *Coccophagus flavifrons*, 362; *C. fraternus*, 96; *C. immaculatus*, 336; *C. lecanii*, 364; *C. lycimnia*, 344, 362, 364; *Coccygomimus aequalis*, 30; *Comperiella bifasciata*, 106, 378; *Compsilura concinnata*, 158; *Cryptus albitarsis*, 24; *Encarsia braziliensis*, 322; *E. formosa*, 320; *E. haldemani*, 322; Encyrtidae, 366; *Ephialtes sanguineipes*, 166; *Eretmocerus haldemani*, 322; *Erynniopsis antennae*, 222; *Eupteromalus peregrinus*, 158; *Eurytoma contractura*, 468; *E. semivenae*, 452; *Glypta* sp., 52; *Grypocentrus albipes*, 184; *Habrolepis dalmanni*, 352; *Helicobia rapax*, 280; *Heterorhabditis heliothidis*, 240; *Homalotylus* sp., 366; *Hormius* spp., 216;

Hypercteina aldrichi, 236; *Idechthis* sp., 252; *Itoplectis conquisitor*, 30, 176, 214; *Lathrolestes nigricollis*, 184; *Lespesia ciliata*, 164; *Lissonota* sp., 262; *Lypha setifacies*, 28; *Lysiphlebus testaceipes*, 312; *Megarhyssa macrurus macrurus*, 494; *Mesostenus thoracicus*, 24, 252; *Metaphycus californicus*, 354; *M. flavus*, 362; *Meteorus hyphantriae*, 166; *M. pulchricornis*, 182; *M. trachynotus*, 28; *Microterys clauseni*, 356; *Nemorilla pyste*, 218; *Omotoma fumiferanae*, 216; *Ooencyrtus* sp., 146; *O. kuvanae*, 138; *Pentacnemus bucculatricis*, 42; *Phaeogenes ater*, 262; *P. gilvilabris*, 214; *P. hariolus*, 28; *Phasganophora sulcata*, 272; *Phorocera* sp., 24; *P. signata*, 262; *Physcus varicornis*, 100; *Platygaster* sp., 436; *Prospaltella* sp., 100, 386; *P. berlesei*, 392; *P. perniciosi*, 378; *Psyllaephagus pachypsyllae*, 452; *Sarcophaga aldrichi*, 168; *Schizonotus sieboldi*, 228; *Scutellista cyanea*, 356; *Spilochalcis odontotae*, 190; *Steinernema feltiae*, 240, 256; *Tachinomyia similis*, 158; *Telenomus* sp., 146, 154; *T. bifidus*, 166; *Tetrastichus* sp., 30, 44, 96; *T. brevistigma*, 222; *T. gallerucae*, 222; *T. malacosomae*, 170; *Tiphia popilliavora*, 236; *T. vernalis*, 236; *Torymus pachypsyllae*, 452; *Trichogramma minutum*, 130, 188, 214; *T. odontotae*, 190; *Trichogrammatomyia* sp., 214; *Trioxys* sp., 292
—microbial: *Aschersonia aleyrodis* (fungus), 322; *Bacillus popilliae* (milky disease; bacterium), 236; *B. thuringiensis* (bacterium), 138, 142, 146, 152, 168, 176, 216, 218; *Beauveria* sp. (fungus), 262; *Empusa fresenii* (fungus), 298; *E. maimaiga* (fungus), 138; *Entomophaga aulicae* (fungus), 138; *E. maimaiga* (fungus), 138; *Entomophthora* sp. (fungus), 168; *E. aphrophora* (fungus), 86; *E. fumosa* (fungus), 324; *Erynia neoaphidis* (fungus), 84; *Herpetomonas* sp. (protozoan), 130; *Massospora cicadina* (fungus), 490; *Pleistophora schubergi* (protozoan), 24; polyhedrosis virus, 146; protozoans, 236; *Xenorhabdus nematophilus* (bacterium), 256; *Zoophthora aphidis* (fungus), 84; *Z. radicans* (fungus), 414
—nematodes: *Heterorhabditis heliothidis*, 240; *Neoaplectana* sp., 280; *N. carpocapsae*, 256; *Steinernema feltiae*, 256
Parectopa robiniella, 190
Parlatoria pittospori, 110
Parthenolecanium sp., 354, 360, 368; *corni*, 98, 364; *fletcheri*, 98, 364; *persicae*, 364; *pruinosum*, 354; *quercifex*, 364
peach bark beetle, 250
peachtree borer, 258, 260
Pealius azaleae, 318
pearleaf blister mite, 486
pear plant bug, 396
pear psylla, 290
pear root aphid, 464
pear sawfly, 130
pear slug, 130
pear thrips, 432
pecan borer, 262
pecan cigar casebearer, 186
pecan leaf casebearer, 212
pecan leaf scorch mite, 474
Pemphigus betae, 460; *populicaulis*, 460; *populitransversus*, 460
peony scale, 372
pepper tree psyllid, 458
Periclista media, 136
periodical cicadas, 13, 490
Periphyllus sp., 302; *lyropictus*, 302
Periploca ceanothiella, 438; *nigra*, 58
Petrova sp., 48, 72; *albicapitana*, 72; *comstockiana*, 72
phaelenopsis mite, 476
Phaeogenes hariolus, 28
Phaeoura mexicanaria, 24
Phasmatidae, 492
Phenacoccus acericola, 342, 346; *aceris*, 324
pheromones, 11, 48, 56, 64, 66, 138, 148, 158, 172, 174, 176, 214, 258, 260, 262, 326, 386, 392
Phigalia titea, 146
Philaenus spumarius, 86, 420
Phloeosinus sp., 66, 70; *canadensis*, 68, 70; *cristatus*, 70; *cupressi*, 70; *dentatus*, 68, 70; *taxodii*, 68, 70
Phloeotribus liminaris, 250
Phomopsis sp. (fungus), 46
Phorocera sp. (fungus), 24
Phryganidia californica, 146
Phyllaphis fagi, 296
Phyllobius intrusus, 240, 244; *oblongus*, 244
Phyllocnistis liriodendronella, 196; *populiella*, 194
Phyllocolpa parva, 468
Phyllocoptes exochordae, 482
Phyllonorycter alnicolella, 196; *blancardella*, 196; *crataegella*, 196; *felinella*, 196; *incanella*, 196; *lucetiella*, 196; *salicifoliella*, 196; *tremuloidiella*, 196; *trinotella*, 196
Phylloxera sp., 460; *caryaecaulis*, 460; *devastatrix*, 460; *notabilis*, 460; *russellae*, 460
Physokermes hemicryphus, 94, 96; *insignicola*, 94; *piceae*, 94, 96
Phytobia amelanchieris, 198; *pruinosa*, 198; *setosa*, 198
Phytocoptella avellanae, 480
Phytomyza ditmani, 206; *glabricola*, 206, 208; *ilicicola*, 206; *ilicis*, 206, 208; *opacae*, 206, 208; *verticillatae*, 206, 208; *vomitoriae*, 206
Phytonemus pallidus, 476
Phytoptus avellanae, 480; *emarginatae*, 482; *pyri*, 486; *tiliae*, 486
Phytoseiidae, 476, 482
Phyxelis melanothrix, 240
Picoides pubescens (woodpecker), 286
pigeon tremex, 132, 494
Pikonema alaskensis, 18
pine bark adelgid, 76, 78
pine bast scale, 92
pine butterfly, 24
pine engraver beetle, 62
pine false webworm, 22
pine leaf adelgid, 76, 78
pine looper, 24
pine needle midge, 44
pine needleminer, 40
pine needle scale, 100, 108
pine needle sheathminer, 38
pine reproduction weevil, 60
pine root collar weevil, 56
pine scale, 108
pine shoot beetle, 64
pine shoot borer, 52
pine shoot moth, 48
pine spittlebug, 86
pine tip moth, 48
pine tortoise scale, 96
pine tube moth, 46
pine tussock moth, 26
pine twig gall scale, 92
Pineus boerneri, 76; *coloradensis*, 76; *floccus*, 76, 112; *pinifoliae*, 76, 78; *similis*, 76, 112; *strobi*, 76; *sylvestris*, 76
pine-webbing sawflies, 18, 22
pine webworm, 22
pinkstriped oakworm, 156
Pinnaspis aspidistrae, 390; *strachani*, 390
Pinyonia edulicola, 116
pinyon needle scale, 92, 94
Pissodes nemorensis, 54, 56; *radiatae*, 56; *schwarzi*, 56; *strobi*, 54; *terminalis*, 54, 56
pitch-eating weevil, 56
pitch mass borer, 72, 260
pitch midges, 44
pitch moths, 72
pitch pine sawfly, 18
pitch pine tip moth, 48
pitch twig moth, 72
pit-making pittosporum scale, 348
pit scales, 310, 348, 352; *Asterolecanium minus*, 352; *A. quercicola*, 352; bamboo, 350; golden oak, 352; holly pit, 352; pit-making pittosporum, 348
pitted ambrosia beetle, 250
Pityogenes sp., 62, 66; *hopkinsi*, 62
Pityophthorus sp., 66; *orarius*, 60
Plagiodera versicolora, 228
Plagiognathus albatus, 400; *punctatipes*, 404
Planococcus citri, 328
plant bugs, 13, 396, 398, 400, 402, 404, 422; ash, 402; boxelder, 398; fourlined, 396; honeylocust, 404; *Lygocoris viburni*, 396; pear, 396; *Plagiognathus punctatipes*, 404; sycamore, 400; tarnished, 398; *Tropidosteptes amoenus*, 402; *T. brooksi*, 402; *T. illitus*, 402; *T. pacificus*, 402; western boxelder, 398
planthoppers, 246, 422; *Acanalonia conica*, 422; *Anormenis chloris*, 422; flatid, 232, 422; *Metcalfa pruinosa*, 232, 422; *Ormenoides venusta*, 422
Plataculus pyramidicus, 484
platanus spider mite, 475
Platycotis vittata, 406, 408
Platygaster sp., 436
Platyphytoptus sabinianae, 122
Plectrodera scalator, 276, 278
Pleistophora schubergi, 24
Pleroneura spp., 20; *brunneicornis*, 20; *lutea*, 20
podocarpus aphid, 294
Podosesia aureocincta, 260–62; *syringae*, 260–62
Poecilocapsus lineatus, 396
Polistes sp., 260
polyhedrosis virus, 146, 168
Polyphagotarsonemus latus, 476
ponderosa pine bark borer, 276
ponderosa pine tip moth, 48
ponderosa pine twig scale, 92, 94
Pontania californica, 468; *proxima*, 468
Popillia japonica, 236
poplar-and-willow borer, 268
poplar borer, 284
poplar clearwing borer, 260
poplar petiolegall aphid, 460
poplar spider mite, 474
poplar tentmaker, 156, 158
poplar twig gall fly, 468
poplar vagabond aphid, 462
potato aphid, 300
potato leafhopper, 414
powdery mildew fungus, 478
powdery pine needle aphid, 84
predators of arthropods
—ants, 496; carpenter, 282, 496; *Crematogaster* spp., 362; *C. clara*, 260
—beetles: *Calosoma sycophanta*, 138; *Conotrachelus buchanani*, 452; *Cybocephalus nipponicus*, 388; *Euclemensia bassettella*, 366; *Plochius amandus*, 162; sapbeetles, 282; *Tenebroides corticalis*, 252. *See also subentry* lady beetles
—birds (woodpeckers), 12, 68, 126, 252, 254, 258, 260, 272, 278, 282–286, 500
—flies, 24, 36; *Leucopis obscura*, 74; syrphid flies, 300, 302, 306
—lacewings, 92, 300–306, 326, 360, 412, 434, 476; *Chrysopa californica*, 412; *C. rufilabris*, 326; *Hemerobius pacificus*, 412; *H. stigmaterus*, 326
—lady beetles, 74, 90, 92, 104, 288, 300–306, 338, 340, 350, 356–362, 366, 370, 380, 386, 392, 434, 476; *Aphidecta obliterata*, 74; *Azya luteipes*, 356; *Chilocorus kuwanae*, 388; *C. stigma* (twicestabbed lady beetle), 104, 362, 366, 370, 382, 386, 392; *Cleis picta*, 92; *Cryptolaemus montrouzieri*, 90; *Diomus pumilio*, 288; *Exochomus childreni*, 392; *Harmonia conformis*, 288; *Hyperaspis bipunctata*, 362; *H. signata*, 362; *Lindorus lophanthae*, 392; *Orcus chalybeus*, 356; *Rhodolia cardinalis*, 338; *Stethorus* sp., 476
—*Laetilia coccidivora* (moth), 356, 362
—mirids: *Deraeocoris aphidiphagus*, 306; *D. nitenatus*, 306; *Lopidea robinea*, 190
—phytoseiid mites, 476, 482, 486; *Amblyseius fallacis*, 476; *Pyemotes tritii*, 30; *P. ventricosus*, 270; *Seius* sp., 486
—true bugs: *Arilus cristatus*, 190; nabid bugs, 200; *Xenotracheliella inimica*, 92. *See also subentry* mirids
—sphecid wasp, 492

Prescott scale, 90, 92, 94
Prionoxystus macmurtrei, 256; *robiniae*, 256
prionids: California prionus, 280; tilehorned prionus, 280
Prionus californicus, 280; *imbricornis*, 280; *laticollis*, 280
Pristiphora erichsonii, 18; *geniculata*, 128
privet mite, 476, 482
privet rust mite, 432, 480
privet thrips, 432
Prociphilus sp., 304; *americanus*, 304; *caryae*, 304
Prodiplosis morrisi, 470
Profenusa canadensis, 188; *platanae*, 188; *thomsoni*, 184
Prosapia bicincta, 420; *bicincta ignipecta*, 420
Proteoteras aesculana, 200; *moffatiana*, 200; *willingana*, 200
Protopulvinaria sp., 354
protozoans, see parasites of arthropods: microbial
prunicola scale, 392
Pseudaonidia duplex, 372; *paeoniae*, 372, 380; *trilobitiformis*, 376
Pseudaulacaspis cockerelli, 374; *pentagona*, 392; *prunicola*, 392
Pseudococcidae, 88, 90, 324–330, 346
Pseudococcus affinis, 88, 324; *comstocki*, 326; *longispinus*, 324; *maritimus*, 88
Pseudophilippia sp., 354; *quaintancii*, 88
Pseudosciaphila duplex, 214
Psilocoris cryptolechiella, 218
Psychidae, 176
psylla, pear, 290
Psylla alba, 290; *americana*, 290; *annulata*, 290; *buxi*, 290; *floccosa*, 306; *pyricola*, 290; *uncatoides*, 288
Psyllidae, see psyllids
psyllids, 288, 290, 450–458; acacia, 288; boxwood, 290; *Calophya californica*, 290; *C. flavida*, 290; *C. nigripennis*, 290; *C. rubra*, 458; *C. triozomima*, 290; *Euphalerus vermiculosus*, 458; *Gyropsylla ilicis*, 454; hackberry, 290, 452; hackberry bud gall, 452; laurel, 458; *Pachypsylla* spp., 450, 452; pear, 290; *Psylla alba*, 290; *P. americana*, 290; *P. annulata*, 290; pepper tree, 458; *Trioza eugenia*, 458; *T. magnoliae*, 456
Pterocomma smithiae, 310
Ptycholoma peritana, 202
Pulvinaria sp., 354; *acericola*, 340, 342, 346; *ericicola*, 342, 346; *floccifera*, 344; *innumerabilis*, 340, 342, 346
Pupillidae (snails), 498
purple mite, 472
purple scale, 322, 380
Putnam scale, 372
Puto albicans, 328; *arctostaphyli*, 328
Pyemotes tritici, 30; *ventricosus*, 270
Pyralidae, 22, 166, 182
pyralid moths, 166, 182
Pyramidobela angelarum, 214
Pyrrhalta decora decora, 224; *luteola*, 222; *viburni*, 224

Quadraspidiotus juglansregiae, 386; *perniciosus*, 386
Quernaspis quercus, 390

rabbits, 12, 500, 502
raspberry bud weevil, 240
redbanded leafhopper, 418
redbanded leafroller, 214
redbanded thrips, 432
red-breasted sapsucker, 500
redheaded ash borer, 278
redheaded pine sawfly, 16, 18
redhumped caterpillar, 156
redhumped oakworm, 156
red imported fire ant, 496
red oak borer, 282
red pine sawfly, 16, 18
red pine scale, 92
red squirrel, 500
red turpentine beetle, 62, 66, 72
red wax scale, 358
reproduction weevil, 60
resin midges, 44. *See also* gall midges; midges
resplendent shield bearer, 208
Resseliella clavula, 436; *liriodendri*, 462; *pinifoliae*, 44
Rhabdophaga strobiloides, 468
Rhadopterus picipes, 242
Rhagoletis sp., 234
Rheumaptera hastata, 146
Rhizotrogus majalis, 236
rhododendron borer, 258
rhododendron gall midge, 470
rhododendron leafhopper, 418
rhododendron stem borer, 288
rhododendron whitefly, 318
Rhopalidae, 398, 422
Rhopalosiphum fitchii, 292
Rhopobota naevana, 206, 208
Rhyacionia buoliana, 48, 50; *bushnelli*, 48; *frustrana*, 48, 50; *montana*, 48; *neomexicana*, 48; *pasadenana*, 48; *rigidana*, 48; *subtropica*, 50; *zozana*, 48
Rhynchaenus ephippiatus, 190; *rufipes*, 190
Rhynchothrips ilex, 434
robin, 154
rodents, 12, 16, 500, 502. *See also* mice
roly-poly galls, 448
Romalea guttata, 490
root borers, 280
root weevils, 240–244; arborvitae weevil, 240, 244; Asiatic oak weevil, 240, 244; black vine weevil, 54, 240, 242; *Calomycterus setarius*, 240, 244; obscure root weevil, 240, 242; *Otiorhynchus rugifrons*, 240; *O. rugosostriatus*, 240; twobanded Japanese weevil, 240, 244; whitefringed beetle, 280; woods weevil, 240
rootworm, cranberry, 242
rose aphid, 308
rose chafer, 236
rose curculio, 238
rose leafhopper, 412
rose midge, 470
rose sawflies, 132
rose scale, 366
roseslug, 132
rosy apple aphid, 296
roundheaded appletree borer, 278
roundheaded borers, 264, 266, 274–288; *Acanthocinus nodosus*, 276; *A. obsoletus*, 276; *A. spectabilis*, 276; cottonwood, 276, 278, 280; dogwood twig, 262, 288; *Elaphidionoides parallelus*, 264; hickory, 282; linden, 276; living beech, 282; locust, 274, 278; oak branch, 282; *Oberea bimaculata*, 262; *O. delongi*, 286; *O. schaumii*, 288; painted hickory, 274, 278; ponderosa pine bark, 276; poplar, 284; redheaded ash, 278; red oak, 282; rhododendron stem, 288; roundheaded appletree, 278; roundheaded oak twig, 266; *Saperda discoidea*, 276; *S. inornata*, 276, 278; *S. lateralis*, 276; *S. populnea*, 276, 278; *S. puncticollis*, 276; sugar maple, 278; thornlimb, 276; white oak, 282, 284
roundheaded oak twig borer, 266
rust mites, 122, 474, 478–480; hemlock, 122; privet, 432, 480
rusty tussock moth, 158, 160

saddled prominent, 154
Saissetia sp., 354; *coffeae*, 360; *miranda*, 358; *neglecta*, 358; *oleae*, 358
San Jose scale, 386
sap beetles, 282
Saperda spp., 254; *calcarata*, 284; *candida*, 278; *discoidea*, 276; *fayi*, 276; *inornata*, 276, 278; *lateralis*, 276; *populnea*, 276, 278; *puncticollis*, 276; *vestita*, 276
sapsuckers, *see* woodpeckers
Saratoga spittlebug, 86
Sarucallis kahawaluokalani, 292
sassafras weevil, 210
satin moth, 158
Saturniidae, 26, 156, 164
sawflies, 13, 16–24, 126, 136, 184–188, 200, 286, 468; alder woolly, 136; birch, 128; blackheaded ash, 134; brownheaded ash, 134; butternut woollyworm, 136; *Caulocampus acericaulis*, 200; dogwood, 126; dusky birch, 128; elm, 136; *Euura* sp., 286, 468; *Macremphytus tarsatus*, 126; *Macrophya punctumalbum*, 136; mountain-ash, 128, 286; *Nematus hispidae*, 134; *N. tibialis*, 134; *Profenusa canadensis*, 188; *P. platanae*, 188; striped alder, 136; willow shoot, 286. *See also* conifer sawflies; leaf-mining sawflies; "slug" sawflies
scale insects: 88–110, 232, 332–394; azalea bark, 336; beech, 332; cottony cypress, 332; cypress bark, 332; *Eriococcus araucariae*, 90; *E. borealis*, 336; European elm, 368; kermes, 364, 366. *See also* armored scales; *Matsucoccus* scales; pit scales; soft scales
Scaphoideus luteolus, 246, 414
Scarabaeidae, 236
scarlet oak sawfly, 130
Schizolachnus piniradiatae, 84
Schizotetranychus sp., 472; *celarius*, 474; *garmani*, 474; *schizopus*, 474; *spireafolia*, 474
Schizura concinna, 150, 156; *ipomoeae*, 156; *unicornis*, 156
Schoene spider mite, 476
Sciopithes obscurus, 240, 242
Sciurus (squirrels): *aberti*, 500; *carolinensis*, 500–502
Sclerroderris canker (fungus), 56
Scolytidae, 60, 62, 64, 66, 68, 70, 248, 250
Scolytus sp., 70; *mali*, 246; *multistriatus*, 246, 248; *quadrispinosus*, 250; *rugulosus*, 250
scurfy scales, 368, 390
Scythropus sp., 54, 58; *californicus*, 58
seagrape borer, 232
seagrape gall midge, 232
Seiridium cardinale (fungus), 60
Seius sp., 486
Selenothrips rubrocinctus, 432
semiochemicals, 13. *See also* pheromones
Semiothisa sexmaculata, 24
Septobasidium pinicola (fungus), 92
sequoia pitch moth, 72
Sesia apiformis, 258; *tibialis*, 258, 260
Sesiidae, 72, 254, 258
Setoptus jonesi, 122
sharpshooters, 418; blue-green, 418
sheathminer, pine needle, 38
shield bearers, 194, 208
shoot borers, *see* tip miners/shoot borers
shothole borer, 250
silk moths, 164
silverspotted skipper, 148
silverspotted tiger moth, 34
Siphoninus phillyreae, 322
Siricidae, 494
Sitka spruce weevil, 54
sixlined ips, 64
six-spotted leafhopper, 418
sixspotted mite, 474
skeletonizers, 1, 164, 212–218; *Acrobasis* spp., 212; apple-and-thorn, 216; birch, 220; *Bucculatrix albertiella*, 220; *B. pomifoliella*, 220; grapeleaf, 164; leaf beetles, 222–228; maple trumpet, 212; oak, 220; pecan leaf casebearer, 212; *Sparganothus* spp., 216; western grapeleaf, 164
skipper, silverspotted, 148
"slug" sawflies, 130, 132; bristly roseslug, 132; curled rose sawfly, 132; pear sawfly, 130; roseslug, 132; scarlet oak sawfly, 130
slugs/snails, 12, 13, 130, 498; banded wood, 498; brown garden, 498; brown-lipped, 498; gray garden, 498; spotted garden, 498; tree-climbing, 498; white garden, 498
small southern pine engraver, 63, 64
smaller European elm bark beetle, 246, 248
small milkweed bug, 398
small southern pine engraver, 64
Smilia camelus, 408

snailcase bagworm, 178
snails, *see* slugs/snails
snowball aphid, 300
snow scales, 390
snowy tree cricket, 492
soft scales: arborvitae soft, 98; barnacle, 356; black, 358; brachyuran wax, 358; buttonbush, 362; calico, 354; candle wax, 358; Caribbean black, 358; Chinese wax, 356, 362; cottony azalea, 342, 346; cottony camellia, 344; cottony cushion, 338; cottony maple, 340, 342, 346; cottony maple leaf, 340, 342, 346; cottony taxus, 344; desert wax, 358; Duges wax, 358; European fruit lecanium, 98, 354, 364; European peach, 364; Fletcher, 98, 364; Florida wax, 356; frosted, 354, 364; globose, 366; hemispherical, 360; hickory lecanium, 364; Indian wax, 356; irregular pine, 94; Japanese wax, 356; lecanium, 354, 364; magnolia, 354, 356; Monterey pine, 94; Nakahara wax, 358; nigra, 360; oak lecanium, 364; pine tortoise, 96; red wax, 358; spruce bud, 96; striped pine, 96; terrapin, 364; tortoise wax, 358; tuliptree, 362; Virginia pine, 96; wax, 356; woolly pine, 88
Solenopsis invicta, 496
solitary oak leafminer, 192
Sonoran tent caterpillar, 170
sourgum scurfy scale, 390
southern cypress bark beetle, 70
southern pine beetle, 64
southern red mite, 475, 476
southwestern pine tip moth, 48
southwestern tent caterpillar, 170
spanworms, 142; Bruce, 146; elm, 144
Sparganothis acerivorana, 216; *directana,* 216; *sulfureana,* 216
spear-marked black moth, 146
speckled green fruitworm, 148
Sphaerolecanium sp., 354; *prunastri,* 366
Sphaerotheca phytoptophila (fungus), 478
sphinx caterpillars, 164
Sphinx drupiferarum, 164
Sphyrapicus varius ruber, 500; *varius varius,* 500
spider mites: 118, 120, 234, 424, 472–476; *Aponychus spinosus,* 474; bamboo spider, 474; boxwood, 475, 476; *Bryobia arborea,* 472; *B. borealis,* 472; *B. kissophila,* 472; *B. lonicerae,* 472; *B. ribis,* 472; *B. ulmophila,* 472; carmine spider, 476; citrus red, 472; clover, 472; *Eotetranychus lewisi,* 474; *E. querci,* 474; *E. tiliarium,* 474; *E. uncatus,* 474; European red, 472; fourspotted spider, 476; Garman spider, 474; honeylocust spider, 472, 474, 476; magnolia spider, 476; maple spider, 475; McDaniel spider, 476; oak spider, 472, 475; *Oligonychus milleri,* 118; *O. newcomeri,* 475; *O. propetes,* 474; *O. subnudus,* 118, 120; *O. viridis,* 475; Pacific spider, 476; pecan leaf scorch, 474; platanus spider, 475; poplar spider, 474; purple, 472; *Schizotetranychus schizopus,* 474; *S. spireafolia,* 474; Schoene spider, 476; sixspotted, 474; southern red, 475, 476; spruce, 98, 118, 120; sycamore spider, 475; *Tetranychus homorus,* 476; *T. hydrangeae,* 476; *T. yuccae,* 476; twospotted spider, 476; willow spider, 474; yellow spider, 474; Yuma spider, 474
Spilonota ocellana, 218
spindlegalls, 212
spined turban galls, 448
spiny elm caterpillar, 152
spiny oakworm, 156
spiny rose gall wasp, 438
spiny witch-hazel gall aphid, 296
spirea aphid, 298
spittlebugs, 86, 420; alder, 420; *Aphrophora canadensis,* 86; *A. permutata,* 86; *Clastoptera arborina,* 86; *C. arizonana,* 86, 420; *C. juniperina,* 86; *C. undulata,* 86; dogwood, 420; meadow, 86, 420; pine, 86; *Prosapia bicincta,* 420; *P. bicincta ignipecta,* 420; Saratoga, 86; western, 86
spotted garden slug, 498
spotted loblolly pine sawfly, 16
spotted pine aphid, 84
spotted tentiform leafminer, 196
spotted tussock moth, 160
spring cankerworm, 142, 144
spruce aphid, 82
spruce bud moth, 28
spruce bud scale, 96
spruce budworm, 28, 500
spruce gall adelgid, 76, 112, 114
spruce needleminer, 32
spruce spider mite, 98, 118, 120, 475
squirrels, 500; gray, 500; red, 500; tassel-eared, 500
Stanford whitefly, 320
Stegophylla essigi, 304; *quercicola,* 304; *quercifolia,* 304
Steinernema feltiae (nematode), 240, 256
Steneotarsonemus pallidus, 476
Stenolechia bathrodyas, 42
Stephanitis pyrioides, 424; *rhododendri,* 424; *takeyai,* 424
Stethorus sp., 476
stick insects, 492–494
Stictocephala bisonia, 406
stinging rose caterpillar, 164
Stomacoccus platani, 334
strawberry root weevil, 240
striped alder sawfly, 136
striped pine scale, 96
Styloxus fulleri, 266
Styrax scurfy scale, 390
subtropical pine tip moth, 50
sugar maple borer, 276, 278
sugar pine scale, 92
Sumitrosis rosea, 190
Susana cupressi, 18
Swaine jack pine sawfly, 18
sweetgum scale, 382
sweetpotato whitefly, 322
sycamore lace bug, 426, 428
sycamore leaf beetle, 400
sycamore leafmining sawfly, 188
sycamore plant bug, 400
sycamore scale, 334
sycamore scurfy scale, 390
sycamore tussock moth, 160
Sylvilagus sp. (rabbit), 502
Symmerista canicosta, 156; *leucitys,* 154, 156
Synanthedon acerni, 260; *acerrubri,* 260; *decipiens,* 260; *exitiosa,* 258, 260; *fatifera,* 260; *fulvipes,* 260; *novaroensis,* 72; *pictipes,* 258, 260; *pini,* 72, 260; *pyri,* 258, 260; *resplendens,* 258; *rhododendri,* 258; *rubrofascia,* 260; *scitula,* 260, 262; *sequoiae,* 72; *viburni,* 260
Syntomeida epilais jucundissima, 162

Taeniothrips inconsequens, 432
Takecallis arundicolens, 312; *arundinariae,* 312
Tamalia coweni, 298
Tamiasciurus hudsonicus, 500
Taphrina coerulescens (fungus), 486
tarnished plant bug, 398
Tarsonemidae, 476
tassel-eared squirrel, 500
tawny emperor butterfly, 152
Taxodiomyia cupressiananassa, 118
Taxomyia taxi, 478
taxus bud mite, 122, 478
taxus mealybug, 88
taxus weevil, 240
tea scale, 394
Teleonemia scrupulosa, 424, 426
tent caterpillars, 168, 170, 270, 500
Tenthredinidae, 16, 17, 128–136, 184–188, 200, 468
tentiform leaf miner, 196
tentmaker, poplar, 156, 158
Tenuipalpidae, 476, 482
Tenuipalpus pacificus, 476
terrapin scale, 364
Tethida cordigera, 134
Tetraleurodes sp., 322; *acacia,* 320; *mori,* 324; *stanfordi,* 320
Tetralopha robustella, 22
Tetranychidae, 118, 120, 472–476
Tetranychus bimaculatus, 476; *canadensis,* 476; *cinnarbarinus,* 476; *homorus,* 476; *hydrangeae,* 476; *magnoliae,* 476; *mcdanieli,* 476; *pacificus,* 476; *schoenei,* 476; *telarius,* 476; *urticae,* 476; *yuccae,* 476
Tetraspinus pyramidicus, 484
Tetrastichus sp., 30, 44, 96; *brevistigma,* 222; *gallerucae,* 222; *malacosomae,* 170
Tettigoniidae, 492
Texas leafcutting ant, 496
Theba pisana, 498
Thecodiplosis piniradiatae, 116; *piniresinosae,* 44; *pinirigidae,* 116; *resinosae,* 116
thornbug, 408
thorn gall, 452
thornlimb borer, 276
Thripidae, *see* thrips
thrips, 13, 432; Cuban laurel, 434; flower, 432; *Frankliniella* sp., 434; greenhouse, 430; pear, 432; privet, 432; redbanded, 432; *Thrips calcaratus,* 432; toyon, 434; western flower, 432
Thrips calcaratus, 432
Thyridopteryx ephemeraeformis, 176, 178
Thysanopyga intractata, 146
Tibicen cinctifer, 490; *linnei,* 490
tiger moths, 34; hickory, 160
tilehorned prionus, 280
Timema californica, 494
Tingidae, *see* lace bugs
tingids, *see* lace bugs
Tinocallis kahawaluokalani, 292
tip miners/shoot borers, 20; *Argyresthia libocedrella,* 42; *A. trifascia,* 42; cypress, 42; European corn borer, 286; European pine shoot moth, 48; Monterey pine tip moth, 48; Nantucket pine tip moth, 48; pine shoot borers, 52; pitch pine tip moth, 48; *Pleroneura* spp., 20; *Proteus aesculana,* 200; *P. moffatiana,* 200; southwestern pine tip moth, 48; western shoot borer, 52; white pine weevil, 54; willow shoot sawfly, 286
tip moths, cypress, 42
Tischeria admirabilis, 196; *badiiella,* 196; *ceanothi,* 196; *citrinipennella,* 196; *discreta,* 196; *immaculata,* 196; *malifoliella,* 196; *mediostriata,* 196
Tischeriidae, 196
Tomicus piniperda, 64
Tomostethus multicinctus, 134
tortoise wax scale, 358
Tortricidae, 28, 32, 38, 72, 172, 206, 212–218, 274
tortrixes: large aspen, 214, 216; orange, 216
Torymus pachypsyllae, 452
Toumeyella sp., 354; *cerifera,* 362; *liriodendri,* 362; *parvicornis,* 96; *pini,* 96; *pinicola,* 94; *virginiana,* 96
Toxoptera aurantii, 232, 302
toyon berry midge, 470
toyon thrips, 434
tree-climbing snails, 498
tree crickets, 13, 492, 494
treehoppers, 404–408; buffalo, 406, 408; *Glossonotus acuminatus,* 406; *Micrutalis calva,* 404; oak, 406; thornbug, 406, 408; twomarked, 410
Tremex columba, 494
Trialeurodes acacia, 320; *merlini,* 322; *stanfordi,* 320; *vaporariorum,* 320, 322; *vittata,* 322
trilobe scale, 376
Trioza alacris, 458; *magnoliae,* 456
Trirhabda bacharidis, 238; *flavolimbata,* 238; *nitidicollis,* 238; *pilosa,* 238
Trisetacus camponodus, 122; *chamaecypari,* 122; *ehmanni,* 122; *gemmavitians,* 122; *quadrisetus,* 122; *thujivagrans,* 122
Tropidosteptes amoenus, 402; *brooksi,* 402; *illitus,* 402; *pacificus,* 402
tube makers: birch, 212; oak, 212
Tuberolachnus salignus, 310

tuliptree aphid, 292
tuliptree leafminer, 210
tuliptree scale, 362
tupelo leafminer, 208
tussock moths, 34, 158, 160
twig girdlers and borers, 260, 264, 266, 286; juniper, 58; Pacific oak, 266, 272
twig pruners, 264
twobanded Japanese weevil, 240, 244
twolined chestnut borer, 270
twolined larch sawfly, 18
twomarked treehopper, 410
twospotted spider mite, 476
Typhlocyba sp., 416; *pomaria*, 412, 416; *ulmi*, 246

uglynest caterpillar, 172
Umbonia crassicornis, 408
Unaspis citri, 390; *euonymi*, 388, 390
unicorn caterpillar, 156
Uresiphita reversalis, 166
Urodus parvula, 148
Utamphorophora crataegi, 314

vagabond gall aphid, 462
Vasates aceriscrumena, 482; *quadripedes*, 482
vedalia beetle, 338
vein pocket gall, 440, 446
Verticillium sp. (fungus), 12
Vespa crabro germana, 258, 494
viburnum aphid, 300
viburnum borer, 260
Vinsonia sp., 354
Virginia pine sawfly, 16
Virginia pine scale, 96
viruses, 12, 34, 138, 146, 464
voles (mice), 12, 500, 502

Wahlgreniella nervata, 298
walkingstick, 13, 492–494
walnut aphid, 292
walnut blister mite, 488
walnut caterpillar, 150, 154
walnut lace bug, 426
walnut purse gall mite, 234, 488
walnut scale, 386
wasps, 494. *See also* parasites of arthropods
wax scales, 356, 358. *See also* scale insects
webbers, 38
webworms: ailanthus, 174; *Athrips rancidella*, 218; boxwood, 166; fall, 166; juniper, 30; mimosa, 180; oak, 172; pine, 22; pine false, 22
webspinning sawflies, 22; *Acantholyda burkei*, 22; *A. erythrocephala*, 22; *A. verticalis*, 22; *A. zappei*, 22; *Cephalcia* spp., 22; *Pamphilius phyllisae*, 136
weevils, 13; arborvitae, 240, 244; Asiatic, 13; Asiatic oak, 240, 244; black vine, 54, 240, 242; *Calomycterus setarius*, 240, 244; citrus root, 240; *Cylindrocopturus eatoni*, 60; *C. furnissi*, 60; deodar, 54, 56; Douglas-fir twig, 60; Engelmann spruce, 54; imported longhorned, 240; ligustrum, 244; lodgepole terminal, 54, 56; Monterey pine, 54, 56; northern pine, 56; obscure root, 240, 242; *Otiorhynchus rugifrons*, 240; *O. rugosostriatus*, 240; pales, 13, 56; pine reproduction, 60; pine root collar, 56; pitch-eating, 56; raspberry bud, 240; root, 240, 242, 244; sassafras, 210; *Scythropus californicus*, 58; Sitka spruce, 54; strawberry root, 240; taxus, 240; twobanded Japanese, 240, 244; white pine, 54; willow flea, 190; woods, 240; yellow poplar, 210; Yosemite bark, 56
western boxelder bug, 398
western flower thrips, 432
western grapeleaf skeletonizer, 164
western hemlock looper, 24
western larch sawfly, 18
western pine beetle, 66
western pine shoot borer, 52
western pine tip moth, 48
western poplar clearwing, 256, 260
western rose chafer, 236
western scurfy scale, 390
western spittlebug, 86
western spruce budworm, 28
western sycamore lace bug, 426
western tent caterpillar, 172
western tussock moth, 160
western willow tingid, 426
wheel bug, 190
white apple leafhopper, 412, 416
whitebanded elm leafhopper, 414
white fir needleminer, 38
whiteflies, 13, 318–324; acacia, 320, 322; *Aleurochiton forbesii*, 324; ash, 322; azalea, 318; barberry, 322; black aleyrodid, 322; citrus, 322; cloudywinged, 322; cottony manzanita, 322; crown, 320; deer brush, 320; grape, 322; greenhouse, 320, 322; Japanese barberry, 322; mulberry, 324; rhododendron, 318; Stanford, 320; sweetpotato, 322; willow, 322; woolly, 322
whitefringed beetle, 280
white garden snail, 498
whiteline leafroller, 214
whitemarked tussock moth, 158, 160
white mealybug, 328
white oak borer, 282, 284
white peach scale, 392
white pine aphid, 84
white pine sawfly, 18
white pine weevil, 54
white prunicola scale, 392
willow and poplar lace bug, 426
willow beaked-gall midge, 468
willow cone gall, 468
willow flea weevil, 190
willow lace bug, 426
willow scurfy scale, 390
willow shoot sawfly, 286
willow spider mite, 474
willow whitefly, 322
winter moth, 146
wisteria scurfy scale, 390
witches'-broom gall, 478
witch-hazel bud gall aphid, 450
witch-hazel cone gall aphid, 450
witch-hazel leaf gall aphid, 450
woodpeckers, 12, 68, 126, 252, 254, 258, 262, 272, 278, 282–286, 500
woods weevil, 240
woolly adelgids, 74
woolly aphids: apple, 306, 316; elm, 306; elm cockscomgall aphid, 464; *Eriosoma crataegi*, 306; *E. pyricola*, 464; oak woolly aphid, 304; pear root aphid, 464; *Prociphilus americanus*, 304; *P. caryae*, 304; *Stegophylla quercifolia*, 304; woolly alder aphid, 304, 306; woolly beech leaf aphid, 296; woolly pine needle aphid, 84
woolly pine scale, 88
woolly whitefly, 322
woollyworm, butternut, 136
wool sower gall, 444
worms, measuring, 24

Xanthogaleruca luteola, 222
Xenorhabdus nematophilus (bacterium), 256
Xyelidae, 20
Xyleborus dispar, 250; *saxeseni*, 250; *sayi*, 250
Xylosandrus compactus, 262; *crassiusculus*, 250; *germanus*, 250

yellow-bellied sapsucker, 500
yellow-billed cuckoo, 152
yellowheaded spruce sawfly, 18
yellownecked caterpillar, 154
yellow poplar scale, 362
yellow poplar weevil, 210
yellow-rumped warbler, 354
yellow scale, 378
yellow spider mite, 474
Yosemite bark weevil, 56
Yponomeuta cagnagella, 174; *malinella*, 174; *padella*, 174
Yponomeutidae, 38, 148, 174
Yuma spider mite, 474

Zale sp., 24
Zeiraphera canadensis, 28
Zelleria haimbachi, 38
Zeuzera pyrina, 254
Zimmerman pine moth, 48, 50
Zoophthora aphidis (fungus), 84
Zygaenidae, 164

Index to Insects by Host Plants

This diagnostic index provides an alphabetical list of plants by their Latin and common names, and the pests associated with each host plant species. Subheadings categorize the plant by its major parts in a topographical arrangement beginning with foliage and ending with roots. When there are many pests, additional subheadings provide common pest groups to narrow the search for the identity of a specific animal. Sometimes the species of plant listed as an insect's host will not appear on the pages referenced, but the reader can infer the association. The Reader's Guide to Identification of Insects and Related Pests (page 10) is an index to the plates.

Abelia: *Callirhopalus bifasciatus* (twobanded Japanese weevil), 244; *Hemiberlesia rapax* (greedy scale), 372

Abies (fir)

—foliage:

caterpillars: *Acleris variana* (eastern blackheaded budworm), 28; *Dasychira pinicola* (pine tussock moth), 26; *D. plagiata*, 26; *Orgyia leucostigma* (whitemarked tussock moth), 158

mites: *Nalepella tsugifoliae* (hemlock rust mite), 122

sawflies: *Pleroneura* sp., 20

—bark, wood, twigs: *Dendroctonus valens* (red turpentine beetle), 62; *Hylobius pales* (pales weevil), 56; *Petrova* sp., 72

birds: *Sphyrapicus* sp. (sapsuckers), 500

—sucking insects: *Adelges piceae* (balsam woolly adelgid), 74

aphids: *Aspidiotus cryptomeriae*, 110; *Cinara curvipes* (bowlegged fir aphid), 82; *Mindarus* sp., 80, 84; *Prociphilus americanus*, 304

scales: *Fiorinia externa* (elongate hemlock scale), 104; *Icerya purchasi* (cottony cushion scale), 338

Abies amabilis (Pacific silver fir): *Adelges piceae* (balsam woolly adelgid), 74; *Lambdina fiscellaria lugubrosa* (western hemlock looper), 24; *Neodiprion tsugae* (hemlock sawfly), 18

Abies balsamea (balsam fir)

—foliage: bud miners, 20; *Dasineura balsamicola*, 116; *Paradiplosis tumifex* (balsam gall midge), 116

caterpillars: *Amorbia humerosana* (whiteline leafroller), 214; *Argyrotaenia velutinana* (redbanded leafroller), 214; *Choristoneura fumiferana* (spruce budworm), 28; *C. occidentalis* (western spruce budworm), 28; *Cingilia catenaria* (chainspotted geometer), 146; *Lambdina fiscellaria* (hemlock looper), 24; *Lymantria dispar* (gypsy moth), 138; *Zeiraphera canadensis* (spruce bud moth), 28

mites: *Nalepella tsugifoliae*, 122

sawflies: *Cephalcia distincta*, 22;

—bark, wood, shoots: *Camponotus* sp. (carpenter ants), 496; *Formica exsectoides* (Allegheny mound ant), 496; *Pleroneura brunneicornis* (balsam shootboring sawfly), 20; shoot borers, 20

—sucking insects: *Adelges piceae* (balsam woolly adelgid), 74, 76; *Aphrophora cribata* (pine spittlebug), 86; *Mindarus abietinus* (balsam twig aphid), 80; *Nuculaspis tsugae*, 102

Abies concolor (white fir): cambium miner, 198; *Choristoneura fumiferana* (spruce budworm), 28; *C. occidentalis* (western spruce budworm), 28; *Cinara curvipes* (bowlegged fir aphid), 82, 84; *Epinotia meritana* (white fir needleminer), 38; *Mindarus abietinus* (balsam twig aphid), 80; *Orgyia pseudotsugata* (Douglas-fir tussock moth), 34; *Pleroneura brunneicornis* (balsam shootboring sawfly), 20; *P. lutea*, 20; sawflies, 20, 22

Abies fraseri (Fraser fir): *Adelges piceae* (balsam woolly adelgid), 74, 76; *Dasineura balsamicola*, 116; *Mindarus abietinus* (balsam twig aphid), 80; *Paradiplosis tumifex* (balsam gall midge), 116

Abies grandis (grand or lowland fir): *Apterona helix* (snailcase bagworm), 178; *Choristoneura fumiferana* (spruce budworm), 28; *C. occidentalis* (western spruce budworm), 34; *Lophocampa argentata* (silverspotted tiger moth), 34; *Mindarus abietinus* (balsam twig aphid), 80; *Orgyia pseudotsugata* (Douglas-fir tussock moth), 34; tiger moths, 34

Abies lasiocarpa (alpine or subalpine fir): *Adelges piceae* (balsam woolly adelgid), 74, 76; *Choristoneura fumiferana* (spruce budworm), 28; *C. occidentalis* (western spruce budworm), 28; *Mindarus abietinus* (balsam twig aphid), 80; *Orgyia pseudotsugata* (Douglas-fir tussock moth), 34

Abies magnifica (red fir): *Epinotia meritana* (white fir needleminer), 38; *Orgyia pseudotsugata* (Douglas-fir tussock moth), 34

Abies procera (noble fir): *Adelges piceae* (balsam woolly adelgid), 76

Abies veitchii (Veitch's silver fir): armored scales, 102; *Nuculaspis tsugae*, 102

Abutilon (flowering maple): *Asynonychus godmani* (Fuller rose beetle), 230; cottony taxus scale, 344; *Niesthrea louisianica* (rhopalid plant bug), 422; *Pulvinaria floccifera* (cottony camellia scale), 344

Acacia

—foliage: *Argyrotaenia citrana* (orange tortrix), 216; *Asynonychus godmani* (Fuller rose beetle), 230; *Selenothrips rubrocinctus* (redbanded thrips), 432

snails: *Bulimulus alternatus mariae*, 498

—bark, twigs: *Oncideres pustulatus* (huisache girdler), 286; *Umbonia crassicornis* (thornbug), 408

birds: *Sphyrapicus* sp. (sapsuckers), 500

—sucking insects: *Acizzia uncatoides* (acacia psyllid), 288; *Clastoptera arizonana*, 86; *Tetraleurodes acaciae* (acacia whitefly), 322; *Umbonia crassicornis* (thornbug), 408

scales: *Aonidiella aurantii* (California red scale), 378; *Aspidiotus nerii* (oleander scale), 374; *Chrysomphalus dictyospermi* (dictyospermum scale), 382; *Hemiberlesia lataniae* (latania scale), 376; *H. rapax* (greedy scale), 372; *Icerya purchasi* (cottony cushion scale), 338; *Pinnaspis aspidistrae* (fern scale), 390; *P. strachani* (lesser snow scale), 390

Acacia cardiophylla: *Clastophora arizonana*, 420

Acacia decurrens (green wattle): *Kunzeana kunzii*, 400; *Sphyrapicus* spp. (sapsuckers), 500

Acacia farnesiana (huisache): *Kunzeana kunzii*, 400; *Oncideres pustulatus* (huisache girdler), 286

Acacia koa: *Acizzia uncatoides* (acacia psyllid), 288

Acacia longifolia (Sydney golden wattle): *Acizzia uncatoides* (acacia psyllid), 288; *Graphocephala atropunctata* (blue-green sharpshooter), 418

Acer (maple)

—foliage: *Heliothrips haemorrhoidalis* (greenhouse thrips), 430; *Rhadopterus picipes* (cranberry rootworm), 242; *Taeniothrips inconsequens* (pear thrips), 432

caterpillars: *Acronicta americana* (American dagger moth), 158; *Anisota senatoria* (orangestriped oakworm), 156; *Automeris io* (io moth), 164; *Choristoneura rosaceana* (obliquebanded leafroller), 216; *Datana ministra* (yellownecked caterpillar), 154; *Dryocampa rubicunda* (greenstriped mapleworm), 156; *Erannis tiliaria* (linden looper), 144; *Hyalophora cecropia* (cecropia moth), 164; *Hyphantria cunea* (fall webworm), 166; *Lithophane antennata* (green fruitworm), 148; *Lymantria dispar* (gypsy moth), 138; *Malacosoma americanum* (eastern tent caterpillar), 168; *M. disstria* (forest tent caterpillar), 168; *Operophtera bruceata* (Bruce spanworm), 146; *Orgyia antiqua* (rusty tussock moth), 158; *O. leucostigma* (whitemarked tussock moth), 158; *Paleacrita vernata* (spring cankerworm), 144; *Phigalia titea* (half-wing geometer), 146; *Proteoteras aesculana*, 200; *Schizura ipomoeae*, 156; *Symmerista leucitys* (orangehumped mapleworm), 154; *Thyridopteryx ephemeraeformis* (bagworm), 176

gall makers: *Acericecis ocellaris*, 462; *Aculops maculatus*, 482; erineum galls, 482; *Eriophyes elongatus*, 482; *E. modestus*, 482; eriophyid spindlegalls, 212, 482; gall midges, 462; Phytoptus sp., 486

leafminers: *Baliosus ruber* (basswood leafminer), 190; *Caulocampus acericaulis* (maple petiole borer), 200; *Phyllonorycter* sp., 196

leafrollers/tiers: *Archips argyrospila* (fruittree leafroller), 172; *Psilocoris cryptolechiella*, 218; *Sparaganothus acerivorana*, 216

mites: *Aceria elongatus*, 482; *A. modestus*, 482; *Aculops maculatus*, 482; erineum galls, 482; eriophyid spindlegalls, 212, 482; *Oligonychus aceris* (maple spider mite), 475; *Phytoptus* sp., 486; *Tetranychus urticae* (common twospotted spider mite), 476

sawflies: *Cimbex americana* (elm sawfly), 136

—bark, wood, twigs: *Agrilus* sp., 272; ambrosia beetle, 286; cambium miners, 198; *Chrysobothris femorata* (flatheaded appletree borer), 270; *Corthylus columbianus* (Colombian timber beetle), 286; *C. punctatissimus* (pitted ambrosia beetle), 250; *Elaphidionoides villosus* (twig pruner), 264; horntails, 494; *Magicicada septendecim* (periodical cicada), 490; *Neoclytus acuminatus* (redheaded ash borer), 278; *Oecanthus angustipennis* (narrowwinged cricket), 492; *Prionoxystus robiniae* (carpenterworm), 256; *Stictocephala bisonia* (buffalo treehopper), 406; *Synanthedon acerni* (maple callus borer), 260; *S. acerrubri* (red maple borer), 260; *Tremex columba* (pigeon tremex), 494; *Xylosandrus compactus* (black twig borer), 262; *X. germanus*, 250; *Zeuzera pyrina* (leopard moth), 254

mammals: mice, 502; rabbits, 502; *Sciurus* sp. (squirrels), 500

—roots: *Prionus laticollis* (broadnecked root borer), 280

Acer (cont.)
—sucking insects: *Anormenis chloris*, 422; *Corythucha pallipes* (birch lace bug), 426; *Edwardsiana rosae* (rose leafhopper), 412; *Metcalfa pruinosa*, 422; *Ormenoides venusta*, 422; *Poecilocapsus lineatus* (fourlined plant bug), 396; psyllids, 290
aphids: *Drepanaphis* sp., 302; *Paraprociphilus tesselatus* (woolly alder aphid), 304; *Periphyllus* sp., 302
mealybugs: *Dysmicoccus wisteriae* (taxus mealybug), 88; *Phenacoccus acericola* (maple mealybug), 342; *P. aceris* (apple mealybug), 324; *Pseudococcus comstocki* (Comstock mealybug), 326
scales: *Aspidiotus nerii* (oleander scale), 374; *Eulecanium cerasorum* (calico scale), 354; *Icerya purchasi* (cottony cushion scale), 338; *Lepidosaphes ulmi* (oystershell scale), 370; *Melanaspis obscura* (obscure scale), 384; *Mesolecanium nigrofasciatum* (terrapin scale), 364; *Pulvinaria acericola* (cottony maple leaf scale), 342; *P. floccifera* (cottony taxus scale), 344; *P. innumerabilis* (cottony maple scale), 340; *Quadraspidiotus juglansregiae* (walnut scale), 386; *Saissetia oleae* (black scale), 358
Acer ginnala (amur maple): *Lygocoris communis* (pear plant bug), 396; *Poecilocapsus lineatus* (fourlined plant bug), 396; *Proteoteras aesculana*, 200
Acer glabrum (Rocky Mountain maple): *Aceria calaceris*, 482
Acer negundo (boxelder)
—foliage: *Asynonychus godmani* (Fuller rose beetle), 230
caterpillars: *Acronicta americana* (American dagger moth), 158; *Archips negundana*, 172, 218; *Coleophora serratella* (birch casebearer), 186; *Halysidota tessellaris* (pale tussock moth), 158; *Hyalophora cecropia* (cecropia moth), 164; *Thyridopteryx ephemeraeformis* (bagworm), 176
mites: *Aceria negundi*, 482, 486
—twigs: *Acanalonia conica*, 422; *Proteoteras willingana* (boxelder twig borer), 200
—sucking insects: *Boisea trivittatus* (boxelder bug), 398; psyllids, 290; *Tetraleurodes mori* (mulberry whitefly), 324
scales: *Aonidiella aurantii* (California red scale), 378; *Melanaspis tenebricosa* (gloomy scale), 384; *Pulvinaria innumerabilis* (cottony maple scale), 340; *Quadraspidiotus juglansregia* (walnut scale), 386
Acer palmatum (Japanese maple): *Popillia japonica* (Japanese beetle), 236; *Pulvinaria floccifera* (cottony taxus scale), 344; *Pulvinaria floccifera* (cottony camellia scale), 344
Acer pensylvanicum (striped maple): *Acericecis ocellaris*, 462
Acer platanoides (Norway maple)
—foliage: *Acronicta americana* (American dagger moth), 158; *Phyllonorycter trinotella*, 196; *Popillia japonica* (Japanese beetle), 236; *Thyridopteryx ephemeraeformis* (bagworm), 176
—sucking insects: *Empoasca fabae* (potato leafhopper), 414; *Tetraleurodes mori* (mulberry whitefly), 324
aphids: *Drepanaphis* sp., 302; *Periphyllus* sp., 302; *P. lyropictus*, 302
Acer pseudoplatanus (sycamore maple): *Pulvinaria acericola* (cottony mapleleaf scale), 342
Acer rubrum (red maple)
—foliage: *Cyrtepistomus castaneus* (Asiatic oak weevil), 244
caterpillars: *Dryocampa rubicunda* (greenstriped mapleworm), 156; *Ennomos subsignarius* (elm spanworm), 144–146; *Epinotia aceriella* (maple trumpet skeletonizer), 212; *Epismus tyrius* (leaf tier), 212; *Paraclemensia acerifoliella* (maple leafcutter), 212; *Phyllonorycter trinotella*, 196; *Proteoterus aesculana*, 200; *P. willingana* (boxelder twig borer), 200; *Sparganothus acerivorana*, 216; *Symmerista leucitys* (orangehumped mapleworm), 154
gall makers: *Acericecis ocellaris*, 462; gall midge, 462; *Vasates quadripedes* (maple bladder gall mite), 482
—bark, wood: *Corthylus columbianus* (Colombian timber beetle), 284, 286; *Phytobia setosa*, 198
birds: *Sphyrapicus* sp. (sapsuckers), 500
—sucking insects: *Alebra albostriella*, 416; *Aleurochiton forbesii* (whitefly), 324; *Stictocephala bisonia* (buffalo treehopper), 406; *Tetraleurodes mori* (mulberry whitefly), 324
scales: *Chionaspis acericola* (maple scurfy scale), 390; *Diaspidiotus liquidambaris*, 382; *Melanaspis tenebricosa* (gloomy scale), 384
Acer saccharinum (silver maple)
—foliage: *Dryocampa rubicunda* (greenstriped mapleworm), 156; *Phyllobius oblongus*, 244; *Rhadopterus picipes* (cranberry rootworm), 242
mites: *Eotetranychus carpini* (yellow spider mite), 474; *Vasates quadripedes* (maple bladder gall mite), 482
—sucking insects: *Boisea trivittatus* (boxelder bug), 398; *Chionaspis acericola* (maple scurfy scale), 390; *Melanaspis tenebricosa* (gloomy scale), 384; *Pulvinaria innumerabilis* (cottony maple scale), 340
aphids: *Paraprociphilus tesselatus* (woolly alder aphid), 304; *Periphyllus* sp., 302; *Pterocomma smithiae*, 310; *Toxoptera aurantii* (black citrus aphid), 302
Acer saccharum (sugar maple)
—foliage: *Acericecis ocellaris*, 462
caterpillars: *Caulocampus acericaulis* (maple petiole borer), 200; *Dryocampa rubicunda* (greenstriped mapleworm), 156; *Epinotia aceriella* (maple trumpet skeletonizer), 212; *Epismus tyrius* (leaf tier), 212; *Heterocampa guttivitta* (saddle prominent), 154; *Paraclemensia acerifoliella* (maple leaf cutter), 212; *Sparganothus acerivorana*, 216; *Symmerista leucitys* (orangehumped mapleworm), 154
mites: *Aceria elongatus*, 482; *Eotetranychus carpini* (yellow spider mite), 474; *Vasates aceriscrumena* (maple spindle gall mite), 482
—bark, wood: *Camponotus* sp. (carpenter ants), 496; *Glycobius speciosus* (sugar maple borer), 276; *Phytobia setosa*, 198
birds: *Sphyrapicus* sp. (sapsuckers), 500
—sucking insects: *Empoasca fabae* (potato leafhopper), 414; *Phenacoccus acericola* (maple mealybug), 342; *Psylla annulata*, 290
scales: *Melanaspis tenebricosa* (gloomy scale), 384
Acer spicatum (mountain maple): *Phyllobus oblongus*, 244
Aesculus (buckeye, horsechestnut)
—foliage
caterpillars: *Archips argyrospila* (fruittree leafroller), 172; *Choristoneura rosaceana* (obliquebanded leafroller), 216; *Ennomos subsignarius* (elm spanworm), 144–146; *Orgyia leucostigma* (whitemarked tussock moth), 158; *Proteoteras aesculana*, 200; *Thyridopteryx ephemeraeformis* (bagworm), 176
mites: *Eotetranychus tiliarium*, 474
—sucking insects: *Corythucha aesculi* (buckeye lace bug), 426; *Empoasca* sp., 416; *Erythroneura gledistia*, 416; *Pseudococcus comstocki* (Comstock mealybug), 326
aphids: *Drepanaphis* sp., 302
scales: *Eulecanium cerasorum* (calico scale), 354; *Icerya purchasi* (cottony cushion scale), 338; *Lepidosaphes ulmi* (oystershell scale), 370; *Periphylla* sp., 302; *Quadraspidiotus juglansregia* (walnut scale), 386
spittlebugs: *Clastoptera obtusa* (alder spittlebug), 420; *C. proteus* (dogwood spittlebug), 420
Aesculus californica (California buckeye): *Archips argyrospila* (fruittree leafroller), 172; *Empoasca* sp., 416; *Icerya purchasi* (cottony cushion scale), 338
Aesculus hippocastanum (European horsechestnut): *Choristoneura rosaceana* (obliquebanded leafroller), 216; *Eotetranychus hicoriae* (pecan leaf scorch mite), 474; *Lepidosaphes ulmi* (oystershell scale), 370; *Orgyia leucostigma* (whitemarked tussock moth), 158; *Quadraspidiotus juglansregiae* (walnut scale), 386; *Tetranychus canadensis* (fourspotted spider mite), 476
African violet: *Phytonemus pallidus* (cyclamen mite), 476
Agathis (Kauri): *Nipaecoccus aurilanatus* (golden mealybug), 90
Aglaonema: *Ceroplastes rubens* (red wax scale), 358
Ailanthus altissima (tree of heaven): *Atteva punctella* (ailanthus webworm), 174
Albizia: *Acizzia uncatoides* (acacia psyllid), 288; *Chionaspis gleditsia*, 390; *Parthenolecanium persicae* (European peach scale), 364; *Toumeyella cerifera* (buttonbush scale), 362
Albizia julibrissin (mimosa, silktree): *Callirhopalus bifasciatus* (twobanded Japanese weevil), 244; *Homadaula anisocentra* (mimosa webworm), 180; *Mesolecanium nigro-fasciatum* (terrapin scale), 364; *Nipaecoccus nipae*, 90; *Oncideres pustulatus* (huisache girdler), 286; *Orgyia leucostigma* (whitemarked tussock moth), 158; *Parlatoria pittospori*, 110; psyllids, 290; *Tetraleurodes acacia* (acacia whitefly), 322; *Umbonia crassicornis* (thornbug), 408
Albizia lebbeck (lebbeck, woman's-tongue): *Umbonia crassicornis* (thornbug), 408
alder, see *Alnus*
black: *Alnus glutinosa*
European: *Alnus cordata*
European green: *Alnus viridis*
mountain: *Alnus tenuifolia*
speckled: *Alnus rugosa*
Aleurites fordii (tung): *Selenothrips rubrocinctus* (redbanded thrips), 432
Algama: *Eriococcus borealis*, 336
alligator pear, see *Persea americana*
almond, see *Prunus amygdalus*
flowering: *Prunus triloba*
Alnus (alder)
—foliage:
beetles: *Calligrapha scalaris* (elm calligrapha), 226; *Chrysomela interrupta*, 226; *C. mainensis* (alder leaf beetle), 224; *Macrohaltica ambiens* (alder flea beetle), 228
caterpillars: *Acronicta dactylina* (alder dagger moth), 160; *Amorbia humerosana* (whiteline leafroller), 214; *Argyrotaenia velutinana* (redbanded leafroller), 214; *Coleophora serratella* (birch casebearer), 186; *Hyphantria cunea* (fall webworm), 166; *Lymantria dispar* (gypsy moth), 138; *Malacosoma californicum pluviale* (western tent caterpillar), 170; *M. disstria* (forest tent caterpillar), 170; *Orgyia antiqua* (rusty tussock moth), 158; *Rheumaptera hastata* (spear-marked black moth), 146
leafminers: *Fenusa dorhnii* (European alder leafminer), 188; *Phyllonorycter alnicolella*, 196; *P. incanella*, 196
leafrollers/tiers: *Apotomis* spp., 214; *Archips negundana*, 172, 214; *A. rosanus*, 218; *Epinotia solandriana*, 218; *Pandemis limitata*, 218
mites: *Eotetranychus carpini* (yellow spider mite), 474; *E. uncatus* (Garman spider mite), 474; erineum galls, 484; eriophyid mites, 484
sawflies: *Arge pictoralis* (birch sawfly), 128; *Cimbex americana* (elm sawfly), 136; *Eriocampa ovata* (alder woolly sawfly), 136; *Hemichroa crocea* (striped alder sawfly), 136
—bark, wood, twigs: *Agrilus anxius* (bronze birch borer), 272; *Chrysobothris mali* (Pacific flatheaded borer), 270; *Cryptorhynchus lapathi* (poplar and willow borer), 268; *Oecanthus angustipennis* (narrowwinged cricket), 492
—sucking insects: *Chionaspis lintneri*, 390; *Clastoptera obtusa* (alder spittlebug), 420; *Corythucha pergandei* (alder lace bug), 426; *Ed-*

wardsiana commisuralis, 416; *Paraprociphilus tesselatus* (woolly alder aphid), 304, 306; *Psylla floccosa* (woolly psyllid), 306; psyllids, 290; *Pulvinaria innumerabilis* (cottony maple scale), 340; *Typhlocyba pomaria* (white apple leafhopper), 416
Alnus cordata (European alder): *Fenusa dorhnii* (European alder leafminer), 188
Alnus glutinosa (black alder): *Acanalonia conica*, 422; *Fenusa dohrnii* (European alder leafminer), 188; *Paraprociphilus tessellatus* (woolly alder aphid), 304
Alnus macrophylla: *Fenusa dohrnii* (European alder leafminer), 188
Alnus rugosa (speckled alder): *Fenusa dohrnii* (European alder leafminer), 188
Alnus tenuifolia (mountain alder): *Clastoptera obtusa* (alder spittlebug), 420
Alnus viridis (European green alder): *Fenusa dohrnii* (European alder leafminer), 188
Aloe: *Aspidiotus nerii* (oleander scale), 374
Aloysia triphylla (verbena): *Icerya purchasi* (cottony cushion scale), 338
Alyxia olivaeformis: *Ceroplastes rubens* (red wax scale), 358
Amelanchier (serviceberry; shadbush)
—foliage: *Bucculatrix pomifoliella*, 220; *Caliroa cerasi* (pear sawfly), 130; *Nymphalis californica* (California tortoiseshell), 152; *Phyllobius oblongus*, 244
  mites: eriophyid leafblister mites, 486; leaf blister, 486; *Oligonychus newcomeri*, 475
—bark, wood, twigs: cambium miners, 198; *Saperda candida* (roundheaded appletree borer), 278; *Scolytus rugulosus* (shothole borer), 250; *Synanthedon pyri* (apple bark borer), 258
—sucking insects: *Aphis fabae* (bean aphid), 300; *Chionaspis lintneri*, 390; *Corythucha cydoniae* (hawthorn lace bug), 426; *Eriosoma americanum* (woolly elm aphid), 306
Amelanchier alnifolia (western serviceberry): *Prociphilus caryae*, 304
Amelanchier arborea (downy serviceberry): *Tetranychus mcdanieli*, 476
Amelanchier canadensis (thicket serviceberry): *Phytobia amelanchieris*, 198
American plane, see *Platanus occidentalis*
Amorpha: *Epargyreus clarus* (silverspotted skipper), 148
Amorpha fructicosa (false indigo): *Odontota dorsalis* (locust leafminer), 190
Andorra: *Contarinia juniperina* (juniper midge), 46
andromeda, see *Pieris japonica*
anemone, bush, see *Carpenteria californica*
antelope brush, see *Purshia tridentata*
Anthurium: *Ceroplastes rubens* (red wax scale), 358
apple, see *Malus*
apricot, see *Prunus armeniaca*
Aralia: *Aphis hederae* (ivy aphid), 308; *Argyrotaenia citrana* (orange tortrix), 216; *Asterolecanium arabidis* (pit-making pittosporum scale), 348; *Ceroplastes rubens* (red wax scale), 358; *Chrysomphalus dictyospermi* (dictyospermum scale), 382; *Saissetia coffeae* (hemispherical scale), 360; *S. oleae* (black scale), 358
aralia, Japanese, see *Fatsia japonica*
Araucaria: *Chrysomphalus dictyospermi* (dictyospermum scale), 382; *Eriococcus araucariae*, 90; *Lindingaspis rossi*, 376
Araucaria bidwelli (Bunya-Bunya): *Nipaecoccus aurilanatus* (golden mealybug), 90
Araucaria excelsa (also see *Araucaria heterophylla*): *Eriococcus araucariae*, 90
Araucaria heterophylla (Norfolk island pine): *Eriococcus araucariae*, 90; *Nipaecoccus aurilanatus* (golden mealybug), 90
arborvitae, see *Thuja*
Arbutus: *Chionaspis ortholobis*, 390; *Prionus californicus* (California prionus), 280; *Pseudococcus maritimus* (grape mealybug), 88
Arbutus menziesii (madrone): *Aleuropleurocelus nigrans* (black aleyrodid), 322; *Coptodisca arbutiella* (madrone shield bearer), 194, 208; *Gelechia panella*, 194; *Hyphantria cunea* (fall webworm), 166; *Malacosoma californicum pluviale* (western tent caterpillar), 170; *Marmara arbutiella*, 194; *Wahlgreniella nervata*, 298
Arbutus unedo (strawberry madrone, strawberry tree): *Coptodisca arbutiella* (madrone shield bearer), 194, 208; *Hemiberlesia rapax* (greedy scale), 372; *Saissetia oleae* (black scale), 358; *Wahlgreniella nervata*, 298
Arctostaphylos (manzanita)
—foliage:
  caterpillars: *Coptodisca arbutiella* (madrone shield bearer), 194, 208; *Gelechia panella*, 194; *Nymphalis californica* (California tortoiseshell), 152; *Orgyia vestuta* (western tussock moth), 160
—sucking insects: *Aleuropleurocelus nigrans* (black aleyrodid), 322; *Puto albicans* (white mealybug), 328; *P. arctostaphyli* (manzanita mealybug), 328; *Trialeurodes merlini* (cottony manzanita whitefly), 322
  aphids: *Aphis fabae* (bean aphid), 300; *Tamalia coweni*, 298; *Wahlgreniella nervata*, 298
  scales: *Aspidiotus nerii* (oleander scale), 374; *Asterolecanium arabidis* (pit-making pittosporum scale), 348; *Chionaspis ortholobis*, 390; *Hemiberlesia rapax* (greedy scale), 372
Arecastrum romanzoffianum (queen palm): *Nipaecoccus nipae* (coconut mealybut), 90
arrowwood, see *Viburnum dentatum*
Artemisia tridentata (sagebrush): *Saissetia oleae* (black scale), 358; *Trirhabda flavolimbata* (baccharis leaf beetle), 238; *T. pilosa*, 238
Ascyra edisonianum: *Toumeyella liriodendri* (tuliptree scale), 362
Ascyra hypericoides: *Toumeyella liriodendri* (tuliptree scale), 362
Ascyra tetrapetalum: *Toumeyella liriodendri* (tuliptree scale), 362
ash, see *Fraxinus*
  American: *Fraxinus americana*
  blue: *Fraxinus quadrangulata*
  green: *Fraxinus pennsylvanica*
  Modesto: *Fraxinus velutina*
  Moraine: *Fraxinus holotricha*
  mountain: *Sorbus*
  prickly: *Zantholyxum*
  red: *Fraxinus pennsylvanica*
  velvet: *Fraxinus velutina*
  white: *Fraxinus americana*
Asimina triloba (pawpaw): *Alebra albostriella* (maple leafhopper), 416
Asparagus sataceus (asparagus fern): *Saissetia oleae* (black scale), 358
aspen (*Populus* in part), 214, 226
—foliage: *Anabrus simplex* (Mormon cricket), 492; *Diapheromera femorata* (walkingstick), 492, 494
  beetles: *Chrysomela tremulae*, 226; *Rhynchaenus rufipes* (willow flea weevil), 190
  caterpillars: *Leucoma salicis* (satin moth), 158; *Lophocampa caryae* (hickory tussock moth), 160; *Malacosoma disstria* (forest tent caterpillar), 168; *Operophtera bruceata* (Bruce spanworm), 146; *Phyllonorycter tremuloidiella* (aspen blotch miner), 196; *Schizura concinna* (redhumped caterpillar), 156
  leafroller/tiers: *Ancylis discigerana*, 218; *Archips agyrospila* (fruittree leafroller), 214; *Choristoneura conflictana* (large aspen tortrix), 214; *C. rosaceana* (obliquebanded leafroller), 216; *Epinotia solandriana*, 218; *Olethreutes cespitana*, 202; *Pseudosciaphila duplex* (aspen leaftier), 214
—bark:
  birds: *Sphyrapicus* sp. (sapsuckers), 500
aspen
  bigtooth: *Populus grandidentata*
  quaking: *Populus tremuloides*
  trembling: *Populus tremuloides*
Aspidistra eliator (iron plant): *Aonidiella aurantii* (California red scale), 378
Atriplex: *Ceroplastes irregularis* (desert wax scale), 358
Aucuba: *Aonidiella citrina* (yellow scale), 378; *Aspidiotus nerii* (oleander scale), 374; *Brevipalpus obovatus* (privet mite), 482; *Chrysomphalus dictyospermi* (dictyospermum scale), 382; *Fiorinia theae* (tea scale), 394; *Hemiberlesia rapax* (greedy scale), 372; *Pseudaulacaspis pentagona* (white peach scale), 392
Aucuba japonica (croton-leaf aucuba): *Acyrthosiphon solani* (foxglove aphid), 310; *Pseudaulacaspis cockerelli* (false oleander scale), 374
Australian bush cherry, see *Syzygium paniculatum*
Australian pine, see *Casuarina*
avocado, see *Persea americana*
azalea
—foliage: *Asynonychus godmani* (Fuller rose beetle), 230; *Callirhopalus bifasciatus* (twobanded Japanese weevil), 244; *Colaspis pseudofavosa*, 224; *Heliothrips haemorrhoidalis* (greenhouse thrips), 430; *Megachile* sp. (leafcutter bees), 494; *Nemocestes incomptus* (raspberry bud weevil, woods weevil), 240; *Otiorhynchus sulcatus* (black vine weevil), 240, 242; *Popillia japonica* (Japanese beetle), 236; *Sciopithes obscurus* (obscure root weevil), 240
  caterpillars: *Archips argyrospila* (fruittree leafroller), 202; *Caloptilia azaleella* (azalea leafminer), 202; *Datana major* (azalea caterpillar), 154; *D. ministra* (yellownecked caterpillar), 154; *Hydria undulata*, 146; *Lithophane antennata* (green fruitworm), 148; *Olethreutes cespitana*, 202; *Ptycholoma peritana*, 202
  mites: *Brevipalpus obovatus* (privet mite), 482; *Oligonychus ilicis* (southern red mite), 475; *Tetranychus urticae* (twospotted spider mite), 474, 476
  sawflies: *Amauronematus azaleae*, 134; *Arge clavicornis*, 134
—bark, wood, twigs: *Corthylus punctatissimus* (pitted ambrosia beetle), 250; *Dendroctonus frontalis* (southern pine beetle), 64; *Oberea myops* (rhododendron stem borer, azalea stem borer), 288; *O. tripunctata* (dogwood twig borer), 262; *Synanthedon rhododendri* (rhododendron borer), 258
—roots: *Callirhopalus bifasciatus* (twobanded Japanese weevil), 244; *Nemocestes incomptus* (raspberry bud weevil, woods weevil), 240; *Otiorhynchus sulcatus* (black vine weevil), 240, 242; *Sciopithes obscurus* (obscure root weevil), 240
—sucking insects: lacebugs, 424; *Metcalfa pruinosa*, 422; *Pealius azaleae* (azalea whitefly), 318; *Poecilocapsus lineatus* (fourlined plant bug), 396; *Stephanitis pyrioides* (azalea lacebug), 424
  leafhoppers: *Graphocephala coccinea* (redbanded leafhopper), 418; *G. fennahi* (rhododendron leafhopper), 418; leafhoppers, 418
  scales: *Aspidiotus nerii* (oleander scale), 374; *Duplaspidiotus claviger* (camellia mining scale), 380; *Eriococcus azaleae* (azalea bark scale), 336; *Hemiberlesia rapax* (greedy scale), 372; *Pseudaonidia paeoniae* (peony scale), 372; *Pulvinaria ericicola* (cottony azalea scale), 342
azalea
  flame: *Rhododendron calendulaceum*
  Pontic: *Rhododendron luteum*
  swamp: *Rhododendron viscosum*

Baccharis: *Pseudococcus affinis* (obscure mealybug), 324
Baccharis halimifolia (eastern baccharis, ground bush): *Trirhabda bacharidis* (baccharis leaf beetle), 238
Baccharis pilularis (coyote bush; dwarf chaparral broom): *Argyrotaenia franciscana*, 218; *Hemiberlesia rapax* (greedy scale), 372; *Trirhabda flavolimbata* (baccharis leaf beetle), 238
baldcypress, see *Taxodium distichum*

balsam fir, see *Abies balsamea*
balsam poplar, see *Populus balsamifera*
Bambusa (bamboo): *Antonina pretiosa* (noxious bamboo mealybug), 330; *Asterolecanium bambusae* (bamboo scale), 350; *Schizotetranychus celarius* (bamboo spider mite), 474; *Takecallis arundicolens* (bamboo aphid), 312
banana shrub, see *Michelia figo*
Barbados cherry, see *Eugenia uniflora*
barberry, see *Berberis*
 Japanese: *Berberis thunbergii*
basswood, see *Tilia americana*
Bauhinia: *Chrysomphalus dictyospermi* (dictyospermum scale), 382; *Tetraleurodes acaciae* (acacia whitefly), 320
Bauhinia purpurea (orchid tree): *Tetraleurodes acaciae* (acacia whitefly), 320
bay, see *Persea*
 loblolly: *Gordonia lasianthus*
 red: *Persea borbonia*
 shore: *Persea borbonia*
 silk: *Persea borbonia* var. *humilis*
 swamp: *Persea borbonia* var. *pubescens*
 sweet: *Magnolia virginiana*
bayberry: *Myrica*
beech: *Fagus*
 American: *Fagus grandifolia*
Berberis: *Asterolecanium arabidis* (pit-making pittosporum scale), 348; *Callirhopalus bifasciatus* (twobanded Japanese weevil), 244; *Ceroplastes ceriferus* (Indian wax scale), 356; *Coryphista meadii*, 144; *Parthenolecanium persicae* (European peach scale), 364; *Trialeurodes vaporariorum* (greenhouse whitefly), 320
Berberis thunbergii (Japanese barberry): *Callirhopalus bifasciatus* (twobanded Japanese weevil), 244
Betula (birch)
—foliage: *Acrobasis betulella* (birch tubemaker), 212; *Anabrus simplex* (Mormon cricket), 492; *Bucculatrix canadensisella* (birch skeletonizer), 220; *Coleophora serratella* (birch casebearer), 186; *Diapheromera femorata* (walkingstick), 492, 494; *Monocesta coryli* (larger elm leaf beetle), 222; *Paraclemensia acerifoliella* (maple leaf cutter), 212; *Popillia japonica* (Japanese beetle), 236; *Rhynchaenus rufipes* (willow flea weevil), 190; *Taeniothrips inconsequens* (pear thrips), 432
 caterpillars: *Anisota senatoria* (orangestriped oakworm moth), 156; *Argyrotaenia velutinana* (redbanded leafroller), 214; *Automeris io* (io moth), 164; *Cingilia catenaria* (chainspotted geometer), 146; *Datana integerrima* (walnut caterpillar), 150; *D. ministra* (yellownecked caterpillar), 154; *Erannis tiliaria* (linden looper), 144; *Heterocampa guttivitta* (saddled prominent), 154; *Hyalophora cecropia* (cecropia moth), 164; *Lophocampa caryae* (hickory tussock moth), 160; *Malacosoma americanum* (eastern tent caterpillar), 168; *M. californicum pluviale* (western tent caterpillar), 170; *M. disstria* (forest tent caterpillar), 168, 170; *Nymphalis antiopa* (mourningcloak butterfly, spiny elm caterpillar), 152; *Orgyia antiqua* (rusty tussock moth), 158; *O. leucostigma* (whitemarked tussock moth), 158; *Rheumaptera hastata* (spear-marked black moth), 146; *Schizura concinna* (redhumped caterpillar), 156; *S. ipomoeae*, 156; *Symmerista leucitys* (orangehumped mapleworm), 154
 leafminers: *Baliosus ruber* (basswood leafminer), 190; *Fenusa pusilla* (birch leafminer), 128, 184; *Messa nana*, 184; *Phyllonorycter* sp., 196; *Profenusa thompsoni*, 184
 leafroller/tiers: *Ancylis discigerana*, 218; *Apotomis* sp., 218; *Archips argyrospila* (fruittree leafroller), 172; *A. negundana*, 172; *Choreutis pariana* (apple-and-thorn skeletonizer), 216; *Choristoneura rosaceana* (obliquebanded leafroller), 216; *Epinotia solandriana*, 218; *Pandemis limitata*, 218; *Psilocoris cryptolechiella*, 218
 mites: *Cecidophyes betulae*, 478; *Eotetranychus querci*, 474; *E. uncatus* (Garman spider mite), 474; *Euceraphis betulae*, 296; *Oligonychus bicolor* (oak spider mite), 475
 sawflies: *Arge clavicornis*, 134; *A. pictoralis* (birch sawfly), 128; *Cimbex americana* (elm sawfly), 136; *Croesus latitarsus* (dusky birch sawfly), 128; *Heterarthrus nemoratus*, 184
—bark, wood, twigs: *Agrilus anxius* (bronze birch borer), 272; ambrosia beetle, 286; cambium miners, 198; *Corthylus columbianus* (Colombian timber beetle), 286; *Magicicada septendecim* (periodical cicada), 490; mice, 502; *Neoclytus acuminatus* (redheaded ash borer), 278; *Paranthrene robiniae* (western poplar clearwing), 256, 260; *Phytobia pruinosa*, 198; *Prionoxystus robiniae* (carpenterworm), 256; rabbits, 502; *Sciurus* sp. (squirrels), 502; *Sphyrapicus* sp. (sapsuckers), 500; *Synanthedon fulvipes*, 260; *S. scitula* (dogwood borer), 262; *Vespa crabro germana* (European hornet), 494; *Xylosandrus germanus*, 250
—sucking insects: *Clastoptera obtusa* (alder spittlebug), 420; *Corythucha pallipes* (birch lace bug), 426; *C. pergandei* (alder lace bug), 426; *Metcalfa pruinosa*, 422; psyllids, 290
 aphids: *Betulaphis* sp., 296; *B. brevipilosa*, 296; *Calaphis* sp., 296; *C. betulaecolens*, 296; *Callipterinella* sp., 296; *C. calliptera*, 296; *Euceraphis* sp., 296; *E. betulae*, 296; *Hamamelistes spinosus* (witch-hazel bud gall aphid), 296, 450; *Hormaphis hamemelidis* (witch hazel leaf gall aphid), 450; *Longistigma caryae* (giant bark aphid), 310
 leafhoppers: *Empoasca fabae* (potato leafhopper), 414; *Graphocephala atropunctata* (blue-green sharpshooter), 418; *Typhlocyba pomaria* (white apple leafhopper), 416
 scales: *Chionaspis floridensis*, 390; *C. lintneri*, 390; *Eulecanium caryae* (hickory lecanium), 364; *Lepidosaphes ulmi* (oystershell scale), 370; *Mesolecanium nigrofasciatum* (terrapin scale), 364; *Parthenolecanium pruinosum* (frosted scale), 354; *P. quercifex* (oak lecanium), 364; *Quadraspidiotus juglansregia* (walnut scale), 386
Betula alba, see *Betula pendula*
Betula alleghaniensis (yellow birch): *Corythucha pallipes* (birch lace bug), 426; *Croesus latitarsus* (dusky birch sawfly), 128; *Heterocampa guttivitta* (saddled prominent), 154; *Phyllobius oblongus*, 244
Betula costata: *Fenusa pusilla* (birch leafminer), 184
Betula davurica: *Fenusa pusilla* (birch leafminer), 184
Betula lenta (black, sweet birch): *Croesus latitarsus* (dusky birch sawfly), 128
Betula maximowiczana: *Agrilus anxius* (bronze birch borer), 272; *Fenusa pusilla* (birch leafminer), 184
Betula nigra (river birch): *Acrobasis betulivorella*, 212; *Agrilus anxius* (bronze birch borer), 272; *Chionaspis acericola* (maple scurfy scale), 390; *Cryptorhynchus lapathi* (poplar-and-willow borer), 268; *Platycotis vittata* (oak treehopper), 406
Betula occidentalis (red, water birch): *Croesus latitarsus* (dusky birch sawfly), 128
Betula papyrifera (paper birch): *Agrilus anxius* (bronze birch borer), 272; *Archips negundana*, 172; *Coleophora serratella* (birch casebearer), 186; *Croesus latitarsus* (dusky birch sawfly), 128; *Fenusa pusilla* (birch leafminer), 184; *Messa nana*, 184
Betula pendula (European white birch): *Agrilus anxius* (bronze birch borer), 272; *Calaphis betulaecolens*, 296; *Corythucha pallipes* (birch lacebug), 426; *Platycotis vittata* (oak treehopper), 406
Betula platyphylla (Japanese white birch): *Agrilus anxius* (bronze birch borer), 272
Betula populifolia (gray birch): *Agrilus anxius* (bronze birch borer), 272; *Croesus latitarsus* (dusky birch sawfly), 128; *Fenusa pusilla* (birch leafminer), 184; *Hamamelistes spinosus* (spiny witch-hazel gall aphid), 296, 450; *Popillia japonica* (Japanese beetle), 236
Betula pubescens: *Calaphis betulaecolens*, 296
Betula pumila: *Cryptorhynchus lapathi* (poplar-and-willow borer), 268
Betula schmitti: *Fenusa pusilla* (birch leafminer), 184
Bignonia (cross vine): *Saissetia coffeae* (hemispherical scale), 360
birch, see *Betula*
bird-of-paradise, see *Strelitzia*
bittersweet, see *Celastrus*
 American: *Celastrus scandens*
black alder, see *Alnus glutinosa*
blackberry: *Aphrophora saratogensis* (Saratoga spittlebug), 86; *Edwardsiana rosae* (rose leafhopper), 412; *Hartigia trimaculata*, 132; horn-tailed wasp, 132
blackgum, see *Nyssa sylvatica*
blackhaw, see *Viburnum prunifolium*
black locust, see *Robinia pseudoacacia*
bladderpod, see *Isomeris arborea*
blueberry, see *Vaccinium*
Boston ivy, see *Parthenocissus tricuspidata*
Bougainvillea: *Aphis craccivora* (cowpea aphid), 308; *A. gossypii* (cotton aphid, melon aphid), 308; *Asciodes gordalis*, 182; *Microcentrum rhombifolium* (broadwinged katydid), 490; *Myzus persicae* (green peach aphid), 300; *Planococcus citri* (citrus mealybug), 328
Bouvardia: *Ceroplastes brachyurus* (brachyuran wax scale), 358
boxelder, see *Acer negundo*
boxwood, see *Buxus*
Brahea edulis (Guadaloupe palm): *Icerya purchasi* (cottony cushion scale), 338
brambles: *Hartigia cressoni*, 132; *Merhynchites bicolor* (rose curculio), 238
Brazilian pepper tree, see *Schinus terebinthifolius*
broom: *Aspidiotus nerii* (oleander scale), 374; *Hemiberlesia rapax* (greedy scale), 372. See also *Cytisus* and *Spartium*
broom, chaparral: *Hemiberlesia rapax* (greedy scale), 372. See also *Baccharis pilularis*
buckbrush, see *Ceanothus cuneatus*
buckeye, see *Aesculus*
buckthorn, see *Rhamnus*
Buddleia (butterflybush): *Aspidiotus nerii* (oleander scale), 374; *Uresiphita reversalis* (genista caterpillar), 166
Buddleia calbillea: *Pyramidobela angelarum*, 214
Bunya-Bunya, see *Araucaria bidwelli*
burning bush, see *Euonymus*
Bursera simaruba (gumbo-limbo, gum-elemi): *Ceroplastes dugesii* (Duges wax scale), 358; *Hemiberlesia rapax* (greedy scale), 372
bush anemone, see *Carpenteria californica*
Butia (palm): *Homaledra sabelella* (palmleaf skeletonizer), 162
butternut, see *Juglans cinerea*
buttonbush, see *Cephalanthus*
Buxus (boxwood)
—foliage: *Datana ministra* (yellownecked caterpillar), 154; *Monarthropalpus flavus* (boxwood leafminer), 204
 mite: *Eurytetranychus buxi* (boxwood mite), 475
 snail: *Helix aspersa* (brown garden snail), 498
—bark: *Vespa crabro germana* (European hornet), 494
—sucking insects: *Aonidiella aurantii* (California red scale), 378; *Aspidiotus nerii* (oleander scale), 374; *Cacopsylla buxi* (boxwood psyllid), 290; *Ceroplastes ceriferus* (Indian wax scale), 356; *Chrysomphalus dictyospermi* (dictyospermum scale), 382; *Hemiberlesia rapax* (greedy scale), 372; *Icerya purchasi* (cottony cushion scale), 338; *Lepidosaphes beckii* (purple scale), 380; *L. gloveri* (Glover scale), 380; *L. ulmi* (oystershell scale), 370; *Pseudococcus comstocki* (Comstock mealybug), 326; *Quadraspidiotus perniciosus* (San Jose scale), 386; *Unaspis euonymi* (euonymus scale), 388

Buxus harlandii (Korean boxwood): *Monarthropalpus flavus* (boxwood leafminer), 204
Buxus microphylla (Japanese boxwood): *Monarthropalpus flavus* (boxwood leafminer), 204
Buxus sempervirens (English boxwood): *Monarthropalpus flavus* (boxwood leafminer), 204

cactus: *Aspidiotus nerii* (oleander scale), 374; *Hemiberlesia rapax* (greedy scale), 372; *Tenuipalpidae* sp. (false spider mites), 476
California bay, see *Umbellularia californica*
California black walnut, see *Juglans hindsii*
California buckeye, see *Aesculus californica*
California Christmas berry, see *Heteromeles arbutifolia*
California coffeeberry, see *Rhamnus californica*
California holly, see *Heteromeles arbutifolia*
California laurel, see *Umbellularia californica*
California-nutmeg tree, see *Torreya californica*
California pepper tree, see *Schinus molle*
Calliandra: *Aleurothrixus floccosus* (woolly whitefly), 322; *Tetraleurodes acacia* (acacia whitefly), 322; *Umbonia crassicornis* (thornbug), 408
Callicarpa: *Pseudaulacaspis pentagona* (white peach scale), 392
Callicarpa americana: *Duplaspidiotus claviger* (camellia mining scale), 380; *Pulvinaria floccifera* (cottony camellia scale, cottony taxus scale), 344
Callistemon: *Chrysomphalus dictyospermi* (dictyospermum scale), 382; *Duplaspidotus claviger* (camellia mining scale), 380; *Fiorinia theae* (tea scale), 394; *Parlatoria pittospori*, 110; *Umbonia crassicornis* (thornbug), 408
Calocedrus (=Libocedrus) decurrens (incense cedar): *Argyresthia arceuthobiella*, 42; *A. libocedrella*, 42; *Carulaspis juniperi* (juniper scale), 106; *Ehrhornia cupressi* (cypress bark mealybug, cottony cypress scale, cypress bark scale), 332
Camellia
—foliage/buds: *Asynonychus godmani* (Fuller rose beetle), 230; *Callirhopalus bifasciatus* (twobanded Japanese weevil), 244; *Nemocestes incomptus* (woods weevil, raspberry bud weevil), 240; *Otiorhynchus sulcatus* (black vine weevil), 240; *Rhadopterus picipes* (cranberry rootworm), 242; *Sciopithes obscurus* (obscure root weevil), 240
mites: *Acaphylla steinwedeni*, 480; *Calcarus carinatus*, 480; *Cosetacus camelliae*, 480; *Oligonychus ilicis* (southern red mite), 475
—bark, wood, twigs: *Nemocestes incomptus* (woods weevil), 240; *Otiorhynchus sulcatus* (black vine weevil), 240; *Rhadopterus picipes* (cranberry rootworm), 242; *Sciopithes obscurus* (obscure root weevil), 240
—sucking insects: *Metcalfa pruinosa*, 422
aphids: *Aphis gossypii* (cotton aphid, melon aphid), 308; *Toxoptera aurantii* (black citrus aphid), 302
scales: *Aspidiotus nerii* (oleander scale), 374; *Ceroplastes ceriferus* (Indian wax scale), 356; *Chrysomphalus dictyospermi* (dictyospermum scale), 382; *Duplaspidiotus tesseratus*, 380; *Fiorinia fioriniae* (palm fiorinia scale), 394; *F. theae* (tea scale), 394; *Hemiberlesia rapax* (greedy scale), 372; *Howardia biclavis*, 380; *Lepidosaphes camelliae*, 370; *Lindingaspis rossi*, 376; *Parasaissetia nigra* (nigra scale), 360; *Pseudaonidia duplex* (camphor scale), 372; *P. paeoniae* (peony scale), 372; *Pulvinaria floccifera* (cottony camellia scale), 344; *Saissetia coffeae* (hemispherical scale), 360; *S. oleae* (black scale), 358; *Unaspis euonymi* (euonymus scale), 388
Camellia japonica: *Duplaspidiotus claviger* (camellia mining scale), 380; *Pinnaspis aspidistrae* (fern scale), 390
Camellia sasangua: *Duplaspidiotus claviger* (camellia mining scale), 380
Camellia sinensis: *Parabemisia myrica* (Japanese bayberry whitefly), 322
camphor tree, see *Cinnamomum camphora*
Canary Island pine, see *Pinus canariensis*
cape jasmine, see *Gardinia augusta*
carob, see *Ceratonia siliqua*
Carpenteria californica (bush anemone): *Asterolecanium arabidis* (pit-making pittosporum scale), 348
Carpinus caroliniana (hornbean, ironwood, blue beech): *Alebra albostriella* (maple leafhopper), 416; *Chionaspis gleditsia*, 390; *Corthylus punctatissimus* (pitted ambrosia beetle), 250; *Corythucha pallipes* (birch lace bug), 426; *Dicera lurida*, 270, 272; *Erythroneura lawsoniana*, 416; *Sciurus carolinensis* (gray squirrel), 502
Carya (hickory, pecan)
—foliage: *Anabrus simplex* (Mormon cricket), 492; *Atta texana* (Texas leafcutting ant), 496; *Cyrtepistomus castaneus* (Asiatic oak weevil), 244; *Diapheromera femorata* (walkingstick), 492, 494; *Eriocampa juglandis* (butternut woollyworm), 136; *Graphognathus* sp. (whitefringed beetle), 280; *Phyllonorycter* sp., 196
caterpillars: *Alsophila pometaria* (fall cankerworm), 142; *Anisota senatoria* (orangestriped oakworm moth), 156; *Datana integerrima* (walnut caterpillar), 150; *D. ministra* (yellownecked caterpillar), 154; *Ennomos subsignarius* (elm spanworm), 144–146; *Erannis tiliaria* (linden looper), 144; *Hyphantria cunea* (fall webworm), 166; *Lithophane antennata* (green fruitworm), 148; *Lophocampa caryae* (hickory tussock moth), 160; *Lymantria dispar* (gypsy moth), 138; *Paleacrita vernata* (spring cankerworm), 144; *Phigalia titea* (half-wing geometer), 146; *Schizura concinna* (redhumped caterpillar), 156
mites: *Eotetranychus hicoriae* (pecan leaf scorch mite), 474; eriophyid mite galls, 484; *Oligonychus bicolor* (oak spider mite), 475; *O. viridis*, 475; *Tetranychus homorus*, 476
rollers/skeletonizers: *Acrobasis exsulella*, 212; *A. juglandis* (pecan leaf casebearer), 212; *Archips argyrospila* (fruittree leafroller), 172; *A. griseus*, 218; *A. rosanus*, 218; *Argyrotaenia juglandana*, 218; *Coleophora laticornella* (pecan cigar casebearer), 186; *Psilocorsis cryptolechiella*, 218
—bark, wood, twigs: *Agrilus* sp., 266, 270; *Chrysobothris femorata* (flatheaded appletree borer), 270; *Conotrachelus aratus*, 234; *Dicera lurida*, 270, 272; *Elaphidionoides parallelus*, 264; *E. villosus* (twig pruner), 264; *Euzophera semifuneralis* (American plum borer), 252; *Goes pulcher* (hickory borer), 282; horntails, 494; *Magicicada septendecim* (periodical cicada), 490; *Megacyllene caryae* (painted hickory borer), 274; *Neoclytus acuminatus* (redheaded ash borer), 278; *Oberea bimaculata*, 262; *Oncideres cingulata* (twig girdler), 264; *Prionoxystus robiniae* (carpenterworm), 256; *Saperda discoidea*, 276; *S. lateralis*, 276; *Scolytus quadrispinosus* (hickory bark beetle), 250; *Stictocephala bisonia* (buffalo treehopper), 406; *Synanthedon scitula* (dogwood borer), 262; *Tremex columba* (pigeon tremex), 494; *Xyleborus saxeseni*, 250; *Xylosandrus campactus* (black twig borer), 262
—roots: *Prionus imbricornus* (tilehorned prionus), 280
—sucking insects: *Acanthocephala femorata*, 424; *Alebra albostriella* (maple leafhopper), 416; *Clastoptera obtusa* (alder spittlebug), 420; *Corythucha ciliata* (sycamore lace bug), 428; *Euthochtha galeator*, 422; *Leptoglossus phyllopus*, 422; *Metcalfa pruinosa*, 422
aphids: *Longistigma caryae* (giant bark aphid), 310; *Melanocollis caryaefoliae* (black pecan aphid), 294
phylloxera: *Phylloxera* sp., 460; *P. caryaecaulis*, 460; *P. devastatrix*, 460; *P. notabilis*, 460; *P. russellae*, 460
scales: *Chionaspis caryae* (hickory scurfy scale), 390; *Eulecanium caryae* (hickory lecanium), 364; *Hemiberlesia rapax* (greedy scale), 372; *Melanaspis obscura* (obscure scale), 384; *Parthenolecanium corni* (European fruit lecanium), 364; *P. quercifex* (oak lecanium), 364; *Quadraspidiotus juglansregiae* (walnut scale), 386
Carya cordiformis: *Toumeyella liriodendri* (tuliptree scale), 362
Carya illinoensis (pecan)
—foliage: *Coleophora laticornella* (pecan cigar casebearer), 186; *Coptodisca* sp., 208; *Monocesta coryli* (larger elm leaf beetle), 222
caterpillars: *Datana integerrima* (walnut caterpillar), 150; *Hyphantria cunea* (fall webworm), 166
mites: *Eotetranychus hicoriae* (pecan leaf scorch mite), 474; *Oligonychus viridis*, 475
rollers/skeletonizers: *Acrobasis exsulella*, 212
—bark, wood, twigs: *Conotrachelus aratus*, 234; *Elaphidionoides villosus* (twig pruner), 264; *Euzophera semifuneralis* (American plum borer), 252; *Goes pulcher* (hickory borer), 282; *Oncideres cingulata* (twig girdler), 264; *Synanthedon scitula* (dogwood borer), 262
—roots: *Prionus imbricornus* (tilehorned prionus), 280
—sucking insects: *Acanthocephala femorata*, 422; leaf-footed bugs, 422, 424; *Metcalfa pruinosa*, 422
aphids: *Longistigma caryae* (giant bark aphid), 310; *Melanocollis caryaefoliae* (black pecan aphid), 292
phylloxera: *Phylloxera devastatrix*, 460; *P. notabilis*, 460; *P. russellae*, 460
scales: *Duplaspidiotus claviger* (camellia mining scale), 380; *Hemiberlesia rapax* (greedy scale), 372; *Icerya purchasi* (cottony cushion scale), 338
Carya ovata (shagbark hickory): *Oligonychus viridis*, 475; *Phylloxera* sp., 460
Cassia: *Aphis craccivora* (cowpea aphid), 308; *Epargyreus clarus* (silver-spotted skipper), 148; *Pinnaspis strachani* (lesser snow scale), 390; *Umbonia crassicornis* (thornbug), 408; *Xylosandrus compactus* (black twig borer), 262
Cassia brachiata: *Tetraleurodes acaciae* (acacia whitefly), 320
Cassia fasciculata: *Toumeyella liriodendri* (tuliptree scale), 362
Cassytha filiformis: *Toxoptera aurantii* (black citrus aphid), 302
Castanea (chestnut)
—foliage: *Amorbia humerosana* (whiteline leafroller), 214; *Bucculatrix ainsliella* (oak skeletonizer), 220; *Callirhopalus bifasciatus* (twobanded Japanese weevil), 244; *Cyrtepistomus castaneus* (Asiatic oak weevil), 240; *Datana ministra* (yellownecked caterpillar), 154; *Phyllonorycter* sp., 196; *Symmerista leucitys* (orange-humped mapleworm), 154
mites: *Eotetranychus hicoriae* (pecan leaf scorch mite), 474; *Oligonychus bicolor* (oak spider mite), 475
—bark, wood, twigs: *Agrilus bilineatus* (twolined chestnut borer), 270; *Elaphidionoides villosus* (twig pruner), 264; *Paranthrene simulans*, 254; *Platycotis vittata* (oak treehopper), 406; *Synanthedon scitula* (dogwood borer), 262; *Zeuzera pyrina* (leopard moth), 254
—roots: *Prionus imbricornus* (tilehorned prionus), 280
—sucking insects: *Empoasca fabae* (potato leafhopper), 414; *Glossonotus acuminatus*, 406; *Longistigma caryae* (giant bark aphid), 310; *Melanaspis obscura* (obscure scale), 384; *Parthenolecanium quercifex* (oak lecanium), 364; *Phenacoccus acercis* (apple mealybug), 324
Castanea dentata (American): *Eulecanium caryae* (hickory lecanium), 364; *Parthenolecanium corni* (European fruit lecanium), 364
Castanea mollissima (Chinese): *Cyrtepistomus castaneus* (Asiatic oak weevil), 240
Castanopsis (chinquapin): *Allokermes* sp., 364, 366; *Hemiberlesia rapax* (greedy scale), 372; *Melanaspis obscura* (obscure scale), 384; *Nanokermes* sp., 364, 366

Casuarina: *Hemiberlesia lataniae* (latania scale), 376; *Icerya purchasi* (cottony cushion scale), 338; *Sphyrapicus* sp. (sapsuckers), 500
Casuarina equisetifolia (horsetail casuarina): *Clastopera undulata*, 86
Catalpa: *Aphis gossypii* (cotton aphid, melon aphid), 308; *Lymantria dispar* (gypsy moth), 138; *Pseudaulacaspis pentagona* (white peach scale), 392; *Ostrinia nubilalis* (European corn borer), 286; *Pseudococcus maritimus* (grape mealybug), 88
Catalpa bungei: *Pseudococcus comstocki* (Comstock mealybug), 326
Catalpa japonica: *Tetraleurodes* sp., 322
Ceanothus
—foliage: *Apterona helix*, 178; *Malacosoma californicum pluviale* (western tent caterpillar), 170; *Schizura ipomoeae*, 156; *Tischeria ceanothi*, 196
—twigs: *Periploca ceanothiella* (ceanothus stem gall moth), 438
—sucking insects: *Aleuropleurocelus nigrans* (black aleyrodid), 322; *Aleurothrixis interrogatonis* (deer brush whitefly), 320; *Pseudococcus longispinus* (longtailed mealybug), 324; *P. maritimus* (grape mealybug), 88
aphids: *Aphis ceanothi*, 298
lacebugs: *Corythucha obliqua* (ceanothus tingid), 426, 428; *Gargaphia angulata* (angulate tingid), 426
psyllids: *Euphalerus vermiculosus*, 458; psyllids, 290
scales: *Aspidiotus nerii* (oleander scale), 374; *Asterolecanium arabidis* (pit-making pittosporum scale), 348; *Chionaspis ortholobis*, 390; *C. sassceri*, 390; *Hemiberlesia rapax* (greedy scale), 372
Ceanothus cordulatus (snowbush): *Nymphalis californica* (California tortoiseshell), 152
Ceanothus cyaneus (San Diego ceanothus): *Periploca ceanothiella* (ceanothus stem gall moth), 438
Ceanothus gloriosus: *Euphalerus vermiculosus*, 458
Ceanothus griseus (carmel ceanothus): *Tischeria immaculata*, 196
Ceanothus griseus horizontalis: *Periploca ceanothiella* (ceanothus stem gall moth), 438
Ceanothus integerrimus (deerbrush): *Aphis ceanothi*, 298; *Periploca ceanothiella* (ceanothus stem gall moth), 438
Ceanothus thyrsiflorus (blue brush): *Periploca ceanothiella* (ceanothus stem gall moth), 438
Ceanothus velutinus (varnishleaf ceanothus): *Keonolla confluens*, 416; *Nymphalis californica* (California tortoiseshell), 152
cedar (unspecified): *Chionaspis pinifoliae* (pine needle scale), 108; *Hemiberlesia rapax* (greedy scale), 372; *Hylobius pales* (pales weevil), 56; *Thyridopteryx ephemeraeformis* (bagworm), 176
atlas: *Cedrus atlantica*
deodar: *Cedrus deodara*
eastern red: *Juniperus virginiana*
incense: *Calocedrus (= Libocedrus) decurrens*
northern white: *Thuja occidentalis*
of Lebanon: *Cedrus libani*
red: *Juniperus* spp.
western red: *Thuja plicata*
Cedrus (cedar): *Chionaspis pinifoliae* (pine needle scale), 108; *Hemiberlesia lataniae* (latania scale), 376; *Hylobius pales* (pales weevil), 56; *Thyridopteryx ephemeraeformis* (bagworm), 176; *Trisetacus quadrisetus juniperinus*, 122
Cedrus atlanticus (Atlas cedar): *Cinara curvipes* (bowlegged fir aphid), 82; *Pissodes nemorensis* (eastern pine weevil), 54
Cedrus deodara (deodar cedar): *Cinara curvipes* (bowlegged fir aphid), 82; *Neodiprion lecontei* (redheaded pine sawfly), 16; *Parlatoria pittospori*, 110; *Pissodes nemorensis* (eastern pine weevil), 54; *Pseudococcus longispinus* (longtailed mealybug), 324; *Saissetia oleae* (black scale), 358; *Thyridopteryx ephemeraeformis* (bagworm), 176
Cedrus libani (cedar of Lebanon): *Icerya purchasi* (cottony cushion scale), 338; *Pissodes nemorensis* (eastern pine weevil), 54, 56
Celastrus (bittersweet): *Unaspis euonymi* (euonymus scale), 388
Celastrus scandens (American bittersweet): *Anormenis chloris*, 422; *Enchenopa binotata* (two-marked treehopper), 410; *Ormenoides venusta*, 422
Celtis (hackberry, sugarberry)
—foliage: *Anabrus simplex* (Mormon cricket), 492; *Atta texana* (Texas leafcutting ant), 496; *Celticecis spiniformis*, 452; *Diapheromera femorata* (walkingsticks), 494
caterpillars: *Asterocampa celtis* (hackberry butterfly), 152; *A. clyton* (tawny emperor), 152; Nymphalis antiopa (spiny elm caterpillar, mourningcloak butterfly), 152; *Sphinx drupiferarum* (cherry sphinx caterpillar), 164
mites: *Eotetranychus matthyssei*, 474; *Oligonychus propetes*, 475; witches'-broom, 230
—bark, wood, twigs: *Agrilus* sp., 272; *Elaphidionoides villosus* (twig pruner), 264; *Megacyllene caryae* (painted hickory borer), 274; *Neoclytus acuminatus* (redheaded ash borer), 278; *Oncideres cingulata* (twig girdler), 264
—sucking insects: *Corythucha celtidis* (hackberry lace bug), 426; *Homalodisca* sp. (leafhopper), 418; *Pseudococcus maritimus* (grape mealybug), 88
psyllids: *Pachypsylla* sp., 450, 452; *P. celtidisgemma* (hackberry bud gall maker), 452; *P. celtidismamma* (hackberry nipple gall maker), 452; *P. celtisvesicula* (hackberry blister gall), 452; psyllids (jumping plant lice), 452
scales: *Chionaspis americana* (elm scurfy scale), 358; *Eriococcus borealis*, 336; *Eulecanium caryae* (hickory lecanium), 364; *Melanaspis obscura* (obscure scale), 384; *M. tenebricosa* (gloomy scale), 384; *Pulvinaria innumerabilis* (cottony maple scale), 340; *Quadraspidiotus juglansregiae* (walnut scale), 386
Celtis laevigata (sugarberry): *Asterocampa celtis* (hackberry butterfly), 152; *A. clyton* (tawny emperor), 152; *Pachypsylla* sp., 450
Celtis occidentalis (hackberry): *Aceria celtis, 478*; *Icerya purchasi* (cottony cushion scale), 338
Cephalanthus (buttonbush): *Toumeyella cerifera* (buttonbush scale), 362; *T. liriodendri* (tuliptree scale), 362
Cephalanthus occidentalis (buttonwillow): *Toumeyella cerifera* (buttonbush scale), 362
Cephalotaxus (plum yew): *Aonidiella taxus* (Asiatic red scale), 110
Ceratonia siliqua (carob): *Aspidiotus nerii* (oleander scale), 374
Cercidiphyllum japonicum: Phenacoccus aceris (apple mealybug), 324
Cercis canadensis (redbud)
—foliage: *Atta texana* (Texas leafcutting ant), 496; *Cyrtepistomus castaneus* (Asiatic oak weevil), 244; *Megachile* sp. (leafcutter bees), 494
caterpillars: *Lithophane antennata* (green fruitworm), 148; *Malacosoma californicum pluviale* (western tent caterpillar), 170; *Orgyia leucostigma* (whitemarked tussock moth), 158; *Parasa indetermina* (stinging rose caterpillar), 164; *Schizura concinna* (redhumped caterpillar), 156
mites: *Eotetranychus matthyssei*, 474; *Tetranychus bimaculatus* (twospotted spider mite), 476
—bark, wood, twigs: *Agrilus* sp., 272; *Elaphidionoides villosus* (twig pruner), 264; *Neoclytus acuminatus* (redheaded ash borer), 278; *Oncideres cingulata* (twig girdler), 264; *Xylosandrus compactus* (black twig borer), 262
—sucking insects: *Acanalonia conica*, 422; *Metcalfa pruinosa*, 422; *Pseudococcus affinis* (obscure mealybug), 324; *Tetraleurodes acaciae* (acacia whitefly), 322; *Trialeurodes vaporariorum* (greenhouse whitefly), 320
leaf/treehoppers: *Enchenopa binotata* (two-marked treehopper), 410; *Erythroneura gleditsia*, 416; *Prosapia bicincta*, 420
scales: *Aspidiotus nerii* (oleander scale), 374; *Hemiberlesia rapax* (greedy scale), 372; *Mesolecanium nigrofasciatum* (terrapin scale), 364; *Parthenolecanium corni* (European fruit lecanium), 364; *Pseudaulacaspis pentagona* (white peach scale), 392
Cercis occidentalis (western redbud): *Schizura concinna* (redhumped caterpillar), 156
Cestrum: *Asterolecanium arabidis* (pit-making pittosporum scale), 348; *Duplaspidiotus claviger* (camellia mining scale), 380; *Hemiberlesia rapax* (greedy scale), 372; *Saissetia oleae* (black scale), 358
Cestrum elegans: *Graphocephala atropunctata* (blue-green sharpshooter), 418
Chaedioria erumpens: *Icerya purchasi* (cottony cushion scale), 338
Chaenomeles (quince)
—foliage: *Bucculatrix pomifoliella*, 220; *Phyllonorycter crataegella*, 196; *Popillia japonica* (Japanese beetle), 236
caterpillars: *Archips argyrospila* (fruit tree leafroller), 218; *Datana ministra* (yellownecked caterpillar), 154; *Malacosoma disstria* (forest tent caterpillar), 170
—bark, wood, twigs: *Elaphidionoides villosus* (twig pruner), 264; *Saperda candida* (roundheaded apple tree borer), 278; *Stictocephala bisonia* (buffalo treehopper), 406
—roots: *Prionus californicus* (California prionus), 280
—sucking insects: *Lygus lineolaris* (tarnished plant bug), 398; *Pseudococcus affinis* (obscure mealybug), 324
aphids: *Aphis pomi* (apple aphid), 300; *Nearctaphis crataegifoliae*, 314
scales: *Aonidiella aurantii* (California red scale), 378; *Ceroplastes ceriferus* (Indian wax scale), 356; *Hemiberlesia rapax* (greedy scale), 372; *Icerya purchasi* (cottony cushion scale), 338; *Quadraspidiotus perniciosus* (San Jose scale), 386; *Sphaerolecanium prunastri* (globose scale), 366. See also *Cydonia*
Chaenomeles japonica (Japanese flowering quince): *Asterolecanium arabidis* (pit-making pittosporum scale), 348; *Hemiberlesia rapax* (greedy scale), 372; *Pseudococcus maritimus* (grape mealybug), 88; *Saissetia oleae* (black scale), 358
Chamaecyparis: *Carulaspis juniperi* (juniper scale), 106; *Phloeosinus* sp., 70; *Phyllobius intrusus* (arborvitae weevil), 240, 244. See also *Cupressus*, *Taxodium*
Chamaecyparis lawsoniana (Lawson cypress): *Argyresthia cupressella*, 42; *A. libocedrella*, 42
Chamaecyparis nootkatensis (Alaska cedar): *Trisetacus chamaecypari*, 122
Chamaedoria erumpens (bamboo palm): *Icerya purchasi* (cottony cushion scale), 338
Chamaerops (palm species): *Diaspis cocois* (cocos scale), 376
chaparral broom, see *Baccharis pilularis*
cherry, see *Prunus*
black: *Prunus serotina*
choke: *Prunus virginiana*
flowering: flowering cherry
hollyleaf: *Prunus ilicifolia*
Japanese flowering: *Prunus serrulata*
laurel: *Prunus laurocerasus*
sour: *Prunus cerasus*
wild: *Prunus serotina*
chestnut, see *Castanea*
Chinese: *Castanea mollissima*
chinaberry, see *Melia azedarch*
Chinese elm, see *Ulmus parvifolia*
chinquapin, see *Castanopsis*
Chionanthus virginicus (fringe tree): *Aulacaspis rosae* (rose scale), 366

Chiranthodendron pentadactylon (monkey-hand tree, handflower tree): *Hemiberlesia lataniae* (latania scale), 376
Choisya ternata (Mexican orange): *Saissetia oleae* (black scale), 358
chokecherry, see *Prunus virginiana*
Christmas berry, see *Heteromeles arbutifolia*
chrysanthemum: *Franklinella* sp., 434
Chrysothamnus nauseosus (rabbitbrush): *Trirhabda nitidicollis,* 238
Cinnamomum (camphor): *Chrysomphalus dictyospermi* (dictyospermum scale), 382; *Euthoracaphis umbellularia* (California laurel aphid), 294; *Hemiberlesia rapax* (greedy scale), 372; *Phenacoccus aceris* (apple mealybug), 324; *Pseudaonidia duplex* (camphor scale), 372
Cinnamomum camphora (camphor tree): *Euthoracaphis umbellulariae* (California laurel aphid), 294; *Urodus parvula* (ermine moth), 148; *Xylosandrus compactus* (black twig borer), 262
Cissus: *Asynonychus godmani* (Fuller rose beetle), 230; *Hemiberlesia rapax* (greedy scale), 372
Cistus: *Asterolecanium arabidis* (pit-making pittosporum scale), 348
Citrofortunella mitis: *Aleurocanthus woglumi* (citrus blackfly), 322
Citrus
—foliage: *Asynonychus godmani* (Fuller rose beetle), 230; grasshoppers, 490; katydids, 490; *Microcentrum rhombifolium* (broadwinged katydid), 490; *Oiketicus abbotii,* 178; *Pachnaeus litus* (citrus root weevil), 240; slugs, 498; snails, 498; *Urodus parvula* (ermine moth), 148
mites: *Brevipalpus lewisi* (citrus flat mite), 476; *B. obovatus* (privet mite), 482; *Eotetranychus lewisi,* 474; *E. sexmaculatus* (six-spotted mite), 474; *E. yumensis* (Yuma spider mite), 474; *Panonychus citri* (citrus red mite), 472
thrips: *Gynaikothrips ficorum* (Cuban laurel thrips), 434; *Heliothrips haemorrhoidalis* (greenhouse thrips), 430
—roots: *Pachnaeus litus* (citrus root weevil), 240
—sucking insects: *Acanthocephala femorata,* 424; *Aleurocanthus woglumi* (citrus blackfly), 322; *Aleurothrixis floccosus* (woolly whitefly), 322; *Dialeurodes citrifolii* (cloudywinged whitefly), 322; *Euthochtha galeator,* 422; *Kunzeana kunzii,* 400; *Leptoglossus phyllopus* (leaf-footed bugs), 422, 424; *Metcalfa pruinosa,* 422; *Parabemisia myricae* (Japanese bayberry whitefly), 322; *Siphoninus phillyreae* (ash whitefly), 322; *Tetraleurodes mori* (mulberry whitefly), 324; whiteflies, 320
aphids: *Aphis citricola* (spirea aphid), 298; *A. craccivora* (cowpea aphid), 308; *Macrosiphum euphorbiae* (potato aphid), 300; *Myzus persicae* (green peach aphid), 300; *Toxoptera aurantii* (black citrus aphid), 302
mealybugs: *Planococcus citri* (citrus mealybug), 328; *Pseudococcus affinis* (obscure mealybug), 324; *P. comstocki* (Comstock mealybug), 326; *P. longispinus* (longtailed mealybug), 324; *P. maritimus* (grape mealybug), 88
scales: *Aonidiella aurantii* (California red scale), 378; *A. citrina* (yellow scale), 378; *Aspidiotus nerii* (oleander scale), 374; *Ceroplastes cirripediformis* (barnacle scale), 356; *C. cistudiformis* (tortoise wax scale), 358; *C. floridensis* (Florida wax scale), 356; *C. rubens* (red wax scale), 358; *C. sinensis* (Chinese wax scale), 356; *Chionaspis sassceri,* 390; *Chrysomphalus aonidium* (Florida red scale), 380; *C. dictyospermi* (dictyospermum scale), 382; *Fiorinia theae* (tea scale), 394; *Hemiberlesia rapax* (greedy scale), 372; *Icerya purchasi* (cottony cushion scale), 338; *Lepidosaphes beckii* (purple scale), 380; *L. gloveri* (Glover scale), 380; *Lindingaspis rossi,* 376; *Pinnaspis aspidistrae* (fern scale), 390; *P. strachani* (lesser snow scale), 390; *Pseudaonidia trilobitiformis* (trilobe scale), 376; *Saissetia coffeae* (hemispherical scale), 360; *S. oleae* (black scale), 358; *Unaspis citri* (citrus snow scale), 390
Citrus sinensis (sweet orange): *Pinnaspis strachani* (lesser snow scale), 390
Clematis: *Alcathoe caudata* (clematis borer), 260; *Chrysomphalus dictyospermi* (dictyospermum scale), 382; *Helix aspersa* (brown garden snail), 498; *Myzus persicae* (green peach aphid), 300
Clethra alnifolia (summersweet; sweet pepperbush): *Oligonychus ilicis* (southern red mite), 475
Cleyera: *Lepidosaphes camelliae,* 370
Clusia: *Toxoptera aurantii* (black citrus aphid), 302
Coccoloba floridana (seagrape): *Ceroplastes nakahari* (Nakahara wax scale), 358
Coccoloba uvifera (seagrape): *Aleurothrixus floccosus* (woolly whitefly), 322; *Ctenodactylomyia watsoni* (gall midge), 232; *Hemiberlesia rapax* (greedy scale), 372; *Hexeris enhydris* (seagrape borer), 232; *Melanaspis coccolobae, 232; Metcalfa pruinosa* (flatid planthopper), 422; *Toxoptera aurantii* (black citrus aphid), 232, 302
Cocos (palm): *Diaspis cocois* (cocos scale), 376; *Pinnaspis aspidistrae* (fern scale), 390
Codiaeum variegatum (croton): *Heliothrips haemorrhoidalis* (greenhouse thrips), 430; *Parasaissetia nigra* (nigra scale), 360; *Pinnaspis aspidistrae* (fern scale), 390; *Planococcus citri* (citrus mealybug), 328; *Saissetia coffeae* (hemispherical scale), 360; *S. oleae* (black scale), 358
Coffea: *Toxoptera aurantii* (black citrus aphid), 302
coffeeberry, see *Rhamnus californica*
Coprosma: *Aonidiella aurantii* (California red scale), 378
Coprosma repens (mirrow-plant): *Graphocephala atropunctata* (blue-green sharpshooter), 418
coral tree, see *Erythrina fusca*
Cornus (dogwood)
—foliage: *Anabrus simplex* (Mormon cricket), 237; *Callirhopalus bifasciatus* (twobanded Japanese weevil), 244; cecidomyid midges, 462; *Cyrtepistomus castaneus* (Asiatic oak weevil), 244; *Diapheromera femorata* (walkingsticks), 492, 494; eyespot galls, 462; *Lygocoris communis* (pear plant bug), 396; *Macremphytus tarsatus,* 126; *M. varianus,* 126; *Odontota dorsalis* (locust leafminer), 87; *Paralellodiplosis subtruncata,* 462
caterpillars: *Choristoneura rosaceana* (obliquebanded leafroller), 216; *Lithophane antennata* (green fruitworm), 148; *Lymantria dispar* (gypsy moth), 138; *Schizura concinna* (redhumped caterpillar), 154
—bark, wood, twigs: *Agrilus* sp., 272; *Corthylus punctatissimus* (pitted ambrosia beetle), 250; *Magicicada septendecim* (periodical cicada), 490; *Neoclytus acuminatus* (redheaded ash borer), 278; *Oberea bimaculata,* 123; *O. tripunctata* (dogwood twig borer), 262; *Odontota dorsalis* (locust leafminer), 190; *Oecanthus angustipennis* (narrowwinged cricket), 492, 494; *O. nigricornis* (blackhorned tree cricket), 492, 494; *Oncideres cingulata* (twig girdler), 264; *Synanthedon scitula* (dogwood borer), 260, 262; *Xyleborus saxeseni,* 250; *Xylosandrus compactus* (black twig borer), 262; *X. germanus,* 250
—roots: *Prionus laticollis* (broadnecked prionus), 280
—sucking insects: *Acanalonia conica,* 422; *Clastoptera proteus* (dogwood spittlebug), 420; *Dysmicoccus wisteriae* (taxus mealybug), 88; *Heliothrips haemorrhoidalis* (greenhouse thrips), 430; *Macrosiphum euphorbiae* (potato aphid), 300; *Metcalfa pruinosa,* 422; *Phenacoccus aceris* (apple mealybug), 324; *Poecilocapsus lineatus* (fourlined plant bug), 396; *Trialeurodes vaporariorum* (greenhouse whitefly), 320
leafhoppers: *Edwardsiana commisuralis,* 416; *E. rosae* (rose leafhopper), 412
scales: *Aspidiotus nerii* (oleander scale), 374; *Chionaspis corni* (dogwood scale), 368; *C. lintneri,* 390; *Eulecanium cerasorum* (calico scale), 352; *Fiorinea theae* (tea scale), 394; *Lepidosaphes ulmi* (oystershell scale), 370; *Melanaspis obscura* (obscure scale), 384; *Pseudaulacaspis pentagona* (white peach scale), 392; *Pulvinaria acericola* (cottony maple leaf scale), 342; *P. innumerabilis* (cottony maple scale), 340; *Quadraspidiotus juglansregiae* (walnut scale), 386; *Q. perniciosus* (San Jose scale), 386
Cornus florida (flowering dogwood): *Duplaspidiotus claviger* (camellia mining scale), 380; *Dysmicoccus wistariae* (taxus mealybug), 88, 326; *Edwardsiana commisuralis,* 416; *Pseudaulacaspis cockerelli* (false oleander scale), 374; *Pulvinaria innumerabilis* (cottony maple scale), 340, 346; *Resseliella clavula* (dogwood club gall midge), 436; *Synanthedon scitula* (dogwood borer), 260, 262; *Tetraleurodes mori* (mulberry whitefly), 324; *Xylosandrus campactus* (black twig borer), 262
Cornus kusa (Korean dogwood): *Synanthedon scitula,* 262
Cornus racemosa (gray dogwood): *Chionaspis corni* (dogwood scale), 368; *Macremphytus tarsatus,* 126
Cornus stolonifera (red-osier dogwood): *Macremphytus tarsatus,*126;*Pandemis pyrusana,* 218
Cornus stricta (swamp dogwood): *Duplaspidiotus claviger* (camellia mining scale), 380
Correa harrisii: *Asterolecanium arabidis* (pit-making pittosporum scale), 348
Corylus (filbert, hazel)
—foliage: *Monocesta coryli* (larger elm leaf beetle), 222
caterpillars: *Dichomeris ligulella,* 30; *Hemileuca maia* (eastern buck moth), 156; *Malacosoma californicum pluviale* (western tent caterpillar), 170; *Rheumaptera hastata* (spear-marked black moth), 146
leafrollers/skeletonizers: *Apotomis* spp., 218; *Archips semiferana* (oak leafroller), 214; *Choristoneura rosaceana* (obliquebanded leafroller), 216; *Coleophora serratella* (birch casebearer), 186; *Pandemis limitata,* 218
mites: *Cecidophyopsis vermiformis,* 480; *Eotetranychus tiliarium,* 474; *Phytoptus avellanae,* 480
sawflies: *Hemichroa crocea* (striped alder sawfly), 136
—bark, wood, twigs: *Corthylus punctatissimus* (pitted ambrosia beetle), 250; *Stictocephala bisonia* (buffalo treehopper), 406; *Synanthedon scitula* (dogwood borer), 262; *Xyleborus dispar,* 250
—sucking insects: *Chionaspis lintneri,* 390; *Corythucha pergandei* (alder lace bug), 426; *Leptoypha costata,* 426; *Phenacoccus aceris* (apple mealybug), 324; psyllids, 290. See hazelnut
Cotoneaster
—foliage: *Acrobasis indiginella* (leaf crumpler), 218; *Caliroa cerasi* (pear sawfly), 130; *Choristoneura rosaceana* (obliquebanded leafroller), 216; slugs, 498; snails, 498
—mites: eriophyid mites, 486; *Oligonychus platani* (platanus spider mite), 475
—bark, wood, twigs: *Saperda candida* (roundheaded appletree borer), 278; *Chrysobothris femorata* (flatheaded appletree borer), 270
—sucking insects: *Corythucha cydoniae* (hawthorn lace bug), 426; *Macrosiphum euphorbiae* (potato aphid), 300; *Phenacoccus aceris* (apple mealybug), 324
scales: *Chrysomphalus dictyospermi* (dictyospermum scale), 382; *Hemiberlesia rapax* (greedy scale), 372; *Lepidosaphes ulmi* (oystershell scale), 370; *Quadraspidiotus perniciosus* (San Jose scale), 386
Cotoneaster acutifolius: *Corythucha cydoniae* (hawthorn lace bug), 426

Cotoneaster buxifolius (dwarf silver-leaf cotoneaster): *Corythucha cydoniae* (hawthorn lace bug), 426
Cotoneaster congestus: Athrips rancidella, 218
Cotoneaster dammeri (bearberry cotoneaster): *Corythucha cydoniae* (hawthorn lace bug), 426
Cotoneaster horizontalis (rock cotoneaster): *Asterolecanium arabidis* (pit-making pittosporum scale), 348; *Athrips rancidella,* 218; *Corythucha cydoniae* (hawthorn lace bug), 426; *Orgyia leucostigma* (whitemarked tussock moth), 158
Cotoneaster lacteus (red clusterberry cotoneaster): *Corythucha cydoniae* (hawthorn lace bug), 426
Cotoneaster microphylla: *Parlatoria pittospori,* 110
Cotoneaster nitens: *Corythucha cydoniae* (hawthorn lace bug), 426
Cotoneaster salicifolius: *Corythucha cydoniae* (hawthorn lace bug), 426
Cotoneaster watereri: *Corythucha cydoniae* (hawthorn lace bug), 426
cottonwood, see *Populus deltoides*
eastern: *Populus deltoides*
plains: *Populus deltoides* var. *occidentalis*
western: *Populus fremontii*
coyote brush, see *Baccharis pilularis*
crabapple (also see crabapple, flowering; *Malus*)
—foliage: *Caliroa cerasi* (pear sawfly), 130; *Coptodisca* sp., 208
caterpillars: *Amorbia humerosana* (whiteline leafroller), 214; *Datana ministra* (yellownecked caterpillar), 154; *Lithophane antennata* (green fruitworm), 148; *Malacosoma americanum* (eastern tent caterpillar), 168; *Orgyia vetusta* (western tussock moth), 160
mites: *Phyllocoptes* sp., 482
rollers/skeletonizers: *Archips argyrospila* (fruit tree leafroller), 172; *Argyrotaenia velutinana* (redbanded leafroller), 214; *Choreutis pariana* (apple-and-thorn skeletonizer), 216; *Sparganothis sulfureana,* 216
—bark, wood, twigs: *Chrysobothris femorata* (flatheaded appletree borer), 270; *Elaphidionoides villosus* (twig pruner), 264; *Neoclytus acuminatus* (redheaded ash borer), 278; *Saperda candida* (roundheaded appletree borer), 278; *Synanthedon pyri* (apple bark borer), 258
—sucking insects: *Corythucha pergandei* (alder lace bug), 426
aphids: *Aphis pomi* (apple aphid), 300; *Macrosiphum euphorbiae* (potato aphid), 300; *M. liriodendri* (tuliptree aphid), 292; *Rhopalosiphum fitchi* (apple grain aphid), 292
leaf/treehoppers: *Stictocephala bisonia* (buffalo treehopper), 406; *Typhlocyba pomaria* (white apple leafhopper), 416
scales: *Parthenolecanium corni* (European fruit lecanium), 364; *P. pruinosum* (frosted scale), 354; *Quadraspidiotus perniciosus* (San Jose scale), 386
crabapple, flowering: *Aphis pomi* (apple aphid), 300; *Bucculatrix pomifoliella,* 220; *Eulecanium cerasorum* (calico scale), 354; *Euzophera semifuneralis* (American plum borer), 252; *Popillia japonica* (Japanese beetle), 236; *Rhopalosiphum fitchii* (apple grain aphid), 292; *Saperda candida* (roundheaded apple tree borer), 278. See also *Malus.*
cranberry, see *Vaccinium*
cranberry bush, European, see *Viburnum opulus*
crape myrtle, see *Lagerstroemia indica*
Crataegus (hawthorn)
—foliage: *Arge clavicornis,* 134; *Caliroa cerasi* (pear sawfly), 130; marginal fold galls, 484; *Monocesta coryli* (larger elm leaf beetle), 222; *Odontota dorsalis* (locust leafminer), 190; *Oecanthus fultoni* (snowy tree cricket), 492; *Profenusa canadensis* (hawthorn leafmining sawfly), 188; *Tischeria malifoliella* (appleleaf trumpet miner), 196
caterpillars: *Amorbia humerosana* (whiteline leafroller), 214; *Archips cerasivorana* (ugly nest caterpillar), 172; *A. rosanus,* 218; *Hyalophora cecropia* (cecropia moth), 164; *Lymantria dispar* (gypsy moth), 138; *Malacosoma disstria* (forest tent caterpillar), 170; *Orgyia vetusta* (western tussock moth), 160; *Schizura concinna* (redhumped caterpillar), 156; *Yponomeuta padella,* 174
mites: *Eotetranychus tiliarium,* 474; eriophyid gall mites, 486; leaf blister, 486; *Oligonychus newcomeri,* 475; *O. propetes,* 475; *Panonychus ulmi* (European red mite), 472
rollers/skeletonizers: *Acrobasis indiginella* (leaf crumpler), 218; *Archips argyrospila* (fruittree leafroller), 172; *Argyrotaenia velutinana* (redbanded leafroller), 214; *Bucculatrix pomifoliella,* 220; *Choreutis pariana* (apple-and-thorn skeletonizer), 216; *Choristoneura rosaceana* (obliquebanded leafroller), 216; *Epinotia aceriella* (maple trumpet skeletonizer), 212; *Sparganothis sulfureana,* 216
—bark, wood, twigs: *Magicicada septendecim* (periodical cicada), 490; *Oberea* sp., 286; *Oecanthus fultoni* (snowy tree cricket), 492; *Saperda candida* (roundheaded appletree borer), 278; *S. fayi* (thornlimb borer), 276; *Scolytus rugulosus* (shothole borer), 250; *Synanthedon pyri* (apple bark borer), 258, 260
—sucking insects: *Anormenis chloris,* 422; *Corythucha bellula,* 426; *C. ciliata* (sycamore lace bug), 426; *C. cydoniae* (hawthorne lace bug), 426; *Frankliniella tritici* (flower thrips), 432; *Ormenoides venusta,* 422; *Phenacoccus aceris* (apple mealybug), 324; *Siphoninus phillyreae* (ash whitefly), 322
aphids: *Aphis pomi* (apple aphid), 300; *Eriosoma crataegi,* 306; *E. lanigerum* (woolly apple aphid), 316; *Nearctaphis crataegifoliae,* 314; *Rhopalosiphum fitchii* (apple grain aphid), 292; *Utamphorophora crataegi* (fourspotted hawthorn aphid), 314
leaf/treehopper: *Alebra albostriella* (maple leafhopper), 416; *Edwardsiana rosae* (rose leafhopper), 412; *Erythroneura gleditsia,* 416; *Stictocephala bisonia* (buffalo treehopper), 406; *Typhlocyba pomaria* (white apple leafhopper), 416
scales: *Chionaspis furfura,* 368; *Eriococcus azaleae* (azalea bark scale), 336; *Mesolecanium nigrofasciatum* (terrapin scale), 364; *Parthenolecanium pruinosum* (frosted scale), 354; *Pulvinaria innumerabilis* (cottony maple scale), 340; *Quadraspidiotus juglansregia* (walnut scale), 386; *Q. perniciosus* (San Jose scale), 386
Crataegus crus-galli (cockspur): *Profenusa·canadensis* (hawthorn leafmining sawfly), 188
Crataegus erecta: *Profenusa canadensis* (hawthorn leafmining sawfly), 188
Crataegus laevigata (English hawthorn): *Eriosoma crataegi,* 306
Crateagus mollis (downy hawthorn): *Profenusa canadensis* (hawthorn leafmining sawfly), 188
Crataegus persimillis: *Profenusa canadensis* (hawthorn leafmining sawfly), 188
Crataegus phaenopyrum (Washington thorn): *Corythucha cydoniae* (hawthorn lace bug), 426
Crataegus stricta: *Caliroa cerasi* (pear sawfly), 130
croton, see *Codiaeum variegatum*
cucumbertree, see *Magnolia acuminata*
Cupressocyparis (Leland-cypress): *Stenolechia bathrodyas,* 42
Cupressus (cypress)
—foliage: *Epinotia subviridis* (cypress leaftier), 38; *Susana cupressi* (cypress sawfly), 18
caterpillars: *Thyridopteryx ephemeraeformis* (bagworm), 176
leafminers: *Argyresthia cupressella,* 42; *A. franciscella,* 42; *A. trifasciae,* 42
—shoots, twigs: cypress tipminers, 42; *Phloeosinus* sp., 66, 70; *Stenolechia bathrodyas,* 42
—sucking insects: *Cinara tujafilina,* 84
scales: *Carulaspis juniperi* (juniper scale), 106; *Chrysomphalus dictyospermi* (dictyospermum scale), 382; *Ehrhornia cupressi* (cottony cypress scale, cypress bark scale, 332; cypress bark mealybug), 332; *Icerya purchasi* (cottony cushion scale), 338; *Lepidosaphes gloveri* (Glover scale), 380
Cupressus arizonica (smooth Arizona cypress): *Ehrhornia cupressi* (cypress bark mealybug), 332
Cupressus arizonica var. *glabra* (smooth Arizona cypress): *Ehrhornia cupressi* (cypress bark mealybug), 332
Cupressus macrocarpa (Monterey cypress): *Argyresthia franciscella* (cypress tip miner), 42; *Coleotechnites juniperella* (leaf/tipminer), 42; *C. stanfordia,* 42; *Ehrhornia cupressi* (cypress bark mealybug), 332; *Laspeyresia cupressana* (cypress bark moth), 60; *Phloeosinus cristatus* (cypress bark beetle), 70; *P. cupressi* (cypress bark beetle), 70; *Susana cupressi* (cypress sawfly), 18
currant, see *Ribes*
Cycas (sago palm): *Aspidiotus nerii* (oleander scale), 162
cyclamen: *Phytonemus pallidus* (cyclamen mite), 476; *Polyphagotarsonemus latus* (broad mite, 476
Cydonia (quince): *Acrobasis indigenella* (leaf crumpler), 218; *Archips argyrospila* (fruittree leafroller), 218; *Corythucha cydoniae* (hawthorn lace bug), 426; *Prionus californicus* (California prionus), 280
cypress: *Cinara tujafilina,* 84; cypress tipminers, 42; *Lepidosaphes gloverii* (Glover scale), 380; *Phloeosinus* sp., 66, 70; *Thyridopteryx ephemeraeformis* (bagworm), 176. See also *Cupressus, Chamaecyparis, Taxodium*
cypress
Arizona: *Cupressus arizonica*
bald: *Taxodium distichum*
Monterey: *Cupressus macrocarpa*
pond: *Taxodium distichum* var. *nutans*
cypress, American: *Ehrhornia cupressi* (cottony cypress scale, cypress bark mealybug, cypress bark scale), 332
Cytisus (brooms): *Chionaspis ortholobis,* 390; *Uresiphita reversalis* (genista caterpillar), 166

daisy: *Franklinella* sp., 434
Daphne: *Aonidiella citrina* (yellow scale), 378; *Aspidiotus nerii* (oleander scale), 374; *Parasaissetia nigra* (nigra scale), 360; *Unaspis euonymi* (euonymus scale), 388
Daubentonia tripetii (glory pea; *Sesbanis punicea*): *Asterolecanium arabidis* (pit-making pittosporum scale), 348
deerweed, see *Lotus scoparius*
Delonix regia (royal poinciana): *Umbonia crassicornis* (thornbug), 408
delphinium: *Magicicada septendecim* (periodical cicada), 492
deodar cedar: *Cedrus deodara*
Deutzia: *Asynonychus godmani* (Fuller rose beetle), 230; *Callirhopalus bifasciatus* (twobanded Japanese weevil), 244; *Caloptilia syringella* (lilac leafminer), 196; *Poecilocapsus lineatus* (fourlined plant bug), 396
Deutzia scabra (fuzzy deutzia): *Asterolecanium arabidis* (pit-making pittosporum scale), 348
Diospyros (persimmon)
—foliage: *Asynonychus godmani* (Fuller rose beetle), 230; *Cyrtepistomus castaneus* (Asiatic oak weevil), 244; *Hyphantria cunea* (fall webworm), 166; *Schizura concinna* (redhumped caterpillar), 156; *Selenothrips rubrocinctus* (redbanded thrips), 432
—bark, wood, twigs: *Agrilus* sp., 272; *Elaphidionoides villosus* (twig pruner), 264; *Euzophera semifuneralis* (American plum borer), 252; *Oncideres cingulata* (twig girdler,), 264
—sucking insects: *Aleuroplatus berbericolus* (barberry whitefly), 322; *Leptoglossus phyllopus*

(leaf-footed bugs), 422–424; *Phenacoccus aceris* (apple mealybug), 324; *Pseudococcus comstocki* (Comstock mealybug), 326; *P. longispinus* (longtailed mealybug), 324
scales: *Ceroplastes dugesii* (Duges wax scale), 358; *Hemiberlesia rapax* (greedy scale), 372; *Parthenolecanium quercifex* (oak lecanium), 364; *Pseudaulacaspis pentagona* (white peach scale), 392
Diplacus longiflorus: *Nipaecoccus aurilanatus* (golden mealybug), 90
Dizygotheca: *Ceroplastes rubens* (red wax scale), 358
dogwood, see *Cornus*
flowering: *Cornus florida*
gray: *Cornus racemosa*
Douglas-fir, see *Pseudotsuga menziesii*
downy serviceberry, see *Amelanchier arborea*
Dracaena: *Asynonychus godmani* (Fuller rose beetle), 230
Duranta: *Saissetia oleae* (black scale), 358
Dutchman's pipe (*Aristolochia*): *Pseudococcus comstocki* (Comstock mealybug), 326
dwarf chaparral broom, see *Baccharis pilularis*

eastern redcedar, see *Juniperus virginiana*
Elaeagnus: *Aspidiotus nerii* (oleander scale), 374; *Chrysomphalus dictyospermi* (dictyospermum scale), 382; *Hemiberlesia rapax* (greedy scale), 372; *Lepidosaphes beckii* (purple scale), 380; *Oligonychus ilicis* (southern red mite), 475; *Panonychus citri* (citrus red mite), 472; *Parthenolecanium persicae* (European peach scale), 364; *Pseudaulacaspis cockerelli* (false oleander scale), 374; *Pseudococcus comstocki* (Comstock mealybug), 326
Elaeagnus pungens (silverberry): *Toxoptera aurantii* (black citrus aphid), 302
elm, see *Ulmus*
American: *Ulmus americana*
Chinese: *Ulmus parvifolia*
Dutch: *Ulmus hollandica*
English: *Ulmus procera*
red: *Ulmus rubra*
Scotch: *Ulmus glabra*
Siberian: *Ulmus pumila*
slippery: *Ulmus rubra*
empress tree, see *Paulownia tomentosa*
English ivy, see *Hedera helix*
English laurel, see *Prunus laurocerasus*
English walnut, see Persian walnut
Ephedra trifurca (joint fir): *Bryobia ephedrae*, 472
Erigonum (umbrella plant): *Aspidiotus nerii* (oleander scale), 374
Eriobotrya japonica (loquat): *Aleuroplatus berbericolus* (barberry whitefly), 322; *Hemiberlesia lataniae* (latania scale), 376; *H. rapax* (greedy scale), 372; *Leptoglossus phyllopus* (leaf-footed bugs), 422; *Oligonychus platani* (platanus spider mite), 475; *Prionoxystus robiniae* (carpenterworm), 256; *Siphoninus phillyreae* (ash whitefly), 322; *Synanthedon scitula* (dogwood borer), 262
Eriophyllum confertiflorum: *Asterolecanium arabidis* (pit-making pittosporum scale), 348
Erythrina fusca (coral tree): *Icerya purchasi* (cottony cushion scale), 338; *Pseudococcus affinis* (obscure mealybug), 324
Eucalyptus: *Aonidiella aurantii* (California red scale), 378; *Aphis craccivora* (cowpea aphid), 308; *A. gossypii* (cotton aphid), 308; *Argyrotaenia citrana* (orange tortrix), 216; *Aspidiotus nerii* (oleander scale), 374; *Chrysomphalus dictyospermi* (dictyospermum scale), 382; *Hemiberlesia rapax* (greedy scale), 372; *Lindingaspis rossi*, 376; *Oligonychus ilicis* (southern red mite), 475; *Prionus californicus* (California prionus), 280; *Pseudococcus longispinus* (longtailed mealybug), 324; *Saissetia oleae* (black scale), 358
Eugenia: *Chrysomphalus dictyospermi* (dictyospermum scale), 382; *Duplaspidiotus claviger* (camellia mining scale), 380; *Oligonychus ilicis* (southern red mite), 475; *Tetraleurodes mori* (mulberry whitefly), 324; *Trioza eugeniae* (eugenia psyllid), 458; *Unaspis euonymi* (euonymus scale), 388
Eugenia myrtifolia: *Ceroplastes nakaharai* (Nakahara wax scale), 358
Eugenia uniflora (Barbados cherry): *Aleurothrixis floccosus* (woolly whitefly), 322
Euonymus (burning bush)
—foliage: *Argyrotaenia citrana* (orange tortrix), 216; *Caloptilia syringella* (lilac leafminer), 196; *Otiorhynchus sulcatus* (black vine weevil), 240, 242
—sucking insects: *Aphis fabae* (bean aphid), 300
scales: *Aonidiella aurantii* (California red scale), 378; *A. citrina* (yellow scale), 378; *Ceroplastes ceriferus* (Indian wax scale), 356; *C. cirripediformis* (barnacle scale), 356; *Chrysomphalus dictyospermi* (dictyospermum scale), 382; *Fiorinia theae* (tea scale), 394; *Hemiberlesia lataniae* (latania scale), 376; *H. rapax* (greedy scale), 372; *Lepidosaphes gloveri* (Glover scale), 380; *L. yanagicola*, 388; *Lindingaspis rossi*, 376; *Parthenolecanium persicae* (European peach scale), 364; *Pulvinaria floccifera* (cottony camellia scale, cottony taxus scale), 344; *P. innumerabilis* (cottony maple scale), 340; *Unaspis euonymi* (euonymus scale), 388
Euonymus alatus (winged burning bush): *Lepidosaphes yanagicola*, 388; *Macrosiphum euphorbiae* (potato aphid), 300; *Pseudococcus comstocki* (Comstock mealybug), 326; *Yponomeuta cagnagella* (euonymus caterpillar), 174
Euonymus europaea (European spindle tree): *Unaspis euonymi* (euonymus scale), 388; *Yponomeuta cagnagella* (euonymus caterpillar), 174
Euonymus japonica (Japanese euonymus): *Aphis gossypii* (cotton, or melon, aphid), 308; *Unaspis euonymi* (euonymus scale), 388; *Yponomeuta cagnagella* (euonymus caterpillar), 174
Euonymus kiautschovica: *Yponomeuta cagnagella* (euonymus caterpillar), 174
European alder, see *Alnus cordata*
European green alder, see *Alnus viridis*
European horse chestnut, see *Aesculus hippocastanum*
European spindle tree, see *Euonymus europeae*
Eurya: *Fiorinia theae* (tea scale), 394
evergreen mock orange, see *Philadelphus mexicanus*
evergreens, broad-leaved: lace bugs, 424; *Otiorhynchus* sp., 240; *Quadraspidiotus juglansregiae* (walnut scale), 386; *Stephanitis* sp., 424
Exochorda racemosa (pearlbush): *Phyllocoptes exochordae*, 482

Fagus (beech)
—foliage: *Atta texana* (Texas leafcutting ant), 496; *Epinotia aceriella* (maple trumpet skeletonizer), 212; *Paraclemensia acerifoliella* (maple leaf cutter), 212; *Phyllonorycter* sp., 196; *Taeniothrips inconsequens* (pear thrips), 432
caterpillars: *Alsophila pometaria* (fall cankerworm), 142; *Automeris io* (io moth), 164; *Datana ministra* (yellownecked caterpillar), 154; *Ennomos subsignarius* (elm spanworm), 144–146; *Halysidota tessellaris* (pale tussock moth), 158; *Heterocampa guttivitta* (saddle prominent), 154; *Lithophane antennata* (green fruitworm), 148; *Lymantria dispar* (gypsy moth), 138; *Operophtera bruceata* (bruce spanworm), 146; *Paleacrita vernata* (spring cankerworm), 142, 144; *Psilocorsis cryptolechiella*, 218; *Symmerista leucitys* (orangehumped mapleworm), 154
beetles: *Brachys aeruginosus*, 190; *Cyrtepistomus castaneus* (Asiatic oak weevil), 244; *Odontota dorsalis* (locust leafminer), 190
mites: *Eotetranychus carpini*, 474; *E. hicoriae* (pecan leaf scorch mite), 474; erineum galls, 482, 484, 486; *Oligonychus bicolor* (oak spider mite), 475; spindle galls, 482
—bark, wood, twigs: *Agrilus bilineatus* (twolined chestnut borer), 270; *Goes pulverulentus* (living beech borer), 282; horntails, 494; *Neoclytus acuminatus* (redheaded ash borer), 278; *Sciurus* sp. (squirrels), 502; *Sphyrapicus* sp. (sapsuckers), 500; *Tremex columba* (pigeon tremex), 494; *Xylosandrus germanus*, 250; *Zeuzera pyrina* (leopard moth), 254
—sucking insects: *Alebra albostriella* (maple leafhopper), 416; *Corythucha pallipes* (birch lace bug), 426; *Fagiphagus imbricator*, 304; *Grylloprociphilus imbricator* (beech blight aphid), 296; *Longistigma caryae* (giant bark aphid), 310; *Phyllaphis fagi* (woolly beech leaf aphid), 296
scales: *Cryptococcus fagisuga* (beech scale), 332; *Eulecanium caryae* (hickory lecanium), 364; *Lepidosaphes ulmi* (oystershell scale), 370; *Melanaspis obscura* (obscure scale), 384; *Pulvinaria innumerabilis* (cottony maple scale), 340; *Saissetia oleae* (black scale), 358
Fagus grandifolia (American beech): *Acalitus fagerinea*, 484; *Corythucha pallipes* (birch lace bug), 426; *Cryptococcus fagisuga*, 332; eriophyid mites, 484; *Symmerista leucitys* (orangehumped mapleworm), 154
Fagus sylvatica (European beech): *Phyllaphis fagi* (woolly beech leaf aphid), 296
false indigo, see *Amorpha fruticosa*
Fatsia japonica (Japanese aralia): *Aphis fabae* (bean aphid), 300; *Hemiberlesia lataniae* (latania scale), 376; *Parasaissetia nigra* (nigra scale), 360
Feijoa: *Asynonychus godmani* (Fuller rose beetle), 230; *Chrysomphalus dictyospermi* (dictyospermum scale), 382
Feijoa sellowiana (pineapple guava): *Asynonychus godmani* (Fuller rose beetle), 230; *Saissetia oleae* (black scale), 358; *Thyridopteryx ephemeraeformis* (bagworm), 176
feltleaf ceanothus, see *Ceanothus arboreus*
ferns: *Aspidiotus nerii* (oleander scale), 374; *Pinnaspis aspidistrae* (fern scale), 390; *Saissetia coffeae* (hemispherical scale), 360
Ficus (fig): *Aleurothrixus floccosus* (woolly whitefly), 322; *Aonidiella aurantii* (California red scale), 378; *Ceroplastes floridensis* (Florida wax scale), 356; *Chrysomphalus dictyospermi* (dictyospermum scale), 382; *Dialeurodes citrifolii* (cloudywinged whitefly), 322; *Duplaspidiotus claviger* (camellia mining scale), 380; *Heliothrips haemorrhoidalis* (greenhouse thrips), 430; *Hemiberlesia rapax* (greedy scale), 372; *Lindingaspis floridana*, 376; *L. rossi*, 376; *Phenacoccus aceris* (apple mealybug), 324; *Pinnaspis aspidistrae* (fern scale), 390; *P. strachani* (lesser snow scale), 390; *Planococcus citri* (citrus mealybug), 328; *Pseudococcus longispinus* (longtailed mealybug), 324; *Saissetia oleae* (black scale), 358; *Toxoptera aurantii* (black citrus aphid), 302
Ficus aurea: *Gynaikothrips ficorum* (Cuban laurel thrips), 434
Ficus benjamina (weeping fig): *Gynaikothrips ficorum* (Cuban laurel thrips), 434; *Pseudococcus longispinus* (longtailed mealybug), 324
Ficus elastica (rubber plant, Indian rubber tree): *Aonidiella citrina* (yellow scale), 378; *Aspidiotus nerii* (oleander scale), 374
Ficus microcarpa (laurel fig, Indian laurel): *Gynaikothrips ficorum* (Cuban laurel thrips), 434
Ficus pumila (climbing fig): *Chrysomphalus dictyospermi* (dictyospermum scale), 382
filbert, see *Corylus*

fir, see *Abies, Pseudotsuga*
alpine or subalpine: *Abies lasiocarpa*
balsam: *Abies balsamea*
Douglas: *Pseudotsuga menziesii*
Fraser: *Abies fraseri*
grand or lowland: *Abies grandis*
joint: *Ephedra trifurca*
noble: *Abies procera*
Pacific silver: *Abies amabilis*
red: *Abies magnifica*
Veitch's silver: *Abies veitchii*
white: *Abies concolor*
Forsythia: *Callirhopalus bifasciatus* (twobanded Japanese weevil), 244; *Graphocephala coccinea* (redbanded leafhopper), 418; *Lithophane antennata* (green fruitworm), 148; *Poecilocapsus lineatus* (fourlined plant bug), 396
flowering almond, see *Prunus triloba*
flowering cherry: *Apiosporina morbosa*, 258; *Chrysobothris femorata* (flatheaded appletree borer), 270; *Elaphidionides villosus* (twig pruner), 264; *Popillia japonica* (Japanese beetle), 236; *Synanthedon pictipes* (lesser peachtree borer), 258; *S. scitula* (dogwood borer), 262. See also *Prunus serrulata*
flowering maple, see *Abutilon*
flowering quince, see *Chaenomeles*
Fraxinus (ash)
—foliage: *Callirhopalus bifasciatus* (twobanded Japanese weevil), 244; *Caloptilia fraxinella*, 196; *C. syringella* (lilac leafminer), 196; *Megachile* sp. (leafcutter bees), 494; *Taeniothrips inconsequens* (pear thrips), 432
caterpillars: *Automeris io* (io moth), 164; *Choristoneura rosaceana* (obliquebanded leafroller), 216; *Ennomos subsignarius* (elm spanworm), 144–146; *Halysidota tessellaris* (pale tussock moth), 158; *Lithophane antennata* (green fruitworm), 148; *Lymantria dispar* (gypsy moth), 138; *Malacosoma americanum* (eastern tent caterpillar), 168; *M. californicum pluviale* (western tent caterpillar), 170; *M. disstria* (forest tent caterpillar), 168; *Oiketicus townsendi*, 176; *Orthosia hibisci* (speckled green fruitworm), 148; *Paleacrita vernata* (spring cankerworm), 144
mites: *Aceria fraxiniflora* (ash flower gall mite), 478; *Oligonychus newcomeri*, 475; *Tetranychus homorus*, 476; *T. urticae* (twospotted spider mite), 476
sawflies: *Macrophya punctumalbum*, 136
—bark, wood, twigs: cambium miner, 198; *Hylesinus* sp., 252; *Magicicada septendecim* (periodical cicada), 490; *Marmara* sp., 198; *Megacyllene caryae* (painted hickory borer), 274; *Neoclytus acuminatus* (redheaded ash borer), 278; *Oecanthus fultoni* (snowy tree cricket), 492, 493; *Podosesia aureocincta* (banded ash clearwing), 260–262; *P. syringae* (lilac borer), 260–262; *Prionoxystus robiniae* (carpenterworm), 256; *Saperda candida* (roundheaded appletree borer), 278; *Stictocephala bisonia* (buffalo treehopper), 406; *Xylosandrus germanus*, 250; *Zeuzera pyrina* (leopard moth), 254
—sucking insects: *Acanalonia conica*, 422; *Corythucha ciliata* (sycamore lace bug), 426, 428; *Leptoypha costata*, 426; *L. minor* (Arizona ash tingid), 426; *Metcalfa pruinosa*, 422; *Prociphilus americanus*, 304; *Siphoninus phillyreae* (ash whitefly), 322; *Trialeurodes vaporariorum* (greenhouse whitefly), 320; *Tropidosteptes amoenus* (ash plant bug), 402; *T. brooski*, 402; *T. illitus*, 402; *T. pacificus*, 402
scales: *Chionaspis acericola* (maple scurfy scale), 390; *C. floridensis*, 390; *C. gleditsia*, 390; *C. kosztarabi*, 390; *C. lintneri*, 390; *C. salicisnigrae*, 390; *Lepidosaphes ulmi* (oystershell scale), 370; *Quadraspidiotus juglansregiae* (walnut scale), 386
Fraxinus americana (white ash): *Asterolecanium arabidis* (pit-making pittosporum scale), 348; *Contarina canadensis* (ash midrib gall midge), 464; *Tethida cordigera* (blackheaded ash sawfly), 134; *Tomostethus multicinctus* (brownheaded ash sawfly), 134; *Tropidosteptes amoenus* (ash plant bug), 402
Fraxinus dipetala (two-petal ash): *Saissetia oleae* (black scale), 358
Fraxinus pennsylvanica (green ash, red ash): *Asterolecanium arabidis* (pit-making pittosporum scale), 348; *Macrophya punctumalbum*, 136; *Podosesia syringae* (ash borer), 260–262; *Prionoxystus robiniae* (carpenterworm), 256; *Tethida cordigera* (blackheaded ash sawfly), 134; *Tomostethus multicinctus* (brownheaded ash sawfly), 134; *Tropidosteptes amoenus* (ash plant bug), 402
Fraxinus quadrangulata (blue ash): *Tropidosteptes amoenus* (ash plant bug), 402
Fraxinus velutina (velvet ash, Modesto ash): *Tropidosteptes illitus*, 402; *T. pacificus*, 402
Fremantia, see *Fremontodendron*
Fremontodendron: *Asterolecanium arabidis* (pit-making pittosporum scale), 348; *Chionaspis sassceri*, 390
fringe tree, see *Chionanthus virginicus*
Fuchsia: *Aculops fuchsiae* (fuchsia gall mite), 486; *Aonidiella aurantii* (California red scale), 378; *Asynonychus godmani* (Fuller rose beetle), 230; *Brevipalpus obovatus* (privet mite), 482; *Heliothrips haemorrhoidalis* (greenhouse thrips), 430; *Hemiberlesia lataniae* (latania scale), 376; *H. rapax* (greedy scale), 372; *Macrosiphum euphorbiae* (potato aphid), 300; *Saissetia oleae* (black scale), 358; *Trialeurodes vaporariorum* (greenhouse whitefly), 320
Fuchsia hybrida: *Graphocephala atropunctata* (blue-green sharpshooter), 418

gallberry, see *Ilex glabra*
Gardenia: *Asynonychus godmani* (Fuller rose beetle), 230; *Ceroplastes cirripediformis* (barnacle scale), 356; *Chrysomphalus dictyospermi* (dictyospermum scale), 382; *Dialeurodes citrifolii* (cloudywinged whitefly), 322; *Duplaspidiotus claviger* (camellia mining scale), 380; *Fiorinia theae* (tea scale), 394; *Lindingaspis rossi*, 376; *Parabemisia myrica* (Japanese bayberry whitefly), 322; *Planococcus citri* (citrus mealybug), 328; *Pseudococcus longispinus* (longtailed mealybug), 324; *Saissetia coffeae* (hemispherical scale), 360
Gardenia augusta (cape jasmine): *Pseudococcus longispinus* (longtailed mealybug), 324; *Toumeyella liriodendri* (tuliptree scale), 362
Garrya: *Icerya purchasi* (cottony cushion scale), 338
Genista (broom): *Aspidiotus nerii* (oleander scale), 374; *Hemiberlesia rapax* (greedy scale), 372
Geranium: *Aulacaspis rosae* (rose scale), 366; *Hemiberlesia rapax* (greedy scale), 372
giant sequoia, see *Sequoiadendron giganteum*
Ginkgo biloba: *Euzophera semifuneralis* (American plum borer), 252; *Pseudococcus maritimus* (grape mealybug), 88
gladiolus: *Franklinella* sp., 434
Gleditsia (honeylocust)
—foliage: *Dasineura gleditchiae* (honeylocust pod gall midge), 466; *Meunieriella aquilonia*, 470; *Neolasioptera brevis*, 470
caterpillars: *Datana integerrima* (walnut caterpillar), 150; *Epargyreus clarus* (silverspotted skipper), 148; *Homadaula anisocentra* (mimosa webworm), 180; *Schizura ipomoeae*, 156; *Sparganothis sulfureana*, 216; *Thyridopteryx ephemeraeformis* (bagworm), 176
mites: *Eotetranychus multidigituli* (honeylocust spider mite), 472
—bark, wood, twigs: *Agrilus* sp., 270; *A. difficilis*, 270; *Elaphidionoides villosus* (twig pruner), 264; *Megacyllene caryae* (painted hickory borer), 274; *Neoclytus acuminatus* (redheaded ash borer), 278; *Oncideres cingulata* (twig girdler), 264; *Xylosandrus germanus*, 240
—sucking insects: *Aphis craccivora* (cowpea aphid), 308; *Diaphnocoris chlorionis* (honeylocust plant bug), 404; *Pseudococcus maritimus* (grape mealybug), 88; *Micrutalis calva*, 404
leafhoppers: *Macropis fumipennis*, 404
scales: *Chionaspis gleditsia*, 390; *Eulecanium caryae* (hickory lecanium), 364; *Parthenolecanium corni* (European fruit lecanium), 364; *P. pruinosum* (frosted scale), 354; *Quadraspidiotus juglansregiae* (walnut scale), 386
Gleditsia triacanthos (honeylocust)
—foliage: *Meunieriella aquilonia*, 470
caterpillars: *Datana ministra* (yellownecked caterpillar), 154; *Epargyreus clarus* (silverspotted skipper), 148; *Homadaula anisocentra* (mimosa webworm), 180; *Lophocampa caryae* (hickory tussock moth), 160; *Schizura ipomoeae*, 156; *Sparganothis sulfureana*, 216; *Thyridopteryx ephemeraeformis* (bagworm), 176
—bark, wood, twigs: *Magicicada septendecim* (periodical cicada), 492; *Megacyllene caryae* (painted hickory borer), 274; *Neolasioptera brevis*, 470; *Ostrinia nubilalis* (European corn borer), 286; *Stictocephala bisonia* (buffalo treehopper), 406; *Xyleborus saxeseni*, 252
—sucking insects: *Diaphnocoris chlorionis* (honeylocust plant bug), 404; *Micrutalis calva*, 404; *Trialeurodes vaporariorum* (greenhouse whitefly), 320
leafhoppers: *Macropis fumipennis*, 404
scales: *Quadraspidiotus juglansregiae* (walnut scale), 386; *Saissetia oleae* (black scale), 358
glory pea, see *Daubentonia tripetii*
golden-chain tree, see *Laburnum anagyroides*
golden raintree, see *Koelreuteria paniculata*
goldenrod, see *Solidago*
gooseberry: *Bryobia ribis* (gooseberry bryobia), 472. See also *Ribes*
Gordonia: *Fiorinia fioriniae* (palm fiorinia scale), 394; *F. theae* (tea scale), 394
Gordonia lasianthus (loblolly bay): *Toumeyella liriodendri* (tuliptree scale), 362
grape, see *Vitis*
Oregon, see *Mahonia aquifolium*
grapefruit: *Aspidiotus nerii* (oleander scale), 374. See also *Citrus*
gray dogwood, see *Cornus racemosa*
Grecian laurel, see *Laurus nobilis*
green ebony, see *Jacaranda acutifolia*
green wattle, see *Acacia decurrens*
Grevillea: *Hemiberlesia lataniae* (latania scale), 376; *Pseudococcus maritimus* (grape mealybug), 88; *Saissetia oleae* (black scale), 358; *Sphyrapicus* sp. (sapsuckers), 500
Grewia: *Saissetia oleae* (black scale), 358
Grossularia: *Eriococcus borealis*, 336
ground bush, see *Baccharis halimifolia*
Guadeloupe palm, see *Brahea edulis*
guava, see *Psidium guajava*
pineapple: *Feijoa sellowiana*
gum
black: *Nyssa sylvatica*
red: *Liquidambar styraciflua*
sour: *Nyssa sylvatica*
sweet: *Liquidambar styraciflua*
gumbo-limbo, see *Bursera simaruba*
gum-elemi, see *Bursera simaruba*
Gymnocladus dioica (Kentucky coffee tree): *Quadraspidiotus juglansregiae* (walnut scale), 386

hackberry, see *Celtis*, *Celtis occidentalis*
Hakea: *Aspidiotus nerii* (oleander scale), 374; *Hemiberlesia rapax* (greedy scale), 372
Hamamelis (witch-hazel)
—foliage: *Camereria hamameliella*, 196; *Phyllonorycter* sp., 196
caterpillars: *Datana ministra* (yellownecked caterpillar), 154; *Malacosoma americanum* (eastern tent caterpillar), 168
rollers/webbers: *Argyrotaenia quercifoliana*, 218
—sucking insects: *Clastoptera obtusa* (alder spit-

tlebug), 420; *Hamamelistes spinosus* (witch-hazel bud gall aphid), 296, 450; *Hormaphis hamamelidis* (witch-hazel leaf gall aphid), 450; *Quadraspidiotus juglansregiae* (walnut scale), 386
handflower tree, see *Chiranthodendron pentadactylon*
hawkweed: *Aphrophora saratogensis* (Saratoga spittlebug), 86
hawthorn, see *Crataegus*
  English: *Crataegus laevigata*
hazel, see *Corylus*
hazelnut: *Anitosa senatoria* (orangestriped oakworm), 156; *Rheumaptera hastata* (spear-marked black moth), 146. See also *Corylus*
heath (*Erica*): *Aspidiotus nerii* (oleander scale), 374; *Hemiberlesia rapax* (greedy scale), 372
Hebe, see *Veronica*
Hedera: *Unaspis euonymi* (euonymus scale), 388
Hedera canariensis (Algerian ivy): *Graphocephala atropunctata* (blue-green sharpshooter), 418
Hedera helix (English ivy): *Aonidiella citrinia* (yellow scale), 378; *Aphis fabae* (bean aphid), 300; *A. hederae* (ivy aphid), 308; *Aspidiotus nerii* (oleander scale), 374; *Asterolecanium arabidis* (pit-making pittosporum scale), 348; *Hemiberlesia lataniae* (latania scale), 376; *H. rapax* (greedy scale), 372; *Lindingaspis rossi*, 376; *Myzus persicae* (green peach aphid), 300; *Parasaissetia nigra* (nigra scale), 360; *Pseudaulacaspis cockerelli* (false oleander scale), 374; *Pseudococcus longispinus* (long-tailed mealybug), 324; *P. maritimus* (grape mealybug), 88; *Pulvinaria floccifera* (cottony camellia scale, cottony taxus scale), 344
Helianthemum nummularium (sun rose): *Asterolecanium arabidis* (pit-making pittosporum scale), 348
hemlock, see *Tsuga*
  Canada: *Tsuga canadensis*
  Carolina: *Tsuga caroliniana*
  eastern: *Tsuga canadensis*
  Japanese: *Tsuga diversifolia*
  mountain: *Tsuga mertensiana*
  western: *Tsuga heterophylla*
Heteromeles arbutifolia (California holly, California Christmas berry, toyon)
—foliage: *Argyrotaenia franciscana*, 218; *Asphondylia photiniae*, 470; *Heliothrips haemorrhoidalis* (greenhouse thrips), 430; *Malacosoma californicum pluviale* (western tent caterpillar), 170; *Orgyia vetusta* (western tussock moth), 160; *Rhynchothrips ilex* (toyon thrips), 434
—sucking insects: *Prociphilus caryae*, 304; *Wahlgreniella nervata*, 298
  lacebugs: *Corythucha bullata* (California Christmas berry tingid), 426, 428; *C. heteromelecola*, 426; *C. incurvata* (California Christmas berry lace bug), 426, 428
  scales: lecanium scales, 364
hetzii juniper, see *Juniperus chinensis* 'Hetzii'
Hibiscus
—foliage: *Asynonychus godmani* (Fuller rose beetle), 230; *Helix aspersa* (brown garden snail), 498; *Trialeurodes vaporariorum* (greenhouse whitefly), 320, 322; *Urodus parvula* (ermine moth), 148
  mite: *Oligonychus ilicis* (southern red mite), 475
—bark, wood, twigs: *Ostrinia nubilalis* (European corn borer), 286
—sucking insects: *Leptoglossus phyllopus* (leaf-footed bugs), 422; *Pseudococcus comstocki* (Comstock mealybug), 326; *P. longispinus* (longtailed mealybug), 324; *Umbonia crassicornis* (thornbug), 408
  aphids: *Aphis gossypii* (cotton, or melon, aphid), 308
  scales: *Parasaissetia nigra* (nigra scale), 360; *Pinnaspis aspidistrae* (fern scale), 390; *P. strachani* (lesser snow scale), 390; *Pseudaulacaspis pentagona* (white peach scale), 392; *Unaspis euonymi* (euonymus scale), 388
Hibiscus calyphyllus: *Acyrthosiphon solani* (foxglove aphid), 310
Hibiscus rosasinensis: *Bemisia tabaci* (sweetpotato whitefly), 322
Hibiscus syriacus (rose of sharon): *Aphis gossypii* (cotton, or melon, aphid), 308; *Niesthrea louisianica* (rhopalid plant bug), 422; *Saissetia oleae* (black scale), 358
hickory, see *Carya*
  shagbark: *Carya ovata*
hog plum, see *Spondias purpurae*
holly, see *Ilex*
  American: *Ilex opaca*
  Burford: *Ilex cornuta* 'Burfordi'
  California: *Heteromeles arbutifolia*
  Chinese: *Ilex cornuta*
  English: *Ilex aquifolium*
  Georgia: *Ilex longipes*
  Japanese: *Ilex crenata*
  mountain: *Nemopanthus*
honeylocust, see *Gleditsia*, *Gleditsia triacanthos*
honeysuckle, see *Lonicera*
hophornbeam, see *Ostrya virginiana*
hoptree, see *Ptelea trifoliata*
hornbeam, see *Carpinus caroliniana*
horsechestnut, see *Aesculus*, *Aesculus hippocastanum*
Howea (palm): *Hemiberlesia lataniae* (latania scale), 376
huckleberry, see *Vaccinium*
huisache, see *Acacia farnesiana*
Hydrangea: *Pulvinaria floccifera* (cottony camellia scale, cottony taxus scale), 344; *Tetranychus hydrangeae*, 476
Hypericum: *Aspidiotus nerii* (oleander scale), 374; *Heliothrips haemorrhoidalis* (greenhouse thrips), 430
Hypericum calycinum (St. Johnswort): *Heliothrips haemorrhoidalis* (greenhouse thrips), 430
Hypericum cistifolium: *Toumeyella liriodendri* (tuliptree scale), 362

Ilex (holly)
—foliage: *Callirhopalus bifasciatus* (twobanded Japanese weevil), 244; *Lymantria dispar* (gypsy moth), 138; *Oligonychus ilicis* (southern red mite), 475; *Phytomyza ilicicola*, 206; *Rhopobota naevana* (holly budmoth, blackheaded fireworm), 206, 208; *Thysanopyga intractata* (holly looper), 146
—bark: cambium miners, 198; mice, 502; *Neoclatus acuminatus* (redheaded ash borer), 278; rabbits, 502; Xylosandrus germanus, 250
—sucking insects: *Metcalfa pruinosa*, 422; *Prosapia bicincta*, 420; *Pseudococcus comstocki* (Comstock mealybug), 326
  scales: *Aonidiella aurantii* (California red scale), 378; *Aspidiotus nerii* (oleander scale), 374; *Ceroplastes brachyurus* (brachyuran wax scale), 358; *C. floridensis* (Florida wax scale), 356; *Chrysomphalus dictyospermi* (dictyospermum scale), 382; *Fiorinia theae* (tea scale), 394; *Hemiberlesia lataniae* (latania scale), 376; *H. rapax* (greedy scale), 372; *Lepidosaphes beckii* (purple scale), 380; *L. camelliae*, 370; *Pseudaonidia paeoniae* (peony scale), 372; *Pulvinaria floccifera* (cottony camellia scale, cottony taxus scale), 344; *Quadraspidiotus juglansregiae* (walnut scale), 386; *Saissetia oleae* (black scale), 358; *Unaspis euonymi* (euonymus scale), 388
Ilex aquifolium (English holly): *Acyrthosiphon solani* (foxglove aphid), 310; *Hemiberlesia rapax* (greedy scale), 372; *Parasaissetia nigra* (nigra scale), 360; *Phytomyza ilicicola* (native holly leafminer), 206, 208; *P. ilicis* (holly leafminer), 206, 208; *P. opacae*, 206, 208; *Pseudaonidia peoniae* (peony scale), 372; *Rhopobota naevana* (holly budmoth), 206, 208; *Thysanopyga intractata* (holly looper), 146; *Toxoptera aurantii* (black citrus aphid), 302
Ilex cornuta (Chinese holly): *Asterolecanium puteanum* (holly pit scale), 352; *Ceroplastes ceriferus* (Indian wax scale), 356; *Fiorinia theae* (tea scale), 394; *Pulvinaria acericola* (cottony maple leaf scale), 342; *Rhadopterus picipes* (cranberry rootworm), 342; *Thysanopyga intractata* (holly looper), 146
Ilex cornuta 'Burfordii' (Burford holly): *Ceroplastes ceriferus* (Indian wax scale), 356; *Fiorinia theae* (tea scale), 394; *Lepidosaphes camelliae*, 370; *Pseudaonidia duplex* (camphor scale), 372; *Rhadopterus picipes* (cranberry rootworm), 242
Ilex crenata (Japanese holly): *Asterolecanium puteanum* (holly pit scale), 352; *Ceroplastes ceriferus* (Indian wax scale), 356; *C. sinensis* (Chinese wax scale), 356; *Oligonychus ilicis* (southern red mite), 475; *Phytomyza ilicicola* (native holly leafminer), 206; *P. opacae*, 206, 208; *Pulvinaria acericola* (cottony maple leaf scale), 342; *Rhadopterus picipes* (cranberry rootworm), 242
Ilex cumicola: *Phytomyza ilicicola* (native holly leafminer), 206; *P. opacae*, 206, 208
Ilex glabra (inkberry, gallberry): *Archips cerasivorana* (uglynest caterpillar), 172; *Melanaspis tenebricosa* (gloomy scale), 384; *Phytomyza glabricola*, 206, 208
Ilex longipes (Georgia holly): *Phytomyza ditmani*, 206
Ilex opaca (American holly): *Asphondylia ilicicola* (hollyberry midge), 470; *Asterolecanium puteanum* (holly pit scale), 352; *Duplaspidiotus claviger* (camellia mining scale), 380; *Phytomyza ilicicola* (native holly leafminer), 206; *P. opacae*, 206, 208; *Pulvinaria acericola* (cottony maple leaf scale), 342; *Tetraleurodes mori* (mulberry whitefly), 324; *Thysanopyga intractata* (holly looper), 146
Ilex serrata (Japanese winterberry): *Phytomyza ditmani*, 206
Ilex verticillata (common winterberry): *Lygocoris communis* (pear plant bug), 396; *Phytomyza verticillatae*, 206, 208; *Pulvinaria floccifera* (cottony camellia scale, cottony taxus scale), 344
Ilex vomitoria (yaupon): *Gyropsylla ilicis*, 454; *Phytomyza vomitoriae*, 206; *Thysanopyga intractata* (holly looper), 146
Impatiens: *Otiorhynchus sulcatus* (black vine weevil), 240
Indian laurel, see *Ficus microcarpa*
Indian rubber tree, see *Ficus elastica*
inkberry, see *Ilex glabra*
iron plant, see *Aspidistra elatior*
ironwood: *Eotetranychus uncatus* (Garman spider mite), 474; *Oligonychus newcomeri*, 475. See also *Carpinus*, *Lyonothamnus*, *Ostrya virginiana*.
Isomeris arborea (bladderpod): *Hemiberlesia rapax* (greedy scale), 372
Itea virginica (sweetspire): *Tetraleurodes mori* (mulberry whitefly), 324
ivy: *Bryobia kissophila* (ivy bryobia), 472; *Lepidosaphes gloveri* (Glover scale), 380
ivy
  Algerian: *Hedera canariensis*
  Boston: *Parthenocissus tricuspidata*
  English: *Hedera helix*
Ixora: *Pseudaonidia trilobitiformis* (trilobe scale), 376; *Toxoptera aurantii* (black citrus aphid), 302
Ixora acuminata: *Ceroplastes nakahara* (Nakahara wax scale), 358

Jacaranda acutifolia (Green ebony): *Umbonia crassicornis* (thornbug), 408
Japanese aralia, see *Fatsia japonica*
Japanese flowering cherry, see *Prunus serrulata*
Japanese quince, see *Chaenomeles*
Japanese winterberry, see *Ilex serrata*
Japanese yew, see *Podocarpus macrophyllus*
Jasminum (jasmine): *Asterolecanium arabidis* (pit-making pittosporum scale), 348; *Pseudococcus longispinus* (longtailed mealybug), 324; *Pulvinaria floccifera* (cottony camellia scale, cottony taxus scale), 344; *Saissetia oleae* (black

Jasminum (jasmine): (*cont.*)
scale), 358; *Unaspis euonymi* (euonymus scale), 388
Jerusalem cherry, see *Solanum pseudocapsicum*
Juglans (walnut)
—foliage: *Acrobasis juglandis* (pecan leaf casebearer), 212; *Argyrotaenia citrana* (orange tortrix), 216; *Datana angusii*, 150, 154; *D. integerrima* (walnut caterpillar), 150; *D. ministra* (yellownecked caterpillar), 154; *Laspeyresia pomonella* (codling moth), 234; *Lophocampa caryae* (hickory tussock moth), 160; *Schizura concinna* (redhumped caterpillar), 150, 156
mites: *Aceria brachytarsus* (walnut purse gall mite), 234, 488; erineum gall mites, 486; *Panonychus ulmi* (European red mite), 472
—fruit: *Rhagoletis* sp. (husk maggots), 234
—bark, wood, twigs: *Agrilus* sp., 272; *Conotrachelus juglandis* (butternut curculio), 234; *Euzophera semifuneralis* (American plum borer), 252; *Neoclatus acuminatus*, (redheaded ash borer), 278; *Oecanthus fultoni* (snowy tree cricket), 492, 494; *Scolytus quadrispinosus* (hickory bark beetle), 250; *Xyleborus dispar*, 252; *Zeuzera pyrina* (leopard moth), 254
—roots: *Prionus californicus* (California prionus), 280
—sucking insects: *Clastoptera obtusa* (alder spittlebug), 420; *Leptoglossus phyllopus* (leaffooted bugs), 422, 424; *Pseudococcus maritimus* (grape mealybug), 88
aphids: *Longistigma caryae* (giant bark aphid), 310; *Myzus persicae* (green peach aphid), 300
scales: *Aonidiella aurantii* (California red scale), 378; *Chionaspis lintneri*, 390; *Eulecanium caryae* (hickory lecanium), 364; *Icerya purchasi* (cottony cushion scale), 338; *Lepidosaphes ulmi* (oystershell scale), 370; *Orgyia vetusta* (western tussock moth), 160; *Parthenolecanium pruinosum* (frosted scale), 354; *Pseudaulacaspis pentagona* (white peach scale), 392; *Quadraspidiotus perniciosus* (San Jose scale), 386; *Toumeyella liriodendri* (tuliptree scale), 362
Juglans ailantifolia (Japanese walnut, Siebold walnut): *Conotrachelus juglandis* (butternut curculio), 234; *Datana integerrima* (walnut caterpillar), 150
Juglans californica (southern California walnut): *Aceria brachytarsus*, 488
Juglans cinerea (butternut)
—foliage: *Datana integerrima* (walnut caterpillar), 150; *D. ministra* (yellownecked caterpillar), 154; *Eriocampa juglandis* (butternut woollyworm), 136; *Lophocampa caryae* (hickory tussock moth), 160; *Lymantria dispar* (gypsy moth), 138; *Schizura concinna* (redhumped caterpillar), 156
mites: *Aceria cinereae*, 488
—bark, wood, twigs: *Conotrachelus juglandis* (butternut curculio), 234; *Enchenopa binotata* (twomarked treehopper), 410; *Megacyllene caryae* (painted hickory borer), 274; *Scolytus quadrispinosus* (hickory bark beetle), 250; *Stictocephala bisonia* (buffalo treehopper), 406
—sucking insects: *Chionaspis furfura*, 368; *Corythucha juglandis* (walnut lace bug), 426
Juglans hindsii (California black walnut): *Aceria brachytarsus* (eriophyid gall mite), 234, 488; *Conotrachelus juglandis* (butternut curculio), 234; *Orgyia vetusta* (western tussock moth), 160
Juglans nigra (black walnut)
—foliage: *Archips argyrospila* (fruittree leafroller), 172; *Coleophora laticornella* (pecan cigar casebearer), 186; *Datana integerrima* (walnut caterpillar), 150; *Eriocampa juglandis* (butternut woollyworm), 136; *Hyphantria cunea* (fall webworm), 166; *Lymantria dispar* (gypsy moth), 138
mites: *Aceria caulis*, 486; erineum gall mites, 486
—fruit: *Conotrachelus juglandis* (butternut curculio), 234; *C. retentus* (black walnut curculio), 234; *Rhagoletis* spp. (husk maggots), 234
—bark, wood, twigs: *Conotrachelus juglandis* (butternut curculio), 234; *Enchenopa binotata* (twomarked treehopper), 410; *Megacyllene caryae* (painted hickory borer), 274
—sucking insects: *Acanalonia conica*, 422; *Anormenis chloris*, 422; *Corythucha juglandis* (walnut lace bug), 426; *Ormenoides venusta*, 422; *Plagiognathus punctatipes* (plant bug), 404
Juglans regia (English, Persian walnut)
—foliage: *Archips argyrospila* (fruittree leafroller), 172; *Datana integerrima* (walnut caterpillar), 150
mites: *Aceria erineus* (walnut blister mite), 488
—fruit: *Conotrachelus juglandis* (butternut curculio), 234; *Laspeyresia pomonella* (codling moth), 234; *Rhagoletis* sp. (husk maggot), 234
—bark, wood, twigs: *Conotrachelus juglandis* (butternut curculio), 234
—sucking insects: *Calaphis juglandis* (duskyveined aphid), 292; *Chromaphis juglandicola* (walnut aphid), 292; *Empoasca fabae* (potato leafhopper), 414; *Eulecanium cerasorum* (calico scale), 354; *Graphocephala atropunctata* (bluegreen sharpshooter), 418; *Hemiberlesia rapax* (greedy scale), 372; *Melanaspis obscura* (obscure scale), 384; *Parthenolecanium pruinosum* (frosted scale), 354; *Pseudococcus maritimus* (grape mealybug), 88; *Quadraspidiotus juglansregiae* (walnut scale), 386; *Saissetia oleae* (black scale), 358
juniper, see *Juniperus*
blue rug or creeping: *Juniperus horizontalis*
California: *Juniperus californica*
Chinese: *Juniperus chinensis*
golden: *Juniperus chinensis* 'Pfitzerana Aurea'
Irish: *Juniperus communis*
Pfitzer: *Juniperus chinensis* 'Pfitzerana Aurea'
Sargent: *Juniperus chinensis* var. *sargentii*
Savin: *Juniperus sabina*
Sierra: *Juniperus occidentalis*
tam: *Juniperus sabina* 'Tamariscifolia'
Juniperus
—foliage: *Argyresthia cupressella* (cypress tip miner), 42; *A. franciscella* (cypress tip miner), 42; *Coleotechnites juniperella*, 42; *Contarinia juniperina* (juniper midge), 46; cypress tip miners, 42; *Dichomeris marginella* (juniper webworm), 30; *Eacles imperialis* (imperial moth), 26; *Epinotia subviridis* (cypress leaftier), 38; *Lophocampa argentata subalpina*, 34; *Lymantria dispar* (gypsy moth), 138; *Stenolechia bathrodyas*, 42; *Susana cupressi* (cypress sawfly), 18; *Thyridopteryx ephemeraeformis* (bagworm), 176
mites: *Oligonychus ununguis* (spruce spider mite), 118, 120; *Trisetacus quadrisetus*, 122; *T. thujivagrans*, 122
—bark, wood, twigs: *Hylobius pales* (pales weevil), 56; *Periploca nigra* (juniper twig girdler), 58; *Phloeosinus* sp., 66, 70; *Phyllobius intrusus* (arborvitae weevil), 240, 244
—sucking insects: *Clastoptera arborina*, 86; *C. juniperina*, 86
aphids: *Cinara sabinae*, 84; *C. tujafilina*, 84; *Mindarus abietinus* (balsam twig aphid), 80
scales: *Carulaspis juniperi* (juniper scale), 106; *C. minima*, 106; *Fiorinia juniperi* (juniper fiorinia scale), 394; *Hemiberlesia rapax* (greedy scale), 372; *Parthenolecanium corni* (European fruit lecanium), 364; *P. fletcheri* (Fletcher scale), 98
Juniperus californica (California juniper): *Ehrhornia cupressi* (cypress bark mealybug, cottony cypress scale), 332
Juniperus chinensis (Chinese juniper): *Argyresthia cupressella*, 42; *Carulaspis juniperi* (juniper scale), 106; *Clastoptera juniperina*, 86; *Contarinia juniperina* (juniper midge), 46; *Dichomeris marginella* (juniper webworm), 30; *Eacles imperialis* (imperial moth), 26; *Lophocampa argentata subalpina*, 34
Juniperus chinensis 'Hetzii': *Contarinia juniperina* (juniper midge), 46
Juniperus chinensis 'Pfitzerana': *Argyresthia cupressella*, 42; *Carulaspis juniperi* (juniper scale), 106
Juniperus chinensis 'Pfitzerana Aurea': *Clastopera juniperina* (western spittlebug), 86; *Dichomeris marginella* (juniper webworm), 30
Juniperus chinensis var. *sargentii*: *Argyresthia cupressella*, 42
Juniperus communis (Irish juniper): *Carulaspis juniperi* (juniper scale), 106; *Saissetia oleae* (black scale), 358; *Trisetacus quadrisetus juniperinus*, 122
Juniperus communis var. *depressa*: *Dichomeris marginella* (juniper webworm), 30
Juniperus communis hibernica (Irish juniper): *Carulaspis juniperi* (juniper scale), 106; *Dichomeris marginella* (juniper webworm), 30
Juniperus horizontalis (blue rug or creeping juniper): *Contarinia juniperina* (juniper midge), 46; *Dichomeris marginella* (juniper webworm), 30
Juniperus occidentalis (Sierra juniper, western juniper): *Argyresthia cupressella*, 42; *Coleotechnites juniperella*, 42; *C. stanfordia*, 42
Juniperus sabina (Savin juniper): *Argyresthia cupressella*, 42; *Carulaspis juniperi* (juniper scale), 106; *Contarinia juniperina* (juniper midge), 46; *Periploca nigra* (juniper twig girdler), 58
Juniperus sabina 'Tamariscifolia': *Argyresthia franciscella*, 42; *Periploca nigra* (juniper twig girdler), 58
Juniperus scopulorum (Rocky mountain juniper): *Argyresthia cupressella*, 42
Juniperus silicicola (southern redcedar): *Trisetacus quadrisetus juniperinus*, 122
Juniperus squamata (scaly-leaved Nepal juniper): *Dichomeris marginella* (juniper webworm), 30
Juniperus stricta: *Dichomeris marginella* (juniper webworm), 30
Juniperus suecica: *Dichomeris marginella* (juniper webworm), 30
Juniperus virginiana (eastern redcedar): *Argyresthia cupressella*, 42; *A. freyella*, 42; *Carulaspis juniperi* (juniper scale), 106; *Cinara tujafilina*, 84; *Contarinia juniperina* (juniper midge), 46; *Lymantria dispar* (gypsy moth), 138; *Phloeosinus canadensis* (northern cedar bark beetle), 68; *P. dentatus* (eastern juniper bark beetle), 68, 70; *Thyridopteryx ephemeraeformis* (bagworm), 176; *Trisetacus quadrisetus juniperinus*), 122; *Xylosandrus germanus*, 250

Kalmia (laurel)
—foliage: *Archips rosanus*, 218; *Spilonota ocellana* (eyespotted bud moth), 218
mites: *Oligonychus ilicis* (southern red mite), 475
—bark, wood, twigs: *Oberea myops* (azalea stem borer, rhododendron stem borer), 288; *O. tripunctata* (dogwood twig borer), 262
—sucking insects: lace bugs, 424; *Metcalfa pruinosa*, 422; *Phenacoccus aceris* (apple mealybug), 324; psyllids, 290
scales: *Chrysomphalus dictyospermi* (dictyospermum scale), 382; *Duplaspidiotus claviger* (camellia mining scale), 380; *Icerya purchasi* (cottony cushion scale), 338; *Parthenolecanium pruinosum* (frosted scale), 354; *Saissetia oleae* (black scale), 358
Kalmia latifolia (mountain laurel): *Callirhopalus bifasciatus* (twobanded Japanese weevil), 244; *Graphocephala coccinea* (redbanded leafhopper), 418; *Lithophane antennata* (green fruitworm), 148; *Stephanitis rhododendri*, 424; *Synanthedon rhododendri* (rhododendron

borer), 258; *Tetraleurodes mori* (mulberry whitefly), 324
Kauri, see *Agathis*
Kentia: *Diaspis cocois* (cocos scale), 376. See also *Howea*
Kentucky coffee tree, see *Gymnocladius dioica*
koa, see *Acacia koa*
Koelreuteria paniculata (golden raintree): *Callirhopalus bifasciatus* (twobanded Japanese weevil), 244; *Pseudaulacaspis pentagona* (white peach scale), 392; *Rhadopterus picipes* (cranberry rootworm), 242; *Xylosandrus compactus* (black twig borer), 262

Laburnum: *Aphis craccivora* (cowpea aphid), 308; *Pseudococcus maritimus* (grape mealybug), 88; *Uresiphita reversalis* (genista caterpillar), 166
Laburnum anagyroides (golden-chain tree): *Odontota dorsalis* (locust leafminer), 190
Lagerstroemia indica (crape myrtle): *Aphis gossypii* (cotton, or melon, aphid), 308; *Duplaspidiotus claviger* (camellia mining scale), 380; *Leptoglossus phyllopus* (leaf-footed bugs), 422–424; *Popillia japonica* (Japanese beetle), 236; *Siphoninus phillyreae* (ash whitefly), 322; *Tinocallis kahawaluokalani*, 292; *Toxoptera aurantii* (black citrus aphid), 302; *Uresiphita reversalis* (genista caterpillar), 166
Lantana (sage): *Bemisia tabaci* (sweetpotato whitefly), 322; *Saissetia coffeae* (hemispherical scale), 360; *Teleonemia scrupulosa* (lantana lace bug), 426; *Trialeurodes vaporariorum* (greenhouse whitefly), 320
larch, see *Larix*
  European: *Larix decidua*
  Siberian: *Larix sibirica*
  western: *Larix occidentalis*
Larix (larch):
—foliage: *Argyrotaenia velutinana* (redbanded leafroller), 214
  caterpillars: *Cingilia catenaria* (chainspotted geometer), 24, 146; *Coleophora laricella* (larch casebearer), 36, 186; *Dasychira plagiata* (pine tussock moth), 26; *Dendroctonus simplex* (eastern larch beetle), 62; *Lambdina fiscillaria* (hemlock looper), 24; *Neodiprion lecontei* (redheaded pine sawfly), 16; *Orgyia leucostigma* (whitemarked tussock moth), 158; *Semiothisa sexmaculata* (larch looper), 24
  mites: *Oligonychus ununguis* (spruce spider mite), 118
  sawflies: *Pristiphora erichsonii* (larch sawfly), 18
—bark, wood, twigs: *Dendroctonus valens* (red turpentine beetle), 62, 66; *Hylobius pales* (pales weevil), 56; *Pissodes schwarzi* (Yosemite bark weevil), 56; *Tomicus piniperda* (pine shoot beetle), 64
  birds: *Sphyrapicus* sp. (sapsuckers), 500
—sucking insects: *Adelges lariciatus* (spruce gall adelgid), 76, 78; *A. laricis*, 76, 78; *A. oregonensis*, 76; adelgids, 76, 78
Larix decidua (European larch): *Adelges laricis*, 78; *Coleophora laricella* (larch casebearer), 11, 36; *Pristiphora erichsonii* (larch sawfly), 16, 18
Larix laricina (tamarack): *Amorbia humerosana* (whiteline leafroller), 214; *Sparganothis sulfureana*, 216
Larix occidentalis (western): *Anoplonyx laricivorus* (western larch sawfly), 18; *A. occidens* (twolined larch sawfly), 18; *Orgyia pseudotsugatae* (Douglas-fir tussock moth), 34
Larix sibirica (Siberian larch): *Oligonychus ununguis* (spruce spider mite), 118
Latania: *Diaspis cocois* (cocos scale), 376; *Homaledra sabelella* (palm leafskeletonizer), 162
laurel, see *Kalmia*
  California: *Umbellularia californica*
  cherry: *Prunus laurocerasus*
  English: *Prunus laurocerasus*
  fig: *Ficus microcarpa*
  Grecian: *Laurus nobilis* (sweetbay)
  mountain: *Kalmia latifolia*
Laurus: *Ceroplastes floridensis* (Florida wax scale), 356; *Chrysomphalus dictyospermi* (dictyospermum scale), 382
Laurus nobilis (Grecian laurel): *Aspidiotus nerii* (oleander scale), 374; *Heliothrips haemorrhoidalis* (greenhouse thrips), 430; *Odontopus calceatus* (yellow poplar weevil), 210; *Trioza alacris* (laurel psyllid), 458
Lavatera (mallow): *Hemiberlesia rapax* (greedy scale), 372
Lawson cypress, see *Chamaecyparis lawsoniana*
lebbeck, see *Albizia lebbeck*
lemon: *Aspidiotus nerii* (oleander scale), 374. See also *Citrus*
Leptospermum laevigatum (tea tree): *Aonidiella aurantii* (California red scale), 378
Leucaena pulverulenta (tepeguaja): *Oncideres pustulatus* (huisache girdler), 286
Leucophyllum: *Teleonemia scrupulosa* (lantana lace bug), 424
Leucophyllum frutescens (Texas sage): *Pinnaspis strachani* (lesser snow scale), 390
Leucothoe: *Stephanitis takeyai*, 424
Libocedrus decurrens: *Argyresthia arceuthobiella*, 42; *A. libocedrella*, 42. See also incense cedar.
Ligustrum (privet)
—foliage: *Archips rosanus*, 218; *Callirhopalus bifasciatus* (twobanded Japanese weevil), 244; *Franklinella* sp., 434; *Otiorhynchus sulcatus* (black vine weevil), 242; snails, 498
  caterpillars: *Caloptilia fraxinella*, 196; *C. syringella* (lilac leafminer), 196
  mites: *Aculus ligustri* (privet rust mite), 480; *Brevipalpus obovatus* (privet mite), 476, 482
  sawflies: *Macrophya punctumalbum*, 136
—bark, wood, twigs: mice, 502; *Podosesia syringae* (lilac borer), 260–262; rabbits, 502
—sucking insects: *Graphocephala coccinea* (redbanded leafhopper), 418; *Leptoglossus phyllopus* (leaf-footed bugs), 422, 424; *Metcalfa pruinosa*, 422; *Siphoninus phillyreae* (ash whitefly), 322
  scales: *Aonidiella aurantii* (California red scale), 378; *A. citrina* (yellow scale), 378; *Aspidiotus nerii* (oleander scale), 374; *Asterolecanium arabidis* (pit-making pittosporum scale), 348; *Chrysomphalus dictyospermi* (dictyospermum scale), 382; *Hemiberlesia rapax* (greedy scale), 372; *Lepidosaphes camelliae*, 370; *L. gloveri* (Glover scale), 380; *Lindingaspis floridana*, 376; *Pseudaonidia paeoniae* (peony scale), 372; *Pseudaulacaspis pentagona* (white peach scale), 392; *P. prunicola* (white prunicola scale), 392; *Quadraspidiotus juglansregiae* (walnut scale), 386; *Q. perniciosus* (San Jose scale), 386; *Saissetia oleae* (black scale), 358; *Unaspis euonymi* (euonymus scale), 388
Ligustrum amurense: *Aculus ligustri* (privet rust mite), 432, 480; *Ochyromera ligustri*, 244
Ligustrum japonicum (Japanese privet): *Duplaspidiotus claviger* (camellia mining scale), 380; *Ochyromera ligustri*, 244
Ligustrum lucidum (glossy privet): *Aonidiella citrina* (yellow scale), 378; *Ochyromera ligustri*, 244
Ligustrum obtusifolium var. *regalianum* (regal privet): *Aculus ligustri* (privet rust mite), 432; *Dendrothrips ornatus* (privet thrips), 432; *Pseudococcus comstocki* (Comstock mealybug), 326
Ligustrum ovalifolium (California privet): *Aculus ligustri* (privet rust mite), 432, 480; *Dendrothrips ornatus* (privet thrips), 432; *Macrophya punctumalbum*, 136; *Popillia japonica* (Japanese beetle), 236; *Pseudococcus comstocki* (Comstock mealybug), 326
Ligustrum vulgare (English privet): *Macrophya punctumalbum*, 136
lilac, see *Syringa*
Limonium perezii (sea lavender): *Graphocephala atropunctata* (blue-green sharpshooter), 418
linden, see *Tilia*
  European: *Tilia vulgaris*
  littleleaf: *Tilia cordata*
  silver: *Tilia tomentosa*
Lindera benzoin (spicebush): *Cyrtepistomus castaneus* (Asiatic oak weevil), 244
Liquidambar styraciflua (American sweetgum, redgum)
—foliage: *Atta texana* (Texas leafcutting ant), 496; *Cyrtepistomus castaneus* (Asiatic oak weevil), 244; *Selenothrips rubrocinctus* (redbanded thrips), 432
  caterpillars: *Hyphantria cunea* (fall webworm), 166; *Malacosoma americanum* (eastern tent caterpillar), 168; *M. disstria* (forest tent caterpillar), 168; *Schizura concinna* (redhumped caterpillar), 156
—bark, wood, twigs: *Camponotus* sp. (carpenter ants), 496; *Elaphidionoides villosus* (twig pruner), 264; *Euzophera semifuneralis* (American plum borer), 252; *Neoclytus acuminatus* (redheaded ash borer), 278; *Xyloborus saxeseni*, 252; *Xylosandrus compactus* (black twig borer), 262; *X. crassiusculus* (ambrosia beetle), 252; *X. germanus*, 250
—sucking insects: *Acanalonia conica*, 422; *Tetraleurodes mori* (mulberry whitefly), 324; *Trialeurodes vaporariorum* (greenhouse whitefly), 320
  scales: *Diaspidiotus liquidambaris* (sweetgum scale), 382; *Eriococcus borealis*, 336; *Eulecanium cerasorum* (calico scale), 354; *Icerya purchasi* (cottony cushion scale), 338; *Melanaspis tenebricosa* (gloomy scale), 384; *Pulvinaria acericola* (cottony mapleleaf scale), 342; *Quadraspidiotus juglansregiae* (walnut scale), 386
Liriodendron tulipifera (tuliptree, yellow poplar)
—foliage: *Cyrtepistomus castaneus* (Asiatic oak weevil), 244; *Odontopus calceatus* (yellow poplar weevil, tuliptree leafminer), 210; *Resseliella liriodendri*, 462
  caterpillars: *Lymantria dispar* (gypsy moth), 138; *Phyllocnistis liriodendronella*, 196
  mites: *Tetranychus magnoliae* (magnolia spider mite), 476
—bark, wood, twigs: *Chrysobothris femorata* (flatheaded appletree borer), 270; *Corthylus columbianus* (Columbian timber beetle), 286; *Euzophora ostricolorella*, 254; *Oecanthus nigricornis* (blackhorned tree cricket), 492, 493; *Xylosandrus crassiusculus* (ambrosia beetle), 286
—sucking insects: *Aphis fabae* (bean aphid), 300; *Illinoia liriodendri* (tuliptree aphid), 292
  scales: *Abgrallaspis liriodendri*, 382; *Lepidosaphes ulmi* (oystershell scale), 370; *Quadraspidiotus juglansregiae* (walnut scale), 386; *Siphoninus phillyreae* (ash whitefly), 322; *Toumeyella liriodendri* (tuliptree scale), 362
Liriope: *Pinnaspis aspidistrae* (fern scale), 390
Litchi chinensis (lychee, lychee nut): *Eriophyes* sp., 486; *Euthochtha galeator*, 422; *Selenothrips rubrocinctus* (redbanded thrips), 432
Lithocarpus (tanbark oak): *Hemiberlesia rapax* (greedy scale), 372; *Quernaspis quercus*, 390; *Tetraleurodes stanfordi* (Stanford whitefly), 320
Lithocarpus densiflorus (tanbark oak, tanoak): *Quernaspis quercus*, 390
Livistona (palm): *Diaspidiotus* sp., 376; *Diaspis cocois* (cocos scale), 376
loblolly bay, see *Gordonia lasianthus*
locust: *Datana ministra* (yellownecked caterpillar), 154; *Diapheromera femorata* (walkingstick), 494; *Glossonotus acuminatus*, 406; *Hemiberlesia rapax* (greedy scale), 372; *Icerya purchasi* (cottony cushion scale), 338; *Lymantria dispar* (gypsy moth), 138. See also *Gleditsia*, *Robinia*
locust
  black: *Robinia pseudoacacia*
  bristly (rose acacia): *Robinia hispida*

London plane, see *Platanus acerifolia*
Lonicera (honeysuckle): *Archips negundana,* 172, 218; *Bryobia lonicerae* (honeysuckle bryobia), 472; *Hemiberlesia rapax* (greedy scale), 372; *Hyadaphis tataricae* (honeysuckle aphid), 314; *Macrosiphum euphorbiae* (potato aphid), 300; *Parthenolecanium persicae* (European peach scale), 364; *Phenacoccus aceris* (apple mealybug), 324; *Pseudaulacaspis pentagona* (white peach scale), 392; *Pulvinaria acericola* (cottony maple leaf scale), 342; *Unaspis euonymi* (euonymus scale), 388; *Uresiphita reversalis* (genista caterpillar), 166
Lonicera involucrata (twinberry): *Paraphytomyza cornigera,* 194
Lonicera korolkowii: *Hyadaphis tataricae* (honeysuckle aphid), 314
Lonicera microphylla: *Hyadaphis tataricae* (honeysuckle aphid), 314
Lonicera tatarica (tatarian honeysuckle): *Hyadaphis tataricae* (honeysuckle aphid), 314
loquat, see *Eriobotrya japonica*
Lotus scoparius (deerweed): *Asterolecanium arabidis* (pit-making pittosporum scale), 348
lupines: *Argyrotaenia franciscana,* 214
lychee, see *Litchi chinensis*
lychee nut, see *Litchi chinensis*
Lyonia fruticosa: *Eriococcus azaleae* (azalea bark scale), 336
Lyonothamnus floribundus (Catalina ironwood): *Icerya purchasi* (cottony cushion scale), 338
Lysimachia: *Mordvilkoja vagabunda* (poplar vagabond aphid), 462

Maclura pomifera (osage orange): *Prionus imbricornis* (tilehorned prionus), 280; *Pulvinaria innumerabilis* (cottony maple scale), 340; *Tetranychus canadensis* (fourspotted spider mite), 476
madrone, see *Arbutus menziesii, Arbutus unedo*
Magnolia
—foliage: *Heliothrips haemorrhoidalis* (greenhouse thrips), 430; *Helix aspersa* (brown garden snail), 498; *Hyphantria cunea* (fall webworm), 166; *Odontopus calceatus* (magnolia leafminer), 210; *Phyllocnistis liriodendronella,* 196
—bark: *Sphyrapicus* sp. (sapsuckers), 500
—sucking insects: *Metcalfa pruinosa,* 422; *Phenacoccus aceris* (apple mealybug), 324; *Pseudococcus affinis* (obscure mealybug), 324
scales: *Aspidiotus nerii* (oleander scale), 374; *Hemiberlesia rapax* (greedy scale), 372; *Icerya purchasi* (cottony cushion scale), 338; *Lepidosaphes gloveri* (Glover scale), 380; *Neolecanium cornuparvum* (magnolia scale), 354; *Saissetia oleae* (black scale), 358; *Toumeyella liriodendri* (tuliptree scale), 354, 362
Magnolia acuminata (cucumber tree): *Neolecanium cornuparvum* (magnolia scale), 354
Magnolia grandiflora (southern magnolia): *Aspidiotus nerii* (oleander scale), 374; *Euzophora ostricolorella,* 254; *Neolecanium cornuparvum* (magnolia scale), 354; *Odontopus calceatus* (magnolia leafminer), 210; *Pseudaulacaspis cockerelli* (false oleander scale), 374; *Tetranychus magnoliae* (magnolia spider mite), 476; *Toumeyella liriodendri* (tuliptree scale), 362
Magnolia macrophylla (bigleaf magnolia): *Tetraleurodes mori* (mulberry whitefly), 324
Magnolia quinquepeta: *Neolecanium cornuparvum* (magnolia scale), 354
Magnolia soulangeana: *Neolecanium cornuparvum* (magnolia scale), 354; *Odontopus calceatus* (magnolia leafminer, yellow poplar weevil), 210; *Toumeyella liriodendri* (tuliptree scale), 362
Magnolia stellata (star magnolia): *Neolecanium cornuparvum* (magnolia scale), 354; *Toumeyella liriodendri* (tuliptree scale), 362
Magnolia tripetala (umbrella tree, umbrella magnolia): *Aspidiotus nerii* (oleander scale), 374; *Hemiberlesia rapax* (greedy scale), 372; *Tetranychus canadensis* (fourspotted spider mite), 476
Magnolia virginiana (sweet bay): *Aonidiella aurantii* (California red scale), 378; *Fiorinia fioriniae* (palm fiorinia scale), 394; *Pinnaspis aspidistrae* (fern scale), 390; *Pseudaulacaspis cockerelli* (false oleander scale), 374; *Rhadopterus picipes* (cranberry rootworm), 242; *Toumeyella liriodendri* (tuliptree scale), 362
Mahonia: *Aleurothrixis interrogatonis* (deer brush whitefly), 320; *Coryphista meadii,* 144
Mahonia aquifolium (Oregon grape): *Aleuroplatus barbericolus* (barberry whitefly), 322; *Hemiberlesia rapax* (greedy scale), 372
mallow, see *Lavatera*
Malpighia: *Fiorinia theae* (tea scale), 394
Malus (apple)
—foliage: *Apterona helix,* 178; *Diapheromera femorata* (walkingstick), 494; *Taeniothrips inconsequens* (pear thrips), 432
beetles: *Asynonychus godmani* (Fuller rose beetle), 230; *Baliosus ruber* (basswood leafminer), 190; *Cyrtepistomus castaneus* (Asiatic oak weevil), 244; *Popillia japonica* (Japanese beetle), 236; *Rhadopterus picipes* (cranberry rootworm), 242; *Rhynchaenus rufipes* (willow flea weevil), 190
caterpillars: *Acronicta americana* (American dagger moth), 158; *Alsophila pometaria* (fall cankerworm), 142; *Amorbia humerosana* (whiteline leafroller), 214; *Amphipyra pyramidoides* (humped green fruitworm), 148; *Datana integerrima* (walnut caterpillar), 150; *D. major* (azalea caterpillar), 154; *D. ministra* (yellownecked caterpillar), 154; *Dichomeris ligulella,* 30; *Erannis tiliaria* (linden looper), 144; *Halysidota tessellaris* (pale tussock moth), 158; *Heterocampa guttivitta* (saddle prominent), 154; *Himella intractata* (fourlined green fruitworm), 148; *Lithophane antennata* (green fruitworm), 148; *L. innominata* (brown fruitworm), 148; *Lophocampa caryae* (hickory tussock moth), 160; *Lymantria dispar* (gypsy moth), 138; *Malacosoma americanum* (eastern tent caterpillar), 168; *M. californicum pluviale* (western tent caterpillar), 170; *M. disstria* (forest tent caterpillar), 168, 170; *Operophtera brumata* (winter moth), 146; *Orgyia antiqua* (rusty tussock moth), 158; *O. leucostigma* (whitemarked tussock moth), 158; *O. vetusta* (western tussock moth), 160; *Orthosia hibisci* (speckled green fruitworm), 148; *Paleacrita vernata* (spring cankerworm), 142, 144; *Parasa indetermina* (stinging rose caterpillar), 164; *Rhopobota naevana* (blackheaded fireworm, holly budmoth), 208; *Schizura concinna* (redhumped caterpillar), 156; *S. ipomoeae,* 156; *Sphinx drupiferarum* (cherry sphinx caterpillar), 164; *Spilonota ocellana* (eyespotted bud moth), 218; *Yponomeuta malinella,* 174; *Y. padella,* 174
miners/skeletonizers: *Acrobasis indigenella* (leaf crumpler), 218; *Bucculatrix pomifoliella,* 220; *Choreutis pariana* (apple-and-thorn skeletonizer), 216; *Coleophora serratella* (birch casebearer), 186; *Coptodisca* sp., 208; *C. splendoriferella* (resplendent shield bearer), 208; *Phyllonorycter* sp., 196; *P. crataegella,* 196; *Tischeria malifoliella* (appleleaf trumpet miner), 196
mites: *Brevipalpus californicus,* 476; *Bryobia arborea* (apple bryobia), 472; *Eotetranychus carpini borealis* (yellow spider mite), 474; *E. uncatus* (Garman spider mite), 474; *Eriophyes pyri* (pear leaf blister mite), 486; false spider mites, 476; *Oligonychus newcomeri,* 475; *Panonychus ulmi* (European red mite), 472; *Tetranychus mcdanieli* (McDaniel spider mite), 476; *T. pacificus* (Pacific spider mite), 476; *T. schoenei* (Schoene spider mite), 476; *T. urticae* (twospotted spider mite), 476
rollers/tiers: *Archips argyrospila* (fruittree leafroller), 172, 218; *A. griseus,* 218; *A. rosanus,* 218; *A. semiferana* (oak leafroller), 214; *Choristoneura rosaceana* (obliquebanded leafroller), 216; *Pandemis pyrusana,* 218; *Sparganothis sulfureana,* 216
sawflies: *Caliroa cerasi* (pear sawfly), 130
—fruit: *Laspeyresia pomonella* (codling moth), 234
—bark, wood, twigs: *Elaphidionoides villosus* (twig pruner), 264; *Euzophera semifuneralis* (American plum borer), 252; horntails, 494; *Magicicada septendecim* (periodical cicada), 490; mice, 502; *Oberea tripunctata* (dogwood twig borer), 262; *Oecanthus angustipennis* (narrowwinged cricket), 492; *O. fultoni* (snowy tree cricket), 492, 494; *Oncideres cingulata* (twig girdler), 264; rabbits, 502; *Saperda candida* (roundheaded appletree borer), 278; *Scolytus mali* (larger shothole borer), 246; *Sphyrapicus* sp. (sapsuckers), 500; *Synanthedon pyri* (apple bark borer), 258, 260; *S. scitula* (dogwood borer), 262; *Tremex columba* (pigeon tremex), 494; *Xyleborus dispar,* 250; *Zeuzera pyrina* (leopard moth), 254
—roots: *Prionus californicus* (California prionus), 280; *P. imbricornus* (tilehorned prionus), 280; *P. laticollis* (broadnecked prionus, broadnecked root borer), 280
—sucking insects: *Aleuroplatus berbericolus* (barberry whitefly), 322; *Boisea trivittatus* (boxelder bug), 398; *Corythucha coelata* (apple lace bug), 426; *C. pergandei* (alder lace bug), 426; *C. salicata* (western willow tingid), 426; *Lygus lineolaris* (tarnished plant bug), 398; *Macrosiphum euphorbiae* (potato aphid), 300; psyllids, 290
aphids: *Aphis craccivora* (cowpea aphid), 308; *A. pomi* (apple aphid), 292, 300; *Dysaphis plantaginea* (rosy apple aphid), 296; *Eriosoma lanigerum* (woolly apple aphid), 316; *Macrosiphum euphorbiae* (potato aphid), 300; *Rhopalosiphum fitchii* (apple grain aphid), 292
leafhoppers: *Edwardsiana rosae* (rose leafhopper), 412; *Empoasca fabae* (potato leafhopper), 414; *Erythroneura lawsoniana,* 416; *Keonolla confluens,* 416; *Stictocephala bisonia* (buffalo treehopper), 406; *Typhlocyba pomaria* (white apple leafhopper), 416
mealybugs: *Phenacoccus aceris* (apple mealybug), 324; *Planococcus citri* (citrus mealybug), 328; *Pseudococcus affinis* (obscure mealybug), 324; *P. comstocki* (Comstock mealybug), 326; *P. longispinus* (longtailed mealybug), 324
scales: *Aonidiella aurantii* (California red scale), 378; *Duplaspidiotus perniciosus* (San Jose scale), 386; *Eulecanium caryae* (hickory lecanium), 364; *E. cerasorum* (calico scale), 354; *Hemiberlesia rapax* (greedy scale), 372; *Icerya purchasi* (cottony cushion scale), 338; *Lepidosaphes ulmi* (oystershell scale), 370; *Mesolecanium nigrofasciatum* (terrapin scale), 364; *Parthenolecanium pruinosum* (frosted scale), 354; *Pulvinaria innumerabilis* (cottony maple scale), 340; *Saissetia oleae* (black scale), 358. See also crabapple.
Malva: *Aphis gossypii* (cotton, or melon, aphid), 308
Mangifera: *Fiorinia theae* (tea scale), 394
Mangifera indica (mango): *Aleurocanthus woglumi* (citrus blackfly), 322; *Aonidiella aurantii* (California red scale), 378; *Ceroplastes rubens* (red wax scale), 358; *Heliothrips haemorrhoidalis* (greenhouse thrips), 430; *Lindingaspis floridana,* 376
mango, see *Mangifera indica*
Manilkara zapota (sapodilla): *Ceroplastes rubens* (red wax scale), 358; *Hemiberlesia rapax* (greedy scale), 372
manzanita, see *Arctostaphylos*
maple, see *Acer*
amur: *Acer ginnala*
flowering: *Abutilon*
boxelder: *Acer negundo*
Japanese maple: *Acer palmatum*
mountain: *Acer spicatum*
Norway: *Acer platanoides*

red: *Acer rubrum*
Rocky Mountain: *Acer glabrum*
silver: *Acer saccharinum*
striped: *Acer pensylvanicum*
sugar: *Acer saccharum*
sycamore: *Acer pseudoplatanus*
Maytenus boaria (mayten): *Parasaissetia nigra* (nigra scale), 360
Melaleuca: *Fiorinia theae* (tea scale), 394
Melia azedarach (chinaberry): *Hemiberlesia rapax* (greedy scale), 372; *Pseudaulacaspis pentagona* (white peach scale), 392
Metopium toxiferum (poisonwood): *Hemiberlesia rapax* (greedy scale), 372
Mexican bush sage, see *Salvia leucantha*
Mexican orange, see *Choisya ternata*
Michelia figo (banana shrub): *Pseudaulacaspis cockerelli* (false oleander scale), 374; *Toumeyella liriodendri* (tuliptree scale), 362
milkweed: *Aphis nerii* (oleander and milkweed aphid), 312; *Lygaeus kalmii* (small milkweed bug), 398
mimosa, see *Albizia julibrissin*
mirrow plant, see *Coprosma repens*
mistletoe: *Aspidiotus nerii* (oleander scale), 374; *Hemiberlesia rapax* (greedy scale), 372
mock orange, see *Choisya ternata, Philadelphus mexicanus, Pittosporum tobira*
Modesto ash: *Tropidosteptes amoenus*, 402; *T. pacificus*, 402
monkey-hand tree, see *Chiranthodendron pentadactylon*
Moraine ash: *Tropidosteptes amoenus*, 402; *T. pacificus*, 402
morning glory: *Hemiberlesia rapax* (greedy scale), 372
Morus (mulberry)
—foliage: *Tetraleurodes mori* (mulberry whitefly), 324
—bark, wood, twigs: *Euzophera semifuneralis* (American plum borer), 252; *Megacyllene caryae* (painted hickory borer), 274; *Phloeotribus liminaris* (peach bark beetle), 350
—sucking insects: *Acanalonia conica*, 422; *Nipaecoccus nipae* (coconut mealybug), 90; *Parabemisia myricae* (Japanese bayberry whitefly), 322; *Phenacoccus aceris* (apple mealybug), 324; *Pseudococcus comstocki* (Comstock mealybug), 326
scales: *Aonidiella aurantii* (California red scale), 378; *Aspidiotus nerii* (oleander scale), 374; *Chrysomphalus dictyospermi* (dictyospermum scale), 382; *Eulecanium caryae* (hickory lecanium), 364; *Hemiberlesia rapax* (greedy scale), 372; *Lepidosaphes gloveri* (Glover scale), 380; *Melanaspis tenebricosa* (gloomy scale), 384; *Mesolecanium nigrofasciatum* (terrapin scale), 364; *Pseudaulacaspis pentagona* (white peach scale), 392; *Pulvinaria floccifera* (cottony camellia scale, cottony taxus scale), 344; *P. innumerabilis* (cottony maple scale), 340
Morus alba (white mulberry): *Aponychus spinosus*, 474
mountain alder, see *Alnus tenuifolia*
mountain ash, see *Sorbus*
mountain holly: *Hemiberlesia rapax* (greedy scale), 372
mulberry, see *Morus*
Myoporum laetum: *Graphocephala atropunctata* (blue-green sharpshooter), 418
Myrica (myrtle, bayberry): *Acyrthosiphon solani* (foxglove aphid), 310; *Phenacoccus aceris* (apple mealybug), 324; *Popillia japonica* (Japanese beetle), 236; *Pseudaonidia paeoniae* (peony scale), 372; *Synanthedon scitula* (dogwood borer), 262
Myrica cerifera (wax myrtle): *Duplaspidiotus claviger* (camellia mining scale), 380; *Parthenolecanium corni* (European fruit lecanium), 364; *Rhadopterus picipes* (cranberry rootworm), 242; *Tetraleurodes mori* (mulberry whitefly), 324
Myrsine: *Aspidiotus nerii* (oleander scale), 374
myrtle: *Aspidiotus nerii* (oleander scale), 374; *Chrysomphalus dictyospermi* (dictyospermum scale), 382; *Coptodisca* sp. (shield bearer), 208; *Hemiberlesia rapax* (greedy scale), 372; *Lepidosaphes gloveri* (Glover scale), 380; *Parasaissetia nigra* (nigra scale), 360; *Synanthedon scitula* (dogwood borer), 262. See also *Myrica, Myrtus, Langerstroemia, Vinca major*
myrtle
crape, see *Langerstroemia indica*
wax, see *Myrica cerifera*
myrtle, wild: *Melanaspis obscura* (obscure scale), 384
Myrtus: *Parasaissetia nigra* (nigra scale), 360; *Saissetia coffeae* (hemispherical scale), 360; *S. oleae* (black scale), 358

Nandina: *Icerya purchasi* (cottony cushion scale), 338
nannyberry, see *Viburnum lentago*
Nemopanthus collinus (mountain holly): *Hemiberlesia rapax* (greedy scale), 372
Nerium oleander (oleander)
—foliage: *Argyrotaenia citrana* (orange tortrix), 216; *Syntomeida epilais jucundissima* (oleander caterpillar), 162
—sucking insects: *Pseudococcus longispinus* (longtailed mealybug), 324
aphids: *Aphis nerii* (oleander and milkweed aphid), 312
scales: *Aspidiotus nerii* (oleander scale), 374; *Ceroplastes floridensis* (Florida wax scale), 356; *Chrysomphalus dictyospermi* (dictyospermum scale), 382; *Lindingaspis rossi*, 376; *Parasaissetia nigra* (nigra scale), 360; *Pseudaonidia trilobitiformis* (trilobe scale), 376; *Pseudaulacaspis cockerelli* (false oleander scale), 374; *Saissetia coffeae* (hemispherical scale), 360; *S. oleae* (black scale), 358
New Jersey tea, see *Ceanothus americanus*
nightshade, see *Solanum*
Norfolk Island pine, see *Araucaria heterophylla*
northern white cedar, see *Thuja occidentalis*
Norway spruce, see *Picea abies*
Nyssa (tupelo): *Antispila nysaefoliella* (tupelo leafminer), 208; *Chionaspis nyssae*, 390; *Malacosoma disstria* (forest tent caterpillar), 168
Nyssa aquatica (water tupelo): *Fusarium solani*, 250; *Malacosoma disstria* (forest tent caterpillar), 168
Nyssa sylvatica (blackgum, black tupelo, sourgum): *Aceria nyssae*, 488; *Antispila nysaefoliella* (tupelo leafminer), 208; eriophyid mites, 484, 488; *Lymantria dispar* (gypsy moth), 138; *Malacosoma americanum* (eastern tent caterpillar), 168; *M. disstria* (forest tent caterpillar), 168; *Pulvinaria acericola* (cottony maple leaf scale), 342; *Synanthedon rubrofascia*, 260

oak, see *Quercus*
black: *Quercus velutina*
blackjack: *Quercus marilandica*
blue: *Quercus douglasii*
burr: *Quercus macrocarpa*
California live: *Quercus agrifolia*
chestnut: *Quercus prinus*
coast live: *Quercus agrifolia*
Engelmann: *Quercus engelmannii*
English: *Quercus robur*
holly: *Quercus ilex*
interior live: *Quercus wislizeni*
live: *Quercus virginiana*
northern red: *Quercus rubra*
Nuttall: *Quercus nuttallii*
Oregon white: *Quercus garryana*
overcup: *Quercus lyrata*
pin: *Quercus palustris*
post: *Quercus stellata*
red: *Quercus falcata*
scarlet: *Quercus coccinea*
scrub: *Quercus ilicifolia, Q. laevis*
southern red: *Quercus falcata*
tanbark: *Lithocarpus, Lithocarpus densiflorus*
tanoak: *Lithocarpus densiflorus*
valley: *Quercus lobata*
water: *Quercus nigra*
white: *Quercus alba*
Olea (olive): *Aspidiotus nerii* (oleander scale), 374; *Aonidiella aurantii* (California red scale), 378; *Chrysomphalus dictyospermi* (dictyospermum scale), 382; *Euzophera semifuneralis* (American plum borer), 252; *Hemiberlesia rapax* (greedy scale), 372; *Lindingaspis rossi*, 376; *Parlatoria pittospori*, 110; *Pseudococcus comstocki* (Comstock mealybug), 326; *Saissetia oleae* (black scale), 358; *Unaspis euonymi* (euonymus scale), 388
Olea europaea (olive): *Hemiberlesia lataniae* (latania scale), 376
oleander, see *Nerium oleander*
yellow: *Thevea peruviana*
olive, see *Olea, Olea europaea*
orange: *Aspidiotus nerii* (oleander scale), 374; *Hemiberlesia rapax* (greedy scale), 372. See also *Citrus*
orange-berry pittosporum: *Pittosporum undulatum*
orange, trifoliate: *Poncirus trifoliata*
orchid: *Diaspis boisduvalii* (boisduval scale), 376; false spider mites, 476; *Lepidosaphes gloverii* (Glover scale), 380; *Tenuipalpus pacificus* (phaelenopsis mite), 476
orchid tree, see *Bauhinia purpurea*
Oregon grape, see *Mahonia aquifolium*
osage orange, see *Maclura pomifera*
Osmanthus: *Aspidiotus nerii* (oleander scale), 374; *Chrysomphalus dictyospermi* (dictyospermum scale), 382; *Duplaspidiotus claviger* (camellia mining scale), 380; *Fiorinia theae* (tea scale), 394
Ostrya virginiana (hophornbeam; ironwood): *Baliosus ruber* (basswood leafminer), 190; *Chionaspis gleditsia*, 390; *C. kosztarabi*, 390; *Corythucha pallipes* (birch lace bug), 426; *Symmerista leucitys* (orangehumped mapleworm), 154
Oxydendrum arboreum (sourwood): *Oberea myops* (rhododendron stem borer, azalea stem borer), 286; *O. tripunctata* (dogwood twig borer), 262

Pachysandra: *Lepidosaphes ulmi* (oystershell scale), 370; *Parthenolecanium fletcheri* (Fletcher scale), 98; *Unaspis euonymi* (euonymus scale), 388
palm
bamboo: *Chamaedoria erumpens*
cocos: *Arecastrum romanzoffianum*
date: *Phoenix dactylifera*
queen: *Arecastrum romanzoffianum*
sago: *Cycas*
palm, coconut: *Homaledra sabelella* (palm leafskeletonizer), 162
palm(s)
—foliage: *Asynonychus godmani* (Fuller rose beetle), 230; *Heliothrips haemorrhoidalis* (greenhouse thrips), 430; *Homaledra sabelella* (palm leafskeletonizer), 162
—bark: *Sphyrapicus* sp. (sapsuckers), 500
—sucking insects: *Nipaecoccus nipae* (coconut mealybug), 90
scales: *Aonidiella aurantii* (California red scale), 378; *A. citrina* (yellow scale), 378; *Aspidiotus nerii* (oleander scale), 374; *Chrysomphalus dictyospermi* (dictyospermum scale), 382; *Diaspis boisduvalii* (boisduval scale), 376; *D. cocois* (cocos scale), 376; *Fiorinia fioriniae* (palm fiorinia scale), 394; *Hemiberlesia lataniae* (latania scale), 376; *H. rapax* (greedy scale), 372; *Lepidosaphes gloverii* (Glover scale), 380; *Lindingaspis floridana*, 376; *L. rossi*, 376; *Parasaissetia nigra* (nigra scale), 360; *Pinnaspis strachani* (lesser snow scale), 390; *Saissetia coffeae* (hemispherical scale), 360; *S. oleae* (black scale), 358. See also *Butia, Chamaerops, Cocos, Howea, Livistonia, Phoenix*

palmetto
cabbage: *Sabal palmetto*
dwarf: *Rhapidophyllum hystrix*
saw: *Serenoa repens*
palmettos: *Homaledra sabelella* (palm leafskeletonizer), 162; *Pseudaulacaspis cockerelli* (false oleander scale), 374
Pandanus utilis (screw pine): *Aonidiella citrina* (yellow scale), 378
Parthenocissus quinquefolia (Virginia creeper): *Albuna fraxini*, 260; *Aspidiotus nerii* (oleander scale), 374; *Erythroneura ziczac*, 416; *Eulecanium cerarosum* (calico scale), 354; *Harrisina americana* (grapeleaf skeletonizer), 164; *Lepidosaphes ulmi* (oystershell scale), 370; *Popillia japonica* (Japanese beetle), 236; *Pulvinaria innumerabilis* (cottony maple scale), 340; *Rhadopterus picipes* (cranberry rootworm), 242; *Saperda puncticollis*, 276
Parthenocissus tricuspidata (Boston ivy): *Aspidiotus nerii* (oleander scale), 374; *Brevipalpus obovatus* (privet mite), 476, 482; *Erythroneura ziczac*, 416; *Eulecanium cerasorum* (calico scale), 354; *Graphocephala atropunctata* (bluegreen sharpshooter), 418; *Icerya purchasi* (cottony cushion scale), 338; *Pseudococcus comstocki* (Comstock mealybug), 326
Passiflora (passion flower, passion vine): *Aonidiella aurantii* (California red scale), 378; *Ceroplastes cistudiformis* (tortoise wax scale), 358; *C. irregularis* (desert wax scale), 358; *Hemiberlesia rapax* (greedy scale), 372
passion flower, passion vine, see *Passiflora*
Paulownia tomentosa (empress tree): *Acyrthosiphon solani* (foxglove aphid), 310; *Pseudococcus comstocki* (Comstock mealybug), 326
pawpaw, see *Asimina triloba*
Paxistima: *Unaspis euonymi* (euonymus scale), 388
pea, glory: *Clianthus* and *Sesbania punicea*
peach, see *Prunus persica*
flowering: *Prunus persica*
pear, *Pyrus communis*
Callery: *Pyrus communis*
pearlbush, see *Exochorda racemosa*
pecan, see *Carya illinoensis*
Pelea: *Toxoptera aurantii* (black citrus aphid), 302
Penstemon: *Argyrotaenia citrana* (orange tortrix), 216; *Asterolecanium arabidis* (pit-making pittosporum scale), 348; *Asynonychus godmani* (Fuller rose beetle), 230
peony: *Franklinella* sp., 434; *Macrodactylus subspinosus* (rose chafer), 236
pepper tree, see *Schinus*
Brazilian: *Schinus terebinthifolius*
California: *Schinus molle*
periwinkle, see *Vinca major*
Persea (bay)
—foliage: *Selenothrips rubrocinctus* (redbanded thrips), 432; *Urodus parvula* (ermine moth), 148
—bark: *Xylosandrus compactus* (black twig borer), 262
—sucking insects: *Nipaecoccus nipae* (coconut mealybug), 90; *Tetraleurodes mori* (mulberry whitefly), 324
aphids: *Aphis gossypii* (cotton, or melon, aphid), 308
scales: *Fiorinia fioriniae* (palm fiorinia scale), 394; *Hemiberlesia lataniae* (latania scale), 376; *H. rapax* (greedy scale), 372
Persea americana (avocado, alligator pear): *Aspidiotus nerii* (oleander scale), 374; *Euthorcaphis umbellulariae* (California laurel aphid), 294; *Heliothrips haemorrhoidalis* (greenhouse thrips), 430; *Hemiberlesia rapax* (greedy scale), 372; *Lindingaspis floridana*, 376; *Parabemisia myrica* (Japanese bayberry whitefly), 322; *Pseudococcus comstocki* (Comstock mealybug), 326; *P. longispinus* (longtailed mealybug), 324
Persea borbonia (redbay): psyllids, 290; *Pulvinaria acericola* (cottony maple leaf scale), 342; *Toumeyella liriodendri* (tuliptree scale), 362; *Trioza magnoliae*, 456; *Urodus parvula* (ermine moth), 148
Persea borbonia var. *humilis* (silkbay): *Trioza magnoliae*, 456
Persea borbonia var. pubescens (swampbay): *Trioza magnoliae*, 456
Persea humilis, see *Persea borbonia* var. *humilis*
Persea littoralis, see *Persea borbonia*
Persea palustris, see *Persea borbonia* var. *pubescens*
persimmon, see *Diospyros*
Philadelphus mexicanus (evergreen mock orange): *Acyrthosiphon solani* (foxglove aphid), 310; *Aphis fabae* (bean aphid), 300; *Asterolecanium arabidis* (pit-making pittosporum scale), 348
Phoenix (palm): *Diaspis cocois* (cocos scale), 376
Phoenix dactylifera (date palm): *Homaledra sabelella* (palm leafskeletonizer), 162
Phoradendron flavescens (mistletoe): *Ceroplastes nakaharai* (Nakahara wax scale), 358
Photinia: *Colaspis pseudofavosa*, 224; *Oligonychus ilicis* (southern red mite), 475; *Rhadopterus picipes* (cranberry rootworm), 242
Photinia arbutifolia, see *Heteromeles arbutifolia*
Photinia villosa: *Pseudococcus comstocki* (Comstock mealybug), 326
Picea (spruce)
—foliage: *Magdalis* sp., 54; *Scythropus* sp., 54
caterpillars: *Acleris variana* (eastern blackheaded budworm), 28; *Amorbia humerosana* (whiteline leafroller), 214; *Argyrotaenia velutinana* (redbanded leafroller), 214; *Choristoneura fumiferana* (spruce budworm), 28; *C. occidentalis* (western spruce budworm), 28; *C. rosaceana* (obliquebanded leafroller), 216; *Cingilia catenaria* (chainspotted geometer), 146; *Coleotechnites piceaella* (orange spruce needleminer), 32; *Dasychira plagiata* (pine tussock moth), 26; *D. pinicola* (pine tussock moth), 26; *Dioryctria abietivorella*, 48; *Endothenia albolineana* (spruce needleminer), 32; *Lambdina athasaria*, 24; *L. fiscellaria* (hemlock looper), 24; *Lymantria dispar* (gypsy moth), 138; *Orgyia antiqua* (rusty tussock moth), 158; *Thyridopteryx ephemeraeformis* (bagworm), 176; *Zeiraphera canadensis* (spruce bud moth), 28
mites: *Nalepella tsugifoliae* (hemlock rust mite), 122; *Oligonychus ununguis* (spruce spider mite), 118, 120; spider mites, 118, 120
sawflies: *Cephalcia* sp., 22; *Gilpinia hercyniae* (European spruce sawfly), 18; *Pikonema alaskensis* (yellowheaded spruce sawfly), 18
—bark, wood, twigs: *Dendroctonus valens* (red turpentine beetle), 62; *Formica exsectoides* (Allegheny mound ant), 496; *Hylobius pales* (pales weevil), 56; *Ips pini* (pine engraver beetle), 62; *Mayetiola piceae*, 114, 116; *Petrova* sp., 72; *Pissodes schwarzi* (Yosemite bark weevil), 56, 112; P. strobi (white pine weevil), 54; *Sciurus* sp. (squirrels), 500, 502; *Synanthedon novaroensis* (Douglas-fir pitch moth), 72; *S. pini* (pitch mass borer), 72, 260; *Tomicus piniperda* (pine shoot beetle), 64
—sucking insects: *Aphrophora cribata* (pine spittlebug), 86; *A. permutata*, 86; *Elatobium abietinum* (spruce aphid), 82; psyllids, 290
adelgids: *Adelges abietis*, 112, 114; *A. cooleyi* (Cooley spruce gall adelgid), 112; *A. lariciatus* (spruce gall adelgid), 112; *A. laricis*, 76, 78, 112; *A. tsugae* (hemlock woolly adelgid), 76; adelgids, 76, 78, 112; *Pineus pinifoliae* (pine leaf adelgid), 78; *P. similis*, 76, 112; *P. strobi* (pine bark adelgid), 76
scales: *Abgrallaspis ithacae* (hemlock scale), 102; *Aspidiotus cryptomeriae*, 110; *Carulaspis minima*, 106; *Chionaspis pinifoliae* (pine needle scale), 108; *Fiorinia externa* (elongate hemlock scale), 104; *Physokermes insignicola* (Monterey pine scale), 94; *P. piceae* (spruce bud scale), 96
Picea abies (Norway spruce): *Adelges abietis* (eastern spruce gall adelgid), 76, 114; *A. lariciatus* (spruce gall adelgid), 76, 78, 112; *Cinara piticornis*, 84; *Endothenia albolineana* (spruce needleminer), 32; *Epinotia nanana* (European spruce needleminer), 32; *Mayetiola piceae*, 114, 116; *Neodiprion lecontei* (redheaded pine sawfly), 16; *Oligonychus ununguis* (spruce spider mite), 118; *Physokermes piceae* (spruce bud scale), 96; *Pineus similis*, 76; *Pissodes strobi* (Engelmann spruce weevil, white pine weevil, Sitka spruce weevil), 54; *Synanthedon pini* (pitch mass borer), 72
Picea englemannii (Engelmann spruce): *Adelges cooleyi* (Cooley spruce gall adelgid), 76, 112; *Cephalcia fascipennis*, 22; *C. provancheri*, 22; *Cinara fornacula* (green spruce aphid), 82; *Endothenia albolineana* (spruce needleminer), 32; *Epinotia nanana* (European spruce needleminer), 32; *Orgyia pseudotsugatae* (Douglas fir tussock moth), 34; *Pineus similis*, 76; *Zeiraphera canadensis* (spruce bud moth), 28
Picea glauca (white spruce)
—foliage: *Zeiraphera canadensis* (spruce bud moth), 28
caterpillars: *Coleotechnites ducharmei*, 32; *Dasychira grisefacta*, 26; *Eacles imperialis* (imperial moth), 26; *Endothenia albolineana* (spruce needleminer), 32; *Epinotia nanana* (European spruce needleminer), 32
sawflies: *Cephalcia fascipennis*, 22; *C. nigra*, 22; *C. provancheri*, 22; *C. semidea*, 22
—bark, wood, twigs: *Mayetiola piceae*, 114, 116; *Synanthedon pini* (pitch mass borer), 72
—sucking insects: *Cinara piticornis*, 84; *Mindarus abietinus* (balsam twig aphid), 80
adelgids: *Adelges abietis* (eastern spruce gall adelgid), 76, 114; *A. cooleyi* (Cooley spruce gall adelgid), 76, 112; *Pineus similis*, 76
Picea glauca var. *albertiana*: *Cinara fornacula* (green spruce aphid), 82; *Oligonychus ununguis* (spruce spider mite), 118, 120; *Physokermes piceae* (spruce bud scale), 96
Picea mariana (black spruce): *Adelges laricis*, 76, 112; *Cephalcia provancheri*, 22; *Cinara fornacula* (green spruce aphid), 82; *Coleotechnites ducharmei*, 32; *Pineus floccus*, 76, 112; *P. pinifoliae* (pine leaf adelgid), 76, 78; *Zeiraphera canadensis* (spruce bud moth), 28
Picea orientalis (Oriental spruce): *Adelges cooleyi* (Cooley spruce gall adelgid), 76, 112
Picea pungens (blue spruce, Colorado blue spruce): *Adelges abietis* (eastern spruce gall adelgid), 114; *A. cooleyi (Cooley spruce gall adelgid), 76, 112; Cephalcia fascipennis*, 22; *Cinara fornacula* (green spruce aphid), 82; *C. piticornis*, 84; *Elatobium abietinum* (spruce aphid), 82; *Endothenea albolineana* (spruce needleminer), 32; *Epinotia nanana* (European spruce needleminer), 32; *Mindarus abietinus* (balsam twig aphid), 80; *Pineus similis*, 76; sawfly, 20; *Synanthedon pini* (pitch mass borer), 72
Picea rubens (red spruce): *Adelges abietis* (eastern spruce gall adelgid), 114; *A. laricis*, 76, 78, 114; *Cephalcia fascipennis*, 22; *C. provancheri*, 22; *Coleotechnites ducharmei*, 32; *Dendroctonus terebrans* (black turpentine beetle), 62; *Endothenia albolineana* (spruce needleminer), 32; *Epinotia nanana* (European spruce needleminer), 32; *Mayetiola piceae*, 114, 116; *Pineus floccus*, 76; *P. pinifoliae* (pine leaf adelgid), 76; *P. similis*, 76; *Sphyrapicus* (sapsuckers), 500
Picea sitchensis (Sitka spruce): *Adelges cooleyi* (Cooley spruce gall adelgid), 76, 112; *Cephalcia provancheri*, 22; *Endothenia albolineana* (spruce needleminer), 32; *Lambdina fiscillaria lugubrosa* (western hemlock looper), 24; *Lophocampa argentata* (silverspotted tiger moth), 34; *Pissodes strobi* (Engelmann spruce weevil, Sitka spruce weevil, white pine weevil), 54; tiger moth, 34
Pieris japonica (andromeda): *Datana major* (azalea caterpillar), 154; *Eriococcus azaleae*

(azalea bark scale), 336; lacebugs, 424; *Oberea tripunctata* (dogwood twig borer), 262; *Pulvinaria acericola* (cottony maple leaf scale), 342; *Stephanitis takeyai*, 424

pine, see *Pinus*
  Aleppo: *Pinus halepensis*
  Apache: *Pinus engelmannii*
  Australian: *Casuarina, C. equisetifolia*
  Austrian: *Pinus nigra*
  Benguet: *Pinus insularis*
  bigcone: *Pinus coulteri*
  Bishop: *Pinus muricata*
  black: *Pinus thunbergiana*
  bristlecone: *Pinus aristata*
  Canary Island: *Pinus canariensis*
  Chihuahua: *Pinus leiophylla* var. *chihuahuana*
  Chinese: *Pinus tabuliformis*
  chir: *Pinus roxburghii*
  cluster: *Pinus pinaster*
  Corsican: *Pinus nigra laricio*
  Coulter: *Pinus coulteri*
  Digger: *Pinus sabiniana*
  eastern white: *Pinus strobus*
  foxtail: *Pinus balfouriana*
  Italian stone: *Pinus pinea*
  jack: *Pinus banksiana*
  Japanese black: *Pinus thunbergiana*
  Japanese red: *Pinus densiflora*
  Jeffrey: *Pinus jeffreyi*
  knobcone: *Pinus attenuata*
  loblolly: *Pinus taeda*
  lodgepole: *Pinus contorta*
  longleaf : *Pinus palustris*
  marsh or pond: *Pinus serotina*
  Mexican: *Pinus patula*
  Mexican pinyon or Mexican stone: *Pinus cembroides*
  Monterey: *Pinus radiata*
  Montezuma: *Pinus montezumae*
  mountain or mugo: *Pinus mugo*
  Norfolk Island: *Araucaria heterophylla*
  pinyon: *Pinus edulis*
  pitch: *Pinus rigida*
  pond: *Pinus serotina*
  ponderosa: *Pinus ponderosa*
  red: *Pinus resinosa*
  sand: *Pinus clausa*
  Scots: *Pinus sylvestris*
  screw, common: *Pandanus utilis*
  scrub (used colloquially): *Pinus virginiana*
  shore: *Pinus contorta* var. *contorta*
  shortleaf: *Pinus echinata*
  slash: *Pinus elliottii*
  spruce: *Pinus glabra*
  sugar: *Pinus lambertiana*
  Swiss mountain: *Pinus mugo*
  Swiss stone: *Pinus cembra*
  table mountain: *Pinus pungens*
  three-needle, Mexican: *Pinus cembroides*
  Torrey: *Pinus torreyana*
  umbrella: *Sciadopitys verticillata*
  Virginia: *Pinus virginiana*
  western white: *Pinus monticola*
  white, eastern: *Pinus strobus*

pineapple guava, see *Feijoa sellowiana*

pine mat, see *Ceanothus diversifolius*

pine, southern (colloquial): *Chionaspis heterophyllae* (pine scale), 108; *Cinara atlantica*, 84; *Dendroctonus terebrans* (black turpentine beetle), 62; *Neodiprion excitans* (blackheaded pine sawfly), 16; *Pissodes nemorensis* (eastern pine weevil), 54, 56

Pinus
—foliage: stick insects, 494
  beetles: *Magdalis* sp., 54; *Scythropus* sp., 54, 58
  caterpillars: *Amorbia humerosana* (whiteline leafroller), 214; *Argyrotaenia citrana* (orange tortrix), 216; *Choristoneura rosaceana* (obliquebanded leafroller), 216; *Cingilia catenaria* (chainspotted geometer), 24; *Coloradia doris* ("Black Hills" pandora moth), 26; *Dasychira* sp. (tussock moths), 26; *D. grisefacta*, 26; *D. manto*, 26; *D. pinicola* (pine tussock moth), 26; *D. plagiata*, 26; *Exoteleia pinifoliella* (pine needleminer), 40; *Lymantria dispar* (gypsy moth), 138; *Orgyia antiqua* (rusty tussock moth), 158; *Sparganothis sulfureana*, 216
  mites: *Oligonychus ununguis* (spruce spider mite), 118, 120; spider mites, 118, 120
  sawflies: *Acantholyda verticalis*, 22; *Cephalcia hopkinsi*, 22; *Diprion similis* (introduced pine sawfly), 16, 18; *Neodiprion abbottii* (Abbott's sawfly), 18; *N. excitans*, 18; *N. lecontei* (redheaded pine sawfly), 16, 18
—bark, wood, twigs: *Cecidomyia resinicola* (pitch midge), 44; *Dioryctria abietivorella*, 48; *Formica exsectoides* (Allegheny mound ant), 496; *Hylastes* sp., 66; *H. radicis* (pine root collar weevil), 56; *Okanagana* sp., 490; *Pissodes nemorensis* (deodar weevil), 56; *P. schwarzi* (Yosemite bark weevil), 56; *Pithophthorus* sp., 66; *Tibicen* sp., 490; *Timema californica*, 494
  bark beetles: bark beetles, 62, 64, 66; *Dendroctonus* sp., 66; *D. brevicomis*, 66; *D. frontalis*, 66; *D. ponderosae* (mountain pine beetle), 66; *D. terebrans* (black turpentine beetle), 62, 66; *D. valens* (red turpentine beetle), 62, 66; *Ips* sp., 62, 64; *I. avulsus* (small southern pine engraver), 64; *I. calligraphus*, 64; *I. grandicollis*, 64; *I. pini* (pine engraver beetle), 62; *I. tridens*, 64; *Pityogenes hopkinsi*, 62; *Pityophthorus* sp., 66; *Tomicus piniperda* (pine shoot beetle), 64; *Xyleborus saxeseni*, 250; *Xylosandrus germanus*, 250
  bark borers: *Acanthocinus nodosus*, 276; *A. obsoletus*, 276; *A. princeps* (ponderosa pine bark borer), 276; *Synanthedon* sp., 72; *S. novaroensis* (Douglas fir pitch moth), 72; *S. pini* (pitch mass borer), 72, 260; *S. sequoia* (sequoia pitch moth), 72
  birds: *Sphyrapicus* (sapsuckers), 500; *S. varius varius* (yellowbellied sapsucker), 500
  rodents: mice, 502; rabbits, 502
—shoot/twig feeders: *Cylindrocopturus* sp., 54; *Dioryctria zimmermani* (Zimmerman pine moth), 48, 50; *Eucosma* sp., 48, 52; *Exoteleia* sp., 48; *Hylobius pales* (pales weevil), 56; *Petrova* sp., 48, 72; *Pissodes strobi* (white pine weevil), 54; *P. terminalis*, 54, 56; *Rhyacionia frustrana* (Nantucket pine tip moth), 48, 50
—sucking insects: *Pineus pinifoliae* (pine leaf adelgid), 76; *P. strobi* (pine bark adelgid), 76; *Pseudococcus comstocki* (Comstock mealybug), 326; psyllids, 290
  aphids: *Prociphilus* sp., 304
  scales: *Aspidiotus cryptomeriae*, 110; *Chionaspis pinifoliae* (pine needle scale), 108; *Fiorinia japonica*, 104; *Icerya purchasi* (cottony cushion scale), 338; *Matsucoccus alabamae* (Alabama pine scale), 92; *Nuculaspis californica* (black pineleaf scale), 100; *Pseudophilippia quaintancii* (woolly pine scale), 88; *Toumeyella parvicornis* (pine tortoise scale), 96; *T. pini* (striped pine scale), 96; *T. virginiana* (Virginia pine scale), 96
  spittlebugs: *Aphrophora canadensis*, 86; *A. parallela* (pine spittlebug), 86; *A. permutata*, 86; *A. saratogensis* (Saratoga spittlebug), 86; *Philaenus spumarius* (meadow spittlebug), 86

Pinus aristata (bristlecone pine): *Matsucoccus acalyptus* (pinyon pine scale), 92

Pinus attenuata (knobcone pine): *Pissodes radiatae* (Monterey pine weevil), 54, 56; *Synanthedon sequoiae* (sequoia pitch moth), 72; *Thecodiplosis piniradiatae* (Monterey pine midge), 116; *Toumeyella pinicola* (irregular pine scale), 94

Pinus balfouriana (foxtail pine): *Matsucoccus acalptus* (pinyon needle scale), 94

Pinus banksiana (jack pine; scrub)
—foliage: *Choristoneura pinus*, 28; *Pissodes terminalis* (lodgepole terminal weevil), 56
  caterpillars: *Coleotechnites canusella*, 40; *Dasychira pinicola* (pine tussock moth), 26; *D. plagiata* (a tussock moth), 26; *Eacles imperialis* (imperial moth), 26; *Exoteleia pinifoliella* (pine needleminer), 40; *Tetralopha robustella* (pine webworm), 22, 24; *Zelleria haimbachi* (pine needle sheathminer), 38
  sawflies: *Cephalcia fulviceps*, 22; *C. marginata*, 22; *Diprion similis* (introduced pine sawfly), 16, 18; *Neodiprion abbotti* (Abbott's sawfly), 18; *N. lecontei* (redheaded pine sawfly), 16; *N. nanulus nanulus* (red pine sawfly), 16; *N. pratti banksianae* (jack pine sawfly), 16; *N. pratti paradoxicus*, 16; *N. sertifer* (European pine sawfly), 16, 18; *N. swainei* (Swaine jack pine sawfly), 18
—bark, wood, twigs, shoots: *Cecidomyia piniinopis*, 44; *C. reeksi*, 44; *Cylindrocopturus eatoni* (pine reproduction weevil), 60; *Dendroctonus ponderosae* (mountain pine beetle), 66; *Eucosma gloriola* (eastern pine shoot borer), 52; *Hylobius pales* (pales weevil), 56; *H. radicis* (pine root collar weevil), 56; *Petrova albicapitana* (northern pitch twig moth), 72; *Pissodes terminalis* (lodgepole terminal weevil), 56; *Synanthedon pini* (pitch mass borer), 72; *S. sequoiae* (sequoia pitch moth), 72
—sucking insects: *Aphrophora cribata* (pine spittlebug), 86; *A. saratogensis* (Saratoga spittlebug), 86; *Schizolachnus piniradiatae* (woolly pine needle aphid), 84
  scales: *Matsucoccus gallicolus* (pine twig gall scale), 92; *Nuculaspis californica* (black pineleaf scale), 100; *Toumeyella parvicornis* (pine tortoise scale), 96

Pinus canariensis (Canary Island pine): *Toumeyella pinicola* (irregular pine scale), 94

Pinus cembroides (three-needled pinyon pine; Mexican pinyon pine): *Lophocampa argentata subalpina*, 34; *Nuculaspis californica* (black pineleaf scale), 100

Pinus clausa (sand pine; scrub): *Cinara cronarti*, 84; *C. piticornus*, 84; *Toumeyella virginiana* (Virginia pine scale), 96

Pinus contorta (lodgepole pine)
—foliage: *Coleotechnites milleri* (lodgepole needleminer), 40; *Coloradia pandora* (pandora moth), 26; *Lophocampa argentata* (silverspotted tiger moth), 34; *Orgyia pseudotsugatae* (Douglas-fir tussock moth), 34; tiger moth, 34
  sawflies: *Acantholyda verticalis*, 22; *Cephalcia californica*, 22; *Neodiprion burkei* (lodgepole pine sawfly), 18
—bark, wood, twigs, shoots: *Cecidomyia piniinopis*, 44; *Dendroctonus ponderosae* (mountain pine beetle), 66; *Eucosma gloriola* (eastern pine shoot borer), 52; *E. sonomana* (western pine shoot borer), 52; *Petrova albicapitana* (northern pitch twig moth), 72; *Pissodes radiatae* (Monterey pine weevil), 56; *P. terminalis* (lodgepole terminal weevil), 54, 56; *Rhyacionia buoliana* (European pine shoot moth), 48, 50; *R. montana*, 48; *Synanthedon sequoiae* (sequoia pitch moth), 72
—sucking insects: *Chionaspis pinifoliae* (pine needle scale), 108; *Matsucoccus bisetosus* (ponderosa pine twig scale), 92, 94; *Nuculaspis californica* (black pineleaf scale), 100; *Schizolachnus piniradiatae* (woolly pine needle aphid), 84; *Toumeyella pini* (striped pine scale), 96

Pinus contorta var. *contorta* (shore pine): *Nuculaspis californica* (black pineleaf scale), 100

Pinus coulteri (bigcone pine, Coulter): *Coloradia pandora* (pandora moth), 26; *Thecodiplosis piniradiata* (Monterey pine midge), 116

Pinus densiflora (Japanese red pine): *Acantholyda erythrocephala* (pine false webworm), 22; *Aphrophora parallela* (pine spittlebug), 86; *Matsucoccus resinosae* (red pine scale), 92; *Neodiprion sertifer* (European pine sawfly), 16, 18; *Synanthedon sequoiae* (sequoia pitch moth), 72

Pinus echinata (shortleaf pine)
—foliage
  caterpillars: *Dasychira manto*, 26; *Lambdina*

Pinus echinata (cont.)
pellucidaria, 24; *Tetralopha robustella* (pine webworm), 22, 24
sawflies: *Diprion similis* (introduced pine sawfly), 16, 18; *Neodiprion abbottii* (Abbott's sawfly), 18; *N. excitans*, 18; *N. lecontei* (redheaded pine sawfly), 16, 18; *N. pratti paradoxicus*, 16; *N. pratti pratti* (Virginia pine sawfly), 16; *N. sertifer* (European pine sawfly), 16, 18; *N. taedae linearis* (loblolly pine sawfly), 16; *N. taedae taedae* (spotted loblolly pine sawfly), 16
—bark, wood, twigs, shoots: *Dendroctonus frontalis* (southern pine beetle), 64; *D. terebrans* (black turpentine beetle), 62, 66; *Hylobius pales* (pales weevil), 56; *Pissodes nemorensis* (eastern pine weevil), 54, 56; *Rhyacionia frustrana* (Nantucket pine tip moth), 48, 50
—sucking insects: *Chionaspis heterophyllae* (pine scale), 108; *Matsucoccus gallicolus* (pine twig gall scale), 92; *Nuculaspis californica* (black pineleaf scale), 100; *Oracella acuta* (mealybug), 88; *Toumeyella pini* (striped pine scale), 96
Pinus edulis (pinyon pine): *Janetiella coloradensis*, 116; *Lophocampa argentata subalpina*, 34; *Matsucoccus acalyptus* (pinyon needle scale), 92, 94; *M. eduli*, 92; *M. monophyllae*, 92; *Nuculaspis californica* (black pineleaf scale), 100; *Pinyonia edulicola*, 116; *Thecodiplosis resinosae*, 116
Pinus elliotti (slash pine): *Aphrophora cribata* (pine spittlebug), 86; *Chionaspis heterophyllae* (pine scale), 108; *Cinara atlantica*, 84; *C. cronarti*, 84; *C. piticornus*, 84; *Dendroctonus frontalis* (southern pine beetle), 64; *D. terebrans* (black turpentine beetle), 62; *Neodiprion excitans* (blackheaded pine sawfly), 18; *N. lecontei* (redheaded pine sawfly), 16, 18; *Rhyacionia frustrana* (Nantucket pine tip moth), 48; *R. rigidana* (pitch pine tip moth, 48; Tetralopha robustella (pine webworm), 22, 24; *Toumeyella parvicornis* (pine tortoise scale), 96; *T. virginiana* (Virginia pine scale), 96
Pinus englemannii (Apache pine): *Dendroctonus frontalis* (southern pine beetle), 66
Pinus flexilis (limber pine): *Matsucoccus paucicatrices* (sugar pine scale), 92
Pinus glabra (spruce pine): *Matsucoccus gallicolus* (pine twig gall scale), 92; *Toumeyella parvicornis* (pine tortoise scale), 96; *T. virginiana* (Virginia pine scale), 96
Pinus halepensis (Aleppo pine): *Chionaspis heterophyllae* (pine scale), 108; *Parlatoria pittospori*, 110; *Synanthedon sequoiae* (sequoia pitch moth), 72; *Toumeyella pinicola* (irregular pine scale), 94
Pinus jeffreyi (Jeffrey pine): *Cecidomyia piniinopis*, 44; *Chionaspis pinifoliae* (pine needle scale), 108; *Coloradia pandora* (pandora moth), 26; *Cylindrocopturus eatoni* (pine reproduction weevil), 60; *Matsucoccus bisetosus* (ponderosa pine twig scale), 92, 94; *M. californicus*, 92; *M. fasciculensis*, (needle fascicle scale), 92; *Nuculaspis californica* (black pineleaf scale), 100; *Orgyia pseudotsugata* (Douglas-fir tussock moth), 34; *Rhyacionia zozana* (ponderosa pine tip moth), 48; *Synanthedon sequoiae* (sequoia pitch moth), 72; *Zelleria haimbachi* (pine needle sheathminer), 38
Pinus lambertiana (sugar pine): *Choristoneura lambertina*, 28; *Coloradia pandora* (pandora moth), 26; *Matsucoccus acalyptus* (pinyon needle scale), 92, 94; *M. paucicatrices* (sugar pine scale), 92; *Nuculaspis californica* (black pineleaf scale), 100; *Orgyia pseudotsugata* (Douglas-fir tussock moth), 34
Pinus monophylla (singleleaf pinyon): *Matsucoccus acalyptus* (pinyon needle scale), 92, 94; *M. monophyllae*, 92
Pinus montana, see *Pinus mugo*
Pinus montezumae (Montezuma pine): *Synanthedon sequoiae* (sequoia pitch moth), 72
Pinus monticola (western white pine): *Acantholyda verticales*, 22; *Lophocampa argentata* (silverspotted tiger moth), 34; *Matsucoccus paucicatrices* (sugar pine scale), 92; *Pineus pinifoliae* (pine leaf adelgid), 76, 78
Pinus mugo (mugho; Swiss mountain pine): *Acantholyda erythrocephala* (pine false webworm), 22; *Chionaspis pinifoliae* (pine needle scale), 108; *Eucosma gloriola* (eastern pine shoot borer), 52; *Eulachnus rileyi*, 84; *Exoteleia pinifoliella* (pine needleminer), 40; *Hylobius radicis* (pine root collar weevil), 56; *Neodiprion sertifer* (European pine sawfly), 16; *Nuculaspis californica* (black pineleaf scale), 100; *Pseudophilippi quaintancii* (woolly pine scale), 88; *Rhyacionia buoliana* (European pine shoot moth), 48, 50; *Tetralopha robustella* (pine webworm), 22, 24; *Toumeyella pinicola* (irregular pine scale), 94
Pinus muricata (Bishop pine): *Pissodes radiatae* (Monterey pine weevil), 54, 56; *Rhyacionia pasadenana* (Monterey pine tip moth), 48; *Scythropus californicus*, 58; *Synanthedon sequoiae* (sequoia pitch moth), 72; *Thecodiplosis piniradiatae* (Monterey pine midge), 116; *Toumeyella pinicola* (irregular pine scale), 94
Pinus nigra (Austrian pine): *Acantholyda erythrocephala* (pine false webworm), 22; *A. zappei*, 22; *Chionaspis pinifoliae* (pine needle scale), 108; *Dioryctria zimmermani* (Zimmerman pine moth), 48, 50; *Diprion similis* (introduced pine sawfly), 16, 18; *Eucosma gloriola* (eastern pine shoot borer), 52; *Eulachnus rileyi*, 84; *Hylobius radicis* (pine root collar weevil), 56; *Neodiprion abbottii* (Abbott's sawfly), 18; *N. lecontei* (redheaded pine sawfly), 16, 18; *Pineus strobi* (pine bark adelgid), 76; *Rhyacionia buoliana* (European pine shoot moth), 48, 50; *Synanthedon pini* (pitch mass borer), 72; *S. sequoiae* (sequoia pitch moth), 72; *Toumeyella parvicornis* (pine tortoise scale), 96
Pinus nigra laricio (Corsican pine): *Hylobius radicis* (pine root collar weevil), 56
Pinus palustris (longleaf; yellow pine): *Chionaspis heterophyllae* (pine scale), 108; *Cinara atlantica*, 84; *Dendroctonus frontalis* (southern pine beetle), 64; *Neodiprion excitans*, 16, 18; *Nuculaspis californica* (black pineleaf scale), 100; *Oracella acuta* (mealybug), 88; *Pissodes nemorensis* (eastern pine weevil), 54, 56; *Pseudophilippia quaintancii* (woolly pine scale), 88; *Rhyacionia frustrana* (Nantucket pine tip moth), 48, 50; *Tetralopha robustella* (pine webworm), 24; *Toumeyella virginiana* (Virginia pine scale), 96
Pinus patula (Mexican yellow pine): *Cinara cronarti*, 84; *C. piticornus*, 84
Pinus pinaster (maritime; cluster pine): *Synanthedon sequoiae* (sequoia pitch moth), 72
Pinus pinea (Italian stone; umbrella pine): *Pseudococcus longispinus* (longtailed mealybug), 324; *Toumeyella pinicola* (irregular pine scale), 94
Pinus ponderosa (ponderosa pine)
—foliage: *Anabrus simplex* (Mormon cricket), 492; *Apterona helix* (snailcase bagworm), 178; *Contarinia coloradensis*, 116; *Cylindrocopturus eatoni* (pine reproduction weevil), 60; *Zelleria haimbachi* (pine needle sheathminer), 38
caterpillars: *Coloradia pandora* (pandora moth), 26; *Dasychira grisefacta*, 26; *Neophasia menapia* (pine butterfly), 24; *Orgyia pseudotsugata* (Douglas-fir tussock moth), 34; *Phaeoura mexicanaria* (pine looper), 24
sawflies: *Acantholyda verticalis*, 22; *Cephalcia californica*, 22
—bark, wood, twigs, shoots: *Acanthocinus spectabilis*, 276; Canonura princeps (ponderosa pine bark borer), 276; *Cecidomyia piniinopis*, 44; *Cylindrocopturus eatoni* (pine reproduction weevil), 60; *Dendroctonus ponderosae* (mountain pine beetle), 66; *Eucosma sonomana* (western pine shoot borer), 52; *Rhyacionia buoliana* (European pine shoot moth), 48, 50; *R. bushnelli* (western pine tip moth), 48; *R. neomexicana* (southwestern pine tip moth), 48; *R. zozana* (ponderosa pine tip moth), 48; *Sciurus aberti* (tassel-eared squirrel), 500
—sucking insects: *Chionaspis pinifoliae* (pine needle scale), 108; *Matsucoccus bisetosus* (ponderosa pine twig scale), 92, 94; *M. californicus*, 92; *M. degeneratus*, 92; *M. fasciculensis* (needle fascicle scale), 92; *M. gallicolus* (pine twig gall scale), 92; *M. secretus*, 92; *M. vexillorum* (Prescott scale), 92, 94; *Nuculaspis californica* (black pineleaf scale), 100; *Philaenus spumarius* (meadow spittlebug), 86; *Schizolachnus piniradiatae* (woolly pine needle aphid), 84
Pinus ponderosa var. *scopulorum* (Rocky Mountain ponderosa pine): *Acanthocinus princeps* (ponderosa pine bark borer), 276; *A. spectabilis*, 276
Pinus pungens (table mountain pine): *Acantholyda erythrocephala* (pine false webworm), 22; *A. zappei*, 22; *Pseudophilippia quaintancii* (woolly pine scale), 88; *Synanthedon sequoiae* (sequoia pitch moth), 72
Pinus radiata (Monterey pine)
—foliage: *Acantholyda burkei*, 22; *Argyresthia pilatella* (Monterey pine needleminer), 40; *Lophocampa argentata sobrina*, 34; *Scythropus californicus*, 58; *Thecodiplosis piniradiatae* (Monterey pine midge), 116
mites: *Oligonychus milleri*, 118; *O. subnudus*, 118, 120
—bark, wood, twigs, shoots: *Cecidomyia resinicoloides* (Monterey pine resin midge), 44; *Dendroctonus valens* (red turpentine beetle), 72; *Exoteleia burkei* (Monterey pine shoot moth), 48; *Pissodes radiatae* (Monterey pine weevil), 54, 56; *Rhyacionia buoliana* (Monterey pine tip moth), 48; *Synanthedon sequoiae* (sequoia pitch moth), 72
—sucking insects: *Aspidiotus nerii* (oleander scale), 374; *Chionaspis pinifoliae* (pine needle scale), 108; *Matsucoccus bisetosus* (ponderosa pine twig scale), 92, 94; *Nuculaspis californica* (black pineleaf scale), 100; *Parlatoria pittospori*, 110; *Physokermes insignicola* (Monterey pine scale), 94; *Schizolachnus piniradiatae* (woolly pine needle aphid), 84; *Toumeyella pinicola* (irregular pine scale), 94
Pinus resinosa (red pine)
—foliage: *Contarinia baeri* (pine needle midge), 44; *Exoteleia pinifoliella* (pine needleminer), 40; *Resseliella pinifoliae* (pine needle midge), 44; *Thecodiplosis piniresinosae*, 44; *T. resinosae*, 116
caterpillars: *Coleotechnites resinosae*, 40; *Dasychira pinicola* (pine tussock moth), 26; *D. plagiata* (tussock moth), 26; *Eacles imperialis* (imperial moth), 26; *Lambdina pellucidaria*, 24; *Tetralopha robustella* (pine webworm), 22, 24; *Zale* sp., 24
sawflies: *Acantholyda erythrocephala* (pine false webworm), 22; *A. zappei*, 22; *Cephalcia frontalis*, 22; *C. fulviceps*, 22; *C. marginata*, 22; *Diprion similis* (introduced pine sawfly), 16, 18; *Gilpinia frutetorum*, 18; *Neodiprion abbottii* (Abbott's sawfly), 18; *N. lecontei* (redheaded pine sawfly), 16, 18; *N. nanulus nanulus* (red pine sawfly), 16, 18; *N. sertifer* (European pine sawfly), 16, 18
—bark, wood, twigs, shoots: *Eucosma gloriola* (eastern pine shoot borer), 52; *Hylobius pales* (pales weevil), 56; *H. radicis* (pine root collar weevil), 56; *Pissodes nemorensis* (eastern pine weevil), 54, 56; *Rhyacionia buoliana* (European pine shoot moth), 48; *R. rigidana* (pitch pine tip moth), 48
—sucking insects: *Aphrophora saratogensis* (Saratoga spittlebug), 86; *Pineus borneri*, 76; *P. coloradensis*, 76
aphids: *Cinara strobi* (white pine aphid), 84; *Eulachnus agilis*, (spotted pine aphid), 84; *E. rileyi*, 84; *Schizolachnus piniradiatae* (woolly pine needle aphid), 84

scales: *Chionaspis pinifoliae* (pine needle scale), 108; *Matsucoccus resinosae* (red pine scale), 92; *Nuculaspis californica* (black pineleaf scale), 100; *Toumeyella parvicornis* (pine tortoise scale), 96; *T. pini* (striped pine scale), 96
Pinus rigida (pitch pine): *Aphrophora cribata* (pine spittlebug), 86; *Cecidomyia piniinopis*, 44; *Contarinia baeri* (pine needle midge), 44; *Eucosma gloriola* (eastern pine shoot borer), 52; *Exoteleia pinifoliella* (pine needleminer), 40; *Hylobius radicis* (pine root collar weevil), 56; *Lambdina pellucidaria*, 24; *Matsucoccus gallicolus* (pine twig gall scale), 92; *Neodiprion abbottei* (Abbott's sawfly), 18; *N. lecontei* (redheaded pine sawfly), 16, 18; *N. pinusrigidae* (pitch pine sawfly), 18; *N. pratti paradoxicus*, 16; *N. sertifer* (European pine sawfly), 16, 18; *Nuculaspis californica* (black pineleaf scale), 100; *Pineus coloradensis*, 76; *Pseudophilippia quaintanceii* (woolly pine scale), 88; *Rhyacionia rigidana* (pitch pine tip moth), 48; *Tetralopha robustella* (pine webworm), 22, 24; *Thecodiplosis pinirigidae*, 116, *Toumeyella pini* (striped pine scale), 96
Pinus sabiniana (Digger pine): *Canonura princeps* (ponderosa pine bark borer), 276; *Matsucoccus bisetosus* (ponderosa pine twig scale), 92, 94; *M. fasciculensis*, 92; *Nuculaspis californica* (black pineleaf scale), 100; *Schizolachnus piniradiatae* (woolly pine needle aphid), 84; *Synanthedon sequoiae* (sequoia pitch moth), 72; *Thecodiplosis piniradiatae* (Monterey pine midge), 116
Pinus serotina (marsh or pond pine): *Cinara cronarti* (black pine aphid), 84; *C. piticornus*, 84; *Matsucoccus gallicolus* (pine twig gall scale), 92
Pinus strobus (eastern white pine)
—foliage: *Argyrotaenia pinatubana* (pine tube moth), 46; *Contarinia baeri* (pine needle midge), 44; *Resseliella pinifoliae*, 44
caterpillars: *Dasychira pinicola* (pine tussock moth), 26; *D. plagiata* (tussock moth), 26; *Eacles imperialis* (imperial moth), 26; *Sparganothis sulfureana*, 216; *Tetralopha robustella* (pine webworm), 22, 24; *Thyridopteryx ephemeraeformis* (bagworm), 176
sawflies: *Acantholyda erythrocephala* (pine false webworm), 22; *A. zappei*, 22; *Cephalcia marginata*, 22; *Diprion similis* (introduced pine sawfly), 16, 18; *Neodiprion lecontei* (redheaded pine sawfly), 16, 18; *N. pinetum* (white pine sawfly), 18
—bark, wood, twigs, shoots: cambium miners, 198; *Cecidomyia piniinopis*, 44; *Eucosma gloriola* (eastern pine shoot borer), 52; *Formica exsectoides* (Allegheny mound ant), 496; *Marmara* sp., 194; *M. fasciella*, 198; *Pissodes nemorensis* (eastern pine weevil), 54; *P. strobi* (Engelmann spruce weevil, Sitka spruce weevil, white pine weevil), 54; *Pityogenes hopkinsi*, 62; *Synanthedon pini* (pitch mass borer), 72; *Tomicus piniperda* (pine shoot beetle), 64
—sucking insects: *Aphrophora cribata* (pine spittlebug), 86; *Matsucoccus macrocicatrices* (Canadian pine scale), 92; *Pineus floccus*, 76; *P. pinifoliae* (pine leaf adelgid), 76, 78; *P. similis*, 76; *P. strobi* (pine bark adelgid), 76
aphids: *Cinara strobi*, (white pine aphid), 84; *Eulachnus rileyi*, 84
Pinus sylvestris (Scots pine)
—foliage: *Contarinia baeri* (pine needle midge), 44
caterpillars: *Choristoneura pinus*, 28; *Eacles imperialis* (imperial moth), 26; *Tetralopha robustella* (pine webworm), 22, 24
mites: *Trisetacus campnotus*, 122; *T. ehmanni*, 122; *T. gemmavitians*, 122
sawflies: *Acantholyda erythrocephala* (pine false webworm), 22; *Diprion similis* (introduced pine sawfly), 16, 18; *Gilpinia frutetorum*, 18; *Neodiprion excitans*, 18; *N. lecontei* (redheaded pine sawfly), 16, 18; *N. sertifer* (European pine sawfly), 16, 18
—bark, wood, twigs, shoots: *Dioryctria zimmermani* (Zimmerman pine moth), 48, 50; *Eucosma gloriola* (eastern pine shoot borer), 52; *Hylobius radicis* (pine root collar weevil), 56; *Pissodes nemorensis* (eastern pine weevil), 54, 56; *Rhyacionia buoliana* (European pine shoot moth), 48; *R. frustrana* (Nantucket pine tip moth), 48; *R. rigidana* (pitch pine tip moth), 48; *Synanthedon pini* (pitch mass borer), 72; *S. sequoiae* (sequoia pitch moth), 72; *Tomicus piniperda* (pine shoot beetle), 64
—sucking insects: *Aphrophora cribata* (pine spittlebug), 86; *Chionaspis pinifoliae* (pine needle scale), 108; *Pineus strobi* (pine bark adelgid), 76; *P. sylvestris*, 76; *Pseudophilippia quaintancii* (woolly pine scale), 88; *Quadraspidiotus juglansregiae* (walnut scale), 386
aphids: *Eulachnus agilis* (spotted pine aphid), 84; *E. rileyi*, 84
scales: *Platyphytoptus sabinianae*, 122; *Steoptus jonesi*, 122; *Toumeyella parvicornis* (pine tortoise scale), 96; *T. pini* (striped pine scale), 96; *Trisetacus campnodus*, 122; *T. ehmanni*, 122; *T. gemmavitians*, 122
Pinus tabuliformis (Chinese pine): *Matsucoccus resinosae* (red pine scale), 92
Pinus taeda (loblolly pine)
—foliage: *Exoteleia pinifoliella* (pine needleminer), 40; *Sparganothis sulfureana*, 216; *Tetralopha robustella* (pine webworm), 22, 24; *Thyridopteryx ephemeraeformis* (bagworm), 176
sawflies: *Neodiprion excitans* (blackheaded pine sawfly), 18; *N. hetricki* (Hetrick's sawfly), 18; *N. lecontei* (redheaded pine sawfly), 16, 18; *N. pratti pratti* (Virginia pine sawfly), 16; *N. taedae linearis* (loblolly pine sawfly), 16; *N. taedae taedae* (spotted loblolly pine sawfly), 16
—bark, wood, shoots: *Dendroctonus frontalis* (southern pine beetle), 64; *D. terebrans* (black turpentine beetle), 62; *Pissodes nemorensis* (eastern pine weevil), 54; *Rhyacionia frustrana* (Nantucket pine tip moth), 48; *R. rigida* (pitch pine tip moth), 48
—sucking insects: *Aphrophora cribata* (pine spittlebug), 86; *Cinara atlantica*, 84; *C. cronarti*, 84; *C. piticornis*, 84; *Oracella acuta* (mealybug), 88
scales: *Chionaspis heterophyllae* (pine scale), 108; *Matsucoccus gallicolus* (pine twig gall scale), 92; *Pseudophilippia quaintancii* (woolly pine scale), 88; *Toumeyella virginiana* (Virginia pine scale), 96
Pinus thunbergiana (Japanese black pine): *Aphrophora cribata* (pine spittlebug), 86; *Dioryctria zimmermani* (Zimmerman pine moth), 48, 50; *Matsucoccus resinosae* (red pine scale), 92; *Rhyacionia buoliana* (European pine shoot moth), 48, 50; *Synanthedon sequoiae* (sequoia pitch moth), 72
Pinus torreyana (Torrey pine): *Synanthedon sequoiae* (sequoia pitch moth), 72
Pinus virginiana (Virginia pine, scrub pine)
—foliage: *Exoteleia pinifoliella* (pine needleminer), 40; *Tetralopha robustella* (pine webworm), 22, 24; *Thyridopteryx ephemeraeformis* (bagworm), 176
sawflies: *Neodiprion lecontei* (redheaded pine sawfly), 16, 18; *N. pratti paradoxicus*, 16; *N. pratti pratti* (Virginia pine sawfly), 16; *N. taedae taedae* (spotted loblolly pine sawfly), 16
—bark, wood, shoots: *Cecidomyia piniinopis*, 44; *Hylobius pales* (pales weevil), 56; *Pissodes nemorensis* (eastern pine weevil), 54, 56; *Rhyacionia frustrana* (Nantucket pine tip moth), 48, 50; *R. rigidana* (pitch pine tip moth), 48; *Synanthedon sequoiae* (sequoia pitch moth), 72
—sucking insects: *Aphrophora cribata* (pine spittlebug), 86; *Eulachnus rileyi*, 84
scales: *Matsucoccus gallicolus* (pine twig gall scale), 92; *Quadraspidiotus juglansregiae* (walnut scale), 386; *Toumeyella pini* (striped pine scale), 96; *T. virginiana* (Virginia pine scale), 16
pinxter azalea, see *Rhododendron periclymenoides*: *Asphondylia azaleae*, 470
pistache, see *Pistacia*
pistachio, see *Pistacia*
Pistacia: *Aonidiella aurantii* (California red scale), 378; *Aspidiotus nerii* (oleander scale), 374; *Leptoglossus clypealis*, 424
Pithecellobium (blackbead): *Umbonia crassicornis* (thornbug), 408
Pittosporum: *Aphis gossypii* (cotton, or melon, aphid), 308; *Argyrotaenia citrana* (orange tortrix), 216; *Aspidiotus nerii* (oleander scale), 374; *Brevipalpus californicus*, 476; broad mites, 476; false spider mites, 476; *Hemiberlesia rapax* (greedy scale), 372; *Icerya purchasi* (cottony cushion scale), 338; *Pinnaspis strachani* (lesser snow scale), 390; *Pseudococcus affinis* (obscure mealybug), 324; *P. longispinus* (longtailed mealybug), 324; *Pulvinaria floccifera* (cottony camellia scale), 344; *Saissetia oleae* (black scale), 358; *Solenopsis* sp. (fire ants), 496; *Toxoptera aurantii* (black citrus aphid), 302
Pittosporum eugenioides (tarata): *Graphocephala atropunctata* (blue-green sharpshooter), 418
Pittosporum tobira (mock orange): *Asterolecanium arabidis* (pit-making pittosporum scale), 348; *Parlatoria pittospori*, 110; *Tetraleurodes mori* (mulberry whitefly), 324
Pittosporum undulatum (victorian box, orange-berry pittosporum): *Parasaissetia nigra* (nigra scale), 360
plane, see *Platanus*
Plantago lanceolata (plantain): *Dysaphis plantaginea* (rosy apple aphid), 296
Platanus (plane, sycamore)
—foliage: *Graphognathus* sp. (whitefringed beetle), 280
caterpillars: *Automeris io* (io moth), 164; *Lymantria dispar* (gypsy moth), 138; *Orgyia antiqua* (rusty tussock moth), 158; *Parasa indetermina* (stinging rose caterpillar), 164; *Phyllonorycter felinella*, 196
beetles: *Neochlamisus platani*, 400
—bark, wood, twigs: ambrosia beetles, 286; *Xylosandrus compactus* (black twig borer), 262; *X. germanus*, 250
—sucking insects: *Graphocephala atropunctata* (blue-green sharpshooter), 418; *Trialeurodes vaporariorum* (greenhouse whitefly), 320
aphids: *Aphis gossypii* (cotton, or melon, aphid), 308
lacebugs: *Corythucha ciliata* (sycamore lace bug), 428
scales: *Chionaspis americana* (elm scurfy scale), 368; *C. platani* (sycamore scurfy scale), 390; *Mesolecanium nigrofasciatum* (terrapin scale), 364; *Parthenolecanium pruinosum* (frosted scale), 354; *P. quercifex* (oak lecanium), 364; *Pulvinaria innumerabilis* (cottony maple scale), 340; *Saissetia oleae* (black scale), 358; *Stomacoccus platani* (sycamore scale), 334
Platanus acerifolia (London plane): *Erythroneura arta*, 416; *E. lawsoniana*, 416; *E. usitata*, 416; *Euzophera semifuneralis* (American plum borer), 252; *Halysidota harrisii* (sycamore tussock moth), 160; *Neochlamisus platani*, 400; *Plagiognathus albatus* (sycamore plant bug), 400
Platanus occidentalis (American plane, sycamore)
—foliage: *Graphognathus* sp. (whitefringed beetles), 280
caterpillars: *Ancylis plantanana*, 218; *Archips rosanus*, 218; *Automeris io* (io moth), 164; *Choristoneura rosaceana* (obliquebanded leafroller), 216; *Ectoedemia platanella*, 196; *Halysidota harrisii* (sycamore tussock moth), 160; *H. tessellaris* (pale tussock moth), 158; *Oiketicus townsendi*, 176; *Orgyia leucostigma* (whitemarked tussock moth), 158; *Thyridopteryx ephemeraeformis* (bagworm), 176

Platanus occidentalis (cont.)
beetles: *Cyrtepistomus castaneus* (Asiatic oak weevil), 244; *Popillia japonica* (Japanese beetle), 236; *Rhadopterus picipes* (cranberry rootworm), 242
mites: *Eotetranychus tiliarium*, 474; *Oligonychus platani* (platanus spider mite), 475
—bark, wood, twigs: *Chrysobothris femorata* (flatheaded appletree borer), 270; *Corthylus columbianus* (Columbian timber beetle), 286; *Euzophera semifuneralis* (American plum borer), 252; *Goes pulverulentus* (living beech borer), 282; horntails, 494; *Stictocephala bisonia* (buffalo treehopper), 406; *Synanthedon resplendens*, 258; *Tremex columba* (pigeon tremex), 494
—sucking insects: *Erythroneura arta*, 416; *E. lawsoniana*, 416; *E. usitata*, 416; *Graphocephala atropunctata* (blue-green sharpshooter), 418; *Plagiognathus albatus* (sycamore plant bug), 400; *Stictocephala bisonia* (buffalo treehopper), 406
aphids: *Drepanaphis* sp., 302; *Longistigma caryae* (giant bark aphid), 310; *Periphyllus* sp., 302
lacebugs: *Corythucha ciliata* (sycamore lace bug), 426, 428; *C. confraterna* (western sycamore lace bug), 426
scales: *Eulecanium caryae* (hickory lecanium), 364; *Lepidosaphes ulmi* (oystershell scale), 370; *Parthenolecanium corni* (European fruit lecanium), 364; *Pulvinaria innumerabilis* (cottony maple scale), 178; *Saissetia oleae* (black scale), 358; *Stomacoccus platani* (sycamore scale), 334
Platanus racemosa (western sycamore): *Ancylis plantanana*, 218; *Archips rosanus*, 218; *Choristoneura rosaceana* (obliquebanded leafroller), 216; *Chrysobothris mali* (Pacific flatheaded borer), 270; *Corythucha confraterna* (western sycamore lace bug), 426; *Halysidota tessellaris* (pale tussock moth), 158; *Orgyia leucostigma* (whitemarked tussock moth), 158; *Pseudococcus affinis* (obscure mealybug), 324; *Profenusa platanae* (sycamore leafmining sawfly), 188; *Synanthedon resplendens*, 258; *Thyridopteryx ephemeraeformis* (bagworm), 176
plum
—foliage: *Dichomeris ligulella*, 30; *Lithophane antennata* (green fruitworm), 148; *Malacosoma americanum* (eastern tent caterpillar), 168; *M. californicum pluviale* (western tent caterpillar), 170; *M. disstria* (forest tent caterpillar), 168, 170; *Pandemis pyrusana*, 218 *Schizura ipomoeae*, 156; *Sphinx drupiferarum* (cherry sphinx caterpillar), 164; *Taeniothrips inconsequens* (pear thrips), 432
—bark: *Scolytus mali* (larger shothole borer), 246; *S. rugulosus* (shothole borer), 250
—sucking insects: *Acanthocephala femorata*, 422, 444; *Boisea trivittatus* (boxelder bug), 398; leaf-footed bugs, 422, 424; *Lygus lineolaris* (tarnished plant bug), 398; *Myzus persicae* (green peach aphid), 300; *Phenacoccus aceris* (apple mealybug), 324; psyllids, 290
scales: *Lepidosaphes ulmi* (oystershell scale), 370; *Melanaspis obscura* (obscure scale), 384; *Pseudaulacaspis pentagona* (white peach scale), 392; *Pulvinaria innumerabilis* (cottony maple scale), 340; *Saissetia oleae* (black scale), 358; *Sphaerolecanium prunastri* (globose scale), 366. See also *Prunus*
plum, hog, see *Spondias mombi*
Plumbago: *Asynonychus godmani* (Fuller rose beetle), 230
Plumeria: *Hemiberlesia rapax* (greedy scale), 372; *Siphoninus phillyreae* (ash whitefly), 322
plum yew, see *Cephalotaxus*.
Podocarpus: *Aonidiella aurantii* (California red scale), 378; *Fiorinia fioriniae* (palm fiorinia scale), 394; *F. japonica*, 104; *Lepidosaphes gloveri* (Glover scale), 380; *Neophyllaphis podocarpi*, 294
Podocarpus elongatus (=*gracilior*): *Parlatoria pittospori*, 110
Podocarpus macrophylla (Japanese yew): *Aonidiella taxus* (Asiatic red scale), 110; *Neophyllaphis podocarpi*, 294
poinsettia: *Aspidiotus nerii* (oleander scale), 374
poisonwood, see *Metopium toxiferum*
pomegranate, see *Punica granatum*
Poncirus trifoliata (trifoliate orange): *Fiorinia theae* (tea scale), 394
pondcypress, see *Taxodium distichum*
pond pine, see *Pinus serotina*
poplar, see *Populus* and aspen
balsam: *Populus balsamifera*
black cottonwood: *Populus trichocarpa*
cottonwood: *Populus deltoides*
Great plains cottonwood: *Populus sargentii*
lombardy: *Populus nigra* 'Italica'
narrowleaf cottonwood: *Populus augustifolia*
plains cottonwood: *Populus deltoides* var. *occidentalis*
tulip: *Liriodendron tulipifera*
western cottonwood: *Populus fremontii*
yellow: *Liriodendron tulipifera*
Populus (poplar)
—foliage: *Oecanthus nigricornis* (blackhorned cricket), 492, 494; *Prodiplosis morrisi*, 470; spindle galls, 482
beetles: *Chrysomela crotchi* (aspen leaf beetle), 226; *C. interrupta*, 226; *C. scripta* (cottonwood leaf beetle), 226; *C. tremulae*, 226; *Plagiodera versicolora* (imported willow leaf beetle), 228
caterpillars: *Archips argyrospila* (fruittree leafroller), 172; *A. rosanus*, 218; *Automeris io* (io moth), 164; *Choristoneura rosaceana* (obliquebanded leafroller), 216; *Cingilia catenaria* (chainspotted geometer), 146; *Hemileuca nevadensis* (Nevada buckmoth), 156; *Hyalophora cecropia* (cecropia moth), 164; *Hydria undulata*, 146; *Ichthyura inclusa* (poplar tentmaker), 158; *Leucoma salicis* (satin moth), 158; *Lymantria dispar* (gypsy moth), 138; *Malacosoma americanum* (eastern tent caterpillar), 168; *M. californicum pluviale* (western tent caterpillar), 170; *M. disstria* (forest tent caterpillar), 170; *M. incurvum* (southwestern tent caterpillar), 170; *Nymphalis antiopa* (spiny elm caterpillar, mourningcloak butterfly), 152; *Orgyia leucostigma* (whitemarked tussock moth), 158; *Rheumaptera hastata* (spear-marked black moth), 146; *Schizura concinna* (redhumped caterpillar), 156
leafminers: *Coptodisca* sp. (shield bearer), 208; *Phyllocnistis* sp. (leafminers), 194; *Phyllonorycter salicifoliella*, 196
mites: *Aceria parapopuli* (poplar bud gall mite), 478; *A. populi*, 478; *Eotetranychus populi* (poplar spider mite), 474; erineum gall, 482; *Tetranychus canadensis* (fourspotted spider mite), 476; *T. urticae* (twospotted spider mite), 476
sawflies: *Cimbex americana* (elm sawfly), 136; *Nematus* sp., 134
—bark, wood, twigs: *Agrilus anxius* (bronze birch borer), 272; *A. liragus* (bronze poplar borer), 272; *Cryptorhynchus lapathi* (poplar-and-willow borer), 268; *Euzophera semifuneralis* (American plum borer), 252; *Gypsonoma haimbachiana* (cottonwood twig borer), 200; *Janus abbreviatus* (willow shoot sawfly), 286; *Oberea delongi*, 288; *O. schaumii*, 288; *Oecanthus nigricornis* (blackhorned cricket), 492, 494; *Oncideres cingulata* (twig girdler), 264; *Paranthrene dollii*, 260; *P. robiniae* (western poplar clearwing), 256, 260; *P. tabaniformis*, 254, 260; *Plectrodera scalator* (cottonwood borer), 276; *Prionoxystus robiniae* (carpenterworm), 256; *Saperda calcarata* (poplar borer), 284; *S. populnea*, 276, 278; *S. vestita* (linden borer), 276; *Sciurus* sp. (squirrels), 500, 502; *Sesia apiformis* (hornet moth), 258, 260; *S. tibialis* (American hornet moth), 258, 260; shoot borers, 286; *Stictocephala bisonia* (buffalo treehopper), 406; twig borers, 286; *Vespa crabro germana* (European hornet), 494; *Zeuzera pyrina* (leopard moth), 254
—roots: *Prionus californicus* (California prionus), 280; *P. imbricornis* (tilehorned prionus), 280; *P. laticollis* (broadnecked prionus), 280; *Sesia apiformis* (hornet moth), 258, 260; *Sesia tibialis* (American hornet moth), 258, 260
—sucking insects: *Lygus lineolaris* (tarnished plant bug), 398; *Pseudococcus comstocki* (Comstock mealybug), 326
aphids: *Mordvilkoja vagabunda* (poplar vagabond aphid), 462; *Pterocomma smithiae*, 310
lacebugs: *Leptoypha minor* (Arizona ash tingid), 426
leafhoppers: *Edwardsiana rosae* (rose leafhopper), 412; *Macropsis graminea*, 404
scales: *Chionaspis lintneri*, 390; *C. salicisnigrae*, 390; *Eriococcus azalae* (azalea bark scale), 336; *E. borealis*, 336; *Lepidosaphes ulmi* (oystershell scale), 370; *Mesolecanium nigrofasciatum* (terrapin scale), 364; *Pulvinaria innumerabilis* (cottony maple scale), 340; *Quadraspidiotus juglansregiae* (walnut scale), 386; *Saissetia oleae* (black scale), 358
Populus angustifolia (narrowleaf cottonwood): *Pemphigus betae*, 460
Populus balsamifera (balsam poplar): *Aceria parapopuli* (poplar bud gall mite), 478; *Agrilus liragus* (bronze poplar borer), 272; *Corythucha elegans* (willow and poplar lacebug), 426; *Janus abbreviatus* (willow shoot sawfly), 286; *Keonolla confluens*, 416
Populus deltoides (eastern cottonwood)
—foliage: *Hyphantria cunea* (fall webworm), 166; *Leucoma salicis* (satin moth), 158; *Phyllonorycter tremuloidiella* (aspen blotch miner), 196; *Prodiplosis morrisi*, 470; *Schizura concinna* (redhumped caterpillar), 156
beetles: *Chrysomela interrupta*, 226; *C. scripta* (cottonwood leaf beetle), 226; *Plagiodera versicolora* (imported willow leaf beetle), 228; *Rhynchaenus ephippiatus*, 190
—bark, wood, twigs: *Aceria parapopuli* (poplar bud gall mite), 478; *Agrilus liragus* (bronze poplar borer), 272; *Cryptorhynchus lapathi* (poplar-and-willow borer), 268; *Gypsonoma haimbachiana* (cottonwood twig borer), 200; *Janus abbreviatus* (willow shoot sawfly), 286; *Plectrodera scalator* (cottonwood borer), 276–280; *Prionoxystus robiniae* (carpenterworm), 256; *Saperda calcarata* (poplar borer), 284
—sucking insects: *Chionaspis furfura*, 368; *Hemiberlesia rapax* (greedy scale), 372; *Pemphigus populicaulis*, 460; *P. populitransversus*, 460
Populus deltoides var. *occidentalis* (= *sargentii*) (plains cottonwood): *Pemphigus populicaulis*, 460; *P. populitransversus*, 460
Populus fremontii (western cottonwood): *Pemphigus populicaulis*, 460; *Saperda populnea*, 278
Populus grandidentata (bigtooth aspen): *Agrilus liragus* (bronze poplar borer), 272; *Corythucha elegans* (willow and poplar lacebug), 426
Populus nigra (Lombardy poplar): *Plagiodera versicolora* (imported willow leaf beetle), 228
Populus nigra 'Italia' (Lombardy poplar): *Plagiodera versicolora* (imported willow leaf beetle), 228
Populus sargentii (Plains cottonwood), see *P. deltoides* var. *occidentalis*
Populus tremuloides (trembling or quaking aspen): *Aceria parapopuli* (poplar bud gall mite), 478; *Agrilus liragus* (bronze poplar borer), 272; *Argyrotaenia velutinana* (redbanded leafroller), 214; *Choristoneura conflictana* (large aspen tortrix), 216, 218; *Corythucha elegans* (willow and poplar lace bug), 426; *Cryptorhynchus lapathi* (poplar-and-willow borer), 268; eriophyid mites, 478; *Hypoxylon mammatum*, 276; *Janus abbreviatus* (willow shoot sawfly), 286; *Malacosoma disstria* (forest tent caterpillar), 168; *Operophtera bruceata* (Bruce spanworm), 146; *Phyllocnistis populiella* (aspen

leafminer), 194; *Saperda fayi* (thornlimb borer), 276; *S. inornata*, 276
Populus trichocarpa (black cottonwood): *Agrilus liragus* (bronze poplar borer), 272; *Paranthrene robiniae* (western poplar clearwing), 256
potato: *Acyrthosiphon solani* (foxglove aphid), 310; *Hemiberlesia lataniae* (latania scale), 376; *Macrosiphum euphorbiae* (potato aphid), 300
prickly ash, see *Zanthoxylum*
primrose: *Asynonychus godmani* (Fuller rose beetle), 230
privet, see *Ligustrum*
California: *Ligustrum ovalifolium*
Regel's: *Ligustrum obtusifolium* var. *regalianum*
prune, see *Prunus*
Prunus (cherry, peach, plum, prune)
—foliage: *Apterona helix* (snailcase bagworm), 178; *Asynonychus godmani* (Fuller rose beetle), 230; *Diapheromera femorata* (walkingstick), 492, 494; erineum galls, 484; *Graphognathus* sp. (whitefringed beetle), 280; *Oecanthus nigrocornis* (blackhorned cricket), 492; spindle galls, 482; *Taeniothrips inconsequens* (pear thrips), 432
caterpillars: *Archips argyrospila* (fruittree leafroller), 172; *A. cerasivorana* (uglynest caterpillar), 172; *Argyrotaenia juglandana*, 218; *Athrips rancidella*, 218; *Choristoneura rosaceana* (obliquebanded leafroller), 216; *Cingilia catenaria* (chainspotted geometer), 146; *Coleophora sacramenta* (California casebearer), 186; *Datana ministra* (yellownecked caterpillar), 154; *Dichomeris ligulella*, 30; *Hemileuca maia* (buckmoth), 156; *Hydria prunivorata* (cherry scallop shell moth), 146; *Lithophane antennata* (green fruitworm), 148; *Malacosoma americanum* (eastern tent caterpillar), 168; M. californicum pluviale (western tent caterpillar), 170; *M. disstria* (forest tent caterpillar), 170; M. incurvum (southwestern tent caterpillar), 170; *Pandemis pyrusana*, 218; *Rhopobota naevana* (blackheaded fireworm), 208; *Schizura concinna* (redhumped caterpillar), 156; *S. ipomoeae*, 156; *Sparganothis sulfureana*, 216; *Sphinx drupiferarum* (cherry sphinx caterpillar), 164; *Spilonota ocellana*, 218; *Stictocephala bisonia* (buffalo treehopper), 406
leafminers: *Baliosus ruber* (basswood leafminer), 190; *Odontota dorsalis* (locust leafminer), 190; *Phyllonorycter crataegella*, 196
mites: *Brevipalpus californicus*, 476; *Eotetranychus carpini* (yellow spider mite), 474; *E. uncatus* (Garman spider mite), 474; erineum galls, 482, 484; *Eriophyes emarginatae*, 482; eriophyid mites, 482, 484; false spider mites, 476; *Oligonychus newcomeri*, 475; *Panonychus ulmi* (European red mite), 472; *Tetranychus mcdanieli* (McDaniel spider mite), 476; *T. pacificus* (Pacific spider mite), 476; *T. schoenei* (Schoene spider mite), 476; *T. urticae* (twospotted spider mite), 476
sawflies: *Caliroa cerasi* (pear sawfly), 130
skeletonizers: *Acrobasis indigenella* (leaf crumpler), 218; *Bucculatrix pomifoliella*, 220; *Choreutis pariana* (apple-and-thorn skeletonizer), 216; *Popillia japonica* (Japanese beetle), 236; *Rhynchaenus rufipes* (willow flea weevil), 190
snails: *Helix aspersa* (brown garden snail), 498
—bark, wood, twigs: cambium miners, 198; *Camponotus* sp. (carpenter ants), 496; *Chrysobothris femorata* (flatheaded appletree borer), 270; *Elaphidionoides villosus* (twig pruner), 264; *Euzophera semifuneralis* (American plum borer), 252; *Goes pulverulentus* (living beech borer), 282; *Oberea tripunctata* (dogwood twig borer), 262; *Oecanthus nigrocornis* (blackhorned cricket), 492, 494; *Phloeotribus liminaris* (peach bark beetle), 250; *Scolytus mali* (larger shothole borer), 246; *S. rugulosus* (shothole borer), 250; *Stictocephala bisonia* (buffalo treehopper), 406; *Synanthedon exitiosa* (peachtree borer), 258, 260; *Xyleborus dispar*, 250; *Xylosandrus crassiusculus* (ambrosia beetle), 250; *Zeuzera pyrina* (leopard moth), 254
—roots: *Graphognathus* sp. (whitefringed beetle), 280; *Prionus californicus* (California prionus), 280;*P. imbricornis* (tilehorned prionus), 280; *P. laticollis* (broadnecked prionus, broadnecked rootborer), 280; *Synanthedon exitiosa*, 260; *S. scitula* (dogwood borer), 260, 262
—sucking insects: *Acanthocephala femorata*, 424; *Aleuroplatus berbericolus* (barberry whitefly), 322; *Anormenis chloris*, 422; *Corythucha associata*, 426; *C. padi* (cherry tingid), 426; *C. pruni* (cherry lace bug), 426, 428; *Leptocoris trivittatus* (boxelder bug), 398; *Leptoglossus phyllopus*, 422, 424; *Lygus lineolaris* (tarnished plant bug), 398; *Metcalfa pruinosa*, 422; *Ormenoides venusta*, 422; psyllids, 290; *Siphoninus phillyreae* (ash whitefly), 322
aphids: *Aphis gossypii* (cotton, or melon, aphid), 308; *A. pomi* (apple aphid), 300; *Myzus persicae* (green peach aphid), 300
leafhoppers: *Alebra albostriella* (maple leafhopper), 416; *Edwardsiana rosae* (rose leafhopper), 412; *Homalodisca* sp., 418; *Keonolla confluens*, 416; *Prosapia bicincta*, 420; *Typhlocyba pomaria* (white apple leafhopper), 416
mealybugs: *Dysmicoccus wistariae* (taxus mealybug), 88; *Phenacoccus aceris* (apple mealybug), 324; *Planococcus citri* (citrus mealybug), 328
scales: *Aspidiotus nerii* (oleander scale), 374; *Chionaspis furfura*, 368; *Duplaspidiotus claviger* (camellia mining scale), 380; *Eulecanium caryae* (hickory lecanium), 364; *Icerya purchasi* (cottony cushion scale), 338; *Melanaspis obscura* (obscure scale), 384; *Mesolecanium nigrofasciatum* (terrapin scale), 364; *Parthenolecanium pruinosum* (frosted scale), 354; *Pseudaulacaspis pentagona* (white peach scale), 392; *P. prunicola* (white prunicola scale), 392; *Pulvinaria innumerabilis* (cottony maple scale), 340; *Saissetia oleae* (black scale), 358; *Sphaerolecanium prunastri* (globose scale), 366; *Unaspis euonymi* (euonymus scale), 388
Prunus amygdalus (almond): *Archips rosanus*, 218; *Boisea rubrolineatus* (western boxelder bug), 398; *Chrysobothris mali* (Pacific flatheaded borer), 270; *Coleophora sacramenta* (California casebearer), 186; *Datana ministra* (yellownecked caterpillar), 154; *Hemiberlesia rapax* (greedy scale), 372; *Icerya purchasi* (cottony cushion scale), 338; *Malacosoma californicum pluviale* (western tent caterpillar), 170; *Panonychus ulmi* (European red mite), 472; *Saissetia oleae* (black scale), 358; *Synanthedon pictipes* (lesser peachtree borer), 258
Prunus armeniaca (apricot): *Icerya puchasi* (cottony cushion scale), 338; *Malacosoma califormicum pluviale* (western tent caterpillar), 170; *Melanaspis obscura* (obscure scale), 384; *Parthenolecanium pruinosum* (frosted scale), 354; *Prionoxystus robiniae* (carpenterworm), 256; *Saissetia oleae* (black scale), 358; *Sphaerolecanium prunastri* (globose scale), 366
Prunus cerasifera (cherry plum): *Sphaerolecanium prunastri* (globose scale), 366; *Synanthedon exitiosa* (peachtree borer), 258
Prunus cerasifera 'Atropurpurea' × *P. pumila* (purpleleaf plum): *Sphaerolecanium prunastri* (globose scale), 366
Prunus cerasus (sour cherry): *Eotetranychus uncatus* (Garman spider mite), 474; *Profenusa canadensis* (hawthorn leafmining sawfly), 188; *Quadraspidiotus juglansregiae* (walnut scale), 386
Prunus cistena (purpleleaf sand cherry): *Synanthedon exitiosa* (peachtree borer), 258
Prunus ilicifolia (hollyleaf cherry): *Paraleucoptera heinrichi*, 198; whiteflies, 198, 320
Prunus laurocerasus (cherry laurel, English laurel): *Chrysomphalus dictyospermi* (dictyospermum scale), 382; *Hemiberlesia rapax* (greedy scale), 372; *Lepidosaphes gloveri* (Glover scale), 380; *Rhadopterus picipes* (cranberry rootworm), 242; *Scolytus rugulosus* (shothole borer), 250; *Trioza alacris* (laurel psyllid), 458
Prunus persica (peach)
—foliage: *Asynonychus godmani* (Fuller rose beetle), 230
caterpillars: *Coleophora sacramenta* (California casebearer), 186; *Datana ministra* (yellownecked caterpillar), 154; *Malacosoma americanum* (eastern tent caterpillar), 168; *M. disstria* (forest tent caterpillar), 168, 170
snail: *Helix aspersa* (brown garden snail), 498
—bark, wood, twigs: *Euzophera semifuneralis* (American plum borer), 252; *Phloeotribus liminaris* (peach bark beetle), 250; *Scolytus rugulosus* (shothole borer), 250; *Stictocephala bisonia* (buffalo treehopper), 406; *Synanthedon pictipes* (lesser peach tree borer), 258
—sucking insects: *Acanthocephala femorata*, 424; *Keonolla confluens*, 416; *Leptoglossus phyllopus* (leaf-footed bugs), 422, 424; *Lygus lineolaris* (tarnished plant bug), 398; *Metcalfa pruinosa*, 422; *Myzus persicae* (green peach aphid), 300; *Pseudococcus comstocki* (Comstock mealybug), 326
scales: *Aulacaspis rosae* (rose scale), 366; *Icerya purchasi* (cottony cushion scale), 338; *Melanaspis obscura* (obscure scale), 352; *Mesolecanium nigrofasciatum* (terrapin scale), 364; *Pseudaulacaspis pentagona* (white peach scale), 392; *Pulvinaria innumerabilis* (cottony maple scale), 340; *Sphaerolecanium prunastri* (globose scale), 366
Prunus serotina (black cherry, wild cherry): *Archips cerasivorana* (uglynest caterpillar), 172; *Corythucha associata*, 426; *C. pruni*, 426, 428; erineum galls, 484; *Hyphantria cunea* (fall webworm), 166; *Malacosoma americanum* (eastern tent caterpillar), 168; *Oberea bimaculata*, 262
Prunus serrulata (Japanese flowering cherry): *Pseudaulacaspis pentagona* (white peach scale), 392; *P. prunicola* (white prunicola scale), 392; *Synanthedon pictipes* (lesser peachtree borer), 258
Prunus triloba (flowering almond): *Archips rosanus*, 218; *Chrysobothris mali* (Pacific flatheaded borer), 270
Prunus virginiana (chokecherry): *Archips cerasivorana* (uglynest caterpillar), 172; *Corythucha padi* (chokecherry tingid), 426; *Halysidota tessellaris* (pale tussock moth), 158; *Sparganothis directana*, 216
Pseudolarix: *Nalepella tsugifoliae* (hemlock rust mite), 122
Pseudotsuga menziesii (Douglas-fir)
—foliage: *Apterona helix* (snailcase bagworm), 178; *Contarinia pseudotsugae*, 44; *Magdalis* sp., 54; *Scythropus* sp., 54; stick insects, 492, 494; *Timema californica*, 494
caterpillars: *Choristoneura fumiferana* (spruce budworm), 28; *C. occidentalis* (western spruce budworm), 28; *Dasychira grisefacta*, 26; *Dioryctria abietivorella*, 48; *Lambdina fiscillaria lugubrosa* (western hemlock looper), 24; *Lophocampa argentata* (silverspotted tiger moth), 34; *Lymantria dispar* (gypsy moth), 138; *Orgyia pseudotsugata* (Douglas-fir tussock moth), 34; tiger moths, 34
mites: *Nalepella* sp., 122; *N. tsugifolia* (hemlock rust mite), 122; *Oligonychus ununguis* (spruce spider mite), 118
—bark, wood, twigs: cambium miners, 198; *Cylindrocopturus* sp., 54; *C. furnissi* (Douglas-fir twig weevil), 60; *Dendroctonus pseudotsugae* (Douglas-fir beetle), 66; *D. valens* (red turpentine beetle), 62; *Eucosma gloriola* (eastern pine shoot borer), 52; *Formica exsectoides* (Allegheny mound ant), 496; *Hylobius pales* (pales weevil), 56; *Pissodes nemorensis* (eastern pine weevil), 56; *P. strobi* (Engelmann spruce weevil, Sitka spruce weevil, white pine weevil), 54; *Pityophthorus orarius*, 60; *Sphyrapicus* sp. (sapsuckers), 500; *Synanthedon novaroensis*

Pseudotsuga menziesii (cont.)
(Douglas-fir pitch moth), 72; *S. sequoiae* (sequoia pitch moth), 72
—sucking insects: *Adelges cooleyi* (Cooley spruce gall adelgid), 76, 112; *Aphrophora cribata*, 86; *A. permutata*, 86
aphids: *Cinara piticornis*, 84; *C. pseudotsugae*, 84; *C. sabinae*, 84
scales: *Chionaspis pinifoliae* (pine needle scale), 108; *Nuculaspis californica* (black pineleaf scale), 100
Psidium (guava): *Ceroplastes floridensis* (Florida wax scale), 356
Psidium guajava (guava): *Aspidiotus nerii* (oleander scale), 374; *Hemiberlesia rapax* (greedy scale), 372; *Saissetia coffeae* (hemispherical scale), 360; *S. oleae* (black scale), 358
Ptelea trifoliata (hoptree): *Enchenopa binotata* (twomarked treehopper), 410
Punica granatum (pomegranate): *Aspidiotus nerii* (oleander scale), 374; *Hemiberlesia rapax* (greedy scale), 372; *Icerya purchasi* (cottony cushion scale), 338; *Saissetia oleae* (black scale), 358
purpleleaf sand cherry, see *Prunus cerasifera* 'Atropurpurea' × *P. pumila*
pussywillow, see *Salix discolor*
Purshia tridentata (antelope brush): *Malacosoma californicum pluviale* (western tent caterpillar), 170
Pyracantha
—foliage: *Acrobasis indigenella* (leaf crumpler), 218; *Choristoneura rosaceana* (obliquebanded leafroller), 216; *Orgyia leucostigma* (whitemarked tussock moth), 158; *O. vetusta* (western tussock moth), 160; *Selenothrips rubrocinctus* (redbanded thrips), 432
mites: *Oligonychus ilicis* (southern red mite), 475; *O. platani* (platanus spider mite), 475; spider mites, 472, 474, 475
—sucking insects: *Corythucha cydoniae* (hawthorn lace bug), 426; lacebugs, 424; *Phenacoccus aceris* (apple mealybug), 324
aphids: *Eriosoma lanigerum* (woolly apple aphid), 316; *Nearctaphis crataegifoliae*, 314; *Utamphorophora crataegi* (fourspotted hawthorn aphid), 314
scales: *Asterolecanium arabidis* (pit-making pittosporum scale), 348; *Ceroplastes ceriferus* (Indian wax scale), 356; *Chionaspis furfura*, 368; *Chrysomphalus dictyospermi* (dictyospermum scale), 382; *Duplaspidiotus claviger* (camellia mining scale), 380; *Eulecanium cerasorum* (calico scale), 354; *Hemiberlesia rapax* (greedy scale), 372; *Quadraspidiotus perniciosus* (San Jose scale), 386; *Sphaerolecanium prunastri* (globose scale), 366
Pyracantha atalantioides: *Corythucha cydoniae* (hawthorn lace bug), 426
Pyracantha angustifolia: *Corythucha cydoniae* (hawthorn lace bug), 426
Pyracantha coccinea: *Corythucha cydoniae* (hawthorn lace bug), 426
Pyracantha crenato-serrata: *Corythucha cydoniae* (hawthorn lace bug), 426
Pyracantha koidzumii: *Corythucha cydoniae* (hawthorn lace bug), 426
Pyrus communis (pear)
—foliage: *Acrobasis indigenella* (leaf crumpler), 218; *Apterona helix* (snailcase bagworm), 178; *Archips argyrospila* (fruittree leafroller), 172; *A. rosanus*, 218; *A. semiferana* (oak leafroller), 214; *Asynonychus godmani* (Fuller rose beetle), 230; *Dichomeris ligulella*, 30; *Lithophane antennata* (green fruitworm), 148; *Malacosoma disstria* (forest tent caterpillar), 168, 170; *Oiketicus townsendi*, 176; *Phyllonorycter* sp., 196; *Schizura concinna* (redhumped caterpillar), 156
mites: *Eotetranychus carpini* (yellow spider mite), 474; *Eriophyes pyri* (pearleaf blister mite), 486; *Oligonychus newcomeri*, 475
—bark, wood, roots: horntails, 494; *Prionoxystus robiniae* (carpenterworm), 256; *Prionus imbricornus* (tilehorned prionus), 280; *Synanthedon pyri* (apple bark borer), 260; *Tremex columba* (pigeon tremex), 494
—sucking insects: *Aonidiella aurantii* (California red scale), 378; *Aphis craccivora* (cowpea aphid), 308; *Aphis pomi* (apple aphid), 300; *Cacopsylla pyricola* (pear psylla), 290; *Eotetranychus carpini* (yellow spider mite), 474; *Eriosoma languinosa* (pear root aphid), 464; *E. pyricola*, 464; *Eulecanium cerasorum* (calico scale), 354; *Glossonotus acuminatus*, 406; *Hemiberlesia rapax* (greedy scale), 372; *Icerya purchasi* (cottony cushion scale), 338; *Lepidosaphes ulmi* (oystershell scale), 370; *Lygocoris communis*, (pear plant bug), 396; *Lygus lineolaris* (tarnished plant bug), 398; *Parthenolecanium corni* (European fruit lecanium), 364; *Pseudococcus comstocki* (Comstock mealybug), 326; *P. maritimus* (grape mealybug), 88; psyllids, 290; *Pulvinaria innumerabilis* (cottony maple scale), 340; *Saissetia oleae* (black scale), 358; *Siphoninus phillyreae* (ash whitefly), 322; *Taeniothrips inconsequens* (pear thrips), 432

queen palm, see *Arecastrum romanzoffianum*
Quercus
—foliage: *Amphibolips confluenta*, 440; *Anabrus simplex* (Mormon cricket), 492; *Atta texana* (Texas leafcutting ant), 496; *Diapheromera femorata* (walkingstick), 492, 494; *Oecanthus angustipennis* (narrowwinged cricket), 492, 494
beetles: *Asynonychus godmani* (Fuller rose beetle), 230; *Baliosus ruber* (basswood leafminer), 190; *Brachys aerosus*, 190; *B. ovatus*, 190; *Cyrtepistomus castaneus* (Asiatic oak weevil), 240, 244; *Odontota dorsalis* (locust leafminer), 190; *Rhadopterus picipes* (cranberry rootworm), 242; *Rhynchaenus rufipes* (willow flea weevil), 190
caterpillars: *Acrobasis minimella* (oak tubemaker), 212; *Acronicta americana* (American dagger moth), 158; *Alsophila pometaria* (fall cankerworm), 142, 270; *Anisota senatoria* (orangestriped oakworm), 156; *A. stigma* (spiny oakworm), 156; *A. virginiensis* (pinkstriped oakworm), 156; *Archips argyrospila* (fruittree leafroller), 172, 218; *A. fervidana* (oak webworm), 214; *A. griseus*, 218; *A. quercifoliana*, 218; *A. rosanus*, 218; *A. semiferana* (oak leafroller), 172, 214, 218; *Argyrotaenia citrana* (orange tortrix), 216; *A. quercifoliana*, 214, 218; *Automeris io* (io moth), 164; *Bucculatrix ainsliella* (oak skeletonizer), 220; *Choristoneura rosaceana* (obliquebanded leafroller), 216; *Cingilia catenaria* (chainspotted geometer), 144, 146; *Coptodisca* sp. (shield bearer), 208; *Croesia albicomana*, 172; *C. semipurpurana*, 172, 218; *Datana integerrima* (walnut caterpillar), 150, 154; *D. ministra* (yellownecked caterpillar), 154; *Dichomeris ligulella*, 30; *Dryocampa rubicunda* (greenstriped mapleworm), 156; *Ennomos subsignarius* (elm spanworm), 144–146; *Erannis tiliaria* (linden looper), 144; *Halysidota tessellaris* (pale tussock moth), 158; *Hemileuca maia* (buck moth), 156; *Leucoma salicis* (satin moth), 158; *Lithophane antennata* (green fruitworm), 148; *Lymantria dispar* (gypsy moth), 138, 140, 270; *Malacosoma americanum* (eastern tent caterpillar), 168; *M. californicum pluviale* (western tent caterpillar), 170; *M. constrictum* (Pacific tent caterpillar), 170; *M. disstria* (forest tent caterpillar), 168, 170, 270; *M. tigris* (Sonoran tent caterpillar), 170; *Operophtera brumata* (winter moth), 146; *Orgyia antiqua* (rusty tussock moth), 158; *O. vetusta* (western tussock moth), 160; *Paleacrita vernata* (spring cankerworm), 142, 144, 270; *Pandemis limitata*, 218; *Parasa indetermina* (stinging rose caterpillar), 164; *Phigalia titea* (half-wing geometer), 146; *Phryganidia californica* (California oakworm), 146; *Phyllonorycter* sp., 196; *Psilocorsis cryptolechiella*, 218; *Schizura ipomoeae*, 156; *Spilonota ocellana*, 218; *Symmerista canicosta* (redhumped oakworm), 156; *Tischeria citrinipennella*, 196; *T. discreta*, 196; *T. mediostriata*, 196; *Urodus parvula* (ermine moth), 148
mites: bud gall mites, 488; *Eotetranychus hicoriae* (pecan leaf scorch mite), 474; *E. querci*, 474; eriophyid mites, 488; *Oligonychus bicolor* (oak spider mite), 475; *O. platani* (platanus spider mite), 475; *O. propetes*, 475
sawflies: *Caliroa quercuscoccineae*, 130; *Periclista* sp., 136
—bark, wood, twigs: *Agrilus* sp., 266; *A. acutipennis*, 266; *A. angelicus* (Pacific oak twig girdler), 266, 272; *A. bilineatus* (twolined chestnut borer), 270; ambrosia beetle, 284; *Chrysobothris femorata* (flatheaded appletree borer), 270; *C. mali* (Pacific flatheaded borer), 270; *Corthylus columbianus* (Columbian timber beetle), 284; *Elaphidionoides parallelus*, 264; *E. villosus* (twig pruner), 264; *Enaphalodes rufulus* (red oak borer), 282; *Goes debilis* (oak branch borer), 282; *G. pulverulentus* (living beech borer), 282; horntails, 494; *Magicicada septendecim* (periodical cicada), 490; *Megacyllene caryae* (painted hickory borer), 274; *Neoclytus acuminatus* (redheaded ash borer), 278; *Oecanthus angustipennis* (narrowwinged cricket), 492, 494; *Oncideres cingulata* (twig girdler), 264; *Paranthrene asilipennis* (oak clearwing moth), 260; *P. simulans*, 260; *Platycotis vittata* (oak treehopper), 406; *Prionoxystus macmurtrei* (little carpenterworm), 256; *P. robiniae* (carpenterworm), 256, 282; *Synanthedon decipiens*, 260; *P. scitula* (dogwood borer), 262; *Tremex columba* (pigeon tremex), 494; *Xylosandrus crassiusculus*, 250; *Zeuzera pyrina* (leopard moth), 254
—roots: *Magicicada septdendecim* (periodical cicada), 490; *Prionus californicus* (California prionus), 280; *P. laticollis* (broadnecked prionus, broadnecked root borer), 280; *Rhadopterus picipes* (cranberry rootworm), 242
—sucking insects: *Alebra albostriella* (maple leafhopper), 416; *Clastoptera obtusa* (alder spittlebug), 420; *Corythucha arcuata* (oak lace bug), 426; *Edwardsiana rosae* (rose leafhopper), 412; *Euthochtha galeator*, 422; *Metcalfa pruinosa*, 422; *Phenacoccus aceris* (apple mealybug), 324; *Stictocephala bisonia* (buffalo treehopper), 406
aphids: *Stegophylla* sp., 304; *S. quercifolia*, 304
scales: *Allokermes kingi*, 366; *Asterolecanium* sp., 352; *A. minus*, 352; *A. quericola*, 352; *A. variolosum* (golden oak scale), 352; *Aonidiella aurantii* (California red scale), 378; *Hemiberlesia rapax* (greedy scale), 372; *Icerya purchasi* (cottony cushion scale), 338; *Melanaspis obscura* (obscure scale), 384; *Nanokermes pubescens*, 366; *Parthenolecanium quercifex* (oak lecanium), 364; pit scales, 352; *Pulvinaria innumerabilis* (cottony maple scale), 340
Quercus agrifolia (coast live oak, California live oak)
—foliage: *Bucculatrix albertiella*, 220; *Dryocosmus dubiosus*, 440, 444; *Malacosoma californicum pluviale* (western tent caterpillar), 170; *Phryganidia californica* (California oakworm), 146
mites: *Aceria mackiei*, 486; erineum galls, 484
—bark, wood, twigs: *Agrilus angelicus* (Pacific oak twig girdler), 266, 272; *Styloxus fulleri*, 266; *Synanthedon resplendens*, 258
—sucking insects: *Atarsaphis agrifoliae*, 296; *Pseudococcus affinis* (obscure mealybug), 324; *Quernaspis quercus*, 390; *Stegophylla essigi*, 304; *Tetraleurodes stanfordi* (Stanford whitefly), 320
Quercus alba (white oak)
—foliage: *Acraspis erinacei* (hedgehog gall), 444; *Acrobasis minimella* (oak tubemaker), 212; *Agromyza viridula*, 206; *Anisota senatoria* (or-

angestriped oakworm), 156; *Cameraria cincinnatiella* (gregarious oak leafminer), 192; *C. hamadryadella* (solitary oak leafminer), 192; *Epinotia aceriella* (maple trumpet skeletonizer), 212; *Macrodiplosis niveipila* (veinfold gall), 446; *Periclista media*, 136; *Symmerista canicosta* (redhumped oakworm), 156; *Tischeria badiiella*, 196
mites: *Aceria triplacis*, 486; *Oligonychus propetes*, 475
—bark, wood, twigs: *Agrilus acutipennis*, 266; A. bilineatus (twolined chestnut borer), 270; *Asterolecanium* sp., 352; *Callirhytis seminator* (wool sower gall), 444; clearwing moths, 254; *Goes tigrinus* (white oak borer), 282; *Paranthrene simulans*, 254
—sucking insects: *Asterolecanium* sp., 352; *A. variolosum* (golden oak scale), 352; *Glossonotus acuminatus*, 406; *Melanaspis obscura* (obscure scale), 384; pit scales, 352
Quercus arizonica (Arizona white oak): *Oligonychus propetes*, 475
Quercus coccinea (scarlet oak): *Agrilus bilineatus* (twolined chestnut borer), 270; *Archips fervidana* (oak webworm), 172, 214; *A. semiferana* (oak leafroller), 214; *Caliroa quercuscoccineae* (scarlet oak sawfly), 130; *Callirhytis quercuspunctata*, 440; *Croesia albicomana*, 172; *C. semipurpurana*, 172; *Enaphalodes rufulus* (red oak borer), 282
Quercus dumosa (California scrub oak): *Antron douglasii* (spined turban gall wasp), 448
Quercus douglasii (blue oak): *Andricus kingi*, 440; *Antron douglasii* (spined turban gall wasp), 448
Quercus engelmannii (Englemann oak): *Agrilus angelicus* (Pacific oak twig girdler), 266; *Andricus fullaway*, 448
Quercus falcata (red oak, southern red oak)
—foliage: *Agromyza viridula*, 206; *Amphibolips confluenta*, 446; *Archips fervidana* (oak webworm), 172, 214; *Camereria hamadryadella* (solitary oak leafminer), 192; *Croesia albicomana*, 172; *C. semipurpurana* (oak leaftier), 172; *Datana major* (azalea caterpillar), 154; *Epinotia aceriella* (maple trumpet skeletonizer), 212; *Macrodiplosis erubescens*, 442; *Oligonychus bicolor* (oak spider mite), 472; "roly-poly" galls, 448
—bark, wood and twigs: *Agrilus* sp., 266; *A. bilineatus* (twolined chestnut borer), 270; *Callirhytis quercuspunctata* (gouty oak gall wasp), 440; clearwing moths, 254; *Goes pulverulentus* (living beech borer), 282; *Paranthrene simulans*, 254; *Prionoxystus robiniae* (carpenterworm), 256
—sucking insects: *Glossonotus acuminatus*, 406; *Melanaspis obscura* (obscure scale), 384; *Nanokermes pubescens* (kermes scale), 364; *Stegophylla quercicola*, 304
Quercus garryana (Oregon white oak): *Andricus kingi*, 440
Quercus ilex (holly oak): *Aspidiotus nerii* (oleander scale), 374
Quercus ilicifolia (bear oak): *Anisota senatoria* (orangestriped oakworm moth), 156; *Archips fervidana* (oak webworm), 172; *Callirhytis cornigera* (horned oak gall wasp), 440, 442; *Macrodiplosis quercusoruca* (vein pocket gall midge), 440, 446; *Stegophylla quercifolia*, 304
Quercus kelloggi (California black oak): *Callirhytis perdens*, 448
Quercus laevis (turkey oak, scrub oak): *Anisota senatoria* (orangestriped oakworm moth), 156; *Archips fervidana* (oak webworm), 172, 214; *Callirhytis cornigera* (horned oak gall wasp), 440, 442; *C. quercuspunctata* (gouty oak gall wasp), 440; *Cameraria hamadryadella* (solitary oak leafminer), 192; *Macrodiplosis quercusoruca* (vein pocket gall midge), 440, 446; *Stegophylla quercifolia*, 304
Quercus lobata (valley oak, California white oak): *Andricus kingi*, 440; *Antron douglasii* (spined turban gall wasp), 448; *Asterolecanium* sp. (an oak pit scale), 352; *Bucculatrix albertiella*, 220; *Neuroterus saltatorius*, 440, 444; *Orgyia vetusta* (western tussock moth), 160; *Phryganidia californica* (California oakworm), 146; *Quernaspis quercus*, 390
Quercus lyrata (overcup oak): *Agrilus acutipennis*, 266; *Goes tigrinus* (white oak borer), 282, 284
Quercus macrocarpa (burr oak): *Agrilus bilineatus* (twolined chestnut borer), 270; *Goes tigrinus* (white oak borer), 282
Quercus marilandica (blackjack oak): *Acrobasis minimella* (oak tubemaker), 212; *Callirhytis cornigera* (horned oak gall wasp), 440, 442; *Cameraria hamadryadella* (solitary oak leafminer), 192
Quercus nigra (water oak): *Callirhytis cornigera* (horned oak gall wasp), 440, 442; *C. quercuspunctata* (gouty oak gall wasp), 440; *Xylosandrus compactus* (black twig borer), 262
Quercus nuttallii (Nuttall oak): clearwing moths, 254; *Paranthrene simulans*, 254; *Prionoxystus robiniae* (carpenterworm), 256
Quercus palustris (pin oak): *Caliroa quercuscoccineae* (scarlet oak sawfly), 130; *Callirhytis cornigera* (horned oak gall wasp), 440, 442; *C. quercuspunctata* (gouty oak gall wasp), 440, 442; *Eotetranychus querci*, 474; *E. uncatus* (Garman spider mite), 474; *Longistigma caryae* (giant bark aphid), 310; *Macrodiplosis erubescens*, 442; *M. quercusoruca* (vein pocket gall midge), 440, 446; *Melanaspis obscura* (obscure scale), 384; *Paranthrene simulans*, 254
Quercus prinus (chestnut oak): *Agrilus bilineatus* (twolined chestnut borer), 270; *Epinotia aceriella* (maple trumpet skeletonizer), 212; *Goes tigrinus* (white oak borer), 282
Quercus robur (English oak): *Camereria hamadryadella* (solitary oak leafminer), 192
Quercus rubra (northern red oak)
—foliage: *Agromyza viridula*, 206; *Archips fervidana* (oak webworm), 172, 214; *A. semiferana* (oak leafroller), 214; *Bucculatrix ainsliella* (oak skeletonizer), 220; *Camereria hamadryadella* (solitary oak leafminer), 192; *Croesia albicomana, 172; C. semipurpurana*, 172; *Ectoedemia platanella*, 196; *Macrodiplosis erubescens*, 446; *M. majalis*, 446; *M. niveipila* (hairy vein pocket gall), 446; *M. qoruca*, 446; *Pamphilus phyllisae*, 136; "roly-poly" galls, 448; *Symmerista leucitys* (orangehumped mapleworm), 154
mite: *Oligonychus bicolor* (oak spider mite), 475
—bark, wood and twigs: *Callirhytis quercuspunctata* (gouty oak gall wasp), 440; *Enaphalodes rufulus* (red oak borer), 282
—sucking insects: *Allokermes kingi*, 366; *Anormenis chloris*, 422; *Eulecanium caryae* (hickory lecanium), 364; *Longistigma caryae* (giant bark aphid), 310; *Ormenoides venusta*, 422; *Stegophylla quercicola* (oak woolly aphid), 304
Quercus stellata (post oak): *Cameraria hamadryadella* (solitary oak leafminer), 192; *Goes tigrinus* (white oak borer), 282
Quercus vaccinifolia: *Contarinia* sp. marginal leaf roll gall), 446
Quercus velutina (black oak): *Agrilus bilineatus* (twolined chestnut borer), 270; *Allokermes kingi*, 366; *Archips fervidana* (oak webworm), 172, 214; *Asterolecanium* sp., 352; *Caliroa quercuscoccineae* (scarlet oak sawfly), 130; *Callirhytis cornigera* (horned oak gall wasp), 440, 442; *C. crypta*, 442; *C. quercuspunctata* (gouty oak gall wasp), 440; *Cameraria hamadryadella* (solitary oak leafminer), 192; *Enaphalodes rufulus* (red oak borer), 282; *Epinotia aceriella* (maple trumpet skeletonizer), 212; *Glossonotus acuminatus*, 406
Quercus virginiana (live oak): *Andricus kingi*, 440; *A. laniger* (live oak woolly leaf gall maker), 440, 446; *Cameraria* sp., 192; *Johnella virginiana*, 486; *Mesolecanium nigrofasciatum* (terrapin scale), 364; *Orgyia leucostigma* (whitemarked tussock moth), 158; *Phryganidia californica* (California oakworm), 146; *Platycotis vittata* (oak treehopper), 406; *Prionoxystus robiniae* (carpenterworm), 256
Quercus wislizeni (interior live oak): *Agrilus angelicus* (Pacific oak twig girdler), 266
quince, see *Chaenomeles*
Japanese flowering: *Chaenomeles japonica*
quince, Japanese: *Hemiberlesia rapax* (greedy scale), 372; *Pseudococcus maritimus* (grape mealybug), 88

rabbitbrush, see *Chrysothamnus nauseosus*
raintree, golden, see *Koelreuteria paniculata*
raspberry, see *Rubus*
redbud, see *Cercis canadensis*
redcedar
eastern: *Juniperus virginiana*
western: *Thuja plicata*
redgum, see *Liquidambar styraciflua*
red-osier dogwood, see *Cornus stolonifera*
redwood, see *Sequoiadendron giganteum*
coast: *Sequoia sempervirens*,
Retinospora, see *Chamaecyparis* and *Thuja*
Rhamnus (buckthorn): *Argyrotaenia quercifoliana*, 218; *Aspidiotus nerii* (oleander scale), 374; *Chionaspis ortholobis*, 390; *Melanaspis tenebricosa* (gloomy scale), 384; *Panonychus ulmi* (European red mite), 472; *Saissetia oleae* (black scale), 358; *Tetraleurodes stanfordi* (Stanford whitefly), 320; *Trialeurodes vaporariorum* (greenhouse whitefly), 320
Rhamnus californica (California buckthorn; California coffeeberry): *Aspidiotus nerii* (oleander scale), 374; *Malacosoma californicum pluviale* (western tent caterpillar), 170; *Orgyia vestuta* (western tussock moth), 160; *Tetraleurodes stanfordi* (Stanford whitefly), 320; *Trialeurodes vittata* (grape whitefly), 322; *T. vaporariorum* (greenhouse whitefly), 320
Rhaphiolepsis indica: *Aleurocanthus woglumi* (citrus blackfly), 322; *Chrysomphalus dictyospermi* (dictyospermum scale), 382; *Lepidosaphes camelliae* (camellia scale), 370; plant bug, 398
Rhapidophyllum hystrix (dwarf palmetto): *Homaledra sabelella* (palm leafskeletonizer), 162
Rhododendron
—foliage: *Heliothrips haemorrhoidalis* (greenhouse thrips), 430
beetles: *Altica ignita* (flea beetle), 228; *Calliropalus bifasciatus* (twobanded Japanese weevil), 244; *Nemocestes incomptus* (woods weevil, raspberry bud weevil), 240; *Otiorhynchus sulcatus* (black vine weevil), 240; *Rhadopterus picipes* (cranberry rootworm), 242; *Sciopithes obscurus* (obscure root weevil), 240, 242; taxus weevil, 240
caterpillars: *Choristoneura rosaceana* (obliquebanded leafroller), 216; *Lithophane antennata* (green fruitworm), 148; *Orgyia antiqua* (rusty tussock moth), 158
mites: *Oligonychus ilicis* (southern red mite), 475
—bark, wood, twigs: *Corthylus punctatissimus* (pitted ambrosia beetle), 250; *Oberea myops* (azalea stem borer, rhododendron stem borer), 288; *O. tripunctata* (dogwood twig borer), 262; *Vespa crabro germana* (European hornet), 494
—roots: *Calliropalus bifasciatus* (twobanded Japanese weevil), 244; *Nemocestes incomptus* (woods weevil, raspberry bud weevil), 240; *Otiorhynchus sulcatus* (black vine weevil), 240; *Prionus laticollis* (broadnecked prionus, broadnecked root borer), 280; *Rhadopterus picipes* (cranberry rootworm), 242; *Sciopithes obscurus* (obscure root weevil), 242; *Synanthedon rhododendri* (rhododendron borer), 258; taxus weevil, 240; *Xylosandrus germanus*, 250
—sucking insects: *Dialeurodes chittendeni* (rhododendron whitefly), 318; *Dysmicoccus wistariae* (taxus mealybug), 88; *Graphocephala coccinea* (redbanded leafhopper), 418; *G. fennahi* (rhododendron leafhopper), 418; lacebugs,

Rhododendron (cont.)
424; *Pseudococcus longispinus* (longtailed mealybug), 324; *Stephanitis rhododendri*, 424
aphids: *Illinoia rhododendri*, 308
scales: *Eriococcus azaleae* (azalea bark scale), 336; *Pulvinaria ericicola* (cottony azalea scale), 342; *P. floccifera* (cottony camellia scale), 344
Rhododendron calendulaceum (flame azalea): *Dendroctonus frontalis* (southern pine beetle), 64; *Nematus lipovskyi*, 134
Rhododendron catawbiense (catawba rhododendron): *Clinodiplosis rhododendri* (rhododendron gall midge), 470; *Graphocephala coccinea* (redbanded leafhopper), 418
Rhododendron edgeworthii: Sciopithes obscurus (obscure root weevil), 242
Rhododendron indicum: Aleurothrixis floccosus (woolly whitefly), 322
Rhododendron luteum (Pontic azalea): *Sciopithes obscurus* (obscure root weevil), 242
Rhododendron maximum (rosebay): *Clinodiplosis rhododendri* (rhododendron gall midge), 470; *Graphocephala coccinea* (redbanded leafhopper), 418
Rhododendron mergeratum: Sciopithes obscurus (obscure root weevil), 242
Rhododendron nudiflora: Pulvinaria ericicola (cottony azalea scale), 342
Rhododendron periclymenoides (pinxter azalea): *Asphondylia azaleae*, 470
rhododendron, rosebay, see *Rhododendron maximum*
Rhododendron trichostomum: *Sciopithes obscurus* (obscure root weevil), 242
Rhododendron viscosum (swamp azalea): *Nematus lipovskyi*, 134
Rhus (sumac): *Alebra albostriella* (maple leafhopper), 416; *Amorbia humerosana* (whiteline leafroller), 214; *Aspidiotus nerii* (oleander scale), 374; *Calophya flavida*, 290; *C. nigripennis*, 290; *Datana ministra* (yellownecked caterpillar), 154; *Euthochtha galeator*, 422; *Parasaissetia nigra* (nigra scale), 360; *Poecilocapsus lineatus* (fourlined plant bug), 396; psyllids, 290; *Pulvinaria immunerablis* (cottony maple scale), 340; *Rhadopterus picipes* (cranberry rootworm), 242; *Saissetia coffeae* (hemispherical scale), 360; *S. oleae* (black scale), 358
Rhus aromatica (fragrant sumac): *Calophya trizomima*, 290
Rhus ovata (sugarbush, sugar sumac): *Calophya californica*, 290
Rhus trilobata (squawbush): *Calophya trizomima*, 290
Ribes (currant, gooseberry): *Aleyrodes amnicola* (willow whitefly), 322; *Aspidiotus nerii* (oleander scale), 374; *Bryobia ribis* (gooseberry bryobia), 472; *Eriococcus borealis*, 336; *Macrosiphum euphorbiae* (potato aphid), 300; *Malacosoma californicum pluviale* (western tent caterpillar), 170; psyllids, 290
Ribes sanguineum (red-flowering currant): *Aphis* sp., 316
Robinia (locust)
—foliage: *Lophocampa caryae* (hickory tussock moth), 160; *Oiketicus townsendi*, 176; sawflies, 134
mites: *Panonychus ulmi* (European red mite), 472; *Tetranychus urticae* (twospotted spider mite), 476
—bark, wood, twigs: *Agrilus* sp., 272; *Megacyllene robiniae* (locust borer), 274; *Stictocephala bisonia* (buffalo treehopper), 406
—sucking insects: *Aphis craccivora* (cowpea aphid), 308; *Erythroneura gleditsia*, 416; *Parthenolecanium pruinosum* (frosted scale), 354
Robinia hispida (rose acacia, bristly locust): *Nematus hispidae*, 134; *N. tibialis*, 134; *Odontota dorsalis* (locust leafminer), 190; sawflies, 134
Robinia pseudoacacia (black locust)
—foliage: *Apion nigrum* (black locust weevil), 224; *Cyrtepistomus castaneus* (Asiatic oak weevil), 244; *Epargyreus clarus* (silverspotted skipper), 148; *Lophocampa caryae* (hickory tussock moth), 160; *Nematus tibialis*, 134; *Obolodiplosis robiniae* (toyon berry midge), 466, 470; *Odontota dorsalis* (locust leafminer), 190; *Parectopa robiniella*, 190; sawflies, 134; *Schizura concinna* (redhumped caterpillar), 156; *Sumitrosis rosea*, 190; *Thyridopteryx ephemeraeformis* (bagworm), 176
mites: *Eotetranychus matthyssei*, 474; *Panonychus ulmi* (European red mite), 472; *Tetranychus urticae* (twospotted spider mite), 476
—bark, wood twigs: *Acanalonia conica*, 422; *Agrilus* sp., 272; *Ecdytolopha insiticiana* (locust twig borer), 274; *Enchenopa binotata* (twomarked treehopper), 410; *Magicicada septendecim* (periodical cicada), 490; *Megacyllene robiniae* (locust borer), 274; *Neoclytus acuminatus* (redheaded ash borer), 278; *Prionoxystus robiniae* (carpenterworm), 256
—sucking insects: *Acanalonia conica*, 422; *Enchenopa binotata* (twomarked treehopper), 410; *Micrutalis calva*, 404; *Pulvinaria innumerabilis* (cottony maple scale), 340; *Quadraspidiotus juglansregiae* (walnut scale), 386; *Saissetia oleae* (black scale), 358; *Stictocephala bisonia* (buffalo treehopper), 406; *Tetraleurodes stanfordi* (Stanford whitefly), 320; *Trialeurodes vaporariorum* (greenhouse whitefly), 320
Rosa (rose)
—foliage: *Apterona helix* (snailcase bagworm), 178; *Dasineura rhodophaga* (rose midge), 470; *Megachile* sp. (leafcutter bees), 494
beetles: *Asynonychus godmani* (Fuller rose beetle), 230; *Callirhopalus bifasciatus* (twobanded Japanese weevil), 244; *Macrodactylus subspinosus* (rose chafer), 236; *M. uniformis* (western rose chafer), 236; *Merhynchites bicolor* (rose curculio), 238; *Nemocestus incomptus* (woods weevil, raspberry bud weevil), 240; *Otiorhynchus sulcatus* (taxus weevil), 240; *Popillia japonica* (Japanese beetle), 236; *Rhadopterus picipes* (cranberry rootworm), 242; *Sciopithes obscurus* (obscure root weevil), 240
caterpillars: *Archips argyrospila* (fruittree leafroller), 172, 218; *A. cerasivorana* (uglynest caterpillar), 172; *A. rosanus*, 218; *Argyrotaenia citrana* (orange tortrix), 216; *Automeris io* (io moth), 164; *Choristoneura rosaceana* (obliquebanded leafroller), 216; *Lithophane antennata* (green fruitworm), 148; *Malacosoma disstria* (forest tent caterpillar), 168, 170; *Orgyia leucostigma* (whitemarked tussock moth), 158; *Pandemis limitata*, 218; *Rheumaptera hastata* (spear-marked black moth), 146; *Schizura concinna* (redhumped caterpillar), 156; *Tischeria admirabilis*, 196
mites: *Panonychus ulmi* (European red mite), 472; *Tetranychus canadensis* (fourspotted spider mite), 476; *T. urticae* (twospotted spider mite), 476
sawflies: *Allantus cinctus* (curled rose sawfly), 132; *Cladius difformis* (bristly roseslug), 132; *Endelomyia aethiops* (roseslug), 132; sawflies, 132
slugs/snails: *Helix aspersa* (brown garden snail), 498
—bark, wood, twigs: cambium miners, 198; *Chrysobothris femorata* (flatheaded appletree borer), 270; *C. mali* (Pacific flatheaded borer), 270; *Diplolepis* sp. (gall wasps), 438; *D. bassetti*, 438; *D. bicolor* (spiny rose gall wasp), 438; *D. dichlorcera* (long rose gall wasp), 438; *D. rosae* (mossyrose gall wasp), 438; *Hartigia cressoni*, 132; *H. trimaculata*, 132; *Oecanthus fultoni* (snowy tree cricket), 492, 494
—roots: *Nemocestus incomptus* (woods weevil, raspberry bud weevil), 240; *Otiorhynchus sulcatus* (taxus weevil), 240; *Rhadopterus picipes* (cranberry rootworm), 242; *Sciopithes obscurus* (obscure root weevil), 240, 242
—sucking insects: *Edwardsiana rosae* (rose leafhopper), 412; *Euthochtha galeator*, 422; *Frankliniella* sp., 432, 434; *F. occidentalis* (western flower thrips), 432; *F. tritici* (flower thrips), 432; *Graphocephala atropunctata* (blue-green sharpshooter), 418; *G. coccinea* (redbanded leafhopper), 418; *G. fennahi* (rhodendron leafhopper), 418; *Leptoglossus phyllopus* (leaf-footed bugs), 422, 424; *Planococcus citri* (citrus mealybug), 328; *Poecilocapsus lineatus* (fourlined plant bug), 396; *Trialeurodes vaporariorum* (greenhouse whitefly), 320; *Typhlocyba pomaria* (white apple leafhopper), 416
aphids: *Macrosiphum euphorbiae* (potato aphid), 300; *M. rosae* (rose aphid), 308; *Wahlgreniella nervata*, 298
scales: *Aonidiella aurantii* (California red scale), 378; *Asterolecanium arabidis* (pitmaking pittosporum scale), 348; *Aspidiotus nerii* (oleander scale), 374; *Aulacaspis rosae* (rose scale), 366; *Chrysomphalus dictyospermi* (dictyospermum scale), 382; *Hemiberlesia lataniae* (latania scale), 376; *Icerya purchasi* (cottony cushion scale), 338; *Parlatoria pittospori*, 110; *Parthenolecanium pruinosum* (frosted scale), 354; *Pinnaspis strachani* (lesser snow scale), 390; *Pulvinaria innumerabilis* (cottony maple scale), 340; *Quadraspidiotus perniciosus* (San Jose scale), 386; *Saissetia oleae* (black scale), 358
rose acacia, *Robinia hispida*
rosebay, see *Rhododendron maximum*
rose of sharon, see *Hibiscus syriacus*
royal poinciana, see *Delonix regia*
Roystonea: *Diaspis cocois* (cocos scale), 376
rubber plant, see *Ficus elastica*
Rubescens: *Fiorinia theae* (tea scale), 394
Rubus (raspberry): *Edwardsiana rosae* (rose leafhopper), 412; *Eotetrachynus carpini* (yellow spider mite), 474; *Hemiberlesia lataniae* (latania scale), 376; *Nemocestes incomptus* (raspberry bud weevil), 240; psyllids, 290; *Schizura ipomoeae*, 156

Sabal palmetto (cabbage palmetto): *Homaledra sabelella* (palm leafskeletonizer), 162; *Lepidosaphes gloveri* (Glover scale), 380; *Litoprosopus futilis* (cabbage palm caterpillar), 162; *Thyridopteryx ephemeraeformis* (bagworm), 176
sage: *Hemiberlesia rapax* (greedy scale), 372. See also *Lantana*, *Salvia*
Mexican bush: *Salvia leucantha*
purple: *Salvia dorrii*
Texas: *Leucophyllum frutescens*
sagebrush, see *Artemesia tridentata*
St. Johnswort, see *Hypericum calycinum*
Salix (willow)
—foliage: *Mayetiola rigidae* (willow beaked-gall midge), 468; *Pontania* sp., 468; *P. proxima*, 468; *Rhabdophaga strobiloides* (willow cone gall midge), 468
beetles: *Calligrapha bigsbyana*, 226; *C. scalaris* (elm calligrapha), 226; *Chrysomela aenicollis*, 226; *C. californica*, 226; *C. crotchi* (aspen leaf beetle), 226; *C. interrupta*, 226; *C. scripta* (cottonwood leaf beetle), 226; *Cyrtepistomus castaneus* (Asiatic oak weevil), 244; *Graphognathus* sp. (whitefringed beetle), 280; *Plagiodera versicolora* (imported willow leaf beetle), 228; *Pyrrhalta decora decora* (gray willow leaf beetle), 224; *Rhynchaenus ephippiatus*, 190; *R. rufipes* (willow flea weevil), 190
caterpillars: *Acronicta americana* (American dagger moth), 158; *Amphipyra pyramidoides*, 148; *Apotomis* sp., 218; *Archips rosanus*, 218; *Argyrotaenia citrana* (orange tortrix), 216; *A. velutinana* (redbanded leafroller), 214; *Automeris io* (io moth), 164; *Choreutis pariana* (apple-and-thorn skeletonizer), 216; *Choristoneura rosaceana* (obliquebanded leafroller), 216; *Cingilia catenaria* (chainspotted geometer), 144; *Coleophora sacramenta* (Cal-

ifornia casebearer), 186; *Datana integerrima* (walnut caterpillar), 150; *Hemileuca maia* (eastern buck moth), 156; *H. nevadensis* (Nevada buck moth), 156; *Hyalophora cecropia* (cecropia moth), 164; *Hydria undulata*, 146; *Hyphantria cunea* (fall webworm), 166; *Leucoma salicis* (satin moth), 158; *Lithophane antennata* (green fruitworm), 148; *Lymantria dispar* (gypsy moth), 138; *Malacosoma americanum* (eastern tent caterpillar), 168; *M. californicum pluviale* (western tent caterpillar), 170; *M. disstria* (forest tent caterpillar), 170; *M. incurvum* (southwestern tent caterpillar), 170; *Nymphalis antiopa* (spiny elm caterpillar), 152; *N. californica* (California tortoiseshell), 152; *Oiketicus townsendi*, 176; *Orgyia antiqua* (rusty tussock moth), 158; *O. vestuta* (western tussock moth), 160; *Orthosia hibisci* (speckled green fruitworm), 148; *Phyllonorycter salicifoliella*, 196; *Rheumaptera hastata* (spear-marked black moth), 146; *Schizura concinna* (redhumped caterpillar), 156

mites: *Eotetranychus carpini* (yellow spider mite), 474; *E. populi* (poplar spider mite), 474; *Schizotetranychus garmani* (willow spider mite), 474; *S. schizopus*, 474

sawflies: *Arge clavicornis*, 134; *A. pictorialis* (birch sawfly), 128; *Cimbex americana* (elm sawfly), 136; *Euura* sp., 468; *Nematus* sp., 134

—bark, wood, twigs, shoots: *Camponotus* spp. (carpenter ants), 496; *Chrysobothris femorata* (flatheaded appletree borer), 270; *Cryptorhynchus lapathi* (poplar and willow borer), 268; *Euura* sp., 468; *E. atra* (sawfly), 286; *E. exiguae*, 468; *E. scoulerianae*, 468; *Janus abbreviatus* (willow shoot sawfly), 286; *Mayetiola rigidae* (willow beaked-gall midge), 468; *Oecanthus fultoni* (snowy tree cricket), 492, 494; *Paranthrene dollii* (poplar clearwing borer), 260; *P. robiniae* (western poplar clearwing), 256, 260; *P. tabaniformis*, 254, 260; *Plectrodera scalator* (cottonwood borer), 276; *Prionoxystus robiniae* (carpenterworm), 256; *Rhabdophaga strobiloides* (willow cone gall midge), 468; *Saperda populnea*, 278; *Sciurus* sp. (squirrels), 502; shoot borers, 286; *Sphyrapicus* sp. (sapsuckers), 500; *Stictocephala bisonia* (buffalo treehopper), 406; *Synanthedon scitula* (dogwood borer), 262; *Vespa crabro germana* (European hornet), 494; *Xylosandrus germanus*, 250; *Zeuzera pyrina* (leopard moth), 254

—sucking insects: *Aleurodes amnicola* (willow whitefly), 322; *Clastoptera obtusa* (alder spittlebug), 420; *Corythucha elegans* (willow and poplar lace bug), 426; *C. mollicula* (willow lace bug), 426; *C. pallipes* (birch lacebug), 426; *C. salicata* (western willow tingid), 426; *C. salicis* (eastern willow tingid), 426; *Erythroneura lawsoniana*, 416; *Graphocephala atropunctata* (blue-green sharpshooter), 418; *Keonolla confluens*, 416; *Pseudococcus affinis* (obscure mealybug), 324; *Psylla alba*, 290; *P. americana*, 290; psyllids, 290; *Stephanitis takeyai*, 424

aphids: *Aphis gossypii* (cotton, or melon, aphid), 308; *A. pomi* (apple aphid), 292; *Longistigma caryae* (giant bark aphid), 310; *Pterocomma smithiae*, 310; *Tuberolachnus salignus* (giant willow aphid), 310

scales: *Aonidiella aurantii* (California red scale), 378; *Chionaspis hamoni*, 390; *C. lintneri*, 390; *C. longiloba*, 390; *C. ortholobis*, 390; *C. salicisnigrae*, 390; *Chrysomphalus dictyospermi* (dictyospermum scale), 382; *Eriococcus azaleae* (azalea bark scale), 336; *E. borealis*, 336; *Eulecanium caryae* (hickory lecanium), 364; *Hemiberlesia lataniae* (latania scale), 376; *H. rapax* (greedy scale), 372; *Icerya purchasi* (cottony cushion scale), 338; *Lepidosaphes gloverii* (Glover scale), 380; *L. ulmi* (oystershell scale), 370; *L. yanagicola*, 388; *Melanaspis obscura* (obscure scale), 384; *Pulvinaria innumerabilis* (cottony maple scale), 340

Salix acutifolia: *Euura atra*, 286
Salix alba (white willow): *Euura atra*, 286; *Janus abbreviatus* (willow shoot sawfly), 286; *Macropsis ocellata*, 404; *Plagiodera versicolora* (imported willow leaf beetle), 228
Salix babylonica (weeping willow): *Euura atra*, 286; *Janus abbreviatus* (willow shoot sawfly), 286; *Plagiodera versicolora* (imported willow leaf beetle), 228; *Pterocomma smithae*, 310
Salix blanda (Wisconsin weeping willow): *Pontania proxima*, 468
Salix caprea (goat pussywillow, French pussywillow): leaf beetles, 226
Salix cinerea: *Euura atra*, 286
Salix discolor (pussywillow): *Cryptorhynchus lapathi* (poplar and willow borer), 268; *Janus abbreviatus* (willow shoot sawfly), 286; leaf beetles, 226; *Mayetiola rigidae* (willow beaked-gall midge), 468; *Popillia japonica* (Japanese beetle), 236
Salix fragilis (crack willow): *Pontania proxima*, 468
Salix interior: *Plagiodera versicolora* (imported willow leaf beetle), 228
Salix lapponum: *Euura atra*, 286
Salix lucida (shining willow): *Plagiodera versicolora* (imported willow leaf beetle), 228
Salix nigra (black willow): *Camponotus* sp. (carpenter ants), 496; *Janus abbreviatus* (willow shoot sawfly), 286; *Plagiodera versicolora* (imported willow leaf beetle), 228; *Rhynchaenus rufipes* (willow flea weevil), 190
Salix purpurea (purple willow): *Euura atra*, 286
Salix viminalis (basket willow): *Euura atra*, 286
saltbush, see *Atriplex*
Salvia (sage): *Asterolecanium arabidis* (pit-making pittosporum scale), 348; *Hemiberlesia rapax* (greedy scale), 372; *Icerya purchasi* (cottony cushion scale), 338
Salvia dorrii (purple sage): *Teleonemia scrupulosa* (lantana lace bug), 424, 426
Salvia leucantha (Mexican bush sage): *Graphocephala atropunctata* (blue-green sharpshooter), 418
San Diego ceanothus, see *Ceanothus cyaneus*
Sapindus (soapberry): *Melanaspis tenebricosa* (gloomy scale), 384
sapodilla, see *Manilkara zapota*
Sassafras albidium: *Acanalonia conica*, 422; *Corthylus punctatissimus* (pitted ambrosia beetle), 250; *Elaphidionoides villosus* (twig pruner), 264; *Lymantria dispar* (gypsy moth), 138; *Mesolecanium nigrofasciatum* (terrapin scale), 364; *Neoclatus acuminatus* (redheaded ash borer), 278; *Odontopus calceatus* (sassafras weevil), 210; *Pandemis limitata*, 218; *Rhadopterus picipes* (cranberry rootworm), 242
Sassafras variifolium (American sassafras, common sassafras): *Euthoracaphis umbellulariae* (California laurel aphid), 294
Schinus (pepper tree): *Argyrotaenia citrana* (orange tortrix), 216; *Aspidiotus nerii* (oleander scale), 374; *Ceroplastes cistudiformis* (tortoise wax scale), 358; *Hemiberlesia rapax* (greedy scale), 372; *Icerya purchasi* (cottony cushion scale), 338; *Leptoglossus phyllopus*, 422; *Pinnaspis strachani* (lesser snow scale), 390; *Saissetia coffeae* (hemispherical scale), 360; *S. oleae* (black scale), 358
Schinus molle (California pepper tree): *Calophya schini* (pepper tree psyllid), 458
Schinus terebinthifolius (Brazilian pepper tree): *Leptoglossus phyllopus*, 422; *Selenothrips rubrocinctus* (redbanded thrips), 432
Sciadopitys verticillata (umbrella pine): *Lepidosaphes gloverii* (Glover scale), 380
screw pine, see *Pandanus utilis*
seagrape, see *Coccoloba uvifera*
sea lavender, see *Limonium perezii*
Sedum: *Hemiberlesia rapax* (greedy scale), 372
Senecio: *Fiorinia theae* (tea scale), 394
Sequoia sempervirens (coast redwood): *Argyresthia cupressella* (cypress tip miner), 42; *Aspidiotus nerii* (oleander scale), 374; *Phloeosinus* sp., 70
Sequoiadendron giganteum (giant sequoia): *Hemiberlesia rapax* (greedy scale), 372
Serenoa repens (saw palmetto): *Homaledra sabelella* (palm leafskeletonizer), 162
serviceberry, shadbush, see *Amelanchier*
shadbush, see *Amelanchier*
shorebay, see *Persea littoralis*
silktree, see *Albizia julibrissin*
silverberry, see *Elaeagnus pungens*
Siberian larch: *Oligonychus ununguis* (spruce spider mite), 118
snowbell, see *Styras*
snowberry, see *Symphoricarpos albus*
snowbush, see *Ceanothus cordulatus*
soapberry, see *Sapindus*
Solanum (nightshade): *Aspidiotus nerii* (oleander scale), 374; *Hemiberlesia rapax* (greedy scale), 372; *Saissetia oleae* (black scale), 358; *Unaspis euonymi* (euonymus scale), 388
Solanum jasminoides: *Saissetia oleae* (black scale), 358
Solanum pseudocapsicum (Jerusalem cherry): *Trialeurodes vaporariorum* (greenhouse whitefly), 320
Solidago (goldenrod): *Ecdytolopha insitiana* (locust twig borer), 174; *Megacyllene robiniae* (locust borer), 274; *Trirhabda flavolimbata* (baccharis leaf beetle), 238
Sophora: *Uresiphita reversalis* (genista caterpillar), 166
Sophora japonica: *Odontota dorsalis* (locust leafminer), 190
Sorbaronia hybrida: *Pristiphora geniculata* (mountain ash sawfly), 128
Sorbus (mountain ash)
—foliage:

caterpillars: *Choreutis pariana* (apple-and-thorn skeletonizer), 216; *Choristoneura rosaceana* (obliquebanded leafroller), 216

mites: *Eriophyes triplacis*, 486; eriophyid gall mites, 484, 486; *Oligonychus newcomeri*, 475; *Panonychus ulmi* (European red mite), 472; *Tetraspinus pyramidicus*, 484

sawflies: *Caliroa cerasi* (pear sawfly), 130; *Pristiphora geniculata* (mountain ash sawfly), 128

—bark, wood, twigs: *Chrysobothris femorata* (flatheaded appletree borer), 270; *Euura atra*, 286; *Euzophera semifuneralis* (American plum borer), 252; *Phloeotribus liminaris* (peach bark beetle), 250; *Saperda candida* (roundheaded appletree borer), 278; *Scolytus rugulosus* (shothole borer), 250; *Synanthedon pyri* (apple bark borer), 258; *S. scitula* (dogwood borer), 262; *Vespa crabro germana* (European hornet), 494

—sucking insects: *Corythucha pallipes* (birch lace bug), 426; *Phenacoccus aceris* (apple mealybug), 324

aphids: *Aphis pomi* (apple aphid), 300; *Eriosoma lanigerum* (woolly apple aphid), 316

scales: *Chionaspis furfura*, 368; *Lepidosaphes ulmi* (oystershell scale), 370; *Parthenolecanium pruinosum* (frosted scale), 354; *Quadraspidiotus juglansregiae* (walnut scale), 386; *Q. perniciosus* (San Jose scale), 386

Sorbus americana (American mountain ash): *Aphis pomi* (apple aphid), 300; *Pristiphora geniculata* (mountain ash sawfly), 128
Sorbus aria (white beam): *Corythucha cydoniae* (hawthorn lace bug), 426
Sorbus aucuparia (European mountain ash): *Pristiphora geniculata* (mountain ash sawfly), 128
Sorbus decora (showy mountain ash): *Pristiphora geniculata* (mountain ash sawfly), 128
sourgum, see *Nyssa aquatica*
sourwood, see *Oxydendrum arboreum*
southern magnolia, see *Magnolia grandiflora*
Spartium (broom): *Asterolecanium arabidis* (pit-making pittosporum scale), 348
speckled alder, see *Alnus rugosa*
spicebush, see *Lindera benzoin*

Spiraea: *Aphis citricola* (spirea aphid), 298; *Callirhopalus bifasciatus* (twobanded Japanese weevil), 244; *Ceroplastes ceriferus* (Indian wax scale), 356; *Choristoneura rosaceana* (obliquebanded leafroller), 216; *Eotetranychus carpini* (yellow spider mite), 474; *Hydria undulata*, 146; *Pseudaulacaspis pentagona* (white peach scale), 392; *Rhadopterus picipes* (cranberry rootworm), 242; *Schizotetranychus spireafolia*, 474
Spondias mombi (hog plum): *Melanaspis obscura* (obscure scale), 384
Spondias purpurea (hog plum): *Melanaspis obscura* (obscure scale), 384
spruce, see *Picea*
Alberta: *Picea glauca* var. *albertiana*
black: *Picea mariana*
blue: *Picea pungens*
Colorado: *Picea pungens*
Colorado blue: *Picea pungens*
Engelmann: *Picea engelmannii*
Norway: *Picea abies*
Oriental: *Picea orientalis*
red: *Picea rubens*
Sitka: *Picea sitchensis*
white: *Picea glauca*
squawbush, see *Rhus trilobata*
star magnolia, see *Magnolia stellata*
Straussia: *Toxoptera aurantii* (black citrus aphid), 302
strawberry tree, see *Arbutus unedo*
Strelitzia (bird-of-paradise): *Hemiberlesia rapax* (greedy scale), 372
Styrax (snowbell): *Chionaspis styracis*, 390; *Stephanitis takeyai*, 424
sugarberry, see *Celtis laevigata*
sugarbush, sugar sumac, see *Rhus ovata*
sumac, see *Rhus*
fragrant, see *Rhus aromatica*
summersweet, see *Clethra alnifolia*
swamp dogwood, see Cornus stricta
sweet bay, see *Magnolia virginiana*. See also *Laurus nobilis*
sweetfern: *Aphrophora saratogensis* (Saratoga spittlebug), 86
sweetgum, see *Liquidambar styraciflua*
sweet orange, see *Citrus sinensis*
sweet pepperbush, see *Clethra alnifolia*
sycamore, see *Platanus occidentalis*
western: *Platanus racemosa*
sydney golden wattle, see *Acacia longifolia*
Symphoricarpos albus (snowberry): *Paraphytomyza cornigera*, 194
Symplocos: *Fiorinia theae* (tea scale), 394
Syringa (lilac)
—foliage: *Callirhopalus bifasciatus* (twobanded Japanese weevil), 244; *Macrophya punctumalbum*, 136; *Ochyromera ligustri*, 244
caterpillars: *Archips rosanus*, 218; *Caloptilia syringella* (lilac leafminer), 196; *Choristoneura rosaceana* (obliquebanded leafroller), 216; *Sphinx drupiferarum* (cherry sphinx caterpillar), 164
—bark, wood, twigs: *Neoclatus acuminatus* (redheaded ash borer, 278; *Oecanthus angustipennis* (narrowwinged cricket), 492, 494; O. fultoni (snowy tree cricket), 492, 494; *Podosesia syringa* (lilac borer), 260–262; *Siphoninus phillyreae* (ash whitefly), 322; *Vespa crabro germana* (European hornet), 494; *Zeuzera pyrina* (leopard moth), 254
scales: *Asterolecanium arabidis* (pit-making pittosporum scale), 348; *Chionaspis lintneri*, 390; *Lepidosaphes ulmi* (oystershell scale), 370; *Pseudaulacaspis pentagona* (white peach scale), 392; *P. prunicola* (white prunicola scale), 392; *Pulvinaria innumerabilis* (cottony maple scale), 340
Syringa vulgaris (common lilac): *Sphinx drupiferarum* (cherry sphinx caterpillar), 164
Syzygium paniculatum (Australian bush cherry): *Graphocephala atropunctata* (blue-green sharpshooter), 418
tamarack, see *Larix laricina*
tanbark oak, see *Lithocarpus*, *Lithocarpus densiflorus*
tarata, see *Pittosporum eugenioides*
tatarian honeysuckle: *Lonicera tatarica*
Taxodium distichum (baldcypress and pondcypress): *Anacamptodes pergracilis* (cypress looper), 26; *Grylloprociphilus imbricator* (beech blight aphid), 296; *Phloeosinus taxodii* (southern cypress bark beetle), 68, 70; *Taxodiomyia cupressiananassa* (cypress twig gall midge), 118; *Xylosandrus germanus*, 250
Taxus (yew)
—foliage: *Nemocestes incomptus* (raspberry bud weevil, woods weevil), 240; *Otiorhynchus ovatus* (strawberry root weevil), 240; *O. rugosostriatus*, 240; *O. rugifrons*, 240; *O. sulcatus* (black vine weevil), 54, 240, 242; *Sciopithes obscurus* (obscure root weevil), 240; *Taxomyia taxi* (artichoke gall midge), 478
mites: *Cecidophyopsis psilaspis* (taxus bud mite), 478; *Nalepella tsugifoliae* (hemlock rust mite), 122
—bark/roots: *Nemocestes incomptus* (raspberry bud weevil, woods weevil), 240; *Oecanthus* sp. (tree crickets), 492; *Otiorhynchus ovatus* (strawberry root weevil), 240; *O. rugosostriatus*, 240; *O. rugifrons*, 240; *O. sulcatus* (black vine weevil), 54, 240, 242; *Sciopithes obscurus* (obscure root weevil), 240
—sucking insects:
mealybugs: *Dysmicoccus wistariae* (taxus mealybug), 88; *Pseudococcus affinis* (obscure mealybug), 88; *P. comstocki* (Comstock mealybug), 326; *P. longispinus* (longtailed mealybug), 324; *P. maritimus* (grape mealybug), 88
scales: *Aspidiotus nerii* (oleander scale), 374; *Chrysomphalus dictyospermi* (dictyospermum scale), 382; *Fiorinia externa* (elongate hemlock scale), 104; *Parthenolecanium fletcheri* (Fletcher scale), 98, 364; *Pseudaulacaspis cockerelli* (false oleander scale), 374; *Pulvinaria floccifera* (cottony camellia scale, cottony taxus scale), 344
Taxus baccata (English yew): *Aonidiella taxus* (Asiatic red scale), 110; *Cecidophyopsis psilaspis* (taxus bud mite), 478
tea tree, see *Leptospermum laevigatum*
Tecoma (trumpet flower): *Hemiberlesia rapax* (greedy scale), 372
tepeguaja, see *Leucaena pulverulenta*
Ternstroemia: *Lepidosaphes camelliae*, 370
Texas sage, see *Leucophyllum frutescens*
Thea, see *Camellia*
Thevea peruviana (yellow oleander): *Palpita flegia*, 162
Thuja (arborvitae)
—foliage: *Argyresthia aureoargentella*, 42; *A. cupressella*, 42; *A. freyella*, 42; *A. thuiella* (arborvitae leafminer), 42; *Coleotechnites thujaella*, 42; cypress tip miners, 42; *Dioryctria abietivorella*, 48, 50; *Epinotia subviridis* (cypress leaftier), 38; *Lambdina athasaria*, 24; *L. fiscellaria* (hemlock looper), 24; *Lymantria dispar* (gypsy moth), 138; *Otiorhynchus* sp., 240; *Susana cupressi* (cypress sawfly), 18; *Thyridopteryx ephemeraeformis* (bagworm), 176
mites: *Oligonychus ununguis* (spruce spider mite), 98, 118, 120; spider mites, 118
—bark, wood, twigs: *Camponotus* sp. (carpenter ants), 496; *Cinara tujafilina*, 84; *Otiorhynchus* sp., 240; *Phloeosinus canadensis* (northern cedar bark beetle), 68, 70; *Phyllobius intrusus* (arborvitae weevil), 240, 244
—sucking insects: *Carulaspis minima*, 106; *Chionaspis pinifoliae* (pine needle scale), 108; *Chrysomphalus dictyospermi* (dictyospermum scale), 382; *Cinara tujafilina*, 84; *Lepidosaphes gloveri* (Glover scale), 380; *Parthenolecanium fletcheri* (Fletcher scale), 98; psyllids, 290
Thuja occidentalis (northern white cedar, arborvitae), see *Thuja*
Thuja plicata (western redcedar): *Argyresthia cupressella*, 42; *Cinara tujafilina*, 84; *Contarinia juniperina* (juniper midge), 46; *Epinotia subviridis* (cypress leaftier), 38; *Lophocampa argentata* (silverspotted tiger moth), 34; spider mites, 118; tiger moths, 34
Thunbergia: *Macrosteles quadrilineatus* (aster leafhopper), 418
Tilia (linden, basswood)
—foliage: *Diapheromera femorata* (walkingstick), 492, 494
beetles: *Baliosus ruber* (basswood leafminer), 190; *Calligrapha scalaris* (elm calligrapha), 226; *Macrodactylus subspinosus* (rose chafer), 236; *Popillia japonica* (Japanese beetle), 236
caterpillars: *Acronicta americana* (American dagger moth), 158; *Alsophila pometaria* (fall cankerworm), 142; *Automeris io* (io moth), 164; *Choristoneura rosaceana* (obliquebanded leafroller), 216; *Datana ministra* (yellownecked caterpillar), 154; *Ennomos subsignarius* (elm spanworm), 144–146; *Erannis tiliaria* (linden looper), 144; *Hyalophora cecropia* (cecropia moth), 164; *Lithophane antennata* (green fruitworm), 148; *Lophocampa caryae* (hickory tussock moth), 160; *Nymphalis antiopa* (spiny elm caterpillar, mourningcloak butterfly), 152; *Orgyia leucostigma* (whitemarked tussock moth), 158; *Paleacrita vernata* (spring cankerworm), 142, 144; *Phyllonorycter* sp., 196; *P. lucetiella*, 196
mites: *Aponychus spinosus*, 474; *Eotetranychus tiliarium*, 474; *E. uncatus* (Garman spider mite), 474; erineum galls, 482, 486; *Eriophyes tiliae*, 486; spindle galls, 482; eriophyid mite galls, 482, 486; *Tetranychus canadensis* (fourspotted spider mite), 476
—bark, wood, twigs: ambrosia beetle, 286; *Corthylus columbianus* (Columbian timber beetle), 286; *Dicera lurida*, 272; *Elaphidionoides villosus* (twig pruner), 264; *Euzophera semifuneralis* (American plum borer), 252; *Neoclytus acuminatus* (redheaded ash borer), 278; *Oncideres cingulata* (twig girdler), 264; *Ostrinia nubilalis* (European corn borer), 286; *Saperda vestita* (linden borer), 276; Xylosandrus germanus, 250
—roots: *Prionus imbricornis* (tilehorned prionus), 280; *P. laticollis* (broadnecked prionus, broadnecked root borer), 280
—sucking insects: *Corythucha juglandis* (walnut lace bug), 426; *C. marmorata*, 426; *Gargaphia tiliae* (basswood lace bug), 426; *Metcalfa pruinosa*, 422; *Phenacoccus aceris* (apple mealybug), 324
aphids: *Eucallipterus tiliae* (linden aphid), 302; *Longistigma caryae* (giant bark aphid), 310
scales: *Lepidosaphes camelliae*, 370; *L. ulmi* (oystershell scale), 370; *Mesolecanium nigrofasciatum* (terrapin scale), 364; *Pulvinaria innumerabilis* (cottony maple scale), 340; *Quadraspidiotus juglansregiae* (walnut scale), 386; *Q. perniciosus* (San Jose scale), 386; *Toumeyella liriodendri* (tùliptree scale), 362
Tilia americana (basswood)
—foliage: *Anabrus simplex* (Mormon cricket), 492; *Cimbex americana* (elm sawfly), 136; *Diapheromera femorata* (walkingstick), 492, 494
beetles: *Baliosus ruber* (basswood leafminer), 190; *Thrips calcaratus* (introduced basswood thrips), 432
caterpillars: *Acronicta americana* (American dagger moth), 158; *Archips rosanus*, 218; *Argyrotaenia velutinana* (redbanded leafroller), 214; *Choristoneura rosaceana* (obliquebanded leafroller), 216; *Ennomos subsignarius* (elm spanworm), 144, 146; *Lymantria dispar* (gypsy moth), 138; *Lophocampa caryae* (hickory tussock moth), 160; *Malacosoma disstria* (forest tent caterpillar), 168; *Paleacrita vernata* (spring cankerworm), 144; *Phyllonorycter lucetiella*, 196; *Symmerista leucitys* (orangehumped mapleworm), 154
mites: *Phytoptus tiliae*, 486

—bark, wood, twigs: *Oncideres cingulata* (twig girdler), 264; *Saperda vestita* (linden borer), 276
—sucking insects: *Alebra albostriella* (maple leafhopper), 416; *Corythucha juglandis* (walnut lace bug), 426; *C. marmorata*, 426; *Gargaphia tiliae* (basswood lace bug), 426
aphids: *Eucallipterus tiliae* (linden aphid), 302
scales: *Lepidosaphes yanagicola*, 388
Tilia cordata (littleleaf linden): *Eucallipterus tiliae* (linden aphid), 302; *Phytoptus tiliae*, 486
Tilia tomentosa (silver linden): *Aponychus spinosus*, 474; *Popillia japonica* (Japanese beetle), 236
Tilia vulgaris (European linden): *Eucallipterus tiliae* (linden aphid), 302; *Saperda vestita* (linden borer), 276
Torreya: *Nalepella tsugifolia* (hemlock rust mite), 122
Torreya californica (California nutmeg tree): *Saissetia oleae* (black scale), 358
toyon, see *Heteromeles arbutifolia*
trifoliate orange, see *Poncirus trifoliata*
trumpet flower, see *Tecoma*
Tsuga (hemlock)
—foliage:
beetles: *Callirhopalus bifasciatus* (twobanded Japanese weevil), 244; *Otiorhynchus sulcatus* (black vine weevil), 240
caterpillars: *Acleris variana* (blackheaded budworm), 28; *Argyrotaenia velutinana* (redbanded leafroller), 214; *Choristoneura rosaceana* (obliquebanded leafroller), 216; *Coleotechnites* sp., 38; *C. apicitripunctella* (green hemlock needleminer), 38; *C. macleodi* (brown hemlock needleminer), 38; *Dasychira plagiata* (pine tussock moth), 26; *D. pinicola*, 26; *Lambdina athasaria*, 24; *L. fiscellaria* (hemlock looper), 24; *Lymantria dispar* (gypsy moth), 138, 140; *Thyridopteryx ephemeraeformis* (bagworm), 176; tussock moths, 26
mites: *Nalepella tsugifoliae* (hemlock rust mite), 122; *Oligonychus ununguis* (spruce spider mite), 118, 120; spider mites, 118, 120
—bark, twigs, roots: *Callirhopalus bifasciatus* (twobanded Japanese weevil), 244; *Hylobius pales* (pales weevil), 56; *Otiorhynchus sulcatus* (black vine weevil), 240; *Sphyrapicus* sp. (sapsuckers), 500
—sucking insects: *Adelges cooleyi* (Cooley spruce gall adelgid), 76, 112; *A. tsugae* (hemlock woolly adelgid), 76, 78; *Aphrophora cribata* (pine spittlebug), 86; *A. permutata*, 86
scales: *Aspidiotus cryptomeriae*, 110; *Ceroplastes ceriferus* (Indian wax scale), 356; *Fiorinia externa* (elongate hemlock scale), 104; *F. japonica* (Japanese elongate hemlock scale), 104; *F. juniperi* (juniper fiorinia scale), 394; *Quadraspidiotus juglansregiae* (walnut scale), 386
Tsuga canadensis (eastern hemlock, Canada hemlock): *Abgrallaspis ithacae* (hemlock scale), 102; *Adelges tsugae* (hemlock woolly adelgid), 76, 78, 112; *Cephalcia distincta*, 22; *Fiorinia externa* (elongate hemlock scale), 104; *Nuculaspis tsugae*, 102
Tsuga caroliniana (Carolina hemlock): *Fiorinia externa* (elongate hemlock scale), 104; *Nuculaspis tsugae*, 102
Tsuga diversifolia (Japanese hemlock): *Fiorinia externa* (elongate hemlock scale), 104; *Nuculaspis tsugae*, 102
Tsuga heterophylla (western hemlock): *Adelges tsugae* (hemlock woolly adelgid), 80; *Dasychira grisefacta*, 26; *Lambdina fiscellaria lugubrosa* (western hemlock looper), 24; *Lophocampa argentata* (silverspotted tiger moth), 34; *Neodiprion tsugae* (hemlock sawfly), 18; *Nuculaspis californica* (black pineleaf scale), 100; *Orgyia pseudotsugata* (Douglas-fir tussock moth), 34; tiger moths, 34
Tsuga mertensiana (mountain hemlock): *Adelges tsugae*, 112
Tsuga sieboldii: *Nuculaspis tsugae*, 102
tuliptree, see *Liriodendron tulipifera*
tung, see *Aleurites fordii*
tupelo, see *Nyssa*
water: *Nyssa aquatica*
twinberry, see *Lonicera involucrata*

Ulmus (elm)
—foliage: *Dasineura ulmi*, 464; spindle galls, 482
beetles: *Brachys aerosus*, 190; *Calligrapha scalaris* (elm calligrapha), 226; *Monocesta coryli* (larger elm leaf beetle), 222; *Odontota dorsalis* (locust leafminer), 190; *Phyllobius oblongus*, 244; *Popillia japonica* (Japanese beetle), 236; *Rhynchaenus rufipes* (willow flea weevil), 190; *Xanthogaleruca luteola* (elm leaf beetle), 222
caterpillars: *Alsophila pometaria* (fall cankerworm), 142; *Amphiphyra pyramidoides* (humped green fruitworm), 148; *Archips argyrospila* (fruittree leafroller), 172; *Automeris io* (io moth), 164; *Coleophora serratella* (birch casebearer), 186; *C. ulmifoliella* (elm casebearer), 186; *Datana ministra* (yellownecked caterpillar), 154; *Ennomos subsignarius* (elm spanworm), 144–146; *Erannis tiliaria* (linden looper), 144; *Halysidota tessellaris* (pale tussock moth), 158; *Hyalophora cecropia* (cecropia moth), 164; *Hyphantria cunea* (fall webworm), 166; *Lithophane antennata* (green fruitworm), 148; *Lophocampa caryae* (hickory tussock moth), 160; *Lymantria dispar* (gypsy moth), 138; *Malacosoma disstria* (forest tent caterpillar), 168; *Nymphalis antiopa* (spiny elm caterpillar, mourningcloak butterfly), 152; *Orgyia leucostigma* (whitemarked tussock moth), 158; *Orthosia hibisci* (speckled green fruitworm), 148; *Paleacrita vernata* (spring cankerworm), 144; *Schizura concinna* (redhumped caterpillar), 156; *S. ipomoeae*, 156; *Sparganothis sulfureana*, 216; *Thyridopteryx ephemeraeformis* (bagworm), 176
mites: *Aceria parulmi*, 482; *Bryobia ulmophila* (elm bryobia), 472; *Eotetranychus matthyssei*, 474; *Eriophyes ulmi*, 464; *Oligonychus bicolor* (oak spider mite), 475; *Panonychus ulmi* (European red mite), 472; *Tetranychus canadensis* (fourspotted spider mite), 476; *T. urticae* (twospotted spider mite), 476
sawflies: *Cimbex americana* (elm sawfly), 136
—bark, wood, twigs: *Elaphidionoides villosus* (twig pruner), 264; *Eutetrapha tridentata*, 276; *Goes pulverulentus* (living beech borer), 282; horntails, 494; *Hylurgopinus rufipes* (native elm bark beetle), 246, 248; *Neoclytus acuminatus* (redheaded ash borer), 278; *Oecanthus fultoni* (snowy tree cricket), 492; *Oncideres cingulata* (twig girdler), 264; *Paranthrene simulans*, 254; *Phloeotribus liminaris* (peach bark beetle), 250; *Prionoxystus robiniae* (carpenterworm), 282; *Sciurus* sp. (squirrels), 500; *Scolytus multistriatus* (European bark beetle), 246, 248; *S. rugulosus* (shothole borer), 250; *Sphyrapicus* sp. (sapsuckers), 500; *Stictocephala bisonia* (buffalo treehopper), 406; *Tremex columba* (pigeon tremex), 494; *Xylosandrus germanus*, 250; *Zeuzera pyrina* (leopard moth), 254
—sucking insects: *Alebra albostriella* (maple leafhopper), 416; *Corythucha pergandei* (alder lace bug), 426; *Edwardsiana rosae* (rose leafhopper), 412; *Metcalfa pruinosa*, 422; *Phenacoccus aceris* (apple mealybug), 324; *Pseudococcus comstocki* (Comstock mealybug), 326; *Scaphoideus luteolus* (whitebanded elm leafhopper), 246, 414; *Typhlocyba pomaria* (white apple leafhopper), 416
aphids: *Colopha ulmicola* (elm cockscombgall aphid), 464; *Eriosoma* sp., 306; *E. americanum* (woolly elm aphid), 306; *E. lanigerum* (woolly apple aphid), 306, 316; *E. crataegi*, 306; *E. rileyi*, 306
scales: *Chionaspis* sp., 368; *C. americana* (elm scurfy scale), 368; *C. furfura*, 368; *Clavaspis* sp., 368; *Diaspidiotus* sp., 368; *Eulecanium caryae* (hickory lecanium), 364; *E. cerasorum* (calico scale), 354; *Gossyparia* sp., 368; *G. spuria* (European elm scale), 368; *Lecanium* sp., 364; *Lepidosaphes* sp., 368; *L. ulmi* (oystershell scale), 370; *L. yanagicola*, 388; *Melanaspis tenebricosa* (gloomy scale), 384; *Parthenolecanium* sp., 364; *P. corni* (European fruit lecanium), 364; *P. persicae* (European peach scale), 364; *P. pruinosum* (frosted scale), 354; *Pulvinaria* innumerabilis (cottony maple scale), 340; *Quadraspidiotus* sp., 368; *Q. juglansregiae* (walnut scale), 386
Ulmus americana (American elm): *Alebra albostriella* (maple leafhopper), 416; *Archips negundana*, 172; *Colopha ulmicola* (elm cockscombgall aphid), 464; *Corythucha ulmi* (elm lace bug), 426; *Eriophyes ulmi*, 464; *Eriosoma americanum* (woolly elm aphid), 306; *E. languinosa* (pear root aphid), 464; *E. lanigerum* (woolly apple aphid), 306, 316; *E. pyricola*, 464; *E. ulmi*, 464; *Fenusa ulmi* (elm leafminer), 186; *Halysidota tessellaris* (pale tussock moth), 158; *Hyphantria cunea* (fall webworm), 166; *Prionoxystus robiniae* (carpenterworm), 256; *Pyrrhalta luteola* (elm leaf beetle), 222; *Scolytus mali* (larger shothole borer), 246; *S. multistriatus* (European elm bark beetle), 246, 248; *Symmerista leucitys* (orangehumped mapleworm), 154; *Xanthogaleruca luteola* (elm leaf beetle), 222
Ulmus carpinifolia (smooth-leaf elm): *Xanthogaleruca luteola* (elm leaf beetle), 222
Ulmus fulva: *Ulmus rubra*
Ulmus glabra (Scotch elm, Camperdown elm): *Fenusa ulmi* (elm leafminer), 186
Ulmus hollandica (Dutch elm): *Xanthogaleruca luteola* (elm leaf beetle), 222
Ulmus laevis (European white elm, Russian elm): erineum galls, 484; *Gossyparia spuria* (European elm scale), 368
Ulmus parvifolia (Chinese elm): *Monocesta coryli* (larger elm leaf beetle), 222; *Pseudaulacaspis pentagona* (white peach scale), 392; *Thyridopteryx ephemeraeformis* (bagworm), 176; *Xanthogaleruca luteola* (elm leaf beetle), 222
Ulmus procera (English elm): *Fenusa ulmi* (elm leafminer), 186; *Xanthogaleruca luteola* (elm leaf beetle), 222
Ulmus pumila (Siberian elm): *Halysidota tessellaris* (pale tussock moth), 158; *Monocesta coryli* (larger elm leaf beetle), 222; *Xanthogaleruca luteola* (elm leaf beetle), 222
Ulmus rubra (slippery elm): *Aponychus spinosus*, 474; *Colopha ulmicola* (elm cockscombgall aphid), 464; *Monocestus coryli* (larger elm leaf beetle), 222; *Scolytus mali* (larger shothole borer), 246;
Ulmus wilsoniana: *Xanthogaleruca luteola* (elm leaf beetle), 222
Umbellularia californica (California bay, California laurel): *Aspidiotus nerii* (oleander scale), 374; *Euthoracaphis umbellulariae* (California laurel aphid), 294; *Hemiberlesia rapax* (greedy scale), 372; *Saissetia oleae* (black scale), 358
umbrella pine, see *Sciadopitys verticillata*
umbrella plant, see *Erigonum*
umbrella tree, umbrella magnolia, see *Magnolia tripetala*

Vaccinium (blueberry, cranberry, huckleberry)
—foliage: *Amorbia humerosana* (whiteline leafroller), 214; *Cingilia catenaria* (chainspotted geometer), 146; *Coptodisca* sp., 208; *Datana major* (azalea caterpillar), 154; *Neochlamisus cribripennis* (blueberry case beetle), 224; *Phigalia titea* (half-wing geometer), 146; *Rhadopterus picipes* (cranberry rootworm), 242; *Rhopobota naevana* (blackheaded fireworm), 208; *Schizura concinna* (redhumped caterpillar), 156; *S. ipomoeae*, 156; *Sparganothis sulfureana*, 216
mites: *Acalitus vaccinii* (blueberry bud mite),

Vaccinium (cont.)
478; *Eotetranychus carpini* (yellow spider mite), 474
—bark, wood, twigs, roots: *Oberea myops* (azalea stem borer, rhododendron stem borer), 288; *O. tripunctata* (dogwood twig borer), 262; *Prionus laticollis* (broadnecked prionus, broadnecked root borer), 280; *Rhadopterus picipes* (cranberry rootworm), 242
—sucking insects: *Anormenis chloris*, 422; *Aspidiotus nerii* (oleander scale), 374; *Clastoptera obtusa* (alder spittlebug), 420; *Hemiberlesia rapax* (greedy scale), 372; *Leptoglossus phyllopus* (leaf-footed bugs), 422, 424; *Mesolecanium nigrofasciatum* (terrapin scale), 364; *Ormenoides venusta*, 422; *Phenacoccus aceris* (apple mealybug), 324; *Pulvinaria ericicola* (cottony azalea scale), 342

verbena, see *Aloysia triphylla*

Veronica: *Asterolecanium arabidis* (pit-making pittosporum scale), 348

Viburnum
—foliage: *Argyrotaenia juglandana*, 218; *Callirhopalus bifasciatus* (twobanded Japanese weevil), 244; *Choristoneura rosaceana* (obliquebanded leafroller), 216; *Cyrtepistomus castaneus* (Asiatic oak weevil), 244; *Gynaikothrips ficorum* (Cuban laurel thrips), 434; *Heliothrips haemorrhoidalis* (greenhouse thrips), 430; *Nemocestes incomptus* (woods weevil, raspberry bud weevil), 240; *Popillia japonica* (Japanese beetle), 236; *Pyrrhalta viburni* (leaf beetle), 224; *Sciopithes obscurus* (obscure root weevil), 240; *Selenothrips rubrocinctus* (redbanded thrips), 432
  mites: *Oligonychus ilicis* (southern red mite), 475
—bark, wood, twigs, roots: *Callirhopalus bifasciatus* (twobanded Japanese weevil), 244; *Enchenopa binotata* (twomarked treehopper), 410; *Metcalfa pruinosa*, 422; *Nemocestes incomptus* (woods weevil, raspberry bud weevil, 240); *Oberea tripunctata* (dogwood twig borer), 262; *Oecanthus nigricornis* (blackhorned tree cricket), 492; *Sciopithes obscurus* (obscure root weevil), 240; *Synanthedon fatifera*, 260; *S. viburni*, 260
—sucking insects: *Acyrthosiphum solani* (foxglove aphid), 310; *Aphis fabae* (bean aphid), 300; *A. hederae* (ivy aphid), 308; *A. viburniphila* (viburnum aphid), 300; *Aspidiotus nerii* (oleander scale), 374; *Ceroplastes rubens* (red wax scale), 358; *Chionaspis lintneri*, 390; *Lepidosaphes ulmi* (oystershell scale), 370; *Neoceruraphis viburnicola* (snowball aphid), 300; *Phenacoccus acericola* (maple mealybug), 346; *Poecilocapsus lineatus* (fourlined plant bug), 396; *Pseudococcus comstocki* (Comstock mealybug), 326; *Pulvinaria* sp., 346; *Selenothrips rubrocinctus* (redbanded thrips), 432

Viburnum acerifolium (mapleleaf viburnum): *Neoceruraphis viburnicola* (snowball aphid), 300; *Pyrrhalta viburni* (leaf beetle), 224
Viburnum dentatum (arrowwood): *Pyrrhalta viburni* (leaf beetle), 224
Viburnum japonicum (Japanese viburnum): *Rhadopterus picipes* (cranberry rootworm), 242
Viburnum lentago (nannyberry): *Enchenopa binotata* (twomarked treehopper), 410; *Lygocoris viburni*, 396; *Pyrrhalta viburni*, 224
Viburnum odoratissimum: thrips, 434
Viburnum opulus (European cranberry bush): *Aphis fabae* (bean aphid), 300; Neoceruraphis viburnicola (snowball aphid), 300; *Pseudococcus* sp., 324; *Pyrrhalta viburni*, 224
Viburnum opulus 'Sterilis': *Neoceruraphis viburnicola* (snowball aphid), 300
Viburnum plicatum: *Neoceruraphis viburnicola* (snowball aphid), 300
Viburnum prunifolium (blackhaw): *Neoceruraphis viburnicola* (snowball aphid), 300
Viburnum rafinesquianum: *Pyrrhalta viburni*, 224
Viburnum rhytidofollium (leatherleaf viburnum): *Aspidiotus nerii* (oleander scale), 374
Viburnum tinus: *Parlatoria pittospori*, 110
Viburnum tomentosum, see *Viburnum plicatum*
Viburnum trilobum (cranberry bush): *Pyrrhalta viburni*, 224
Vinca: *Aspidiotus nerii* (oleander scale), 374; *Chrysomphalus dictyospermi* (dictyospermum scale), 382; *Myzus persicae* (green peach aphid), 300
Vinca major (myrtle, periwinkle): *Acyrthosiphon solani* (foxglove aphid), 310; *Graphocephala atropunctata* (blue-green sharpshooter), 418; *Macrosteles quadrilineatus* (aster leafhopper), 418
Virginia creeper, see *Parthenocissus quinquefolia*
Vitex: *Aspidiotus nerii* (oleander scale), 374; *Asterolecanium arabidis* (pit-making pittosporum scale), 348
Vitis (grape): *Aspidiotus nerii* (oleander scale), 374; *Coptodisca* sp., 208; *Erythroneura lawsoniana*, 416; *Harrisina americana* (grapeleaf skeletonizer), 164; *H. brillians* (western grapeleaf skeletonizer), 164; *Hemiberlesia rapax* (greedy scale), 372; *Melanaspis obscura* (obscure scale), 384; *Ochyromera ligustri*, 244; *Parthenolecanium persicae* (European peach scale), 364; *P. pruinosum* (frosted scale), 354; *Saissetia oleae* (black scale), 358

walnut, see *Juglans*
  butternut: *Juglans cinerae*
  black: *Juglans nigra*
  English: *Juglans regia*
  Japanese: *Juglans ailantifolia*
  Persian: *Juglans regia*
  Siebold: *Juglans ailantifolia*
wattle, see *Acacia*
wavy-leaf ceanothus, see *Ceanothus foliosus*
wax myrtle, see *Myrica cerifera*
Weigela: *Callirhopalus bifasciatus* (twobanded Japanese weevil), 244; *Poecilocapsus lineatus* (fourlined plant bug), 396; *Popillia japonica* (Japanese beetle), 236; *Pseudococcus comstocki* (Comstock mealybug), 326
Weigela florida: *Asterolecanium arabidis* (pit-making pittosporum scale), 348
western hemlock, see *Tsuga heterophylla*
western redcedar, see *Thuja occidentalis*
western serviceberry, see *Amelanchier alnifolia*
western sycamore, see *Platanus racemosa*
white beam, see *Sorbus aria*
willow, see *Salix*
Wisteria: *Chionaspis wisteriae*, (wisteria scurfy scale), 390; *Elaphidionoides villosus* (twig pruner), 264; *Enchenopa binotata* (twomarked treehopper), 410; *Epargyreus clarus* (silverspotted skipper), 148; *Eulecanium cerasorum* (calico scale), 354; *Odontota dorsalis* (locust leafminer), 190; *Popillia japonica* (Japanese beetle), 236; *Pseudococcus comstocki* (Comstock mealybug), 326
Wisteria floribunda (Japanese wisteria): *Aphis craccivora* (cowpea aphid), 308
witch-hazel, see *Hamamelis*
woman's-tongue, see *Albizia lebbeck*

yaupon, see *Ilex vomitoria*
yellow poplar, see *Liriodendron tulipifera*
yew, see *Taxus*
  Japanese: *Podocarpus macrophyllus*
Yucca: *Aspidiotus nerii* (oleander scale), 374; *Diaspis boisduvalii* (boisduval scale), 376; *Hemiberlesia lataniae* (latania scale), 376; *Tetranychus yuccae*, 476

Zamia: *Saissetia coffeae* (hemispherical scale), 360
Zanthoxylum (prickly ash): *Parthenolecanium quercifex* (oak lecanium), 364
Zelkova serrata (Japanese zelkova): *Eulecanium cerasorum* (calico scale), 354; *Xanthogaleruca luteola* (elm leaf beetle), 222